ELECTRONIC CIRCUITS
Discrete and Integrated

McGraw-Hill Series in Electrical Engineering

Stephen W. Director, Carnegie-Mellon University
Consulting Editor

Networks and Systems
Communications and Information Theory
Control Theory
Electronics and Electronic Circuits
Power and Energy
Electromagnetics
Computer Engineering and Switching Theory
Introductory and Survey
Radio, Television, Radar, and Antennas

Previous Consulting Editors

Ronald M. Bracewell, Colin Cherry, James F. Gibbons, Willis W. Harman, Hubert Heffner, Edward W. Herold, John G. Linvill, Simon Ramo, Ronald A. Rohrer, Anthony E. Siegman, Charles Susskind, Frederick E. Terman, John G. Truxal, Ernst Wever, and John R. Whinnery

Electronics and Electronic Circuits

Stephen W. Director, Carnegie-Mellon University
Consulting Editor

ELECTRONIC CIRCUITS
Discrete and Integrated

Second Edition

Donald L. Schilling

Professor of Electrical Engineering
The City College of the City University of New York

Charles Belove

Professor of Electrical Engineering and Computer Science
New York Institute of Technology

McGraw-Hill Book Company

New York   St. Louis   San Francisco   Auckland   Bogotá   Düsseldorf
Johannesburg   London   Madrid   Mexico   Montreal   New Delhi
Panama   Paris   São Paulo   Singapore   Sydney   Tokyo   Toronto

ELECTRONIC CIRCUITS: Discrete and Integrated

Copyright © 1979, 1968 by McGraw-Hill, Inc. All rights reserved.
Printed in the United States of America. No part of this publication may be
reproduced, stored in a retrieval system, or transmitted, in any form or by any means,
electronic, mechanical, photocopying, recording, or otherwise, without the prior writte:
permission of the publisher.

4567890 FGRFGR   83210

This book was set in Times Roman.
The editors were Julienne V. Brown and Madelaine Eichberg;
the cover was designed by Rafael Hernandez;
the production supervisor was Dennis J. Conroy.
The drawings were done by Santype Ltd.
Fairfield Graphics was printer and binder.

Library of Congress Cataloging in Publication Data

Schilling, Donald L
    Electronic circuits, discrete and integrated.

    (Electrical and electronic engineering)
    Includes bibliographies and index.
    1. Electronic circuits. I. Belove, Charles, joint author. II. Title.
TK7867.S33 1979   621.381'53   78–12142
ISBN 0–07–055294–0

To our wives
ANNETTE and GOLDA
for their assistance and patience

CONTENTS

PREFACE

In the preface to the first edition, the authors predicted that advances in micro-electronics and integrated circuits might make the then current thinking obsolete within a decade. This has indeed been the case, and circuit design is now largely a process of selecting the appropriate combination of integrated circuits. These circuit blocks must be properly interfaced and operated within their rated capacities so that their interconnection will solve the problem at hand.

Specification sheets for integrated circuits often include circuit diagrams. These diagrams show the elements (transistors, resistors, and diodes) on the IC chip and are included so that the engineer can intelligently design the necessary interfacing circuits and use the chip most efficiently. In the design of large-scale integrated (LSI) circuits the design engineer must indicate to the chip manufacturer how he or she wishes to interconnect the thousands of transistors on the chip. This requires an understanding of the operation of bipolar junction and field-effect transistors.

The present edition is designed to prepare the student to use the new techniques effectively. It is intended as a beginning text in electronic circuits for upper-sophomore or junior-level engineering and physics students. The background acquired will provide the student with a sufficient depth of understanding to handle circuit problems and to be able to comprehend new devices as they become available. Upon completion of the course, students should be sufficiently prepared to function competently both in industry and in more sophisticated senior- and graduate-level courses.

The first edition of this text found wide acceptance in engineering schools throughout the world and in many industry short courses. This second edition is based on the same philosophy of teaching, which is that practical circuit design seldom makes direct use of device physics, using terminal properties instead. The authors present physical theory descriptively to enhance the discussion of these terminal properties. The knowledge of the practical use of semiconductor devices will prepare the way for a subsequent course in semi-conductor physics.

It is assumed that the student has a background in linear passive-circuit theory, which includes a thorough grounding in the use of Kirchoff's laws in dc and ac circuits. This, along with a knowledge of simple power calculations, is all that is required for the most of the text. In Chapters 9 and 10 some knowledge of the complex-frequency plane and the concept of frequency response will be helpful. The elements of boolean algebra and logic functions are included for students who may not have had a previous course in digital logic.

The prime objective of this text is to provide insight into the analysis and design of electronic circuits. As noted previously, this has evolved today into a process of combining ICs. In order to carry out this process intelligently and to be able to create combinations of ICs which are unconventional, the designer must understand the internal operation of the ICs. Chapters 1 to 8 cover transistor circuits at low frequencies. Design procedures and the most useful models are presented. The emphasis is on the graphical approach, in which the authors believe very strongly. The important concepts of dc and ac load lines, along with large-signal and small-signal analyses, are introduced in Chapter 1 in connection with diode circuits. When the bipolar junction transistor is introduced in Chapter 2 and the field-effect transistor in Chapter 3, the student has these techniques firmly established. Bias stability is covered in Chapter 4. The type of bias arrangement employed in integrated circuits is introduced in addition to the standard techniques. Chapter 5 introduces low-frequency power amplifiers. In Chapter 6, the small-signal low-frequency behavior of the various transistor configurations is studied, using the hybrid model. Here the technique of impedance reflection is introduced as a shortcut in the analysis of complicated circuits. The influence of the bias point on the small-signal behavior is also considered. In Chapter 7 multiple-transistor circuits are examined, with emphasis on circuit configurations (such as the difference amplifier and operational amplifiers) which lend themselves naturally to integrated-circuit fabrication techniques. In Chapter 8, applications of operational amplifiers are covered.

The frequency-response of RC-coupled amplifiers at low and high frequencies is considered in Chapter 9, along with tuned narrowband amplifiers and the transistor switch. The principles and advantages of feedback are considered in Chapter 10. Gain, sensitivity, and impedances are studied, and examples stress the type of feedback amplifiers which utilize linear integrated circuits. Chapter 10 also considers the frequency response of feedback amplifiers. The stability problems associated with integrated circuits are considered, and frequency-compensation techniques are studied. A section on practical transistor oscillators is also included in this chapter.

The second part of the book deals with digital circuits and begins with Chapter 11, which introduces the logic analysis techniques required for subsequent chapters. Logic gates are considered in Chapter 12. Those logic families which have withstood the test of time (TTL, ECL, and CMOS) are covered along with methods for interfacing between different types and interpretation of manufacturers data sheets. Chapter 13 covers flip-flops, which are used in

Chapter 14 to form shift registers and counters. Also included in Chapter 14 are discussions of arithmetic circuits and digital filters, both of which utilize the gates and flip-flops discussed previously.

Chapter 15 considers sample-and-hold circuits and digital-to-analog and analog-to-digital converters along with timing circuits (astable and monostable multivibrators) constructed using ICs.

Finally, Chapter 16 discusses the fabrication of integrated circuits and provides an introduction to the recently developed technique known as integrated injection logic (I^2L).

Every effort has been made to use practical parameter values in the numerous illustrative examples presented throughout the text. Typical manufacturers' specifications are given, so that the student gains an idea of the practical range of parameter values for the various devices. Numerous homework problems are included, ranging from drill in analysis to difficult designs. Appendixes B and C include standard resistor and capacitor values and manufacturers' data sheets for various devices, which can be used for the design problems and as future references. A solutions manual is available from the publisher.

The authors would like to acknowledge the excellent review of the entire manuscript by Dr. Edward Nelson, who also prepared the solutions manual. Our thanks also go to Mrs. Joy Rubin who typed the entire manuscript.

Donald L. Schilling
Charles Belove

NOTATION

The symbols for currents and voltages at the terminals of active devices have subscripts which indicate the pertinent terminal for currents or terminal pair for voltages. In addition, uppercase and lowercase symbols and subscripts are used to distinguish between quiescent values, total values, and incremental values. The International System of Units is used throughout.

EXAMPLES

$$I_{BQ}, I_{CQ}, V_{CEQ} = \text{quiescent-point value}$$
$$I_B, I_C, V_{CE} = \text{dc value, with signal}$$
$$i_B, i_C, i_E, v_{CE} = \text{total instantaneous value}$$
$$i_b, i_c, i_e, v_{ce} = \text{instantaneous value of time-varying component (zero average)}$$
$$I_b, I_e, V_{ce} = \text{rms value of sinusoidal component}$$
$$I_{bm}, I_{em}, V_{cem} = \text{max (peak) value of time-varying component of variable}$$
$$V_{BB}, V_{CC}, V_{DD} = \text{supply voltages}$$

GRAPHICAL ILLUSTRATION OF NOTATION

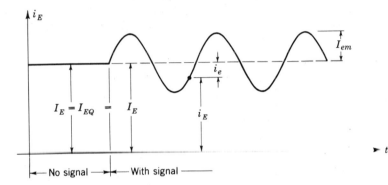

Sinusoidal signal, no distortion

$$i_E(t) = I_E + I_{em} \sin \omega t = I_{EQ} + I_{em} \sin \omega t$$
$$i_e(t) = I_{em} \sin \omega t$$

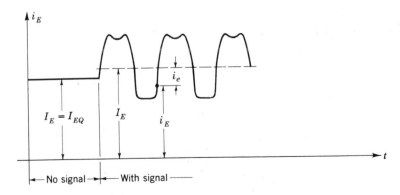

Signal, with distortion

$$i_E(t) = I_E + i_e(t)$$

DIODE-CIRCUIT ANALYSIS

INTRODUCTION

The diode is the simplest of the nonlinear devices with which this text is concerned. In this chapter we discuss circuit characteristics and applications of the junction diode, the Zener diode, and the Schottky diode. Graphical techniques are emphasized throughout the chapter because they provide a visual picture of circuit operation and often yield insights not readily obtained from purely algebraic treatments. These graphical techniques include a thorough treatment of dc and ac load lines as applied to both small and large signals. Although these methods are not often used in the analysis of diode circuits, their introduction at this point serves to establish them firmly in the student's repertoire. Later, when transistors are involved, the additional problems encountered will be more easily solved as a result of the experience gained with diodes.

1.1 NONLINEAR PROPERTIES; THE IDEAL DIODE

Students usually begin their study of circuits by considering models of linear elements, the simplest of these being the resistor. The volt-ampere (vi) characteristic of the ideal resistor is described by such a simple relation—Ohm's law—that we sometimes lose sight of its graphical interpretation. The linear character of the resistance is evident in Fig. 1.1-1. The vi characteristic of the *ideal* diode is shown in Fig. 1.1-2. The nonlinear character of the diode is clearly evident here. When the source voltage v_i is positive, i_D is positive and the diode is a short circuit ($v_D = 0$), while when v_i is negative, i_D is zero and the diode is an open circuit

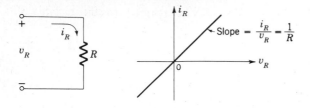

Figure 1.1-1 The resistance element and its vi characteristic.

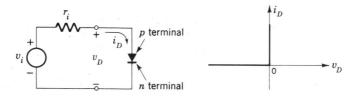

Figure 1.1-2 The ideal diode and its vi characteristic.

$(v_D = v_i)$. The diode can be thought of as a switch controlled by the polarity of the source voltage. The switch is closed for positive source voltages and open for negative source voltages.

Another way to look at this element is to note that the diode conducts current only from p to n (Fig. 1.1-2) and conduction takes place only when the source voltage is positive. The diode does not conduct when the source voltage is negative.

Physical diodes have inherent characteristics and limitations which cause them to differ from the ideal. These are discussed in succeeding sections. For present purposes, the diodes are considered to be ideal.

The following examples illustrate some of the operations on signals which are often achieved with simple diode circuits.

Example 1.1-1: Half-wave rectifier or clipping circuit One of the principal applications of the diode is in the production of a dc voltage from an ac supply, a process called *rectification*. A useful by-product of rectification consists of signals at frequencies which are integral multiples of the supply frequency. A typical *half-wave-rectifier* circuit is shown in Fig. 1.1-3. (*a*) The source voltage is sinusoidal, $v_i = V_{im} \cos \omega_0 t$, where $V_{im} = 10$ V. Find and sketch the waveform of the load voltage. Find its average (dc) value. (*b*) Repeat part (*a*) if $v_i = -5 + 10 \cos \omega_0 t$.

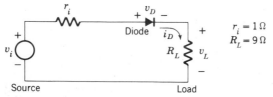

$r_i = 1\,\Omega$
$R_L = 9\,\Omega$

Figure 1.1-3 Half-wave-rectifier circuit for Example 1.1-1.

SOLUTION (a) Kirchhoff's voltage law (KVL) applied to the circuit of Fig. 1.1-3 yields

$$v_i = i_D r_i + v_D + i_D R_L \qquad \text{or} \qquad i_D = \frac{v_i - v_D}{r_i + R_L}$$

This equation contains two unknowns, v_D and i_D. They in turn are related by the diode vi characteristic. The solution for i_D or v_D thus requires "substitution" of the vi curve into the equation. This can be done in the following way. The diode characteristic indicates that only positive current in the reference direction can flow in this circuit. This requires that $v_i > v_D$. However, when the diode is conducting, $v_D = 0$, so that current flows in the positive direction only when $v_i > 0$.

When v_i is negative, current flow should be opposite to the reference direction; but the diode cannot conduct in this direction; so $i_D = 0$ when $v_i < 0$.

This discussion can be summarized by drawing two circuits, one of which holds for $v_i > 0$ and one for $v_i < 0$, as shown in Fig. 1.1-4. Using the circuits shown in this figure, the unknowns v_D and i_D can be found. Thus the diode current i_D is

$$i_D = \begin{cases} \dfrac{V_{im}}{r_i + R_L} \cos \omega_0 t & \text{when } v_i > 0 \\ 0 & \text{when } v_i < 0 \end{cases}$$

and the load voltage v_L is

$$v_L = i_D R_L$$

The load voltage v_L and signal voltage v_i are sketched in Fig. 1.1-5. Note that the current waveform has the same shape as the load voltage v_L. This is a *half-wave-rectified* sine wave. Its average value is obtained by dividing the area by the period, 2π.

$$V_{L,\,\text{dc}} = \frac{1}{2\pi} \int_{-\pi/2}^{\pi/2} (V_{Lm} \cos \omega_0 t)\, d(\omega_0 t) = \frac{V_{Lm}}{\pi} = \frac{9}{\pi} = 2.86 \text{ V} \qquad (1.1\text{-}1)$$

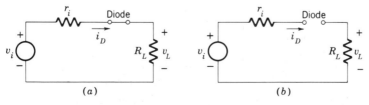

(a)   (b)

Figure 1.1-4 Conducting and nonconducting states of the diode rectifier: (a) $v_i > 0$; (b) $v_i < 0$.

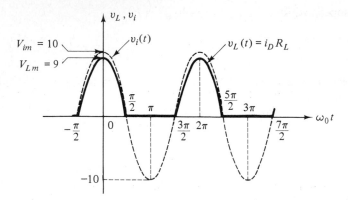

Figure 1.1-5 Waveforms in the rectifier circuit of Example 1.1-1.

The Fourier-series expansion for $v_L(t)$ can be shown to be (see Prob. 1.1-20)

$$v_L(t) = V_{Lm}\left(\frac{1}{\pi} + \frac{1}{2}\cos \omega_0 t + \frac{2}{3\pi}\cos 2\omega_0 t - \frac{2}{15\pi}\cos 4\omega_0 t + \cdots\right) \quad (1.1\text{-}2)$$

This expression shows clearly that the effect of the diode has been to generate not only the dc term and one at the same frequency as the source but also terms at harmonic frequencies not present in the source voltage.

If the circuit is to produce a dc voltage, the dc component must be separated from the harmonics by filtering $v_L(t)$. This can be accomplished by using a simple passive filter like those shown in Fig. 1.1-6. Let us assume that one of these filters, say the RC filter shown in Fig. 1.1-6a, is placed across resistor R_L in Fig. 1.1-3 and R is adjusted so that $R \gg R_L$ to assure negligible loading. Then, if the RC product is adjusted so that $RC = 100/\omega_0$, the amplitude of the output voltage V_{on} at the frequency $n\omega_0$ is

$$V_{on} = \frac{V_{Ln}}{\sqrt{1 + (n\omega_0 RC)^2}} \approx \frac{V_{Ln}}{100n} \quad \text{when } n \geq 1$$

where V_{Ln} is the amplitude of the load voltage at the frequency $n\omega_0$, for example, from (1.1-2), $V_{L2} = 2V_{Lm}/3\pi$.

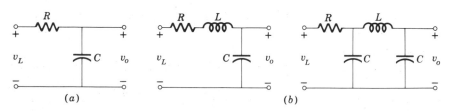

Figure 1.1-6 Passive power-supply filters.

If we use superposition, the output voltage is

$$v_o(t) \approx V_{Lm}\left(\frac{1}{\pi} + \frac{1}{200}\sin \omega_0 t + \frac{1}{300\pi}\sin 2\omega_0 t - \frac{1}{3000\pi}\sin 4\omega_0 t + \cdots\right)$$

Thus the output voltage consists of a dc voltage V_{Lm}/π and a small ripple voltage v_r, where

$$v_r = V_{Lm}\left(\frac{1}{200}\sin \omega_0 t + \frac{1}{300\pi}\sin 2\omega_0 t - \cdots\right)$$

The ratio of the rms value of the ripple voltage to the dc voltage is a measure of the effectiveness of the filter in separating the dc voltage from the harmonics. For the RC filter of this example

$$(v_r)_{\text{rms}} \equiv \left|\frac{1}{2\pi}\int_0^{2\pi}[v_r(\omega_0 t)]^2 \, d(\omega_0 t)\right|^{1/2}$$

$$= \frac{V_{Lm}}{\sqrt{2}}\sqrt{\frac{1}{(200)^2} + \frac{1}{(300\pi)^2} + \cdots} \approx \frac{V_{Lm}}{280}$$

and
$$\frac{(v_r)_{\text{rms}}}{V_{L,\,dc}} \approx \frac{\pi}{280} \approx 0.011$$

Thus the rms ripple is approximately 1 percent of the dc voltage in the output.

More complicated filters, such as the LC or CLC filters shown in Fig. 1.1-6b, yield a much smaller rms ripple, which can be calculated approximately using the above method (see Probs. 1.1-4 and 1.1-5.)

In the design of a practical power supply we do not usually set the filter resistance R to be much larger than R_L. Actually R_L is often made infinite by removing it from the circuit since this results in an increased output voltage. In this example, we have assumed $R_L \ll R$ to allow a linear analysis. A typical practical power-supply circuit is the peak-detector circuit which will be discussed in Example 1.1-3.

(b) The waveform of v_i is sketched in Fig. 1.1-7. In this case a negative bias has been added to the signal. The waveform for v_L is obtained by noting that current will flow only when v_i is positive. The exact time $\pm t_1$ at which the current flow starts and stops is found by setting $v_i = 0$; then

$$-5 + 10 \cos \omega_0 t_1 = 0$$

from which $\qquad \cos \omega_0 t_1 = 0.5 \qquad$ and $\qquad \omega_0 t_1 = \pm \frac{\pi}{3}$

From the symmetry of the cosine function, the diode is seen to conduct current when

$$2\pi n - \frac{\pi}{3} \leq \omega_0 t \leq 2\pi n + \frac{\pi}{3}$$

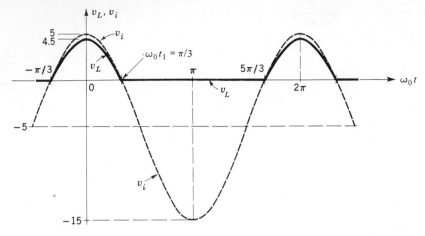

Figure 1.1-7 Rectifier-circuit waveforms with bias added to signal.

Thus the load voltage is

$$
v_L =
\begin{cases}
-4.5 + 9 \cos \omega_0 t & 2\pi n - \dfrac{\pi}{3} \le \omega_0 t \le 2\pi n + \dfrac{\pi}{3} \\[2ex]
0 & 2\pi n + \dfrac{\pi}{3} \le \omega_0 t \le 2\pi n + \dfrac{5\pi}{3}
\end{cases}
$$

The average value of v_L is found as before.

$$
\begin{aligned}
V_{L,\,\text{dc}} &= \frac{1}{2\pi} \int_{-\pi/3}^{\pi/3} (-4.5 + 9 \cos \omega_0 t)\, d(\omega_0 t) \\[2ex]
&= (-4.5)(\tfrac{1}{3}) + \frac{9}{\pi} \sin \frac{\pi}{3} \\[2ex]
&= -1.5 + \frac{9}{\pi} \frac{\sqrt{3}}{2} \approx 0.98 \text{ V}
\end{aligned}
$$

///

Example 1.1-2: The full-wave rectifier The ripple voltage in the half-wave rectifier is primarily due to the signal component at the fundamental frequency ω_0. The *full-wave rectifier* yields a load voltage with the lowest ripple-frequency term at $2\omega_0$, and, in addition, the dc component is doubled. This type of circuit, one form of which is shown in Fig. 1.1-8, is therefore more efficient for the production of a dc voltage with low ripple and is the basic rectifier circuit for most dc power supplies.

The circuit operation can be explained qualitatively if the ideal transformer is eliminated by redrawing Fig. 1.1-8 as shown in Fig. 1.1-9*a*. In this figure the transformer is seen to reflect the ac source from the primary into the center-tapped secondary circuit. When v_i is positive, D_1 is a short circuit and

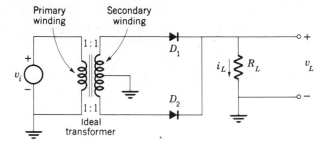

Figure 1.1-8 Full-wave rectifier.

D_2 is an open circuit. When v_i is negative, D_1 is an open circuit and D_2 is a short circuit. In each case the load current i_L is in the same positive direction, as shown in Fig. 1.1-9a, and since one or the other of the diodes D_1 or D_2 is a short circuit on each alternate half-cycle, the load voltage can be written $v_L = |v_i|$. The current and voltage waveforms are shown in Fig. 1.1-9b.

The Fourier series for v_L can be shown to be (see Prob. 1.1-21)

$$v_L = V_{Lm}\left(\frac{2}{\pi} + \frac{4}{3\pi}\cos 2\omega_0 t - \frac{4}{15\pi}\cos 4\omega_0 t + \cdots\right) \qquad (1.1\text{-}3)$$

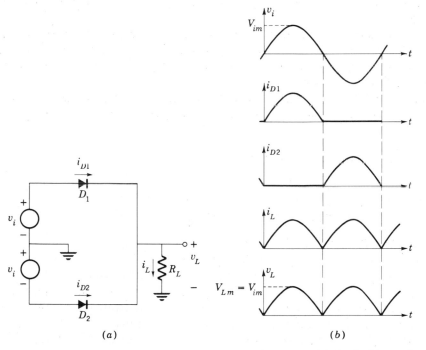

(a)   (b)

Figure 1.1-9 Full-wave-rectifier equivalent circuit and waveforms: (a) circuit; (b) waveforms.

The dc component is $(2/\pi)V_{Lm}$, which is twice the dc value obtained using the half-wave rectifier. If v_L is passed through the RC filter of Fig. 1.1-6a with $\omega_0 RC = 100$ as before, the output ripple voltage becomes

$$v_r = \frac{4V_{Lm}}{3\pi}\left(\frac{1}{200}\sin 2\omega t - \frac{1}{2000}\sin 4\omega t + \cdots\right)$$

and the rms ripple is

$$(v_r)_{\text{rms}} \approx \frac{V_{Lm}}{210\pi}$$

The ratio of ripple voltage to dc voltage is

$$\frac{(v_r)_{\text{rms}}}{V_{L,\,\text{dc}}} \approx \frac{1}{420} \approx 0.0024$$

which is considerably less than that obtained using a half-wave rectifier. ///

Almost all electronic circuits require dc power for their operation, and since ac power is usually available, some sort of rectifier-filter combination will be found in most electronic equipment. Examples 1.1-1 and 1.1-2 have presented the most basic rectifier-filter combinations used to produce a dc voltage from an ac supply. These basic circuits suffer from several drawbacks, which make them unsuitable for many applications. The first of these is the variation of the dc load voltage with load current. This is measured by a quantity called the *regulation*, which is defined as

$$\text{Regulation} = \frac{\text{no-load voltage} - \text{full-load voltage}}{\text{full-load voltage}}$$

An ideal supply would provide a dc voltage which is constant and independent of the load current, i.e., zero regulation. This is equivalent to saying that the supply would have zero output resistance as seen from the load terminals. However, in practical circuits, the diode resistance, which was neglected in the examples, and the filter-circuit resistance are not negligible and result in a finite output resistance. If the output resistance is equal to the load resistance, the full-load voltage is equal to one-half of the no-load voltage, and the regulation is 100 percent. A second disadvantage of this basic rectifier circuit when used as a dc supply is that the dc output is directly proportional to the magnitude of the ac supply voltage. Since most ac power lines do not maintain an absolutely constant voltage, the dc output will vary proportionally. For many applications, this variation cannot be tolerated, even though it may be relatively small. A third disadvantage of the above circuit is that even the small ripple voltage found in the examples is often more than can be tolerated for proper operation of sophisticated electronic circuits.

Many techniques exist for overcoming the effects mentioned above. Manufacturers of rectifiers provide data and handbooks[1]† which contain complete information on the design of power supplies of many varieties. These handbooks usually contain the latest *state-of-the-art* information and provide an important source of information for the design engineer. Several techniques for improving power-supply performance are discussed in Secs. 1.10 and 8.8.

Example 1.1-3: The peak detector The dc component in the output of the full-wave rectifier is only about 64 percent of the peak voltage of the input sinusoid. The *peak* (or *envelope*) *detector* provides a dc output comparable to the peak value of the input voltage and is therefore used as a power supply. This circuit (Fig. 1.1-10a) is also used in AM receivers to detect the *envelope* of the *amplitude-modulated* carrier waveform.

The operation of the peak detector is most easily explained by letting $v_i = V_{im} \sin \omega_0 t$ and assuming that the load resistor R_L is infinite. Then, during the first quarter-cycle of the input waveform the diode acts like a short circuit, and the capacitor will therefore follow v_i, as shown in Fig. 1.1-10b. When $\omega_0 t = \pi/2$, the capacitor will have charged to $v_L = V_{im}$. When v_i decreases, the capacitor voltage cannot decrease because with R_L infinite the capacitor discharge current must flow through the diode in the reverse direction. Since current cannot flow through the diode in the reverse direction, the capacitor cannot discharge. The load voltage v_L therefore remains at the peak value V_{im} until the peak value of v_i is increased.

Capacitors are never perfect, and their departure from the ideal can best be modeled by a parallel resistance such as R_L in Fig. 1.1-10a. For the case when R_L is not infinite the output voltage of the peak detector appears as shown in Fig. 1.1-10c. From this figure we see that again $v_L = v_i$ during the first quarter-cycle since the diode acts as a short circuit. However, when v_i decreases, the voltage across the capacitor also decreases even though the diode is off, since there is a path for the discharge current to flow, i.e., through R_L. The output voltage between times t_1 and t_2 (see Fig. 1.1-10c) decays exponentially following the equation

$$v_L = V_{im} \epsilon^{-(t-t_1)/R_L C} \qquad t_1 \le t \le t_2 \qquad (1.1\text{-}4)$$

Between times t_2 and t_3 the diode again appears to be a short circuit, and the capacitor voltage follows the input. For this circuit the peak-to-peak ripple is

$$V_{Lr,\,p\text{-}p} = v_L(t_1) - v_L(t_2)$$
$$= V_{im}\left(\epsilon^{-(t_1-t_1)/R_L C} - \epsilon^{-(t_2-t_1)/R_L C}\right)$$
$$= V_{im}\left(1 - \epsilon^{-(t_2-t_1)/R_L C}\right) \qquad (1.1\text{-}5)$$

† A superior number refers to a citation in the References at the end of the chapter.

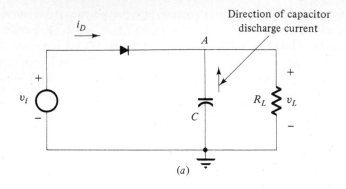

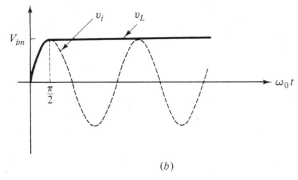

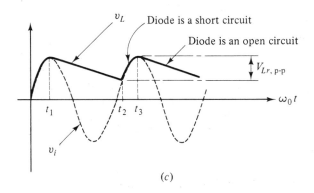

Figure 1.1-10 The half-wave peak detector: (a) circuit; (b) waveform for $R_L = \infty$; (c) waveform for discharge time constant = $R_L C$.

If $R_L C \gg t_2 - t_1$, as is usually the case, we can use the expansion $\epsilon^{-x} \approx 1 - x$, which is valid for $x \ll 1$. Then

$$V_{Lr,\ \text{p-p}} \approx V_{im}\left[1 - \left(1 - \frac{t_2 - t_1}{R_L C}\right)\right] \approx \frac{V_{im}(t_2 - t_1)}{R_L C} \qquad (1.1\text{-}6)$$

For the half-wave-rectifier peak detector shown in Fig. 1.1-10a the duration of the discharge part of the cycle, $t_2 - t_1$, will be very nearly equal to the period of the input sine wave. Thus $t_2 - t_1 \approx 1/f_0$, and the ripple is

$$V_{Lr, \text{p-p}} \approx \frac{V_{im}}{f_0 R_L C} \tag{1.1-7}$$

The dc component of the load voltage is approximately

$$V_{L, \text{dc}} \approx V_{im} - \tfrac{1}{2}V_{Lr, \text{p-p}} \approx V_{im}\left(1 - \frac{1}{2f_0 R_L C}\right) \tag{1.1-8}$$

Note that when R_L (or C) approaches infinity, $V_{L, \text{dc}}$ approaches V_{im}, as noted previously. For a full-wave-rectifier peak detector, the ripple is halved.

When a peak detector is used in a voltmeter, for example, R_L is made as large as possible so that a minimum of ripple will be present and v_L will be very close to the peak value of the input voltage.

In AM systems, where the peak detector is called an envelope detector (or demodulator), the problems are a little different. Here the time constant in the peak detector must be chosen so that the circuit *can* follow the time-varying envelope which carries the desired information.

As an example, consider that the modulating signal (information) is

$$m(t) = 0.5 \cos \omega_m t \tag{1.1-9}$$

This is modulated onto a high-frequency carrier by passing it through a circuit which yields as output

$$v_i(t) = V_{im}[1 + m(t)] \cos \omega_0 t = v_e(t) \cos \omega_0 t \tag{1.1-10}$$

where $v_e(t) = V_{im}[1 + m(t)]$ is the *envelope*. In this equation, ω_0 is the carrier angular frequency, and usually $\omega_m \ll \omega_0$. The modulated carrier is sketched in Fig. 1.1-11a. The envelope detector is used to recover the modulating signal $m(t)$ by following the slow variations in $m(t)$ and ignoring the much more rapid variations of the high-frequency carrier.

In order to make clear what is happening, we have sketched in Fig. 1.1-11b a portion of the waveform of v_i expanded to show the details of the detection process. The signal at the output of the envelope detector is shown in solid lines and is designated as v_L. We see that in order for v_L to follow the envelope of v_i the capacitor voltage must be able to change by an amount equal to the maximum change of the envelope in the time interval required by the carrier to go through one complete cycle. This time is designated $T(= 1/f_0)$.

The rate of change of the envelope is

$$\frac{d}{dt}v_e(t) = V_{im}\frac{dm}{dt} = -V_{im}(0.5\omega_m) \sin \omega_m t$$

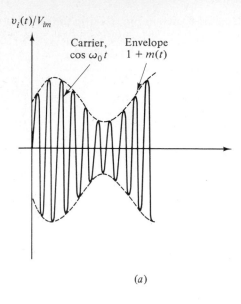

$v_i(t)/V_{im}$

Carrier, $\cos \omega_0 t$ Envelope $1 + m(t)$

(a)

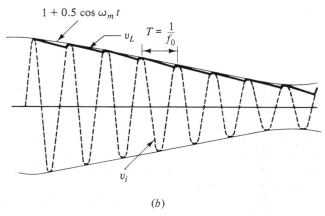

$1 + 0.5 \cos \omega_m t$

v_L $T = \dfrac{1}{f_0}$

v_i

(b)

Figure 1.1-11 Envelope detector: (a) AM waveform; (b) expanded view of detector output.

Its maximum magnitude occurs when $m(t) = 0$ and is

$$\left| \frac{dv_e(t)}{dt} \right|_{\text{max}} = 0.5\omega_m V_{im} \tag{1.1-11}$$

In order to find the maximum amount that the envelope can decrease in a time equal to one period of the carrier $(T = 1/f_0)$ we replace $dv_e(t)/dt$ by $\Delta v_e/\Delta t$ and set $\Delta t = T$. Then, over one period of the carrier

$$\Delta v_e = 0.5\omega_m T V_{im} \tag{1.1-12}$$

When $m(t) = 0$, the detector output begins to decay exponentially from the value V_{im} (see Fig. 1.1-11b) toward zero, according to the relation

$$v_L(t) = V_{im} \epsilon^{-t/R_L C} \tag{1.1-13}$$

After one cycle of the carrier (time T) the diode again conducts, and the total change in $v_L(t)$ is

$$\Delta v_L = V_{im} - V_{im} \epsilon^{-T/R_L C} \tag{1.1-14}$$

Since $T \ll R_L C$, we can again use the expansion $\epsilon^{-x} \approx 1 - x$ ($x \ll 1$). This yields

$$\Delta v_L \approx V_{im} \frac{T}{R_L C} \tag{1.1-15}$$

As noted previously, the change in output voltage must be at least equal to the maximum change in the envelope over one cycle of the carrier. The conditions for this to be true are found by equating (1.1-12) and (1.1-15), yielding

$$R_L C = \frac{2}{\omega_m} = \frac{1}{\pi f_m} \tag{1.1-16}$$

If the $R_L C$ time constant exceeds $1/\pi f_m$, the capacitor discharge will be too slow and the detector output v_L will not follow the envelope; if $R_L C$ is less then $1/\pi f_m$, v_L will follow the envelope but excessive ripple may result. In practice the modulation usually contains a band of frequencies, and a compromise must be made. Usually, the highest modulation frequency would be used in Eq. (1.1-16).　　　　///

Example 1.1-4: Clamping circuit The *clamping circuit* shown in Fig. 1.1-12a is similar in operation to the peak detector of Fig. 1.1-10a. In fact, the two circuits are identical if R_L in the clamping circuit is set equal to infinity. To explain the circuit operation qualitatively, assume that R_L is infinite and that the diode is ideal. Noting how the diode is connected, we observe that the output voltage v_L can never be greater than the reference voltage V_R. The instant that v_L tries to exceed V_R the diode will turn ON, becoming a short circuit and forcing the condition $v_L = V_R$. Then capacitor C must charge to the voltage $V_R - V_{im}$, as shown in Fig. 1.1-12b. With R_L infinite capacitor C cannot discharge, and v_L is

$$v_L = (V_R - V_{im}) + V_{im} \sin \omega_0 t$$

This waveform is shown in Fig. 1.1-12b. We say that the output voltage v_L is *clamped* to V_R.

In practical systems the input waveform is not sinusoidal but has a varying amplitude, as shown in Fig. 1.1-12c. In order to accommodate the changes in amplitude, we include resistor R_L so that the capacitor can discharge and maintain the clamping at V_R.　　　　///

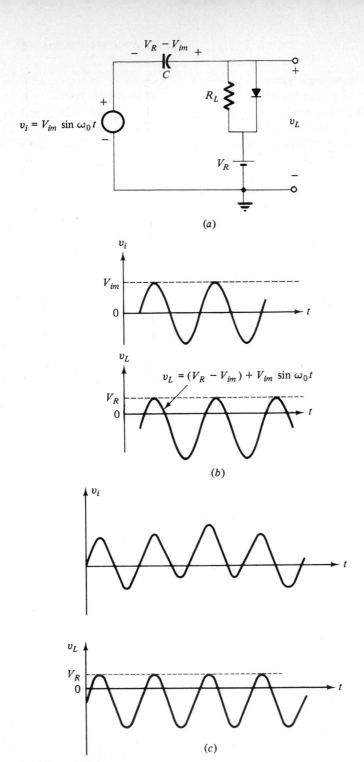

Figure 1.1-12 Clamping circuit: (*a*) circuit; (*b*) waveforms for sine-wave input; (*c*) waveforms for input with varying amplitude.

Example 1.1-5: Diode logic gates Diodes can be used to form logic *gates*, which perform some of the logical operations required in digital computers. The two operations most easily implemented with diodes are called the AND and OR operations. The circuit for a three-input OR gate is shown in Fig. 1.1-13*a*. The input voltages v_1, v_2, and v_3 and the output voltage v_L can take on only one of two values, 0 or 5 V. This OR gate operates according to the following definition. *The output of the OR gate becomes 5 V if one or more inputs are 5 V.* Referring to the circuit, we see that if, for example, $v_1 = 5$ V while $v_2 = v_3 = 0$ V, then D_1 is a short circuit and $v_L = 5$ V. D_2 and D_3 are reverse-biased. A little thought will show that the circuit obeys the definition for all combinations of input voltages. It is called an OR gate because $v_L = 5$ V if $v_1 = 5$ V OR $v_2 = 5$ V OR $v_3 = 5$ V. Of course, if $v_1 = v_2 = v_3 = 0$ V, then $v_L = 0$ V.

A three-input diode AND gate is shown in Fig. 1.1-13*b*. Here the following definition can be written. *The output of the AND gate becomes 5 V only when all the inputs are equal to 5 V.* Referring to the circuit, we see that if, for example, $v_1 = 0$ V, then D_1 will become forward-biased and thus a short circuit. This leads to the condition $v_L = 0$ V regardless of the values of v_2 and v_3. When all three inputs are 5 V, each of the diodes will be OFF and the output will be 5 V.

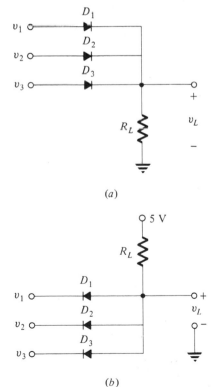

(a)

(b)

Figure 1.1-13 Diode logic gate: (*a*) OR gate; (*b*) AND gate.

In practice diodes are not ideal and when drawing current act more like a 0.7-V battery than like a short circuit. If real diodes are used in the OR gate of Fig. 1.1-13a, we see that if $v_1 = 5$ V, then $v_L = 4.3$ V. If this gate output is now the input to another OR gate, the output of the second gate will be 3.6 V. Thus, a cascade of seven gates would yield an output of only 0.1 V. Ideally, digital circuits will recognize only two voltages, such as 0 and 5 V, but in practice any voltage less than, say, 2.5 V is recognized as 0 V. When a diode gate is used with a transistor, the problem of signal attenuation illustrated above is eliminated since the transistor amplifies the signal. As a result of the attenuation present in diode gates, diode logic is not used in practice.   ///

1.2 AN INTRODUCTION TO SEMICONDUCTOR-DIODE THEORY[2, 3]

A brief qualitative discussion of the basic concepts governing current flow in a semiconductor diode is presented in this section. No attempt is made to be rigorous or to derive equations. The interested student should refer to any of the many texts on solid-state physics to fill in the details.

The basic material used in the construction of most diodes and transistors today is silicon. Formerly, germanium was used extensively, but it is fast disappearing. Silicon is a semiconductor; i.e., at room temperature very few electrons exist in the conduction band of the silicon crystal. Since the current is proportional to the number of electrons in motion, the current is small; hence the material has a high resistance. The conduction and valence bands of pure silicon are shown in Fig. 1.2-1.

At 0 K (absolute zero), all the electrons are at their lowest possible energy levels. At room temperature, an occasional electron has enough energy to escape from the valence band and move to the conduction band, as shown by the dot in Fig. 1.2-1. The vacancy left by the electron is shown as a circle, or *hole*. If an electric field is applied to the material as shown in Fig. 1.2-2, the electron moves toward the positive battery terminal, as expected. An electron in the valence band can also move toward the positive battery terminal if it has enough energy to take it from its energy level to the energy level of the hole. When this electron does escape into the hole, it leaves a hole behind. Thus it appears that the hole moves to

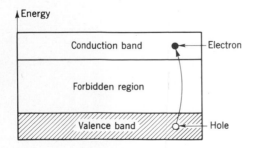

Figure 1.2-1 Energy bands in silicon at room temperature.

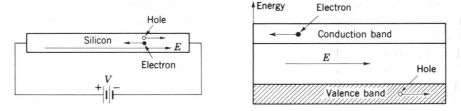

Figure 1.2-2 Electron and hole motion in silicon with applied electric field.

the right, toward the negative battery terminal. The net current is therefore the sum of the current due to the electron motion in the conduction band and the current due to hole motion. We refer to hole, or positive-charge, motion rather than to electron motion in the valence band to avoid confusion with the electron motion in the conduction band. The conventional current, due to the flow of electrons, and the hole current are of course in the direction of the electric field.

It should be noted that the electron moves more rapidly to the positive terminal than the hole moves toward the negative terminal since the probability of an electron having the energy required to move to an empty state in the conduction band (which is almost empty) is much greater than the probability of an electron having the energy required to move to an empty state in the valence band (which is almost filled). Thus the current due to electron flow in the conduction band is greater than the hole current in silicon. However, the net current is small, and hence the material is a semiconductor.

To construct a diode, we take silicon and add atoms of another element, such as boron. This process is called *doping*. Boron is called an *acceptor* material because it is able to accept electrons from the valence band of the silicon. At room temperature, the electrons from the valence band of the silicon fill the acceptor space of the boron, as shown in Fig. 1.2-3, since the probability of the valence electron's having sufficient energy at room temperature to bridge the small gap is very high. The result is that an extremely large number of holes exist. When an electric field is applied across this doped silicon, the hole current is very high and

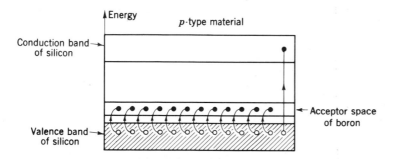

Figure 1.2-3 Energy bands for silicon with boron added.

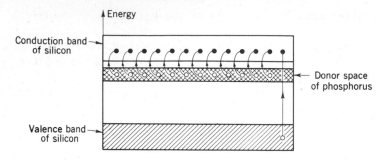

Figure 1.2-4 Energy bands for silicon with phosphorus added.

the material is now a good conductor. This is called *p-type* material. Note that in *p-type* material conduction is due primarily to the motion of holes.

We then take another piece of silicon and add atoms of another element, such as phosphorus. The phosphorus is called a *donor* material since it is able to donate electrons to the conduction band of the silicon. It therefore donates (at room temperature) all its electrons to the conduction band of the silicon, as shown in Fig. 1.2-4. Now the current flow, when an electric field is applied, is due primarily to electron flow. This doped material is called *n type*.

The diode consists of *p-* and *n-*type material fabricated as shown schematically in Fig. 1.2-5. The *junction* between the *n* and *p* materials is the basis for the name *junction diode*. Figure 1.2-5*a* shows a forward-biased diode and its circuit symbol. Holes from the *p* region flow into the *n* region, while electrons from the *n* region flow into the *p* region. A small voltage *V* is sufficient to yield a high current.

Figure 1.2-5*b* shows a reverse-biased diode. Electrons in the *p* region now flow into the *n* region, and holes in the *n* region flow into the *p* region. The current flow is hence very small because of the small number of charges in motion. If V_r is increased beyond the diode's *breakdown voltage*, the diode current increases considerably for a very small change in V_r. This is called the *Zener region*, and the breakdown is commonly called a *Zener* or *avalanche breakdown*. The avalanche and Zener-breakdown mechanisms differ. Avalanche breakdown occurs at high reverse voltages, and Zener breakdown occurs at small reverse voltages but the

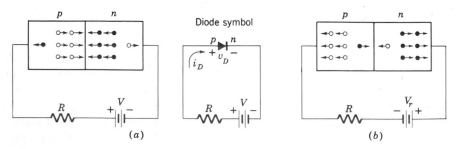

Figure 1.2-5 The junction diode: (*a*) forward bias; (*b*) reverse bias.

effect on the circuit is the same. No distinction is made in this book between these two processes.

The avalanche-breakdown process can be thought of as a moving electron colliding with a fixed electron, knocking it free; these two electrons free two more electrons; etc. This results in a large current flow in this region.

A physical analysis of the diode (neglecting the Zener region) shows that the current and voltage are related by

$$i_D = I_o(\epsilon^{qv_D/mkT} - 1) \tag{1.2-1}$$

where i_D = current through diode, A

v_D = voltage across diode, V

I_o = reverse saturation current, A

q = electron charge = 1.6×10^{-19} C

k = Boltzmann's constant = 1.38×10^{-23} J/K

T = absolute temperature, K

m = empirical constant between 1 and 2 (in this text we shall assume for simplicity that $m = 1$)

At room temperature (300 K)

$$V_T \equiv \frac{kT}{q} \approx 25 \text{ mV} \tag{1.2-2}$$

Equation (1.2-1) states that if v_D is negative with magnitude much greater than kT/q, the current i_D is the reverse saturation current $-I_o$. This reverse current $-I_o$ is a function of material, geometry, and temperature. If, however, v_D is positive and much larger than kT/q, the forward current is

$$i_D \approx I_o \epsilon^{qv_D/kT} = I_o \epsilon^{v_D/V_T} \tag{1.2-3}$$

Equation (1.2-1) is sketched in Fig. 1.2-6a, for germanium and silicon.

The actual characteristic of a typical diode differs from the exponential curve because of various effects. At relatively large forward currents the ohmic resistance of the contacts and the semiconductor material effectively increases the forward resistance. In the reverse direction, surface leakage, which is the current along the surface of the silicon rather than through the junction between the p- and n-type regions, effectively decreases the reverse resistance. At large reverse voltages avalanche breakdown takes over. Other effects come into play over various portions of the diode characteristic, but for most practical purposes they are negligible. The curves of Fig. 1.2-6a, when plotted to a suitable scale, appear to *turn on* at approximately 0.2 V for germanium and 0.7 V for silicon. (See Fig. 1.2-6b.)

For large-signal applications, the diode is often considered to behave in accordance with the straight-line approximations of Fig. 1.2-6c. Such approximations are called *piecewise-linear curves.*

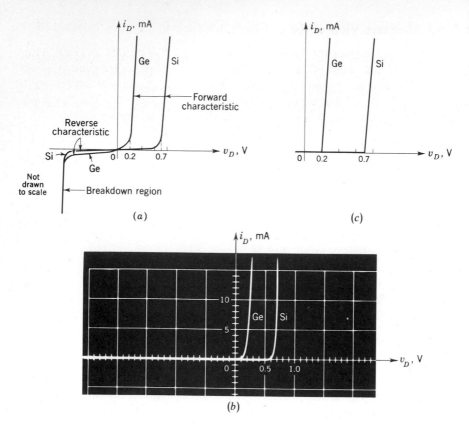

Figure 1.2-6 Diode characteristics: (a) actual characteristic; (b) oscillogram of diode characteristics; (c) straight-line (piecewise-linear) characteristics.

Considering the reverse characteristic, it is interesting to note that while germanium has a 0.2-V break in its forward characteristic as compared with 0.7 V for silicon, the reverse characteristics, shown in Fig. 1.2-7, indicate that at the same reverse diode voltage V_r, the silicon diode draws considerably less current than the germanium diode.

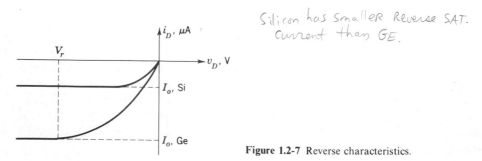

Silicon has smaller Reverse SAT. Current than GE.

Figure 1.2-7 Reverse characteristics.

1.2-1 An Alternate View of the vi Characteristic

Equation (1.2-3) describes the forward characteristic of the diode. A useful relationship can be found from this equation by considering the currents and voltages at two different operating points. Assume that current i_{D1} flows with corresponding diode voltage v_{D1}. If the current changes to i_{D2}, we wish to find the new diode voltage v_{D2}. We assume $v_D \gg kT/q$. Then (1.2-3) yields the two equations

$$i_{D1} = I_o \, \epsilon^{v_{D1}/V_T} \tag{1.2-4a}$$

$$i_{D2} = I_o \, \epsilon^{v_{D2}/V_T} \tag{1.2-4b}$$

Dividing these two equations, we get

$$\frac{i_{D1}}{i_{D2}} = \epsilon^{(v_{D1} - v_{D2})/V_T} \tag{1.2-4c}$$

Taking natural logarithms of both sides yields the relation of interest:

$$v_{D1} - v_{D2} = V_T \ln \frac{i_{D1}}{i_{D2}} \tag{1.2-4d}$$

The usefulness of this equation can be seen by considering that the current ratio $i_{D1}/i_{D2} = 10$. Then

$$v_{D1} - v_{D2} = 25 \ln 10 \approx 60 \text{ mV}$$

Thus, if we know that a particular diode has a voltage of 0.7 V when the current is 0.5 mA, the voltage will be 0.76 V when the current is 5 mA. If the current ratio is 2, corresponding to a doubling of the current, then $v_{D1} - v_{D2} = 25 \ln 2 = 17$ mV, an often negligible amount.

It is the small change in voltage for a very large change in current that makes the characteristics shown in Fig. 1.2-6a appear to be almost vertical lines after turn-on. For example, in Fig. 1.2-6b, the diode current range is 1 to 15 mA, while the voltage drop across the silicon diode remains approximately 0.7 V. If the diode were used in an application requiring perhaps 100 mA rather than 10 mA, the current scale shown in Fig. 1.2-6c would be multiplied by a factor of 10. Accordingly the approximate turn-on voltage would appear to be increased by about 60 mV to 0.76 V. Thus, on a linear current scale the diode characteristic does not appear to be exponential but piecewise-linear, as in Fig. 1.2-6c.

1.2-2 Piecewise-linear Equivalent Circuit

In many applications the actual diode characteristic can be approximated by a piecewise-linear characteristic, as shown in Fig. 1.2-8a. When this is the case, the diode can be replaced by an equivalent circuit consisting of a battery V_F in series with an ideal diode, as shown in Fig. 1.2-8b. In accordance with the discussion of Sec. 1.2-1, we take cognizance of the fact that the value of V_F employed is a function of the current level. However, for most of our applications we shall assume that $V_F = 0.7$ V.

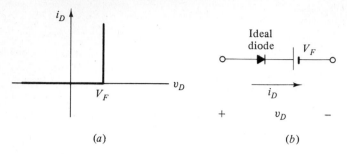

Figure 1.2-8 Piecewise-linear equivalents: (a) vi characteristic; (b) circuit.

For example, in the half-wave rectifier circuit shown in Fig. 1.1-3 if the diode were replaced by the piecewise-linear equivalent circuit of Fig. 1.2-8b, we would find that the diode current is

$$i_D = \begin{cases} \dfrac{1}{r_i + R_L}(V_{im}\cos\omega_0 t - V_F) & V_{im}\cos\omega_0 t \geq V_F \\ 0 & V_{im}\cos\omega_0 t < V_F \end{cases} \quad (1.2\text{-}5)$$

1.3 ANALYSIS OF SIMPLE DIODE CIRCUITS; THE DC LOAD LINE

In this section we consider the behavior of basic circuits consisting of independent sources, diodes, and resistors. The circuit to be analyzed is the simple half-wave rectifier shown in Fig. 1.3-1. This circuit was analyzed in Example 1.1-1, assuming an ideal diode. In this section the actual diode characteristics are taken into account.

The idea behind the graphical analysis is based on two simple facts:

1. The behavior of the diode is completely characterized at low frequencies by its vi characteristic, which is usually available graphically in the manufacturer's specifications or can be easily measured.
2. The other elements of the circuit, being linear, can be replaced by a Thevenin equivalent circuit as seen from the diode terminals.

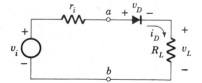

Figure 1.3-1 Half-wave rectifier.

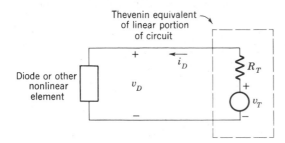

Thevenin equivalent
of linear portion
of circuit

Diode or other
nonlinear
element

Figure 1.3-2 General circuit containing a nonlinear element.

Let us consider the corresponding two parts of the circuit, as shown pictorially in Fig. 1.3-2. The terminal relations for the two parts can be written

Nonlinear element:
$$i_D = f(v_D) \qquad\qquad (1.3\text{-}1)$$

Thevenin equivalent:
$$v_D = v_T - i_D R_T \qquad\qquad (1.3\text{-}2)$$

We have two equations with two unknowns, v_D and i_D. When the two parts of the circuit are connected, these two relations are satisfied simultaneously and the circuit will operate at the point given by the solution of the equations. This solution can be arrived at analytically if the functional form of the vi characteristic of the nonlinear element is known. If, for example, the element is a silicon diode, (1.2-1) can be used as the nonlinear relation and the solution found. Because of the exponential nature of (1.2-1), it is clear that this is not going to be a routine calculation and may in fact involve considerable labor. In some cases this may be justifiable, but in most it is not, for two reasons: the accuracy *required* in most cases is not great, so that simpler or approximate methods are justified; and the accuracy *achievable* by a detailed calculation is often meaningless because the behavior of most diodes differs from the theoretical characteristic given by (1.2-1), and in fact large variations will often be encountered in batches of diodes of the same type.

Problems of this type are most often solved graphically by plotting (1.3-1) and (1.3-2) on the same set of axes. The intersection of the two resulting curves gives the operating point of the circuit. A typical plot is shown in Fig. 1.3-3. The straight-line characteristic of the Thevenin circuit (usually called the *dc load line*) is drawn for a dc Thevenin voltage of 1.5 V and a Thevenin resistance of 50 Ω. The intersection of the dc load line (1.3-2) and the diode characteristic (1.3-1) gives the operating point (often called the *quiescent* or Q point) for these conditions. The intersection of these two curves occurs at the point Q_1, where $v_D \approx 0.7$ V, $i_D \approx 15$ mA. If the Thevenin voltage v_T changes to 2 V, the load line shifts horizontally 0.5 V to the right as shown and the operating point moves to Q_2 on the graph. As long as R_T remains constant, any change in v_T is accounted for by a simple horizontal shift of the load line.

Thus, if v_T is sinusoidal, that is, $V_{Tm} \sin \omega t$, the *current* waveform can be found by choosing several points on the sine wave and drawing the corresponding load lines to find the resulting currents. This technique is illustrated in Fig. 1.3-4 for

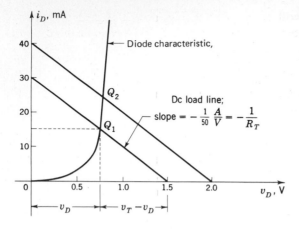

Figure 1.3-3 Typical diode characteristic and load lines.

$V_{Tm} = 1.5$ V. The nonlinear nature of the diode characteristic causes the current waveform to be distorted, i.e., to have a shape different from v_T. This waveform should be compared with that of Fig. 1.1-5, which shows the result when an ideal diode is used.

Returning to the original circuit, Fig. 1.3-1, we observe that for this simple case the Thevenin voltage v_T is the same as the source voltage v_i and $R_T = r_i + R_L$, so that all one need do is multiply the current in Fig. 1.3-4 by R_L to

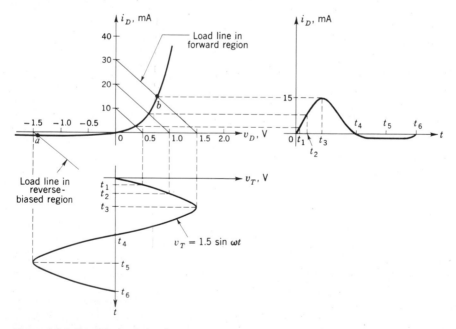

Figure 1.3-4 Graphical solution for current when sinusoidal voltage is applied.

obtain a plot of the response voltage v_L. Note that the operating path of the diode with this signal applied lies between points a and b, as shown on the diode vi characteristic.

1.4 SMALL-SIGNAL ANALYSIS; THE CONCEPT OF DYNAMIC RESISTANCE

The total peak-to-peak variation (swing) of the signal is often a small fraction of its *dc* component, hence the name *small signal*. When this condition occurs, a combined graphical-analytical approach can be employed which greatly simplifies the analysis. This approach is illustrated using the circuit of Fig. 1.3-1 with a dc voltage added to v_i, so that

$$v_T = V_{dc} + v_i = V_{dc} + V_{im} \sin \omega t \qquad \text{where } V_{im} \ll V_{dc} \qquad (1.4\text{-}1)$$

The voltage V_{dc} is usually called the *bias voltage*.

The technique used here is based on the fact that the inequality in (1.4-1) forces the circuit to operate over a very small region of its possible operating range. For most practical purposes the diode characteristic can then be considered linear in this region and the diode replaced by a resistance. The resulting linear circuit is then amenable to standard circuit-analysis techniques.

Keeping in mind that the circuit is to be linearized, we first determine the operating point for $v_T = V_{dc}$, that is, $V_{im} = 0$. This is the quiescent (Q) point. The procedure here is precisely the same as that used in Sec. 1.3, and the pertinent graph is repeated in Fig. 1.4-1 for $V_{dc} = 1.5$ V and $r_i + R_L = 50$ Ω.

The construction necessary to determine the exact current waveform as in Sec. 1.3 (Fig. 1.3-4) is shown on the graph. Only that portion of the diode characteristic lying between points a and b is of importance in determining the response. If the characteristic is reasonably linear between these points, it can be replaced by a straight line for purposes of calculating the ac component. In order to focus attention on the ac response, we construct a new set of axes with their origin at the Q point. The variables associated with these axes are then

Current: $\qquad\qquad\qquad\qquad i_d = i_D - I_{DQ}$ $\qquad\qquad\qquad\qquad\qquad$ (1.4-2a)

Voltage: $\qquad\qquad\qquad\qquad v_d = v_D - V_{DQ}$ $\qquad\qquad\qquad\qquad\qquad$ (1.4-2b)

That portion of the graph (Fig. 1.4-1) of interest is shown in Fig. 1.4-2, with the new variables drawn on expanded scales. The operating path ab is assumed to be linear, and passes through the origin. This is equivalent to replacing the diode by a resistance of value equal to the inverse slope of line ab. This is called the *dynamic resistance* r_d of the diode, and can be found by evaluating the inverse slope of the diode characteristic at the Q point. Hence

dynamic
resistance $\qquad\qquad r_d = \dfrac{\Delta v_D}{\Delta i_D}\bigg|_{Q\text{ point}}$ $\qquad\qquad\qquad\qquad$ (1.4-3)

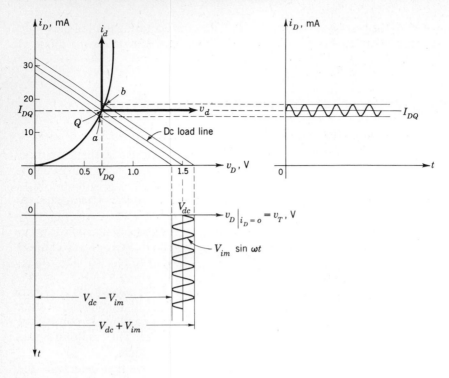

Figure 1.4-1 Graphical determination of load current.

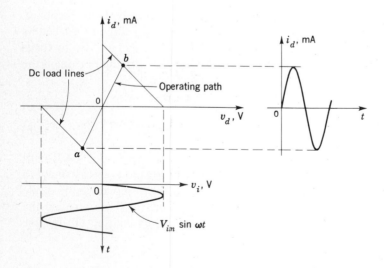

Figure 1.4-2 Graphical interpretation of auxiliary variables.

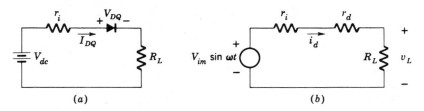

(a)   (b)

Figure 1.4-3 Diode circuit considered as two separate circuits: (a) circuit for calculating dc operating point; (b) circuit for calculating small-signal ac component.

Once r_d has been determined, any of the circuit variables (for small-signal ac operation only) can be calculated by simple application of Ohm's law.

The original circuit can be considered to be two separate circuits, as shown in Fig. 1.4-3. Figure 1.4-3a is used to find I_{DQ} and V_{DQ} (the quiescent operating point) and Fig. 1.4-3b to find i_d and v_d (the small-signal components). The total diode current and voltage can then be found using (1.4-2a) and (1.4-2b).

The development which led to the equivalent circuits of Fig. 1.4-3 can also be carried through analytically, using a Taylor-series expansion of the diode characteristic at the Q point. This is an often used engineering approximation. The diode vi characteristic is given by

$$i_D = f(v_D) \tag{1.3-1}$$

For small signals and no distortion

$$i_D = I_{DQ} + i_d \quad \text{and} \quad v_D = V_{DQ} + v_d \tag{1.4-4}$$

where

$$|i_d| \ll I_{DQ} \quad \text{and} \quad |v_d| \ll V_{DQ}$$

Then (1.3-1) becomes

$$I_{DQ} + i_d = f(V_{DQ} + v_d) \tag{1.4-5}$$

Taylor's series, from which $f(x + \Delta x)$ can be found, given $f(x)$, is

$$f(x + \Delta x) = f(x) + \Delta x f'(x) + \text{higher-order terms} \tag{1.4-6}$$

We neglect the higher-order terms and identify x with V_{DQ} and Δx with v_d, so that

$$i_D = I_{DQ} + i_d \approx f(V_{DQ}) + v_d \frac{di_D}{dv_D}\bigg|_{Q\,\text{point}} \tag{1.4-7}$$

Noting that $f(V_{DQ}) = I_{DQ}$, we see that this simplifies to

$$i_d \approx v_d \frac{di_D}{dv_D}\bigg|_{Q\,\text{point}} \tag{1.4-8}$$

and finally

$$\frac{v_d}{i_d} \approx \frac{dv_D}{di_D}\bigg|_{Q\,\text{point}} \approx \frac{\Delta v_D}{\Delta i_D}\bigg|_{Q\,\text{point}} = r_d \tag{1.4-9}$$

Kirchhoff's voltage law (KVL) for the circuit of Fig. 1.3-1 is

$$v_T = v_D + i_D R_T \qquad \text{where } R_T = r_i + R_L \qquad (1.4\text{-}10)$$

When we substitute the small-signal definitions (1.4-1) and (1.4-4) into (1.4-10), we obtain

$$V_{dc} + v_i = V_{DQ} + v_d + I_{DQ} R_T + i_d R_T \qquad (1.4\text{-}11)$$

Since, for the assumed small-signal conditions with no distortion, v_i, v_d, and i_d are all zero-average time-varying signals and V_{dc}, V_{DQ}, and I_{DQ} are constants, (1.4-11) can be separated into a dc and an ac equation

$$V_{dc} = V_{DQ} + I_{DQ} R_T \qquad (1.4\text{-}12)$$

and

$$v_i = v_d + i_d R_T \qquad (1.4\text{-}13)$$

Finally, using (1.4-9) in (1.4-13), we get

$$v_i = i_d (r_d + R_T) \qquad (1.4\text{-}14)$$

Equations (1.4-12) and (1.4-14) describe the dc and ac equivalent circuits of Fig. 1.4-3. The dc calculation is performed graphically, using the diode characteristic, while the small-signal analysis is carried out using Ohm's law, with r_d evaluated from the diode characteristic at the Q point.

Calculation of r_d An analytic expression for the dynamic resistance of a silicon diode in the forward direction can be found by differentiating the diode equation (1.2-1), inverting the result, and evaluating r_d at the operating point as follows:

$$i_D = I_o(\epsilon^{v_D/V_T} - 1) \approx I_o \epsilon^{v_D/V_T} \qquad (1.2\text{-}1)$$

$$\frac{di_D}{dv_D} = \frac{1}{V_T} I_o \epsilon^{v_D/V_T} \approx \frac{i_D}{V_T}$$

and

$$r_d = \frac{dv_D}{di_D}\bigg|_{Q \text{ point}} \approx \frac{V_T}{I_{DQ}} \approx \frac{25 \text{ mV}}{I_{DQ}} \qquad \begin{array}{l} \text{at } T = 300 \text{ K} \\ I_{DQ} \text{ in A} \end{array} \qquad (1.4\text{-}15)$$

Typically, the dynamic resistance of a junction diode operating at 1 mA dc is 25 Ω.

Note that the preceding analysis is valid only if the operating portion of the diode characteristic (a to b in Fig. 1.4-1) can be considered a straight line. (If this assumption is not met, the current and voltage waveforms will be distorted.) The final results, the load current and voltage, are found by superimposing the responses of the circuits of Fig. 1.4-3a and b:

$$i_D = I_{DQ} + i_d = I_{DQ} + \frac{V_{im}}{r_i + r_d + R_L} \sin \omega t$$

and

$$v_L = R_L i_D$$

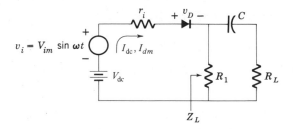

Figure 1.4-4 Diode circuit with a reactive element.

Reactive elements When small-signal conditions hold, it is a simple matter to take into account reactive elements such as the RC filter shown in Fig. 1.4-4. Clearly, the capacitor can have no effect on the operating point, and so the dc calculation is unaltered. Also, the slope of the diode characteristic $1/r_d$ at the Q point does not change, and to determine the ac current and diode voltage one can use Ohm's law to obtain

$$I_{dm} = \frac{V_{im}}{|r_i + r_d + Z_L|} \tag{1.4-16}$$

where I_{dm} and V_{im} denote the peak current and voltage amplitude, and Z_L is the complex impedance. This is illustrated in the example which follows.

Example 1.4-1 A junction diode is used in the circuit of Fig. 1.4-4 with

$$V_{dc} = 1.5 \text{ V} \qquad V_{im} = 20 \text{ mV} \qquad r_i = 10 \text{ } \Omega \qquad R_1 = 90 \text{ } \Omega$$

$$R_L = 200 \text{ } \Omega \qquad C = 100 \text{ } \mu\text{F} \qquad \omega = 10^4 \text{ rad/s}$$

Find the voltage across R_L.

Solution To find the Q point, the dc load line is drawn through the point $v_D - 1.5$ V, with slope $-1/(r_i + R_1) = -0.01$. The intersection of this load line and the diode characteristic occurs at 7.5 mA and 0.75 V (Fig. 1.4-5). The location of the Q point on the curve and the size of the ac signal indicate that small-signal theory will be applicable.

From (1.4-15)

$$r_d = \frac{25 \times 10^{-3}}{7.5 \times 10^{-3}} = 3.3 \text{ } \Omega$$

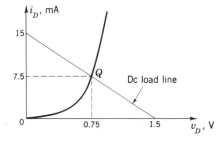

Figure 1.4-5 Graphical evaluation of Q point for the circuit of Fig. 1.4-4.

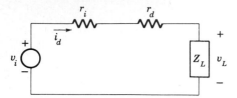

Figure 1.4-6 Small-signal equivalent circuit.

Since $|X_c| = 1/\omega C \approx 1\ \Omega$, the capacitor is seen to have negligible impedance in comparison with R_L. Taking this into account, we have $Z_L \approx R_1 \| R_L = 62\ \Omega$.† The small-signal circuit analogous to Fig. 1.4-3b takes the form shown in Fig. 1.4-6. Using the values obtained for r_d and Z_L, we get

$$V_{L,\ dc} = 0$$

and $\qquad V_{Lm} = I_{dm}|Z_L| = \dfrac{V_{im}|Z_L|}{|r_i + r_d + Z_L|} \approx \dfrac{(20)(62)}{10 + 3.3 + 62} \approx 17\ \text{mV} \qquad$ ///

1.5 SMALL-SIGNAL ANALYSIS; THE AC LOAD LINE

The circuit of Fig. 1.4-4 was described in Sec. 1.4 by a combination of graphical and analytical methods. By a simple extension of the dc-load-line concept it is possible to perform both the dc and ac analyses graphically as long as the reactance of the capacitor is negligible. This procedure leads to the concept of the *ac load line*, which, while not often used in practice for diode circuits, is often used to analyze and design transistor circuits. Since the concept is easier to grasp in terms of the simple diode, we introduce it at this point.

For the circuit of Fig. 1.4-4, the dc load line and Q point are obtained as shown in Fig. 1.5-1. The dc conditions are not influenced by that part of the circuit consisting of the capacitor C and load resistance R_L because of the dc blocking action of the capacitor. The slope of the dc load line is therefore determined by the resistance $r_i + R_1$. When an ac signal is present (assume that the capacitor acts as a short circuit at the frequencies involved), the effective resistance as seen by the diode is $r_i + (R_1 \| R_L)$, which is the negative of the inverse slope of the ac load line. In order to draw this ac load line, we need only one point, since the slope is known. The point where the ac signal is zero is the easiest to obtain. This is simply the Q point. Thus the ac load line is drawn through the Q point with a slope $-1/[r_i + (R_1 \| R_L)]$, as shown in Fig. 1.5-1.

As the signal varies with time, the ac load line moves back and forth to define the operating path for the diode. Compare this with Fig. 1.4-1, where the dc load line moves back and forth. The difference between these two lies in the fact that the ac impedance is *not* the same as the dc resistance seen by the diode for the circuit being considered in this section.

The amplitude of the ac component of current is found using the graphical

---

† The notation $R_1 \| R_L$ is an abbreviation for "R_1 in parallel with R_L" and is used throughout this text.

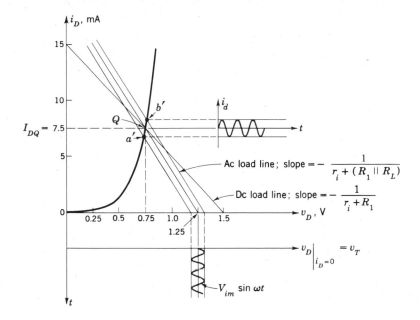

Figure 1.5-1 Graphical solution of the circuit of Fig. 1.4-4.

construction shown in Fig. 1.5-1. The operating path is along the segment $a'b'$ of the diode characteristic. This procedure will yield results identical with the analytical results obtained in (1.4-16) as long as segment $a'b'$ is approximately linear. As in Sec. 1.4, it is conceptually useful to superimpose a set of i_d-v_d axes on the curves of Fig. 1.5-1. This is left as an exercise.

The equations for the dc and ac load lines can be obtained simultaneously from the circuit of Fig. 1.4-4 by using KVL. We have

$$V_{dc} + v_i = i_D r_i + v_D + I_{DQ} R_1 + i_d(R_1 \| R_L) \tag{1.5-1}$$

Setting $i_D = i_d + I_{DQ}$ and $v_D = v_d + V_{DQ}$ and making the reasonable assumption that superposition applies yields

$$V_{dc} = I_{DQ}(r_i + R_1) + V_{DQ} \qquad \text{dc-load-line equation} \tag{1.5-2a}$$

and
$$v_i = i_d(r_i + (R_1 \| R_2)) + v_d \qquad \text{ac-load-line equation} \tag{1.5-2b}$$

1.6 THE DIODE ARRAY

The *diode array* has many applications, among which are the analog switch, multiplier, and phase detector. A schematic of a typical commercially available diode array (LM3019) is shown in Fig. 1.6-1. It consists of a diode quad and two isolated diodes, all manufactured at one time on the same silicon chip. The details of fabrication of such integrated circuits (ICs) will be discussed in Chap. 16.

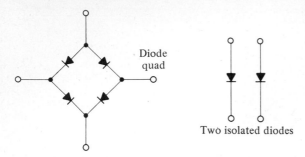

Diode
quad

Two isolated diodes

Figure 1.6-1 A diode array.

However, we may note that the diodes in the IC are all manufactured simultaneously and therefore have almost identical characteristics, while the characteristics of discrete diodes, selected at random, may differ widely. For example, the diodes in the array will have a maximum voltage difference of 5 mV between any two when the current through them is 1 mA. If two discrete diodes of the same type were selected at random, the voltage difference might be 25 to 50 mV.

We first discuss the analog switch shown in Fig. 1.6-2. The function of this circuit is to produce an output voltage v_L proportional to the input analog voltage v_i when the control voltage $v_C = V_{on}$ and have $v_L = 0$ when $v_C = V_{off}$. Thus, in this application the array is acting like an ordinary switch which is turned on and off by the control voltage.

To see that the circuit performs as expected let $v_C = V_{on}$. Then with $v_i = 0$ all four diodes are on, and each has a voltage drop V_F across it. Hence, the voltage drop from a to b is zero. V_{on} is adjusted so that the diodes remain on, even when v_i

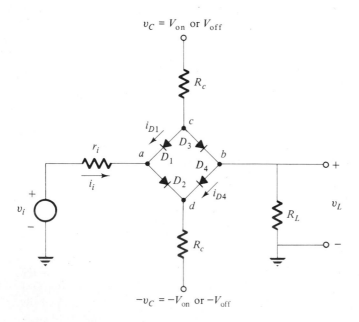

Figure 1.6-2 A four-diode analog switch.

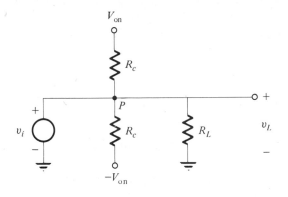

Figure 1.6-3 Equivalent circuit when $v_C = V_{on}$.

is not zero. If $r_i = 0$, it is clear that $v_L = v_i$ when $v_C = V_{on}$. However, even if r_i is not zero, v_L is proportional to v_i.

The voltage V_{off} is chosen to be sufficiently small to ensure that when $v_C - V_{off}$, all four diodes are off and therefore conduct no current. Hence, the current in R_L is zero and $v_L = 0$ V independent of v_i.

Quantitative analysis In practice the analog switch is driven from an op-amp (see Chap. 8) which has an output impedance much less than 1 Ω. Thus, it is reasonable to assume that $r_i = 0$ Ω.

A quantitative analysis of the circuit of Fig. 1.6-2 is rather involved unless we approximate each diode by the piecewise-linear equivalent circuit of the ideal diode described in Fig. 1.1-2. If now $v_C = V_{on}$, the voltage drop across each diode is zero and the resulting circuit is shown in Fig. 1.6-3. Note that points $a, b, c,$ and d in Fig. 1.6-2 now merge to a single point p and

$$v_L = v_i \tag{1.6-1}$$

Calculation of maximum allowable input voltage V_{im} The circuit ceases to operate properly when v_i increases to the point where $v_i = V_{im}$, a voltage sufficiently positive to cause D_1 and D_4 to turn off so that i_{D1} and i_{D4} are zero. This results in a complete opening of the diode bridge, thereby disconnecting the input v_i from the load R_L. The resulting equivalent circuit is shown in Fig. 1.6-4. Let us assume that the diodes are just cut off so that the diode voltages v_{ac} and v_{bd} are both equal to zero. Then, considering voltage drops with respect to ground, we have

$$v_a = v_d = V_{im} \tag{1.6-2a}$$

and
$$v_c = v_b = v_L = \frac{V_{on} R_L}{R_c + R_L} \tag{1.6-2b}$$

Setting $v_{ac} = v_a - v_c$ equal to zero yields

$$V_{im} = V_{on} \frac{R_L}{R_c + R_L} \tag{1.6-3}$$

Recall that this is the value of V_{im} at which diodes D_1 and D_4 are just cut off.

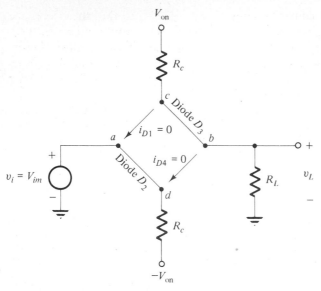

Figure 1.6-4 Equivalent circuit when $v_i = V_{im}$ and diodes D_1 and D_4 are at the edge of cutoff.

If $R_c \gg R_L$, (1.6-3) reduces to

$$V_{im} \approx \frac{R_L}{R_c} V_{on} \tag{1.6-4a}$$

However, if the diode quad is connected to the input of a noninverting op-amp (Sec. 8.2) the input impedance R_L of which is much greater than R_c, we shall have

$$V_{im} \approx V_{on} \tag{1.6-4b}$$

In either case if v_i is less than the value of V_{im} given in (1.6-3), the diode switch will operate to give $v_L = v_i$. If v_i exceeds V_{im}, the output voltage v_L will remain fixed at the value given in (1.6-2b). [It is readily shown that (1.6-3) also results if we set $v_i = -V_{im}$, in which case diodes D_2 and D_3 are off.]

The results obtained using the above analysis are only approximately correct since the equivalent circuit used to represent the diodes is not exact. For example, consider the diodes to be represented by the piecewise-linear equivalent circuit of Fig. 1.2-8. Then, referring to Fig. 1.6-2, we see that with $v_i = V_i$, a positive constant voltage, the currents in diodes D_1 and D_3 differ, D_3 having the larger current. The reason for this difference in currents is as follows. The current flow in the four diodes can be found using superposition. There are two sources v_C and v_i. v_C causes current in all the diodes D_1 to D_4 in such direction as to turn them on. Source v_i, however, supplies a current which tends to decrease the total current in diodes D_1 and D_4 and increase the total current in diodes D_2 and D_3. As a result diodes D_2 and D_3 carry larger currents than diodes D_1 and D_4. Hence, the voltage drop V_{F3} across D_3 is greater than the voltage drop V_{F1} across D_1. As a result v_{ba} is not equal to zero but is a nonlinear function of v_i. We neglect this second-order effect.

Determination of V_{off} The control voltage V_{off} is selected so that the diodes are operating in the reverse region (so as not to conduct current) but not beyond their Zener breakdown. If the diodes were reverse-biased beyond their Zener-breakdown voltage, they would conduct reverse currents which might be large enough to cause physical damage. Diodes used in ICs typically have a Zener breakdown when the reverse voltage is about $V_{ZB} = 6$ V.

In order to determine the permissible range of $v_C = V_{off}$ we note that to keep the diodes D_3 and D_4 off we must have $V_{cd} = V_c - V_d < 2(0.7) = 1.4$ V. In this mode of operation there is no current flowing in either resistor R_c; hence $2V_{off} = V_{cd} < 1.4$ V and $V_{off} < 0.7$ V.

When the diodes are cut off, there is a possibility that they may suffer a Zener breakdown if $v_i = V_{im}$ is too large. Since no current flows through r_i when the diodes are cut off, $V_a = V_{im}$, so that $V_{ac} = V_{im} - V_{off}$. To ensure that the diodes do not break down we set $V_{off} > V_{im} - V_{ZB}$, where V_{ZB} is the Zener-breakdown voltage of the diodes. Hence V_{off} must lie in the range

$$V_{im} - V_{ZB} < V_{off} < 0.7 \text{ V}$$

For example, if $V_{im} = 5$ V, we might choose $V_{off} = 0$ V. If this choice is made, D_2 will turn on when v_i exceeds 0.7 V and D_1 will turn on when v_i becomes less than -0.7 V. However, v_L will remain equal to 0 V independent of v_i because D_3 and D_4 are off and therefore the gate will remain cut off.

Example 1.6-1: Six-diode analog switch A six-diode analog switch using the full six-diode array circuit is shown in Fig. 1.6-5. The function of this circuit is

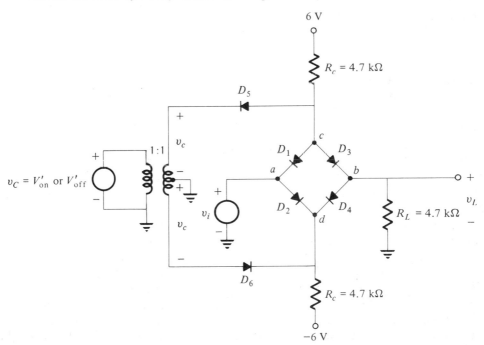

Figure 1.6-5 A six-diode analog switch.

the same as that of the four-diode analog switch shown in Fig. 1.6-2; i.e., when $v_C = V'_{on}$, the load voltage v_L is proportional to v_i, while when $v_C = V'_{off}$, the load voltage v_L is zero. Find (a) v_L as a function of v_i when $v_C = V'_{on}$, (b) the maximum allowable V_{im}, and (c) the minimum value of V'_{on} and the allowable range of V'_{off}.

SOLUTION (a and b) The transformer provides the means for switching v_C from V'_{on} to V'_{off} and back again. This was not shown in the circuit of Fig. 1.6-2 although a transformer could be used. The circuit is designed so that when $v_C = V'_{on}$, diodes D_5 and D_6 are *off* and diodes D_1 through D_4 are *on* providing that v_i is less than the value V_{im} given by (1.6-3). When we use (1.6-3) with $V_{on} = 6$ V the maximum V_{im} can be shown to be $V_{im} = 3$ V, and for $|v_i| < V_{im}$

$$v_L \approx v_i \tag{1.6-5}$$

(c) Since v_L will never exceed $V_{Lm} = V_{im} = 3$ V, the voltage at point c will not exceed $3 + 0.7 \approx 3.7$ V. Therefore diodes D_5 and D_6 will be *off* whenever $v_C = V'_{on} > 3.7 - V_{D5} = 3.7 - 0.7 = 3$ V. Diodes D_1 through D_4 will be *off* whenever the voltage drop from c to d in Fig. 1.6-5 is less than 1.4 V. Using KVL, we have

$$v_{cd} = V_{F5} + 2v_C + V_{F6} < 1.4$$

Assuming that $V_{F5} = V_{F6} = 0.7$ V, diodes D_1 through D_4 will be off if $v_C = V'_{off} < 0$ V. Since we do not want diodes D_1 through D_4 to be reverse-biased by more than the Zener-breakdown voltage V_{ZB}, we have

$$-V_{D1} = V_{im} - V'_{off} - V_{F5} < 6 \text{ V}$$

Hence
$$V'_{off} > V_{im} - 6.7 \text{ V}$$

Thus the permissible range of V'_{off} is

$$V_{im} - 6.7 \text{ V} < V'_{off} < 0 \text{ V} \qquad /// \tag{1.6-6}$$

Example 1.6-2: Use of the six-diode switch as a multiplier Show that the six-diode switch described in Example 1.6-1 can be used as a multiplier.

SOLUTION The circuit of a typical diode array multiplier is the same as the diode switch shown in Fig. 1.6-5. If we let v_i be a sine wave, such as $v_i = \sin 2\pi f_i t$, and v_C be another sinusoidal waveform, such as $v_C = \cos 2\pi f_c t$, then, if we assume that the switch turns on and off when v_C goes through zero, the control signal will switch v_i on and off at a rate equal to f_c. This is illustrated in Fig. 1.6-6a, where we have chosen $f_c < f_i$. In order to explain the multiplication process using this array, the switching operation is best represented as shown in Fig. 1.6-6b. Here the sinusoidal control voltage of frequency f_c is replaced by the square-wave switching function $S(t)$ of frequency f_c, which is either 1 or 0 V. When $S(t)$ is 1 V, $v_L = v_i(t)$. When

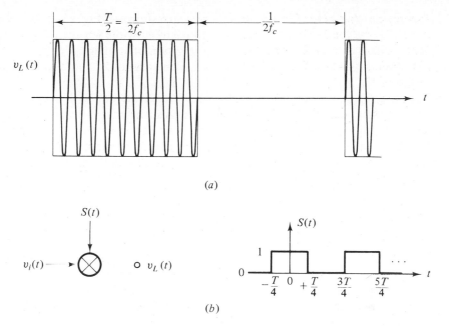

(a)

(b)

Figure 1.6-6 The diode switch as a multiplier: (a) $v_L(t)$ when $v_i = \sin 4\pi \times 10^7 t$ and $v_C = \sin 2\pi \times 10^6 t$; (b) equivalent representation of diode switch.

$S(t) = 0$ V, $v_L = 0$ V. Since $S(t)$ is periodic, it can be represented by the Fourier series

$$S(t) = \frac{1}{2} + \sum_{n-1}^{\infty} \frac{\sin n\pi/2}{n\pi/2} \cos 2\pi n f_c t \qquad (1.6\text{-}7)$$

Thus $v_L(t)$ is

$$v_L(t) = S(t)v_i(t) = \tfrac{1}{2} \sin 2\pi f_i t + \sum_{n=1}^{\infty} \frac{\sin n\pi/2}{n\pi/2} \sin 2\pi f_i t \cos 2\pi n f_c t \qquad (1.6\text{-}8)$$

Equation (1.6-8) shows that $v_L(t)$ consists of the original signal v_i, the product of the two signals, and products of $v_i(t)$ and harmonics of the original control signal. Hence if the switch is followed by a tuned circuit with center frequency f_i and bandwidth $B = 2f_c$, the voltage at the output of the filter will be $v'_L(t) = K \sin 2\pi f_i t \cos 2\pi f_c t$. Thus, the filter output voltage $v'_L(t)$ is proportional to the product of $v_C(t)$ and $v_i(t)$. ///

The circuit of Fig. 1.6-6 has several names, depending upon the design objective. For example, if we intend that $v_L(t) = S(t)v_i(t)$, the circuit is called a *multiplier*. If $f_i = f_c$ but there is a phase shift θ between $v_i(t)$ and $S(t)$, then if $v_L(t)$ is low-pass-filtered, the output of the filter is proportional to $\sin \theta$. The circuit is now called a *phase detector*. If the output voltage $v_L(t)$ is followed by a tuned circuit centered at $f_i + nf_c$, we call the circuit a *mixer* and say that we have shifted or heterodyned the frequency f_i to the new frequency $f_i + nf_c$.

1.7 FUNCTION GENERATION

The technique of function generation involves approximating the vi characteristic of a nonlinear device with connected straight-line segments and then utilizing diodes, resistors, and constant-voltage and constant-current sources to construct an equivalent circuit which realizes the piecewise-linear curve. The equivalent circuit is then amenable to standard circuit-analysis methods, especially for small signals, and graphical procedures are thus avoided.

An application of the use of function generators is to the solution of problems in which a nonlinear differential equation representing a physical system is to be solved using an analog computer. Nonlinear terms are approximated by piecewise-linear curves and synthesized. The complete problem is then solved electronically, and the result is plotted automatically or displayed on an oscilloscope.

Consider the three-segment approximation to the nonlinear characteristic shown in Fig. 1.7-1a as an example of the synthesis of a piecewise-linear circuit. Note that any nonlinear curve can be approximated as closely as desired by using one or more straight-line segments. For practical purposes, the number of line segments is kept as small as possible. The curve of Fig. 1.7-1b, consisting of only two segments, both semi-infinite, can be used to approximate the curve when the accuracy required permits the large error shown. It is synthesized as the circuit of Fig. 1.7-2a. The operation of the circuit is easily explained if one thinks of the input voltage as a dc source which changes slowly from a large negative value to a large positive value. Recall that the ideal diode will conduct only when the voltage $v_T - V_3$ is positive (forward bias) and is an open circuit when $v_T - V_3$ is negative (reverse bias); no current will flow unless $v_T \geq V_3$; thus $v_T < V_3$ gives segment 1 of the curve. Now when $v_T = V_3$ (the breakpoint), a change of state occurs. As soon as v_T becomes greater than V_3, the diode conducts, appearing as a short circuit. Thus, for all values of $v_T > V_3$, the circuit consists of resistor R_3 in series with voltage V_3, which furnishes segment 2 of the curve.

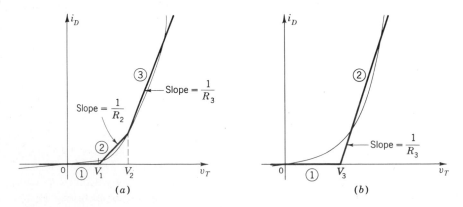

Figure 1.7-1 Piecewise-linear approximations: (a) three segments; (b) two segments.

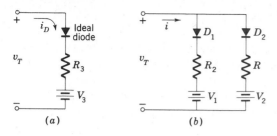

Figure 1.7-2 Piecewise-linear circuits: (a) synthesis of Fig. 1.7-1b; (b) synthesis of Fig. 1.7-1a.

This result can be used to synthesize the three-segment curve of Fig. 1.7-1a. The circuit of Fig. 1.7-2a synthesizes segments 1 and 2 (with R_3 replaced by R_2, and V_3 replaced by V_1). A breakpoint must still be provided at V_2, and the slope of the curve in region 3 must be $1/R_3$. Note that this slope is greater than that of segment 2, so that the equivalent resistance over this range must be less. This suggests that a parallel circuit be added to achieve a reduction in resistance. The circuit of Fig. 1.7-2b will produce the desired result. The left-hand branch is the same as the one previously considered and yields segments 1 and 2 of the curve. The right-hand branch will have no effect until $v_T = V_2$, where D_2 conducts. For $v_T > V_2$, D_2 and D_1 are both short circuits, and the resistance of the circuit is $R\|R_2$. In order to match the desired slope $1/R_3$,

$$\frac{1}{R_3} = \frac{1}{R_2} + \frac{1}{R} \qquad \text{or} \qquad R = \frac{R_2 R_3}{R_2 - R_3}$$

This completes the circuit.

It is important to distinguish between the piecewise-linear equivalent circuits under discussion here and the small-signal linear equivalent discussed in Sec. 1.4. The resistance values may differ considerably, because in the small-signal case slopes are measured at a particular operating point, whereas for the piecewise-linear case they are values averaged over relatively large ranges. The piecewise-linear circuit can be used to calculate *total* currents and voltages, while the small-signal equivalent circuit is restricted to small variations about the operating point.

Example 1.7-1 Synthesize a circuit to yield a piecewise-linear approximation to the function $y = \ln x$ for the range $0 < y < 3$.

SOLUTION A plot of $x = e^y$ for $0 < y < 3$ is shown in Fig. 1.7-3. A three-segment approximation will suffice, with breakpoints at

$$i = \begin{cases} 0 & v_T = 1 \\ 1.61 & v_T = 5 \end{cases}$$

and segment 3 must pass through the point $i = 3$, $v_T = 20$.

One possible synthesis procedure and circuit are described below. Referring to the piecewise-linear curve of Fig. 1.7-3, for v_T less than 1 V, the current

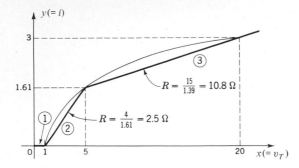

Figure 1.7-3 The function $x = e^y$ and its piecewise-linear approximation.

i is zero, while from 1 to 5 V the slope is that of a 2.5-Ω resistance. This set of conditions is provided by the circuit of Fig. 1.7-4.

Above 5 V the resistance increases to 10.8 Ω. When synthesizing a circuit for the curve of Fig. 1.7-1a, it was seen that if the resistance decreased after a breakpoint, parallel circuits should be added. Reasoning this way, we try a circuit in series with the circuit of Fig. 1.7-4 to increase the resistance. This second circuit must be a short circuit for voltages less than 5 V (currents less than 1.61 A). A possible configuration is shown in Fig. 1.7-5. Observe that for $i < 1.61$ A, D_2 is a short circuit and the 1.61 A from the current source flows through D_2. When $i = 1.61$ A, the current i_{D_2}, in D_2, becomes zero, and when i exceeds 1.61 A, D_2 becomes an open circuit. Then, since D_1 is a short circuit, KVL yields

$$v_T = (i - 1.61)(8.3) + 2.5i + 1 \qquad \begin{matrix} i > 1.61 \text{ A} \\ v_T > 5 \text{ V} \end{matrix}$$

The piecewise-linear slope in this range is

$$\frac{\Delta v_T}{\Delta i} = 8.3 + 2.5 = 10.8 \ \Omega$$

as required.

Several practical problems arise when attempting to construct the circuit of Fig. 1.7-5 using real diodes and resistances because a real diode has a nonzero break voltage and an equivalent series resistance. In addition, the equivalent resistance *decreases* with current, which produces a characteristic with curvature opposite that desired.

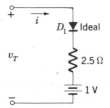

Figure 1.7-4 Circuit for realizing segments 1 and 2 of Fig. 1.7-3.

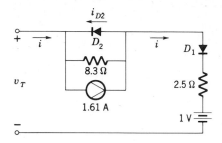

Figure 1.7-5 Complete circuit for Example 1.7-1.

To minimize these problems, we usually *scale* the variables, so that milliamperes, rather than amperes, flow in the circuit. To see how this transformation is made, refer to the original equation and rewrite it as

$$x = \epsilon^y$$

Now substitute

$$x = av_T \quad \text{and} \quad y = bi$$

Then

$$v_T = \frac{1}{a} \epsilon^{bi}$$

If $b = 10^3$, the current is given in milliamperes. By similarly adjusting the constant a the voltage can be scaled independently. Of course, all resistor values change accordingly. If $a = 1$ and $b = 10^3$, all resistors in the circuit of Fig. 1.7-5 are multiplied by 10^3 and currents in the new circuit are 10^{-3} times as large as those in the original circuit. Voltages throughout the new circuit are the same as in the original. In general, voltage levels should be chosen high enough to ensure that the diode voltage drops will not cause large errors.

A scaled version of the circuit of Fig. 1.7-5 is shown in Fig. 1.7-6. In this circuit the 1-V supply is eliminated, and D_1 is replaced by a silicon and germanium diode in series. The sum of the break voltages of these two diodes is approximately equal to 1 V, and their combined internal resistance is much less than 2.5 kΩ at these current levels, so that a reasonably straight-line characteristic would result between 1 and 1.61 mA. The current source is replaced by a 13.5-V battery in series with the 8.3-kΩ resistor.   ///

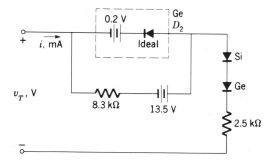

Figure 1.7-6 Practical version of the circuit of Fig. 1.7-5.

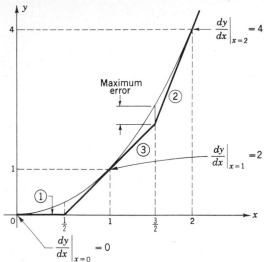

Figure 1.7-7 Piecewise-linear approximation to $y = x^2$.

Example 1.7-2 Using piecewise-linear circuits, find the simultaneous solution of the equations

$$y = x^2 \qquad 0 < x < 2$$

and

$$y = x \qquad 0 < x < 2$$

The solution is, of course, $x = 1$. Let us see how this solution can be obtained electronically.

SOLUTION First represent the equation $y = x^2$ by the three line segments shown in Fig. 1.7-7. A straightforward procedure for choosing the piecewise-linear segments is to divide up the interval into equal parts, as shown. We start with segments 1 and 2, the slopes being the slopes at the points considered (in this case $x = 0$ and $x = 2$). If the error between the piecewise-linear and the true nonlinear equation is too great, add a segment 3 as shown. The slope of segment 3 is the slope at $x = 1$. If the error is still too great, two more segments can be added at $x = \frac{1}{2}$ and $x = \frac{3}{2}$, etc. It must be noted that even with three segments the error could be reduced by simply placing segment 3 as a chord rather than as a tangent to the curve. (Since calculations are simplified by using a tangent, we do not worry about the reduced error here.)

Now let

$$y = i \text{ (mA)} \qquad \text{and} \qquad x = v \text{ (V)}$$

Then the circuit for $y = x^2$ can be synthesized as shown in Fig. 1.7-8.

The circuit for $y = x$ ($y = i$ in milliamperes; $x = v$ in volts) is simply a 1-kΩ resistance. To find the simultaneous solution to the equations we force

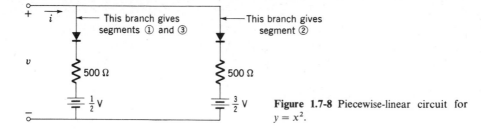

Figure 1.7-8 Piecewise-linear circuit for $y = x^2$.

$x (= v)$ to be the same in both circuits by connecting them in parallel, as shown in Fig. 1.7-9.

The voltage $v = V_1$ is then varied until $i = i_1$. At this point $v = V_1 (= x)$ and $i = i_1 (= y)$ represent the simultaneous solution of the equations. For this simple case $v = V_1 = 1$ V and $i = i_1 = 1$ mA are seen to yield the solution.

An alternative method involves connecting the 1-kΩ resistance in series with the $y = x^2$ network, thus forcing the current y to be the same in both circuits. The input voltage is then varied until the voltage $(= x)$ across each circuit is the same. ///

Example 1.7-3 Use the result of Example 1.7-2 to solve the nonlinear differential equation

$$\frac{dx}{dt} + x^2 = f(t) \qquad \text{with} \qquad \begin{matrix} x(0) - 0 \\ f(t) = 4 \end{matrix}$$

SOLUTION As in Example 1.7-2, we set x to be analogous to v, so that the circuit of Fig. 1.7-8 can be used for the second term of the equation. This yields a current in mA equal to $v^2 (= x^2)$. The first term can be represented by a 1000-μF capacitor, in which the current is

$$i = C\frac{dv}{dt} \qquad \text{or} \qquad y = i \text{ (mA)} = \frac{dv}{dt} = \frac{dx}{dt}$$

Thus we have currents analogous to the two variable terms of the differential equation. This suggests that we use KCL to add the two currents to a 4-mA constant current source, as shown in Fig. 1.7-10.

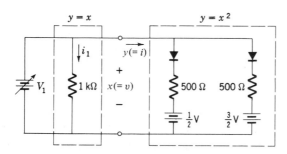

Figure 1.7-9 Connection for simultaneous solution of $y = x^2$ and $y = x$.

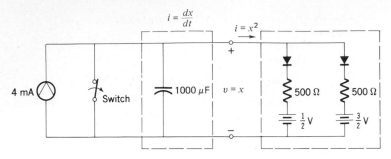

Figure 1.7-10 Circuit for solving $dx/dt + x^2 = 4$.

If the switch is opened at $t = 0$, with the capacitor uncharged, so that $v(0) = x(0) = 0$, then an oscilloscope placed across the circuit will record the solution $v(t) = x(t)$.

The reader will note that a nonzero initial condition can be accounted for by placing an initial charge on the capacitor. A different $f(t)$ would simply require a current source having the same waveform as $f(t)$.   ///

1.8 DIODE CAPACITANCE

In this section we describe certain effects which occur in diodes and which lead to capacitance elements in the circuit model for the diode. The size of these capacitances depends on the magnitude and polarity of the voltage applied to the diode as well as the type of junction formed during the manufacturing process. The actual capacitance is nonlinear but is usually approximated as a linear element.

Reverse bias, the transition capacitance Consider the reverse-biased pn-junction diode shown in Fig. 1.8-1a. When the diode operates in this mode, the holes in the p region and the electrons in the n region each move away from the junction, thereby forming a *depletion region* from which the charge carriers have been removed. The effective length L of the depletion region becomes larger as the reverse voltage V_R increases since the electric field increases proportionally with V_R.

Since the electrons and holes have moved away from the junction, the depletion region established in the p material becomes negatively charged while the depletion region established in the n material becomes positively charged. The junction with reverse bias therefore behaves like a capacitor, whose capacitance theoretically varies inversely with the voltage drop V_{NP} from N to P. Actually the capacitance C_R is inversely proportional to the $\frac{1}{2}$ or $\frac{1}{3}$ power of V_{NP}, depending on whether the device has an alloy junction or a grown junction. In a high-speed diode this capacitance is rather small, usually less than 5 pF. In high-current rectifier diodes, it can be as large as 500 pF.

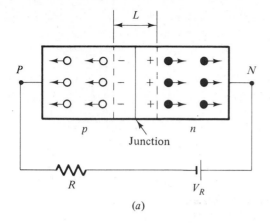

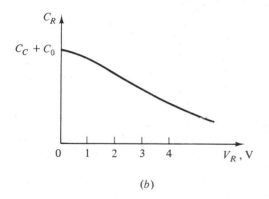

(a)

(b)

Figure **1.8-1** Diode capacitance, reverse bias: (a) pictorial view; (b) typical variation of capacitance with applied reverse voltage.

Varicap or varactor diodes are manufactured specifically to operate in the reverse mode. They can be designed for a capacitance of up to several hundred picofarads if required. An application of the use of such diodes is in frequency modulation (FM) circuits, where a reverse-biased diode is placed in parallel with an inductor. The resonant frequency of the resulting tuned circuit can be changed by varying V_R. Hence, if V_R is a voice signal, for example, the resonant frequency will be proportional to the voice-signal amplitude; i.e., the frequency will be modulated. Many FM systems are constructed using this principle.

An equation which relates the transition capacitance across a reverse-biased diode to the diode voltage V_R is

$$C_R \approx C_C + \frac{C_0}{(1 + 2V_R)^n} \tag{1.8-1}$$

where C_C = capacitance due to diode case
C_0 = capacitance of diode when $V_R = 0$
n = either $\frac{1}{2}$ or $\frac{1}{3}$

A graph of the diode capacitance as a function of V_R is shown in Fig. 1.8-1b. The nonlinear nature of C_R is usually neglected, and a constant value is used in calculations.

Forward bias, the storage capacitance When the diode is forward-biased, the depletion width L decreases and the transition capacitance increases relative to the capacitance value found in a reverse-biased diode. However, under the forward-bias condition a much larger capacitance effect occurs, which is modeled by an element called the *storage*, or *diffusion*, *capacitance*.

In Sec. 1.2 we described the mechanism of current flow as electrons moving from one hole (vacancy) to another. Let us assume that the average time that it takes for an electron to move between vacancies is τ s (τ is the average time considering both electron flow in the conduction band and electron flow in the valence band). Then the average current flow is $I_D = Q/\tau$, where Q is the average charge. However, from the diode equation (1.2-3) we have

$$I_D = \frac{Q}{\tau} = I_o \epsilon^{V_D/V_T} \qquad (1.8\text{-}2)$$

If we define the storage capacitance C_S as $C_S = dQ/dV_D$, we easily find that

$$C_S = \frac{I_D \tau}{V_T} \qquad (1.8\text{-}3)$$

Thus, the capacitance is directly proportional to the forward diode current and can be quite large. For example if $\tau = 1$ ns and $I_D = 1$ mA, then $C_S = 40$ pF. It is this capacitance that limits the switching speed in logic circuits using junction devices.

1.9 SCHOTTKY DIODES[4]

A relatively new component used in integrated circuits is the Schottky diode, which is formed by bonding a metal, such as platinum, to n-type silicon. These devices have negligible charge storage and are finding more and more use in high-speed switching applications.

A metal, e.g., platinum, acts as an acceptor material for electrons when bonded to n-type silicon. Thus, when the metal is connected to the n-type silicon, electrons initially diffuse from the silicon into the metal. As shown in Fig. 1.9-1a, this diffusion results in the n material's being depleted of electrons near the junction and therefore attaining a positive potential. When it becomes large enough, this positive voltage inhibits further diffusion of electrons. On the other hand, when a sufficiently large positive voltage is applied externally across the diode, as shown in Fig. 1.9-1b, the electrons in the n region see a positive potential in the metal side of the junction and electron flow resumes.

The reader should note the distinction being made between the *rectifying contact* described above and an *ohmic contact*, which is made to connect a p or n

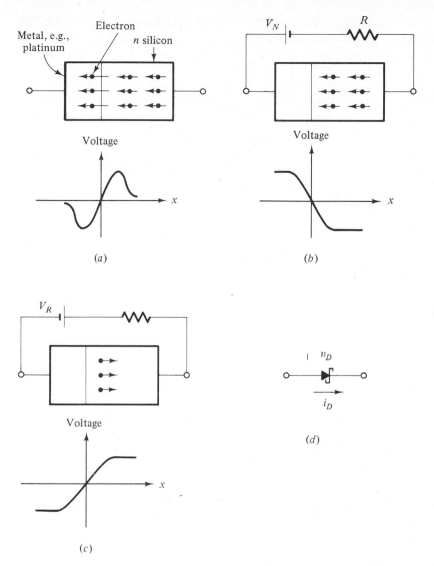

Figure 1.9-1 Schottky diodes: (*a*) potential distribution after initial diffusion; (*b*) potential distribution after application of a positive voltage; (*c*) potential distribution after application of a negative voltage; (*d*) circuit symbol.

region to an external circuit. In a rectifying contact, negligible current flows until V_N exceeds a certain minimum voltage $V_\gamma \cdot V_\gamma$ is the voltage needed to flatten the voltage curve shown in Fig. 1.9-1*a* (in a *pn* silicon diode the voltage V_γ is approximately 0.65 V). A small increase in the voltage V_N above V_γ produces a very large change in current. When the voltage applied to the diode is reversed so that the *n* material is made positive with respect to the platinum (or *p* material), the voltage on the n side of the junction increases (see Fig. 1.9-1c) beyond the level indicated in Fig. 1.9-1a and there is no current flow.

When an ohmic contact is made, there is no initial diffusion of electrons across the junction since the two materials are such that the density and energy of the electrons on both sides of the junction are the same.

It is interesting to note that there is a metal–n-silicon junction formed even in a pn diode since a metal wire must be attached to the n material using aluminum in order to make connection to the external circuit. To prevent this n-type silicon-aluminum connection from behaving like a diode the n-type silicon is doped so that it has a surplus of electrons at the end to be bonded to the metal. This excess-electron region is called an n^+ region. After the initial diffusion of electrons to the acceptor metal the n and n^+ regions take on the same characteristics as the metal and the contact becomes ohmic; i.e., it acts like a low resistance for voltages of either polarity.

When the Schottky diode is operated in the forward mode, current is due to electrons moving from the n-type silicon across the junction and through the metal. Since electrons move relatively unimpeded through metal, the recombination time τ [see Eqs. (1.8-2) and (1.8-3)] is very small, typically being of the order of 10 ps. This is several orders of magnitude less than that found using pn silicon diodes. For example, using (1.8-3), we find, with $I_D = 1$ mA, that $C_S = 0.4$ pF.

The circuit symbol for a Schottky diode is shown in Fig. 1.9-1d. The diode has a vi characteristic similar to that of an ordinary pn silicon diode except that the forward break voltage of the diode is $V_F \approx 0.3$ V.

1.10 ZENER DIODES

Zener, or breakdown, diodes are semiconductor pn junction diodes with controlled reverse-bias properties which make them extremely useful in many applications, especially as voltage-reference devices. A typical vi characteristic is shown in Fig. 1.10-1.

The forward characteristic is similar to that of the standard semiconductor diode. The reverse characteristic, however, exhibits a region in which the terminal voltage is almost independent of the diode current, as discussed in Sec. 1.2. The Zener voltage of any particular diode is controlled by the amount of doping applied in the manufacturing process. Typical values range from about 2 to 200 V, with power-handling capabilities up to 100 W.

In most applications, the Zener diode operates in this reverse-biased region. A typical application is the simple voltage-regulator circuit shown in Fig. 1.10-2. When this circuit is properly designed, the load voltage V_L remains at an essentially constant value, equal to the nominal Zener voltage, even though the input voltage V_{dc} and the load resistance R_L may vary over a wide range. The operation of the circuit can be explained qualitatively in terms of the vi characteristic of Fig. 1.10-1. If the input voltage increases, the diode tends to maintain a constant voltage across the load so that the voltage drop across r_i must increase. The

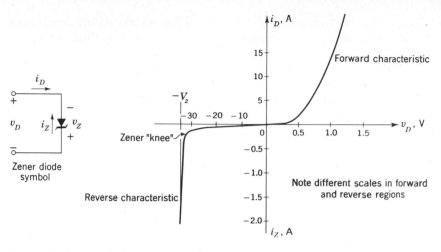

Zener diode
symbol

Reverse characteristic

Forward characteristic

Zener "knee"

Note different scales in forward
and reverse regions

Figure 1.10-1 Zener-diode circuit symbol and vi characteristic.

resultant increase in I_i flows through the diode, while the load current remains constant.

Now let the input voltage remain constant but decrease the load resistance. This requires an increase in load current. This extra current cannot come from the source since the drop in r_i, and therefore the source current, will not change as long as the diode is within its regulating range. The additional load current will, of course, come from a decrease in the Zener-diode current.

Both these regulating actions depend on the operation of the diode below the knee of the curve, where the diode voltage remains nearly constant, and the range of the regulation will depend on the value of r_i in this particular circuit. (Temperature characteristics will be discussed in Sec. 1.11.)

Example 1.10-1 A 7.2-V Zener diode is used in the circuit of Fig. 1.10-2, and the load current is to vary from 12 to 100 mA. Find the value of r_i required to maintain this load current if the supply voltage $V_{dc} = 12$ V.

SOLUTION For a shunt regulator such as this, an empirical factor of 10 percent of the maximum load current is used as the minimum Zener-diode current. Thus the minimum Zener-diode current for the stated conditions must be at least 10 mA. Applying Ohm's law to the circuit, we get

$$r_i = \frac{V_{dc} - V_L}{I_Z + I_L}$$

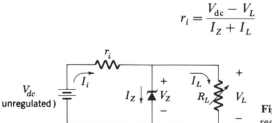

Figure 1.10-2 Zener-diode voltage regulator.

The voltage across r_i must remain at $12 - 7.2 = 4.8$ V over the regulating range. The minimum Zener current will occur when the load current is maximum, so that

$$r_i = \frac{V_{dc} - V_L}{I_{z, min} + I_{L, max}} = \frac{V_{dc} - V_L}{(1 + 0.1)I_{L, max}} = \frac{4.8}{0.11} = 43.5 \ \Omega$$

As the load current decreases with r_i set at this value, the Zener-diode current will increase, their sum remaining constant at 110 mA.

Note that if the load resistance should accidentally become open-circuited so that $I_L = 0$ and $I_Z = 110$ mA, the Zener diode must be capable of dissipating

$$P_z = (7.2)(110 \times 10^{-3}) \approx 0.8 \ \text{W}$$

to protect it from possible destruction due to excessive power dissipation.

Let us choose a 1-W Zener diode which has a voltage of 7.2 V when drawing 10 mA and a dynamic resistance r_d of 2 Ω. We shall calculate the output-voltage variation across the load, using graphical methods.

The vi characteristic for this diode is shown in Fig. 1.10-3. Also shown is the maximum power curve

$$P_Z = \frac{1}{T} \int_{-T/2}^{T/2} v_z i_z \, dt = V_{ZQ} I_{ZQ} = 1 \ \text{W}$$

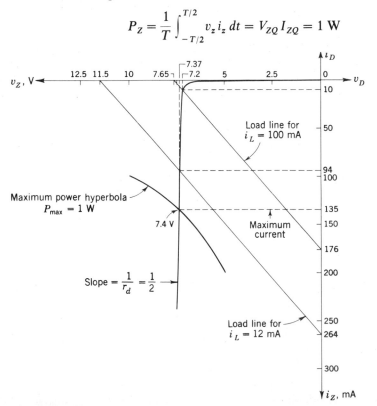

Figure 1.10-3 Zener-diode characteristic and load lines for Example 1.10-1.

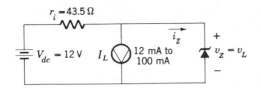

Figure 1.10-4 The circuit of Fig. 1.10-2 redrawn for clarity.

This is the equation of the hyperbola sketched on the vi characteristic. Operation beyond this hyperbola will result in diode power dissipation in excess of its maximum rating of 1 W.

The dc-load-line equations can be found by redrawing Fig. 1.10-2 as shown in Fig. 1.10-4. From this circuit we find the load-line equation

$$v_Z = 12 - 43.5(I_L + i_Z) \tag{1.10-1}$$

Thus
$$v_Z + i_Z(43.5) \approx \begin{cases} 11.5 & I_L = 12 \text{ mA} \\ 7.65 & I_L = 100 \text{ mA} \end{cases} \tag{1.10-2}$$

The two load lines of (1.10-2) are plotted in Fig. 1.10-3. From the graph, it is evident that the diode reverse voltage v_L and hence the load voltage v_L vary from 7.2 V when $I_L = 100$ mA to 7.37 V when $I_L = 12$ mA. Note that if the Zener diode were not present, the load voltage would vary (keeping $r_i = 43.5 \, \Omega$) from 7.65 V when $I_L = 100$ mA to 11.5 V when $I_L = 12$ mA. Thus the Zener diode is seen to have provided voltage regulation with respect to changes in load current. This occurs since the impedance seen by the load is small ($2 \, \Omega$) compared with the load resistance $R_L > (7.2 \text{ V})/(100 \text{ mA}) = 72 \, \Omega$.   ///

Manufacturers specify Zener diodes according to their Zener voltage and maximum power dissipation. The current drawn by the diode at its nominal Zener voltage is called the *test current* I_{ZT}. Typically the maximum power rating of the diode is 4 times the power dissipation at the Zener voltage, i.e.,

$$P_{Z,\,\text{max}} = 4I_{ZT}V_Z \tag{1.10-3}$$

Another parameter of interest to us is the current at the knee of the Zener characteristic. This *knee current* I_{Zk} is found to be approximately constant for a specified maximum dissipation and is independent of the rated Zener voltage. A listing of Zener-diode specifications is presented in Appendix C.

For the Zener diode required in Example 1.10-1, which we have characterized as a 7.2-V 1-W device, Fig. C.1-1 indicates that $I_{Zk} = \frac{1}{4}$ mA and, from (1.10-3), that $I_{ZT} = 35$ mA. Note that the minimum current expected to flow in the diode is 10 mA, which is much greater than I_{Zk}, and also that I_{ZT} lies between the minimum and maximum values of current expected. Thus our choice of diode is reasonable.

Example 1.10-2: Ripple reduction with the Zener regulator The 7.2-V Zener diode is used in a circuit similar to that shown in Fig. 1.10-2, with an ac ripple

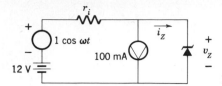

Figure 1.10-5 Zener regulator for Example 1.10-2.

voltage added to the unregulated dc voltage. As stated before, these voltages are typical of the output of a dc power supply. The load draws a current of 100 mA. The output of the unregulated supply can be represented as

$$v_T = 12 + 1 \cos \omega t \tag{1.10-4}$$

Find the source resistance r_i for proper operation and the peak-to-peak value of the ripple voltage present across the load.

SOLUTION The load-line equation for this problem is easily obtained if we first redraw Fig. 1.10-2 as shown in Fig. 1.10-5. Thus

$$v_Z = 12 + 1 \cos \omega t - r_i(0.1 + i_Z)$$

which is rearranged to yield

$$v_Z + r_i i_Z = 12 - 0.1 r_i + 1 \cos \omega t \tag{1.10-5}$$

The problem can now be solved analytically or graphically using the diode characteristic shown in Fig. 1.10-3. We use the graphical technique since it gives insight into the problem. The characteristic shown in Fig. 1.10-3 is redrawn in Fig. 1.10-6.

The range of possible values of source resistance r_i will be bounded because of the minimum and maximum allowable Zener currents. Thus, if $i_{Z, \min} = 10$ mA (which is 10 percent of the maximum current), then $v_{Z, \min} = 7.2$ V and from (1.10-5)

$$r_i < \left(\frac{12 - 7.2 + \cos \omega t}{0.1 + 0.01}\right)_{\min} = \frac{3.8}{0.11} = 34.5 \ \Omega \tag{1.10-6}$$

A lower bound on r_i is obtained by using the maximum diode voltage and current ratings. Thus, using (1.10-5) and

$$i_{Z, \max} = 135 \text{ mA} \qquad v_{Z, \max} \approx 7.4 \text{ V}$$

we find that the source resistor r_i must exceed

$$r_i > \left(\frac{12 - 7.4 + \cos \omega t}{0.1 + 0.135}\right)_{\max} = \frac{5.6}{0.235} \approx 24 \ \Omega \tag{1.10-7}$$

We arbitrarily choose $r_i = 32 \ \Omega$. The load-line equation (1.10-5) can now be plotted in Fig. 1.10-6 and the output ripple voltage determined. From the figure, the peak-to-peak ripple voltage across the load is $7.33 - 7.22 = 0.11$ V, compared with a 2-V input peak-to-peak ripple voltage.

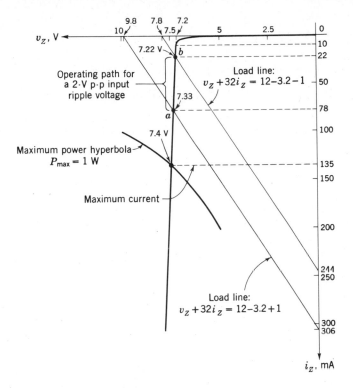

Figure 1.10-6 Zener characteristic and load lines for Example 1.10-2.

The output ripple can be calculated more easily by assuming that the circuit is linear for the 2-V input ripple voltage. (The validity of this assumption is borne out by the ripple operating path shown in Fig. 1.10-6.) This allows us to use superposition and consider the ac portion of the response separately. The ac circuit is shown in Fig. 1.10-7. In the circuit $r_d = 2\ \Omega$ represents the dynamic resistance of the diode, while the load is shown as a

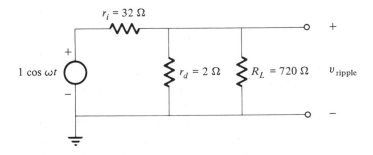

Figure 1.10-7 Ac circuit for ripple calculation.

720-Ω resistor [(7.2 V)/(100 mA)]. Clearly, we can neglect the load resistance, so that using the voltage-divider formula, we can find the output ripple as

$$v_{\text{ripple}} \approx \frac{2}{2+32}(2) = 0.118 \text{ V peak to peak}$$

This agrees well with the value found previously.   ///

1.11 TEMPERATURE EFFECTS

In designing diode circuits, especially at high power levels, the temperature dependence of the diode characteristics must be taken into account. This dependence results in a variation of the forward voltage V_F, the Zener voltage V_Z, and the reverse saturation current I_o. Furthermore, a significant increase in temperature is often followed by an increase in power dissipation, which causes the temperature of the diode to increase still further. This is called *thermal runaway*, and unless the heat developed within the diode case can be dissipated, physical damage to the diode may result. Some of the more important temperature effects are described in this section.

Diode voltage When a diode is operated in the forward direction, an increase in temperature results in a decrease in voltage. A relation which holds when the current is maintained constant is

$$v_D(T_1) - v_D(T_0) = -k(T_1 - T_0) \qquad i_D = \text{const} \qquad (1.11\text{-}1)$$

where T_0 = room temperature-25°C
  T_1 = new temperature of diode, °C
  $v_D(T_0)$ = diode voltage at room temperature, V
  $v_D(T_1)$ = diode voltage at T_1, V
  k = temperature coefficient, V/°C

A plot of k as a function of the forward diode voltage at room temperature is given in Fig. 1.11-1, which indicates that the temperature coefficient varies significantly with the diode voltage. However, it is standard engineering practice to assume that k is a constant. The values usually assumed are approximations found from Fig. 1.11-1 at the typical diode voltages, $v_D(T_0) = 0.7$ V for silicon, 0.3 V for Schottky, and 0.2 V for germanium. They are

$$k \approx \begin{cases} 2.5 \text{ mV/°C} & \text{germanium} \\ 2 \text{ mV/°C} & \text{silicon} \\ 1.5 \text{ mV/°C} & \text{Schottky} \end{cases} \qquad (1.11\text{-}2)$$

For example, if a silicon diode has a voltage drop of 0.7 V at 25°C with a diode current of 1 mA and the temperature of the diode is increased to 125°C, then if the diode current is maintained at 1 mA, (1.11-1) indicates that the diode voltage will decrease to 0.5 V. If the temperature had decreased to −75°C, the diode voltage would have increased to 0.9 V at 1 mA.

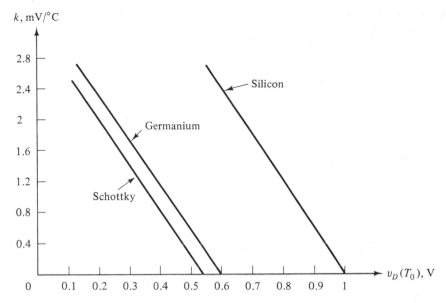

Figure 1.11-1 Temperature coefficients of diodes as a function of diode voltage at room temperature.

Reverse saturation current The reverse saturation current I_o is also a function of temperature. It is found in practice that in the neighborhood of room temperature, I_o approximately doubles for an increase in temperature of 10°C (similarly I_o will halve for a decrease in temperature of 10°C).

Zener voltage The change of the Zener voltage V_Z as a result of a change of temperature is found to be proportional to the value of the Zener voltage as well as to the change in temperature. This variation is usually expressed as a *temperature coefficient* (TC), where

$$\text{TC} = \frac{\Delta V_Z / V_Z}{\Delta T} \times 100\%/°C \qquad (1.11\text{-}3)$$

It is found that when V_Z is approximately 6 V, the TC is zero and V_Z is independent of temperature. If V_Z exceeds about 6 V, the TC is positive, and if V_Z is less than 6 V, the TC is negative. A typical TC ≈ 0.1 percent, i.e., the Zener voltage increases $0.001V_Z$ for each 1°C temperature rise.

The effect of heat dissipation on diode operation The practical problem of the maximum allowable heat dissipation in a *pn* junction diode will be considered in this section. The dissipation of electric power as heat in the diode causes the junction temperature to rise. This temperature rise must be kept within acceptable limits, or the diode will suffer physical damage. The thermal problem is easily handled by a simple thermal analog of Ohm's law, in which current is replaced by power, voltage by temperature, and electric resistance by thermal resistance θ.

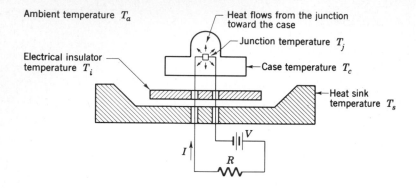

Figure 1.11-2 Diode mounted on heat sink.

Figure 1.11-2 shows a diode mounted on a heat sink. The diode is insulated electrically (not thermally) from the large heat sink. The operation of the system is as follows. With no electrical connection made to the diode, the junction temperature T_j will be the same as the ambient temperature T_a. When a signal is applied, power will be dissipated in the diode, causing the junction temperature to rise. The heat produced flows outward to the case and is then carried by conduction from the diode case to the heat sink. The heat sink has a large surface area, from which it radiates the heat to the surrounding environment.

If the power dissipated at the junction is constant and within the power-handling capabilities of the diode, after a sufficient time has elapsed, the system will reach a state of thermal equilibrium. Each of the elements will, in general, be at a different temperature. A good approximation is that the temperature rise is in proportion to the power dissipated at the junction. The constant of proportionality is called the *thermal resistance* θ.

An increase in the junction temperature above the case temperature is related to the power dissipated by the equation

$$T_j - T_c = \theta_{jc} P_j \qquad (1.11\text{-}4)$$

where $T_j - T_c$ = rise in junction temperature above case temperature, °C
$\qquad P_j$ = electric power dissipated at junction, W
$\qquad \theta_{jc}$ = thermal resistance between junction and case, °C/W

The thermal resistance is a function of the construction of the diode and case and is usually specified by the manufacturer.

Consider a diode in a case without any mounting, as shown in Fig. 1.11-3. The case and ambient temperatures differ by an amount equal to the product of P_j and the thermal resistance between the case and ambient θ_{ca}. Thus the increase in case temperature above ambient temperature is given by

$$T_c - T_a = \theta_{ca} P_j \qquad (1.11\text{-}5)$$

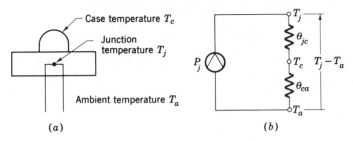

Figure 1.11-3 Diode and electrical analog of thermal system: (a) diode without mounting; (b) electrical analog.

The circuit shown in Fig. 1.11-3b represents an electrical analog of the thermal system of Fig. 1.11-3a, with the following analogs derived from (1.11-4) and (1.11-5):

Temperature drop $T_j - T_a$	Voltage drop
Power dissipation P_j	Constant-current generator
Thermal resistance $\theta_{jc} + \theta_{ca}$	Electric resistance

It is seen from Fig. 1.11-3 and (1.11-4) and (1.11-5) that

$$T_j = P_j\theta_{jc} + P_j\theta_{ca} + T_a \qquad (1.11\text{-}6)$$

This result can immediately be extended to cover the heat-conduction system of Fig. 1.11-2

$$T_j = P_j(\theta_{jc} + \theta_{cs} + \theta_{sa}) + T_a \qquad (1.11\text{-}7)$$

Here the thermal resistances not previously defined are

θ_{cs} = case-to-heat-sink (includes insulator) thermal resistance

θ_{sa} = heat-sink-to-ambient thermal resistance

In a practical design problem, use of this equation is based on the following conditions:

1. The maximum allowable junction temperature is furnished by the manufacturer. Typical values are about 100°C for germanium diodes and 150 to 200°C for silicon diodes.
2. The ambient temperature is an uncontrolled variable depending on the environment in which the equipment will ultimately be operated.
3. The power dissipated at the junction depends on the electric system and for time-varying currents and voltages is given by

$$P_j = \frac{1}{T}\int_0^T v_D(t)i_D(t)\, dt$$

For dc operation this is simply

$$P_j = V_D I_D$$

4. Once a particular diode has been chosen to meet electrical specifications, its thermal resistance θ_{jc} will be fixed. This figure is usually provided by the manufacturer.

Noting these facts and (1.11-6), we see that the case-to-ambient thermal resistance is the only variable available for adjustment in order to maintain the junction temperature at a safe value.

Solving (1.11-6) yields

$$\theta_{ca} = \frac{T_{j,\,max} - T_a}{P_j} - \theta_{jc} \qquad (1.11\text{-}8)$$

This expression is used to determine the maximum θ_{ca}, given all the factors on the right. Equation (1.11-7) can, of course, be used to determine any of the variables involved.

Example 1.11-1 A 50-W silicon Zener diode is to dissipate 10 W in a particular circuit. The maximum allowable junction temperature is 175°C, the ambient temperature is 50°C, and $\theta_{jc} = 2.4$°C/W. Find the maximum thermal resistance which can be provided between case and ambient so that the junction temperature will not exceed 175°C.

SOLUTION From (1.11-8)

$$\theta_{ca} = \frac{175 - 50}{10} - 2.4 = 10.1°\text{C/W}$$

A typical heat sink with mica insulator for this type of diode has $\theta_{ca} = 3.2$°C/W, which would be adequate for this application. If this heat sink is used, the actual junction temperature can be calculated from (1.11-6)

$$T_j = (10)(2.4 + 3.2) + 50 = 106°\text{C} \qquad ///$$

Derating curves The power rating of a diode is usually specified for an ambient temperature of 25°C. The maximum power that the diode can dissipate is determined by the junction temperature. Thus the power rating must be decreased as the ambient temperature rises to keep the junction temperature within a safe limit. The manufacturer usually provides derating curves which can be used to determine the maximum allowable power dissipation for a given case temperature.

A typical derating curve is shown in Fig. 1.11-4. It is seen that at case temperatures less than T_{co}, the diode can dissipate its maximum allowable power. At case temperatures exceeding T_{co}, the maximum allowable dissipation is decreased until

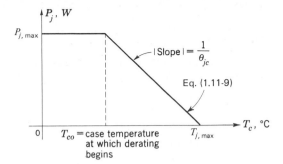

Figure 1.11-4 Diode derating curve.

$T_c = T_{j,\,max}$. At this maximum temperature the diode cannot dissipate any power without exceeding the maximum allowable junction temperature.

Using (1.11-4), we see that the rate of decrease of power with increased case temperature is θ_{jc}. Referring to Fig. 1.11-4, we note that when $P_j = 0$, $T_c = T_{j,\,max}$. Referring to (1.11-4) and replacing T_j by $T_{j,\,max}$, we have

$$T_c = T_{j,\,max} - \theta_{jc}P_j \tag{1.11-9}$$

Equation (1.11-9) represents the sloping portion of the derating curve. Again referring to Fig. 1.11-4, we see that when $T_c = T_{co}$, $P_j = P_{j,\,max}$. Substituting in (1.11-9) and solving for θ_{jc}, we have

$$\theta_{jc} = \frac{T_{j,\,max} - T_{co}}{P_{j,\,max}} \tag{1.11-10}$$

The case-to-ambient thermal resistance θ_{ca} can be determined graphically using (1.11-5) and the derating curve of Fig. 1.11-4, as illustrated in the following example.

Example 1.11-2 An 8-W Zener diode (derating curve shown in Fig. 1.11-5) is to dissipate 3.5 W in a particular circuit. The ambient temperature is 100°C. (a) Find θ_{ca} so that the transistor will not overheat. (b) Find T_c for these conditions. (c) If an infinite heat sink were available, how much power could this diode dissipate?

SOLUTION (a) We begin by plotting (1.11-5) on the derating curve. The two known points on (1.11-5) are $T_c = T_a = 100°C$ when $P_j = 0$ and the point formed by $P_j = 3.5$ W lying on the derating curve. Referring to (1.11-5), we see that when $T_c = 0$,

$$P_j = \frac{-T_a}{\theta_{ca}}$$

Thus from Fig. 1.11-5 we have $-T_a/\theta_{ca} \approx -11$, and therefore

$$\theta_{ca} \approx \tfrac{100}{11} \approx 9°C/W$$

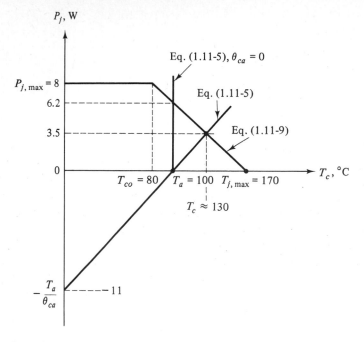

Figure 1.11-5 Graphical technique to find θ_{ca}. Two values of θ_{ca} are illustrated.

(b) The intersection of (1.11-5) and the derating curve at the operating power of 3.5 W yields the maximum allowable case temperature to avoid overheating. From Fig. 1.11-5, $T_c \approx 130°C$.

(c) If $T_a = 100°C$ and $\theta_{ca} = 0$ (an infinite heat sink), (1.11-5) can be plotted as a vertical line, as shown in Fig. 1.11-5. The intersection of this line and the derating curve yields the maximum allowable power dissipation. From the figure this is $P_j \approx 6.2$ W.   ///

1.12 MANUFACTURERS' SPECIFICATIONS

1.12-1 The Diode Rectifier

The diode characteristics commonly specified by manufacturers are listed as follows (see Figs. 1.12-1 to 1.12-3):

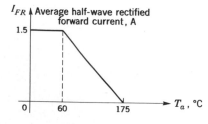

Figure 1.12-1 Current-derating curve for 1N566 silicon diode.

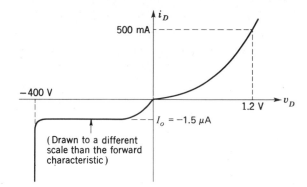

Figure 1.12-2 Forward and reverse diode characteristics.

Type: Silicon diode 1N566 (values are given at 25°C)
1. Peak inverse voltage (PIV) = − 400 V
2. Maximum reverse current I_o (at PIV) = 1.5 μA
3. Maximum dc forward voltage = 1.2 V at 500 mA
4. Average half-wave-rectified forward current I_{FR} = 1.5 A
5. Maximum junction temperature $T_{j,\,max}$ = 175°C
6. Current derating curve as given in Fig. 1.12-1

Characteristics 1 to 3 are most easily explained by referring to Fig. 1.12-2.

Characteristic 1 The PIV is the maximum allowable negative voltage that can be applied to the diode before it will break down. This voltage is therefore somewhat less than the Zener voltage, but in practice we do not refer to the Zener voltage of a diode which is intended to be used as a rectifier.

Characteristic 2 The maximum reverse saturation current I_o is 1.5 μA. Thus, when the diode is used in a rectifying circuit, the maximum negative current through it is 1.5 μA.

Characteristic 3 If a dc current of 500 mA is passed through the diode, the voltage drop across it will not exceed 1.2 V.

Characteristics 4 to 6 can be explained by referring to Figs. 1.12-3 and 1.12-1.

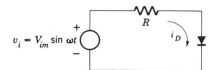

Figure 1.12-3 Half-wave diode rectifier.

Characteristic 4 When the input voltage v_i in Fig. 1.12-3 is sinusoidal, as indicated in the figure, the average half-wave-rectified forward current is

$$I_{FR} = \frac{V_{im}}{\pi R}$$

Characteristics 5 and 6 Figure 1.12-1 indicates that if the ambient temperature is less than 60°C,

$$I_{FR} = \frac{V_{im}}{\pi R} \leq 1.5 \text{ A}$$

At temperatures greater than 60°C, this maximum current is linearly derated until the ambient temperature is the same as the maximum junction temperature of 175°C, at which point $I_{FR} = 0$.

In designing diode circuits, good engineering practice dictates the use of a 10 to 20 percent safety factor on all published maximum ratings, to take variations between units into account.

1.12-2 The Zener Diode

Type: 18-V Zener diode 1N2816
1. Nominal reference voltage $V_{ZT} = 18$ V
2. Tolerance = 5%
3. Maximum dissipation (at 25°C) = 50 W
4. Test current $I_{ZT} = 700$ mA
5. Dynamic impedance at I_{ZT}, $R_{ZT} = 2 \ \Omega$
6. Knee current $I_{Zk} = 5$ mA
7. Dynamic impedance at I_{Zk}, $R_{Zk} = 80 \ \Omega$
8. Maximum junction temperature = 150°C
9. Temperature coefficient TC = 0.075%/°C

The characteristics are easily explained using Figs. 1.12-4 to 1.12-6. Figure 1.12-4 illustrates items 1, 2, and 4. Notice that an 18-V Zener diode may exhibit a "nominal" voltage anywhere from 17.1 to 18.9 V because of the 5 percent tolerance. The test current indicates the nominal operating region. Item 3, the maximum dissipation, gives the maximum permissible power which the diode can dissipate at room temperature. The dynamic impedance, item 5, is the slope of the reverse-biased diode characteristic measured at the test current I_{ZT}. The maximum junction temperature and maximum dissipation are related by the power-derating curve shown in Fig. 1.12-5. θ_{jc} is the thermal resistance (≈ 2.4°C/W for this diode), usually specified by the manufacturer. The temperature T_{co} is the maximum case temperature at which full rated power can be dissipated.

The temperature coefficient, item 9, is defined in (1.11-3) as

$$\text{TC}(\%) = \frac{\Delta V_Z / V_{ZT}}{\Delta T} \qquad \%/°C$$

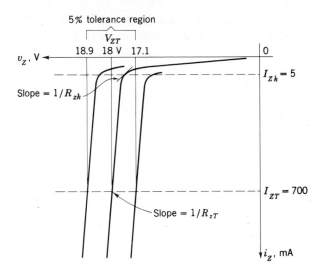

Figure 1.12-4 Reverse characteristic for 1N2816 Zener diode.

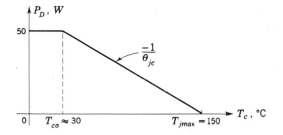

Figure 1.12-5 Derating curve for 1N2816 Zener diode.

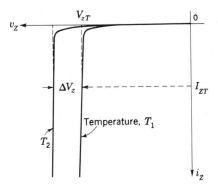

Figure 1.12-6 Zener-diode characteristic illustrating the temperature coefficient.

In Sec. 1.11 we pointed out that the TC is positive when V_{ZT} exceeds approximately 6 V and negative if V_{ZT} is less than 6 V. For this 18-V 50-W diode, a temperature rise of 50°C would result in a change in Zener voltage of

$$\frac{\Delta V_Z}{V_{ZT}} = (0.075\%/°C)(50°C) = 3.75\%$$

Thus

$$\Delta V_Z = (0.0375)(18) = 0.67 \text{ V} \qquad \text{and} \qquad V_{ZT} \approx 18.7 \text{ V}$$

when the temperature rises 50°C.

REFERENCES

1. " Silicon Rectifier Handbook " and " Silicon Zener Diode and Rectifier Handbook," Motorola Inc., Semiconductor Products Division, Phoenix, Ariz.
2. A. S. Grove, " Semiconductor Physics," Wiley, New York, 1967.
3. Motorola Inc., Engineering Staff, " Integrated Circuits," McGraw-Hill, New York, 1965.
4. D. Hamilton and W. Howard, " Basic Integrated Circuit Engineering," pp. 24, 269, 488, McGraw-Hill, New York, 1975.

PROBLEMS†

1.1-1 For the circuit of Fig. P1.1-1a let $r_i = 100$ Ω and $R_L = 600$ Ω. (a) Sketch v_L as a function of time t in milliseconds for v_i, as shown in Fig. P1.1-1b. Repeat part (a) if $v_i(t)$ is (b) sinusoidal (1-V peak) and (c) triangular (1-V peak).

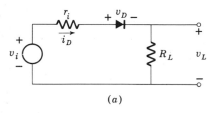
(a)

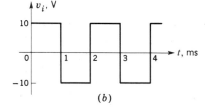
(b)

Figure P1.1-1

1.1-2 For the circuit of Fig. P1.1-2a (a) sketch v_L as a function of time in milliseconds for v_i, as shown in Fig. P1.1-2b.

(b) Repeat part (a) if $v_i(t)$ is sinusoidal, triangular (1-V peak).

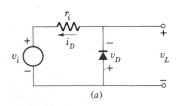
(a)

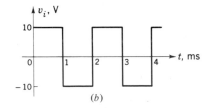
(b)

Figure P1.1-2

† The first two digits of the problem number indicate the text section to which the problem applies.

1.1-3 For the circuit of Fig. P1.1-3a (a) sketch $v_L(t)$ if $R_b = 100$ kΩ, $r_i = R_L = 1$ kΩ, and v_i is as shown in Fig. P1.1-3b.

  (b) Repeat part (a) if $v_i(t)$ is sinusoidal, triangular (1-V peak).

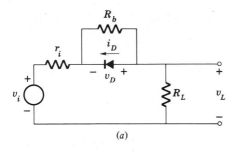
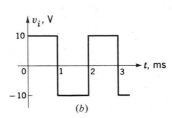

(a) (b)

Figure P1.1-3

1.1-4 The output voltage given by (1.1-3) is passed through the LC filter shown in Fig. P1.1-4. Let $LC = 10^4/\omega_0^2$ and $RC = 100\sqrt{2}/\omega_0$.

  (a) Find $v_o(t)$. Compare with the result obtained using the RC filter.

  (b) Calculate the rms ripple voltage.

  (c) Calculate the ratio of the rms ripple voltage to the dc voltage.

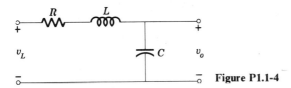

Figure P1.1-4

1.1-5 Repeat Prob. 1.1-4 using the pi filter shown in Fig. P1.1-5.

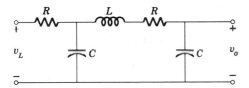

Figure P1.1-5

1.1-6 The circuit shown in Fig. P1.1-6 is a *clamping* circuit. Find $v_L(t)$ when $v_i(t) = A \cos \omega_0 t$.

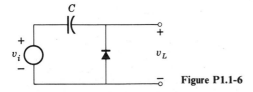

Figure P1.1-6

1.1-7 The circuit shown in Fig. P1.1-7 is a more practical version of the clamping circuit of Fig. P1.1-6 since it includes the diode resistances. Sketch $v_L(t)$ when v_i is the square wave shown. Assume that $R_b = 1 \text{ k}\Omega$, $r_d = 50 \ \Omega$, and $r_d C \ll T$, while $R_b C \gg T$.

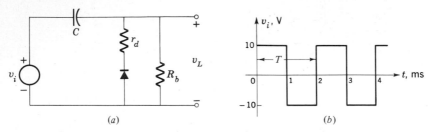

(a)   (b)

Figure P1.1-7

1.1-8 The circuit shown in Fig. P1.1-8 is a diode full-wave-rectifier bridge. Sketch $v_L(t)$ when $v_i(t)$ is sinusoidal.

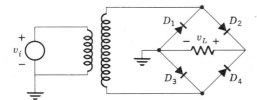

Figure P1.1-8

1.1-9 The full-wave-rectifier circuit of Fig. 1.1-8 is complicated because of the presence of resistors in series with D_1 and D_2, as shown in Fig. P1.1-9. Sketch $v_L(t)$ when v_i is sinusoidal with frequency ω_0. Assume that $R = 0.2R_L$.

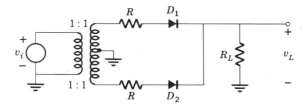

Figure P1.1-9

1.1-10 The input to the half-wave peak detector shown in Fig. 1.1-10a is a square wave with voltage levels $+5$ and -5 V and a period of 1 μs. Calculate the required $R_L C$ time constant if the decay of v_L from the maximum value of $+5$ V is to be less than 5 mV. This decay is called the *tilt*.

1.1-11 The input to the circuit shown in Fig. 1.1-10a is

$$v_i(t) = 5[1 + m(t)] \cos \omega_0 t$$

where $m(t)$ is a square wave with voltage levels $+1$ and -1 V and a period of 1 ms. The frequency $f_0 = 1$ MHz. Sketch the voltage v_L when $R_L C = 0.1$, 1, and 10 μs. Which value results in v_L most closely reproducing the square wave?

1.1-12 In Fig. 1.1-10 let $v_i = 10 \cos 200\pi t$, $R_L = 1 \text{ k}\Omega$, and $C = 100 \ \mu$F. Calculate v_L (dc and rms ripple).

1.1-13 The circuit of Fig. P1.1-13 is a full-wave peak-rectifier power supply, where D_1 and D_2 are ideal diodes.

    (a) Sketch v_L.

    (b) Calculate the dc voltage and the ripple at v_L.

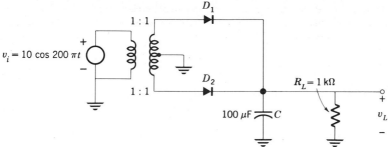

Figure P1.1-13

1.1-14 Specify n, C, and the ripple in the peak-detector power supply of Fig. P1.1-14 so that it will deliver 30 V dc to the 100-Ω load with a ripple voltage equal to or less than 10 percent of the dc voltage.

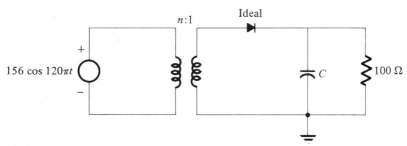

Figure P1.1-14

1.1-15 The input to the diode clamping circuit of Fig. 1.1-12a is the waveform shown in Fig. P.1.1-15. Assume that $V_R = 10$ V.
 (a) What is the dc level of the waveform?
 (b) What is the dc voltage across the capacitor?
 (c) Sketch v_L.

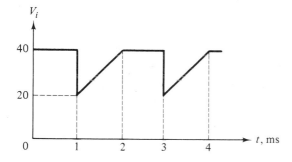

Figure P1.1-15

1.1-16 The peak detector of Fig. 1.1-10 can be analyzed in an alternate fashion which allows the inclusion of additional filters.
 (a) Refer to Fig. 1.1-10c and assume small ripple $V_{Lr} \ll V_{im}$. Use the vi equation for the capacitor

$$v_L = \frac{1}{C} \int_{t_1}^{t_2} i \, dt + V_{im}$$

and the fact that $i \approx I_{DC} = -V_{im}/R_L$ to calculate V_{Lr} and obtain (1.1-6).

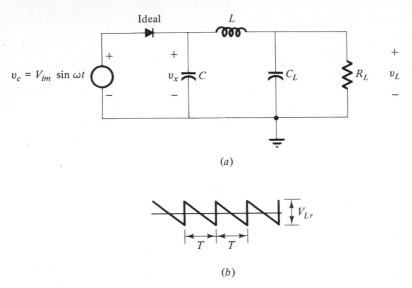

(a)

(b)

Figure P1.1-16

(b) Consider the circuit in Fig. P1.1-16a. Sketch v_x assuming that the diode is ideal and the ripple is small.

(c) Use the technique in part (a) to calculate the peak-to-peak ripple at v_x.

(d) The ripple at v_x can be approximated by the waveform in Fig. P1.1-16b. Use a Fourier series

$$v_x = V_1 \sin (\omega t + \varphi_1) + V_2 \sin (2\omega t + \varphi_2) + \cdots$$

where $V_1 = V_{Lr}/\pi$ and $V_2 = V_1/2$. State the conditions that are necessary for $v_L = V_L \sin (\omega t + \varphi_{1L})$.

(e) Use the results of part (d) to show that

$$V_L \approx \frac{V_{Lr}}{\pi} \frac{1}{\omega^2 L C_L}$$

1.1-17 In Fig. P1.1-16 let $v_i = 10 \sin 200\pi t$, $R_L = 1 \text{ k}\Omega$, $C = C_L = 100 \text{ }\mu\text{F}$, and $L = 10 \text{ H}$.

(a) Sketch v_x and calculate the peak-to-peak ripple. Check the assumption of small ripple.

(b) Sketch the waveshape expected at v_L. What is the peak-to-peak ripple?

1.1-18 For the circuit of Fig. P1.1-18 (a) sketch v_x and calculate the peak-to-peak ripple using the assumptions in Prob. 1.1-16.

(b) Write the Fourier series for the ripple at v_x as in part (d) of Prob. 1.1-16 and state the conditions for v_L to be sinusoidal.

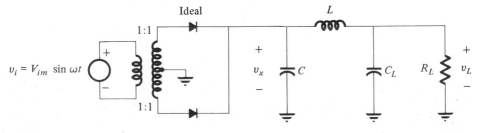

Figure P1.1-18

(c) Show the advantage of full-wave rectification by calculating the peak-to-peak ripple at v_L and comparing with the result in part (e) of Prob. 1.1-16.

1.1-19 Design a full-wave peak-detector power supply using the CLC_L filter of Fig. P1.1-18 to deliver 30 V dc from a 120-V rms 60-Hz source, into a 100-Ω load at less than 0.1 percent ripple. Determine the transformer turns ratio n and L and C_L. Assume that losses in the coil are zero, that $C = 500$ μF, and that 20 percent ripple at v_x is sufficiently small to allow use of the approximations stated in Prob. 1.1-16.

1.1-20 Verify (1.1-2).

1.1-21 Verify (1.1-3).

1.2-1 Sketch i_D/I_o as given by (1.2-1) as a function of the diode voltage v_D using semilog paper. Assume $T = 300$ K (room temperature).

1.2-2 Repeat Prob. 1.2-1 for $T = 500$ and 200 K. Discuss these results.

1.2-3 A silicon diode has a reverse saturation current $I_o = 1$ nA. Sketch i_D versus v_D. Do not use semilog paper. A germanium diode having the same power-dissipation capability as the silicon diode has an $I_o = 100$ μA. On the same axes sketch i_D versus v_D for the germanium diode.

(a) Determine the turn-on voltages.

(b) Find approximate straight-line (piecewise-linear) representations for the two diodes. Use two straight lines in the forward region to obtain results similar to those shown in Fig. 1.2-6b.

1.2-4 Calculate the i_{D1}/i_{D2} that results from changing the diode voltage by 50, 100, and 200 mV. Assume $T = 300$ K.

1.2-5 Calculate the change in the diode voltage for a current ratio of 100 and 1000.

1.2-6 Repeat Prob. 1.1-1 using the piecewise-linear model of Fig. 1.2-8.

1.3-1 For the circuit of Fig. P1.3-1, sketch i_D when v_T is a square wave having an average value of zero and a peak-to-peak voltage of 2 V. Obtain the answer analytically and graphically.

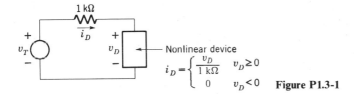

$$i_D = \begin{cases} \dfrac{v_D}{1\text{ k}\Omega} & v_D \geq 0 \\ 0 & v_D < 0 \end{cases}$$

Figure P1.3-1

1.3-2 For the circuit of Fig. P1.3-2a sketch i_D when v_T is sinusoidal with a peak value of 4 V (use the graphical technique). The nonlinear device characteristic is shown in Fig. P1.3-2b. Use $R = r_i = 1000$ Ω, $R_L = 500$ Ω, and $v_{DC} = 4$ V.

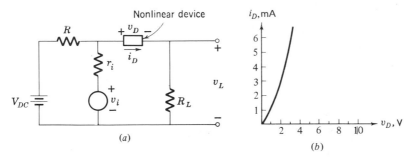

(a)

(b)

Figure P1.3-2

1.3-3 (a) Redraw Fig. P1.3-2a and obtain the Thevenin circuit (v_T and R_T).

(b) Let $R = 1$ kΩ, $r_i = 1.5$ kΩ, $R_L = 1.4$ kΩ, $V_{dc} = 5$ V, and $v_i(t) = 10 \sin \omega_0 t$ V. Plot the dc load line when $\omega_0 t = 0$, $\pm \pi/3$, and $\pm \pi/2$.

(c) Sketch $v_L(t)$. *Hint:* $v_L = R_L i_D$.

1.3-4 In the circuit of Fig. P1.3-4 calculate $v_L(t)$ using the characteristic of Fig. 1.2-8 with $V_F = 0.7$ V.

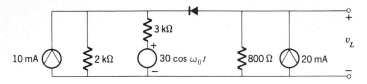

Figure P1.3-4

1.3-5 Assuming that diodes D_1 and D_2 are identical in the circuit of Fig. P1.3-5, find v_L.

$$i_{D1} = \begin{cases} 2 \times 10^{-3} v_{D1}^2, & v_{D1} \geq 0 \\ -I_0 & v_{D1} < 0 \end{cases}$$

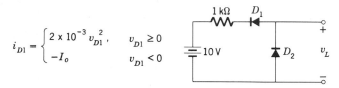

Figure P1.3-5

1.3-6 For the limiter circuit shown in Fig. P1.3-6a, where diodes D_1 and D_2 are described by Fig. 1.2-8 with $V_F = 0.7$ V, find(a) $v_L(t)$ with v_i as shown in Fig. P1.3-6b and(b) $v_L(t)$ with $v_i(t) = 10 \cos \omega_0 t$.

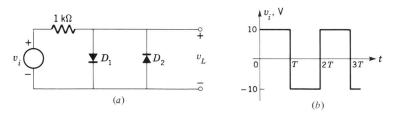

Figure P1.3-6

1.4-1 In Fig. P1.4-1, the diode vi characteristic is given by

$$i_D = 10^{-6}(\epsilon^{qv_D/kT} - 1) \qquad \text{where } T = 300 \text{ K}$$

(a) Obtain the Thevenin circuit, R_T, and v_T.
(b) Determine the quiescent current in the diode.
(c) Calculate r_d.
(d) Calculate $v_L(t)$.

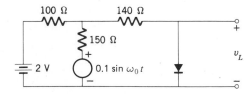

Figure P1.4-1

1.4-2 The diode in Fig. P1.4-2 is characterized by (1.2-1) where $I_o = 10^{-9}$ A and $T = 300$ K. Find i_D.

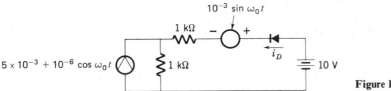

$10^{-3} \sin \omega_0 t$

1 kΩ

$5 \times 10^{-3} + 10^{-6} \cos \omega_0 t$ 1 kΩ i_D 10 V

Figure P1.4-2

1.4-3 (a) Find the Thevenin equivalent circuit, R_T, and v_T for the circuit in Fig. P1.4-3.
(b) Sketch $i_D(t)$.
(c) Find the average current through the diode.

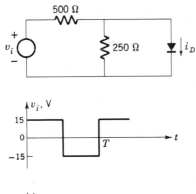

500 Ω

v_i 250 Ω i_D

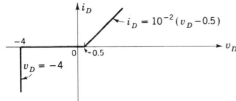

v_i, V

15

0

T

-15

t

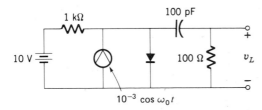

i_D

$i_D = 10^{-2}(v_D - 0.5)$

-4

0 -0.5 v_D

$v_D = -4$

Figure P1.4-3

1.4-4 The diode vi characteristic is given by

$$i_D = I_o\left(e^{qv_D/kT} - 1\right)$$

(a) Expand this equation in a Taylor series about the point I_D, V_D.
(b) Show that if $v_d \ll kT/q$, all terms beyond the first two in the expansion can be dropped; that is, $i_D \approx I_D + v_d/r_d$.

1.4-5 (a) Determine the quiescent diode current in Fig. P1.4-5. The diode vi characteristic is given in Prob. 1.4-1.

1 kΩ   100 pF

10 V   100 Ω v_L

$10^{-3} \cos \omega_0 t$ **Figure P1.4-5**

(b) Calculate r_d.

(c) Find $v_L(t)$ when $\omega_0 = 10^6$, 10^8, 10^{10} rad/s.

1.5-1 (a) For the circuit of Fig. P1.5-1, obtain the equation of the dc load line and plot it on the vi characteristic. Obtain the quiescent current.

(b) Obtain the equation of the ac load line and plot it on the vi characteristic.

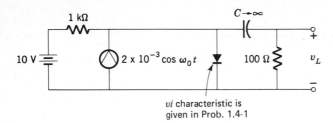

vi characteristic is
given in Prob. 1.4-1

Figure P1.5-1

1.5-2 The diode circuit of Fig. P1.5-2 has a signal voltage $v_i = 0.1 \cos \omega_0 t$.

(a) Find the quiescent diode voltage and diode current.

(b) Construct the ac load line.

(c) Determine the dynamic resistance of the diode.

(d) Calculate v_L.

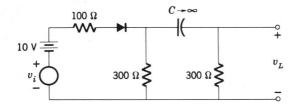

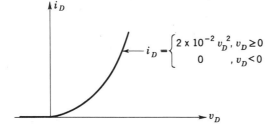

$$i_D = \begin{cases} 2 \times 10^{-2} \, v_D^2, & v_D \geq 0 \\ 0 & , v_D < 0 \end{cases}$$

Figure P1.5-2

1.6-1 In Fig. 1.6-2 let $r_i = 10 \, \Omega$, $R_c = 10 \, \text{k}\Omega$, $R_L = 5.1 \, \text{k}\Omega$, and $v_i = 5 \sin \omega_0 t$. The control signal is shown in Fig. P1.6-1, and the diodes are ideal.

(a) Determine an acceptable V_1 and V_2 for the control signal.

(b) For the values in part (a) sketch v_L if $\omega = 4\omega_0$.

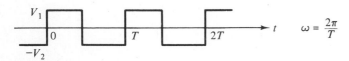

$$\omega = \frac{2\pi}{T}$$

Figure P1.6-1

1.6-2 In Fig. 1.6-2 let $r_i = 10\ \Omega$, $R_L = 1\ \text{k}\Omega$, $v_i = V_{im} \sin \omega_0 t$, $V_{on} = 15\ \text{V}$, and $V_{off} = 0$.

    (a) Calculate V_{im}.

    (b) Find a suitable value for R_c.

1.6-3 Use the piecewise model of Fig. 1.2-8 in Fig. 1.6-2 and show that the maximum allowable input voltage is

$$V_{im} = (V_{on} - V_F) \frac{R_L}{R_c + R_L}$$

1.6-4 The six-diode analog switch of Fig. 1.6-5 has a control signal $v_C(t)$ as shown in Fig. P1.6-1, where $V_1 = V_2$. Let $v_i = \sin \omega_0 t$.

    (a) Determine an acceptable range of square-wave amplitude ($V_1 = V_2$) so that the switch functions properly.

    (b) Sketch $v_L(t)$ if $\omega = \omega_0/5$ and show that v_L is the product of v_i and a square wave $S_1(t)$.

1.6-5 $S_1(t)$ in Prob. 1.6-4 can be written as a Fourier series by noting that in (1.6-7) $f_c = f = 1/T$ and that t is replaced by $t - T/4$. Express $v_L(t)$ as a sum of harmonics.

1.6-6 (a) In the circuit of Fig. 1.6-5 let $v_i = V_{im} \sin \omega_0 t$. Show that for the control signal to be a symmetrical square wave of amplitude V_1 and the switch to function properly, we must have

$$V_{im} \le \frac{5.3 R_L}{R_c + R_L} = 2.65\ \text{V} \qquad \text{where } V_F - 0.7\ \text{V}$$

    (b) Find the square-wave amplitude for V_1 corresponding to the result in part (a).

1.6-7 The six-diode array of Fig. 1.6-1 is used as a shunt gate in Fig. P1.6-7. Diodes D_5 and D_6 provide the control source with a symmetric load. Assume that all diodes are ideal and the control signal is a symmetrical square wave (see Fig. P1.6-1 and let $V_1 = V_2$).

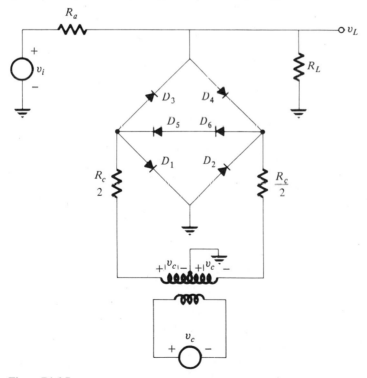

Figure P1.6-7

(*a*) Draw the equivalent circuit seen by $v_i - v_L$ when the diodes D_1 to D_4 conduct.

(*b*) If $v_i = V_{im} \sin \omega_0 t$ and $\omega = 2\omega_0$ (see Fig. P1.6-1), sketch v_L assuming that the circuit is functioning properly.

(*c*) Show that the current in D_3 is given by

$$i_{D3} = \frac{V_1}{R_c} - \frac{v_i}{2R_a}$$

Hint: Assume a balanced bridge and use superposition.

(*d*) What is the maximum value of V_{im} allowed if $V_1 = 5$ V, $R_c = 10$ kΩ, $R_L = 1$ kΩ, and $R_a = 1$ kΩ?

1.7-1 Obtain a two-segment piecewise-linear approximation to the equation

$$y = t^2 \qquad 0 \le t \le 3$$

such that the absolute value of the maximum difference between the function y and its approximation $\hat{y}$, at any time t, is always less than 0.5.

1.7-2 Using one resistor (1 kΩ), one battery (1 V), and one ideal diode, synthesize circuits to realize each of the four basic vi characteristics shown in Fig. P1.7-2.

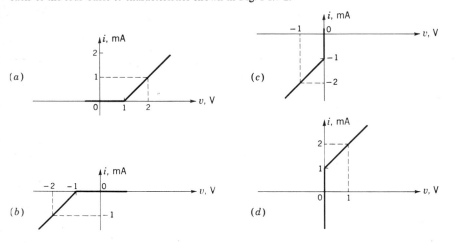

Figure P1.7-2

1.7-3 Obtain a three-segment piecewise-linear approximation to the function $y = \ln x$ for the range $0 < y < 3$ that will yield a maximum error $|\hat{y} - y|_{\max}$ which is less than that given in Fig. 1.7-3. Use a trial-and-error procedure. How small can you make $|\hat{y} - y|_{\max}$?

1.7-4 Scale the circuit of Example 1.7-1 so that the currents are in microamperes and the voltages are in volts.

1.7-5 Design a circuit using resistors and ideal diodes to synthesize the vi characteristic shown in Fig. P1.7-5.

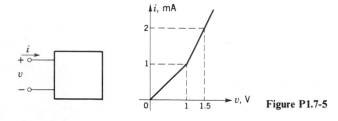

Figure P1.7-5

1.7-6 Design a circuit using resistors and ideal diodes to synthesize the vi characteristic in Fig. P1.7-6.

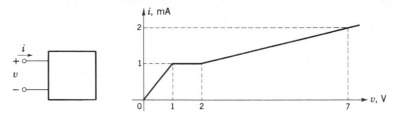

Figure P1.7-6

1.7-7 Design a circuit to solve the nonlinear differential equation

$$\frac{dy}{dt} + \sin y = 0.8 \qquad y(0) = 0 \qquad t \ge 0 \qquad -\frac{\pi}{2} \le y \le \frac{\pi}{2}$$

All currents should be in milliamperes and voltages in volts.

1.7-8 Solve Example 1.7-3 when $x(0) = 1$.

1.10-1 The Zener diode in Fig. P1.10-1 has a fixed voltage drop of 18 V across it as long as i_z is maintained between 200 mA and 2 A.

    (a) Find r_i so that V_L remains at 18 V while V_{dc} is free to vary from 22 to 28 V.

    (b) Find the maximum power dissipated by the diode.

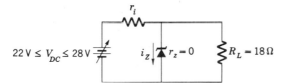

Figure P1.10-1

1.10-2 A 10-V Zener diode is used to regulate the voltage across a variable-load resistor as shown in Fig. P1.10-2. The input voltage v_i varies between 13 and 16 V. The load current i_L varies between 10 and 85 mA. The minimum Zener current is 15 mA.

    (a) Calculate the maximum value of r_i.

    (b) Calculate the maximum power dissipated by the Zener diode using this value of r_i.

Figure P1.10-2

1.10-3 An unregulated supply varies between 20 and 25 V and has a 10-Ω internal impedance. A 10-V Zener diode is to regulate this voltage for use in a tape recorder. The recorder draws 30 mA while recording and 50 mA when playing back. The Zener diode has a resistance of 10 Ω when the Zener current is 30 mA. The knee of the Zener characteristic occurs at 10 mA. In addition, the Zener diode can dissipate a maximum power of 800 mW.

    (a) Find r_i so that the diode regulates continuously (assume $r_i \gg 10\ \Omega$).

    (b) Find the maximum peak-to-peak output ripple.

1.10-4 The Zener characteristic is usually due to an avalanche breakdown. This is described approximately by the equation

$$i_Z = \frac{I_o}{(1 - v_Z/V_o)^n} \qquad v_Z \le V_o \qquad \frac{i_Z}{I_o} \le 100$$

(a) Sketch the $v_Z i_Z$ characteristic for $n = 4$ and $n = 10$. *Hint:* Plot i_Z/I_o versus v_Z/V_o to obtain a normalized characteristic.

(b) Obtain the knee of the curve.

(c) Sketch a piecewise-linear approximation to the curve using two straight lines.

(d) The slope of the straight-line curve below the knee is called the *Zener resistance* r_z. Calculate r_z. Note that r_z continually changes.

1.11-1 Obtain an electrical analog of the thermal system of Fig. 1.11-3a in which temperature is analogous to current. What are P_j and θ_{jc} analogous to?

1.11-2 A silicon Zener diode is rated at 15 W. The maximum allowable junction temperature is 200°C, the ambient temperature is 25°C, and $\theta_{jc} = 2.4°C/W$.

(a) If an infinite heat sink is provided, find the maximum allowable diode power rating.

(b) What is the junction temperature of the diode?

(c) What is the case temperature of the diode?

1.11-3 Repeat Prob. 1.11-2 if the heat sink has a thermal resistance θ_{ca} of 2°C/W.

1.11-4 A diode can dissipate 20 W at temperatures less than 50°C. The maximum junction temperature is 200°C. Find θ_{jc}.

1.11-5 Repeat Prob. 1.11-2 if $T_{j,\,max} = 100°C$ and $\theta_{ca} = 3°C/W$.

1.11-6 A 50-W Zener diode (derating curve shown in Fig. P1.11-6) is to dissipate 6 W in a particular circuit. The ambient temperature is 85°C.

(a) Find θ_{ca} so that the transistor will not overheat.

(b) Find T_c for these conditions.

(c) If an infinite heat sink were available, how much power could this transistor dissipate?

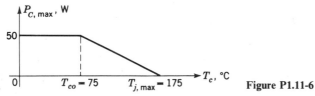

Figure P1.11-6

1.12-1 The reverse saturation current of the diode in Fig. P1.12-1 is 1 μA. Its peak inverse voltage is 500 V. Find r_i so that the PIV is not exceeded.

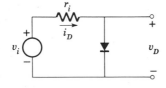
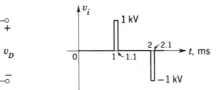

Figure P1.12-1

1.12-2 The maximum average half-wave-rectified forward current through the diode is 1 A in Fig. P1.12-2.

(a) Find R_L so that this value is not exceeded.

(b) If $I_o = 1 \ \mu A$, find the minimum required PIV to prevent diode breakdown.

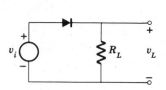 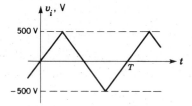

Figure P1.12-2

1.12-3 At a test current of 100 mA a Zener diode has a nominal voltage of 20 V. It has a 2 percent tolerance. Calculate the range of operating test voltages.

1.12-4 The 1N2816 Zener diode in Fig. P1.12-4 is to be used to maintain a fairly constant dc voltage across a load which varies from 10 to 100 Ω. The input voltage varies from 80 to 100 V.

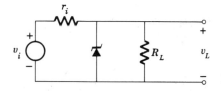

Figure P1.12-4

(a) Calculate r_i assuming $V_{zT} - 18$ V and $R_{zT} \approx 2 \ \Omega$.

(b) Calculate v_L taking into account the 5 percent tolerance and the fact that the circuit operates over the temperature range 0 to 25°C.

1.12-5 Show that the PIV rating of the diodes in the full-wave-rectifier circuit of Fig. 1.1-9 should be greater than $2V_{im}$. Assume ideal diodes.

1.12-6 Show that the PIV rating of the diodes in the full-wave bridge-rectifier circuit of Fig. P1.1-8 should be greater than V_{im}. Assume ideal diodes.

INTRODUCTION TO TRANSISTOR CIRCUITS

INTRODUCTION

In the preceding chapter the *pn* junction diode and some of its physical and circuit properties were considered. In this chapter attention is focused on the device which has caused a veritable revolution in the electronics field since the early 1950s. This device is, of course, the transistor. As we shall see, the basic junction transistor consists essentially of two *pn* junctions placed back to back, so that much of the theory of Chap. 1 applies with minor modifications.

Conceptually, the transistor is a device that acts as a current amplifier, and in this and succeeding chapters its properties, uses, and limitations are studied.

2.1 CURRENT-FLOW MECHANISM IN THE JUNCTION TRANSISTOR[1]

Here, the basic mechanism of current flow, as viewed from the terminals of the device, is considered. The junction transistor consists of two *pn* junctions placed back to back, as shown pictorially in Fig. 2.1-1*a*. The circuit symbol used to represent the *pnp* transistor is shown in Fig. 2.1-1*b*. Note that the transistor in the figure has its semiconductor materials arranged *p-n-p*; hence the name *pnp* transistor. The alternative arrangement, *npn*, is discussed in Sec. 2.2.

Physically, the transistor consists of three parts, emitter, base, and collector, the base region being very thin. Its operation can be explained qualitatively as follows.

PNP

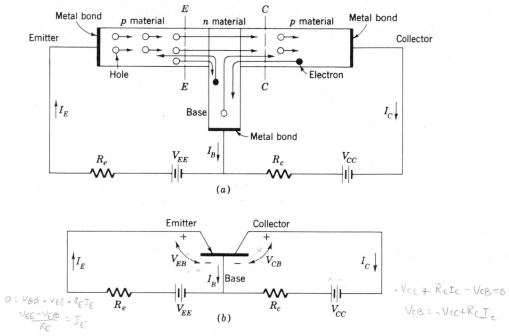

$$0 = V_{BB} - V_{EE} + R_E I_E$$
$$\frac{V_{EE} - V_{EB}}{R_E} = I_E$$

$$-V_{cc} + R_c I_c - V_{cB} = 0$$
$$V_{cB} = -V_{cc} + R_c I_c$$

Figure 2.1-1 The *pnp* junction transistor: (*a*) pictorial representation; (*b*) circuit symbol and reference directions for the common-base connection.

The battery V_{EE} forward-biases the emitter-base *pn* junction, causing the emitter to inject holes into the *n* material. Most of the holes travel across the narrow base region, through the second junction, into the right-hand negatively biased *p* region (the collector). A small amount of these holes (approximately 1 percent) is caught in the *n* material and is collected by the base. Electrons in the base material flow into the emitter as shown.

While the emitter-base junction represents a forward-biased diode with its characteristic properties of low impedance and low voltage drop, the collector-base junction is reverse-biased because of the polarity of V_{CC}. This is essentially a reverse-biased diode, and the collector-base impedance is thus very high.

The current measured in the emitter circuit (the flow of charge across boundary *EE* per unit time), called the *emitter current* I_E, is positive into the material.†
The current measured in the collector circuit (the flow of charge across boundary *CC* per unit time) is called the collector current I_C. This current consists of two terms, the predominant term representing the percentage of emitter current reaching the collector. The percentage depends almost solely on the construction of the

---

† Reference directions for the transistor currents are often taken as positive flowing *into* the base, emitter, and collector. In this text, the convention to be used is that currents are positive in the direction in which an ammeter would indicate positive current when the transistor is biased for ordinary linear operation.

transistor (the size and shape of the material and the doping of the emitter) and can be considered a constant for a particular transistor. The constant of proportionality is defined as α,‡ so that the major part of the collector current is αI_E. Typical values of α range from 0.90 to 0.99.

The second term represents the current flow through the reverse-biased collector-base junction when $I_E = 0$. This current is called I_{CBO} (it was called I_o in the diode), and, as expected, it is relatively quite small. Since both currents flow out of the collector, this direction is defined as the positive collector-current direction, and

$$I_C = \alpha I_E + I_{CBO} \tag{2.1-1}$$

Applying Kirchhoff's current law (KCL) to the transistor of Fig. 2.1-1, and noting the indicated current directions, we have

$$I_E = I_B + I_C \tag{2.1-2}$$

Substituting (2.1-1) into (2.1-2), we find the base current to be

$$I_B = (1 - \alpha)I_E - I_{CBO} = \frac{1 - \alpha}{\alpha} I_C - \frac{I_{CBO}}{\alpha} \tag{2.1-3}$$

The symbol β is used to represent the ratio $\alpha/(1 - \alpha)$, which arises continually in the study of transistors. Equations (2.1-1) to (2.1-3) describe the transistor in terms of its terminal currents.§ In the remainder of this section we show that in normal operation the transistor can be considered to be two isolated pn junctions, one forward-biased and one reverse-biased. The analysis of transistor circuits can then be carried out using the techniques developed for pn junction diodes in Chap. 1.

2.1-1 The Emitter-Base Junction

When we apply Kirchhoff's voltage law (KVL) to the emitter-base loop in Fig. 2.1-1b, the emitter current is

$$I_E = \frac{V_{EE} - V_{EB}}{R_e} \tag{2.1-4}$$

where V_{EB} is the voltage across the forward-biased emitter-base junction. The vi characteristic of the emitter-base junction is shown in Fig. 2.1-2a and b. The reader will observe that this characteristic should also depend on the state of the collector junction and the operating temperature. However, in normal operation, the collector-voltage dependence is negligible, and is often omitted. Further discussion of the dependence of the emitter current on the collector-base voltage is postponed to Sec. 2.2, and temperature effects are considered in Chap. 4. With this simplification, the analysis of the emitter-base circuit is exactly the same as the analysis of the diode circuits discussed in Secs. 1.3 and 1.4. In particular, the dc

‡ The symbol h_{FB} is often used instead of α.
§ The symbol h_{FE} is often used instead of β.

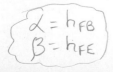

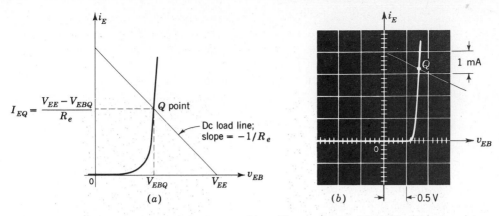

Figure 2.1-2 (a) Emitter-base vi characteristic; (b) oscillogram of TI 179 6508 silicon transistor emitter-base vi characteristic.

load line can be drawn as shown in Fig. 2.1-2. As with the diode, a voltage *threshold*, or *break*, exists, where V_{EBQ} is approximately 0.7 V for silicon. The transistor with the vi characteristic shown in Fig. 2.1-2b is operating at $V_{EBQ} = 0.8$ V when $V_{EE} = 4$ V and $R_e = 1$ kΩ. Notice that v_{EB} varies between 0.7 and 0.9 V, depending on the quiescent emitter (or base) current.

Drawing on our experience with diodes, we can linearize the emitter-base circuit by replacing the emitter-base diode by a piecewise-linear equivalent. The corresponding model and vi characteristic are shown in Fig. 2.1-3. It must be remembered that when operating near the break voltage, the exact curve of Fig. 2.1-2 should be used.

The resistance r_d represents the slope of the characteristic at the Q point; thus, since the diode equation (1.4-15) applies,

$$r_d = \frac{V_T}{I_{EQ}} \approx \frac{25 \times 10^{-3}}{I_{EQ}} \ \Omega \qquad \text{at room temperature} \qquad (2.1\text{-}5)$$

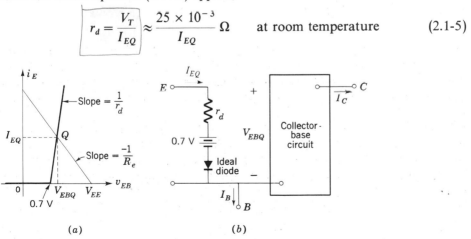

Figure 2.1-3 Piecewise-linear transistor input circuit: (a) vi characteristic; (b) piecewise-linear equivalent.

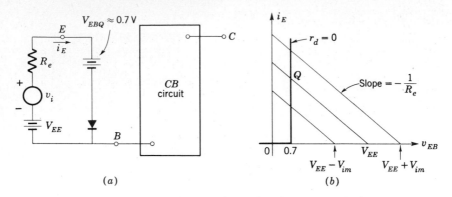

Figure 2.1-4 Common-base circuit with signal applied: (*a*) equivalent circuit; (*b*) graphical representation.

where $V_T = kT/q$. This resistance is usually relatively small, and consequently the impedance looking into the emitter-base circuit will be small.

In Chaps. 2 to 5, which are concerned with large-signal behavior, this small "input" impedance is neglected, that is, $r_d \approx 0$. The discussion is also restricted to silicon transistors, so that, for all cases, $V_{EBQ} \approx 0.7$ V. These simplifications permit us to focus more shaply on the central problems considered in these chapters. Later on, when facility has been gained in handling this simplified model, additional effects are examined.

To illustrate the use of the model, assume that an oscillator is inserted in series with V_{EE}, as shown in Fig. 2.1-4*a*. The ac signal is

$$v_i = V_{im} \cos \omega t \tag{2.1-6}$$

and from Fig. 2.1-4*b* we must have

$$V_{EE} - V_{im} > V_{EBQ} = 0.7 \text{ V} \tag{2.1-7}$$

so that the junction is always forward-biased and operating past the break. This places an upper bound on the peak signal V_{im} for linear operation for any given V_{EE}. Thus

$$i_E = \frac{V_{EE} + V_{im} \cos \omega t - V_{EBQ}}{R_e} \tag{2.1-8}$$

and noting that $i_E = I_{EQ} + i_e$, we get

$$I_{EQ} = \frac{V_{EE} - V_{EBQ}}{R_e} \tag{2.1-9}$$

and

$$i_e = \frac{V_{im}}{R_e} \cos \omega t \tag{2.1-10}$$

The assumptions inherent in (2.1-8) are that the *vi* characteristic of the junction, as shown in Fig. 2.1-4*b*, can be considered a vertical straight line ($r_d \ll R_e$) and that inequality (2.1-7) is satisfied.

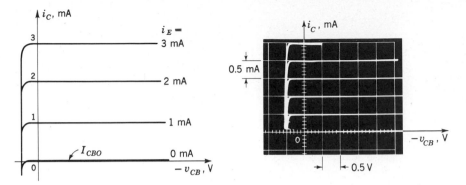

Figure 2.1-5 Common-base output characteristics.

2.1-2 The Collector-Base Junction

In order to complete the model of Fig. 2.1-4a, an equivalent-circuit model is needed for the collector-base junction. Perhaps the easiest way to find this model is to consider the common-base output *vi* characteristics, as shown in Fig. 2.1-5.

For values of $V_{CB} < 0.5$ V these curves can be considered as a family of straight lines which obey the relation

$$I_C = \alpha I_E + I_{CBO} \qquad (2.1\text{-}1)$$

This leads to the equivalent circuit of Fig. 2.1-6a. The current source αI_E is a *dependent* (or *controlled*) source since the current from the source depends on the emitter current I_E. Thus, the current source αI_E is the mechanism by which changes in the emitter current are transmitted to the collector circuit. This type of source is always present in models of active elements.

If we restrict the discussion to silicon transistors, for which $I_{CBO} \ll \alpha I_E$ at normal operating temperatures, the model reduces to that of Fig. 2.1-6b, where the ideal diode in the emitter circuit has also been omitted. This is permissible for linear operation if we remember that we are operating past the break at all times,

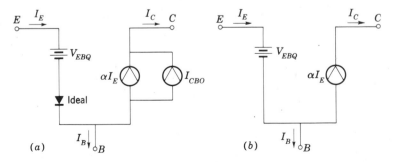

Figure 2.1-6 Common-base equivalent circuits: (a) basic model; (b) simplified model.

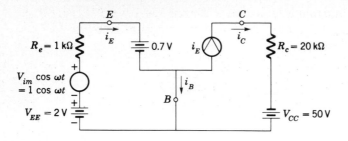

Figure 2.1-7 Common-base equivalent circuit for Example 2.1-1.

so that the piecewise-linear model remains valid. In addition, in order to avoid the nonlinear region to the left of the i_C axis of the output characteristic of Fig. 2.1-5, it is necessary that $v_{CB} < 0.5$ V at all times.

The complete large-signal model of Fig. 2.1-6b can be used for most low-frequency large-signal calculations, as illustrated by the example which follows.

Example 2.1-1 In the circuit of Fig. 2.1-1b, $\alpha \approx 1$, $I_{CBO} \approx 0$, $V_{EE} = 2$ V, $R_e = 1\,\text{k}\Omega$, $V_{CC} = 50$ V, $R_c = 20\,\text{k}\Omega$, and a 1-V peak sinusoidal source is connected in series with V_{EE}. Find i_E and v_{CB}.

SOLUTION The complete equivalent circuit takes the form shown in Fig. 2.1-7. For the emitter-base circuit, the junction is forward-biased as long as

 $V_{im} < 1.3$ V, and

$$i_E = \frac{V_{EE} - 0.7 + V_{im} \cos \omega t}{R_e} = 1.3 + 1.0 \cos \omega t \qquad \text{mA} \quad (2.1\text{-}11)$$

For the collector-base circuit KVL yields

knowing $i_C \approx i_E$ $v_{CB} = -V_{CC} + i_C R_c \approx -V_{CC} + i_E R_c \qquad\qquad (2.1\text{-}12)$

Substituting (2.1-11) into (2.1-12) gives

$$v_{CB} = -V_{CC} + \frac{V_{EE} - 0.7 + V_{im} \cos \omega t}{R_e} R_c$$

$$= -V_{CC} + I_{EQ} R_c + \frac{R_c}{R_e} V_{im} \cos \omega t \qquad (2.1\text{-}13)$$

and substituting numerical values gives

$$v_{CB} = -50 + (1.3)(20) + \tfrac{20}{1} \cos \omega t$$

$$= -24 + 20 \cos \omega t \quad \text{V}$$

The collector-base junction is always reverse-biased ($v_{CB} < 0.5$ V), so that the linear model of Fig. 2.1-7 is valid. Note that the transistor amplifies the input ac voltage and the resulting voltage gain A_v is

$$A_v = \frac{V_{cbm}}{V_{im}} = \frac{20}{1} = 20 \qquad ///$$

2.2 CURRENT AMPLIFICATION IN THE TRANSISTOR[1]

In the preceding section a much abbreviated description of the mechanism of current flow in the junction transistor was presented. It was just enough to allow us to extract some of the terminal relations for linear operation. In this section the discussion is extended to show how current amplification is achieved. Briefly, the process can be explained this way. Referring to (2.1-1) to (2.1-3), we see that if $\alpha \approx 1$ and I_{CBO} is small, a change in emitter current i_E produces a change of approximately the same amount in collector current i_C and a much smaller change in base current (a factor of $1 - \alpha$). To achieve current amplification, the change is initiated in the base current rather than in the emitter current. This causes the collector current and the emitter current to change by a factor of approximately $\alpha/(1 - \alpha) = \beta$.

This result assumes that $\beta = \alpha/(1 - \alpha)$ does not vary with base current. Thus, neglecting I_{CBO}, we have

$$i_C = \beta i_B \tag{2.2-1a}$$

A change in i_B results in a change in i_C such that

$$\frac{\Delta i_C}{\Delta i_B} = \beta + \frac{\Delta \beta}{\Delta i_B} i_B \tag{2.2-1b}$$

This ratio is called the *small-signal current-amplification factor* h_{fe}. Therefore

$$h_{fe} = \beta + \frac{\Delta \beta}{\Delta i_B} i_B \tag{2.2-2}$$

If $(\Delta\beta/\Delta i_B) i_B$ is small compared with β, then

$$h_{fe} \approx \beta \equiv h_{FE} \tag{2.2-3}$$

Small signal current amplification

Figure 2.2-1 shows a typical variation of β and h_{fe} with collector current. Note that in a typical operating range of 1 to 100 mA, h_{fe} is approximately equal to β and is relatively independent of changes in collector current. The assumption that $\beta = h_{fe} =$ constant is often made to simplify analysis. We make this assumption throughout this text.

$\beta = h_{FE}$

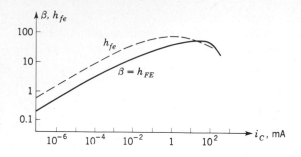

Figure 2.2-1 Common-emitter large- and small-signal current gains.

The explanation presented above is, of course, far from complete. In order to obtain a quantitative picture, consider the circuit of Fig. 2.2-2, which shows a *pnp* transistor in what is known as the *common-emitter* (CE) *configuration*.

In this circuit V_{BB}, V_{CC}, and R_c are adjusted so that, with no signal present, the base-emitter junction is forward-biased and the collector-base junction is reverse-biased, as in the common-base configuration. Writing KVL around the base-emitter loop yields for the base current

$$i_B = \frac{V_{BB} + V_{im} \cos \omega t + V_{BEQ}}{R_b} \tag{2.2-4}$$

where the assumption has been made that the base-emitter junction is operating in its linear range with $V_{BB} - V_{im} \gg V_{BEQ}$; that is, $r_d \approx 0$.

Since linear operation with no distortion is assumed, i_B will consist of both a dc and an ac component

$$\overset{\text{Total}}{i_B} = \overset{DC}{I_{BQ}} + \overset{AC}{i_b} \tag{2.2-5}$$

where

$$I_{BQ} \approx \frac{V_{BB} + V_{BEQ}}{R_b} \quad (DC \text{ comp.}) \tag{2.2-6}$$

and

$$i_b = \frac{V_{im}}{R_b} \cos \omega t \quad (AC \text{ comp.}) \tag{2.2-7}$$

and the collector current, from (2.2-1a), is

$$i_C = I_{CQ} + i_c = \beta(I_{BQ} + i_b) \tag{2.2-8}$$

Common Emitter

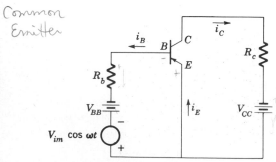

Base-emitter : forward biased
Collector-base : reversed biased

$i_c R_c$ ~

Figure 2.2-2 Basic transistor amplifier.

PNP

$R_b i_b - V_{BB} - V_{im} \cos \omega t + V_{EB} = 0$

$V_{EB} = - V_{BE}$

$i_b = \dfrac{V_{BB} + V_{BE} + V_{im} \cos \omega t}{R_b}$

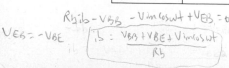

**Table 2.2-1 Relation-
ship between α and β**

α	$\beta = \alpha/(1 - \alpha)$
0.95	19
0.98	49
0.99	99
0.995	199

The small-signal current amplification is then

$$A_i = \frac{i_c}{i_b} \equiv h_{fe} = \beta \qquad \text{Current Gain} \qquad (2.2\text{-}9)$$

This is independent of the external circuit and is a property of the transistor alone, subject to the assumptions made in the derivation.

β is approximately constant for an individual transistor, although it does vary with temperature and slightly with the collector current. Hence the β and α used are average values. Another transistor of the same type may have a β differing by a factor of 3 or more to 1, from the first transistor. This large variability is caused by very small changes in α, as can be seen from Table 2.2-1.

The npn transistor Current flow in the pnp transistor has been considered in the preceding sections. Now let us briefly investigate current flow in the npn transistor, shown in Fig. 2.2-3.

The npn transistor has its bias voltages V_{EE} and V_{CC} reversed from those in the pnp transistor. This is necessary to forward-bias the emitter-base junction and reverse-bias the collector-base junction. Since the emitter current is mainly due to

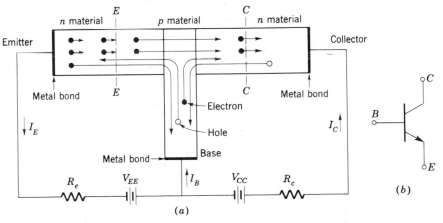

Figure 2.2-3 The npn junction transistor: (a) pictorial representation; (b) circuit symbol.
common base

electrons moving from the emitter to the collector, the positive direction of emitter current is out of the emitter, as shown in Fig. 2.2-3. The collector and base currents flow into the transistor, and

$$I_C = \alpha I_E + I_{CBO} \tag{2.2-10}$$

$$I_B = (1 - \alpha)I_E - I_{CBO} \tag{2.2-11}$$

as before.

In Figs. 2.1-1 and 2.2-3 the arrow identifies the emitter and indicates the type of transistor. A *pnp* transistor has emitter current flowing into the transistor; hence the arrow points in, as shown in Fig. 2.1-1. In Fig. 2.2-3 the arrow points out, indicating the direction of positive emitter current in an *npn* transistor.

The common-emitter characteristic For either type of transistor, the actual terminal relations are nonlinear and of course have the exponential form of the junction-diode equation. Since there are three terminals to deal with, the characteristics can be displayed in a variety of ways. Probably the most useful curves are

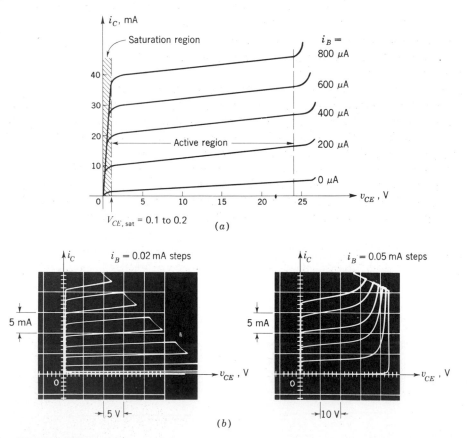

Figure 2.2-4 Common-emitter output characteristics of *npn* transistor: (*a*) typical characteristics; (*b*) oscillograms of TI179 6508 common-emitter output characteristics. Note the different scales and the avalanche breakdown for $V_{CE} > \approx 30$ V.

the common-emitter static characteristics shown in Fig. 2.2-4 for a typical *npn* low-power silicon transistor.

Figure 2.2-4 shows the common-emitter output (or collector) *vi* characteristic. The shaded region shown at the left of the figure is called the *saturation* region. The collector-emitter saturation voltage which defines this region is typically 0.1 to 0.2 V for low-power (less than 1 W) transistors and may be as large as 1 to 2 V for high-power transistors. A more detailed description of the operation of a transistor in saturation is given in Example 2.2-1.

Referring to Fig. 2.2-4, we observe that in the saturation region an increase in base current does not result in a proportionate increase in collector current. Thus, in the design of linear amplifiers, the saturation region is avoided. However, when a transistor is used as a logic gate or switch, the transistor is often switched between cutoff (no emitter current) and saturation in order to take advantage of the fact that a transistor in saturation dissipates little power.

For large values of v_{CE} (about 30 V for the transistor of Fig. 2.2-4b), an avalanche breakdown takes place similar to that described in Sec. 1.2 for the diode. The collector-base breakdown voltage is related to the collector-emitter breakdown voltage by the equation $BV_{CEO} \approx BV_{CBO}/\sqrt[n]{\beta}$, where n varies between 2 and 4 for silicon.

In the region between saturation and breakdown, called the *active region*, the relation between base and collector currents is given by (2.1-3), and it is in this region that linear amplification is obtained.

The active region also has upper and lower bounds on the collector current. The upper bound is the maximum collector current that can flow without causing physical damage to the transistor. This is always specified by the manufacturer. The lower bound is called collector *cutoff*, below which essentially no emitter current flows. This is usually taken to be zero.

Example 2.2-1 The behavior of the transistor in the saturation region becomes important in the design of switching circuits. To illustrate this, consider the circuit of Fig. 2.2-5, with $V_{CC} = 10$ V, $R_b = 10$ kΩ, and $R_c = 1$ kΩ. The transistor has $\beta = 100$, $V_{BE} = +0.7$ V, and a saturation voltage $V_{CE,\,sat} \approx 0.1$ V. Find the operating conditions when (*a*) $V_{BB} = 1.5$ V and (*b*) 10.7 V.

SOLUTION (*a*) For $V_{BB} = 1.5$ V application of KVL around the base-emitter loop yields

$$-V_{BB} + I_B R_b + V_{BE} = 0$$

$$I_B = \frac{V_{BB} - V_{BE}}{R_b} = \frac{1.5 - 0.7}{10^4} = 0.08 \text{ mA}$$

$$I_C = \beta I_B = (100)(0.08) = 8 \text{ mA} \qquad Q \text{ pt}$$

$$I_E \approx I_C = 8 \text{ mA}$$

$$V_{CE} = V_{CC} - I_C R_c = 10 - (8)(1) = 2 \text{ V}$$

$I_C = \beta I_B$

$I_E = I_B(1+\beta)$

useful eq'ns

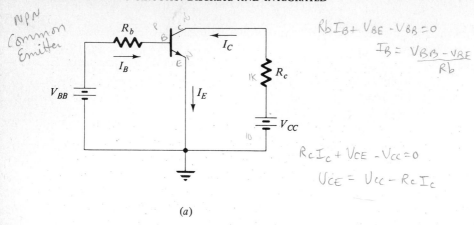

NPN
Common
Emitter

$R_b I_B + V_{BE} - V_{BB} = 0$

$$I_B = \frac{V_{BB} - V_{BE}}{R_b}$$

$R_c I_c + V_{CE} - V_{cc} = 0$

$V_{CE} = V_{cc} - R_c I_c$

(a)

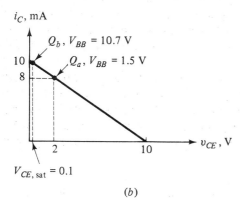

(b)

Figure 2.2-5 Example 2.2-5: (a) circuit; (b) load line.

Thus the transistor is operating within the active region ($V_{CE} > V_{CE, \text{sat}}$) as shown on the load line of Fig. 2.2-5b.

(b) For $V_{BB} = 10.7$ V,

$$I_B = \frac{10.7 - 0.7}{10^4} = 1 \text{ mA}$$

If the basic relation $I_C = \beta I_B$ were to hold here, we should have $I_C = 100$ mA and $V_{CE} = 10 - 100 = -90$ V, an impossible situation. Thus the transistor is in saturation, and

$$V_{CE} = V_{CE, \text{sat}} \approx 0.1 \text{ V}$$

The collector current is

in saturation →

$$I_C = \frac{V_{CC} - V_{CE, \text{sat}}}{R_c} = \frac{10 - 0.1}{1 \text{ k}\Omega} = 9.9 \text{ mA}$$

and

$$I_E = I_C + I_B = 10.9 \text{ mA}$$

Note that the effective β in this particular saturation condition is $I_C/I_B = 9.9$.

///

MISTAKE S/B Rc

2.2-1 Ebers-Moll Transistor Model

The transistor model shown in Fig. 2.1-6 is a linearized model which does not take into account the exponential nature of the pn-junction vi characteristic. It assumes that a pn junction acts like a 0.7-V battery which is either ON or OFF, depending on the input-signal magnitude. In the Ebers-Moll model, discussed below, pn-junction diodes with exponential characteristics in conjunction with appropriate controlled sources are used to represent the transistor. It is found that agreement between the characteristics of this model and experimental measurements is quite good. The Ebers-Moll model also accurately predicts transistor behavior in the saturation region and in the reverse direction, a region which is completely omitted from the linearized model.

The circuit for the Ebers-Moll model can be constructed by considering the linearized common-base equivalent circuit shown in Fig. 2.1-6 for a pnp transistor operating in the normal direction. This circuit is redrawn in Fig. 2.2-6a for an npn

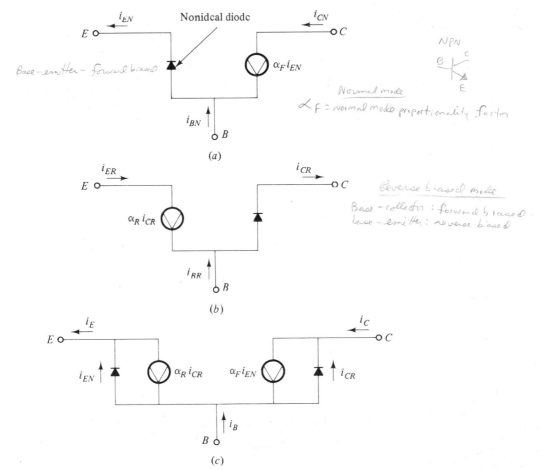

Figure 2.2-6 Ebers-Moll model: (*a*) equivalent circuit of an npn transistor operating in the normal mode; (*b*) the reverse-mode equivalent circuit; (*c*) the complete Ebers-Moll equivalent circuit.

transistor. The ideal diode in the linearized version has been replaced by a non-ideal diode which represents the actual *pn* junction, and the subscript N has been added to the currents to indicate the normal mode of operation. α_F is the normal-mode proportionality factor (current gain) between emitter and collector currents.

When the transistor is operated in the reverse direction, i.e., with the base-collector junction forward-biased and the base-emitter junction reverse-biased, the model takes the form shown in Fig. 2.2-6*b*, where the subscript R has been added to indicate operation in the reverse direction. Note that this circuit is identical in form to the circuit for normal operation except that emitter and collector terminals are interchanged. As in the linearized model, the controlled source is the element which provides the transistor action.

The complete Ebers-Moll model combines both normal and reverse circuits, as shown in Fig. 2.2-6*c*. In the model the total currents are

$$i_E = i_{EN} - i_{ER} = i_{EN} - \alpha_R i_{CR} \tag{2.2-12a}$$

$$i_C = i_{CN} - i_{CR} = \alpha_F i_{EN} - i_{CR} \tag{2.2-12b}$$

$$i_B = i_{BN} + i_{BR} \tag{2.2-12c}$$

In order to express these currents in terms of the terminal voltages we proceed as follows. In Fig. 2.2-6*a* i_{EN} is the current flowing in the base-emitter diode: thus,

$$i_{EN} = I_{EO}(\epsilon^{v_{BE}/V_T} - 1) \tag{2.2-13a}$$

In Fig. 2.2-6*b* i_{CR} is the current in the base-collector diode; thus

$$i_{CR} = I_{CO}(\epsilon^{v_{BC}/V_T} - 1) \tag{2.2-13b}$$

where I_{EO} and I_{CO} are the diode saturation currents. Substituting (2.2-13) into (2.2-12b), we have

$$i_C = i_{CN} - i_{CR}$$
$$i_C = \alpha_F I_{EO}(\epsilon^{v_{BE}/V_T} - 1) - I_{CO}(\epsilon^{v_{BC}/V_T} - 1) \tag{2.2-14a}$$

Using (2.2-12a), we get

$$i_E = i_{EN} - i_{ER}$$
$$i_E = I_{EO}(\epsilon^{v_{BE}/V_T} - 1) - \alpha_R I_{CO}(\epsilon^{v_{BC}/V_T} - 1) \tag{2.2-14b}$$

Since $i_B = i_E - i_C$, we have

$$i_B = (1 - \alpha_F)I_{EO}(\epsilon^{v_{BE}/V_T} - 1) + (1 - \alpha_R)I_{CO}(\epsilon^{v_{BC}/V_T} - 1) \tag{2.2-14c}$$

In practice, transistors are not constructed in the symmetrical form indicated in Fig. 2.1-1*a*, which would result in current gains $\alpha_F \approx \alpha_R$. Typically, they are designed so that $\alpha_F \approx 1$ and $\alpha_R \approx 0.01$. In addition, the saturation currents are related to the current gains by Einstein's relation

$$\alpha_F I_{EO} = \alpha_R I_{CO} \tag{2.2-15}$$

usually $I_{CO} > I_{EO}$

From the typical values of α_F and α_R given above we see that I_{CO} will usually be significantly larger than I_{EO}.

Saturation region The Ebers-Moll equations are useful for determining circuit conditions when a transistor is operating in the saturation region. In this region both the base-emitter junction and the base-collector junction are forward-biased, so that currents i_{EN} and i_{CR} are flowing simultaneously. Assume that v_{BE} and v_{BC} both exceed $V_T(\approx 25 \text{ mV})$ by a sufficient amount for us to neglect the -1 terms in (2.2-14). Then dividing i_C by i_B yields

$$\frac{i_C}{i_B} = \frac{\alpha_F I_{EO} \epsilon^{v_{BE}/V_T} - I_{CO} \epsilon^{v_{BC}/V_T}}{(1 - \alpha_F)I_{EO} \epsilon^{v_{BE}/V_T} + (1 - \alpha_R)I_{CO} \epsilon^{v_{BC}/V_T}} \qquad (2.2\text{-}16)$$

Using (2.2-15) and noting that $v_{CE} = v_{BE} - v_{BC}$, we see that (2.2-16) can be manipulated to yield, after some algebra,

$$\frac{i_C}{i_B} = h_{FE} \frac{\epsilon^{v_{CE}/V_T} - 1/\alpha_R}{\epsilon^{v_{CE}/V_T} + h_{FE}/h_{FC}} \qquad (2.2\text{-}17a)$$

where $h_{FE} = \alpha_F/(1 - \alpha_F)$, $h_{FC} = \alpha_R/(1 - \alpha_R)$, and $h_{FE} \gg h_{FC}$. If we wish to express the collector-emitter voltage in terms of i_C/i_B, (2.2-17a) can be rearranged to yield

$$v_{CE} = V_T \ln \frac{1/\alpha_R + (i_C/i_B)/h_{FC}}{1 - (i_C/i_B)/h_{FE}} \qquad (2.2\text{-}17b)$$

Equation (2.2-17a) is plotted in Fig. 2.2-7 for a typical transistor. In this plot the characteristics have been normalized by letting the abscissa be v_{CE}/V_T and the ordinate be i_C/i_B.

If we define the *edge* of saturation as the point where $i_C/i_B = 0.9h_{FE}$, we find v_{CE} at this point from (2.2-17b) as follows:

$$v_{CE} = V_{CE,\text{ sat}} = V_T \ln \frac{(1 + h_{FC})/h_{FC} + 0.9h_{FE}/h_{FC}}{1 - 0.9}$$

$$v_{CE} = V_T \ln 10 \frac{1 + h_{FC} + 0.9h_{FE}}{h_{FC}}$$

Typically $0.9h_{FE} \gg 1 + h_{FC}$, so that this simplifies to

$$V_{CE,\text{ sat}} \approx V_T \ln \frac{9h_{FE}}{h_{FC}} = V_T\left(2.2 + \ln \frac{h_{FE}}{h_{FC}}\right) \qquad (2.2\text{-}17c)$$

for example, if $h_{FE} = 100$ and $h_{FC} = 0.01$, then $V_{CE,\text{ sat}} \approx 285 \text{ mV}$.

If we reconsider part (b) of Example 2.2-1 using (2.2-17b) with $h_{FE} = 100$ and $h_{FC} = 0.01$, we find that if $I_C/I_B = 9.9$, then $v_{CE} \approx 0.18 \text{ V}$ rather than 0.1 V. Hence, $I_C \approx 9.8 \text{ mA}$. However, if $h_{FE} = 100$ and $h_{FC} = 0.1$, rather than 0.01, then $v_{CE} = 0.12 \text{ V}$, a value which is closer to our assumption of 0.1 V. Note, however, that the value of the collector current is not altered significantly since the collector-

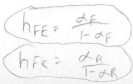

$$h_{FE} = \frac{\alpha_F}{1 - \alpha_F}$$

$$h_{FC} = \frac{\alpha_R}{1 - \alpha_R}$$

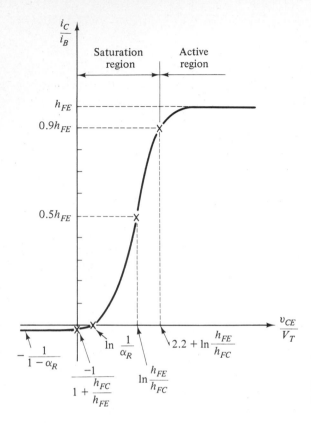

Figure 2.2-7 Normalized vi characteristic of a transistor ($h_{FE} \gg h_{FC}$).

emitter voltage in saturation is small compared with the supply voltage V_{CC}. Figure 2.2-7 shows that even if V_{BB} [part (b) of Example 2.2-1] is increased indefinitely so that i_B becomes very large, the collector-emitter voltage will remain positive and approach the limit

$$v_{CE} \to V_T \ln \frac{1}{\alpha_R} \quad \text{as} \quad \frac{i_C}{i_B} \to 0 \qquad (2.2\text{-}17d)$$

Hence, with $\alpha_R = 0.01$, v_{CE} approaches 0.115 V; and with $\alpha_R = 0.1$, v_{CE} approaches 0.057 V.

In some circuits, e.g., the emitter-follower (Sec. 2.7), and TTL logic gates (Sec. 12.2), the collector current can be made negative. In these cases Fig. 2.2-7 shows that the collector-emitter voltage can become zero or negative. Furthermore, referring to Fig. 2.2-4, the reader is left with the impression that when $i_C = 0$, v_{CE} is also equal to zero. Figure 2.2-7 clearly shows that such is not the case and that when $i_C = 0$, $v_{CE} = V_T \ln (1/\alpha_R)$, a positive voltage.

Many of the circuits studied in this text consider the transistor to be operating in the active region. When such is the case and when no ambiguity will arise, we shall represent α_F by the symbol α. We shall also assume that $V_{CE,\text{sat}} \approx 0.2$ V for most applications.

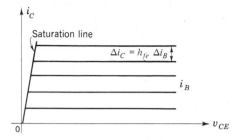

Figure 2.2-8 Idealized output characteristic for *npn* transistor.

At this juncture, we wish to point out that the Ebers-Moll model provides very useful mathematical relationships that apply to a transistor under any operating conditions. However, the equations are obviously unwieldy and hardly suitable for routine analysis or design. Fortunately, there are approximations which we can make to simplify the model so that circuit calculations are reasonably easy while maintaining sufficient accuracy for most applications. The linearized model discussed in Sec. 2.1 is an example of this simplification. Other models, useful in certain other regions of operation, will be presented in subsequent sections.

A useful model for the collector circuit The vertical distance between adjacent curves in Fig. 2.2-4 obeys the relation $\Delta i_C = h_{fe} \, \Delta i_B$ (2.2-1*b*). However, h_{fe} is not absolutely constant over the full range of i_C, as shown in Fig. 2.2-1. For many applications it is sufficient to consider that the output characteristics are a set of uniformly spaced horizontal straight lines, as in Fig. 2.2-8; that is, $h_{fe} = h_{FE} =$ constant. It is extremely important to note that while the transistor *vi* characteristic can be used to obtain insight into the operation of the device, it should not be used to derive any quantitative information regarding the variation of i_C with i_B because the h_{fe} of transistors of the same type may differ considerably.

The assumption of uniformly spaced horizontal lines on the output characteristic permits the collector circuit to be replaced by a controlled current source, as shown in Fig. 2.2-9. This model does not implicitly account for the saturation or breakdown regions and is valid only in the active region; it is identical to the output circuit shown in Fig. 2.2-6*a* except that the source is controlled by the base current rather than the emitter current.

We can go one step farther and take into account the nonzero slope of the actual curves by including a resistance R_o in parallel with the current source, as shown in Fig. 2.2-10*a*. For a silicon transistor this resistor is typically 100 kΩ. Because of its large size, it is often neglected (R_o is often considered to be infinite).

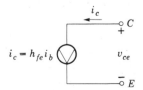

Figure 2.2-9 Small-signal controlled-current source model of idealized *npn* transistor valid for the active region.

resistor included to take into account slope of lines

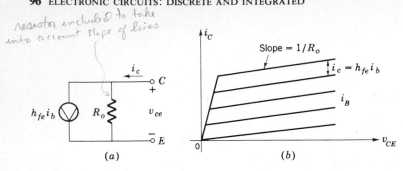

(a)

(b)

Figure 2.2-10 Piecewise linear *vi* characteristic and small-signal equivalent circuit of *npn* transistor, including R_o: (*a*) equivalent circuit; (*b*) piecewise linear *vi* characteristic.

The corresponding output *vi* characteristic is shown in Fig. 2.2-10*b*. It should be noted that the circuit of Fig. 2.2-10 is the small-signal equivalent of the output circuit and is valid only in the active region (this equivalent circuit will be studied in detail in Chap. 6).

The remainder of this chapter assumes that the transistor is a linear device (some of the nonlinear characteristics of the transistor are studied in subsequent chapters). In addition, unless otherwise stated, it is assumed that silicon transistors are used, that $|V_{BE}| = 0.7$ V, and that I_{CBO} is negligibly small and can be neglected.

2.2-2 The Schottky Transistor

When a transistor is saturated, excess charge is stored in the base-emitter and base-collector junctions. Referring to Sec. 1.8, we see that if we were to attempt to suddenly reverse-bias a diode, thereby driving the diode to cutoff, we would first have to discharge the diode storage capacitance C_s. Since the base-emitter and base-collector junctions of a transistor can be considered to be forward-biased diodes, we expect that in order to drive a saturated transistor to cutoff we must

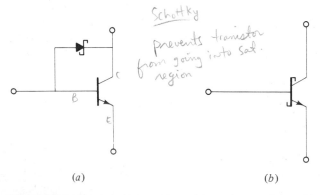

Schottky prevents transistor from going into Sat. region

(a)

(b)

Figure 2.2-11 The Schottky transistor: (*a*) regular transistor with base-collector junction shunted by Schottky diode; (*b*) circuit symbol.

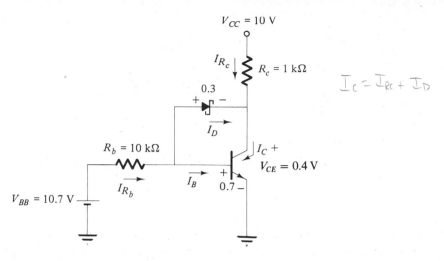

Figure 2.2-12 Effect of the Schottky diode on the quiescent operation of a transistor amplifier.

first remove the excess charge stored in the base-emitter- and base-collector-diode capacitances. A Schottky diode connected between base and collector serves to speed the transition from forward bias to cutoff since in a Schottky diode $C_s \approx 0$. The Schottky diode can also be used to prevent a transistor from saturating. Such a configuration, consisting of a Schottky diode and a transistor, is shown in Fig. 2.2-11a and is called a *Schottky transistor*. The circuit symbol for a Schottky transistor is shown in Fig. 2.2-11b. When the Schottky diode is placed across the transistor, the base-collector voltage of the transistor cannot become more positive than 0.3 V, the forward voltage of the Schottky diode. Since $v_{CE} = v_{BE} - v_{BC} \approx 0.7 - 0.3 = 0.4$ V, we see that the collector-emitter voltage of the transistor will always be greater than or equal to 0.4 V. When $v_{CE} - 0.4$ V in Fig. 2.2-7, the transistor is not in the saturation region; i.e., for reasonable values of h_{FE} and h_{FC}, $i_C/i_B \approx h_{FE}$ when $v_{CE} \geq 0.4$ V. For example, consider that a Schottky transistor is used in part (b) of Example 2.2-1 (Fig. 2.2-5). The resulting circuit is shown in Fig. 2.2-12. Since $V_{BB} = 10.7$ V and $V_{BE} = 0.7$ V, $I_{R_b} = 1$ mA. However, the collector current I_C is now given by

$$I_C = I_{R_c} + I_D \qquad (2.2\text{-}18a)$$

Since the diode voltage is 0.3 V, the collector-emitter voltage $V_{CE} = 0.4$ V and $I_{R_c} = 9.6$ mA. The diode current I_D is

$$I_D = I_{R_b} - I_B = 1 \text{ mA} - \frac{I_C}{h_{FE}} \qquad (2.2\text{-}18b)$$

Substituting (2.2-18b) into (2.2-18a) yields

$$I_C = 9.6 \text{ mA} + 1 \text{ mA} - \frac{I_C}{h_{FE}} \qquad (2.2\text{-}18c)$$

Therefore $$I_C \approx 10.5 \text{ mA}$$

The base current is therefore $I_B \approx 105 \ \mu\text{A}$, and the diode current is $I_D \approx 0.9 \text{ mA}$. The current delivered by the 10-V power supply is $I_{R_c} = 9.6 \text{ mA}$.

2.3 GRAPHICAL ANALYSIS OF TRANSISTOR CIRCUITS

In Chap. 1 the analysis of diode circuits was discussed in terms of three general methods, the graphical method, the piecewise-linear approximation, and the small-signal linear analysis using *incremental*, or *dynamic*, parameters. All three methods are used for the analysis of transistor circuits. In this and the following sections a combination of the graphical and piecewise-linear methods is used to design a common-emitter amplifier to obtain maximum symmetrical variation in the collector current. We also determine the maximum power dissipated by the transistor, the maximum power dissipated in the load resistor, and the power furnished by the supply V_{CC}.

The basic amplifier A basic transistor amplifier in the common-emitter configuration is shown in Fig. 2.3-1. The resistors R_1, R_2, R_L, and R_e and the supply voltage V_{CC} are to be chosen so that the transistor operates linearly and a maximum peak-to-peak swing in i_C is possible. Resistors R_1 and R_2 form a voltage divider across the V_{CC} supply. The function of this network is to provide *bias* conditions which ensure that the emitter-base junction is operating in the proper region. Figure 2.3-1 can be simplified by obtaining a Thevenin equivalent circuit for R_1, R_2, and V_{CC}, as shown in Fig. 2.3-2a. The Thevenin conversion can be made in either direction, using

$$V_{BB} = \frac{R_1}{R_1 + R_2} V_{CC} \tag{2.3-1a}$$

and

$$R_b = \frac{R_1 R_2}{R_1 + R_2} \tag{2.3-1b}$$

or

$$R_1 = \frac{R_b}{1 - V_{BB}/V_{CC}} \tag{2.3-1c}$$

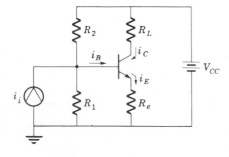

Common emitter

Figure 2.3-1 Basic common-emitter amplifier.

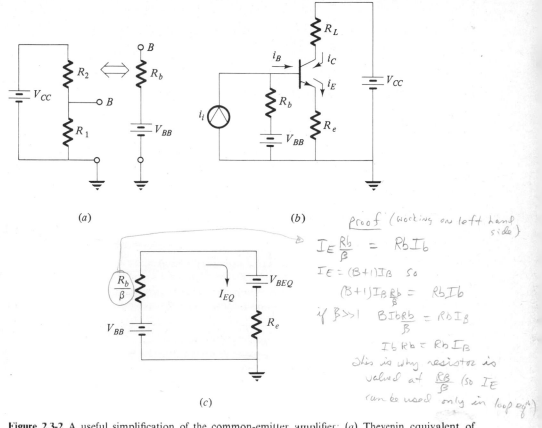

(a)

(b)

(c)

Proof (working on left hand side)

$$I_E \frac{Rb}{\beta} = Rb I_b$$

$$I_E = (\beta+1)I_B \quad So$$

$$(\beta+1)I_B \frac{Rb}{\beta} = Rb I_b$$

if $\beta \gg 1 \quad \beta I_b \frac{bRb}{\beta} = Rb I_B$

$$I_b \, Rb = Rb \, I_B$$

this is why resistor is valued at $\frac{RB}{\beta}$ *(so* I_E *can be used only in loop eq^n)*

Figure 2.3-2 A useful simplification of the common-emitter amplifier: (a) Thevenin equivalent of bias circuit; (b) simplified amplifier; (c) dc bias circuit.

and

$$R_2 = R_b \frac{V_{CC}}{V_{BB}} \tag{2.3-1d}$$

The resulting simplified circuit of the common-emitter amplifier is shown in Fig. 2.3-2b. Note that due to the dc supply voltages V_{CC} and V_{BB}, dc currents I_{BQ}, I_{CQ}, and I_{EQ} flow in the transistor and that due to the input current i_i, the small signal currents i_b, i_c, and i_e also flow. Hence, the total current flow in the transistor is

$$i_B = I_{BQ} + i_b \tag{2.3-2a}$$

$$i_C = I_{CQ} + i_c \tag{2.3-2b}$$

and

$$i_E = I_{EQ} + i_e \tag{2.3-2c}$$

Since i_i causes the collector current to change, the collector-emitter voltage also changes and

$$v_{CE} = V_{CEQ} + v_{ce} \tag{2.3-2d}$$

The operation of the amplifier is determined by first setting $i_i = 0$ so that dc conditions prevail and then calculating the quiescent operating point. Next, with i_i present, we determine the load line which describes the current-voltage variation in the circuit.

We begin by letting $i_i = 0$. Then

$$V_{CC} = v_{CE} + i_C R_L + i_E R_e \tag{2.3-3a}$$

Since $i_C = \alpha i_E \approx i_E$, (2.3-3a) becomes

$$V_{CC} \approx v_{CE} + i_C(R_L + R_e) \tag{2.3-3b}$$

The second equation is obtained using KVL around the emitter-base circuit. Again with $i_i = 0$,

$$V_{BB} - i_B R_b = v_{BE} + i_E R_e \tag{2.3-3c}$$

Since

$$(\text{Eq } 2.2\text{-}11) \quad i_B = i_E(1 - \alpha)$$

(handwritten:) $iE = iB + iC$
$iE = ib + \beta ib$
$iE = ib(1+\beta)$ $ib = \dfrac{iE}{1+\beta}$

(2.3-3c) becomes

(handwritten:) $\beta = \dfrac{\alpha}{1-\alpha}$

$$V_{BB} - v_{BE} = i_E[R_e + (1 - \alpha)R_b]$$

and

$$i_E = \frac{V_{BB} - v_{BE}}{R_e + (1 - \alpha)R_b} \tag{2.3-4a}$$

Since $1 - \alpha = 1/(\beta + 1) \approx 1/\beta$ for $\beta \gg 1$, which is usually the case, and since no signal is present, so that $v_{BE} = V_{BEQ}$, (2.3-4a) becomes

$$I_{EQ} = \frac{V_{BB} - V_{BEQ}}{R_e + R_b/\beta} \tag{2.3-4b}$$

(handwritten:) $\beta = \dfrac{I_C R_b}{V_{BB} - V_{BEQ} - I_{CQ} R_E}$

or

$$I_{EQ} R_e + I_{EQ}\frac{R_b}{\beta} = V_{BB} - V_{BEQ} \tag{2.3-4c}$$

This equation exactly describes the circuit of Fig. 2.3-2c, called the *dc bias circuit*. This circuit is useful as a conceptual aid in bias calculations because it is easily remembered. Its use will be illustrated in the examples.

Differences in individual transistors can cause $1/(\beta + 1)$ to change by a factor of 3 or more to 1, thus changing the dc emitter current. Consequently, circuits are designed so that

(handwritten:) Desirable to have $\rightarrow$

$$R_e \gg \frac{R_b}{\beta + 1} \approx \frac{R_b}{\beta} \tag{2.3-5a}$$

to eliminate variations in i_E due to variations in β. As a rule of thumb it is usually sufficient to have

$$R_e = \frac{10R_b}{\beta} \quad \text{or} \quad R_b = \frac{\beta R_e}{10} \tag{2.3-5b}$$

Using this rule of thumb we see that (2.3-4a) becomes

$$I_{CQ} \approx I_{EQ} \approx \frac{V_{BB} - V_{BEQ}}{1.1R_e} \tag{2.3-6}$$

where $V_{BEQ} \approx 0.7$ V.

(handwritten:) $V_{CEQ} = V_{CC} - I_E(R_C + R_E)$
$\approx I_C$

Equations (2.3-3b) and (2.3-6) can be solved algebraically for $v_{CE} = V_{CEQ}$

$$V_{CEQ} = V_{CC} - (V_{BB} - 0.7)(0.9)\left(1 + \frac{R_L}{R_e}\right) \qquad (2.3\text{-}7)$$

Equations (2.3-6) and (2.3-7) give the dc (quiescent) operating conditions for the transistor circuit of Fig. 2.3-2. We now turn to a graphical analysis in order to determine I_{CQ} (the quiescent collector current) to permit a maximum swing in collector current. Before attempting this it will be helpful to discuss the interpretation of some of the equations which have been written, in terms of the diode analysis of Chap. 1 and the transistor theory of Sec. 2.2. Consider (2.3-3b), which describes the operation of the collector-emitter loop. This has exactly the same form as that describing the diode circuit of Sec. 1.3, with the important difference that the junction is reverse-biased here. The variables v_{CE} and i_C are also related by the vi characteristic given in Fig. 2.2-4. Thus (2.3-3b) represents a dc load line, and we proceed to solve graphically, exactly as for the diode. This load line is plotted in Fig. 2.3-3a.

The load line defines the operating path of this circuit. When v_{CE} is less than 0.1 to 0.2 V (Fig. 2.2-4), the transistor is said to be *saturated*, as discussed

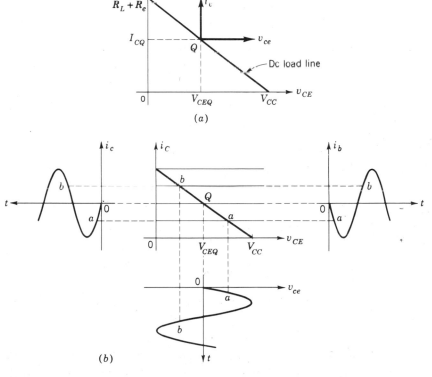

Figure 2.3-3 Graphical analysis: (*a*) load line; (*b*) waveforms.

previously. When the collector current becomes zero ($v_{CE} = V_{CC}$ in this example), the transistor is said to be at *cutoff*. Note that

$$i_C = I_{CQ} + i_c \quad \text{and} \quad v_{CE} = V_{CEQ} + v_{ce}$$

Hence, as in Chap. 1, a set of i_c-v_{ce} axes passing through the Q point can be drawn, as shown in the figure.

If the variation of i_b is known, values of v_{ce} and i_c can be found by using the graphical construction of Fig. 2.3-3b. This need not be done unless the nonlinearity of the collector characteristics is to be taken into account.

Because of the variability of β, the insertion of the constant i_B curves onto the vi characteristic is meaningful quantitatively only if they were measured using the transistor being considered. The results then obtained from Fig. 2.3-3b would probably not be valid if we replaced the measured transistor with a different one of the same type but having a different value of β. For this reason the base-current i_B curves are usually omitted when performing a graphical analysis. Figure 2.3-3b is of importance since it illustrates the phase relationships between i_b, i_c, and v_{ce}. Thus, as i_b increases, i_c increases and v_{ce} decreases. This is clearly seen by referring to points a and b on the waveforms of Fig. 2.3-3b.

2.3-1 Maximum Symmetrical Swing

One of the first decisions involved in designing a common-emitter amplifier is the choice of the Q-point location. In Fig. 2.3-3a the Q point is placed so that the load line is bisected. This will allow a maximum symmetrical variation in collector current, a condition known as *maximum symmetrical swing*. If the input-signal excursions are large enough for the operating point to move over an appreciable portion of the load line, clearly this is a design condition which will ensure linear operation over a maximum range of input signal. In this section we show how to design the bias network so that the Q point is placed in the center of the load line for maximum symmetrical swing.

There are many conditions under which maximum symmetrical swing is neither necessary nor desirable. For example, if the signal excursion covers only a small portion of the load line, we may choose the Q-point location on the basis of other constraints. A very practical problem is to minimize quiescent power-supply current drain. In order to satisfy this requirement we would place the Q point as close to cutoff as possible. Such Q-point placement will be discussed in the next section.

Since the collector current can vary from zero to approximately $V_{CC}/(R_L + R_e)$ (neglecting the small saturation voltage), a quiescent current

$$I_{CQ} = \frac{V_{CC}/2}{R_L + R_e} \qquad (2.3\text{-}8)$$

FOR MAX SYM SWING

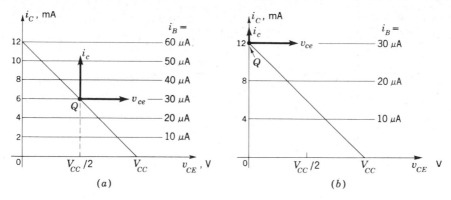

Figure 2.3-4 Effect of β variation on the Q point: (a) $\beta_1 = 200$; (b) $\beta_2 = 400$.

will yield the maximum symmetrical collector-current swing. If a sinusoidal base current is applied to produce this swing, the total collector current will be

$$i_c = I_{CQ} + i_c \qquad (2.3\text{-}9)$$

where

$$I_{CQ} = \frac{V_{cc}/2}{R_L + R_e} \qquad (2.3\text{-}10)$$

and

$$i_c = \frac{V_{cc}/2}{R_L + R_e} \cos \omega t = I_{cm} \cos \omega t \qquad (2.3\text{-}11)$$

Limitations on the amount of current that can flow safely, the collector voltage, and the power that can be dissipated in the transistor are discussed in Chap. 5.

It is interesting to observe the effect of variations in β on the response of the amplifier circuit of Fig. 2.3-2b. This is accomplished with the aid of Fig. 2.3-4a, which shows a nominal quiescent point at

for max sym. swing

$$V_{CEQ} = \frac{V_{cc}}{2} \qquad I_{CQ} = 6 \text{ mA} \qquad \text{and} \qquad I_{BQ} = 30 \ \mu\text{A}$$

If we now replace the original transistor by one with twice the β, that is, $\beta_2 = 2\beta_1$, and adjust R_1 and R_2 so that the quiescent base current remains at 30 μA, the Q point will shift (see Fig. 2.3-4b) to

$$V_{CEQ} = 0 \qquad I_{CQ} = 12 \text{ mA} \qquad I_{BQ} = 30 \ \mu\text{A}$$

Thus, with one transistor, we have a maximum collector-voltage swing, while the other transistor is in saturation and hence there is no swing. This is the basic reason for requiring the inequality in (2.3-5a) to be met so that the transistor is biased with a constant-emitter rather than a constant-base current. The important question of how to arrange the input circuit to achieve the desired Q point is postponed until later.

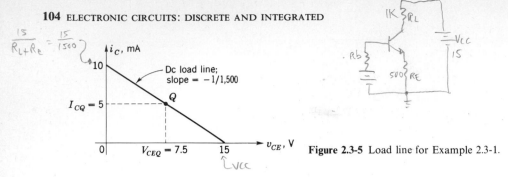

$$\frac{15}{R_L + R_e} = \frac{15}{1500}$$

Figure 2.3-5 Load line for Example 2.3-1.

Example 2.3-1 In the circuit of Fig. 2.3-2, $V_{CC} = 15$ V, $R_L = 1$ kΩ, and $R_e = 500\ \Omega$. Determine the maximum symmetrical swing in collector current and the Q point.

SOLUTION The dc load line is plotted in Fig. 2.3-5. In order to obtain the maximum symmetrical swing, choose the quiescent collector current in the center of the load line. Thus

$$I_{CQ} = 5 \text{ mA} \qquad \text{and} \qquad V_{CEQ} = 7.5 \text{ V}$$

The peak-to-peak collector current can reach 10 mA. ///

Example 2.3-2 Find the Q point for maximum symmetrical collector-current swing in the circuit of Fig. 2.3-6.

SOLUTION The dc-load-line equation for the circuit is

$$9 \approx V_{CEQ} + I_{CQ}(1000 + 200)$$

where it has been assumed that $200I_{EQ} \approx 200I_{CQ}$, that is, $\alpha \approx 1$. The dc-load-line equation is then plotted as shown. When the collector-emitter voltage is zero (saturation), maximum collector current flows (7.5 mA). Thus, to achieve a maximum symmetrical swing, the Q point should be set at 3.75 mA, so that the peak current will be 3.75 mA. Note that the result assumes that the saturation voltage is zero. The reader can readily verify that the solution is not altered significantly by assuming that saturation occurs when $v_{CE, \text{sat}} = 0.2$ V.

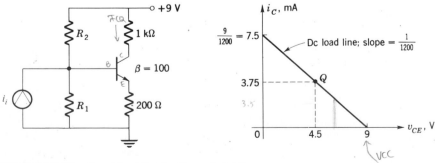

Figure 2.3-6 Circuit and load line for Example 2.3-2.

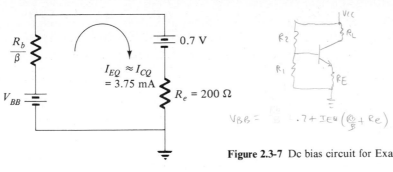

$$V_{BB} = \quad .7 + I_{EU}\left(\tfrac{R_b}{B} + R_e\right)$$

Figure 2.3-7 Dc bias circuit for Example 2.3-3.

If the Q point were set at 3.5 mA, the maximum symmetrical swing would be reduced to 3.5 mA; if the Q point were set at 4 mA, the maximum symmetrical swing would again be 3.5 mA. ///

Example 2.3-3 Find R_1 and R_2 in Example 2.3-2 to set the Q point at $I_{CQ} = 3.75$ mA, $V_{CEQ} = 4.5$ V.

SOLUTION The dc bias circuit is shown in Fig. 2.3-7. In order to stabilize the quiescent current against β variations we use the rule of thumb given in (2.3-5b)

$$R_b = \frac{\beta R_e}{10} - \frac{(100)(200)}{10} - 2\ \text{k}\Omega \qquad R_b = \frac{R_1 R_2}{R_1 + R_2}$$

Then from the dc bias circuit $V_{BB} = I_E R_E + I_E R_b + .7$

$$V_{BB} = (3.75)(200 + \tfrac{2000}{100}) \times 10^{-3} + 0.7 = 1.525\ \text{V}$$

When we know V_{BB} and R_b, R_1 and R_2 can be determined from (2.3-1c) and (2.3-1d)

$$R_1 = \frac{R_b}{1 - V_{BB}/V_{CC}} = \frac{2\ \text{k}\Omega}{1 - 1.525/9} \approx 2.41\ \text{k}\Omega$$

and

$$R_2 = \frac{V_{CC}}{V_{BB}} R_b = \frac{9}{1.525}(2\ \text{k}\Omega) \approx 11.8\ \text{k}\Omega$$

As a practical matter, standard resistors would be used. Thus

$$R_1 = 2.2\ \text{k}\Omega \qquad \text{and} \qquad R_2 = 12\ \text{k}\Omega$$

These standard values for R_1 and R_2 in turn yield

$$V_{BB} = \frac{2.2}{14.2}(9) \approx 1.4\ \text{V} \qquad \text{and} \qquad R_b = \frac{(2.2)(12)}{14.2} \approx 1.9\ \text{k}\Omega$$

The resulting quiescent current is then [Eq. (2.3-6)]

$$I_{CQ} \approx \frac{1.4 - 0.7}{(1.1)(200)} = 3.18\ \text{mA}$$

Thus, because of the use of standard resistors, the maximum swing is 3.18 mA rather than 3.75 mA. Since saturation has been neglected, we should expect the actual maximum swing to be somewhat less than 3.18 mA. ///

2.3-2 Arbitrary Q-Point Placement

As noted previously, the Q point may be located anywhere on the dc load line by properly choosing bias resistors R_1 and R_2. The dc bias circuit can be used to advantage as an aid in visualization during this design process. Once R_b has been chosen to provide the desired bias stability, all the information required can be obtained from the dc bias circuit. A typical problem is solved in the following example.

Example 2.3-4 In the circuit of Fig. 2.3-6 find new values for R_1 and R_2 which will minimize quiescent power-supply current drain. Assume that the input signal is such that the maximum expected collector-current swing will be 1 mA peak to peak about the Q point.

SOLUTION In order to provide a safety margin, we allow 0.25 mA above cutoff to the lowest expected point on the collector-current swing. Thus, as shown on the dc load line in Fig. 2.3-8, the Q point is located at $I_{CQ} = 0.75$ mA. The

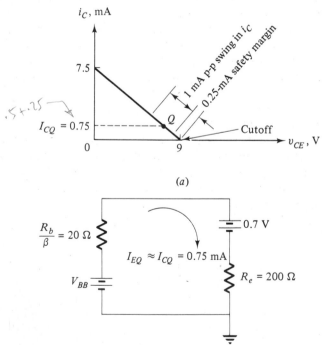

(a)

(b)

Figure 2.3-8 Load line and dc bias circuit for Example 2.3-4.

resulting dc bias circuit is also shown in the figure. The factor-of-10 rule of thumb for bias stability has been used to choose R_b. From the dc bias circuit

$$V_{BB} = (0.75)(220)(10^{-3}) + 0.7 = 0.865 \text{ V}$$

Next we use (2.3-1c) and (2.3-1d)

$$R_1 = \frac{R_b}{1 - V_{BB}/V_{CC}} = \frac{2 \text{ k}\Omega}{1 - 0.865/9} = 2.2 \text{ k}\Omega$$

$$R_2 = \frac{V_{CC}}{V_{BB}} R_b = \frac{9}{0.865}(2 \text{ k}\Omega) = 20.8 \text{ k}\Omega$$

Choosing standard values, we would use $R_1 = 2.2 \text{ k}\Omega$ and $R_2 = 20 \text{ k}\Omega$. (The next smaller standard value is chosen for R_2 in order not to decrease our safety margin. Why?)   ///

Quick estimates of quiescent conditions in transistor circuits It is often useful to be able to make a quick estimate of the quiescent current in a transistor amplifier. With the dc bias circuit this is easily done by neglecting R_b/β. Then

$$I_{CQ} \approx \frac{V_{BB} - 0.7}{R_e}$$

With R_e known, V_{BB} is found from the R_1 and R_2 voltage-divider ratio. Consider Example 2.3-4, where $R_1 = 2.2 \text{ k}\Omega$ and $R_2 = 20 \text{ k}\Omega$. Then

$$V_{BB} \approx \frac{2.2}{2.2 + 20}(9) = 0.892 \text{ V} \qquad \text{and} \qquad I_{CQ} \approx \frac{0.892 - 0.7}{0.2} = 0.96 \text{ mA}$$

The design value for I_{CQ} was 0.75 mA, but since we neglected R_b/β, we would expect our estimate to be high.

2.4 POWER CALCULATIONS[2]

Power calculations need to be made in a transistor-amplifier design for several reasons. One is that amplifiers must accept signals at a given power level and deliver them to a load at another (usually much higher) power level. Another reason is that the elements in a transistor amplifier all have limitations on the amount of power they can dissipate or supply. Transistors are rated according to maximum allowable collector dissipation; if this rating is exceeded, the transistor may be damaged. Resistors are also rated according to maximum power dissipation, typical values being 0.1, 0.5, 1, 2, and 10 W. Since size and cost are proportional to power rating, we always try to use the lowest power rating consistent with the actual power we expect the resistor to dissipate. Finally the dc power supply, which provides V_{CC}, is capable of supplying only a finite amount of power which cannot be exceeded. In this section we calculate the power supplied by the

V_{CC} supply, the power dissipated in R_L and R_e, and the power dissipated in the transistor.

The average power supplied to or dissipated by any linear or nonlinear device is simply

$$P = \frac{1}{T} \int_0^T V(t)I(t)\, dt \tag{2.4-1}$$

where V = total voltage across device
I = total current through device
T = period of any periodic time-varying part of V (or I)

If

$$V = V_{av} + v(t) \tag{2.4-2a}$$

and $$I = I_{av} + i(t) \tag{2.4-2b}$$

where V_{av} and I_{av} are average values and $v(t)$ and $i(t)$ are time-varying components having an average value of zero, then

$$P = \frac{1}{T} \int_0^T [V_{av} + v(t)][I_{av} + i(t)]\, dt$$

$$= \frac{1}{T} \int_0^T V_{av} I_{av}\, dt + \frac{1}{T} \int_0^T V_{av} i(t)\, dt$$

$$+ \frac{1}{T} \int_0^T I_{av} v(t)\, dt + \frac{1}{T} \int_0^T v(t)i(t)\, dt$$

$$P = V_{av} I_{av} + \frac{1}{T} \int_0^T v(t)i(t)\, dt \tag{2.4-3a}$$

since

$$\frac{1}{T} \int_0^T i(t)\, dt = \frac{1}{T} \int_0^T v(t)\, dt = 0 \tag{2.4-3b}$$

by definition of $i(t)$ and $v(t)$. Equation (2.4-3a) shows that the average power supplied (or dissipated) by a device consists of the sum of the power in the dc (average) terms and the power in the ac terms.

Average power dissipated in load Let us now turn our attention to the transistor circuit of Fig. 2.3-2b. The ac power dissipated in the load $P_{L,\,ac}$ is

$$P_{L,\,ac} = \frac{1}{T} \int_0^T i_c^2 R_L\, dt \tag{2.4-4a}$$

If we assume that i_c is sinusoidal

$$i_c = I_{cm} \cos \omega t \tag{2.4-4b}$$

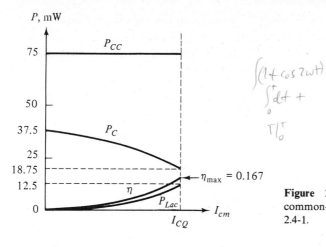

$\int (1 + \cos 2\omega t)$

$\int_{0}^{t} dt +$

$T\big|_{0}^{T}$

$\cos^{2}\theta = \frac{1}{2}(1 + \cos 2\theta)$

Figure 2.4-1 Power relations in the common-emitter amplifier of Example 2.4-1.

the ac power dissipated in the load becomes

$$P_{L, ac} = \frac{1}{T} \int_{0}^{T} R_L I_{cm}^2 \cos^2 \omega t \; dt$$

$$= R_L \frac{1}{T} \int_{0}^{T} \frac{I_{cm}^2}{2} (1 + \cos 2\omega t) \; dt = \frac{I_{cm}^2 R_L}{2} \qquad (2.4\text{-}4c)$$

Equation (2.4-4c) is plotted in Fig. 2.4-1 as a function of I_{cm} for the amplifier of Example 2.3-1. Since the power increases parabolically with I_{cm}, maximum ac power is dissipated in the load when I_{cm} is a maximum. If the quiescent collector current is chosen for maximum swing,

$$I_{cm, max} = I_{CQ}$$

The maximum average power dissipated in the load is then

$$(P_{L, ac})_{max} = \frac{I_{CQ}^2 R_L}{2} \qquad (2.4\text{-}5a)$$

Using (2.3-10), we have

$$(P_{L, ac})_{max} = \frac{V_{CC}^2 R_L}{8(R_L + R_e)^2} \qquad (2.4\text{-}5b)$$

To maximize the power dissipated in the load resistor, the emitter resistor is made much smaller than the load resistor; that is, $R_L \gg R_e$. Then the peak-to-peak ac voltage swing across R_L is approximately V_{CC}, and

$$(P_{L, ac})_{max} \approx \frac{V_{CC}^2}{8R_L} \qquad (2.4\text{-}5c)$$

It must be noted that decreasing R_e requires a decrease in R_b [Eq. (2.3-5)], which in turn results in a decrease in current gain (Chap. 6). There is a practical limit to how small R_e can be made for a given degree of stability against β and temperature variations. This is discussed in Chap. 4.

Average power delivered by the supply The average power delivered by the supply is

In general

$$P_{CC} = \frac{1}{T}\int_0^T V_{CC} i_C \, dt = \frac{1}{T}\int_0^T V_{CC}[I_{CQ} + i_c(t)] \, dt = \boxed{V_{CC}I_{CQ}} \quad (2.4\text{-}6a)$$

The supplied power is seen to be a constant, independent of signal power for the distortionless conditions that have been assumed. Under the above conditions, when I_{CQ} is chosen for maximum swing,

$$I_{CQ} = \frac{V_{CC}}{2(R_L + R_e)} \qquad \boxed{\text{For max swing}} \qquad (2.4\text{-}6b)$$

Hence $\qquad P_{CC} = \frac{V_{CC}^2}{2(R_L + R_e)} \approx \frac{V_{CC}^2}{2R_L} \qquad$ when $R_L \gg R_e \qquad (2.4\text{-}6c)$

Average power dissipated in the collector The power dissipated in the collector P_C is

but $v_{CE} = V_{CC} - (R_L + R_e)i_C$

$$P_C = \frac{1}{T}\int_0^T v_{CE} i_C \, dt = \frac{1}{T}\int_0^T [V_{CC} - (R_L + R_e)i_C]i_C \, dt$$

$$= \frac{1}{T}\int_0^T V_{CC} i_C \, dt - (R_L + R_e)\frac{1}{T}\int_0^T i_C^2 \, dt \qquad (2.4\text{-}7a)$$

The first term is recognized as the power delivered by the supply P_{CC}. The second integral represents the dc and ac power dissipated in the load P_L and in the emitter resistor P_E. This result should be intuitively obvious, since the power delivered by the supply must equal the sum of all the other power components:

$$P_{CC} = P_C + P_L + P_E \qquad (2.4\text{-}7b)$$

Equation (2.4-7a) can be evaluated after performing the integration

$$\frac{1}{T}\int_0^T i_C^2 \, dt = \frac{1}{T}\int_0^T (I_{CQ} + I_{cm}\cos\omega t)^2 \, dt = I_{CQ}^2 + \frac{I_{cm}^2}{2} \qquad (2.4\text{-}7c)$$

Thus $\qquad P_C = P_{CC} - (R_L + R_e)I_{CQ}^2 - (R_L + R_e)\frac{I_{cm}^2}{2} \qquad (2.4\text{-}7d)$

The power dissipated in the collector [Eq. (2.4-7d)] is plotted in Fig. 2.4-1. It is seen that the collector dissipation is a maximum when no signal is present.

$$P_{C,\,max} = P_{CC} - (R_L + R_e)I_{CQ}^2 = \frac{V_{CC}^2}{4(R_L + R_e)} \approx \frac{V_{CC}^2}{4R_L} \qquad (2.4\text{-}8)$$

✳ Max. Collector Dissipation is when No A.C. Signal is Present

In most low-power transistor circuits the power dissipated in the input circuit is small, so that P_C represents the total dissipation internal to the transistor. The maximum value of P_C is always specified by the manufacturer, and this value must not be exceeded if the junction temperature is to be kept within safe limits. This is discussed in detail in Chap. 5.

It is interesting and important to note that the power supply furnishes only dc power. However, the useful power dissipated in the load is ac power, i.e., dissipation resulting from an ac signal present in the load. This ac load power is generated in the transistor amplifier and appears as the third term of (2.4-7d). Thus increasing the ac current increases the ac generated power. This decreases the collector dissipation and increases the power delivered to the load (and to the emitter resistor, in this case).

The maximum collector dissipation is twice the maximum power that can be delivered to the load. Thus this device is a very inefficient power amplifier. Ideally, one would like no power to be dissipated if there were no signal present. For example, consider using this amplifier in the receiver of an intercom system. The receiver is always on, yet we do not want to dissipate power in the receiver unless a voice signal is present. Use of the amplifier above would not be economical in this application. The class B amplifier studied in Chap. 5 is a circuit which dissipates almost no power unless a signal is present.

Efficiency The ratio of the ac power dissipated in the load resistor to the power delivered by the supply is defined as the efficiency η of the amplifier

efficiency

$$\eta = \frac{P_{L,\,ac}}{P_{CC}} = \frac{I_{cm}^2(R_L/2)}{V_{CC}^2/2R_L} = \frac{I_c^2(R_L/2)}{V_{cc}(I_{CQ})} \tag{2.4-9a}$$

The maximum efficiency occurs when the signal is maximum, since P_{CC} is constant and $P_{L,\,ac}$ increases with increasing current. Then

Max efficiency = 25%

$$\eta_{max} = \frac{V_{CC}^2/8R_L}{V_{CC}^2/2R_L} = 0.25 \qquad R_L \gg R_e \tag{2.4-9b}$$

This type of amplifier is extremely inefficient and is not often used to amplify or deliver large amounts of power. It is, however, used extensively, as will be seen in Chap. 5, to amplify current at low power levels (under 500 mW).

The ratio of the maximum collector power to the maximum ac load power is

$$\frac{P_{C,\,max}}{(P_{L,\,ac})_{max}} \approx \frac{V_{CC}^2/4R_L}{V_{CC}^2/8R_L} = 2 \tag{2.4-10}$$

Thus, using this type of amplifier, to obtain 1 W dissipation in the load requires a transistor capable of handling 2 W of collector dissipation. This is extremely wasteful, and this ratio can be significantly reduced (to $\frac{1}{5}$), using a class B amplifier. This configuration is discussed in detail in Chap. 5.

Example 2.4-1 For the circuit of Example 2.3-1 calculate the power supplied by the collector supply, the power dissipated in the load and emitter resistors, the power dissipated in the transistor, and the efficiency of operation.

SOLUTION The power furnished by the collector supply is

$$P_{CC} = V_{CC}I_{CQ} = (15)(5 \times 10^{-3}) = 75 \text{ mW}$$

The power dissipated in the load and emitter resistors (assuming a sinusoidal signal) is

$$P_L + P_E \approx I_{CQ}^2(R_L + R_e) + \frac{1}{T}\int_0^T i_c^2(t)(R_L + R_e)\, dt$$

$$= (R_L + R_e)\left(I_{CQ}^2 + \frac{I_{cm}^2}{2}\right)$$

and

$$(P_L + P_E)_{\text{max}} = (1.5 \times 10^3)\left[(5 \times 10^{-3})^2 + \frac{(5 \times 10^{-3})^2}{2}\right] = 56.25 \text{ mW}$$

The power dissipated in the transistor varies with ac collector current. Using (2.4-7d)

$$P_C = V_{CC}I_{CQ} - (R_L + R_e)I_{CQ}^2 - (R_L + R_e)\frac{I_{cm}^2}{2}$$

$$= [V_{CC} - (R_L + R_e)I_{CQ}]I_{CQ} - (R_L + R_e)\frac{I_{cm}^2}{2}$$

$$= V_{CE}I_{CQ} - (R_L + R_e)\frac{I_{cm}^2}{2}$$

$$= (37.5 \times 10^{-3}) - \tfrac{1500}{2}I_{cm}^2$$

Notice that maximum collector dissipation occurs when there is no ac signal.

$$P_{C,\text{max}} = 37.5 \text{ mW}$$

The efficiency is

$$\eta = \frac{P_{L,\text{ac}}}{P_{CC}} = \frac{\frac{1}{2}I_{cm}^2 R_L}{V_{CC}I_{CQ}} = \frac{I_{cm}^2 \times 10^3}{(2)(15)(5 \times 10^{-3})} = \frac{10^6}{150}I_{cm}^2$$

The efficiency is 16.7 percent at $I_{cm,\text{max}}$. This value is less than 25 percent [Eq. 2.4-9)] because R_e is not negligible for this amplifier. The results are plotted as a function of I_{cm} in Fig. 2.4-1.　　　///

2.5 THE INFINITE BYPASS CAPACITOR

The emitter resistor R_e is required to obtain the desired dc quiescent emitter-current stability. However, the inclusion of R_e causes a decrease in amplification at nonzero frequencies. Thus R_e is an element which we need at its full value for dc but would prefer to have as a short circuit for ac. This can be accomplished by connecting a capacitor in parallel with R_e, as shown in Fig. 2.5-1. In this section, an infinite capacitance is assumed, to avoid frequency effects; then the emitter is at ground potential for all ac signals. The effect of finite capacitance on the response of the amplifier is determined in Chap. 9.

The quiescent point for this circuit is found as before, since at dc the capacitor acts as an open circuit. The dc load line is shown in Fig. 2.5-2.

Now we are faced with a situation similar to that of Sec. 1.5, where both dc and ac conditions had to be satisfied. For ac signals the collector-emitter circuit impedance is not $R_L + R_e$, as in Sec. 2.3, but simply R_L, because the capacitor effectively short-circuits R_e at all signal frequencies. Thus we must construct an ac load line with a slope $-1/R_L$. The ac load line will then be the operating path for ac signals and must pass through the Q point, because when the ac signal is reduced to zero, the operating path must reduce to the Q point. Let us locate the Q point so as to obtain a maximum symmetrical swing. The situation is illustrated graphically in Fig. 2.5-3.

The equation of the ac load line is

$$i_c + \frac{v_{ce}}{R_L} = 0$$

AC load
Line Eqⁿ

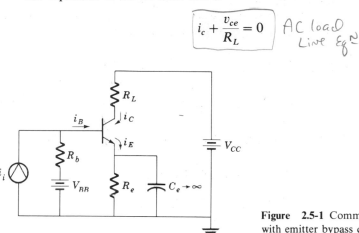

Figure **2.5-1** Common-emitter amplifier with emitter bypass capacitor.

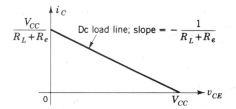

Figure **2.5-2** Dc load line for the common-emitter amplifier.

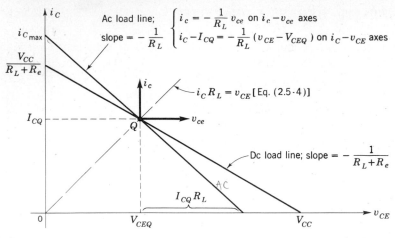

Figure 2.5-3 Dc and ac load lines for the amplifier of Fig. 2.5-1.

which can be written

$$i_C - I_{CQ} = -\frac{1}{R_L}(v_{CE} - V_{CEQ}) \tag{2.5-1}$$

The maximum value of i_C occurs when $v_{CE} = 0$ $(I_{CQ} + i_c)R_C + I_{CQ}R_E + V_{CE} = V_{CC}$ ⓔ '

$$\boxed{i_{C,\,max} = I_{CQ} + \frac{V_{CEQ}}{R_L}} \tag{2.5-2}$$

AC load line value $I_{CQ}(R_E + R_C) + V_{CEQ} = V_{CC}$ ⓔ 2
subtracting ⋯⋯
$-V_{CEQ} + i_c R_C + V_{CE} = 0$

To obtain maximum symmetrical swing the Q point should bisect the ac load line so that   make $V_{CE} = 0$

$$i_{C,\,max} = 2I_{CQ} \tag{2.5-3}$$

$-V_{CEQ} = -i_c R_C$
$i_C = \frac{V_{CEQ}}{R_L}$

Substituting in (2.5-2) gives   but $I_c = I_{CQ} + i_c$

$$2I_{CQ} = I_{CQ} + \frac{V_{CEQ}}{R_L}$$ $i_{C\,max} = I_{CQ} + \frac{V_{CEQ}}{R_L}$

and $$I_{CQ} = \frac{V_{CEQ}}{R_L} \tag{2.5-4}$$

The Q point (2.5-4) lies on the line $i_C R_L = v_{CE}$, which passes through the origin. Its intersection with the dc load line yields the Q point for maximum symmetrical swing, as shown in Fig. 2.5-3. The design procedure is to first construct the dc load line, then the line $i_C R_L = v_{CE}$ through the origin, to determine the Q point. The ac load line is then drawn through the Q point with slope $-1/R_L$. The operating conditions must be chosen so that the maximum collector dissipation is not exceeded.

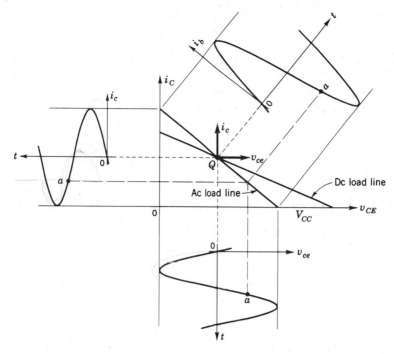

Figure 2.5-4 Waveforms in the common-emitter amplifier of Fig. 2.5-1 when adjusted for maximum swing.

It should be noted that the optimum Q point can also be obtained analytically by substituting (2.5-4) into the dc-load-line equation

$$V_{CC} = V_{CEQ} + I_{CQ}(R_L + R_e) \qquad (2.5\text{-}5)$$

Then
$$I_{CQ} \text{ (for maximum symmetrical swing)} = \frac{V_{CC}}{2R_L + R_e} \qquad (2.5\text{-}6a)$$

This important design equation can be put into an easily remembered form by noting that the ac load in the collector-emitter circuit is $R_{ac} = R_L$ while the dc load is $R_{dc} = R_L + R_e$. When these definitions are used, (2.5-6a) becomes

$$I_{CQ} = \frac{V_{CC}}{R_{ac} + R_{dc}} \qquad \text{Max swing} \qquad (2.5\text{-}6b)$$

and
$$V_{CEQ} = \frac{V_{CC}}{2 + R_e/R_L} = \frac{V_{CC}}{1 + R_{dc}/R_{ac}} \qquad (2.5\text{-}6c)$$

Now that the Q point and ac load line are specified, the current and voltage waveforms can be sketched. Figure 2.5-4 shows that as i_B increases, i_C increases and v_{CE} decreases. Thus i_b is in phase with i_c, and v_{ce} is 180° out of phase with both i_b and i_c.

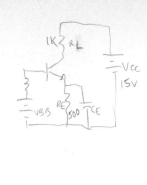

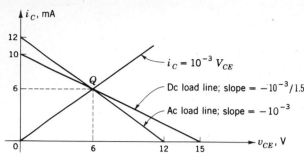

Figure 2.5-5 Load lines for Example 2.5-1.

Example 2.5-1 In Fig. 2.5-1, $V_{CC} = 15$ V, $R_L = 1$ kΩ, and $R_e = 500$ Ω. Find the maximum possible collector swing and the Q point.

SOLUTION The dc and ac load lines are shown in Fig. 2.5-5. This result can also be obtained analytically. We use (2.5-6b) with $R_{ac} = 1$ kΩ, and $R_{dc} = 1.5$ kΩ:

$$I_{CQ} = \frac{15}{2.5 \text{ k}\Omega} = 6 \text{ mA} \quad \text{and} \quad V_{CE} = \frac{15}{1 + 1.5/1} = 6 \text{ V}$$

The maximum peak collector swing is therefore

$$I_{cm} = 6 \text{ mA} \qquad\qquad ///$$

Example 2.5-2 Using the optimum Q point found in Example 2.5-1, calculate R_1 and R_2.

SOLUTION We use the factor-of-10 rule of thumb for bias stability so that

$$R_b = \frac{\beta R_e}{10} = \frac{(100)(0.5 \text{ k}\Omega)}{10} = 5 \text{ k}\Omega$$

Then from the dc bias circuit of Fig. 2.5-6b

$$V_{BB} = (6 \text{ mA})(0.55 \text{ k}\Omega) + 0.7 = 4 \text{ V}$$

$$R_1 = \frac{R_b}{1 - V_{BB}/V_{CC}} = \frac{5 \text{ k}\Omega}{1 - \frac{4}{15}} = 6.8 \text{ k}\Omega$$

and

$$R_2 = R_b \frac{V_{CC}}{V_{BB}} = (5 \text{ k}\Omega)(\tfrac{15}{4}) = 18.75 \text{ k}\Omega$$

Using standard values, we have

$$R_1 = 6.8 \text{ k}\Omega \qquad R_2 = 18 \text{ k}\Omega \qquad\qquad ///$$

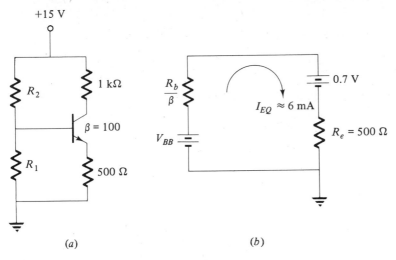

(a)   (b)

Figure 2.5-6 Example 2.5-2: (a) circuit; (b) dc bias circuit.

2.6 THE INFINITE COUPLING CAPACITOR

Quite often the load resistor must be ac-coupled to the transistor so that dc current will not flow through the load. This is usually accomplished by inserting a *coupling* capacitor between the collector and load, as shown in Fig. 2.6-1. This capacitor serves to *block* dc currents while permitting currents at signal frequencies to pass. The effect of this capacitor on the frequency characteristics of the amplifier is studied in Chap. 9. In this section maximum swing conditions are found when C_c is assumed to be infinite.

The dc-load-line equation is

$$V_{CC} = i_C(R_c + R_e) + v_{CE} \tag{2.6-1}$$

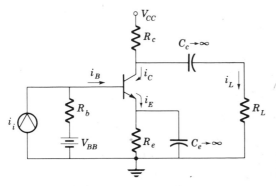

Figure 2.6-1 Amplifier with ac-coupled load.

and, noting that R_L and R_c are *in parallel* as seen from the collector terminal, we have the ac-load-line equation

$$i_C - I_{CQ} = -\frac{R_L + R_c}{R_L R_c}(v_{CE} - V_{CEQ}) \tag{2.6-2}$$

The resistance which determines the slope of the ac load line is the resistance seen by the collector at signal frequencies, $R_{ac} = R_L \| R_c$.

Since (2.6-2) is similar to (2.5-1), the same procedure can be used to determine the Q point for maximum swing. Thus the intersection of (2.6-1) and the straight line

$$i_C = \frac{R_c + R_L}{R_c R_L}v_{CE} = \frac{v_{CE}}{R_{ac}} \tag{2.6-3}$$

determines the Q point for maximum symmetrical swing. (The graphical interpretation of these three equations is the same as the curves shown in Fig. 2.5-3.) The quiescent current can be calculated by combining (2.6-1) and (2.6-3). This yields

$$V_{CC} = i_C\left(R_c + R_e + \frac{R_L R_c}{R_L + R_c}\right) = i_C(R_{dc} + R_{ac})$$

and at the Q point

$$I_{CQ} = \frac{V_{CC}}{R_c + R_e + R_L R_c/(R_L + R_c)} = \frac{V_{CC}}{R_{dc} + R_{ac}} \tag{2.6-4}$$

The maximum sinusoidal ac collector current with these bias conditions is

$$i_c = \frac{V_{CC}}{R_c + R_e + R_L R_c/(R_L + R_c)}\cos \omega t \tag{2.6-5}$$

and the maximum current in the load R_L is

$$i_L = \frac{R_c}{(R_c + R_L)}\frac{V_{CC}}{(R_c + R_e + R_L R_c/(R_L + R_c))}\cos \omega t \tag{2.6-6}$$

Note that the analysis above was performed algebraically, without reference to the previously described graphical technique. This is possible because of our assumption that the transistor is a *linear* amplifier over the range of voltages and currents of interest. The graphical technique is recommended in conjunction with the analysis because of the insight it provides. For example, in a design problem, a glance at the appropriate diagram often immediately indicates the effects of parameter changes, which can frequently become obscured in an equation.

Example 2.6-1 In Fig. 2.6-1, $V_{CC} = 15$ V, $R_c = 1$ kΩ, $R_e = 500$ Ω, and $R_L = 1$ kΩ. Find the Q point and the maximum symmetrical collector-current swing.

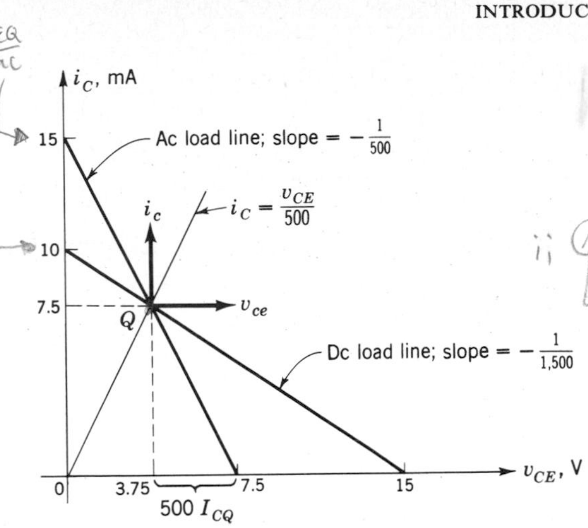

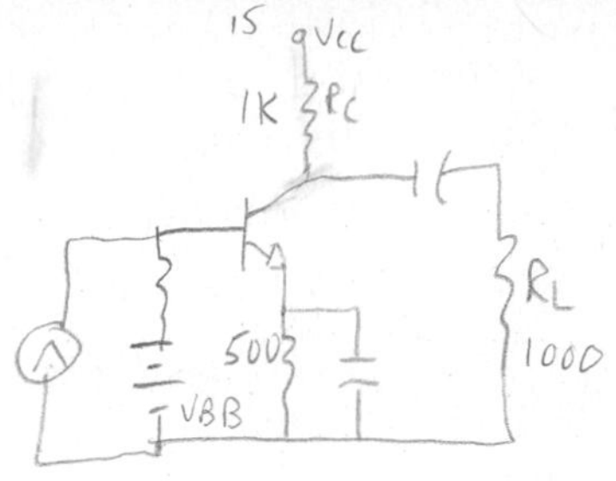

Figure 2.6-2 Load lines for Example 2.6-1.

SOLUTION The dc and ac load lines are plotted in Fig. 2.6-2. The Q point is most easily obtained from (2.6-4) with $R_{ac} = 500\ \Omega$ and $R_{dc} = 1.5\ k\Omega$

$$I_{CQ} = \frac{V_{CC}}{R_{ac} + R_{dc}} = \frac{15}{0.5 + 1.5} = 7.5 \text{ mA}$$

and
$$V_{CEQ} = I_{CQ} R_{ac} = (7.5)(0.5) = 3.75 \text{ V}$$

The maximum peak sinusoidal ac collector current is $I_{cm} = 7.5$ mA. ///

Example 2.6-2 If in Example 2.6-1 the emitter resistor is unbypassed, find the Q point, the maximum peak ac collector-current swing, and R_1 and R_2 (Fig. 2.6-3a).

SOLUTION As a result of the input current i_i the base current, collector current, and emitter current each consist of a small-signal current proportional to i_i as well as the dc quiescent current. Since the emitter resistor is

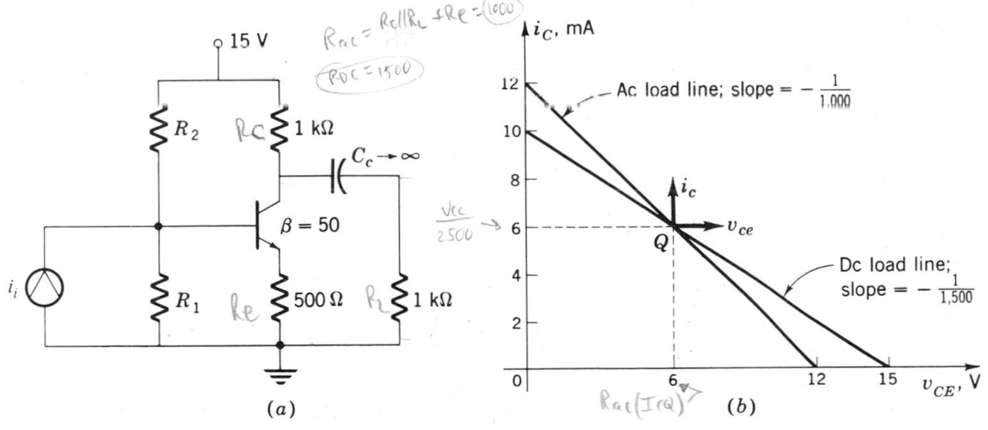

Figure 2.6-3 Circuit and load lines for Example 2.6-2: (a) circuit; (b) load lines.

unbypassed, the small-signal emitter current flows through R_e. The quiescent current can be obtained either graphically or by solving (2.6-1) and (2.6-3) simultaneously as follows: substitute into (2.6-1) to get

$$15 = v_{CE} + 1500i_C$$

Since the ac load seen by the collector is $(R_c \| R_L) + R_e$, (2.6-3) becomes $v_{CE} = 1000i_C$. Thus $I_{CQ} = 6\,\text{mA}$ and the maximum peak ac collector-current swing is $I_{cm} = 6\,\text{mA}$. The bias resistors R_1 and R_2 are calculated as before using the dc bias circuit.

Thus,

$$R_b = \frac{\beta R_e}{10} = \frac{(50)(500)}{10} = 2.5\ \text{k}\Omega$$

Then

$$V_{BB} = +(6)(0.55) + 0.7 = 4\ \text{V}$$

and $\quad R_2 = (2500)(\tfrac{15}{4}) = 9.4\ \text{k}\Omega \quad$ and $\quad R_1 = \dfrac{2500}{1 - \tfrac{4}{15}} = 3.41\ \text{k}\Omega$

With standard resistors, $R_2 = 10\ \text{k}\Omega$ and $R_1 = 3.3\ \text{k}\Omega$. ///

2.7 THE EMITTER FOLLOWER

The circuit shown in Fig. 2.7-1a represents the common-collector, or emitter-follower (EF), configuration. The term *follower* refers to the fact that the output voltage follows the signal voltage quite closely, as will be seen.

As before, the bias network is replaced by its Thevenin equivalent to obtain the circuit of Fig. 2.7-1b. The dc load line for this circuit has a slope $-1/R_e$ and is shown in Fig. 2.7-2.

For this circuit, application of KVL to the collector-emitter loop yields the simple equation

$$V_{CC} = v_{CE} + v_E$$

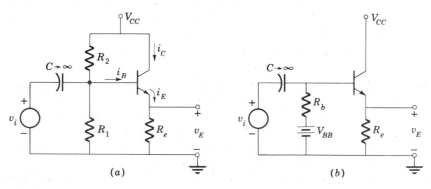

Figure 2.7-1 (a) Emitter-follower circuit; (b) simplified circuit.

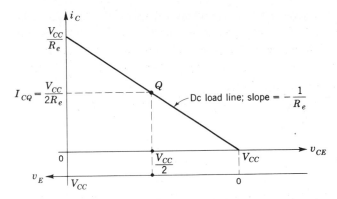

Figure 2.7-2 Load line and auxiliary output-voltage scale for emitter follower.

Therefore
$$v_E = V_{CC} - v_{CE} \tag{2.7-1}$$

Since V_{CC} is the point at which the dc load line intersects the v_{CE} axis, we can easily construct a v_E scale as shown in Fig. 2.7-2. Now, if the Q point is placed in the center of the load line (at $I_{CQ} = V_{CC}/2R_e$), the output swing will be symmetrical, varying from $v_E = 0$ to $v_E = V_{CC}$.

Writing KVL around the base-emitter loop, we have

$$v_B = v_{BE} + v_E \tag{2.7-2}$$

If the time variation of v_{BE} is negligible, we have

$$V_{BB} \approx V_E + 0.7$$

and
$$v_i \approx v_e \tag{2.7-3}$$

so that the output voltage follows the signal.

Usually, the load resistor is ac-coupled to the emitter, as shown in Fig. 2.7-3a.

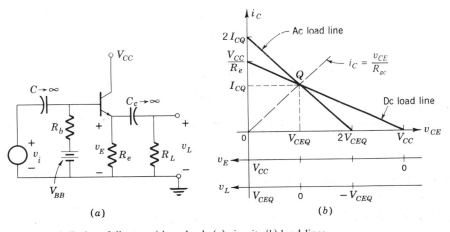

Figure 2.7-3 Emitter follower with ac load: (*a*) circuit; (*b*) load lines.

The corresponding dc and ac load lines are sketched in Fig. 2.7-3b. As before, the maximum symmetrical output swing occurs when the Q point is placed at the point where the equation

$$i_C = \frac{R_e + R_L}{R_e R_L} v_{CE} = \frac{v_{CE}}{R_{ac}} \qquad (2.7\text{-}4)$$

intersects the dc load line. The ac load line is drawn through this point, with slope $-1/R_{ac}$. The peak ac voltage swing across the load is V_{CEQ}, and the peak collector-current swing is I_{CQ}.

Notice that we performed the same calculations when considering the common-emitter amplifier circuits of Secs. 2.5 and 2.6 as when handling the emitter follower. The reason is that approximately the same current flows in the collector and emitter circuits. Thus, as far as the collector-emitter circuit is concerned, R_e and R_c are interchangeable. The difference is, of course, that when the output is taken from the emitter, $v_e \approx v_i$, while when the output is taken from the collector, $i_L \approx h_{fe} i_i$. The small-signal amplification and impedance characteristics of this device are discussed in detail in Chap. 6.

Example 2.7-1 Consider the circuit of Fig. 2.7-4. Find the values of R_1 and R_2 which will permit a maximum symmetrical swing in the output.

SOLUTION The design of the $R_1 - R_2$ bias network for the emitter follower parallels that for the common-emitter amplifier. The quiescent current is found from (2.6-4) (why does this equation apply here?):

For Max Swing

$$I_{CQ} = \frac{V_{CC}}{R_{dc} + R_{ac}} = \frac{21}{(1.5\,\text{k}\Omega \| 1\,\text{k}\Omega) + 1.5\,\text{k}\Omega} = 10\,\text{mA}$$

The Q point and the load lines are shown in Fig. 2.7-5a. Students should be certain they understand how the v_L scale was constructed. From the curves, v_L is seen to have a maximum peak-to-peak ac swing of 12 V.

For bias stability we choose

$$R_b = \frac{\beta Re}{10} = \frac{(100)(1.5\,\text{k}\Omega)}{10} = 15\,\text{k}\Omega$$

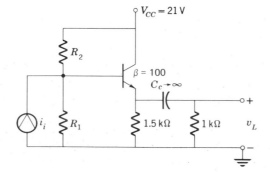

Figure 2.7-4 Emitter follower with ac load and bias network.

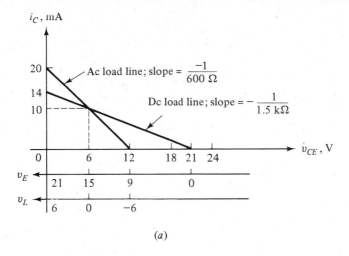

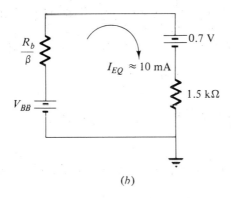

(b)

Figure 2.7-5 Example 2.7-1: (a) load lines; (b) dc bias circuit.

From the dc bias circuit

$$V_{BB} = (10)(1.65) + 0.7 = 17.2 \text{ V}$$

Then

$$R_1 = \frac{R_b}{1 - V_{BB}/V_{CC}} = \frac{15}{1 - 17.2/21} = 82.9 \text{ k}\Omega$$

and

$$R_2 = R_b \frac{V_{CC}}{V_{BB}} = 15 \frac{21}{17.2} = 18.3 \text{ k}\Omega$$

Standard values to be used would be

$$R_1 = 82 \text{ k}\Omega \qquad R_2 = 18 \text{ k}\Omega \qquad ///$$

We are often required to build a single amplifier and are not concerned with Q-point changes with β variation due to transistor replacement. In this case R_1

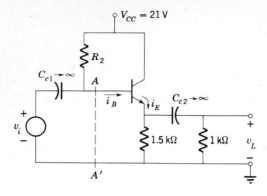

Figure 2.7-6 Emitter follower with base-injection bias.

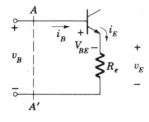

Figure 2.7-7 Dc portion of emitter follower of Fig. 2.7-6.

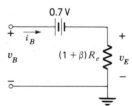

Figure 2.7-8 Equivalent circuit looking into the base terminal.

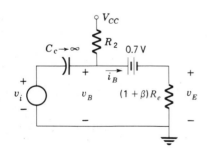

Figure 2.7-9 Complete dc equivalent circuit for base-injection bias.

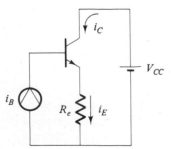

Figure 2.7-10 The emitter follower for Example 2.7-2.

can usually be eliminated. Figure 2.7-6 shows the emitter follower of Fig. 2.7-4 with R_1 removed. This is called *base-injection* bias.

Let us find R_2 so that v_L can undergo a maximum symmetrical swing. This could be done using the dc bias circuit, but instead we follow a somewhat different procedure. We first determine the Thevenin equivalent circuit looking into the base at AA'. The dc portion of the circuit as seen from AA' is shown in Fig. 2.7-7.

Application of KVL yields

$$v_B = V_{BE} + i_E R_e$$

or
$$v_B = 0.7 + (1 + \beta)i_B R_e \tag{2.7-5}$$

The circuit described by (2.7-5) is shown in Fig. 2.7-8. This is the desired Thevenin equivalent circuit looking into the base of the transistor.

Now return to Fig. 2.7-6. Making use of the Thevenin equivalent, we insert R_2 and V_{CC}, as shown in Fig. 2.7-9.

In this example $V_{CC} = 21$ V, $V_{BQ} = 15.7$ V, $R_e = 1500$ Ω, and $\beta = 100$. Thus

$$I_{BQ} = \frac{V_E}{R_e(1 + \beta)} = \frac{15}{(1500)(101)} \approx 100 \ \mu A$$

and
$$R_2 = \frac{V_{CC} - V_{BQ}}{I_{BQ}} = \frac{21 - 15.7}{10^{-4}} = 53 \ k\Omega$$

(A standard value of 56 kΩ would be used in practice.)

Note that if $R_2 = 53$ kΩ and the transistor is replaced with another, where $\beta = 50$ instead of 100 so that $R_e(1 + \beta) \approx 75$ kΩ, then

$$V_{EQ} = \frac{(V_{CC} - 0.7)[R_e(1 + \beta)]}{R_2 + R_e(1 + \beta)} \approx 12 \ V$$

instead of 15 V. The maximum peak-to-peak symmetrical swing under these conditions will be reduced accordingly.

An interesting result of this analysis is that the emitter resistor as viewed from the base terminal appears as $R_e(1 + \beta)$, a much larger resistance. In the dc bias circuits used up to this example we found that the base resistor R_b as viewed from the emitter terminal appeared as $R_b/(1 + \beta)$, a much smaller resistance. These points will be explored to advantage in Chap. 6, where small-signal equivalent circuits are considered.

2.7-1 Emitter Follower in Saturation

The basic emitter follower is shown in Fig. 2.7-10. When operation is in the active region, an increase in the base current i_B causes the collector and the emitter currents to increase and the collector-emitter voltage to decrease. Such operation continues until the transistor leaves the active region and enters the saturation region (refer to Fig. 2.2-7). As i_B increases further, driving the transistor deep into saturation, the collector current stops increasing and then decreases, eventually becoming zero and then negative. When $i_C = 0$, the base current equals the emitter current, and from Fig. 2.2-7,

$$v_{CE} = V_T \ln \frac{1}{\alpha_R} \qquad i_C = 0 \qquad (2.7\text{-}6)$$

When i_C is negative, the transistor is deep in saturation, and for sufficiently large values of i_B, v_{CE} can be made zero and even negative. Referring to Fig. 2.2-7, we see that

$$i_C = \frac{-i_B}{1 + h_{FC}/h_{FE}} \qquad \text{when } v_{CE} = 0 \qquad (2.7\text{-}7)$$

Now since

$$i_C + i_B = i_E \qquad (2.7\text{-}8)$$

we find, substituting (2.7-7) into (2.7-8), that the ratio of i_C to i_E is

$$\frac{i_C}{i_E} = -\frac{h_{FE}}{h_{FC}} \qquad (2.7\text{-}9)$$

Thus, when $v_{CE} = 0$, the magnitude of i_C is significantly greater than the magnitude of i_E.

As i_B increases further, the ratio of the magnitude of i_C/i_B also increases and approaches

$$\frac{i_C}{i_B} = -1 \qquad (2.7\text{-}10)$$

Substituting (2.7-10) into (2.2-17b), we find that as i_B increases, the collector-emitter voltage approaches

$$v_{CE} = -V_T \ln \frac{h_{FE} + 1}{h_{FE}} \qquad (2.7\text{-}11)$$

Some numerical results are given in the example which follows.

Example 2.7-2: The emitter follower in saturation In Fig. 2.7-10 let $R_e = 10 \text{ k}\Omega$, $V_{CC} = 3$ V, $h_{FE} = 50$, and $h_{FC} = 0.1$. (a) Find the value of i_B which will produce $v_{CE} = 0$ and determine i_C and i_E for this condition. (b) Calculate i_E, i_B, and v_{CE} when $i_C = 0$. (c) Calculate v_{CE} when i_B becomes extremely large.

SOLUTION (a) When $v_{CE} = 0$ V, $i_E = V_{CC}/R_e = 0.3$ mA. From (2.7-7) we have for $v_{CE} = 0$ V

$$\frac{i_C}{i_B} = -\frac{1}{1 + h_{FC}/h_{FE}} = -\frac{1}{1 + 1/500} \approx -1 + 2 \times 10^{-3}$$

Since $i_B + i_C = i_E$, we have

$$i_B \left(1 + \frac{i_C}{i_B} \right) = i_E$$

Hence
$$i_B = \frac{0.3 \times 10^{-3}}{2 \times 10^{-3}} = 150 \text{ mA}$$

Therefore
$$i_C \approx -150 \text{ mA}$$

Note that the collector current is negative.

(b) When $i_C = 0$,

$$v_{CE} = V_T \ln (1/\alpha_R) = V_T \ln [(1 + h_{FC})/h_{FC}] = V_T \ln 11 \approx 62 \text{ mV}.$$

The emitter current (which is equal to the base current since $i_C = 0$) is then

$$i_E = i_B = \frac{V_{CC} - v_{CE}}{R_e} \approx 0.3 \text{ mA}$$

Note that a change in collector current from -150 to 0 mA results in a negligible change in i_E since the transistor remains in saturation.

(c) Using (2.7-11) yields

$$v_{CE} = -V_T \ln \frac{h_{FE} + 1}{h_{FE}} = -25 \ln 1.02 \text{ mV} \approx -0.5 \text{ mV} \qquad ///$$

REFERENCES

1. A. S. Grove, "Semiconductor Physics," Wiley, New York, 1967; Motorola, Inc. Engineering Staff, "Integrated Circuits," McGraw-Hill, New York, 1965.
2. D. J. Hamilton and W. Howard, "Basic Integrated Circuit Engineering," McGraw-Hill, New York, 1975.

PROBLEMS

A sketch of the pertinent load lines should be included in all cases, and the problems should be solved graphically wherever possible.

2.1-1 The emitter-base junction of a *pnp* silicon transistor in the common-base configuration can be represented approximately as a 0.5-V battery in series with a 10-Ω resistance and an ideal diode (Figs. 2.1-1 and 2.1-3). Find V_{EBQ} for $R_e = 1000 \ \Omega$ and 10 kΩ and $V_{EE} = 6$ V.

2.1-2 Repeat Example 2.1-1 with $V_{im} = 2$ V.

2.2-1 The dc current-amplification factor β for a hypothetical transistor is given by

$$\beta = 100 i_C \epsilon^{-6(i_C - 0.1)^2}$$

(a) Calculate h_{fe}.

(b) Plot h_{fe} and β as a function of i_C.

2.2-2 Rewrite the Ebers-Moll equations in terms of the terminal currents and show that

$$i_C = \alpha_F i_E - (1 - \alpha_F \alpha_R) I_{CO}(\epsilon^{V_{BC}/V_T} - 1)$$

$$i_E = \alpha_R i_C + (1 - \alpha_F \alpha_R) I_{EO}(\epsilon^{V_{BE}/V_T} - 1)$$

2.2-3 Plot i_C versus v_{CB} using the results in Prob. 2.2-2 if $\alpha_F = 0.99$, $\alpha_R = 0.01$, and $I_{CO} = 1\,\mu A$.

2.2-4 (a) Show, using (2.1-3), that

$$I_C = \beta I_B + \frac{I_{CBO}}{1 - \alpha}$$

The above equation is valid only in the active region. At high collector-emitter voltages the equation becomes

$$I_C = \beta I_B + \frac{I_{CBO}}{1 - \dfrac{\alpha}{1 - (V_{CE}/BV_{CBO})^n}}$$

(b) Find the value of V_{CE}, where I_C becomes infinite. This voltage is called the *common-emitter breakdown voltage* BV_{CEO}; n varies between 2 and 4 in silicon.

(c) Find the breakdown region, i.e., the region where I_C greatly increases. Let $I_B = 1$ mA, $\beta = 100$, $I_{CBO} = 0.1\,\mu A$, $BV_{CBO} = 20$ V, and $n = 3$.

2.2-5 Verify Eq. (2.2-17).

2.2-6 In Fig. 2.2-5a let $V_{CC} = 10$ V, $R_C = 500\,\Omega$, $R_b = 10$ kΩ, $V_{BB} = 5.7$ V, $h_{FE} = 100$, and $h_{FC} = 0.01$. Assuming $V_{BE} = 0.7$ V even though the transistor may be saturated, find v_{CE}.

2.2-7 In the circuit of Fig. P2.2-7 find I_B to obtain the same v_{CE} as in Prob. 2.2-6.

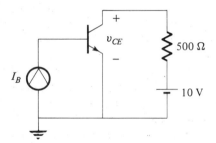

Figure P2.2-7

2.2-8 In Fig. P2.2-8 estimate (a) I_C at saturation and (b) I_B at saturation; (c) find v_{CE} and recheck your result in part (a).

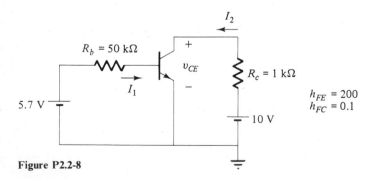

Figure P2.2-8

2.2-9 Assume the transistor in Fig. P2.2-8 is a Schottky transistor as described in Sec. 2.2-2. Use the model of Fig. 2.2-12 and calculate (a) I_1 and I_2 and (b) the base current I_B, the diode current, I_D, and the collector current I_C.

2.3-1 Find the Q point of the amplifier of Fig. P2.3-1 for (a) $R_b = 1$ kΩ and (b) $R_b = 10$ kΩ.

(handwritten annotations)
$(V_{CEQ}, I_{Cα})$

V_{CE}, I_C

$$\beta = \frac{\alpha}{1-\alpha} = \frac{.99}{1-.99} = 99$$

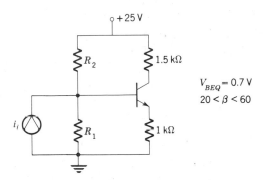

Figure P2.3-1

(handwritten)
$$I_{EQ} = \frac{V_{BB} - V_{BEQ}}{R_e + R_b/\beta}$$

$$V_{CEQ} = V_{cc} - I_{EQ}(R_e + R_c)$$

$V_{BEQ} = 0.7$ V
$\alpha = 0.99$
$I_{CBO} = 0$
$\beta = 99$

2.3-2 For the amplifier of Fig. P2.3-2 find R_1 and R_2 so that $V_{CEQ} \approx 5$ V. The quiescent current I_{CQ} must vary by no more than 10 percent as β varies from 20 to 60.

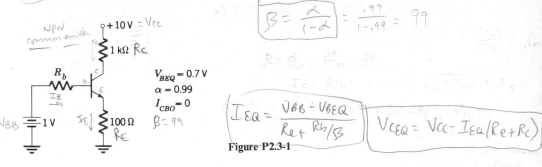

$V_{BEQ} = 0.7$ V
$20 < \beta < 60$

Figure P2.3-2

2.3-3 For the amplifier of Fig. P2.3-2 ($\beta = 100$) find new values of R_1 and R_2 which will permit a maximum symmetrical swing in i_C. *(handwritten: $R_{ac} = 2500$ $R_{DC} = 2500$ $I_{CQ} = \frac{25}{5000} = 5$ MA)*

2.3-4 For the amplifier of Fig. P2.3-2 ($\beta = 100$) find new values of R_1 and R_2 which will minimize power supply current drain. Assume an input signal such that the maximum collector current swing will be 10 mA peak to peak about the Q point.

2.3-5 (a) Find the Q point for the circuit of Fig. P2.3-5.

(b) Another transistor of the same type is to be plugged into the same circuit. What is the minimum β that the new transistor may have if the quiescent collector current is not to change by more than 10 percent?

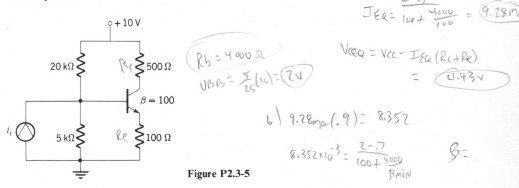

Figure P2.3-5

(handwritten annotations)
$R_b = 4000$ Ω
$V_{BB} = \frac{5}{25}(10) = (2V)$

$$I_{EQ} = \frac{2 - .7}{100 + \frac{4000}{100}} = 9.28\ mA$$

$V_{CEQ} = V_{cc} - I_{EQ}(R_c + R_e)$
$= 4.43$ V

b) $9.28\ mA\ (.9) = 8.352$

$$8.352 \times 10^{-3} = \frac{2 - .7}{100 + \frac{4000}{\beta_{min}}}$$

$\beta =$

2.4-1 For the amplifier of Prob. 2.3-2 calculate the quiescent power (*a*) supplied by the battery and (*b*) dissipated in R_1, R_2, R_e, R_c, and at the collector junction.

2.4-2 In the circuit at Fig. P2.4-2 (*a*) find V_{BB} for maximum symmetrical collector swing. Calculate the efficiency under this condition.

(*b*) Repeat part (*a*), assuming that the transistor saturates at $V_{CE,sat} = 2$ V.

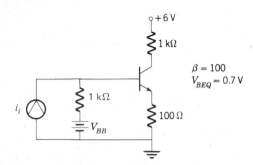

$\beta = 100$
$V_{BEQ} = 0.7$ V

Figure P2.4-2

2.4-3 In the circuit of Fig. P2.4-2 $V_{BB} = 1.2$ V and $\beta = 20$. Find the maximum possible symmetrical collector swing and the efficiency. $I_{EQ} = \frac{V_{BB} - V_{BEQ}}{R_e + R_b/\beta}$

2.5-1 (*a*) Find R_1 and R_2 so that $I_{CQ} = 10$ mA in Fig. P2.5-1.

(*b*) Find the maximum symmetrical collector swing possible with these values of R_1 and R_2.

(*c*) Draw the dc and ac load lines labeling all points of intersection.

(*d*) Sketch the maximum undistorted waveforms for i_C and v_{CE} with i_i sinusoidal as shown.

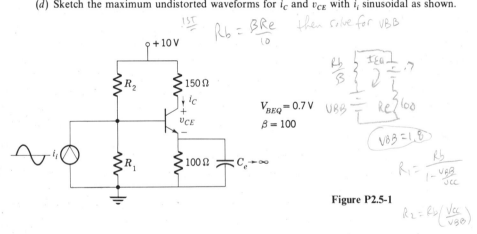

$R_b = \frac{\beta R_e}{10}$ then solve for V_{BB}

$V_{BEQ} = 0.7$ V $V_{BB} = R_e / 100$
$\beta = 100$

$V_{BB} = 1.8$

$R_1 = \frac{R_b}{1 - \frac{V_{BB}}{V_{CC}}}$

Figure P2.5-1

$R_2 = R_b \left(\frac{V_{CC}}{V_{BB}} \right)$

2.5-2 In Fig. P2.5-1 let $R_1 = 1.5$ kΩ and $R_2 = 10$ kΩ. Repeat Prob. 2.5-1.

2.5-3 In Fig. P2.5-1 let $R_1 = 10$ kΩ and $R_2 = 1.5$ kΩ. Repeat Prob. 2.5-1.

2.5-4 In Fig. P2.5-1 find R_1 and R_2 for maximum symmetrical collector-current swing. Specify the Q point for this condition and repeat parts (*c*) and (*d*) of Prob. 2.5-1.

2.5-5 In Fig. P2.5-1 the maximum required collector-current swing is 10 mA peak to peak. In order to reduce the current demand on the power supply, I_{CQ} is to be as small as possible. Specify the Q point and the required values of R_1 and R_2 assuming that the transistor is cut off at $i_C = 0$.

2.6-1 Find R_1, R_2, R_c, and R_e so that 4 mA peak ac current can flow in the 100-Ω load in the circuit of Fig. P2.6-1. Note that the solution is not unique.

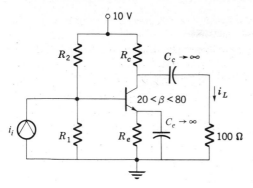

Figure P2.6-1

2.6-2 (a) Refer to Fig. P2.6-2. Find R_1 and R_2 for maximum symmetrical load current swing.
(b) Draw the dc and ac load lines.
(c) Sketch the maximum undistorted i_L.

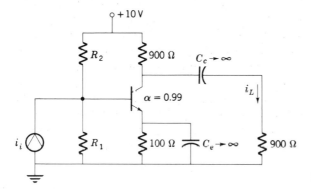

Figure P2.6-2

2.6-3 Find R_c in Fig. P2.6-3 for maximum symmetrical output voltage v_L.

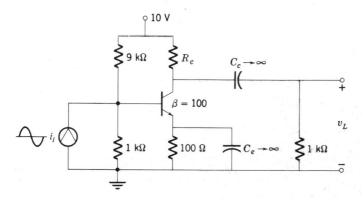

Figure P2.6-3

2.7-1 (*a*) Find the maximum possible symmetrical swing at v_L in Fig. P2.7-1.
(*b*) Sketch the waveform at v_L assuming a sinusoidal input.

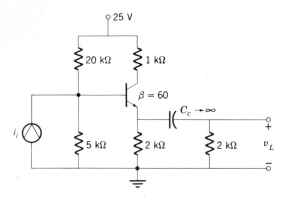

Figure P2.7-1

2.7-2 Find the Q point and the maximum symmetrical v_L in Fig. P2.7-2. Show all load lines.

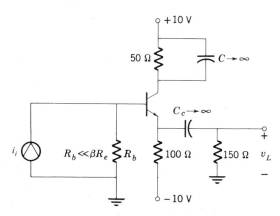

Figure P2.7-2

2.7-3 In Fig. P2.7-3 find R_2 so that $I_{CQ} = 5$ mA. Sketch the maximum undistorted v_L at this value of I_{CQ}.

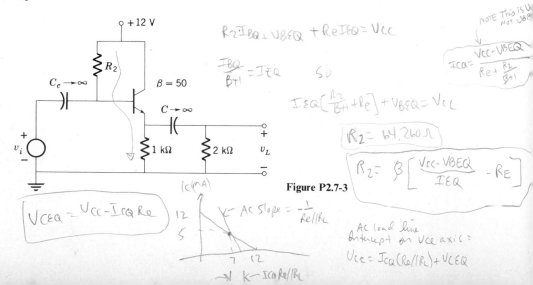

Figure P2.7-3

$R_2 I_{BQ} + V_{BEQ} + R_E I_{EQ} = V_{CC}$

$\dfrac{I_{BQ}}{B+1} = I_{EQ}$ So

$I_{EQ}\left[\dfrac{R_2}{B+1} + R_e\right] + V_{BEQ} = V_{CC}$

$\boxed{R_2 = 64,260\,\Omega}$

$\boxed{R_2 = B\left[\dfrac{V_{CC} - V_{BEQ}}{I_{EQ}} - R_E\right]}$

NOTE This is V_{BE}
NOT V_{BE}

$I_{CQ} = \dfrac{V_{CC} - V_{BEQ}}{R_e + \dfrac{R_2}{B+1}}$

$\boxed{V_{CEQ} = V_{CC} - I_{CQ} R_e}$

$i_C \,(mA)$
12
5
7   12
$\rightarrow$ k $I_{CQ} R_e / R_L$

$\leftarrow$ AC slope $= -\dfrac{1}{R_e \| R_L}$

AC load line
intercept on V_{CC} axis:
$V_{ce} = I_{CQ}(R_e \| R_L) + V_{CEQ}$

2.7-4 Refer to Fig. P2.7-4.

(a) If $v_i = 0$, find v_C.

(b) If $v_i = -3$ V dc, find v_C.

(c) Find v_i for $v_C = 2.5$ V.

(d) Find v_C and v_i for cutoff and saturation.

(e) Plot v_C versus v_i for $-6 < v_i < 6$. This is the *transfer characteristic* of the amplifier.

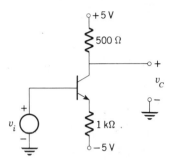

Figure P2.7-4

2.7-5 (a) Plot the CB characteristics i_C versus v_{CB} for the transistor in Fig. P2.7-5.

(b) Find the Q point and plot with dc and ac load lines on the CB characteristics.

(c) Find the maximum symmetrical v_L.

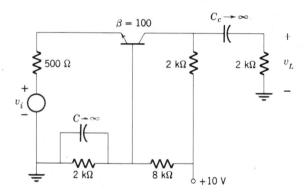

Figure P2.7-5

2.7-6 In Fig. 2.7-10 let $R_e = 1$ kΩ, $V_{CC} = 10$ V, $h_{FE} = 200$, and $h_{FC} = 0.01$.

(a) Find i_B so that $v_{CE} = 0$. Determine i_C and i_E for this condition.

(b) Calculate v_{CE} as $i_B \to \infty$.

(c) Calculate v_{CE}, i_E, i_B when $i_C = 0$.

THREE

THE FIELD-EFFECT TRANSISTOR

INTRODUCTION

Another type of transistor that is particularly suitable for use in integrated circuits is the *field-effect transistor* (FET), available in two types, the junction field-effect transistor (JFET) and the metal-oxide-semiconductor field-effect transistor (MOSFET). In very broad terms, the FET differs from the junction transistor discussed in Chap. 2 in that it is a voltage-sensitive device which has an extremely high input impedance (as high as 10^{14} Ω) and a relatively high output impedance. The FET finds use in both digital and analog circuits as an amplifier or switch.

The importance of the FET is a consequence of four of its properties. The first is its physical size; the MOSFET is so small compared with the bipolar junction transistor (BJT) that it occupies only 20 to 30 percent of the chip area taken up by a typical BJT. Thus, MOSFETs can be packed quite densely on an IC chip, and they are widely used for *large-scale integration* (LSI). The second property exhibited by MOSFETs is that over a portion of their operating range they act like voltage-controlled resistance elements and occupy much less area on a chip than the corresponding IC resistor. The third property is the extremely high input resistance. This means that the time constant of the input circuit is long enough to enable the charge stored on the small input capacitance to remain sufficiently long for the device to be useful as a storage element in digital circuits. The fourth property is its ability to dissipate high power and switch large currents in several nanoseconds. This is much faster than that possible using the BJT. Having the first three of these properties available in one device means that many different circuit functions can be included on one silicon chip containing only MOSFETs. The fourth characteristic enables the FET to be used as a high-frequency high-power switch.

Another advantage of MOSFETs in digital circuits, when used in the complementary connection (CMOS), is that the quiescent power dissipation is essentially zero at low frequencies. In this chapter we shall discuss the operation of the FET as an amplifier and as a switch and, where appropriate, compare it with the BJT.

3.1 INTRODUCTION TO THE THEORY OF OPERATION OF THE JFET

The JFET is shown schematically and with its circuit symbol in Fig. 3.1-1. The device consists of a thin layer of n-type material with two ohmic contacts, the *source S* and the *drain D*, along with two rectifying contacts, called the *gates G*. The conducting path between the source and the drain is called the *channel*. Also available are p-type JFETs, in which the channel is p type and the gates are n type.

To understand the operation of this device, assume the source and the gates to be at ground potential. Now let us study the effect of placing a small positive potential on the drain. Since there is a positive voltage between drain and source, electrons will flow from the source to the drain (current flow is from drain to source). Note that negligible current flows between source (or drain) and gate since the diode formed by the channel-to-gate junction is reverse-biased. The amount of current flowing from the drain to the source depends initially on the drain-to-source potential difference v_{DS} and the resistance of the n material in the channel between the drain and the source. This resistance is a function of the doping of the n material and the channel width, length, and thickness.

As the drain potential v_{DS} is increased, the channel voltage increases, and since the gates are fixed at 0 V, the pn diode formed by the gate-channel junction is further reverse-biased. Let us see what happens to the channel when this occurs. Figure 3.1-2 shows a reverse-biased diode. Initially, the holes in the p material flow toward the negative terminal of the battery, and the electrons in the n material flow toward the positive terminal of the battery. This results in the formation of a central region of length l, which is void of free charges (holes and electrons). Since the region contained within l has been depleted of free charges, it is called the *depletion region*. As the reverse voltage is increased, the free charges (holes and

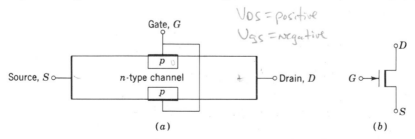

Figure 3.1-1 The JFET: (a) schematic; (b) circuit symbol, n channel.

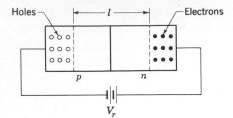

Figure 3.1-2 A reverse-biased diode.

electrons) move farther from the junction and the effective separation length l increases.

This result is directly applicable to the JFET under consideration. Sketches of the space-charge regions in the FET for several values of v_{DS} are shown in Fig. 3.1-3. We see in Fig. 3.1-3a that as v_{DS} increases, the depletion region increases, causing narrowing of the channel. The channel area decreases, resulting in an increase in channel resistance; hence the rate of increase of current per unit increase in v_{DS} decreases. This decrease is shown in Fig. 3.1-4.

When $v_{DS} = V_{p0}$, the depletion regions on each side of the channel join together, as shown in Fig. 3.1-3b. The voltage V_{p0} is called the *pinch-off* voltage† since it pinches off the channel connection between drain and source. In Fig. 3.1-3c the drain voltage v_{DS} is larger than the pinch-off voltage. In this region the depletion area thickens. However, we find that the potential at point a remains essentially at the pinch-off voltage V_{p0}. Thus the current i_{DS} remains almost

† The first subscript p represents pinch-off, while the second subscript 0 means that $v_{GS} = 0$ V.

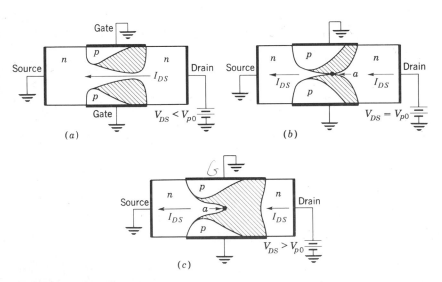

Figure 3.1-3 The JFET below, at, and above pinch-off. The shaded areas indicate depletion regions. (a) $V_{DS} < V_{p0}$; (b) $V_{DS} = V_{p0}$; (c) $V_{DS} > V_{p0}$.

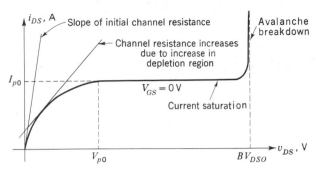

Figure 3.1-4 Ideal JFET characteristic ($V_{GS} = 0$).

constant as v_{DS} increases above V_{p0}. This is called saturation in the FET and is shown in Fig. 3.1-4.

Note that the drain current i_{DS} increases rapidly as v_{DS} increases toward V_{p0}. Above V_{p0}, the current tends to level off at I_{p0}, and then rises slowly. When v_{DS} equals the breakdown voltage BV_{DSO} an avalanche breakdown occurs (Sec. 1.10) and the current again rises rapidly.

Let us now consider holding the drain-source voltage fixed and varying the gate-source voltage. As the gate-source voltage is made negative, the pn junction is reverse-biased, increasing the depletion region between the gate and the source. This decreases the channel width, increasing the channel resistance. The current i_{DS} therefore decreases. When the gate voltage is made positive, the depletion region decreases until, for large positive gate voltages, the channel opens. Then the pn junction between gate and source becomes forward-biased and current flows from the gate to the source. The n-type JFET is usually operated so that the gate-to-source potential is either negative or slightly positive to avoid gate-to-source current.

Summarizing, we see that varying the gate voltage varies the channel width, and hence the channel resistance. This in turn varies the current from drain to source, i_{DS}. We note that it is the gate *voltage* variation which varies i_{DS}; thus the FET is a *voltage-sensitive* device, compared with the junction transistor, which is a current-sensitive device.

A typical set of vi output characteristics for the JFET is shown in Fig. 3.1-5 with gate-to-source voltage as the parameter. The pinch-off voltage for the JFET is 5 V when $v_{GS} = 0$, and the drain current at this point is 10 mA. Notice that as the gate potential decreases, the pinch-off voltage also decreases. The drain-source voltage at which pinch-off occurs is approximately given by the equation

$$v_{DS} \text{ (at pinch-off)} = V_p = V_{p0} + v_{GS} \qquad (3.1\text{-}1)$$

Thus, when $v_{GS} = 0$, $V_p = V_{p0}$, as expected. For the vi characteristics shown in Fig. 3.1-5, the pinch-off voltage is zero when $v_{GS} = -5$ V. At this negative potential no drain current flows.

$V_{DS} \text{ (at pinch off)} = V_{P0} + V_{GS}$

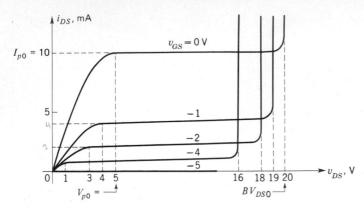

Figure 3.1-5 JFET *vi* characteristic.

The breakdown voltage is also a function of the gate-to-source voltage. This variation is given by

$$BV_{DS} \approx BV_{DS0} + v_{GS} \qquad (3.1\text{-}2)$$

where BV_{DS0} is the breakdown voltage for $v_{GS} = 0$ (for this set of characteristics $BV_{DS0} = 20$ V) and BV_{DS} is the breakdown voltage for an arbitrary v_{GS}.

Since pinch-off and breakdown of the channel are directly dependent on the voltage drop across the channel-gate junction, it is interesting to rewrite Eqs. (3.1-1) and (3.1-2) in terms of v_{DG} rather than v_{GS}, since the drain-to-gate voltage is larger than the gate-to-source voltage and therefore produces a larger depletion region. First, we see that (3.1-1) becomes

$$v_{DG} = v_{DS} - v_{GS} = V_{p0} \qquad \text{at pinch-off} \qquad (3.1\text{-}3)$$

From this equation we can interpret V_{p0} as the voltage needed between the drain and the gate to cause pinch-off. Also note that v_{DG} is independent of v_{GS} at pinch-off.

The expression for the breakdown voltage given in (3.1-2) can also be modified. If we call the drain-to-gate voltage needed for breakdown BV_{DG}, then

$$BV_{DG} = BV_{DS} - v_{GS} = BV_{DS0} \qquad (3.1\text{-}4)$$

Thus, the breakdown voltage measured between drain and gate is equal to the breakdown voltage measured when the gate-source voltage is zero and is not a function of v_{GS}.

At drain-to-source potentials between pinch-off and breakdown, called the *saturation region* since the current has saturated and does not change appreciably as a function of v_{DS}, the drain current can be approximated as

$$i_{DS} = I_{p0} \left[1 + \frac{3v_{GS}}{V_{p0}} + 2\left(-\frac{v_{GS}}{V_{p0}} \right)^{3/2} \right] \qquad v_{GS} < 0 \qquad (3.1\text{-}5)$$

Normalized I_{p0}

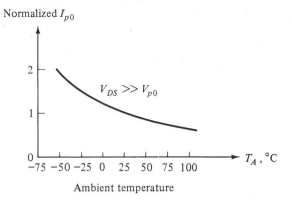

Figure 3.1-6 Normalized pinch-off current as a function of temperature.

We see from (3.1-5) that when $v_{GS} = 0$, $i_{DS} = I_{p0}$, and when $v_{GS} = -V_{p0}$, $i_{DS} = 0$. These results and (3.1-5) are independent of v_{DS}.

At drain-to-source potentials between 0 V and pinch-off the FET is said to be operating in the *linear region* since, as seen from Fig. 3.1-5, the drain-to-source current is approximately proportional to v_{DS} as in a "linear" resistor. The current equation in this region of operation is given below in (3.2-2).

The current I_{p0} can be shown to be inversely proportional to the $\frac{3}{2}$ power of the temperature

$$I_{p0} \propto T^{-3/2} \tag{3.1-6a}$$

The dependence is illustrated in Fig. 3.1-6, in which the current $I_{p0}(T_A)$ is normalized with respect to $I_{p0}(T_A = 25°C)$.

The pinch-off voltage V_{p0} is also a function of temperature in much the same way as the base-emitter voltage of a BJT:

$$\Delta V_{p0} = -k\ \Delta T \qquad k = 2\ \text{mV/°C} \tag{3.1-6b}$$

3.2 INTRODUCTION TO THE THEORY OF OPERATION OF THE MOSFET

The operation of the MOSFET is similar to the operation of the JFET. There are, however, basic differences which result in the MOSFET's having lower capacitance and higher input impedance than the JFET.

An n-channel MOSFET (Fig. 3.2-1) consists of a p-type substrate into which two n^+ regions have been diffused. These two regions form the *source* and the *drain*. No channel is actually fabricated in this device, which should be compared with the JFET (Fig. 3.1-1).

The gate is formed by covering the region between the drain and the source with a silicon dioxide layer, on top of which is deposited a metal plate. (It is this gate formation of a *metal*, *oxide*, and *semiconductor* that results in the name MOSFET.)

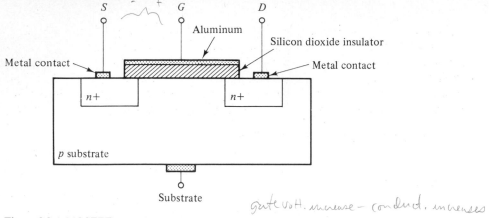

Figure 3.2-1 MOSFET.

gate volt. increase — conduct. increases

The MOSFET is usually operated with a positive gate-source potential. This is called the *enhancement mode* of operation. When the gate is positive, an *n*-type channel is *induced* between the source and the drain (Fig. 3.2-2) as a result of electrons leaving the source and drain and being attracted toward the gate by its positive potential. As the gate voltage is increased the number of electrons attracted to the channel beneath the gate increases, thereby increasing the conductivity of the channel. As the drain voltage is increased, a depletion region forms, constricting the channel. The channel impedance increases until, when $v_{DS} = V_p$, the channel pinches off and the channel impedance becomes infinite. (Actually, as with the JFET, the impedance never really becomes infinite; 100 kΩ

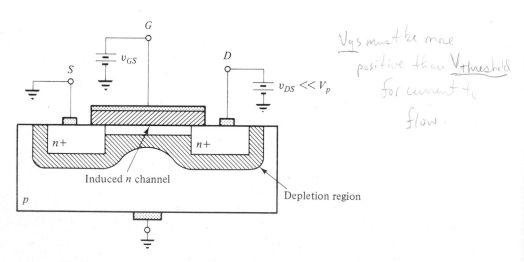

V_{gs} must be more positive than $V_{threshold}$ for current to flow.

Figure 3.2-2 The MOSFET below pinch-off.

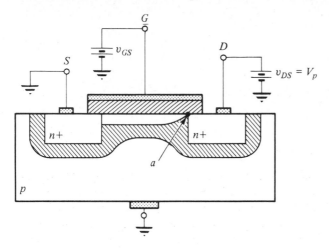

Figure 3.2-3 The MOSFET at pinch-off.

is a typical value.) The depletion region at pinch-off is shown schematically in Fig. 3.2-3. Further increases in drain-source potential result in only a slight increase in the drain-source current.

When the drain-source potential exceeds the pinch-off voltage, a depletion region forms between the drain and the channel, as shown in Fig. 3.2-4. Point a, which denotes the pinch-off point, moves only slightly toward the source. Note the similarity to the JFET shown in Fig. 3.1-3.

Now consider varying the gate-source potential, keeping the drain-source potential fixed and above pinch-off. Increasing the gate voltage increases the

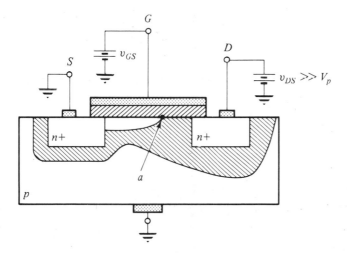

Figure 3.2-4 The MOSFET above pinch-off.

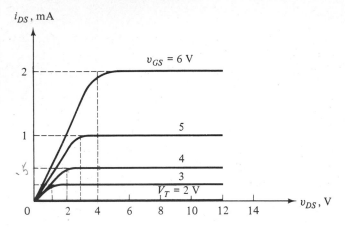

Figure 3.2-5 MOSFET *vi* characteristic.

conductivity of the channel and thereby increases the current. Thus, the drain-source *current* is modulated by the gate-source *voltage*.

The basic operation of the MOSFET and the JFET are therefore seen to be similar. Above pinch-off, increases in the drain voltage do not result in a proportional increase in drain current, and the drain *current* is proportional only to changes in the gate *voltage*.

To the circuit-design engineer the most significant difference between the JFET and the enhancement-mode MOSFET is that in the operating range of the MOSFET v_{GS} is always positive. Since v_{GS} must be positive to produce a channel, and since no current can flow until the channel is formed, current can flow only when v_{GS} exceeds a positive voltage called the *threshold voltage* V_{TN}.

The *vi* characteristics of a typical MOSFET are shown in Fig. 3.2-5. Note that when $v_{GS} = V_{TN} = 2$ V, the current $i_{DS} = 0$ for all values of v_{DS}. However, when v_{GS} is greater than V_{TN}, current flows. Here we see that the drain-source voltage at pinch-off is

$$v_{DS} \text{ (at pinch-off)} = v_{GS} - V_{TN} \qquad (3.2\text{-}1a)$$

Thus, when $v_{GS} = V_{TN}$, pinch-off occurs at $v_{DS} = 0$ V and no current can flow. Note that (3.2-1a) can be written

$$v_{GD} \text{ (at pinch-off)} = V_{TN} \qquad (3.2\text{-}1b)$$

We see from this equation that pinch-off is determined by the gate-to-drain voltage. Referring to Fig. 3.2-3, we observe that the point *a* where pinch-off occurs lies between the gate and drain. Thus, the potential at point *a* is a function of v_{GD}. Equation (3.2-1b) indicates that the potential at *a* is sufficiently large to cause pinch-off when $v_{GD} = V_{TN}$.

The equations for the current i_{DS} below and above pinch-off are:
Below pinch-off (linear region):

$$i_{DS} = k_n[2(v_{GS} - V_{TN})v_{DS} - v_{DS}^2] \qquad v_{DS} < v_{GS} - V_{TN} \qquad (3.2\text{-}2a)$$

For linear range

Above pinch-off (saturation):

$$i_{DS} = k_n(v_{GS} - V_{TN})^2 \qquad v_{DS} \geq v_{GS} - V_{TN} \tag{3.2-2b}$$

The constant k_n is given by

$$k_n = \frac{\mu\epsilon}{2t}\frac{W}{L} \tag{3.2-2c}$$

where μ = mobility of carriers in channel (electrons in NMOS)
ϵ = dielectric constant of oxide under gate
t = thickness of oxide
W = channel width
L = channel length

$$K_n = \frac{Amp}{V^2}.$$

The dimensions of k_n are amperes per volt squared, and typically k_n lies between 10^{-3} and 10^{-2}. A FET designed to act as a low resistance has a large width-to-length (W/L) ratio† and therefore has a large k_n while a FET designed to act as a high resistance has a small W/L ratio and hence a small k_n.

It is interesting to compare the equation for current in the MOSFET [Eq. (3.2-2b)] with the equation for current flow in the JFET given in (3.1-5). To make this comparison we normalize (3.2-2b) by letting $I_{p0} = k_n V_{TN}^2$ and $V_{p0} = -V_{TN}$, so that

$$\frac{i_{DS}}{I_{p0}} = \left(1 + \frac{v_{GS}}{V_{p0}}\right)^2 \tag{3.2-3}$$

The comparison, shown graphically in Fig. 3.2-6, indicates that the MOSFET and

† Resistance is proportional to the ratio of length to cross-sectional area. Since the width W is proportional to the area and the length L to the channel length, R is proportional to L/W.

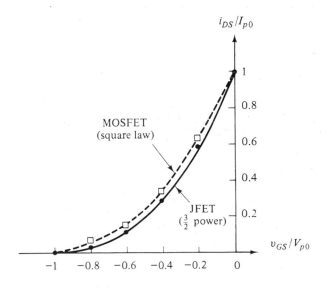

Figure 3.2-6 Current-gate-voltage characteristic of the MOS and junction FET showing similarity of the two characteristics.

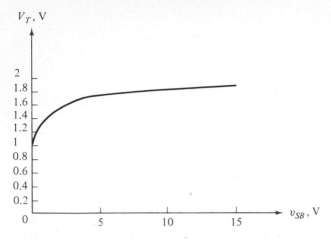

Figure 3.2-7 Variation of threshold voltage due to changes in the source-substrate voltage V_{SB}.

JFET characteristics above pinch-off are almost identical. Accordingly throughout the remainder of this text we shall assume that the MOSFET and the JFET follow the square-law characteristic.

Substrate The p region of the NMOSFET is called the *substrate*. Referring to Fig. 3.2-4, we see that the source, substrate, and drain form an *npn* BJT. To ensure that no BJT action occurs we must be certain that the BJT is cut off. Hence in Fig. 3.2-4 the substrate is connected to ground. In general, the substrate in an NMOSFET is always connected to the most negative potential available in the circuit.

The substrate voltage affects the threshold voltage V_T and the MOSFET drain-current–gate-voltage characteristic. Figure 3.2-7 shows the variation of threshold voltage V_T with changes in the source-to-substrate voltage v_{SB} for a typical MOSFET. Theoretically, it can be shown that

$$v_{TN}(v_{SB}) \approx V_{TN}(0) + K\sqrt{v_{SB}} \qquad (3.2\text{-}4)$$

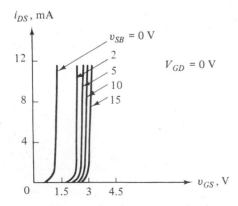

Figure 3.2-8 Drain-source current as a function of gate-source voltage for several values of source-substrate voltage. Curves are drawn for $V_{GD} = 0$ V.

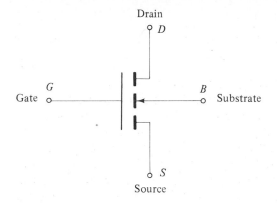

Drain

G
Gate

B
Substrate

S
Source

Figure 3.2-9 Circuit symbol for an NMOSFET explicitly showing the substrate terminal.

where $v_{TN}(v_{SB})$ is the threshold voltage at the source-substrate voltage v_{SB}, $V_{TN}(0)$ is the threshold voltage at $v_{SB} = 0$, and K is a constant of proportionality depending on the width, length, doping, etc., employed in the fabrication of the FET. Typical values of K are between 0.25 and 2.

The drain-source current is also affected by changing the substrate voltage because changing the source-substrate voltage changes the threshold voltage. For example, if the MOSFET is in saturation,

$$i_{DS} = k[v_{GS} - v_{TN}(v_{SB})]^2 \approx k[v_{GS} - V_{TN}(0) - K\sqrt{v_{SB}}]^2 \qquad (3.2\text{-}5)$$

Experimental results obtained for a typical NMOSFET, connected so that $v_{GD} = 0$, are shown in Fig. 3.2-8. Note that each curve has the same shape except for the translation caused by the changes in the threshold voltage.

We shall represent both the n-channel JFET and MOSFET by the circuit symbol shown in Fig. 3.1-1b. When it is important in a schematic drawing to show the substrate voltage explicitly, we shall employ the FET circuit symbol shown in Fig. 3.2-9.

3.3 REVERSIBILITY OF DRAIN AND SOURCE

The FET is usually manufactured so that the drain and source can be interchanged without any noticeable change in the vi characteristic. Many manufacturers even advertise that their FETs offer "symmetrical" operation. We shall therefore adopt the convention that in an n-channel FET the *source* is the terminal which is the "source of the electron flow," while the *drain* is the terminal into which the "electrons drain."

Figure 3.3-1 shows a FET connected between a source of voltage v_i and a load resistance R_L. The gate voltage V_{GG} is adjusted so that current flows in the FET. If v_i is positive, current flows from point a to point b and the voltage drop $V_{ab} > 0$. In this case point a is the drain and point b is the source. When v_i is negative, current flow is from point b to point a and the voltage drop $V_{ba} > 0$. Now point a is the source and point b is the drain.

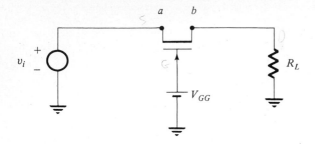

Figure 3.3-1 A FET connected between a source and a load to illustrate that point a is the drain when $V_a > V_b$ and that point b is the drain when $V_b > V_a$.

It should be noted that the reversibility of the transistor terminals is unique to the FET. If the terminals of the BJT are reversed, so that in an *npn* transistor the emitter-to-collector voltage is positive, the *vi* characteristics change dramatically since the relation between the base current and collector current $(i_C = h_{FE} i_B)$ becomes a relation between base current and reverse emitter current $(i_{ER} = h_{FC} i_B)$. The reason for this is that the BJT is not symmetrical: the emitter surface is much smaller than the collector surface, and the emitter doping is much greater than the collector doping.

3.4 *p*-CHANNEL FET

A *p*-channel JFET is shown in Fig. 3.4-1*a*, and the circuit symbol is shown in Fig. 3.4-1*b*. In a *p*-channel FET the source is positive with respect to the drain. Here the source is the *source of holes* which flow through the channel to the *drain*.

To control the flow of holes from source to drain we must establish a depletion region between the gate and the channel in order to achieve pinch-off.

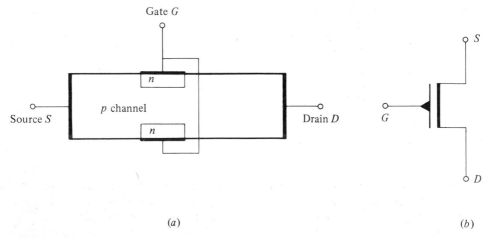

(*a*)   (*b*)

Figure 3.4-1 *p*-channel JFET: (*a*) schematic view; (*b*) circuit symbol.

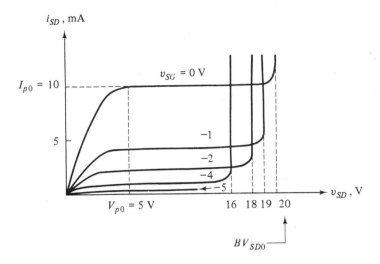

Figure 3.4-2 *p*-channel JFET *vi* characteristics.

This can be done by making the source-to-gate voltage v_{SG} negative, thereby reverse-biasing the *pn* diode formed by the channel and the gate.

A typical set of *p*-channel JFET *vi* characteristics is shown in Fig. 3.4-2. Note the similarity between these characteristics and those shown for the *n*-channel JFET in Fig. 3.1-5. The curves are identical except that i_{SD}, v_{SD}, and v_{SG} replace, respectively, i_{DS}, v_{DS}, and v_{GS}.

A PMOSFET is shown schematically in Fig. 3.4-3a, and its *vi* characteristic is shown in Fig. 3.4-3b. The circuit symbol is identical to that of the NMOSFET except that the arrowhead is reversed in Fig. 3.1-1b or Fig. 3.2-9. In Fig. 3.4-3a the voltages have been adjusted so that the device is operating below pinch-off in the linear region. The *substrate voltage* is always the most positive potential in the circuit so that the *pn* diodes formed between the substrate and the drain, source, and channel are reverse-biased. With the polarities shown for V_{GG} and V_{DD} we achieve this end by grounding the substrate.

The gate voltage $-V_{GG}$ is less than the source voltage, so that holes from the source are attracted toward the gate, thereby forming the channel. The drain is also at a lower potential than the source, so that holes from the source are drawn through the channel to the drain.

Pinch-off occurs in a PMOSFET when the drain-to-gate voltage is equal to a threshold voltage V_{TP}. At this value the potential of the *p* channel at point *a* is sufficiently negative with respect to the substrate voltage near point *a* for the depletion region that forms to cut off the channel.

In Fig. 3.4-3b we again use the convenient coordinates i_{SD}, v_{SD}, and v_{SG}, which are all positive. Note that current flows only when $v_{SG} > V_{TP}$.

The equations for the current i_{SD} of a PMOSFET as a function of v_{SG} and v_{SD} are:

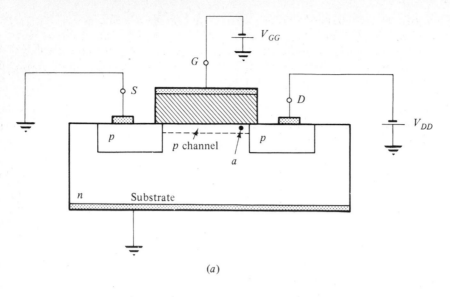

(a)

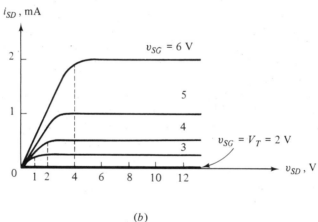

(b)

Figure 3.4-3 The PMOSFET: (a) schematic; (b) vi characteristic.

Below pinch-off (linear region):

$$i_{SD} = k_p[2(v_{SG} - V_{TP})v_{SD} - v_{SD}^2] \qquad v_{SD} < v_{SG} - V_{TP} \qquad (3.4\text{-}1a)$$

Above pinch-off (saturation):

$$i_{SD} = k_p(v_{SG} - V_{TP})^2 \qquad v_{SD} \geq v_{SG} - V_{TP} \qquad (3.4\text{-}1b)$$

The values of k_p for a PMOSFET having the same geometry as an NMOSFET are approximately one-third as large as k_n, while the values of V_{TP} are comparable to those found for V_{TN}.

As before, we shall assume that the current equations for a JFET are the same as for a MOSFET, so that for a p-channel JFET $I_{p0} = k_p V_{TP}^2$ and $V_{p0} = -V_{TP}$.

Since the values of V_{TP} are comparable to the values of V_{TN}, we shall omit the N and P subscripts and refer to the threshold voltage as V_T except where some ambiguity may arise.

3.5 DEPLETION-MODE MOSFET $U_{Thres} = negative$

The MOSFET can also be fabricated to operate in the *depletion mode*. For an NMOSFET a channel is present even when the gate-to-source voltage is zero. In this case v_{GS} must become negative in order to cut the device off so that $i_{DS} = 0$ for all v_{DS}. Similarly, a PMOSFET must have v_{SG} negative to achieve cutoff. Thus, in a depletion-mode FET the threshold voltage V_T is negative. Typical values of V_T for a depletion-mode FET lie between $V_T = -4$ V and $V_T = -10$ V.

3.5-1 Comparison of the Three Types of MOSFET

The various types of FET can profitably be compared on the basis of their *transfer characteristics*, which are plots of the output (drain-source current) versus the input (gate-source voltage). In the normal operating region between pinch-off or turn-on and breakdown, the drain current is nearly independent of drain-source voltage. The transfer characteristic for each device in this region is then approximately a single curve, as shown in Fig. 3.5-1.

Figure 3.5-1a shows the transfer characteristic for a JFET. From the curve we see that there is considerable drain current for $v_{GS} = 0$. The drain current is controlled by applying a negative v_{GS}. If we wish to use an n-channel device like this as an amplifier, we require a negative bias voltage between gate and source so that the applied signal will vary the drain current above and below the value set by the bias. Manufacturers often specify typical values of I_{p0} and V_{p0}.

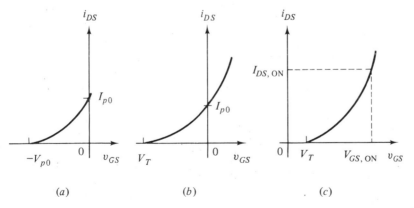

Figure 3.5-1 Transfer characteristics: (a) JFET; (b) depletion-mode n-channel MOSFET; (c) enhancement-mode n-channel MOSFET.

Figure 3.5-1*b* shows the transfer characteristic of a depletion-mode *n*-channel MOSFET. Here both depletion and enhancement regions are present; there is appreciable drain current for $v_{GS} = 0$ but typically less than for the JFET. Drain current begins to flow only when v_{GS} becomes greater than the negative threshold voltage V_T. For this device manufacturers specify V_T and I_{p0}.

Figure 3.5-1*c* shows the transfer characteristic for the *n*-channel enhancement MOSFET. Here, there is no drain current for $v_{GS} = 0$; drain current just begins to flow when a positive voltage equal to the threshold voltage V_T is applied. After that the characteristic is similar to that of the depletion-mode MOSFET for $v_{GS} > V_T$. In practice the manufacturer will specify V_T and a particular value of $I_{DS, ON}$ corresponding to a specified value of $V_{GS, ON}$.

3.6 THE MOSFET INVERTER

The FET can be used as a linear amplifier or as a logic gate. In either mode of operation the output signal has a polarity opposite that of the input signal. A typical MOSFET *inverter* is shown in Fig. 3.6-1*a*. NMOS is shown here although PMOS can, of course, be employed.

The MOSFET inverter shown in the figure consists of two transistors rather than a single transistor and a resistance load, as in the BJT inverter. Transistor T_2 acts like a resistive load, the ratio of the resistance of the load to the resistance of the driver T_1 being proportional to the ratio k_{n1}/k_{n2}, which we shall define as λ.

The input-output characteristics of the FET inverter take on different forms depending not only on $\lambda = k_{n1}/k_{n2}$ but also on the relative values of the supply voltages V_{GG} and V_{DD}.

Input-output characteristics Three different input-output characteristics are possible:

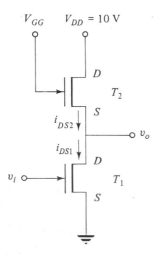

Figure 3.6-1 MOSFET inverter.

1. T_2 operating above pinch-off
2. T_2 operating below pinch-off
3. T_2 a depletion-mode device

The resulting characteristics are shown for these three cases in Fig. 3.6-2. The curves were obtained using (3.2-2a) and (3.2-2b) to represent transistors T_1 and T_2 in the appropriate regions. The details of the calculations are left for the problems.

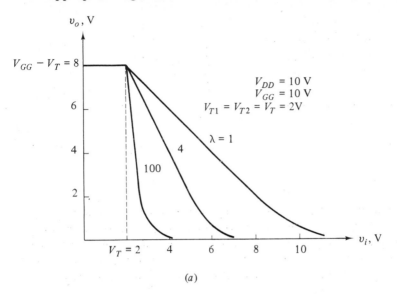

(a)

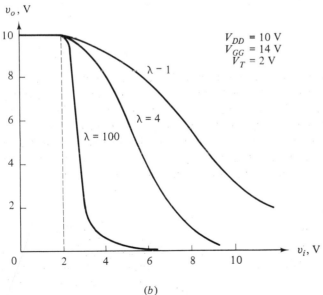

(b)

Figure 3.6-2 MOSFET inverter characteristics: (a) $V_{DD} \geq V_{GG} - V_T$; T_2 operating above pinch-off; (b) $V_{DD} < V_{GG} - V_T$; T_2 below pinch-off;

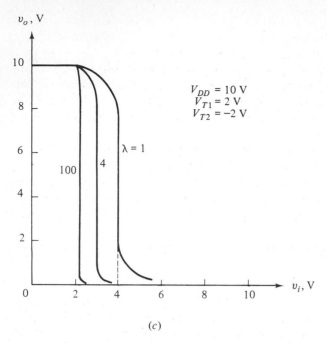

(c)

Figure 3.6-2 (c) $V_{GS} = 0$ (gate tied to source); T_2 a depletion-mode device.

Figure 3.6-2a to c should be compared on the basis of the intended application; i.e., is the inverter to be used as a logic element? If so, we are interested in a *low state* and a *high state*. Or is the inverter to be used as a linear amplifier? In this case we would like a linear relationship to exist between v_o and v_i.

If the inverter is to be used as a logic element, we would like the output v_o to switch from a high voltage to a low voltage when the input v_i increases an infinitesimal amount above $V_T = 2$ V. From Fig. 3.6-2c we see that if T_2 is a depletion-mode device, we obtain an extremely abrupt transition for any λ. We also note that the transitions are more abrupt than found in Fig. 3.6-2a or b. Unfortunately, with present technology only enhancement-mode FETs are fabricated in IC form; depletion-mode FETs are found only in discrete form.

As noted above, a logic inverter requires an abrupt transition. The transition obtained when T_2 operates above pinch-off (Fig. 3.6-2a) is more abrupt than when T_2 operates below pinch-off (Fig. 3.6-2b). The circuit of Fig. 3.6-2a is also a convenient choice since a single power supply can be used; i.e., we can let $V_{GG} = V_{DD}$.

If the inverter is to be used as a linear amplifier, again we would select the characteristics shown in Fig. 3.6-2a since with T_2 operating above pinch-off the characteristic is linear over a reasonable range of values of input voltage v_i. For example, with $\lambda = 100$, the relationship between output and input is

$$v_o = -10v_i \qquad (3.6\text{-}1)$$

for values of v_i between 2 and 2.5 V. Thus, if the input to the NMOS inverter is adjusted so that the quiescent output voltage is 5 V, the inverter can swing ± 3 V with respect to the Q point. Note that for this configuration the gain $A = v_o / v_i = \sqrt{\lambda}$.

There are problems associated with using the NMOS inverter as a logic element. First, the dimensions of the load transistor greatly exceed those of the driver transistor. For example, if

$$\lambda = \frac{(W/L)_D}{(W/L)_L} = 100$$

then assuming that T_1 and T_2 have the same width, the length of the load transistor T_2 is 100 times the length of the driver transistor T_1. In IC design, where chip "real estate" is hard to come by, this ratio is unrealistic, and we must search for another way to design this inverter. Second, the transition region of v_i is quite large, extending (for $\lambda = 100$) from $v_i \approx 2$ V to $v_i \approx 4$ V (see Fig. 3.6-2a). As the input signal crosses this region, power is dissipated in the FET. This power must be supplied by the V_{DD} supply.

3.7 COMPLEMENTARY-SYMMETRY MOS

A complementary-symmetry FET (CMOS) connected as an inverter is fabricated as shown in Fig. 3.7-1a. It consists of one PMOSFET and one NMOSFET. Both are enhancement-mode devices and are designed so that $k_p = k_n$. Other connections between the PMOSFET and the NMOSFETs are employed when CMOS is used in different applications to be discussed later.

The circuit diagram of the CMOS inverter is shown in Fig. 3.7-1b. Here T_2 is the PMOSFET and T_1 is the NMOSFET. The two drains are connected together so that current can flow from the V_{SS} supply to ground. The circuit on the right shows the substrate connections.

The operation of the inverter is as follows. When $v_i < V_{T1}$, transistor T_1 is OFF and, as is usually the case, $v_{SG2} = V_{SS} - v_i > V_{T2}$, so that transistor T_2 is ON. With T_1 OFF no current can flow in T_2 even though it is ON. Hence from Fig. 3.4-3b or from (3.4-1a) we see that with $i_{SD2} = 0$ we shall have $v_{SD2} = 0$. Thus, the output of the inverter is $v_o = V_{SS}$. As the input voltage v_i increases above V_{T1}, both T_1 and T_2 turn ON and the output voltage decreases. Finally, when v_i increases sufficiently for T_2 to go OFF, that is, $V_{SS} - v_i < V_{T2}$, then, since T_1 is ON, the output voltage $v_o = 0$ V.

Calculation of the input-output characteristic of the CMOS inverter The input-output characteristic of a typical CMOS inverter is shown in Fig. 3.7-2. It is drawn for a circuit in which $V_{SS} = 10$ V, $V_{T1} = V_{T2} = 3$ V, and $k_p = k_n$. Note that when v_i is less than 3 V, T_1 is OFF, T_2 is ON, and $v_o = 10$ V; while when v_i is greater than 7 V, T_2 is OFF, T_1 is ON, and $v_o = 0$ V.

In the region from A to D both T_1 and T_2 are ON. In region AB, T_1 is in saturation while T_2 is in the *linear* range of operation since $v_o > v_i - V_{T1}$ and

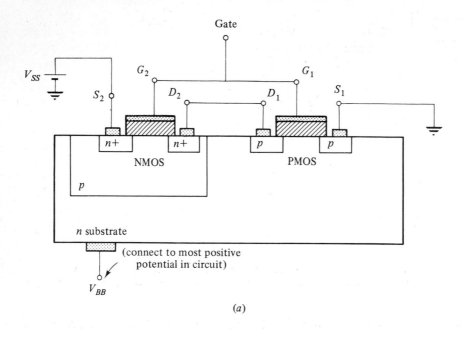

(a)

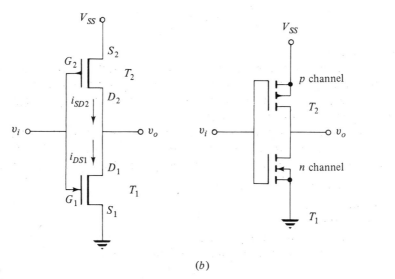

(b)

Figure 3.7-1 CMOS inverter: (a) pictorial view; (b) circuit symbols.

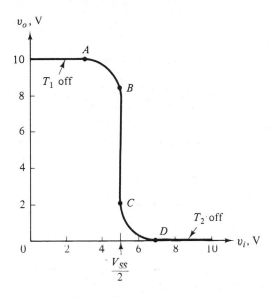

Figure 3.7-2 Input-output characteristic of a CMOS inverter.

$V_{SS} - v_o < V_{SS} - v_i - V_{T2}$ [see (3.2-2) and (3.4-1)]. Thus, the characteristic between A and B is found by equating (3.2-2b) and (3.4-1a) as follows:

$$k_n(v_i - V_{T1})^2 - k_p[2(V_{SS} - v_i - V_{T2})(V_{SS} - v_o) - (V_{SS} - v_o)^2] \quad (3.7\text{-}1)$$

By symmetry we see that in region CD, T_2 is in saturation while T_1 is in the linear region. Hence the characteristic in region CD is found by equating (3.2-2a) and (3.4-1b)

$$k_n[2(v_i - V_{T1})v_o - v_o^2] = k_p(V_{SS} - v_i - V_{T2})^2 \quad (3.7\text{-}2)$$

In region BC both transistors T_1 and T_2 are in saturation. The characteristic in this region is found by equating (3.2-2b) and (3.4-1b). The result is

$$k_n(v_i - V_{T1})^2 = k_p(V_{SS} - v_i - V_{T2})^2 \quad (3.7\text{-}3)$$

With $k_n = k_p$ and $V_{T1} = V_{T2}$ we have

$$v_i = \frac{V_{SS}}{2} = 5 \text{ V} \quad (3.7\text{-}4)$$

The characteristic in region BC is a vertical line, indicating that the output voltage changes abruptly as v_i moves across the value $V_{SS}/2$. In practical circuits, of course, the transition is not abrupt but is very steep, signifying high gain.

3.8 THE FET SWITCH

A FET can be used as a linear switch similar to the diode quad discussed in Sec. 1.6. However, it is much slower since the Schottky-diode switch can change state in less than 0.1 ns while the FET switch requires more than 1 ns to go from

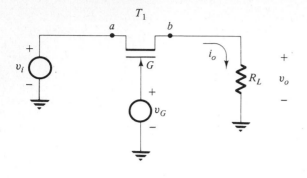

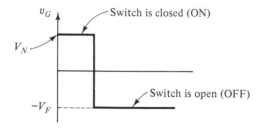

Figure 3.8-1 A FET switch. Point a is the drain when v_i is positive, and point b is the drain when v_i is negative.

ON to OFF. Most FET switches require 10 ns or more to switch states, and they are therefore used to switch low-frequency signals.

The FET switch does not act as a zero resistance when it turns ON but as a small resistance, so that the circuit of Fig. 3.8-1 is equivalent to a resistive voltage divider. For the output voltage v_o to be an undistorted scaled replica of the input voltage v_i the equivalent resistance of the FET must be linear, i.e., independent of the input voltage v_i. This is accomplished by choosing the available parameters so that the FET is operating well below pinch-off (see Fig. 3.2-5), where its characteristic can be considered a straight line.

We now show that the CMOS switch is completely linear but the FET switch only approximately so.

A typical FET switch is shown in Fig. 3.8-1. v_G is a control voltage which turns the switch ON or OFF. When v_G is a suitably large positive voltage V_N, T_1 turns ON and the voltage drop v_o across R_L will be proportional to v_i for suitable values of V_N and R_L. For this to be true T_1 must act like a linear resistor when it is ON. When v_G is a negative voltage $-V_F$, T_1 turns OFF, no current can flow, and the output voltage is zero.

Let us assume that with $v_G = V_N$, T_1 is operating below pinch-off. Then, when v_i is positive, we have $v_{DS} = v_i - v_o$ and $v_{GS} = V_N - v_o$ and the load current is, from (3.2-2a),

$$i_o = k_n[2(V_N - v_o - V_T)(v_i - v_o) - (v_i - v_o)^2] \tag{3.8-1}$$

This equation can be rearranged to read

$$i_o = k_n[(v_i - v_o)^2 + 2(V_N - V_T - v_i)(v_i - v_o)] \tag{3.8-2a}$$

When v_i is negative, the source and drain are interchanged and we have $v_{DS} = v_o - v_i$ and $v_{GS} = V_N - v_i$. In addition, the current in the FET reverses, and therefore (3.2-2a) becomes

$$i_o = -k_n[2(V_N - V_T - v_i)(v_i - v_o) + (v_i - v_o)^2] \qquad (3.8\text{-}2b)$$

Note that (3.8-2a) and (3.8-2b) are identical equations even though i_o is positive in (3.8-2a) and negative in (3.8-2b).

Referring to (3.8-2), we see that the relation between the current through the FET i_o and the voltage drop across the FET $v_i - v_o$ is parabolic rather than linear. Thus the equivalent resistance $R_{FET} = v_{DS}/i_{DS} = (v_i - v_o)/i_o$ is a nonlinear function of the voltage drop across the FET $v_i - v_o$ and the input voltage v_i.

Even though R_{FET} is nonlinear, its range of values is small, typically 50 to 500 Ω. This is true for all values of v_i provided that the gate voltage $V_N \gg V_T + v_i$. Thus, by choosing the load resistance $R_L \gg 500 \ \Omega$ we are assured that the output voltage v_o will be a good approximation of the input voltage v_i. For example, suppose we design a FET switch such that the voltage drop across the switch is 1 percent of the voltage drop across the load resistor R_L. Then if V_N is large, since the voltage drop across the FET $v_i - v_o$ is small, we can neglect the term $(v_i - v_o)^2$ in (3.8-2). The FET resistance is then

$$R_{FET} \approx \frac{v_i - v_o}{i_o} \approx \frac{1}{2k_n(V_N - V_T - v_i)} \approx \frac{1}{2k_n(V_N - V_T)} \qquad (3.8\text{-}3)$$

if $k_n = 10^{-3}$, $V_T = 2$ V, and $V_N = 12$ V, then $R_{FET} \approx 50 \ \Omega$. If $R_L > 5000 \ \Omega$, the actual variation of R_{FET} will not result in a large distortion of the output voltage.

It should also be noted that if the load impedance $R_L \approx 0 \ \Omega$, so that $v_o = 0$ V, (3.8-2a) and (3.8-2b) reduce to $i_o - 2k_n (V_N - V_T)v_i$ and therefore i_o is directly proportional to the input voltage v_i. In this case R_{FET} is still given by (3.8-3). This mode of operation occurs when the FET switch drives an op-amp (see Sec. 15.1).

Switch control voltage To turn the switch on, so as to ensure that the switch resistance R_{FET} will be small compared with the load impedance R_L, we saw that the required condition is

$$V_N \gg V_T + (v_i)_{max} \qquad (3.8\text{-}4a)$$

To ensure that the FET switch is OFF we note that if $v_i > 0$ and $i_o = 0$, we must have $v_{GS} = -V_F < V_T$. However, if $v_i < 0$, the drain and source reverse and $i_o = 0$ when $v_{GS} = -V_F - (v_i)_{min}$, which must be less than V_T. Thus, the condition required to ensure that the FET switch is OFF is

$$-V_F < V_T + (v_i)_{min} \qquad (3.8\text{-}4b)$$

For example, if v_i varies in the range ± 5 V, so that $(v_i)_{max} = +5$ V and $(v_i)_{min} = -5$ V and $V_T = 2$ V, we must have $V_N \gg 7$ V and $-V_F < -3$ V for proper operation.

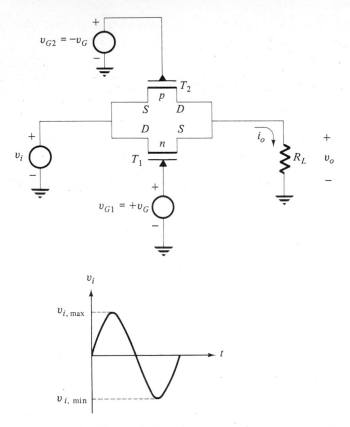

Figure 3.8-2 A CMOS switch. Drain and source terminals are shown for $v_i > 0$. When $v_i > 0$, these terminals are reversed.

Resistance of a CMOS switch A CMOS switch, shown in Fig. 3.8-2, consists of an NMOSFET in parallel with a PMOSFET. The substrate connections have been omitted from the figure for simplicity, but it should be remembered that, if possible, the p substrate of the NMOSFET should be connected to the most negative voltage in the circuit and the n substrate of the PMOSFET should be connected to the most positive voltage in the circuit.

When $v_i \geq 0$, the drain and source terminals are as shown in the figure. When $v_i < 0$, the drain and source terminals reverse.

To turn the CMOS switch ON, v_G is made a large positive voltage V_N, and to turn the switch OFF v_G is made a large negative voltage $-V_F$. If V_N is sufficiently large, T_1 and T_2 are both below pinch-off. Note that when $v_i > 0$, $v_{DS}(T_1) = v_{SD}(T_2) = v_i - v_o$, $v_{GS}(T_1) = V_N - v_o$, and $v_{SG}(T_2) = v_i - V_N$, so that the output current is

$$i_o = k_n[2(V_N - v_o - V_{T1})(v_i - v_o) - (v_i - v_o)^2]$$
$$+ k_p[2(v_i + V_N - V_{T2})(v_i - v_o) - (v_i - v_o)^2] \qquad (3.8\text{-}5)$$

The same equation results if $v_i < 0$. If we set $k = k_n = k_p$ and $V_T = V_{T1} = V_{T2}$, this equation can be simplified to

$$i_o = 4k(V_N - V_T)(v_i - v_o) \tag{3.8-6}$$

The effective resistance of the CMOS switch is then

$$R_{CMOS} = \frac{1}{4k(V_N - V_T)} \tag{3.8-7}$$

Hence the CMOS resistance is linear and independent of v_i. It can readily be shown that (3.8-7) is valid for v_i positive or negative.

It is interesting to observe that T_1 and T_2 act as resistors in parallel. When $v_i = 0$, the two resistances are the same. As v_i increases, the resistance of T_1 increases (since v_{GS1} decreases) and the resistance of T_2 decreases (since v_{SG2} increases). The variation is such as to maintain the equivalent parallel resistance approximately constant.

To ensure that T_1 and T_2 remain ON simultaneously and that the voltage drop across the FET switch $v_i - v_o$ will be small when $v_G = V_N$ it is necessary that $v_{GS}(T_1) = V_N - (v_i)_{max} \geq V_T$ and $v_{SG}(T_2) = (v_i)_{min} - V_N \geq V_T$.

This result is true for $v_i > 0$; however, the same result occurs when $v_i < 0$ since $v_i \approx v_o$ when T_1 and T_2 are ON. Hence T_1 is ON when

$$V_N \geq V_T + (v_i)_{max} \tag{3.8-8a}$$

and T_2 is ON when

$$V_N \geq V_T - (v_i)_{min} \tag{3.8-8b}$$

For example, if $-5 \text{ V} \leq v_i \leq 5 \text{ V}$ and $V_T = 2 \text{ V}$, we must have $V_N \geq 7 \text{ V}$.

To turn T_1 OFF requires that $v_G = -V_F$ and $v_{GS}(T_1) = -V_F - (v_i)_{min} \leq V_T$. Hence

$$-V_F \leq V_T + (v_i)_{min} \tag{3.8-9a}$$

while to turn T_2 OFF requires that $v_{SG}(T_2) = (v_i)_{max} - V_F \leq V_T$. Thus

$$V_F \geq (v_i)_{max} - V_T \tag{3.8-9b}$$

Using the same example as before, with $-5 \text{ V} \leq v_i \leq 5 \text{ V}$ and $V_T = 2 \text{ V}$, we find that we must have $-V_F \leq -3 \text{ V}$.

Example 3.8-1 A FET switch in which the FET is in parallel with the load is shown in Fig. 3.8-3a and the equivalent voltage divider circuit in Fig. 3.8-3b. The FET is characterized by the values $k_n = 10^{-3}$ and $V_T = 2 \text{ V}$. The input voltage v_i varies between -5 and $+5 \text{ V}$. The gate voltage used to turn the switch ON is $v_G = V_N$, and the voltage used to turn the switch OFF is $v_G = -V_F$. If $R_1 = 10 \text{ k}\Omega$, (a) show that the resistance of the FET when used in this configuration is approximately constant if the output voltage $v_o \ll 2(V_N - V_T)$. (b) Assuming that R_{FET} is constant, calculate R_{FET}. (c) Using

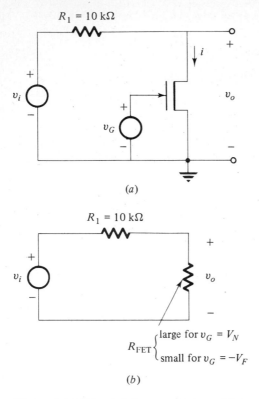

(a)

(b)

Figure 3.8-3 FET switch for Example 3.8-1: (a) circuit; (b) equivalent circuit.

the value of R_{FET} found in part (b), find v_o and the maximum value of v_o. (d) Determine V_N so that $(v_o)_{\text{max}} \ll 2(V_N - V_T)$ and also to be sure that $v_o \le 0.01 v_i$ when the switch is ON. Find R_{FET} for this value of V_N, and (e) find V_F so that the gate can be turned OFF.

SOLUTION (a and b) From (3.2-2a) and Fig. 3.8-3 we find that for v_i positive or negative

$$i = k_N[2(V_N - V_T)v_o - v_o^2]$$

Thus if $v_o \ll 2(V_N - V_T)$, we can neglect the term in v_o^2 and get

$$R_{\text{FET}} \approx \frac{v_o}{i} = \frac{1}{2k_N(V_N - V_T)} = \frac{500}{V_N - V_T}$$

(c) When the voltage-divider formula is used, the output voltage v_o becomes

$$v_o = v_i \frac{R_{\text{FET}}}{R_{\text{FET}} + R_1} = \frac{v_i}{1 + 20(V_N - V_T)}$$

and since $(v_i)_{max} = +5$ V,

$$(v_o)_{max} = \frac{5}{1 + 20(V_N - V_T)}$$

(d) We wish to ensure that when the switch is ON,

$$(v_o)_{max} = \frac{5}{1 + 20(V_N - V_T)} \ll 2(V_N - V_T)$$

Solving the inequality, we find that we must choose V_N so that $V_N - V_T \gg 0.35$ V.

In addition, to ensure that $v_o/v_i \le 0.01$ we set

$$1 + 20(V_N - V_T) \ge 100$$

from which we find that we must choose V_N so that

$$V_N - V_T \ge 5 \text{ V}$$

This is a much stronger condition. With $V_T = 2$ V we must have $V_N \ge 7$ V. With $V_N = 7$ V, $R_{FET} = 100 \, \Omega$.

(e) To ensure that the FET remains OFF for all values of v_i consider the worst-case condition, where $(v_i)_{min} = -5$ V. Then, from (3.8-4b) we must choose V_F so that

$$-V_F < V_T + (v_i)_{min}$$

which yields

$$-V_F < 2 - 5 = -3 \text{ V} \qquad\qquad ///$$

3.9 TEMPERATURE EFFECTS IN MOSFETS

The threshold voltage V_T varies with temperature in the same way as the voltage drop across a diode, i.e.,

$$\frac{\Delta V_T}{\Delta T} = -2 \text{ mV/°C} \qquad\qquad (3.9\text{-}1)$$

If $V_T = 2$ V at 25°C, then at 125°C, $\Delta T = 100$°C and

$$\Delta V_T = (-2 \text{ mV/°C})(100\text{°C}) = -200 \text{ mV}$$

Then $V_T(125\text{°C}) = V_T(25\text{°C}) + \Delta V_T = 2 - 0.2 = 1.8$ V.

The effect of temperature on diode voltage (or the base-emitter voltage of a junction transistor) is much more severe than on FET threshold voltage since the nominal diode voltage is only 0.7 V and a 0.2-V change therefore represents a change of 29 percent with respect to the nominal value.

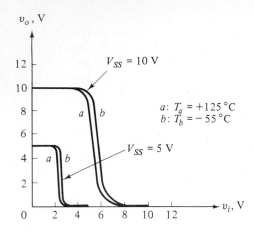

Figure 3.9-1 Input-output characteristic of a CMOS inverter as a function of supply voltage and temperature.

The resistance of an FET is also temperature-dependent. Typically, R_{FET} increases by 0.7 percent for each degree Celsius the temperature is raised. Such an increase can be represented by

$$R_2 = R_1 \epsilon^{K \, \Delta T}$$

where K is found by noting that if $\Delta T = 1°C$, $R_2/R_1 = 1.007$. Hence

$$1.007 = \epsilon^K \quad \text{and} \quad K = \ln 1.007 = 0.007$$

Thus, if $R_{FET} = 50 \, \Omega$ at 25°C and the temperature is increased to 125°C, R_{FET} will increase to

$$R_2 = 50\epsilon^{(0.007)(100)} = 100 \, \Omega$$

A CMOS inverter characteristic showing a typical temperature variation is given in Fig. 3.9-1. The effect of temperature on the input-output characteristic is negligible for this application.

3.10 THE CHARGE-COUPLED DEVICE

The charge-coupled device (CCD) is used as a *delay line* for analog signals and as a shift register (see Sec. 14.1) for digital signals. Basically, the CCD is a MOSFET constructed with many gates, often several thousand. A pictorial sketch is shown in Fig. 3.10-1a, which shows a p-channel MOSFET with an input voltage v_i applied to the source and a negative supply voltage $-V_{DD}$ applied to the drain through a load resister.

The delay-line action is obtained by appropriately adjusting the voltages on the gates. This causes charge to leave the source and to arrive at the drain after a specified delay which depends on the number of gates and the variation of the gate voltage.

The CCD receives the voltage samples $v_i(kT)$ of an input analog voltage or of an input binary signal. Assume that v_i is positive at a time $t = 0$ and that the voltage v_{G1} at gate 1 is made negative, so that v_{SG1} exceeds the threshold voltage V_T. Holes from the p-type source then flow into the minichannel created under gate 1. This is shown in Fig. 3.10-1b. To stop the flow of charge, the input voltage is reduced so that v_{SG1} is less than V_T. The holes now remain under gate 1 and will not migrate farther since gate 1 is now at the most negative potential in the device.

To cause the charge to move toward the drain we abruptly decrease the voltage at gate 2 and slowly increase the voltage at gate 1. The result is illustrated in Fig. 3.10-1c at time $t = \Delta T$. Finally, at $t = T$, all the holes have moved into the *well* (it is common practice to call the minichannel a *potential well* or *well* since it represents the region of minimum potential) under gate 2. The voltage at gate 1 is now $v_{G1} = 0$ V. To move the charge farther toward the drain we repeat the process by abruptly decreasing the voltage at gate 3 while slowly increasing the voltage at gate 2. This is shown in Fig. 3.10-1d. Similarly, each T we move the charge from gate to adjacent gate until the charge arrives under gate G_N. The voltages at gate G_N and the drain are illustrated in Fig. 3.10-1e. Note that the voltage at gate G_N is less than that at the drain. However, this difference is less than V_T to ensure that holes do not flow from the drain to gate G_N. When the voltage at gate G_N rises above that at the drain, the holes flow into the drain, causing a pulse of voltage to appear across the load R_L.

If the time for the holes to move between adjacent gates is T, the time delay observed between the time that the charge was injected into gate 1 of the CCD and the time that the voltage pulse is observed at the load is $t_D = NT$.

In practice it is required to delay a sequence of voltages rather than a single voltage. Referring to Fig. 3.10-1a to d, we see that a second voltage cannot be applied to the source until gate G_2 has returned to 0 V, that is, until G_2 has transferred all its charge to G_3. For if the potential at G_1 dropped abruptly while G_2 was transferring charge to G_3, some charge from G_2 would flow backward into G_1 since the potential at G_1 (as well as the potential at G_3) would be less than the potential at G_2. For this reason the voltage at G_1 is made to drop abruptly at the same time as the voltage at G_4 drops abruptly. Such a timing system is called a

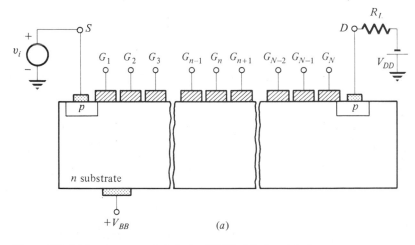

Figure 3.10-1 The charge-coupled device (CCD): (a) pictorial view.

Voltage

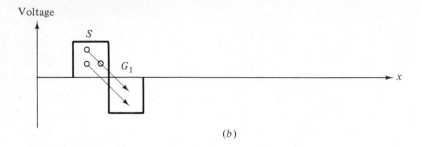

(b)

Voltage

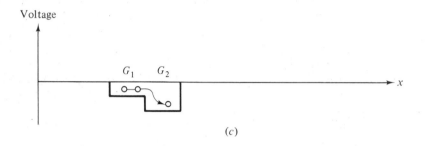

(c)

Voltage

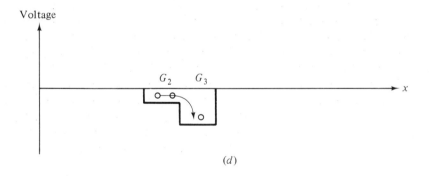

(d)

Voltage

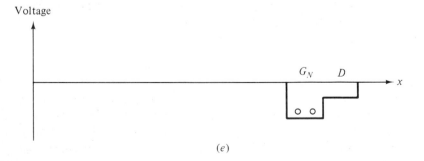

(e)

Figure 3.10-1 (*continued*) (*b*) Voltages at time $t = 0$ showing holes moving from the source to gate 1; (*c*) voltages at $t = \Delta T$; (*d*) voltages at $t = T + \Delta T$; (*e*) voltages at gate G_N and drain.

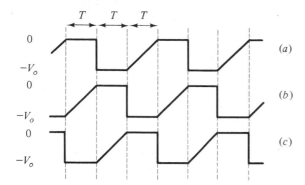

Figure 3.10-2 Three-phase gate-voltage waveforms: (*a*) G_1, G_4, G_7, ... ; (*b*) G_2, G_5, G_8, ... ; (*c*) G_3, G_6, G_9, ...

three-phase system since the voltages at G_1, G_4, G_7, etc., change identically, the voltages at G_2, G_5, G_8, etc., change identically, and the voltages at G_3, G_6, G_9, etc., change identically. Typical gate-voltage waveforms are shown in Fig. 3.10-2.

Noise in a CCD One might infer from the above discussion that the time T between shifts can be made arbitrarily long. Such is not the case since holes in the *n*-type substrate tend to flow into the well created under the low-potential gate. These holes constitute extraneous charge, or *noise*, which cannot be distinguished from the charge we are intentionally moving into the well. In practice as long as the signal charge is shifted at a rate $1/T > 1\,\text{kHz}$ ($T < 1$ ms) and the total delay is less than approximately 1 s, the noise due to the extraneous charge does not seriously affect the operation of the circuit. This constraint is particularly important when analog signals are to be delayed using the CCD. In this case such noise deteriorates the quality of the delayed signal.

3.11 INPUT PROTECTION IN THE MOSFET

If the gate-to-source voltage in a MOSFET exceeds approximately 100 V, *breakdown* (rupture of the silicon dioxide layer beneath the gate) will occur. This can result in permanent damage since excessive current will flow.

Since the gate of a FET is one plate of an almost perfect capacitor, charge introduced at the gate will remain stored on this capacitor plate and not leak off. Stray electrostatic charge is easily capable of developing enough voltage on this capacitor to cause breakdown. For example, a person walking across a typical laboratory floor can generate static voltages as high as 10 kV under suitable conditions. If this person touches the input terminal of a FET device, the energy stored in body capacitance (typically 300 pF) will be sufficient to transfer enough voltage to the FET to cause breakdown.

In order to prevent breakdown, manufacturers construct a diode *protection circuit* at the FET input. One type of protection circuit using two diodes D_1 and D_2 and a resistor R_s (typically 250 Ω to 1.5 kΩ) is shown in Fig. 3.11-1. Diode D_3

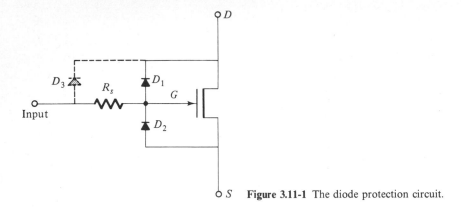

Figure 3.11-1 The diode protection circuit.

is formed as a result of the fabrication process used to construct R_s; it does not contribute to the protection of the gate.

Diode D_1 protects the gate from large positive input voltages (it clamps the gate to the voltage at the drain terminal). Diode D_2 protects the gate from excessive negative input voltages by clamping it to the voltage at the source terminal. The limits on the gate voltage due to this clamping action are

$$v_S - 0.7 < v_G < 0.7 + v_D \qquad (3.11\text{-}1)$$

For the typical range of source and drain voltages, that is, ± 20 V, this is sufficient to prevent breakdown.

3.12 POWER FET

The JFET and MOSFET can be constructed with large surface areas so that they can transmit large currents and dissipate high power. Typically, such devices can pass 2 A or more, dissipate 50 W, and have breakdown voltages greater than 200 V. Furthermore, these high-power FETs have switching speeds as low as 5 ns. In addition, the power needed to drive the FET switch is extremely low, compared with the power needed to drive the BJT, since the FET gate draws a very small current.

A comparison of a power BJT and a power JFET and MOSFET is shown in Table 3.12-1. Note that the BJT has higher maximum ratings but is also significantly slower than the FET. An excellent feature of the MOSFET is that it can be driven from a standard CMOS amplifier. For example, a typical circuit is shown in Fig. 3.12-1a. Here amplifier C_1 is a CMOS amplifier connected so that the output can swing from 0 to 10 V. Transistor T_1 is a power MOSFET selected so that when $v_{GS}(T_1) = 10$ V, the drain source current is 2 A. The power MOSFET is cut off when $v_{GS} < V_T \ (\approx 2$ V).

Since the power MOSFET dissipates a significant amount of power, heat sinking must be employed; this topic will be discussed in Sec. 4.6.

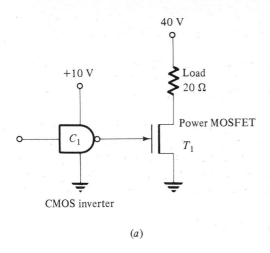

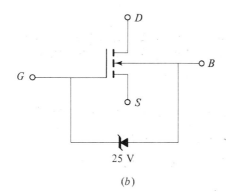

Figure 3.12-1 The power FET: (*a*) CMOS inverter driving a high-power MOSFET; (*b*) Zener-diode protection for a power FET.

Manufacturers protect the power FET against excessive static charge by placing a 25-V zener diode between gate and substrate, as shown in Fig. 3.12-1*b*. In practice the substrate of the power transistor is connected internally to the case and is grounded so that the Zener diode bypasses any voltage transients outside the range -0.7 to $+25$ V.

Table 3.12-1 Comparison of power switching transistors†

Transistor type	Absolute maximum ratings			Drive current, 1-A output	Drive voltage, V	Average switching time, ns
	Power, W	Current, A	Voltage, V			
BJT 2N6308	175	8	350	~ 250 mA	~ 1	1300
JFET 2SK60	63	5	170	~ 100 μA	25	—
MOSFET VMP-1	35	2	60	< 1 μA	10	5

† Courtesy of *Electronic Design*, Apr. 26, 1976, p. 67.

REFERENCES

1. L. J. Sevin, "Field Effect Transistors," McGraw-Hill, New York, 1965.

PROBLEMS

3.1-1 Using (3.1-5), find i_{DS} when

$$v_{GS} = V_{GSQ} + \epsilon \cos \omega t \qquad \text{where } \epsilon \ll V_{p0}$$

Show that i_D consists of an ac term and a quiescent term. Comment on the ratio ϵ/V_{p0} with regard to a linear model representing (3.1-5).

3.2-1 Assuming that

$$v_{GS} = V_{GSQ} + \epsilon \cos \omega_0 t$$

and using (3.2-2b), show that the MOSFET can "amplify" signals. Explain your answer in terms of the ratio ϵ/V_{TN}.

3.2-2 Plot i_{DS} versus v_{DS} from (3.2-2) using v_{GS} as the parameter. Let $k_n = \frac{3}{16}$ mA/V^2 and $V_{TN} = 1$ V. Do this for $v_{GS} = 4, 5$, and 6 V.

3.2-3 Use (3.1-5) and (3.2-3) to verify the graph of Fig. 3.2-6.

3.6-1 (a) For the MOSFET in Fig. P3.6-1 show that the transistor is operating above pinch-off, i.e., saturated.

(b) Find i_x as a function of v_{DS} and plot assuming that $k_n = \frac{3}{16} \times 10^{-3}$ A/V^2 and $V_{TN} = 1$ V. Hint: $i_x = i_{DS}$. Why?

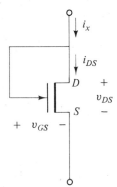

Figure P3.6-1

3.6-2 The MOSFET load T_2 in Fig. P3.6-2a is characterized by the curves shown in Fig. P3.6-2b.

(a) From the result in part (b) of Prob. 3.6-1, $i_{DS2} = k_{n2}(v_{DS2} - V_{TN2})^2$. Modify this equation to obtain i_{DS1} as a function of v_{DS1}.

(b) The result in part (a) is a nonlinear load line. Plot the load line on the curves of Fig. 3.6-2b if $V_{TN2} = 1$ V and $k_{n2} = \frac{3}{16} \times 10^{-3}$ A/V^2.

(c) Plot v_o versus v_i and compare with the curve in Fig. 3.6-2a. (The curves in Fig. P3.6-2b correspond to $k_{n1} = \frac{3}{16} \times 10^{-3}$ A/V^2 and $V_{TN1} = 1$ V, so that $\lambda = 1$.)

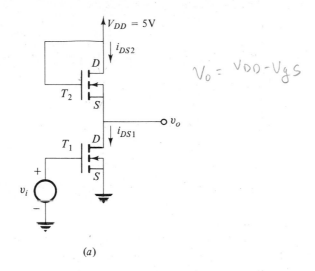

$$V_o = V_{DD} - V_{gs}$$

(a)

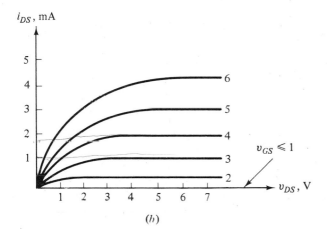

(b)

Figure P3.6-2

3.6-3 The MOSFET load T_2 in Fig. P3.6-2a has a channel width-to-length ratio that is one-fourth the value in Prob. 3.6-2, so that $k_{n2} = \frac{3}{64} \times 10^{-3}$ A/V^2. The threshold voltage V_{TN2} is assumed to be 1 V.

(a) Plot the load line on the curve of Fig. P3.6-2a. (Refer to Prob. 3.6-2.)

(b) Plot v_o versus v_i and compare with Fig. 3.6-2a ($\lambda = 4$).

3.6-4 (a) For the MOSFET inverter of Fig. P3.6-2a use the analytical expressions (3.2-2a) and (3.2-2b) and show that the transfer characteristic is given by the graph in Fig. P3.6-4. Note that best switching action is indicated by a large value for λ.

(b) Write the equation describing the linear region.

(c) Indicate the states of the MOSFETs in each region.

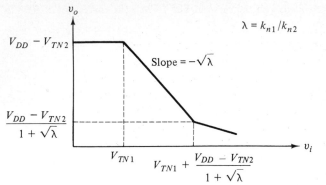

Figure P3.6-4

3.6-5 For the MOSFET in Fig. P3.6-5, described by the curves of Fig. 3.2-5, (a) draw the load line on the curves and (b) plot v_o versus v_G by choosing $v_{GS} = 2, 3, 4, 5, 6$ V. *Hint:* Find i_{DS} from the load line and solve for v_G and v_o.

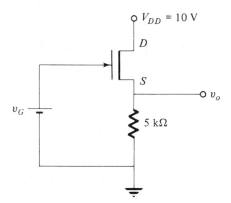

Figure P3.6-5

3.6-6 For the JFET inverter in Fig. P3.6-6, described by the curves of Fig. 3.1-5, (a) draw the load line on the curves and (b) plot v_o versus v_G by choosing $v_{GS} = 0, -1, -2, -4, -5$ V.

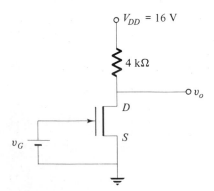

Figure P3.6-6

3.6-7 Repeat Prob. 3.6-4 for Fig. 3.6-1 and Fig. 3.6-2a.

3.6-8 Find the transfer characteristics for the MOSFET inverter of Fig. 3.6-1 and Fig. 3.6-2b using the technique of Prob. 3.6-4. Verify that the top of the curve is described by

$$\lambda(v_i - 2)^2 = (12 - v_o)^2 - 4$$

3.6-9 Find the transfer characteristics for the MOSFET inverter of Fig. 3.6-1 and Fig. 3.6-2c using the technique of Prob. 3.6-4. Verify that when both T_1 and T_2 are above pinchoff

$$v_i = \frac{2}{\sqrt{\lambda}} + 2$$

3.7-1 (a) For the CMOS inverter of Fig. 3.7-1b use the analytical expressions (3.2-2) and (3.4-1) and show that the transfer characteristic is given by the graph in Fig. P3.7-1.

(b) Indicate the states of the MOSFETs constituting the CMOS inverter in regions A and B.

(c) Let $V_{SS} = 10$ V and $V_{TP} = V_{TN} = 3$ V and compare with Fig. 3.7-2. Using these values, find the equation relating v_o and v_i in region A.

(d) Let $V_{SS} = 10$ V, $V_{TP} = 4$ V, and $V_{IN} = 2$ V. Sketch the transfer curve v_o versus v_i.

(e) Repeat part (d) for $V_{SS} = 10$ V, $V_{TP} = 2$ V, and $V_{TN} = 4$ V.

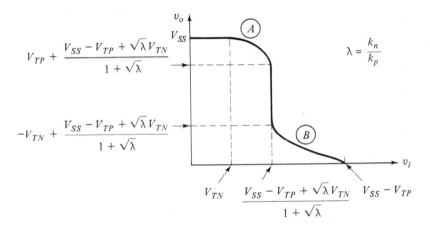

Figure P3.7-1

3.7-2 For the CMOS inverter of Fig. 3.7-1b (a) find the equivalent resistance of the NMOS transistor in the LOW state assuming that $v_{DS1} \ll 2(v_i - V_{TN})$ and $v_i \approx V_{SS}$.

(b) Find the equivalent resistance of the PMOS transistor in the HIGH state assuming that $v_{SD2} \ll 2(V_{SS} - v_i - V_{TP})$ and $v_i \approx 0$.

(c) Use the results of parts (a) and (b) to calculate the equivalent resistances for $k_n = k_p = 10^{-3}$ A/V^2, $V_{TN} = V_{TP} = 3$ V, and $V_{SS} = 10$ V.

3.8-1 The purpose of the diode in Fig. P3.8-1 is to improve the switch-control voltage limit in the OFF case when the load represents a short such as an op-amp input (10 Ω is used here).

(a) If $V_N = 15$ V, estimate R_{FET} and show that the diode is always open with the switch ON. Assume a diode cut-in voltage of 0.65 V.

(b) Find V_F to keep the switch OFF and compare with (3.8-4b).

(c) Sketch i_L if $\omega = 2\pi/T = \omega_0/4$.

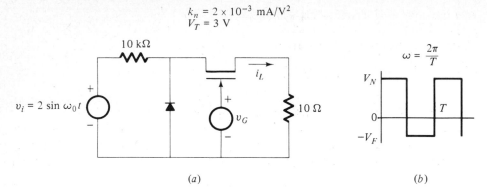

$$k_n = 2 \times 10^{-3} \text{ mA/V}^2$$
$$V_T = 3 \text{ V}$$

(a)

(b)

Figure P3.8-1

3.8-2 In Fig. 3.8-1 let $k_n = 1.5 \times 10^{-3}$ A/V², $V_T = 2$ V, $R_L = 6.8$ kΩ, and $v_i = \sin \omega_0 t$ V.
(a) Find V_F to keep the switch OFF.
(b) Determine V_N so that $R_{FET} < R_L/10$. Is this sufficient to keep the switch ON?
(c) Sketch v_L if the period of the control signal is equal to $T_o/2$.

3.8-3 The CMOS switch in Fig. 3.8-2 is characterized by the values $k_n = k_p = 2 \times 10^{-3}$ A/V², $V_T = 3$ V, and $R_L = 1$ kΩ. The input signal ranges between -8 and 6 V.
(a) Design the control signal v_G so that the switch functions properly.
(b) Calculate the value of R_{CMOS} for the design in part (a).
(c) Adjust the control signal for $R_{CMOS} < 10$ Ω.

3.8-4 The CMOS switch in Fig. P3.8-4 is characterized by $k_n = k_p = 1$ mA/V², $V_T = 2$ V, $R_L = 10$ kΩ, $R_s = 5$ kΩ, and $v_{in} = 6 \sin \omega_0 t$ V.
(a) Determine a satisfactory value for V_F.
(b) Show that the switch will remain ON as long as

$$V_N \geq V_T + \frac{(V_{in})_{max}(R_L + R_{CMOS})}{R_{CMOS} + R_S + R_L} \quad \text{and} \quad V_N \geq V_T - \frac{(V_{in})_{min}(R_{CMOS} + R_L)}{R_{CMOS} + R_S + R_L}$$

and find V_N assuming R_{CMOS} negligible. Check this assumption by calculating R_{CMOS} and comparing it to R_L.
(c) Repeat parts (a) and (b) if $R_L = 1$ kΩ and $R_s = 500$ Ω.

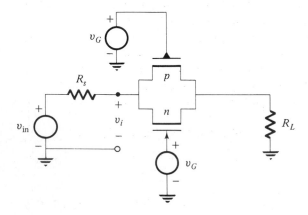

Figure P3.8-4

3.8-5 Design the control signal in Fig. 3.8-3 so that $v_o \leq 0.02$ V when the switch is ON. The signal ranges from $-4 \leq v_i < 7$ V. Use the values in Example 3.8-1.

3.8-6 Repeat Prob. 3.8-5 if a 10-kΩ load is placed across the output v_o in Fig. 3.8-3.

FOUR

BIAS STABILITY

INTRODUCTION

In the practical design of transistor circuits, the quiescent operating point Q is carefully established to ensure that the transistor will operate over a specified range, that linearity (and perhaps a maximum linear swing) will be achieved, and that $P_{C, \max}$ will not be exceeded. Once a design has been completed, it is necessary to check for quiescent-point variations due to temperature changes and possible unit-to-unit amplifier-parameter variations. These variations must be kept within acceptable limits as set by the specifications.

Among the independent parameters which can cause a shift of the Q point of a junction transistor are the following:

1. The wide variation in the current amplification β (often 5 to 1 or more) for a particular transistor type
2. Variation in the collector cutoff current I_{CBO} due to its dependence on temperature
3. Variations in the quiescent base-emitter voltage V_{BEQ} due to its dependence on temperature
4. Variations in the supply voltages due to imperfect regulation
5. Variations in the circuit resistances due to tolerance and/or temperature effects

Some of these parameters, e.g., temperature effects, are of importance for all designs while others, e.g., resistor tolerance and β variation, are more important when we are concerned with a production run of a number of identical amplifiers.

In a FET, the threshold voltage and the current parameter k are functions of temperature. These parameters also vary somewhat from unit to unit as a result of differences in fabrication.

In this chapter we discuss all these factors, along with methods for minimizing the effect of their changes on the Q point.

4.1 QUIESCENT-POINT VARIATION DUE TO UNCERTAINTY IN β

When transistors first appeared, engineers used common-emitter characteristics like those shown in Fig. 4.1-1 to establish the Q point for an amplifier. A suitable dc load line was drawn, as shown, and the Q point was established at some base current I_{BQ}. The input circuit was designed to maintain the quiescent base current at this value. A problem arose, however, when the amplifier was mass-produced. Since the β of various transistors of the same type may be subject to a typical variation of $5:1$, the quiescent collector current is subject to the same variation (if I_{BQ} is held constant). This had the effect of changing the scale of i_C, so that the Q point, with I_{BQ} fixed, could be at saturation in one circuit, where the β was high, or at cutoff in another, where the β was low.

When we studied the common-emitter amplifier in Sec. 2.3, we found that the quiescent collector current could be stabilized against unit-to-unit β variation by using an emitter resistor and maintaining a certain relation between the base and emitter-circuit resistances. This relation is given by the inequality

$$R_b \ll \beta R_e \tag{4.1-1}$$

We shall now show that when this inequality is satisfied, the Q point is essentially independent of the transistor characteristics. The circuit employed is shown in Fig. 4.1-2. It should be noted that while Fig. 4.1-2 can be thought of as a common-emitter circuit, it also represents the dc circuit for the common-base configuration, and if we let R_c equal zero, the circuit represents the emitter-follower configuration. Thus Fig. 4.1-2 and all the results which follow apply equally well to all three configurations.

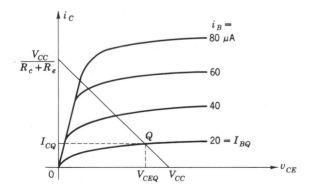

Figure 4.1-1 Common-emitter characteristics.

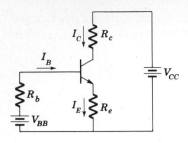

Figure 4.1-2 Common-emitter circuit.

Using results obtained in Sec. 2.3, we have:

Collector current: $\qquad\qquad I_C = \beta I_B + (\beta + 1)I_{CBO}$ $\qquad\qquad$ (4.1-2)

Collector circuit using KVL: $V_{CC} = I_C R_c + I_E R_e + V_{CE}$ $\qquad$ (4.1-3)

Base circuit using KVL: $\quad V_{BB} = I_B R_b + V_{BE} + I_E R_e$ $\qquad$ (4.1-4)

From (2.1-3) $\qquad\qquad\qquad I_B = \dfrac{I_E}{\beta + 1} - I_{CBO}$ $\qquad\qquad\qquad$ (4.1-5)

Combining (4.1-4) and (4.1-5), we obtain

$$V_{BB} = V_{BE} - I_{CBO} R_b + I_E\left(R_e + \frac{R_b}{\beta + 1}\right) \qquad (4.1\text{-}6)$$

Now when we use

$$I_C = \frac{\beta}{\beta + 1} I_E + I_{CBO} \qquad (4.1\text{-}7)$$

in (4.1-6), the quiescent collector current becomes

$$I_{CQ} = \frac{\beta(V_{BB} - V_{BE}) + (\beta + 1)I_{CBO}(R_e + R_b)}{(\beta + 1)R_e + R_b} \qquad (4.1\text{-}8)$$

The quiescent collector-emitter voltage V_{CEQ} can be obtained from (4.1-3) and (4.1-8). These equations can be simplified considerably by making three practical assumptions:

$$\alpha = \frac{\beta}{\beta + 1} \approx 1 \qquad \text{since } \beta \gg 1 \qquad (4.1\text{-}9a)$$

and since I_{CBO} is very small in silicon transistors, we shall assume that (4.1-7) becomes

$$I_C \approx \frac{\beta}{\beta + 1} I_E \approx I_E \qquad (4.1\text{-}9b)$$

We shall further assume that in the numerator of (4.1-8)

$$I_{CBO}(R_e + R_b) \ll \frac{\beta}{\beta + 1}(V_{BB} - V_{BE}) \approx V_{BB} - V_{BE} \qquad (4.1\text{-}9c)$$

With these assumptions (4.1-3) becomes

$$V_{CC} \approx V_{CE} + I_C(R_c + R_e) \tag{4.1-10}$$

and (4.1-8) becomes

$$I_{CQ} \approx \frac{V_{BB} - V_{BE}}{R_e + R_b/(\beta + 1)} \tag{4.1-11a}$$

If the inequality (4.1-1) is satisfied, (4.1-11a) simplifies to

if $R_b \ll R_e$ over β $I_{CQ} \approx \dfrac{V_{BB} - V_{BE}}{R_e} \approx \dfrac{V_{BB} - 0.7}{R_e}$ $\tag{4.1-11b}$

since V_{BE} is assumed equal to 0.7 V for silicon units.

The location of the Q point as given by (4.1-11b) is seen to be independent of β when the inequality of (4.1-1) is satisfied.

A design procedure can now be prescribed.

1. Choose a suitable dc load line and Q point on the basis of such considerations as available supply voltage V_{CC}, desired current swing, desired quiescent power dissipation, etc. The slope of the dc load line then fixes $R_c + R_e$, and the intercept on the v_{CE} axis fixes V_{CC}.
2. V_{BB}, R_b, and either R_c or R_e remain to be determined, and considerable latitude is allowed. Equation (4.1-11b) can be used to determine V_{BB} once a suitable value of R_e has been chosen. For a given value of R_e, (4.1-1) represents an upper bound on R_b. In Chap. 6 we show that R_b should be as large as possible so that current gain is not lost through attenuation in the input circuit. With this requirement in mind, we assume that a factor of 10 will satisfy the inequality and write

$$\frac{R_b}{\beta_{min} + 1} = \frac{R_e}{10} \tag{4.1-12a}$$

from which

$$R_b \approx \frac{\beta_{min} R_e}{10} \tag{4.1-12b}$$

This fixes R_b, and since V_{BB} and V_{CC} are known, the practical bias circuit of Fig. 2.3-1 can be determined using (2.3-1c) and (2.3-1d).

Example 4.1-1 In the circuit of Fig. 4.1-2, let $V_{CC} = 10$ V, $I_{CQ} = 10$ mA, $V_{CEQ} = 5$ V, $R_c = 400\ \Omega$, and $40 \leq \beta \leq 120$. Find suitable values for (a) R_e and (b) R_b. (c) With fixed values of R_e, R_b, and V_{BB}, the quiescent current will vary as β varies over the total indicated range. Determine V_{BB} so that when $\beta = 40$ and 120, the magnitudes of the changes in the quiescent current about the nominal value of 10 mA are equal. Also calculate the maximum variation in the collector current.

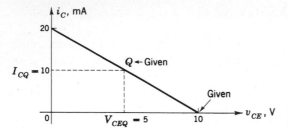

Figure 4.1-3 Load lines for Example 4.1-1.

SOLUTION The specifications give enough information to determine the load line and locate the Q point shown in Fig. 4.1-3.

(*a*) From the load line

$$R_c + R_e = \frac{10}{20 \times 10^{-3}} = 500 \ \Omega$$

and since $R_c = 400 \ \Omega$,

$$R_e = 100 \ \Omega$$

(*b*) From (4.1-12*b*)

$$R_b \approx \frac{\beta_{min} R_e}{10} = \frac{(40)(100)}{10} = 400 \ \Omega$$

(*c*) To calculate the effect of β on the Q point, we use (4.1-11*a*) and assume that $\beta \approx \beta + 1$. Then

$$I_{CQ} \approx \frac{V_{BB} - 0.7}{R_e + R_b/\beta}$$

When $\beta = 40$,

$$I_{CQ} = \frac{V_{BB} - 0.7}{100 + 400/40} = 10 \ \text{mA} - \Delta I_{CQ}$$

When $\beta = 120$,

$$I_{CQ} = \frac{V_{BB} - 0.7}{100 + 400/120} = 10 \ \text{mA} + \Delta I_{CQ}$$

Solving for V_{BB} and ΔI_{CQ} yields

$$V_{BB} \approx 1.76 \ \text{V} \qquad \text{and} \qquad \Delta I_{CQ} \approx 0.33 \ \text{mA}$$

Thus a 3 : 1 variation in β produces a negligible shift of the Q point. ///

4.2 THE EFFECT OF TEMPERATURE ON THE Q POINT

In the preceding section the variation of Q-point position with respect to unit-to-unit variations in β and a biasing arrangement which minimized these variations were discussed. Another important cause of Q-point variation is the transistor operating temperature. In this section we study the variation of the Q point due to the dependence of I_{CBO} and V_{BE} on temperature.

We begin the analysis with the exact expression for the quiescent collector current [Eq. (4.1-8)]. This expression can be simplified by noting that usually $\beta \gg 1$, so that $R_e + R_b/(\beta + 1) \approx R_e$ because we have stabilized against variations of β as in Sec. 4.1. Thus (4.1-8) becomes   uncompensated

$$I_{CQ} \approx \frac{V_{BB} - V_{BE}}{R_e} + I_{CBO}\left(1 + \frac{R_b}{R_e}\right) \tag{4.2-1}$$

This is the desired relation between collector current and the two temperature-dependent variables I_{CBO} and V_{BE}.

In the preceding section we neglected I_{CBO} and used (4.1-11b) to find the quiescent collector current. The assumptions leading to (4.1-11b) are valid at room temperature. Equation (4.2-1) is now used to investigate the validity of (4.1-11b) at elevated temperatures.

The base-emitter voltage V_{BE} is found to decrease linearly with temperature according to the relation

$$\Delta V_{BE} = V_{BE2} - V_{BE1} = -k(T_2 - T_1) \tag{4.2-2}$$

where $\qquad k \approx 2 \text{ mV/°C} \qquad T \text{ in °C}$

The reverse saturation current I_{CBO} approximately doubles for every 10°C temperature rise. This fact is embodied in the formula

$$I_{CBO2} = I_{CBO1}\epsilon^{K(T_2 - T_1)} \tag{4.2-3}$$

where, since $\epsilon^{0.7} \approx 2$,

$$K \approx 0.07/°C \qquad T \text{ in °C}$$

Typically, the I_{CBO} of low-power transistors at room temperatures is 1 μA or less.

The variation of I_{CQ} with temperature can be found from (4.2-1). Assuming that only V_{BE} and I_{CBO} vary, we have

$$\frac{\Delta I_{CQ}}{\Delta T} = -\frac{1}{R_e}\frac{\Delta V_{BE}}{\Delta T} + \left(1 + \frac{R_b}{R_e}\right)\frac{\Delta I_{CBO}}{\Delta T} \tag{4.2-4}$$

From (4.2-2), with $\Delta T = T_2 - T_1$,

$$\frac{\Delta V_{BE}}{\Delta T} = -k \tag{4.2-5}$$

and from (4.2-3)

$$\frac{\Delta I_{CBO}}{\Delta T} = \frac{I_{CBO2} - I_{CBO1}}{\Delta T} = \frac{I_{CBO1}(\epsilon^{K\,\Delta T} - 1)}{\Delta T} \tag{4.2-6}$$

Substituting (4.2-5) and (4.2-6) into (4.2-4), we get

$$\frac{\Delta I_{CQ}}{\Delta T} = \frac{k}{R_e} + \left(1 + \frac{R_b}{R_e}\right) I_{CBO1} \frac{\epsilon^{K\,\Delta T} - 1}{\Delta T} \tag{4.2-7}$$

from which

$$\Delta I_{CQ} = \frac{k\,\Delta T}{R_e} + \left(1 + \frac{R_b}{R_e}\right) I_{CBO1}(\epsilon^{K\,\Delta T} - 1) \tag{4.2-8}$$

In the example to follow, some typical values will be calculated.

Example 4.2-1 Consider the circuit of Fig. 4.1-2, with $R_b = 400\ \Omega$, $R_e = 100\ \Omega$, and $I_{CQ} = 10$ mA, at room temperature (25°C). Calculate the change in I_{CQ} if the temperature increases to 55°C.

SOLUTION Substituting the given values in (4.2-8) with $\Delta T = 30$°C gives

$$\Delta I_{CQ} = \frac{(2 \times 10^{-3})(30)}{100} + (1 + 4)(I_{CBO1})(7.2) = (0.6 \times 10^{-3}) + 36 I_{CBO1}$$

A typical value for a low-power silicon transistor is

$$I_{CBO1} = 0.1\ \mu\text{A}$$

Therefore $\qquad \Delta I_{CQ} = (0.6 + 0.0036) \times 10^{-3} \approx 0.6$ mA $\qquad$ ///

We see from this example that the shift in I_{CQ} for the silicon transistor would be negligible for most applications. Note also that most of the current change is due to the change in base-emitter voltage V_{BE}. This is often the case, and so we tend to neglect the effect of changes in I_{CBO}.

4.3 STABILITY-FACTOR ANALYSIS

The method of analysis presented in this section is often used in engineering practice. Briefly stated, the problem is the following. Given a physical variable (in our case, I_{CQ}), what change will it undergo when the variables on which it depends (in our case, I_{CBO}, V_{BE}, β, V_{CC}, etc.) change by prescribed (usually small) amounts? This type of analysis goes under various names, e.g., sensitivity analysis, variability analysis, and stability-factor analysis. All these methods are based on the assumption that, for *small* changes, the variable of interest is a linear function of the other variables and can be expressed in the form of a total differential. For our case, we write

$$I_{CQ} = I_{CQ}(I_{CBO}, V_{BE}, \beta, \ldots) \tag{4.3-1}$$

Then the total differential is

$$dI_{CQ} = \frac{\partial I_{CQ}}{\partial I_{CBO}} dI_{CBO} + \frac{\partial I_{CQ}}{\partial V_{BE}} dV_{BE} + \frac{\partial I_{CQ}}{\partial \beta} d\beta + \cdots \tag{4.3-2}$$

Now we define stability factors

$$S_I = \frac{\Delta I_{CQ}}{\Delta I_{CBO}} \tag{4.3-3a}$$

$$S_V = \frac{\Delta I_{CQ}}{\Delta V_{BE}} \tag{4.3-3b}$$

$$S_\beta = \frac{\Delta I_{CQ}}{\Delta \beta} \tag{4.3-3c}$$

If the changes in the independent variables I_{CBO}, V_{BE}, β, etc., are small,

$$S_I \approx \frac{\partial I_{CQ}}{\partial I_{CBO}} \qquad S_V \approx \frac{\partial I_{CQ}}{\partial V_{BE}} \qquad S_\beta \approx \frac{\partial I_{CQ}}{\partial \beta} \tag{4.3-4a}$$

and $\quad \Delta I_{CQ} \approx dI_{CQ} \qquad \Delta I_{CBO} \approx dI_{CBO} \qquad \Delta V_{BE} \approx dV_{BE} \qquad \Delta \beta \approx d\beta \quad$ (4.3-4b)

Therefore $\qquad \Delta I_{CQ} \approx S_I \Delta I_{CBO} + S_V \Delta V_{BE} + S_\beta \Delta \beta + \cdots \tag{4.3-5}$

From this relation we can easily find ΔI_{CQ} for changes in any of the independent variables if the changes are small enough for our assumption that the increment ΔI is approximately equal to the differential dI to be valid. If the assumption is not valid, we must calculate the actual increment. (This will usually be the case when variations in β are considered.†)

Equation (4.1-8) is the complete relation between I_{CQ} and the variables of interest for the conventional common-emitter amplifier. Taking the partial derivatives as indicated by (4.3-4a), we find, assuming $R_e \gg R_b/(\beta + 1)$,

$$S_I = \frac{\partial I_{CQ}}{\partial I_{CBO}} = \frac{R_e + R_b}{R_e + R_b/(\beta + 1)} \approx 1 + \frac{R_b}{R_e} \quad = S_I \tag{4.3-6a}$$

and $\qquad S_V = \frac{\partial I_{CQ}}{\partial V_{BE}} = \frac{-\beta}{(\beta + 1)R_e + R_b} \approx -\frac{1}{R_e} \quad = S_V \tag{4.3-6b}$

---

† If I_{CQ} is a linear function of a variable x, then $\Delta I_{CQ}/\Delta x = dI_{CQ}/dx$ even when Δx is large (since the slope of a straight line is a constant). However, if I_{CQ} is a nonlinear function of x, $\Delta I_{CQ}/\Delta x$ does not represent the slope of the curve except in the limit where $\Delta x \to dx$.

To find S_β we must calculate the actual increment because of the large change involved. Since $V_{BB} - V_{BE} \gg I_{CBO}(R_e + R_b)$ in the active region, the I_{CBO} terms will be neglected in (4.1-8). Thus

$$I_{CQ} \approx \frac{\beta(V_{BB} - V_{BE})}{R_b + (\beta + 1)R_e} \tag{4.3-7}$$

Now let β_2 and β_1 represent the upper and lower limits, respectively, on β, with I_{CQ2} and I_{CQ1} the corresponding collector currents. Next, we form the ratio

$$\frac{I_{CQ2}}{I_{CQ1}} = \frac{\beta_2}{\beta_1} \left(\frac{R_b + (\beta_1 + 1)R_e}{R_b + (\beta_2 + 1)R_e} \right) \tag{4.3-8}$$

Unity is subtracted from both sides of (4.3-8) and the result manipulated to yield

$$\frac{I_{CQ2} - I_{CQ1}}{I_{CQ1}} = \frac{\Delta I_{CQ}}{I_{CQ1}} = \frac{\Delta\beta(R_b + R_e)}{\beta_1[R_b + (\beta_2 + 1)R_e]} \tag{4.3-9}$$

where $\qquad \Delta I_{CQ} = I_{CQ2} - I_{CQ1} \qquad$ and $\qquad \Delta\beta = \beta_2 - \beta_1$

Therefore $\qquad S_\beta \equiv \dfrac{\Delta I_{CQ}}{\Delta\beta} = \dfrac{I_{CQ1}}{\beta_1} \left(\dfrac{R_b + R_e}{R_b + (\beta_2 + 1)R_e} \right) \tag{4.3-10}$

Thus, finally,

$$\Delta I_{CQ} \approx \left(1 + \frac{R_b}{R_e} \right) \Delta I_{CBO} - \frac{1}{R_e} \Delta V_{BE}$$
$$+ \frac{I_{CO1}}{\beta_1} \left(\frac{R_b + R_e}{R_b + (\beta_2 + 1)R_e} \right) \Delta\beta + \cdots \tag{4.3-11a}$$

It should be noted that since the change in β is large, the total change ΔI_{CQ}, in I_{CQ}, cannot always be determined as in (4.3-11a). In the present case, the stability factors S_I and S_V do not contain β; hence (4.3-11a) is correct. If, however, we have, say, $R_e = 0$, S_I and S_V as given by (4.3-6) would contain β. Since β can have any value from β_1 to β_2, we cannot know which value of β to insert in the stability factors. In such a case the total change in I_{CQ} must be obtained directly; i.e.,

$$\Delta I_{CQ} = I_{CQ}(I_{CBO_2}, V_{BE_2}, \beta_2) - I_{CQ}(I_{CBO_1}, V_{BE_1}, \beta_1) \tag{4.3-11b}$$

where the independent variables are chosen to maximize ΔI_{CQ} in order to provide a worst-case condition.

The increments in I_{CBO} and V_{BE} can be related directly to temperature by making use of (4.2-2) and (4.2-3). Thus

$$\Delta V_{BE} = -k\,\Delta T \tag{4.2-2}$$

and $\qquad \Delta I_{CBO} = I_{CBO1}(\epsilon^{K\,\Delta T} - 1) \tag{4.2-3}$

Note that it is relatively easy to take into account other factors which might affect I_{CQ}. For example, if changes in I_{CQ} due to changes in V_{CC} and R_e are also to be found, we write

$$\Delta I_{CQ} = S_I \, \Delta I_{CBO} + S_V \, \Delta V_{BE} + S_\beta \, \Delta\beta + S_{Vcc} \, \Delta V_{CC} + S_{R_e} \, \Delta R_e \quad (4.3\text{-}12)$$

where
$$S_{Vcc} \approx \frac{\partial I_{CQ}}{\partial V_{CC}} \quad \text{and} \quad S_{R_e} \approx \frac{\partial I_{CQ}}{\partial R_e} \quad (4.3\text{-}13)$$

for small changes of I_{CQ}.

Equation (4.3-12) shows that the total change ΔI_{CQ}, in the quiescent current is proportional to the changes in each of the independent variables and to their stability factors. Thus, to design for small ΔI_{CQ}, we design to *minimize* the stability factors. In the usual application of (4.3-12), extreme accuracy is not required, and it is customary to use this type of analysis for relatively large changes (20 percent or more) and to exercise engineering judgment in the interpretation of the results. The examples which follow illustrate typical orders of magnitude for these quantities.

Example 4.3-1 Find V_{BB}, R_b, and R_e for the amplifier shown in Fig. 4.3-1, so that i_C can swing by at least ± 5 mA. β varies from 40 to 120, and V_{BEQ} is between 0.6 and 0.8 V. The collector-emitter saturation voltage $V_{CE, \text{sat}}$ is 0.1 V.

SOLUTION To have a peak swing of ± 5 mA requires that the quiescent current I_{CQ} be bounded by I_{CQ1} and I_{CQ2}, as shown on Fig. 4.3-2. The point Q_1 ensures that i_C can swing -5 mA before cutoff, while Q_2 ensures that i_C can swing $+5$ mA before saturation. If we assume cutoff at $i_C = 0$, $I_{CQ1} = 5$ mA. From Fig. 4.3-2, I_{CQ2} is seen to be given by the intersection of the dc and ac load lines.

dc load line:
$$V_{CC} = V_{CEQ2} + I_{CQ2}(R_c + R_e)$$

ac load line:
$$V_{CEQ2} - V_{CE, \text{sat}} = R_c(I_{CQ2} + 5 \times 10^{-3} - I_{CQ2})$$

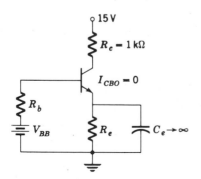

Figure 4.3-1 Circuit for Example 4.3-1.

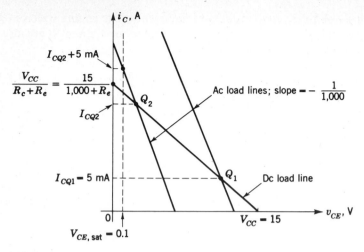

Figure 4.3-2 Load lines for Example 4.3-1.

Eliminating V_{CEQ2}, we get

$$I_{CQ2} = \frac{V_{CC} - 5 \times 10^{-3}R_c - V_{CE,\,sat}}{R_c + R_e}$$

$$= \frac{15 - 5 - 0.1}{1000 + R_e} \approx \frac{10}{1000 + R_e}$$

Therefore the quiescent collector current must satisfy the inequality

$$I_{CQ1} \le I_{CQ} \le I_{CQ2}$$

or $$5 \text{ mA} \le I_{CQ} \le \frac{10}{1000 + R_e} \text{ mA}$$

Solving, we find that R_e must be less than 1 kΩ in order to ensure that the Q point will be between Q_1 and Q_2.

The next step in the solution is to investigate the effect of the expected β and V_{BEQ} variations. The pertinent relation is

$$I_{CQ} = \frac{\beta(V_{BB} - V_{BE})}{R_b + (\beta + 1)R_e} \tag{4.3-7}$$

From (4.3-11a)

$$\Delta I_{CQ} \approx -\frac{1}{R_e}\Delta V_{BE} + \frac{I_{CQ1}}{\beta_1}\frac{R_b + R_e}{R_b + (\beta_2 + 1)R_e}\Delta\beta$$

The worst case will occur when $\Delta V_{BE} = -0.2$ V and $\Delta \beta = 120 - 40 = 80$

Then
$$\Delta I_{CQ} \approx \frac{0.2}{R_e} + \frac{I_{CQ1}}{40} \frac{(R_b + R_e)(80)}{R_b + 120R_e}$$

$$\approx \frac{0.2}{R_e} + \frac{I_{CQ1}}{60}\left(1 + \frac{R_b}{R_e}\right) \qquad 120R_e \gg R_b$$

Clearly, a wide range of values of R_b, R_e, and V_{BB} will maintain the Q point between the limits Q_1 and Q_2. If we choose

$$R_e = 500 \ \Omega \qquad R_b = 2000 \ \Omega \qquad V_{BB} = 3.7 \text{ V}$$

the lower limit on I_{CQ} will occur when $V_{BE} = 0.8$ V and $\beta = 40$. From (4.3-7)

$$I_{CQ, \text{ min}} = \frac{(40)(3.7 - 0.8)}{2000 + (40)(500)} \approx 5.3 \text{ mA}$$

and
$$\Delta I_{CQ} = \frac{0.2}{500} + \frac{5.3}{60} \times 10^{-3}\left(1 + \frac{2000}{500}\right) \approx 0.8 \text{ mA}$$

Thus
$$I_{CQ, \text{ max}} = 5.3 + 0.8 = 6.1 \text{ mA}$$

The load lines shown in Fig. 4.3-3 illustrate the possible Q-point variation under these conditions. The collector current is seen to be capable of $+5$ mA swing for the specified range of β and V_{BE}. ///

Example 4.3-2 (a) For the circuit of Fig. 4.3-4, find I_{CQ} at room temperature using the nominal values given on the figure. (b) Find ΔI_{CQ} for the indicated tolerances on V_{CC}, R_e, and β. The ambient temperature ranges from 25 to 125°C.

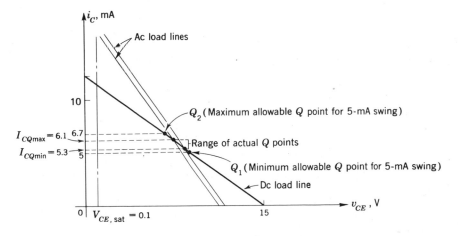

Figure 4.3-3 Q-point variation for Example 4.3-1.

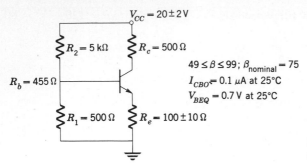

$V_{CC} = 20 \pm 2\,V$

$R_2 = 5\ k\Omega$ $R_c = 500\ \Omega$

$R_b = 455\ \Omega$

$R_1 = 500\ \Omega$ $R_e = 100 \pm 10\ \Omega$

$49 \le \beta \le 99;\ \beta_{nominal} = 75$
$I_{CBO} = 0.1\ \mu A$ at $25°C$
$V_{BEQ} = 0.7\,V$ at $25°C$

Figure 4.3-4 Circuit for Example 4.3-2.

SOLUTION (a) From (4.1-8), the nominal quiescent current if I_{CBO} is neglected is

$$
\begin{aligned}
I_{CQ} &= \frac{\beta(V_{BB} - V_{BEQ})}{(\beta + 1)R_e + R_b} \\[2mm]
&= \frac{\beta\{[R_1/(R_1 + R_2)]V_{CC} - V_{BEQ}\}}{(\beta + 1)R_e + R_b} \\[2mm]
&= \frac{75[(0.5/5.5)20 - 0.7]}{(76)(100) + 455} \\[2mm]
&= 10.6\ \text{mA}
\end{aligned}
$$

(b) To find ΔI_{CQ} we use (4.3-5). The required stability factors are S_I, S_V, S_β, $S_{V_{cc}}$, and S_{R_e}. From (4.3-6a)

$$
S_I = \frac{R_e + R_b}{R_e + R_b/(\beta + 1)} \approx \frac{100 + 455}{100} = 5.5\ \text{mA/mA}
$$

and since $\beta R_e \gg R_b$, the approximations in (4.3-6) can be used. From (4.3-6b)

$$
S_V \approx -\frac{1}{R_e} = -0.01\ \text{A/V} = -10\ \text{mA/V}
$$

With $\beta_1 = 49$ the corresponding I_{CQ1} is found to be $I_{CQ1} = 10.2$ mA. Then from (4.3-10)

$$
S_\beta \approx \frac{10.2}{49} \times 10^{-3}\,\frac{455 + 100}{455 + 10,000} \approx 0.01\ \text{mA per unit change in } \beta
$$

To find $S_{V_{cc}}$ we refer to part (a). Then

$$
\begin{aligned}
S_{V_{CC}} &= \frac{\partial I_{CQ}}{\partial V_{CC}} = \frac{\beta}{(\beta + 1)R_e + R_b}\left(\frac{R_1}{R_1 + R_2}\right) \approx \frac{R_1}{(R_1 + R_2)R_e} \\[2mm]
&= \frac{500}{(5500)(100)} \approx +0.9\ \text{mA/V}
\end{aligned}
$$

To find S_{R_e}, we apply the definition to (4.1-8)

$$S_{R_e} = \frac{\partial I_{CQ}}{\partial R_e} \approx \frac{-(V_{BB} - V_{BE})}{[R_e + R_b/(\beta + 1)]^2} \approx \frac{-1.1}{(100)^2}$$

$$\approx -10^{-4} \text{ A}/\Omega = -0.1 \text{ mA}/\Omega$$

Thus all the stability factors are determined. The next step in the solution is to find ΔI_{CBO}, ΔV_{BE}, $\Delta \beta$, ΔV_{CC}, and ΔR_e.

Using (4.2-3), we have

$$\Delta I_{CBO} = I_{CBO1}(\epsilon^{K \, \Delta T} - 1) = 0.1 \times 10^{-6}(\epsilon^{(0.07)(100)} - 1) \approx 0.1 \text{ mA}$$

and from (4.2-2)

$$\Delta V_{BE} = -k \, \Delta T = -(2 \times 10^{-3})(100) = -200 \text{ mV}$$

From the specifications

$$\Delta \beta = 50 \qquad \Delta V_{CC} = 4 \text{ V} \qquad \Delta R_e = 20 \text{ }\Omega$$

The *worst* possible Q-point shift from the minimum value will be

$$\Delta I_{CQ} = |S_I \, \Delta I_{CBO}| + |S_V \, \Delta V_{BE}| + |S_\beta \, \Delta \beta| + |S_{V_{CC}} \, \Delta V_{CC}| + |S_{R_e} \, \Delta R_e|$$

$$= (5.5)(0.1) + (10)(0.2) + (0.01)(50) + (0.9)(4) + (0.1)(20)$$

$$= 0.55 + 2 + 0.5 + 3.6 + 2 \text{ mA}$$

and
$$\Delta I_{CQ} \le 8.65 \text{ mA}$$

The maximum possible variation from the nominal value of 10.6 mA will then be about one-half of this figure, or $\approx \pm 4.3$ mA.

The large variation with temperature, due primarily to variations in V_{BE}, caused approximately 2.55 mA of the total 8.65 mA quiescent-current variation. To reduce the variations due to V_{CC} and R_e, a regulated supply could be used, so that $\Delta V_{CC} = 0.1$ V instead of 4 V, and a 1 percent rather than a 10 percent resistor employed, so that $\Delta R_e = \pm 1$ Ω. Then the above calculations can be modified to yield

$$\Delta I_{CQ} \le \pm 1.67 \text{ mA}$$

Even this change, approximately 20 percent of the nominal I_{CQ}, is not insignificant. This analysis procedure is therefore extremely important since the information enables the designer to determine whether the required peak collector-current swing can be obtained. If not, a redesign is required.   ///

4.4 TEMPERATURE COMPENSATION USING DIODE BIASING

In preceding sections we saw that changes in ambient temperature can result in significant variation in quiescent collector current. This variation is due primarily to the base-emitter voltage V_{BE}, which is a function of temperature (the effect of I_{CBO} is usually negligible).

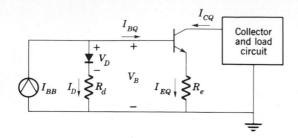

Figure 4.4-1 Diode-biasing: simplified circuit.

One method of reducing this current variation becomes obvious when one considers the stability factor S_V of (4.3-6b). If R_e is increased, S_V decreases and ΔI_{CQ} is decreased. However, this also reduces the quiescent current. Thus this scheme of minimizing S_V cannot be carried very far.

Single diode compensation An alternative method of reducing the base-emitter voltage variation is to use diode compensation. To understand this technique, consider the circuit of Fig. 4.4-1.

In this circuit, the bias is supplied by a constant-current source I_{BB}. If the diode is chosen to match the base-emitter junction characteristic of the transistor, so that their temperature dependence is the same, we shall have

$$\frac{\Delta V_D}{\Delta T} = \frac{\Delta V_{BE}}{\Delta T} \qquad (4.4\text{-}1)$$

The bias supply current is constant, so that

$$I_{BB} = I_D + I_{BQ} = I_D + \frac{I_{EQ}}{\beta + 1} = \text{const} \qquad (4.4\text{-}2)$$

The base voltage is found from the circuit to be

$$V_B = V_D + I_D R_d = V_{BEQ} + I_{EQ} R_e \qquad (4.4\text{-}3)$$

Thus, from (4.4-2) and (4.4-3), the quiescent emitter current is

$$I_{EQ} = \frac{V_D - V_{BEQ} + I_{BB} R_d}{R_e + R_d/(\beta + 1)} \qquad (4.4\text{-}4)$$

Since I_{BB} is constant,

$$\frac{\Delta I_{EQ}}{\Delta T} = \frac{\Delta V_D/\Delta T - \Delta V_{BE}/\Delta T}{R_e + R_d/(\beta + 1)} = 0 \qquad (4.4\text{-}5)$$

Thus I_{EQ} is insensitive to temperature variations.

Note also from (4.4-4) that the emitter current I_{EQ} is relatively independent of variations in β if

$$R_e \gg \frac{R_d}{\beta + 1} \qquad (4.4\text{-}6)$$

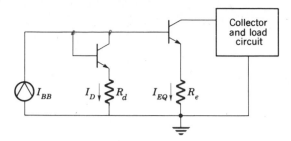

Figure 4.4-2 Diode biasing with a transistor connected as a diode.

Clearly, the degree of temperature stabilization depends on the matching of the external diode to the base-emitter diode of the transistor. In using discrete components care should be used in selecting the diode. However, in ICs, diodes are formed from transistors by connecting the base to the collector, as shown in Fig. 4.4-2. These IC transistors are usually closely matched because of the nature of the manufacturing process, and so the temperature compensation is quite good.

It should be noted that since (4.4-5) is independent of the value of R_d, perfect compensation can theoretically be achieved if $R_d = 0$. However, in such a circuit, $V_D > V_{BE}$, so that the current supplied to the bias circuit exceeds the current in the transistor. This approach results in increased power dissipation.

Example 4.4-1 A diode biasing circuit in which the bias current source I_{BB} is replaced by a battery in series with a resistor is shown in Fig. 4.4-3. Determine the effect of temperature changes on the quiescent emitter current.

SOLUTION We begin our solution to the problem by obtaining the Thevenin equivalent circuit of the biasing network. To do this we represent the diode by a battery of voltage V_D. The resulting equivalent circuit is then shown in Fig. 4.4-4. Using KVL around the base-emitter loop and noting that $(\beta + 1)I_{BQ} = I_{EQ}$, we find the quiescent emitter current

$$I_{EQ} \approx \frac{(V_{CC}R_d + V_D R_b)/(R_b + R_d) - V_{BE}}{R_e}$$

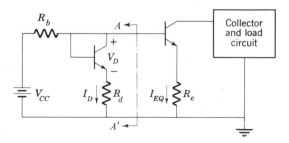

Figure 4.4-3 Practical diode-biasing circuit.

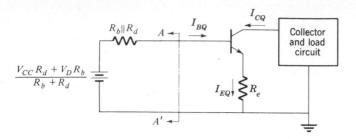

Figure 4.4-4 Equivalent circuit.

where we have assumed $R_e \gg (R_b \| R_d)/(\beta + 1)$. This current can be set at the desired value by adjusting R_b and R_d. The change in I_{EQ} due to a change in temperature is

$$\frac{\partial I_{EQ}}{\partial T} = \frac{1}{R_e} \frac{R_b}{R_b + R_d} \frac{\partial V_D}{\partial T} - \frac{\partial V_{BEQ}}{\partial T}$$

But

$$\frac{\partial V_D}{\partial T} = \frac{\partial V_{BEQ}}{\partial T} = -k \qquad \text{where } k \approx 2 \text{ mV/°C}$$

Therefore

$$\boxed{\frac{\partial I_{EQ}}{\partial T} = +\frac{k}{R_e} \frac{1}{1 + R_b/R_d}} \quad \text{With compens.}$$

We saw in Sec. 4.3 that without diode compensation, if I_{CBO} is neglected,

$$\boxed{\left| \frac{\partial I_{EQ}}{\partial T} = +\frac{k}{R_e} \right|} \quad \text{without compens.}$$

Thus diode compensation, when employed as above, reduces sensitivity to temperature. If, for example, $R_b = 2.5$ kΩ and $R_d = 250$ Ω, then

$$\frac{\partial I_{EQ}}{\partial T} = \frac{k}{11 R_e}$$

and temperature effects are reduced by a factor of 11, compared with an amplifier stage without diode stabilization.   ///

Two diode compensation Example 4.4-1 illustrates the fact that biasing using a single diode cannot result in perfect temperature compensation. However, when two or more diodes are used, as in Fig. 4.4-5, $\Delta I_{EQ}/\Delta T$ can be made equal to zero so that we have perfect compensation. For some applications it is desirable to make $\Delta I_E/\Delta T$ a specific positive or negative value. Such adjustment of the temperature variation of I_{EQ} is possible using the circuit of Fig. 4.4-5a.

In order to determine the conditions necessary to make $\Delta I_{EQ}/\Delta T = 0$ we first obtain the equivalent circuit of the biasing network, as shown in Fig. 4.4-5b. Then,

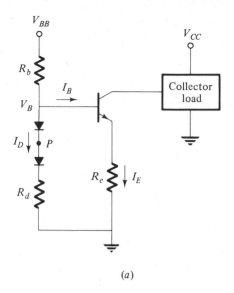

(a)

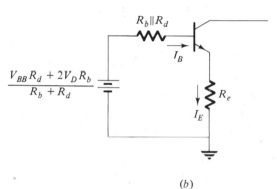

Figure 4.4-5 Biasing with two diodes:
(b)   (a) circuit; (b) equivalent circuit.

using KVL around the base-emitter loop and assuming that $R_e \gg (R_b \| R_d)/(\beta + 1)$, we find I_{EQ}

$$I_{EQ} = \frac{(V_{BB} R_d + 2V_D R_b)/(R_b + R_d) - V_{BE}}{R_e} \qquad (4.4\text{-}7)$$

Since $\Delta V_D/\Delta T = \Delta V_{BE}/\Delta T$, the quiescent emitter current is independent of variations of temperature if

$$\frac{\Delta I_{EQ}}{\Delta T} = \frac{2R_b/(R_b + R_d) \, \Delta V_D/\Delta T - \Delta V_D/\Delta T}{R_e} = 0 \qquad (4.4\text{-}8)$$

This is satisfied when

$$R_b = R_d \qquad (4.4\text{-}9)$$

We have therefore shown that in the circuit of Fig. 4.4-5 with $R_b = R_d$ the emitter current is independent of temperature. The following example illustrates the design of a two-diode biased transistor circuit.

Example 4.4-2 The two-diode biasing circuit of Fig. 4.4-5 is used to obtain temperature compensation of the emitter current I_E. If $I_E R_e = 2$ V, find V_{BB}. If $I_E = 1$ mA, determine a suitable value for $R_b = R_d$.

SOLUTION Neglecting the base current I_B in comparison with I_D, we note that the voltage at point P is equal to $V_{BB}/2$ (since $R_b = R_d$). We observe that the voltage at P is equal to the emitter voltage if the diode current $I_D = I_E$. Therefore,

$$V_P = \frac{V_{BB}}{2} = I_E R_e$$

Hence $V_{BB} = 4$ V. In addition we have

$$I_D R_d = V_P - V_{D2} = \frac{V_{BB}}{2} - 0.7 = 1.3 \text{ V}$$

With $I_D = I_E = 1$ mA $\qquad R_b = R_d = 1.3$ kΩ

(A standard value would be 1.2 kΩ.) If, to save power, we decide to make I_D much less than I_E the above design procedure would not be valid since V_{D1} would be less than V_{BE}. A suitable design procedure for this case is given in Prob. 4.4-6. ///

In Example 4.4-2 we found $V_{BB} = 4$ V. Typically this value of V_{BB} is less than the collector voltage V_{CC}. V_{BB} and R_b are then obtained from V_{CC} using a resistive voltage-divider network (see Fig. 2.3-2).

4.5 BIAS STABILITY IN THE FET

The vi characteristics of the FET change with temperature, like those of the BJT. In addition, characteristics of FETs of the same type vary not unlike the large β variation found in BJTs of the same type. In this section we discuss several types of bias circuits which control the bias variation in the JFET and MOSFET. Although n-channel devices are used throughout, the same techniques can be used to bias p-channel FETs.

Biasing the JFET A JFET amplifier is shown in Fig. 4.5-1. We wish to bias this stage at a prescribed nominal value of quiescent drain-to-source voltage V_{DSQ}. The variation of the Q point with FET-parameter variation is to remain within prescribed limits. To accomplish this we must bias the JFET to ensure that the variation of quiescent drain-to-source current falls within prescribed limits since changes in I_{DSQ} are reflected directly in V_{DSQ}.

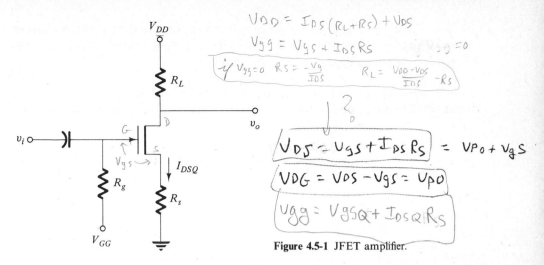

Handwritten annotations:

$$V_{DD} = I_{DS}(R_L + R_S) + V_{DS}$$

$$V_{gg} = V_{gs} + I_{DS} R_S \qquad V_{gg} = 0$$

if $V_{gg} = 0$ $R_S = -\dfrac{V_g}{I_{DS}}$ $\qquad R_L = \dfrac{V_{DD} - V_{DS}}{I_{DS}} - R_S$

$$\boxed{V_{DS} = V_{gs} + I_{DS} R_S} = V_{P0} + V_{gs}$$

$$\boxed{V_{DG} = V_{DS} - V_{gs} = V_{P0}}$$

$$\boxed{V_{gg} = V_{gSQ} + I_{DSQ} R_S}$$

Figure 4.5-1 JFET amplifier.

The transfer characteristics of Fig. 4.5-2 show the worst-case variation of drain-source current as a function of gate-source voltage for the particular FET type. The transfer curves are drawn assuming operation in the saturation region, so that (3.2-3) applies:

$$i_{DS} = I_{p0}\left(1 + \frac{v_{GS}}{V_{p0}}\right)^2 \qquad (3.2\text{-}3)$$

Two curves are drawn representing the worst-case values of I_{p0} and V_{p0}; these values are provided by the manufacturer for each particular FET type. The characteristics of any FET of this type will then lie somewhere between these two curves, with the nominal values approximately in the center of the region.

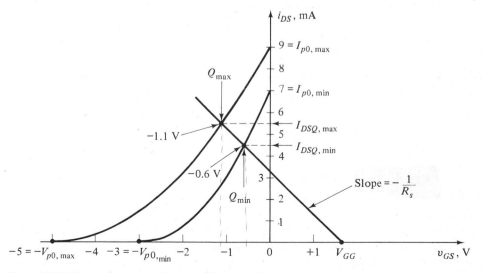

Figure 4.5-2 Worst-case transconductance characteristics.

The curves are used in the following way to determine V_{GG} and R_s for the circuit of Fig. 4.5-1. First we select a nominal quiescent current and the maximum allowable deviation from this nominal value. Typically we may allow a ± 10 percent variation in I_{DSQ}. Hence $I_{DSQ, \text{max}} = 1.1 I_{DSQ}$ and $I_{DSQ, \text{min}} = 0.9 I_{DSQ}$. These values are marked on the two transconductance curves at points Q_{max} and Q_{min}, as shown in Fig. 4.5-2, and the corresponding values of V_{GSQ} are found from the curves.

Referring to Fig. 4.5-1, we have

$$V_{GG} = V_{GSQ} + I_{DSQ} R_s \tag{4.5-1}$$

The straight line representing this equation must pass through the points Q_{max} and Q_{min}, as shown in Fig. 4.5-2. The intersection of this line and the v_{GS} axis yields V_{GG}, while the slope of the line is $-1/R_s$.

The design of the bias circuit is now complete. When V_{DSQ} and V_{DD} are known, the load resistor R_L can readily be calculated. Example 4.5-1 illustrates the design procedure.

Example 4.5-1 Design the bias circuit for the JFET amplifier shown in Fig. 4.5-1 by finding V_{GG} and R_s. $V_{p0, \text{max}} = 5$ V, $V_{p0, \text{min}} = 3$ V, $I_{p0, \text{max}} = 9$ mA, $I_{p0, \text{min}} = 7$ mA. $I_{DSQ, \text{nom}} = 5$ mA, and a ± 10 percent variation is permitted. If the nominal value of $V_{DSQ} = 6$ V and $V_{DD} = 15$ V, find the variation in V_{DSQ} and determine R_L.

SOLUTION The worst-case transconductance curves are given in Fig. 4.5-2. $I_{DSQ, \text{max}}$ intersects the top curve at Q_{max}, where $V_{GS} = -1.1$ V. $I_{DSQ, \text{min}}$ intersects the lower curve at $v_{GS} = -0.6$ V (Q_{min}). Equation (4.5-1) is plotted in Fig. 4.5-2 to go through points Q_{max} and Q_{min}. The source resistance R_s is then found to be

$$R_s = \frac{\Delta V_{GS}}{I_{DSQ, \text{max}} - I_{DSQ, \text{min}}} = \frac{1.1 - 0.6}{1 \times 10^{-3}} = 500 \ \Omega$$

The bias voltage V_{GG} is found to be $V_{GG} = 1.65$ V. With $V_{DD} = 15$ V and V_{DSQ} nominally set at 6 V, we have

$$R_L + R_s = \frac{V_{DD} - V_{DSQ}}{I_{DSQ, \text{nom}}} = \frac{15 - 6}{5 \times 10^{-3}} = 1800 \ \Omega$$

Since $R_s = 500 \ \Omega$, $R_L = 1.3$ kΩ (a standard value would be 1.2 kΩ). Referring to Fig. 4.5-1, we see that

$$\Delta V_{DSQ} = -\Delta I_{DSQ}(R_L + R_s)$$

Since $\Delta I_{DSQ} = \pm 0.5$ mA, the change in drain-to-source voltage is $\Delta V_{DSQ} = \pm 0.9$ V (± 15 percent from the nominal value compared with ± 5 percent specified for I_{DSQ}).   ///

Typical: for $\pm 10\%$ I_{DSQ} changes

$I_{DSQ_{MAX}} = 1.1 I_{DSQ}$
$I_{DSQ_{min}} = .9 I_{DSQ}$

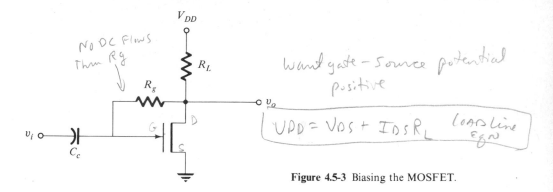

Figure 4.5-3 Biasing the MOSFET.

Biasing the MOSFET An NMOSFET amplifier is shown in Fig. 4.5-3. The gate bias is provided from the drain voltage through resistor R_g. Since no dc current can flow into the gate of the MOSFET, no dc current flows in R_g and the dc voltage at the gate is equal to the dc voltage at the drain, i.e.,

$$V_{GS} = V_{DS} \tag{4.5-2}$$

This bias circuit is similar to that used to bias the JFET in the following sense. In the JFET circuit of Fig. 4.5-1 any increase in drain current causes an increase in the source voltage, and therefore the gate-source voltage becomes more negative. This tends to reduce the current, thereby reducing the current increase which started the cycle. This sequence of events is called *negative feedback* (see Sec. 10.1). In the MOSFET circuit shown in Fig. 4.5-3 an increase in I_{DS} causes V_{DS} to fall. Since $V_{DS} = V_{GS}$, the gate voltage also falls, thereby tending to reduce the original increase in current. Example 4.5-2 illustrates quantitatively the stabilizing influence of the bias circuit.

Example 4.5-2 Figure 4.5-4 shows the worst-case maximum and minimum *vi* transfer characteristics for a particular NMOSFET which is biased using the circuit shown in Fig. 4.5-3. Since $V_{DS} = V_{GS}$, the FET is operating in the saturation region, so that the current i_{DS} is independent of the drain-source voltage and varies with the square of the gate-source voltage as in (3.2-2b). The worst-case equations are

$$i_{DS} = k_{max}(v_{GS} - V_{T,\,min})^2 \tag{4.5-3a}$$

and
$$i_{DS} = k_{min}(v_{GS} - V_{T,\,max})^2 \tag{4.5-3b}$$

Equation (4.5-3a) yields the largest i_{DS} for a given value of v_{GS} while (4.5-3b) produces the smallest i_{DS}. The characteristic drawn with a dashed line has the parameters $V_T = 4$ V and $k_n = 2 \times 10^{-3}$ A/V^2, and that drawn using a solid line has the parameters $V_T = 5$ V and $k_n = 10^{-3}$ A/V^2. $V_{DD} = 15$ V and $R_L = 1.5$ kΩ. Find the maximum possible variation of I_{DSQ} and V_{DSQ}.

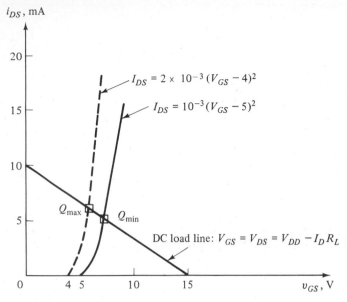

Figure 4.5-4 Vi characteristics of a MOSFET for worst-case parameter variations.

SOLUTION The load-line equation

$$V_{DD} = V_{DS} + I_{DS}R_L$$

is plotted on the same axes as the vi transfer characteristics as shown in Fig. 4.5-4. The quiescent points Q_{max} and Q_{min} are at the intersection of the two vi curves and the load-line equation. In this example Q_{max} and Q_{min} are read from the graph as

Q_{max}: $I_{DSQ} = 6.2$ mA $V_{DSQ} = 5.7$ V

Q_{min}: $I_{DSQ} = 5.3$ mA $V_{DSQ} = 7.1$ V

Thus, the nominal Q point is at the average values $I_{DSQ} \approx 5.7$ mA and $V_{DSQ} \approx 6.5$ V. The maximum variation of I_{DSQ} is therefore $\Delta I_{DSQ} = \pm 0.5$ mA, or approximately 10 percent of the nominal value of I_{DSQ}, while $\Delta V_{DSQ} = \pm 0.8$ V, which is about 11 percent of the nominal value of V_{DSQ}.   ///

The above example shows that feedback biasing produces a Q point which is stable despite large parameter variations. It is left as a problem to show that if feedback was not employed and the FET was biased as in Fig. 4.5-5, variations of the amount shown in Fig. 4.5-4 would produce substantially larger variations in the position of the Q point.

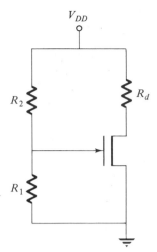

Figure 4.5-5 FET biased without feedback.

4.6 ENVIRONMENTAL THERMAL CONSIDERATIONS IN TRANSISTOR AMPLIFIERS

In this section we consider the effect of power dissipation and ambient temperature on the operation of transistor circuits. Since most power-amplifier circuits employ the BJT, we shall focus on this device. FET power amplifiers are discussed in Sec. 4.6-1.

The practical design of transistor circuits almost always involves thermal as well as electrical considerations because the maximum average power that the transistor can dissipate is limited by the temperature the collector-base junction can withstand. Thus all circuit designs should include or be followed by a calculation of thermal conditions to ensure that the maximum allowable junction temperature has not been exceeded. The maximum allowable operating temperature for silicon is in the range 150 to 200°C; at higher temperatures the transistor will suffer physical damage. The average power dissipated in the collector circuit P_c is equal to the average of the product of the collector current and the collector-base voltage, and the maximum allowable average collector power is specified by the manufacturer. This rating can be exceeded momentarily provided the transistor does not have sufficient time to heat up to the point where it burns out.

The analysis of the thermal situation in a transistor is the same as for the junction diode considered in Sec. 1.11. The typical physical configuration shown in Fig. 4.6-1 is described exactly by (1.11-7), and the discussion in connection with that equation carries over directly. Information on the thermal resistances θ_{jc} and θ_{ca} is usually provided by the transistor manufacturer.

Example 4.6-1 A silicon transistor has the thermal ratings

$$T_{j,\,\text{max}} = 150°C \qquad \theta_{jc} = 0.7°C/W$$

$\theta_{jc},\ \theta_{ca}$ Thermal resistances

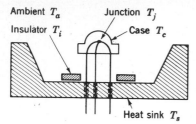

Ambient T_a Junction T_j
Insulator T_i Case T_c
Heat sink T_s

Figure 4.6-1 Transistor and heat sink.

$T_{jMAX} = 150°C$

$\theta_{jc} = .7°C/W$

Find (*a*) the power this transistor could dissipate if the case could be maintained at 50°C regardless of the junction temperature; (*b*) the power that could be dissipated with an ambient temperature of 50°C and a heat sink having $\theta_{ca} = 1°C/W$.

SOLUTION (*a*) For these conditions

$$P_j = \frac{T_j - T_c}{\theta_{jc}} = \frac{150 - 50}{0.7} \approx 143 \text{ W}$$

(*b*) Using (1.11-6), we get

$$P_j = \frac{T_j - T_a}{\theta_{jc} + \theta_{ca}} = \frac{150 - 50}{0.7 + 1} \approx 59 \text{ W}$$

Thus an "infinite" heat sink as in (*a*) permits the transistor to dissipate over twice the power permitted when using the "real" heat sink specified in (*b*).

///

Derating curves The variation of maximum collector dissipation with case temperature is an important characteristic supplied by the transistor manufacturer. A typical curve is shown in Fig. 4.6-2. At case temperatures less than T_{co} the transistor can dissipate its maximum allowable power. At case temperatures exceeding T_{co} the maximum allowable collector dissipation is decreased, as shown.

The collector dissipation in this region is given by

$$\frac{T_{j,\text{max}} - T_{co}}{P_{C,\text{max}}} = \frac{T_{j,\text{max}} - T_c}{P_C} \tag{4.6-1}$$

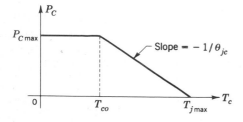

Slope $= -1/\theta_{jc}$

Figure 4.6-2 Derating curve.

Since

$$T_j - T_c = P_c \theta_{jc} \tag{4.6-2}$$

we see that

$$\theta_{jc} = \frac{T_{j,\,max} - T_{co}}{P_{C,\,max}} \tag{4.6-3}$$

Example 4.6-2 A high-power silicon transistor can dissipate 150 W as long as the case temperature is less than 45°C. Above this temperature the collector power is linearly derated, as shown in Fig. 4.6-2. The maximum junction temperature is 120°C. The amplifier is to be capable of operating at very high ambient temperatures, reaching 80°C. Determine the maximum power this transistor can dissipate and the thermal resistance of the heat sink and insulator required to avoid having the junction temperature exceed its maximum allowable value.

SOLUTION $T_j = T_{j,\,max} = 120°C$, and θ_{jc} can be found using (4.6-3):

$$\theta_{jc} = \frac{120 - 45}{150} = 0.5°C/W \quad \text{thermal resistance}$$

To avoid exceeding the maximum junction temperature we need

$$\frac{T_j - T_a}{\theta_{jA}} = P_c \qquad T_{j,\,max} = T_{a,\,max} + P_c\theta_{ja} \qquad \boxed{\theta_{jA} = \theta_{jc} + \theta_{cA}}$$

Thus $\qquad 120 - 80 = 40 = P_c\theta_{ja} - P_c(\theta_{jc} + \theta_{ca}) - P_c(0.5 + \theta_{ca})$

At this point engineering judgment must be used. It is clear that with an infinite heat sink $P_C = 80$ W. However, infinite heat sinks cannot be bought. A very good heat sink (with insulator) has a thermal resistance of 0.5°C/W. Thus

$$P_{C,\,max} = 40 \text{ W}$$

Observe that the 150-W transistor could dissipate only 40 W because of the high-ambient-temperature requirement. ///

4.6-1 Thermal Conditions Using the Power FET

In Sec. 3.12 we discussed some of the characteristics of a power FET. In this section we consider the effect of temperature increases on the operation of this device.

The FET switch, when ON, has an effective resistance R_{FET} which increases with temperature at the rate of approximately 0.7 percent per Celsius degree. The relation between resistance and temperature can be written as

$$R_{FET}(T_2) = R_{FET}(T_1)\epsilon^{0.007(T_2 - T_1)} \tag{4.6-4}$$

where T_2 and T_1 are the two temperatures of interest. (Note that $\epsilon^{0.007} \approx 1.007$, so that when $T_2 - T_1 = 1°C$, the resistance has increased 0.7 percent.)

The power dissipated by the FET in the ON position is

$$P_F = I_{DS}^2 R_{FET} \qquad (4.6\text{-}5)$$

where I_{DS} is the direct current flowing in the switch. Since power is being dissipated in the FET, its temperature will rise, so that R_{FET} will increase, causing the power dissipation to increase further. The relation between the power dissipated and the temperature increase ΔT is given by (see Sec. 1.11)

$$\Delta T = T_j - T_a = \theta_{ja} P_F \qquad (4.6\text{-}6a)$$

Substituting (4.6-5) into (4.6-6a) gives

$$\Delta T = T_j - T_a = \theta_{ja} I_{DS}^2 R_{FET}(T_j) \qquad (4.6\text{-}6b)$$

Equation (4.6-4) relates $R_{FET}(T_2 = T_j)$ to $R_{FET}(T_1 = T_a)$. Making the substitution in (4.6-6b) yields

$$\Delta T\, \epsilon^{-0.007\Delta T} = \theta_{ja} I_{DS}^2 R_{FET}(T_a) \qquad (4.6\text{-}7)$$

In Fig. 4.6-3 we have plotted $\Delta T\, \epsilon^{-0.007\Delta T}$ versus ΔT.

Equation (4.6-7) and Fig. 4.6-3 can be used to determine the thermal resistance of the heat sink needed in a given design, as illustrated in Example 4.6-3.

Example 4.6-3 A power FET has a maximum junction temperature of 175°C and a thermal resistance $\theta_{jc} = 4°C/W$. When the transistor is ON, the drain-source current is 1 A and the FET ON resistance at an ambient temperature of 25° is $R_{FET}(25°C) = 10\ \Omega$. If the junction temperature is to be 125°C or less, calculate (a) the thermal resistance required of the heat sink θ_{ca}, (b) the FET resistance when $T_j = 125°C$ and the power dissipated in the FET at $T_j = 25$ and 125°C assuming that I_{DS} remains at 1 A. (c) Under what conditions can we assume that I_{DS} remains constant even though R_{FET} changes?

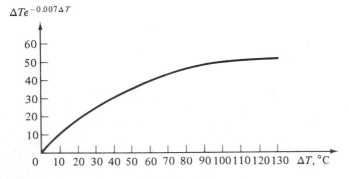

Figure 4.6-3 Plot of $\Delta T\, \epsilon^{-0.007\,\Delta T}$ versus ΔT.

SOLUTION (a) From (4.6-7) we have

$$\theta_{ja} = \theta_{jc} + \theta_{ca} = \frac{\Delta T \, \epsilon^{-0.007\Delta T}}{I_{DS}^2 R_{FET}(T_a)}$$

Using Fig. 4.6-3 with $\Delta T = 100°C$, we find

$$\Delta T \, \epsilon^{-0.007\Delta T} \approx 50$$

Hence
$$\theta_{ja} \approx \frac{50}{(1)^2(10)} = 5°C/W$$

Since $\theta_{jc} = 4°C/W$, the thermal resistance of the heat sink must be less than 1°C/W.

(b) Using (4.6-4), we have

$$R_{FET}(125°C) = 10 \, \epsilon^{(0.007)(100)} = 20 \, \Omega$$

Hence $P_F(25°C) = 10$ W and $P_F(125°C) = 20$ W.

(c) We can assume that I_{DS} remains constant while R_{FET} increases from 10 to 20 Ω if the load resistance greatly exceeds the 20 Ω FET resistance. Thus, if the FET switch is driving a 100-Ω load, the current change due to the increase in R_{FET} is negligible. ///

4.7 MANUFACTURERS' SPECIFICATIONS FOR HIGH-POWER ($P_{C,\,max} > 1$ W) TRANSISTORS

In this section, some common specifications provided by transistor manufacturers are discussed. The specifications given below are for an *npn* silicon diffused-junction power transistor.

Transistor type: 2N2015, silicon *npn*
A. Maximum thermal resistance $\theta_{jc} = 0.7°C/W$
B. Maximum collector dissipation with infinite heat sink at 25°C $P_C = 150$ W
C. Maximum junction temperature $T_{j,\,max} = 140°C$
D. Absolute maximum ratings at 25°C:
   1. $I_C = 10$ A
   2. $I_B = 6$ A
   3. Breakdown voltage
     a. Collector-base $(BV_{CBO}) = 100$ V
     b. Emitter-base $(BV_{EBO}) = 10$ V
     c. Collector-emitter $(BV_{CEO}) = 50$ V
E. Maximum I_{CBO} at maximum V_{CB} at 25°C = 50 μA
F. Current amplification β at $V_{CE} = 4$ V and $I_C = 5$ A: $15 < \beta \leq 50$

Explanation of symbols The maximum thermal resistance θ_{jc}, the maximum collector dissipation $P_{C,\,max}$, and the maximum junction temperature $T_{j,\,max}$, discussed in Secs. 1.11 and 4.6, can be summarized using the power-derating curve

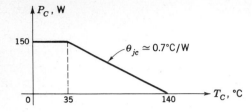

Figure 4.7-1 Derating curve.

shown in Fig. 4.7-1. Thus, if an infinite heat sink is employed ($\theta_{ca} = 0$), 150 W can be dissipated as long as the case temperature (which equals the ambient temperature for an infinite heat sink) is less than 35°C. If the case temperature increases above this value, the allowable power dissipation in the transistor decreases as shown.

The absolute maximum ratings specified indicate upper bounds on the current and voltage capability of the transistor. Thus a collector current of 10 A should never be exceeded. A base current of 6 A (making sure that $I_C < 10$ A) should never be exceeded. The collector-base and collector-emitter voltages should not exceed 100 or 50 V, respectively. This is called the breakdown voltage. When this voltage is exceeded, the junction breaks down (as in a Zener diode), and avalanche current multiplication results in a vi characteristic as shown in Fig. 1.10-1. Note that I_{CBO} of 50 μA is possible. This is not very large if we note that 10 A can flow in the collector. High-power transistors normally have higher relative collector-base currents than low-power transistors.

We see that the current-amplification factor β is only 15 to 50. This is due to construction problems. In a high-power transistor the base region is widened to increase the breakdown voltage, and as a result β decreases.

PROBLEMS

4.1-1 The silicon transistor to be used in the circuit of Fig. P4.1-1 has a β that varies from 50 to 200. If $V_{BB} = 3$ V and $R_e = 200$ Ω, find the variation in Q point for $R_b = 1$ kΩ and $R_b = 10$ kΩ.

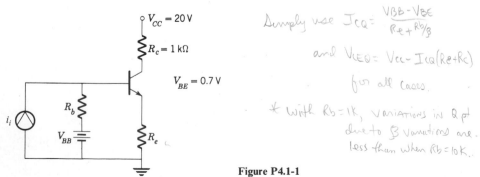

Simply use $I_{CQ} = \dfrac{V_{BB} - V_{BE}}{R_e + R_b/\beta}$

and $V_{CEO} = V_{CC} - I_{CQ}(R_e + R_C)$

for all cases.

* with $R_b = 1K$, variations in Q pt due to β variations are less than when $R_b = 10K$.

Figure P4.1-1

4.1-2 The transistor of Prob. 4.1-1 is to be used in the circuit of Fig. P4.1-2. Find the variation in quiescent current as β varies from 50 to 200.

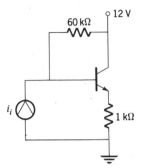

Figure P4.1-2

4.1-3 The amplifier shown in Fig. P4.1-3 is to be designed to have a maximum symmetrical swing. If β varies from 50 to 150 for this type of transistor, find V_{BB}, R_e, and the maximum swing in i_C.

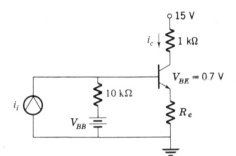

Figure P4.1-3

careful

4.2-1 In the circuit of Fig. P4.1-1 $V_{BB} = 3$ V, $I_{CBO} = 0.1$ μA, $R_b = 1$ kΩ, and $R_e = 200$ Ω. Find the variation in quiescent current as the temperature varies from 25 to 175°C. Assume $\beta = 100$.

4.2-2 In the circuit of Fig. P4.1-2 $I_{CBO} = 10$ μA and $\beta = 100$. Find the variation in quiescent current as the temperature varies from 25 to 175°C and from 25 to -55°C.

4.2-3 Calculate v_C as the temperature varies from 25 to 100°C in Fig. P4.2-3. Assume that T_1 and T_2 are identical silicon transistors with $I_{CBO} = 1$ μA and $\beta = 20$.

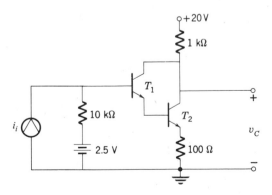

Figure P4.2-3

4.2-4 The amplifier shown in Fig. P4.2-4 is to be designed to have a maximum symmetrical swing. The temperature range lies between -55 and $+125°C$. $I_{CBO} = 0.1~\mu A$, $\beta \to \infty$, ΔI_{CQ} is to be less than 1 mA. Find (a) V_{BB} and R_e and (b) the maximum swing.

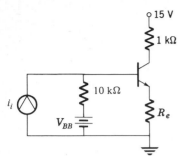

Figure P4.2-4

4.3-1 Using the stability factor S_β, find values for R_e and R_b in the circuit of Fig. P4.1-1 such that the voltage across R_c will not vary by more than ± 0.5 V as β varies from 50 to 200. The quiescent current is to be about 10 mA.

4.3-2 In the circuit of Fig. P4.3-2

$$50 < \beta < 200 \qquad 25°C < T < 75°C$$

$$V_{CC} = 6~V \pm 0.2~V \qquad I_{CBO} = 0.01~\mu A \text{ at } 25°C$$

Find the quiescent current, all the pertinent stability factors, and the worst-case quiescent-current shift.

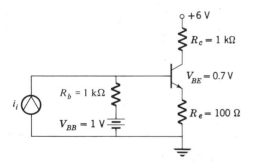

Figure P4.3-2

4.3-3 (a) The circuit of Fig. P4.3-3 illustrates the use of collector feedback to bias a transistor. Show that

$$I_{CQ} = \frac{\beta[V_{CC} - V_{BEQ} + (R_c + R_b)I_{CBO}]}{(\beta + 1)R_c + R_b}$$

(b) Find S_I, S_V, and S_β.

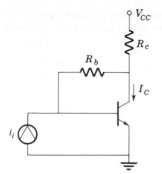

Figure P4.3-3

4.3-4 In Fig. P4.3-3, $V_{CC} = 20$ V and $R_c = 1$ kΩ. Find R_b so that $I_{CQ} \approx 10$ mA. Calculate S_I, S_V, and S_β. Determine the Q-point shift for $50 < \beta < 200$ and $25°C < T < 100°C$, with $I_{CBO}(25°C) = 1$ μA.

4.3-5 (a) An emitter resistor R_e is added to the circuit of Fig. P4.3-3. Show that

$$I_{CQ} = \frac{\beta(R_e + R_b + R_c)I_{CBO} + \beta(V_{CC} - V_{BEQ})}{R_b + (\beta + 1)(R_e + R_c)}$$

(b) Find S_I, S_V, and S_β.

4.3-6 Using the results of Prob. 4.3-5, find the Q-point shift when $50 < \beta < 200$, $25°C < T < 100°C$, and $I_{CBO} = 1$ μA at room temperature. Find R_b such that $I_{CQ} = 10$ mA when $V_{CC} = 20$ V, $R_c = 800$ Ω, and $R_e = 200$ Ω.

4.3-7 The amplifier shown in Fig. P4.3-7 is to be operated over the temperature range -25 to $75°C$. The range of β for the transistor used lies between 100 and 300. The transistor has $I_{CBO} = 0.1$ μA and $V_{BE} = 0.7$ V at room temperature. If $R_1 \| R_2 \geq 1$ kΩ, find the maximum possible symmetrical swing. Specify R_1, R_2, and R_e.

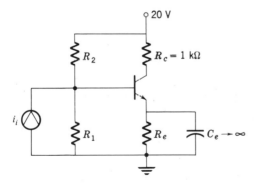

Figure P4.3-7

$I_{CE} = I_{CEN}$

4.4-1 The amplifiers shown in Fig. P4.4-1 are to operate over the temperature range 25 to 90°C.

   (a) Calculate the change in quiescent current for the uncompensated and compensated amplifiers and compare.

   (b) Calculate V_{BB} for the amplifier in Fig. P4.4-1b.

$\Delta T = 65$

$$\Delta I_{CQ} = \frac{K\Delta T}{Re} + \left(1 + \frac{Rb}{Re}\right) I_{CBO}\left(e^{K\Delta T} - 1\right)$$

I_{CBO} typical $= .1 \mu A$

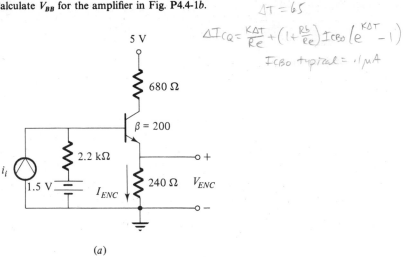

(a)

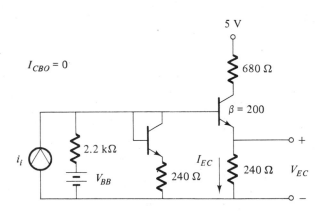

(b)

Figure P4.4-1 (b) Compensated; (a) uncompensated.

4.4-2 Determine R_2, R_1, and R_d in Fig. P4.4-2 for maximum symmetrical swing. Assume $I_D = I_E$ and $\beta = 250$. *Hint:* Take a Thevenin equivalent, obtaining V_{BB} and R_b as in Fig. 4.4-5a.

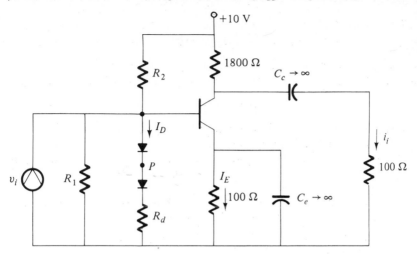

Figure P4.4-2

4.4-3 A commonly used IC connection to approximate several diodes in series is shown in Fig. P4.4-3a. Use the transistor model in Fig. P4.4-3b and prove that the circuit of Fig. P4.4-3c results. *Hint:* Write $v_D = i_{R2} R_2 + i_{R1} R_1$ and solve for i_{R2} and i_{R1} in terms of i_n.

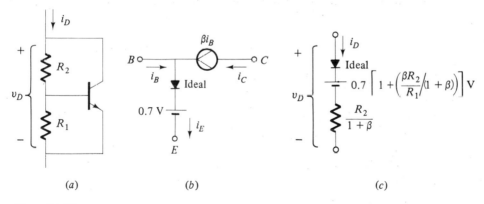

(a)   (b)   (c)

Figure P4.4-3

4.4-4 The circuit of Fig. 4.4-3 with $R_d = R_c = 0$ is used quite often in IC biasing and is referred to as a *current mirror*. This is redrawn in Fig. P4.4-4. Assuming matched transistors, show that

$$I_{EQ} = \frac{(V_{BB} - 0.7)/R_b}{1 + 1/(1 + \beta_2)}$$

Find S_1.

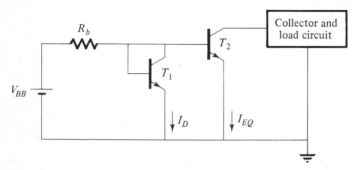

Figure P4.4-4

4.4-5 By inserting a resistor R_e into the emitter circuit of T_2 in Fig. P4.4-4, small values of I_{EQ} can be obtained

(*a*) Assuming matched transistors and $I_D \gg I_{EQ}$, show that

$$R_e = \frac{25 \times 10^{-3}}{I_{EQ}} \ln \frac{I_D}{I_{EQ}} \qquad I_D = \frac{V_{BB} - 0.7}{R_b}$$

Hint: Use (1.2-4*d*).

(*b*) Given $V_{BB} = 6$ V and $R_b = 2$ kΩ, calculate R_e so that $I_{EQ} = 10$ μA.

4.4-6 (*a*) Refer to Fig. 4.4-5*b*. Show that when $R_b = R_d$,

$$I_E = \frac{V_{BB}/2 - 25 \times 10^{-3} \ln (I_E/I_D)}{R_e + R_b/2(\beta + 1)}$$

(*b*) As in Example 4.4-2, let $I_E = 1$ mA and $R_e = 2$ kΩ. Assume $\beta = 100$, and let $I_D = 0.01$ mA to save power. Use a rule of thumb similar to (2.3-5*b*) and calculate V_{BB} and R_b.

4.4-7 In Fig. 4.4-3 use n diodes in place of the single diode. Show that $\partial I_{EQ}/\partial T = 0$ when $R_b = R_d/(n-1)$.

4.5-1 A JFET characterized by the curves in Fig. 3.1-5 is used in the circuit of Fig. 4.5-1. Let $V_{DD} = 16$, $V_{GG} = 0$, $V_{GS} = -2$ V, and $I_{DS} = 2$ mA. $VDG = VDS-VgS=VpO$ $VDS = Vpo+VgS$
(*a*) Find V_{DS}.
(*b*) Calculate R_d and R_s. $VDD=(RD+RS)IDSQ + VDS$

4.5-2 A MOSFET characterized by the curves of Fig. 3.2-5 is used in the circuit of Fig. 4.5-1. Let $V_{DD} = 12$ V, $V_{GS} = 4$ V, and $I_{DS} = 0.5$ mA. $VDS = Vgs-VTN = 4-2=(2V)$ $p.of 142$
(*a*) Find V_{DS}
(*b*) Calculate R_d and V_{GG}. → $Vgg = Vgs + IDSRs = 4+.5mA(2000) = (5v)$ $Ro given in class = 10k$

4.5-3 The MOSFET characterized in Fig. 3.2-5 is used in the circuit of Fig. 4.5-3. Let $V_{DD} = 12$ V, $R_g = 1$ MΩ, $V_{GS} = 4$ V, and $I_{DS} = 0.5$ mA. Find V_{DS} and R_L.

Given
$Ri = Rd = 2K$

Given
$Ro = 10K$
$Rs = 2K$

4.5-4 The JFET 2N4223 (characteristics in Appendix C) used in the circuit of Fig. P4.5-4, is biased at $V_{GS} = -2$ V and $V_{DS} = 10$ V. If $V_{DD} = 16$ V, find (a) R_{s1} and R_{s2} and (b) v_o.

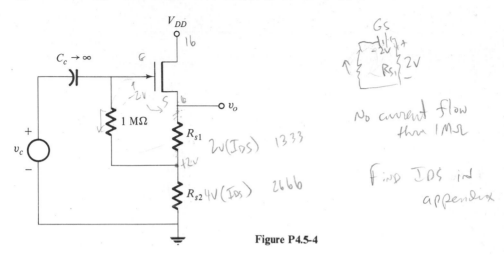

Handwritten annotations:

GS

R_s 2V

No current flow
thru 1 MΩ

2v(I_{DS}) 1333

R_{s2} 4V(I_{DS}) 2666

Find I_{DS} in
appendix

Figure P4.5-4

4.5-5 The MOSFET described in Fig. 4.5-4 is used in the circuit of Fig. 4.5-5. Let $V_{DD} = 15$ V and $R_D = 1.5$ kΩ.

(a) Let $R_1 \| R_2 = 1$ MΩ. Find R_1 and R_2 so that $V_{GSQ} = 5.8$ V.

(b) For the result in part (a) calculate the variation in I_{DSQ} and compare with the variation using feedback as in Example 4.5-2.

4.5-6 Repeat Example 4.5-1 if $I_{DSQ, \text{nom}} = 3$ mA and a 10 percent variation is permitted. Assume a nominal value of $V_{DSQ} = 8$ V and $V_{DD} = 20$ V.

4.5-7 In Fig. 4.5-3, $R_g = 1$ MΩ, $R_L = 3.3$ kΩ, and $V_{DD} = 10$ V. The MOSFET is characterized by (4.5-3a) and (4.5-3b), which are plotted in Fig. 4.5-4. Find the maximum possible variation in I_{DSQ} and V_{DSQ}.

4.6-1 A silicon power transistor has the thermal ratings

$$P_{C, \text{max}} = 200 \text{ W} \qquad T_{j, \text{max}} = 175°C \qquad \theta_{jc} = 0.7°C/W$$

$T_A = 25°C$

(a) Find the power that this transistor could dissipate if the case could be maintained at room temperature (25°C). $T_A = 25°C$

(b) The transistor is mounted directly to a flat aluminium heat sink which has $\theta_{sa} = 8°C/W$. The direct mounting results in $\theta_{cs} = 0.2°C/W$. Find the maximum allowable dissipation.

4.6-2 (a) The heat sink of part (b) of Prob. 4.6-1 is used with the transistor in that problem, but the transistor is electrically insulated from the sink with a mica washer, so that θ_{cs} increases to 2°C/W. Find the maximum allowable dissipation.

(b) Find the temperature of the case and heat sink.

4.6-3 (a) The transistor of Prob. 4.6-1 is used with a large finned heat sink, which has $\theta_{sa} \approx 0.9°C/W$. Find the maximum allowable dissipation if the transistor is mounted directly or if it is electrically insulated by the mica washer, as in Prob. 4.6-2.

(b) Find the temperature of the case and of the heat sink.

4.6-4 Repeat Example 4.6-3 for a power FET having a maximum junction temperature of 160°C and a thermal resistance of 5°C/W. When the FET is ON, the drain-source current is 0.8 A and the ON resistance at an ambient temperature of 25°C is $R_{FET}(25°C) = 12$ Ω. The junction temperature is to be 125°C or less.

AUDIO-FREQUENCY LINEAR
POWER AMPLIFIERS

INTRODUCTION

Several major problems associated with amplifiers required to furnish large amounts of power are discussed in this chapter. The aim in most applications is to furnish the required power as economically as possible while meeting other specifications, which may include limitations on size, weight, dc supply voltage, distortion, etc. The designer often has to make several compromises along the way in order to achieve an optimum design. Often, the transistors are driven to the limits of their useful operating range, and careful design is required to ensure that physical damage, due to excessive heating, does not occur.

Power amplifiers are classified according to the portion of the input sine-wave cycle during which load current flows (Fig. 5.1). To achieve low-distortion amplification of audio-frequency signals, only class A would seem to apply. However, if one uses *complementary-symmetry* or a *push-pull* arrangement, described in Secs. 5.3 and 5.4, class AB and class B amplifiers can also yield essentially linear amplification. Class C power amplifiers are used extensively at radio frequencies where tuned circuits remove the distortion resulting from the nonlinear operation of the circuit.

In this chapter Q-point placement and power relations in the most commonly used class A and B audio-frequency BJT circuits are studied.

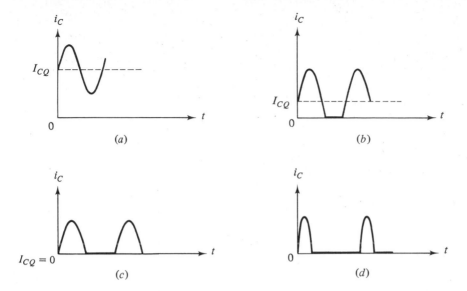

Figure 5.1 (*a*) Class A: current flows for 360° (full-cycle operation); (*b*) class AB: current flows for more than one half-cycle but less than full cycle; (*c*) class B: current flows for one half-cycle; (*d*) class C: current flows for less than one half-cycle.

5.1 THE CLASS A COMMON-EMITTER POWER AMPLIFIER

In Chap. 2 it was observed that a large amount of power is dissipated in the collector resistor R_c because of the quiescent collector current I_{CQ}. This results in a maximum operational efficiency of only 25 percent. Thus, when 1 W of signal power is to be dissipated in the load under maximum signal conditions, 4 W must be furnished continuously by the dc power supply. In this section it is shown that replacing R_c with a large inductor (often called a *choke*) increases the maximum efficiency to 50 percent.

Replace Rc by a choke → increases maximum efficiency from 25 to 50%.

5.1-1 *Q*-Point Placement

A class A circuit with choke coupling is shown in Fig. 5.1-1*a*. The circuit is designed so that all capacitors are essentially short circuits and the inductor is essentially an open circuit at signal frequencies. At dc the capacitors are, of course, open circuits, while the inductor is a short circuit. For simplicity the inductor is assumed to have no internal resistance.

To determine the Q point, apply KVL around the collector circuit, including only dc voltage drops. The dc-load-line equation for this amplifier is then

$$V_{CC} = v_{CE} + i_c R_e \qquad (5.1\text{-}1)$$

$$V_{ce} = V_{cc} - i_c R_e$$

$V_{cc} = V_{ce} + i_c R_e$

$V_{ce} = V_{cc} - i_c R_e$

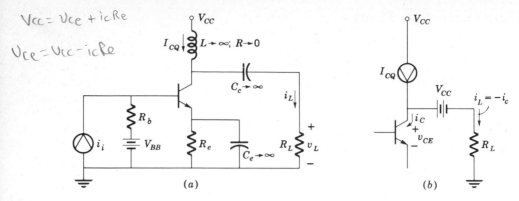

Figure 5.1-1 Inductor-coupled power amplifier: (a) circuit; (b) equivalent circuit.

The emitter resistor is kept as small as possible in order to minimize bias-circuit power loss while maintaining adequate Q-point stability. Thus the dc load line is almost vertical, as shown in Fig. 5.1-2.

Applying KVL around the collector circuit (Fig. 5.1-1a), including only ac voltage drops, yields the ac-load-line equation

$$v_{ce} = -i_c R_L = i_L R_L \qquad (5.1\text{-}2a)$$

which can be written

$$i_C - I_{CQ} = -\frac{1}{R_L}(v_{CE} - V_{CEQ}) \qquad (5.1\text{-}2b)$$

To place the Q point for maximum symmetrical swing, we use

For maximum
Symmetrical Swing

$$\boxed{I_{CQ} = \frac{V_{CC}}{R_{ac} + R_{dc}}} \qquad (2.5\text{-}6b)$$

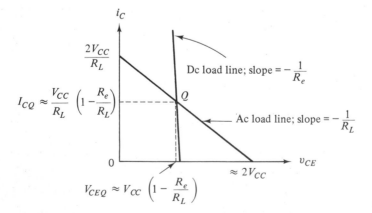

Figure 5.1-2 Power-amplifier load lines.

For this circuit $R_{ac} = R_L$ and $R_{dc} = R_e$, so that   FOR MAX SWING

$$I_{CQ} = \frac{V_{CC}}{R_L + R_e} = \boxed{\frac{V_{CC}/R_L}{1 + R_e/R_L} = I_{CQ}} \qquad (5.1\text{-}3a)$$

The collector-emitter voltage at the Q point is found by setting $v_{CE} = V_{CEQ}$ and $i_C = I_{CQ}$ in (5.1-1) and using the value of I_{CQ} given in (5.1-3a). The result is

$$V_{CEQ} = V_{CC}\frac{R_L}{R_L + R_e} = \boxed{\frac{V_{CC}}{1 + R_e/R_L} = V_{CEQ}} \qquad (5.1\text{-}3b)$$

MAX Swing

Usually $R_L \gg R_e$, so that (5.1-3a) and (5.1-3b) reduce to

$$I_{CQ} \approx \frac{V_{CC}}{R_L}\left(1 - \frac{R_e}{R_L}\right) \approx \frac{V_{CC}}{R_L} \quad \Big\{ \text{if } R_L \gg R_e \qquad (5.1\text{-}3c)$$

and

$$V_{CEQ} \approx V_{CC}\left(1 - \frac{R_e}{R_L}\right) \approx V_{CC} \qquad (5.1\text{-}3d)$$

The ac load line passes through the Q point with slope $-1/R_L$, as shown in Fig. 5.1-2. Note that the maximum collector-current swing is from 0 to $2I_{CQ}$ as v_{CE} swings from $2V_{CC}$ to 0. Note, too, that v_{CE} is limited by the saturation voltage of the transistor to a minimum voltage of $V_{CE,\,sat}$. To simplify calculations, the saturation voltage is assumed to be zero. Its effect, however, is considered in the examples.

It is interesting to consider how the collector-emitter voltage can become twice the supply voltage. Since the inductance is very large, no ac current will flow through it, and for purposes of analysis, it can be replaced by a constant current source of strength I_{CQ}. Since the capacitive reactance is very small, no ac voltage will appear across the capacitor, and it can be replaced by a battery of voltage V_{CC}, the voltage to which it is charged when no signal is present. With these two substitutions, the collector-load circuit takes the equivalent form shown in Fig. 5.1-1b. Assume that a sinusoidal signal is present, and consider an instant of time when $i_C = 0$. At that instant, $i_L = I_{CQ}$, so that $v_{CE} = V_{CC} + i_L R_L$. From (5.1-3), $I_{CQ} R_L = i_L R_L = V_{CC}$, so that $v_{CE} = 2V_{CC}$. This establishes the upper limit on v_{CE}. When the signal reverses polarity, $i_C = 2I_{CQ}$. Then i_L must equal $-I_{CQ}$ so that $v_{CE} = 0$, establishing the lower limit.

5.1-2 Power Calculations

With the amplifier biased as in (5.1-3), the currents and voltages of interest are (the voltage across the emitter resistor is neglected for simplicity)

$$i_C = I_{CQ} + i_c = \frac{V_{CC}}{R_L} + i_c \qquad (5.1\text{-}4a)$$

$$i_L = -i_c \qquad (5.1\text{-}4b)$$

$$i_{supply} = i_L + i_C = I_{CQ} = \frac{V_{CC}}{R_L} \qquad (5.1\text{-}4c)$$

From Fig. 5.1-1b we have

$$v_{CE} = V_{CC} - i_c R_L \tag{5.1-5a}$$

and

$$v_L = +i_L R_L = -i_c R_L \tag{5.1-5b}$$

If the signal current is sinusoidal

$$i_i = I_{im} \sin \omega t \tag{5.1-6a}$$

then

$$i_c = I_{cm} \sin \omega t \tag{5.1-6b}$$

It should be noted that the *maximum* peak value of ac collector current is I_{CQ}, so that

$$i_{c,\,max} = I_{CQ} \sin \omega t \tag{5.1-7a}$$

Therefore

$$I_{cm} \leq I_{CQ} \tag{5.1-7b}$$

The power supplied, the power dissipated in the collector and the load, and the efficiency are found in the same manner as in Sec. 2.4. The results are given in (5.1-8) to (5.1-11) and are plotted in Fig. 5.1-3.

Supplied power

$$P_{CC} = V_{CC} I_{CQ} \approx \frac{V_{CC}^2}{R_L} \tag{5.1-8}$$

which is constant and essentially independent of signal current as long as the distortion is negligible.

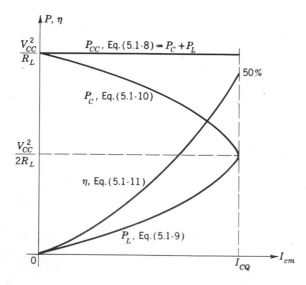

Figure 5.1-3 Variation of power and efficiency with collector current.

Power transferred to load

$$P_L = \frac{I_{Lm}^2 R_L}{2} = \frac{I_{cm}^2 R_L}{2} \tag{5.1-9a}$$

since $i_L = -i_c$, $I_{Lm} = -I_{cm}$.

The *maximum* average power dissipated by the load occurs when

When $\boxed{I_{cm} = I_{CQ}}$

Thus

$$P_{L,\,max} = \frac{I_{CQ}^2 R_L}{2} = \frac{V_{CC}^2}{2R_L} \tag{5.1-9b}$$

Collector dissipation

$$P_C = P_{CC} - P_L = \frac{V_{CC}^2}{R_L} - \frac{I_{cm}^2 R_L}{2} \tag{5.1-10a}$$

Thus the minimum power dissipated in the collector occurs when the maximum power is dissipated by the load:

$$P_{C,\,min} = \frac{V_{CC}^2}{2R_L} \qquad P_C \text{ minimum} \tag{5.1-10b}$$

The maximum power dissipated by the collector occurs when no signal is present.

$$P_{C,\,max} - \frac{V_{CC}^2}{R_L} = V_{CEQ}I_{CQ} \qquad P_C \text{ maximum} \tag{5.1-10c}$$

Efficiency The efficiency of operation of the inductor-coupled amplifier for a sinusoidal signal is

$$\eta = \frac{P_L}{P_{CC}} = \frac{I_{cm}^2 (R_L/2)}{V_{CC}I_{CQ}} = \frac{1}{2}\left(\frac{I_{cm}}{I_{CQ}}\right)^2 \tag{5.1-11a}$$

Thus maximum efficiency occurs at maximum signal current. Then

$$\eta_{max} = \tfrac{1}{2} = 50\% \tag{5.1-11b}$$

The efficiency of operation has been doubled by using an inductor rather than a resistor R_c in the dc collector circuit.

The variations of supply, load and collector powers and efficiency are plotted in Fig. 5.1-3 as a function of collector current for sinusoidal signals. Note that as the load power increases, the collector dissipation decreases, their sum remaining constant ($P_{CC} = P_C + P_L$). Also note that P_C is a maximum when no signal is present.

Figure of merit A useful figure of merit for a power amplifier is the ratio of maximum transistor collector dissipation to maximum power dissipated in the load. Using (5.1-10c) and (5.1-9b) or Fig. 5.1-3, we have

$$\frac{P_{C,\,max}}{P_{L,\,max}} = 2 \tag{5.1-12}$$

This is the same result obtained in Sec. 2.4. Thus, if $P_{L,\,max} = 25$ W, the collector junction must be capable of dissipating at least 50 W. In Chap. 4 it was found that in order to operate at high ambient temperatures, transistors often must be derated. Thus, to dissipate 25 W might require a transistor with an allowable collector dissipation of 100 W if the ambient temperature is high.

5.1-3 The Maximum-Dissipation Hyperbola

Once the maximum power to be delivered to the load and the anticipated temperature range are known, the power rating of the transistor can be determined and the transistor can be selected. The maximum allowable collector dissipation is, as stated above, usually less than the maximum rating specified for the transistor.

In addition to having the specified power rating, the transistor must be able to handle currents as high as $2I_{CQ}$ (Fig. 5.1-2) and a collector-emitter voltage up to $2V_{CC}$. It must also have an operating frequency at least as high as the signal frequency. These ratings are generally supplied by the transistor manufacturer (Sec. 4.7).

Design and transistor specifications will, in general, include

$$
\begin{aligned}
&i_{C,\,max} \geq 2I_{CQ} \\
&BV_{CEO} \geq 2V_{CC} \\
\text{and} \qquad &P_{C,\,max} = V_{CEQ}I_{CQ}
\end{aligned}
\qquad (5.1\text{-}10c)
$$

These maximum ratings place bounds on the permissible operating region of the transistor, as in Fig. 5.1-4. The figure shows that for safe operation the Q point must lie on or below the hyperbola

$$v_{CE}i_C = P_{C,\,max}$$

This hyperbola represents the locus of all operating points at which the collector dissipation is exactly $P_{C,\,max}$.

The ac load line, having a slope $-1/R_L$, must pass through the Q point,

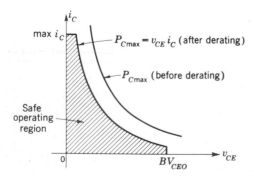

Figure 5.1-4 Maximum-collector-dissipation hyperbola.

intersect the v_{CE} axis at a voltage less than BV_{CEO}, and intersect the i_C axis at a current less than max i_C; that is,

$$2V_{CC} \leq BV_{CEO} \tag{5.1-13a}$$

$$2I_{CQ} \leq i_{C,\,max} \tag{5.1-13b}$$

In order to obtain a maximum symmetrical swing we must have

$$I_{CQ} = \frac{1}{R_L} V_{CEQ} \tag{2.5-4}$$

Combining (2.5-4) and (5.1-10c) gives the Q point at

$$I_{CQ} = \sqrt{\frac{P_{C,\,max}}{R_L}} \quad \begin{cases} \text{For} \\ \text{max Swing} \end{cases} \tag{5.1-14a}$$

and

$$V_{CEQ} = \sqrt{P_{C,\,max} R_L} \tag{5.1-14b}$$

It is interesting to note that at the Q point the slope of the hyperbola is

$$\frac{\partial i_C}{\partial v_{CE}} = -\frac{I_{CQ}}{V_{CEQ}} = -\frac{1}{R_L} \tag{5.1-15}$$

Thus the slope of the ac load line is the same as the slope of the hyperbola, and the ac load line is tangent to the hyperbola at the Q point when maximum symmetrical swing is obtained.

Example 5.1-1 A transistor having the following specifications is used in the circuit of Fig. 5.1-1:

$$P_{C,\,max} \text{ (after derating)} = 4 \text{ W} \qquad BV_{CEO} = 40 \text{ V} \qquad i_{C,\,max} = 2 \text{ A}$$

The load resistance R_L is 10 Ω. Determine the Q point so that maximum power is dissipated by the load resistor. Specify the required supply voltage V_{CC}.

SOLUTION The permissible operating range is shown in Fig. 5.1-5. The Q point is obtained by plotting

$$i_C = \frac{1}{R_L} v_{CE} = \frac{v_{CE}}{10} \tag{2.5-4}$$

on the vi characteristic and finding the intersection of this equation with the $P_{C,\,max}$ hyperbola. The intersection is the desired Q point, and the ac load line is tangent to the hyperbola at this point.

The Q point is obtained graphically, as in Fig. 5.1-5, or analytically, using (5.1-14a) and (5.1-14b):

$$I_{CQ} = \sqrt{\tfrac{4}{10}} \approx 0.63 \text{ A} \qquad \text{and} \qquad V_{CEQ} = \sqrt{(4)(10)} \approx 6.3 \text{ V}$$

These results are seen to verify those obtained graphically.

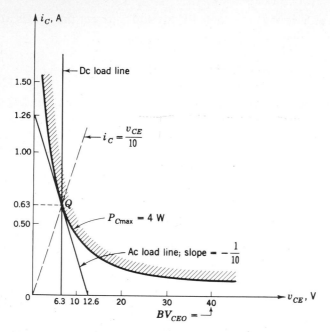

Figure 5.1-5 Load lines and maximum-dissipation hyperbola for Example 5.1-1.

The supply voltage V_{CC} is equal to V_{CEQ} (if we neglect the voltage drop in R_e). Thus

$$V_{CC} = 6.3 \text{ V}$$

and the maximum v_{CE} is equal to 12.6 V and is less than the breakdown voltage. The maximum collector current i_C is equal to 1.26 A and is less than the maximum allowable current.

The maximum power delivered to the load is then

$$P_{L,\max} = \frac{I_{CQ}^2 R_L}{2} = \frac{(0.63)^2(10)}{2} = 2 \text{ W} \qquad (5.1\text{-}9b)$$

Selecting R_e, R_b, and V_{BB} It should be noted that the quiescent operating point is obtained using the standard techniques of Chap. 2. From Fig. 5.1-1a, R_b is chosen according to the rule given in (2.3-5b)

$$R_b = \tfrac{1}{10}\beta R_e$$

In addition, R_e is chosen to be small, so that its power dissipation is negligible. For example, we could choose $R_e = 1 \ \Omega$, so that

$$P_{R_e} = I_{CQ}^2 \times 1 = 0.4 \text{ W} \ll P_{C,\max} = 4 \text{ W}$$

If $\beta = 40$,

$$R_b \approx 4 \ \Omega$$

The base supply voltage is $V_{BB} \approx .7 + I_{CQ}(R_e)$

$$V_{BB} \approx 0.7 + (0.63)(1) = 1.33 \text{ V}$$

The Q point is found to be shifted slightly if the effect of R_e is included in the design. This results in an increase in the V_{CC} required.

Let us consider the same problem with the maximum collector-current rating changed to max $i_C = 1$ A. The previous solution results in $i_{C,\,max} = 1.26$ A, which exceeds the new maximum allowable current. If the Q point is left unchanged, the peak safe ac current swing is reduced to 0.37 A. If the signal is sinusoidal, the maximum ac current is then

$$i_C = 0.37 \sin \omega t \qquad A$$

and the maximum power dissipated in the load now becomes

$$P_{L,\,max} = (\tfrac{1}{2})(0.37)^2(10) \approx 0.69 \text{ W}$$

Since the 10-Ω load is fixed, the slope of the ac load line is fixed. However, if the load line is moved so that it intersects the i_C axis at $i_{C,\,max} = 1$ A and the Q point placed at

$$I_{CQ} = 0.5 \text{ A} \qquad V_{CEQ} = V_{CC} = 5 \text{ V}$$

the maximum ac collector current is

$$i_C = 0.5 \sin \omega t \qquad A$$

and the maximum power in the load becomes

$$P_{L,\,max} = (\tfrac{1}{2})(0.5)^2(10) = 1.25 \text{ W}$$

In either case the power delivered to the load is far below the available power of 2 W because we are unable to compensate for the reduced maximum allowable collector current. In the next section the transformer-coupled power amplifier is considered. Using this circuit, we can place the Q point without regard to the actual load resistance by making use of the impedance-transforming property of the transformer. The problem faced above, of not being able to deliver maximum available power to the load, is thus alleviated.

///

Example 5.1-2 An *npn* silicon transistor is used in the circuit of Fig. 5.1-6a with an 8-Ω load resistor. The maximum ratings of this transistor are

$$P_{C,\,max} = 24 \text{ W (after derating)} \qquad BV_{CEO} = 80 \text{ V} \qquad V_{CE,\,sat} = 2 \text{ V}$$

Determine the maximum attainable swing and the maximum power dissipated by the load.

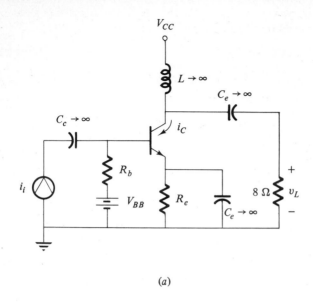

(a)

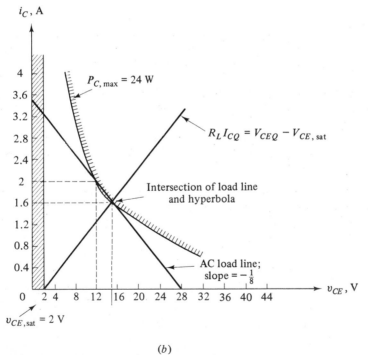

(b)

Figure 5.1-6 Example 5.1-2: (a) circuit; (b) operating conditions.

SOLUTION The amplifier circuit and the vi characteristic are shown in Fig. 5.1-6. Maximum power is dissipated by the load when the collector-current swing is a maximum. The ac-load-line equation is

$$R_L(i_C - I_{CQ}) = -(v_{CE} - V_{CEQ})$$

When i_C is a maximum, $i_C = 2I_{CQ}$ (Sec. 2.5) and $v_{CE} = V_{CE, \text{sat}}$. Then

$$R_L I_{CQ} = V_{CEQ} - V_{CE, \text{sat}} \tag{5.1-16}$$

To avoid exceeding the maximum average collector dissipation, we set

$$\boxed{I_{CQ} V_{CEQ} = P_{C, \text{max}}} \tag{5.1-17}$$

Combining (5.1-16) and (5.1-17) and solving the resulting equation for I_{CQ} and V_{CEQ} yields

$$I_{CQ} - -\frac{V_{CE, \text{sat}}}{2R_L} + \sqrt{\frac{P_{C, \text{max}}}{R_L} + \left(\frac{V_{CE, \text{sat}}}{2R_L}\right)^2} \tag{5.1-18a}$$

and

$$V_{CEQ} = \frac{V_{CE, \text{sat}}}{2} + \sqrt{P_{C, \text{max}} R_L + \left(\frac{V_{CE, \text{sat}}}{2}\right)^2} \tag{5.1-18b}$$

The Q point can, of course, also be obtained graphically from the intersection of (5.1-16) and (5.1-17), as shown in Fig. 5.1-6b. The Q point is found to be

$$I_{CQ} \approx 1.6 \text{ A} \quad \text{and} \quad V_{CEQ} \approx 15 \text{ V}$$

Thus the collector supply voltage V_{CC} is 15 V, neglecting the drop across R_e.

Note that the ac load line crosses the maximum-power hyperbola. This does not mean that the average power dissipated in the collector circuit exceeds $P_{C, \text{max}}$. Since maximum collector dissipation occurs when there is no signal and under this condition the collector dissipation is $P_{C, \text{max}}$ [Eq. (5.1-17)], maximum collector dissipation is not exceeded.

The maximum peak ac collector current is 1.6 A, and the maximum average power dissipated by the load resistor is

$$P_{L, \text{max}} = \frac{\frac{1}{2}I_{CQ}^2 R_L}{2} = (\tfrac{1}{2})(1.6)^2(8) = 10.2 \text{ W}$$

Note that the maximum efficiency, neglecting the loss in R_e, is only

$$\eta_{\text{max}} - \frac{P_{L, \text{max}}}{P_{CC}} - \frac{10.2}{(15)(1.6)} = 42\% \qquad ///$$

5.2 TRANSFORMER-COUPLED AMPLIFIER

A class A transformer-coupled amplifier is shown in Fig. 5.2-1. In the analysis presented below the transformer is assumed to be ideal. This implies that

$$v_c = Nv_L \tag{5.2-1}$$

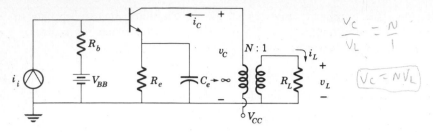

Figure 5.2-1 Transformer-coupled power amplifier.

and
$$Ni_c = -i_L \qquad (5.2\text{-}2)$$

Multiplying (5.2-1) and (5.2-2) gives
$$v_c(-i_c) = v_L i_L \qquad (5.2\text{-}3)$$

Dividing (5.2-1) by (5.2-2), we get
$$\frac{v_c}{-i_c} = N^2 \frac{v_L}{i_L} = N^2 R_L \equiv R_L' \qquad (5.2\text{-}4)$$

Therefore the ac impedance R_L' seen looking into the transformer is N^2 times the load resistance R_L.

The dc-load-line equation for this amplifier is the same as for the inductor-coupled amplifier
$$V_{CC} = v_{CE} + i_E R_e \approx v_{CE} + i_c R_e \qquad (5.2\text{-}5)$$

R_e is again made small. The ac load line can be obtained directly from (5.2-4), noting that $v_{ce} = v_c$. Thus the slope of the ac load line is
$$\frac{i_c}{v_{ce}} = -\frac{1}{R_L'} \qquad (5.2\text{-}6)$$

The dc and ac load lines are plotted in Fig. 5.2-2. If R_e is chosen so that
$$R_e \ll R_L'$$

the quiescent current for maximum symmetrical swing is

FOR MAX SYMM SWING
$$I_{CQ} \approx \frac{V_{CC}}{R_L'}\left(1 - \frac{R_e}{R_L'}\right) \qquad (5.2\text{-}7a)$$

and the quiescent collector-emitter voltage, found by substituting (5.2-7a) into (5.2-5), is
$$V_{CEQ} \approx V_{CC}\left(1 - \frac{R_e}{R_L'}\right) \qquad (5.2\text{-}7b)$$

Thus the impedance-transformation ratio provides the extra freedom required to set the Q point for maximum power transfer to the load.

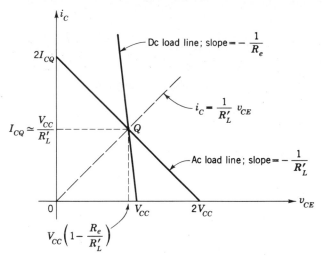

Figure 5.2-2 Load lines for transformer-coupled amplifier.

5.2-1 Power Calculations

The power calculations presented below are identical with those made in Sec. 5.1, with R_L replaced by R'_L. The signal i_i is sinusoidal; thus

$$i_c = I_{cm} \sin \omega t \qquad (5.2\text{-}8)$$

Supplied power

$$P_{CC} = V_{CC} I_{CQ} = \frac{V_{CC}^2}{R'_L} \qquad (5.2\text{-}9)$$

Power transferred to load When the signal is sinusoidal, the load current is also sinusoidal (neglecting distortion):

$$i_L = I_{Lm} \sin \omega t \qquad (5.2\text{-}10)$$

Then

$$P_L = \frac{I_{Lm}^2}{2} R_L \qquad (5.2\text{-}11)$$

Using (5.2-2), we have

$$I_{Lm} = N I_{cm} \qquad (5.2\text{-}12)$$

Therefore

$$P_L = \frac{I_{cm}^2}{2} R'_L \qquad (5.2\text{-}13a)$$

and

$$P_{L,\,\text{max}} = \frac{I_{CQ}^2}{2} R'_L = \frac{V_{CC}^2}{2R'_L} \qquad (5.2\text{-}13b)$$

Collector dissipation

$$P_C = \frac{V_{CC}^2}{R_L'} - \frac{I_{cm}^2}{2} R_L' \qquad (5.2\text{-}14a)$$

and the maximum collector dissipation is, with no signal present,

$$P_{C,\,max} = \frac{V_{CC}^2}{R_L'} = V_{CEQ} I_{CQ} \qquad (5.2\text{-}14b)$$

Efficiency The efficiency of operation is also unchanged.

$$\eta = \frac{1}{2}\left(\frac{I_{cm}}{I_{CQ}}\right)^2 \qquad (5.2\text{-}15a)$$

and

$$\eta_{max} = 50\% \qquad (5.2\text{-}15b)$$

Figure of merit The transistor figure of merit remains

$$\frac{P_{C,\,max}}{P_{L,\,max}} = 2 \qquad (5.2\text{-}16)$$

From the above equations it is seen that the transformer performs a single function over and above that of the inductor-capacitor coupling circuit shown in Fig. 5.1-1a, that is, transforming the load impedance. This factor provides the flexibility needed to place the ac operating path in the optimum position. It has been assumed in the above analysis that this transformation takes place without any power loss. In practice, however, the transformer is not ideal, and power losses occur in it, reducing the load power and the efficiency of operation. In addition, audiofrequency transformers are always of the iron-core variety and hence may be heavy and space-consuming, and they often introduce distortion.

Example 5.2-1 Using the transistor of Example 5.1-1, where $i_{C,\,max} = 1$ A, with transformer coupling to the 10-Ω load, redesign the amplifier for maximum power transfer to the load. Specify the required supply voltage, the power dissipated in the load, and the transformer turns ratio N.

SOLUTION The vi characteristic showing the permissible operating region is presented in Fig. 5.2-3. The quiescent point which will provide maximum power transfer to the load can be obtained graphically or analytically using (5.1-14).

$$I_{CQ} = \sqrt{\frac{P_{C,\,max}}{N^2 R_L}} = \sqrt{\frac{0.4}{N^2}} = \frac{0.63}{N} \text{ A} \qquad (5.2\text{-}17a)$$

and

$$V_{CEQ} = \sqrt{P_{C,\,max} N^2 R_L} = 6.3N \text{ V} \qquad (5.2\text{-}17b)$$

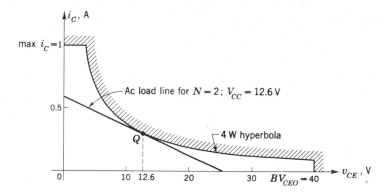

Figure 5.2-3 Operating region and optimum load line for Example 5.2-1.

Thus, using a transformer, the Q point may be chosen almost arbitrarily, as long as

$$2I_{CQ} = \frac{1.26}{N} < 1 = i_{C,\,max} \qquad (5.2\text{-}18a)$$

and
$$2V_{CEQ} = 12.6N < 40 = BV_{CEO} \qquad (5.2\text{-}18b)$$

These inequalities set bounds on the turns ratio; i.e.,

$$1.26 < N < 3.17$$

When faced with the problem of choosing from a range of Q points, other considerations must be taken into account. For example, it is good practice to use as little current as possible, since the greater the current capability of the dc supply, the greater the size and cost. Often a specified supply voltage must be used, and this determines V_{CC} (and hence V_{CEQ}). A very important consideration is the availability of a transformer with the proper turns ratio. One common turns ratio is $N = 2$. If a transformer having this turns ratio is chosen,

$$I_{CQ} \approx 0.32 \text{ A} \qquad \text{and} \qquad V_{CEQ} = 12.6 \text{ V} \approx V_{CC}$$

Then
$$P_{L,\,max} = (\tfrac{1}{2})(0.32)^2(2)^2(10) = 2 \text{ W}$$

The load line is shown in Fig. 5.2-3 for these conditions. ///

5.3 CLASS B PUSH-PULL POWER AMPLIFIERS

As found in the previous sections, the maximum attainable efficiency in class A operation is 50 percent, because the peak ac collector current never exceeds the quiescent collector current. In the class B amplifier, the dc collector current is less than the peak ac current. Thus less collector dissipation results, and the efficiency

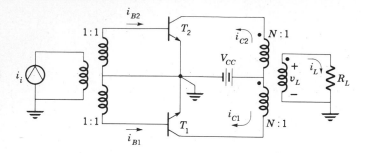

Figure 5.3-1 Push-pull amplifier.

increases. The class B push-pull amplifier shown in Fig. 5.3-1 has a maximum efficiency of 78.5 percent, an improvement of 28.5 percent over the class A amplifiers discussed in Secs. 5.1 and 5.2.

Let us first examine the operation of this circuit, assuming ideal transistors, to determine the upper limits on efficiency and power output. Circuit operation can be explained in terms of the waveforms shown in Fig. 5.3-2. The center-tapped input transformer supplies two base currents of equal amplitudes but 180° out of phase (Fig. 5.3-2b and c). On the first half-cycle, i_{B1} is zero, and because T_1 is biased at cutoff, i_{C1} is zero as in Fig. 5.3-2d. However, in this same interval, i_{B2} is positive, T_2 conducts, and the collector current i_{C2} is shown in Fig. 5.3-2e. Thus one transistor is cut off while the other is conducting. On the second half-cycle, the roles are reversed: T_2 is cut off, and T_1 conducts. When T_2 conducts, the current shown in Fig. 5.3-2e flows through the upper half of the primary winding, and the resulting time-varying flux in the transformer core induces a voltage in the secondary winding. This voltage, in turn, produces the first half-cycle of current through the load in Fig. 5.3-2f. When T_1 conducts, the current i_{C1} induces a flux in the core in a direction opposite to the flux of the previous half-cycle, resulting in the second half-cycle of load current. The final load current under these ideal conditions is thus directly proportional to the signal current i_i. It is seen from Fig. 5.3-1 that the load current i_L is related to the individual currents by

$$i_L = N(i_{C1} - i_{C2}) \tag{5.3-1}$$

If the circuit of Fig. 5.3-1 were used in practice, the load current would be extremely distorted near the zero crossing, as seen from the oscillogram of the load current shown in Fig. 5.3-2g. This effect, called *crossover distortion*, is due to the base-emitter voltage v_{BE} being zero when no signal is applied. However, linear operation of the transistor begins only when i_B is positive enough for v_{BE} to exceed the cutin voltage, which is assumed to be 0.65 V for silicon. This distortion is shown in dotted lines in Fig. 5.3-2d to f, and is clearly visible in the oscillogram of Fig. 5.3-2g.

To eliminate this distortion, base-emitter junctions are biased at approximately 0.65 V. The result is then class AB rather than class B operation, although it

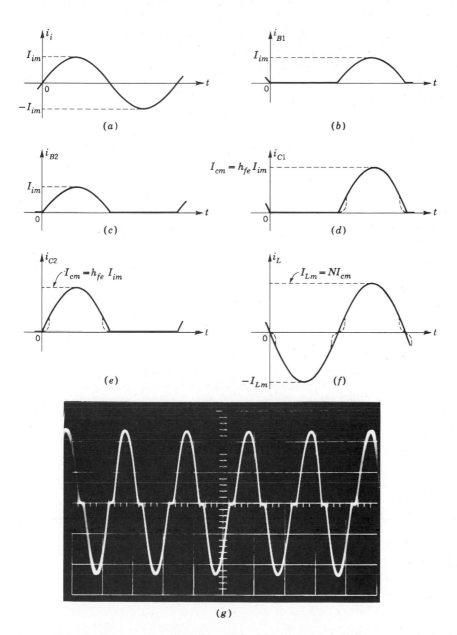

Figure 5.3-2 Waveforms in the push-pull amplifier: (*a*) input current; (*b*) base current in T_1; (*c*) base current in T_2; (*d*) collector current in T_1; (*e*) collector current in T_2; (*f*) load current; (*g*) oscillogram of load-current waveform, illustrating crossover distortion.

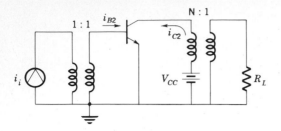

Figure 5.3-3 One-half of class B push-pull stage.

is so nearly class B that it is usually called simply class B. This bias is called the *turn-on bias*. In practice, one often allows crossover distortion and relies on the transformer and internal and stray capacitances to filter it out.

5.3-1 Load-Line Determination

Since each transistor operates in a symmetrical fashion and only half the time, we need study the operation of only one of the transistors. Consider T_2 as shown in Fig. 5.3-3. This circuit enables us to describe the operation of the amplifier. The dc load line is a vertical line, $v_{CE} = V_{CC}$, and the ac load line has a slope

$$\frac{i_C}{v_{CE}} = -\frac{1}{R'_L} \qquad \text{Ac load line slope} \qquad (5.3\text{-}2)$$

as shown in Fig. 5.3-4. During the time that T_2 is off, $i_{C2} = 0$ and

$$v_{CE2} = V_{CC} + Nv_L$$

varies from V_{CC} (which occurs when $v_{CE1} = V_{CC}$, and hence $Nv_L = 0$) to $2V_{CC}$ (which occurs when $v_{CE1} = 0$, and hence $Nv_L = V_{CC}$). Thus, while the transistor is off, the ac load line is horizontal, with $i_{C2} = 0$.

The maximum value of both i_{C1} and i_{C2} (Figs. 5.3-2*d* and *e* and 5.3-4) is

$$I_{cm} = \frac{V_{CC}}{R'_L} \qquad (5.3\text{-}3)$$

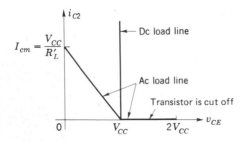

Figure 5.3-4 Load lines for class B stage.

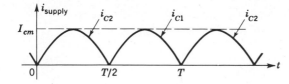

Figure 5.3-5 Power-supply current waveform.

5.3-2 Power Calculations

Assume that the signal current is sinusoidal,

$$i_i = I_{im} \sin \omega t$$

Supplied power The power delivered by the supply is

$$P_{CC} = V_{CC} \frac{1}{T} \int_{-T/2}^{T/2} [i_{C1}(t) + i_{C2}(t)] \, dt \qquad (5.3\text{-}4a)$$

The current $i_{C1} + i_{C2}$ is the current flowing through the supply. From Fig. 5.3-2d and e, this is a full-wave-rectified current, as shown in Fig. 5.3-5. The average value of a full-wave-rectified sine wave is $2/\pi$ times its peak value. Thus

$$\frac{1}{T} \int_{-T/2}^{T/2} (i_{C1} + i_{C2}) \, dt = \frac{2}{\pi} I_{cm}$$

The supplied power P_{CC} is then

$$P_{CC} = \frac{2}{\pi} V_{CC} I_{cm} \qquad (5.3\text{-}4b)$$

Its maximum value is (Fig. 5.3-4)

$$P_{CC,\,\text{max}} = \frac{2}{\pi} V_{CC} \frac{V_{CC}}{R'_L} = \frac{2 V_{CC}^2}{\pi R'_L} \quad \text{Max} \qquad (5.3\text{-}4c)$$

Power transferred to load The power transferred to the load is

$$P_L = \tfrac{1}{2} I_{Lm}^2 R_L = \tfrac{1}{2} I_{cm}^2 N^2 R_L = \tfrac{1}{2} I_{cm}^2 R'_L \qquad (5.3\text{-}5a)$$

Its maximum value is

$$P_{L,\,\text{max}} = \frac{V_{CC}^2}{2R'_L} \qquad (5.3\text{-}5b)$$

Power dissipated in the collector The power dissipated in the collectors of transistors T_1 and T_2 totals

$$2P_C = P_{CC} - P_L$$

Using (5.3-4b) and (5.3-5a), we get

$$2P_C = \frac{2}{\pi} V_{CC} I_{cm} - \frac{R'_L I_{cm}^2}{2} \qquad (5.3\text{-}6)$$

The maximum value of collector dissipation $P_{C,\,max}$ is found by differentiating P_C with respect to I_{cm} and setting the result equal to zero

$$2\frac{dP_C}{dI_{cm}} = \frac{2}{\pi}V_{CC} - R'_L I_{cm} = 0 \tag{5.3-7}$$

The collector current at which the collector dissipation is a maximum is then

$$I_{cm} = \frac{2}{\pi}\frac{V_{CC}}{R'_L} \tag{5.3-8}$$

and combining (5.3-6) and (5.3-8) gives the maximum collector dissipation

$$2P_{C,\,max} = \frac{2}{\pi}V_{CC}\frac{2V_{CC}}{\pi R'_L} - \frac{R'_L}{2}\left(\frac{2}{\pi}\frac{V_{CC}}{R'_L}\right)^2 = \frac{2}{\pi^2}\frac{V_{CC}^2}{R'_L} \tag{5.3-9}$$

The power dissipated in each collector is then

$$P_{C,\,max} = \frac{1}{\pi^2}\frac{V_{CC}^2}{R'_L} \approx 0.1\frac{V_{CC}^2}{R'_L} \tag{5.3-10}$$

Efficiency The efficiency of operation η is calculated from (5.3-4b) and (5.3-5a).

$$\eta = \frac{P_L}{P_{CC}} = \frac{\frac{1}{2}R'_L I_{cm}^2}{(2/\pi)V_{CC}I_{cm}} = \frac{\pi}{4}\frac{I_{cm}}{V_{CC}/R'_L} \tag{5.3-11}$$

Since the maximum attainable collector current is V_{CC}/R'_L, the maximum attainable efficiency is

$$\eta_{max} = \frac{\pi}{4} \approx 78.5\% \tag{5.3-12}$$

The supplied power, the load power, the collector dissipation, and the efficiency are plotted in Fig. 5.3-6. These results should be compared with those obtained using the class A amplifier of Sec. 5.1, which are shown in Fig. 5.1-3.

Figure of merit The transistor utilization figure of merit for the class B push-pull amplifier is

$$\frac{P_{C,\,max}}{P_{L,\,max}} = \frac{V_{CC}^2/\pi^2 R'_L}{V_{CC}^2/2R'_L} = \frac{2}{\pi^2} \approx \frac{1}{5} \tag{5.3-13}$$

Note the improvement by a factor of 10 in the figure of merit achieved over the class A amplifier. Thus, if $P_{L,\,max}$ is to be 25 W, each collector must be rated at only 5 W. Although it is true that the circuit requires two transistors and two transformers, the lower power rating of each of the transistors means that they will take up less space and require significantly less heat sinking than one high-power transistor. In this case a single-transistor class A amplifier would require 50 W of power-dissipation capability.

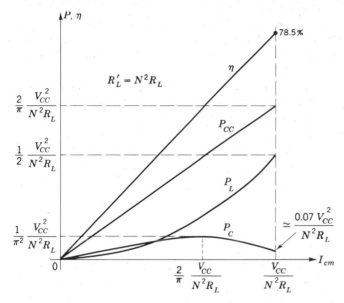

Figure 5.3-6 Power and efficiency variation in the class B push-pull amplifier.

Another major advantage of class B operation is that the no-signal current drain from the battery is zero, while in class A operation the no-signal current is the same as the full-load current.

It is important to keep in mind that the efficiency and collector dissipation ratings are derived for sinusoidal signals and are theoretical maxima, which can only be approached in actual practice.

Example 5.3-1 Design a push-pull class B amplifier to achieve maximum power output to a 10-Ω load. Use two transistors with ratings as in Example 5.1-1 ($i_{C,\,max} = 1$ A). Specify V_{CC}, N, and a bias network to eliminate crossover distortion. Calculate the power output and efficiency.

SOLUTION For convenience, the transistor ratings are repeated.

$$P_{C,\,max} = 4 \text{ W} \qquad BV_{CEO} = 40 \text{ V} \qquad i_{C,\,max} = 1 \text{ A}$$

The maximum power output is given by (5.3-5b) and (5.3-3)

$$\boxed{P_{L,\,max} = \frac{V_{CC}^2}{2R_L'} = \frac{V_{CC} I_{cm}}{2}}$$

Thus the power output can be increased by increasing V_{CC} and I_{cm}. However, V_{CC} and I_{cm} cannot be increased indefinitely. The transistor ratings establish upper bounds on V_{CC} and I_{cm} as follows:

$$V_{CC} \leq \tfrac{1}{2} BV_{CEO} = 20 \text{ V} \qquad I_{cm} \leq i_{C,\,max} = 1 \text{ A}$$

and, using (5.3-13),

$$P_{L,\,\text{max}} = \frac{V_{CC}I_{cm}}{2} \leq 5P_{C,\,\text{max}} = 20 \text{ W}$$

The Q point is chosen to drive the transistor to its rated maximum i_C and BV_{CEO}. Thus

$$V_{CC} = 20 \text{ V} \quad \text{and} \quad I_{cm} = 1 \text{ A}$$

Then

$$P_{L,\,\text{max}} = 10 \text{ W}$$

The turns ratio N is found as follows. Since

$$I_{cm} = \frac{V_{CC}}{N^2 R_L}$$

we have

$$N^2 = 2 \quad \text{and} \quad N = 1.414$$

Note that the two transistors in push-pull provide 5 times the power that a single transistor can supply to the given load resistance without exceeding maximum ratings (see Example 5.2-1).

Figure 5.3-7 shows the vi characteristic of each transistor and the ac load line. Note that the ac load line intersects the maximum average collector-dissipation curve. This means that the instantaneous power *can* exceed the maximum average power. Our power restriction is simply that the *average* collector dissipation be less than the maximum allowable average collector dissipation.† Note that we do not optimize the design of a class B push-pull amplifier by making the ac load line tangent to the $P_{C,\,\text{max}}$ hyperbola.

---

† If a very low frequency signal is being amplified, the time during which the maximum-dissipation curve is exceeded may be long enough to result in the transistor's overheating. The lowest frequency which can be amplified depends on the thermal time constant of the transistor.

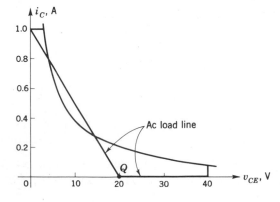

Figure 5.3-7 Load line for Example 5.3-1.

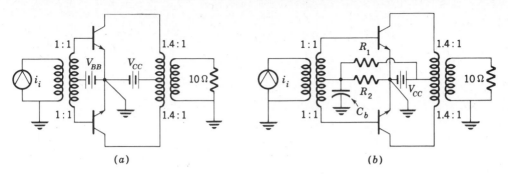

Figure 5.3-8 Push-pull amplifier with bias supply: (a) battery supply; (b) resistance-divider supply.

To complete the design, a bias network must be selected. Figure 5.3-8a shows the push-pull amplifier with a separate bias supply. V_{BB} is adjusted to the turn-on voltage of the transistor, which is approximately 0.7 V for silicon.

In practice, the V_{CC} supply with a suitable voltage divider is used rather than a separate supply, as shown in Fig. 5.3-8b. R_1 and R_2 are chosen so that the base-emitter drop is about 0.7 V. A silicon diode is often used in place of R_2, since the voltage drop across a silicon diode operating above its break voltage is similar to the quiescent base-emitter voltage required to turn the transistor on. ///

5.3-3 Direct-coupled Push-Pull Amplifier

A push-pull amplifier with direct-coupled output is shown in Fig. 5.3-9. If we compare this circuit with the push-pull amplifier shown in Fig. 5.3-1, we see that in both circuits an input transformer is used to couple the input current i_i into the base of T_1 or T_2, depending on the direction of current flow. However, the circuit of Fig. 5.3-9a requires no output transformer.

The operation of the circuit can be explained by noting that when $i_i(t)$ is positive, T_1 is ON and T_2 is OFF (see Fig. 5.3-9b). In this case the collector current $i_{C1}(t)$ flows into the load R_L, and $i_L(t) = i_{C1}(t)$. Hence, $i_L(t) = h_{fe1} i_i(t)$ when $i_i(t) > 0$. Next consider what happens when $i_i(t)$ is negative. In this case, Fig. 5.3-9b shows that T_2 is ON and T_1 is OFF. Now the load current $i_L(t) = -i_{C2}(t) = -h_{fe2} i_{B2}(t) = -h_{fe2}(-i_i) = h_{fe2} i_i(t)$. The resulting load current, due to a full cycle of input current, is shown in Fig. 5.3-9b.

Note that the effect of omitting the output transformer, used in the circuit shown in Fig. 5.3-1, is to cause the collector impedance of the ON transistor to be R_L rather than $N^2 R_L$. This is illustrated in the following example.

Example 5.3-2 Repeat Example 5.3-1 using the circuit of Fig. 5.3-9.

SOLUTION In order to provide maximum power to the load resistor we must supply the load with the maximum possible current. The transistor used in

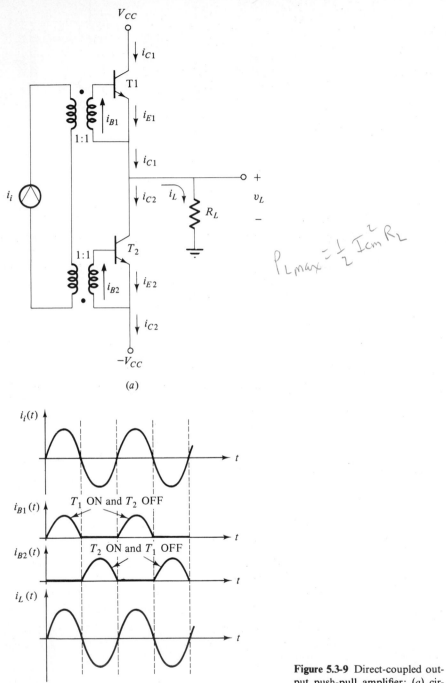

$$P_{Lmax} = \frac{1}{2} I_{cm}^2 R_L$$

(a)

(b)

Figure 5.3-9 Direct-coupled output push-pull amplifier: (a) circuit; (b) waveforms.

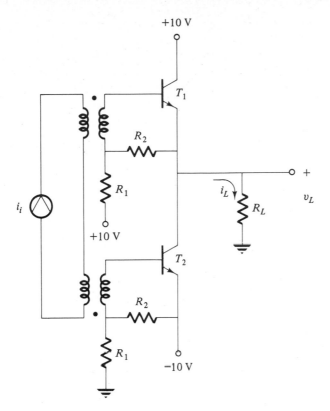

Figure 5.3-10 Direct-coupled push-pull amplifier with bias networks to eliminate crossover distortion.

Example 5.3-1 can supply a maximum current of 1 A. Since the push-pull amplifier shown in Fig. 5.3-9a is connected directly to load R_L and not transformer-coupled to the load, the maximum power that can be dissipated in the load is

$$P_{L,\,max} = \tfrac{1}{2}I_{cm}^2 R_L = (\tfrac{1}{2})(1)^2(10) = 5 \text{ W}$$

This value of power output is one-half the maximum load power found in Example 5.3-1. To minimize the power dissipated in T_1 and T_2 we choose the minimum possible value for V_{CC} so that the transistor swings from saturation to cutoff. Since $R_L = 10 \ \Omega$ and $I_{cm} = 1$ A, we find $V_{CC} = V_{Lm} = R_L I_{cm} = 10$ V. To avoid the practical problem of saturation ($V_{CE,\,sat} = 2$ V) the actual V_{CC} employed would be 12 V. In order to eliminate crossover distortion the base-emitter junctions are forward-biased by approximately 0.65 V, as shown in the final push-pull circuit of Fig. 5.3-10. ///

An advantage of the direct-coupled push-pull amplifier is that no output transformer is required. The input transformer can also be omitted and replaced by a *phase-splitting circuit* (Sec. 6.11). Since transformers are large, heavy, and expensive, their elimination from the circuit results in considerable savings. A disadvantage is that we cannot obtain maximum power transfer to the load since we no longer have available the impedance-transforming property of the transformer. A second disadvantage is the need for two power supplies.

5.4 AMPLIFIERS USING COMPLEMENTARY SYMMETRY

Figure 5.4-1 illustrates a type of push-pull class B amplifier which employs one *pnp* and one *npn* transistor and requires no transformers. This type of amplifier uses *complementary symmetry*. Its operation can be explained by referring to the figure. When the signal voltage is positive, T_1 (the *npn* transistor) conducts, while T_2 (the *pnp* transistor) is cut off. When the signal voltage is negative, T_2 conducts while T_1 is cut off. The load current is

$$i_L = i_{C1} - i_{C2} \tag{5.4-1}$$

The load-line and output-circuit power relations for this amplifier are the same as for the conventional class B amplifier of Sec. 5.3. Some advantages of the circuit are that the transformerless operation saves on weight and cost and balanced push-pull input signals are not required. Disadvantages are the need for both positive and negative supply voltages and the problem of obtaining pairs of transistors matched closely enough to achieve low distortion.

Example 5.4-1 A complementary-symmetry amplifier uses transistors having the same specifications as those of Example 5.3-1:

$$P_{c,\,\text{max}} = 4 \text{ W} \qquad BV_{CEO} = 40 \text{ V} \qquad i_{C,\,\text{max}} = 1 \text{ A}$$

Find V_{CC} and the maximum power which can be delivered to a 10-Ω load.

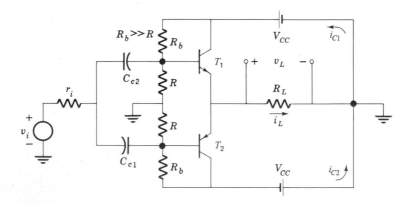

Figure 5.4-1 Complementary-symmetry amplifier.

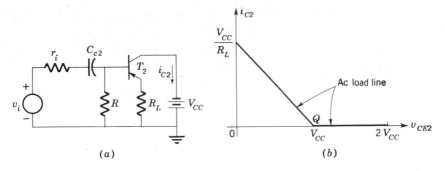

Figure 5.4-2 Circuit and load line for T_2 of the complementary-symmetry emitter-follower circuit: (*a*) circuit; (*b*) load line.

SOLUTION Each transistor is essentially a class B emitter follower. Consider T_2. Its equivalent circuit during conduction and the corresponding load line and operating path are shown in Fig. 5.4-2. Since the peak value of i_{c2} cannot exceed 1 A,

$$\frac{V_{CC}}{R_L} = 1 \text{ A}$$

Therefore

$$V_{CC} = 10 \text{ V}$$

Note that the $2V_{CC}$ swing is less than the collector-emitter breakdown voltage, as required. The load current is

$$i_L = i_{C1} - i_{C2}$$

If v_i is sinusoidal, i_L is also sinusoidal, with a maximum peak current

$$I_{Lm} = \frac{V_{CC}}{R_L}$$

Thus

$$i_L = I_{Lm} \sin \omega t = \frac{V_{CC}}{R_L} \sin \omega t$$

The maximum power into the load is

$$P_{L, \text{max}} = \frac{V_{CC}^2}{2R_L} = 5 \text{ W}$$

The push-pull amplifier of Sec. 5.3-1 could deliver 10 W because the transformer doubled the effective load impedance. The complementary-symmetry amplifier and the direct-coupled push-pull amplifier described in Sec. 5.3-3 do not employ an output transformer and therefore do not transform the load impedance.

We can get some idea of the signal voltage required to drive this amplifier by assuming that $r_i \ll R$. Then, because the circuit is basically an emitter follower, the maximum signal $v_i = V_{im} \sin \omega t$ must have a peak value of V_{CC}, that is, $V_{im} = V_{CC}$, in order to transfer the full 5 W to the load. ///

SUMMARY

The basic class A and push-pull class B power amplifiers have been considered, and procedures have been presented for establishing preliminary circuit designs. In all cases, the circuit design of a power amplifier must be accompanied by a thermal design which will ensure that the junction temperature is maintained within safe limits. The push-pull class B configurations yield significantly higher efficiency and lower power-supply drain. Thus, for a given power output, much smaller and lighter transistors are required. In particular, the complementary-symmetry arrangement is very attractive because of the simplicity of the circuit and the availability of matched pairs of transistors.

In practice, a preliminary power-amplifier design is usually done on paper and final adjustments made in the laboratory where power output, distortion, efficiency, and temperature stability are easily measured.

PROBLEMS

5.1-1 For the circuit of Fig. P5.1-1 (a) sketch the dc and ac load lines and (b) find the maximum undistorted v_{CE}, i_C, and i_L.

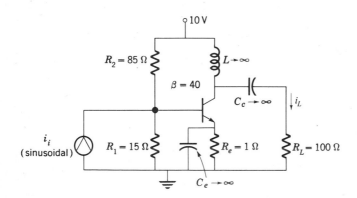

Figure P5.1-1

(c) Calculate the maximum power dissipated in the load, the total power delivered by the V_{CC} supply, the power dissipated at the collector, and the efficiency.

5.1-2 The circuit shown in Fig. P5.1-2 is a class A power amplifier which must supply a maximum undistorted power of 2 W to the 10-Ω load. The *minimum* necessary transistor ratings are to be specified. Find I_{CQ}, P_{CC}, and η. Also specify $P_{C,\,max}$, $v_{CE,max}$, and $i_{C,\,max}$ for the transistor. Neglect R_e and bias-circuit losses.

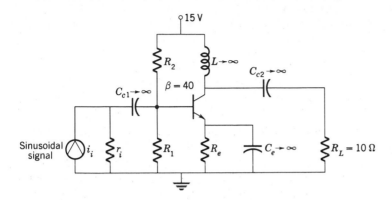

Figure P5.1-2

5.1-3 In Example 5.1-2 assume that $i_{C,\,max} = 2.0$ A. Determine the maximum attainable swing and the maximum power dissipated in the load.

5.1-4 Repeat Prob. 5.1-1 if the transistor collector-emitter saturation voltage is 1 V.

5.2-1 The transistor ratings are to be specified for class A operation in Fig. P5.2-1. The maximum required load power is 2 W. Neglect R_e and bias-circuit losses.

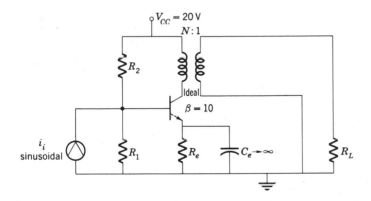

Figure P5.2-1

    (a) Find P_{CC}, assuming that the amplifier is designed for maximum efficiency.
    (b) Find I_{CQ}.
    (c) Specify the $i_{C,\,max}$, $v_{CE,\,max}$, and $P_{C,\,max}$ ratings for the transistor.
    (d) If $R_L = 6.25$ Ω, find the turns ratio N.

5.2-2 Repeat Prob. 5.2-1, assuming that the transformer efficiency is 75 percent.

5.2-3 In Prob. 5.2-1 the maximum power of $P_L = 2$ W is obtained when $i_i = I_{im} \sin \omega t$. Find P_L when

$$i_i = \frac{I_{im}}{2} \sin \omega t + \frac{I_{im}}{2} \sin 3\omega t$$

5.2-4 Repeat Prob. 5.2-1 assuming $V_{CE,\,sat} = 1$ V. Include the effect of losses in the emitter and bias circuits. Assume that $R_e = 1\ \Omega$ and $R_1 = 10\ \Omega$.

5.2-5 In the emitter-coupled class A power amplifier shown in Fig. P5.2-5, $P_{C,\,max} = 100$ W. Find R_b, V_{BB}, and N so that maximum power can be transmitted to the load. Also find P_{CC}, $P_{L,\,max}$, P_C, and η.

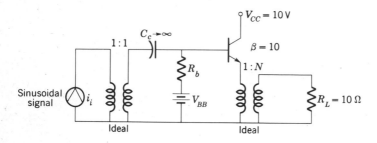

Figure P5.2-5

5.2-6 In the circuit of Fig. P5.2-6 find N so that maximum power can be dissipated in the load. Calculate $P_{L,\,max}$, $P_{C,\,max}$, and P_{CC}. Include the effect of losses in the bias and emitter circuits.

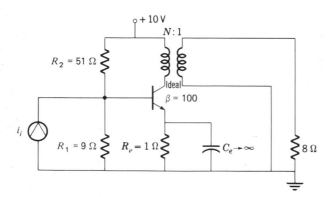

Figure P5.2-6

5.2-7 In Example 5.2-1 calculate N and P_{Lm} if $V_{CC} = 9$ V. Compare with the result in the example, where $V_{CC} = 12.6$ V.

5.2-8 Repeat Prob. 5.2-7 if $V_{CC} = 18$ V.

5.3-1 For the class B push-pull amplifier shown in Fig. P5.3-1, calculate the maximum values of i_C, i_L, v_{CE}, P_L, P_C, and P_{CC}. Plot P_{CC}, P_L, and P_C versus i_C for the range $0 \le i_C \le i_{C,\,max}$.

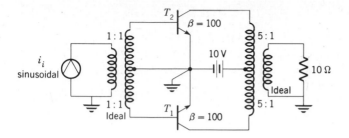

Figure P5.3-1

5.3-2 Design a class B push-pull amplifier to deliver 10 W to a 10-Ω load, using transistors which have $BV_{CEO} = 40$ V. Specify $P_{C,\,max}$ for each transistor, V_{CC}, and the required N. $R_L' = 20\Omega$ $V_{CC} = 20$ $N = 1.44$

5.3-3 Repeat Prob. 5.3-1 with the transformer in the emitter circuit. $P_{C\,max} = 2WATT$

5.3-4 Transistors having $BV_{CEO} = 50$ V and $P_{C,\,max} = 1$ W are available. Design a push-pull class B amplifier using these transistors, if $V_{CC} = 22.5$ V.

    (a) Specify the reflected load resistance, and find the maximum power output. $Find\ I_{CM} = \frac{V_{CC}}{R_L'} = .444$

    (b) Find the input-current swing required if $\beta \approx 50$. $I_{Bmax} = \frac{I_{Cmax}}{B} = 8.8mA$

5.3-5 A loudspeaker rated at 8 Ω and 500 mW is to be driven by a class B push-pull amplifier. The power supply is 9 V. The transistors to be used have $V_{CE,\,sat} = 1$ V. Select a suitable value of N and find P_{CC} and P_C when 500 mW is being dissipated in the load.

5.3-6 Transistors T_1 and T_2 are nonlinear in the circuit of Fig. P5.3-6, so that

$$i_{C2} = 10i_{B2} + i_{B2}^2 \qquad \text{and} \qquad i_{C1} = 10i_{B1} + i_{B1}^2$$

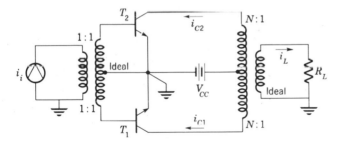

Figure P5.3-6

If

$$i_i = \cos \omega_0 t$$

find i_L. Note the distortion resulting from the nonlinearities.

5.3-7 Design a direct-coupled output push-pull amplifier as in Fig. 5.3-10 to achieve maximum output to an 8 Ω load. The transistor has ratings $P_{C,\,max} = 6$ W, $BV_{CEO} = 50$ V, $i_{C,\,max} = 1$ A, and $h_{FE} = 100$.

5.4-1 For the complementary-symmetry amplifier shown in Fig. P5.4-1 (a) calculate the maximum power dissipated in the load, (b) the maximum power furnished by both the V_{CC} and $-V_{EE}$ supplies; and (c) the maximum power dissipated by each transistor.

    (d) Plot P_L, P_{CC}, and P_C versus I_{cm} in the range $0 < I_{cm} < 2$ A.

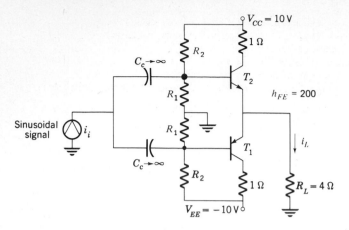

Figure P5.4-1

(e) Design the bias network to eliminate crossover distortion. Choose $R_1 \| R_2 = 10 \text{ k}\Omega$ and adjust for 0.65 V.

(f) Sketch the maximum undistorted v_{CE2}, i_{C2}, and i_L.

5.4-2 For the complementary-symmetry amplifier shown in Fig. P5.4-2 (a) calculate R_x to eliminate crossover distortion at 0.65 V.

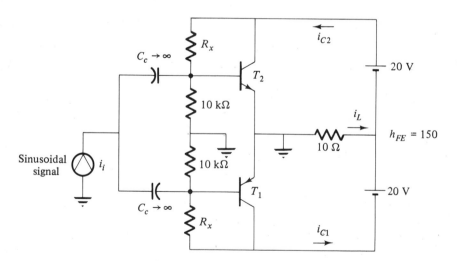

Figure P5.4-2

(b) Sketch the maximum undistorted sinusoidal waveform for v_{CE2}, i_{C2}, and i_L.

(c) Determine P_{out}, P_{CC}, and η for maximum output current.

(d) Determine the maximum collector dissipation and the maximum collector-current amplitude yielding this dissipation.

5.4-3 The circuit in Fig. P5.4-3 eliminates crossover distortion using diode biasing.
  (a) Calculate the current in D_1, D_2, T_1, T_2 when $i_i = 0$. Assume symmetry.
  (b) When $i_i > 0$, which transistor turns off? Repeat for $i_i < 0$.
  (c) Sketch the maximum undistorted v_{CE2}, i_{C2}, and i_L.
  (d) Determine P_L, P_{CC}, and η for maximum output current.
  (e) Determine the maximum collector dissipation and the maximum collector-current amplitude yielding this dissipation.

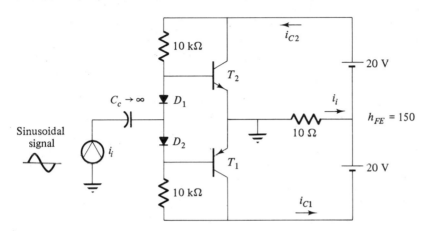

Figure P5.4-3

SIX

SMALL-SIGNAL LOW-FREQUENCY
ANALYSIS AND DESIGN

INTRODUCTION

Up to this point, the graphical approach has been emphasized as a conceptual aid in the solution of problems in analysis and design. This approach was found useful when considering dc biasing and power amplifiers, where large signals are encountered. In this chapter, the response of transistor circuits to *small signals* is studied.

When the collector-emitter or drain-source voltage and current swings are very small, the transistor is considered to be linear and can be replaced, for purposes of analysis, by a *small-signal equivalent-circuit model*. The graphical approach is abandoned, and this linear model can be analyzed using standard network analysis techniques, e.g., loop or node equations, in order to determine the small-signal response. To simplify response calculations, the technique of *impedance reflection* is introduced. With this technique, complicated circuits can often be simplified to the point where they can be solved by inspection.

The small-signal equivalent circuits developed in this chapter are assumed to be independent of frequency. Frequency response is considered in detail in Chap. 9.

In practice, the design of small-signal current or voltage amplifiers is broken down into two essentially separate parts. The first consists of setting the dc bias, i.e., finding a suitable Q point; here the graphical method is used. The second part involves gain and impedance calculations at signal frequencies; here the small-signal equivalent circuit is used. These two parts are not completely independent because, as will be seen, the values of some of the components in the equivalent circuit depend on the Q point.

6.1 THE HYBRID PARAMETERS; BIPOLAR JUNCTION TRANSISTOR

The elements of the equivalent circuit for the BJT can be developed from the internal physics of the device or from its terminal properties. The latter approach is employed here since it is more general and has many conceptual advantages.

When analyzing or designing a transistor amplifier, attention is focused on two pairs of terminals, input and output. Thus use can be made of the applicable results of two-port network theory. There are six possible pairs of equations relating input and output quantities, which can be used to define completely the terminal behavior of the two-port network shown in Fig. 6.1-1a. These six pairs of equations involve impedance, admittance, hybrid, and chain parameters, all of which are interrelated. For most BJT work, the *hybrid*, or *h*, parameters are the most useful because they are easily measured and therefore most often specified in manufacturers' data sheets, and they provide quick estimates of circuit performance.

The standard form for the hybrid equations is

$$v_1 = h_{11}i_1 + h_{12}v_2 \tag{6.1-1a}$$

$$i_2 = h_{21}i_1 + h_{22}v_2 \tag{6.1-1b}$$

In these equations, the independent variables are the input current i_1 and the output voltage v_2. The voltage and current variables are understood to represent small variations about the quiescent operating point. Note that the small-signal-current directions shown in Fig. 6.1-1b are into the network. This differs from the convention employed to describe dc and total currents in the transistor.

In transistor circuit theory, the numerical subscripts are exchanged for letters which identify the physical nature of the parameter

$$v_1 = h_i i_1 + h_r v_2 \tag{6.1-2}$$

$$i_2 = h_f i_1 + h_o v_2 \tag{6.1-3}$$

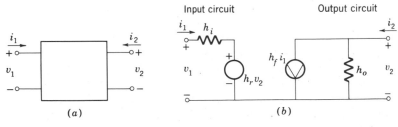

Figure 6.1-1 The two-port network: (a) general two-port; (b) h-parameter equivalent circuit.

In the equivalent circuit of Fig. 6.1-1b, the *input* circuit is derived from (6.1-2), using KVL, and the *output* circuit from (6.1-3), using KCL. The physical meaning of the *h* parameters can be obtained from the defining equations or from the circuit. For example, (6.1-2) indicates that h_i is dimensionally an impedance. From the circuit of Fig. 6.1-1b it is seen to be the input impedance with the output short-circuited $(v_2 = 0)$. The subscript *i* thus stands for *input*. Similarly, h_r is dimensionless and represents the *reverse* open-circuit voltage ratio.

Terminal definitions for all four parameters are as follows:

$$h_i = \left. \frac{v_1}{i_1} \right|_{v_2 = 0} = \text{short-circuit input impedance} \qquad (6.1\text{-}4)$$

$$h_r = \left. \frac{v_1}{v_2} \right|_{i_1 = 0} = \text{open-circuit reverse voltage gain} \qquad (6.1\text{-}5)$$

$$h_f = \left. \frac{i_2}{i_1} \right|_{v_2 = 0} = \text{short-circuit forward current gain} \qquad (6.1\text{-}6)$$

$$h_o = \left. \frac{i_2}{v_2} \right|_{i_1 = 0} = \text{open-circuit output admittance} \qquad (6.1\text{-}7)$$

The equivalent circuit of Fig. 6.1-1b is extremely useful for several reasons: (1) it isolates the input and output circuits, their interaction being accounted for by the two controlled sources; (2) the two parts of the circuit are in a form which makes it simple to take into account source and load circuits. The input circuit is a Thevenin equivalent and the output circuit a Norton equivalent.

The circuit and definitions of (6.1-4) to (6.1-7) also suggest small-signal methods of measurement for the various parameters. For example, (6.1-6) indicates that h_f can be measured by placing an ac short circuit (a large capacitor) across the output (so that $v_2 = 0$), applying a small ac current to the input, and then measuring the current ratio. Note that the dc conditions must be maintained so that the *h* parameters can be determined with respect to a specified *Q* point. We show below that the *h* parameters are each a function of the *Q* point.

We now find the equivalent circuits for the common-base, common-emitter, and common-collector (emitter-follower) configurations, using the *h* parameters wherever practicable. It should be noted that these parameters must all be evaluated at the *Q* point. The parameters are in general different for each configuration and are distinguished by adding an identifying letter as a second subscript. Thus, for example, h_{ob} is the output admittance for the CB configuration.

Often, the manufacturer will give the CB *h* parameters, and the designer may need the CE parameters. The conversion from one set to another can be made using simple circuit-analysis methods, to be described later.

6.2 THE COMMON-EMITTER CONFIGURATION

In this section we find the small-signal equivalent-circuit model for the common-emitter configuration shown in Fig. 6.2-1a. In Chaps. 2 and 5, where large-signal behavior was considered, it was assumed that the time variation of v_{BE} was negligible compared with the signal. Thus the base-emitter circuit was represented by a battery (≈ 0.7 V). In this chapter, which is concerned with *small* signals, this assumption, which would imply that h_{ie} and h_{re} are zero, is *not* made.

The load current i_L, in the circuit of Fig. 6.2-1a, contains a dc as well as a small-signal ac component. Since linear operation is assumed, the ac and dc components can be treated separately, using superposition. Thus the batteries and the capacitor can be replaced by short circuits, yielding the small-signal ac circuit of Fig. 6.2-1b.

Now, of the four hybrid parameters, three can be disposed of rather quickly. The reverse voltage gain h_{re} is usually negligible and is omitted. The output admittance h_{oe} can be written using (6.1-7)

$$h_{oe} = \left. \frac{i_c}{v_{ce}} \right|_{i_b = 0} \tag{6.2-1a}$$

where i_c and v_{ce} are defined as small variations about the nominal operating point. Thus the parameter h_{oe} is simply the slope of the collector characteristic at the Q point, as shown in Fig. 6.2-2a

$$h_{oe} = \left. \frac{\Delta i_C}{\Delta v_{CE}} \right|_{Q \text{ point}} \tag{6.2-1b}$$

The output admittance h_{oe} can be calculated from the Ebers-Moll equations. Consider

$$\frac{i_C}{i_B} = h_{FE} \left(\frac{\epsilon^{v_{CE}/V_T} - 1/\alpha_R}{\epsilon^{v_{CE}/V_T} + h_{FE}/h_{FC}} \right) \tag{2.2-17a}$$

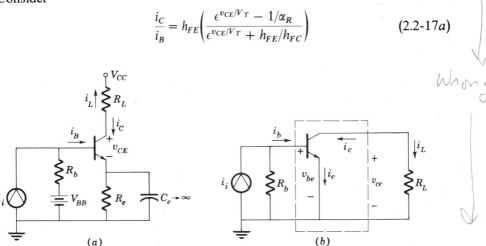

(a)   *(b)*

Figure 6.2-1 The common-emitter configuration: *(a)* complete circuit; *(b)* small-signal circuit.

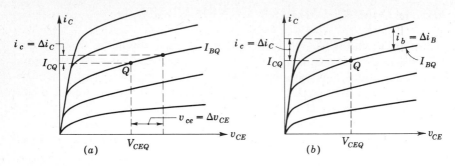

Figure 6.2-2 Estimating h_{oe} and h_{fe} from the vi characteristic. (a) h_{oe}; (b) h_{fe}.

Differentiating i_C with respect to v_{CE} and evaluating the result at $i_B = I_{BQ}$ yields

$$h_{oe} = \frac{\partial i_C}{\partial v_{CE}}\bigg|_{i_B = I_{BQ}}$$

Wrong

$$= \frac{I_{BQ}\, h_{FE}}{V_T}\, \epsilon^{v_{CE}/V_T} \left[\frac{\dfrac{h_{FE}}{h_{FC}} + \dfrac{1}{\alpha_R}}{(\epsilon^{v_{CE}/V_T} + h_{FE}/h_{FC})^2} \right] \tag{6.2-1c}$$

Since $1/\alpha_R \ll h_{FE}/h_{FC}$ and $I_{CQ} = h_{FE}\, I_{BQ}$, (6.2-1c) reduces to

$$h_{oe} = \frac{h_{FE}\, I_{CQ}}{h_{FC}\, V_T} \left[\frac{\epsilon^{v_{CE}/V_T}}{(\epsilon^{v_{CE}/V_T} + h_{FE}/h_{FC})^2} \right] \tag{6.2-1d}$$

The output admittance can be further simplified if we assume

$$\epsilon^{v_{CE}/V_T} \gg h_{FE}/h_{FC} \tag{6.2-1e}$$

Then

$$h_{oe} \approx \frac{h_{FE}\, I_{CQ}}{h_{FC}\, V_T}\, \epsilon^{-v_{CE}/V_T} \tag{6.2-1f}$$

Hence h_{oe} is directly proportional to the quiescent collector current and varies exponentially with the collector-emitter voltage.

Numerical values for h_{oe} can be obtained from the vi characteristic if it is available. For most transistors, h_{oe} has a value less than 10^{-4} S,† and since it is in parallel with a load resistance R_L, it can be neglected as long as R_L is less than 1 or 2 kΩ, which is very often the case.

The short-circuit current gain h_{fe} is obtained by setting $R_L = 0$. Then, from (6.1-6),

$$h_{fe} = \frac{i_c}{i_b}\bigg|_{Q \text{ point}} = \frac{\Delta i_C}{\Delta i_B}\bigg|_{Q \text{ point}} \tag{6.2-2}$$

---

† S is the abbreviation for siemens, the SI unit which has replaced the mho as the unit of conductance (named after the German electrical engineer Werner von Siemens).

This parameter can also be obtained from the vi characteristic, as shown in Fig. 6.2-2b. Figure 2.2-1 shows that h_{fe} is approximately equal to h_{FE} and is a function of the quiescent current. Throughout the remainder of this text, it will be assumed, unless otherwise stated, that $h_{fe} = h_{FE}$. The symbol h_{fe} will be used.

Finally, h_{ie} is calculated using (6.1-4).

$$h_{ie} = \frac{v_{be}}{i_b}\bigg|_{v_{ce}=0} \tag{6.2-3a}$$

Refer to the forward-biased junction diode, which is seen looking into the base-emitter terminals. The small-signal ratio v_{be}/i_b represents the dynamic resistance of that junction evaluated at the Q point. This resistance was determined in (2.1-5). Making use of that result, we write

$$h_{ie} = \frac{v_{be}}{i_b}\bigg|_{Q\ \text{point}} = \frac{V_T}{I_{BQ}} \approx h_{fe}\frac{V_T}{I_{CQ}} \approx h_{fe}\frac{V_T}{I_{EQ}} \tag{6.2-3b}$$

At room temperature $V_T \approx 25$ mV, so that a transistor with $h_{fe} = 100$ and $I_{CQ} = 10$ mA would have an input impedance

$$h_{ie} \approx 250\ \Omega$$

Note that h_{fe} may vary by $3:1$ for the same transistor type. If the h_{fe} of this transistor varied from 50 to 150, the range of h_{ie} would be

$$125\ \Omega < h_{ie} < 375\ \Omega$$

When designing transistor circuits we generally use the nominal value of h_{fe}, and for this example the nominal $h_{ie} = 250\ \Omega$. However, we must always keep in mind the possible variation of this resistance.

To sum up the results for the common-emitter configuration, the equivalent circuit is shown in three successively simplified versions in Fig. 6.2-3. The simple version of Fig. 6.2-3c is easily remembered and serves adequately for most calculations.

Let us return to the amplifier circuit of Fig. 6.2-1b and insert the equivalent circuit in place of the transistor, as shown in Fig. 6.2-4. The important quantities are the input and output impedance and the current gain. These are easily calculated directly from the circuit. For the current gain

$$\frac{i_b}{i_i} = \frac{R_b}{R_b + h_{ie}} = \frac{1}{1 + h_{ie}/R_b} \tag{6.2-4}$$

and

$$i_L = -i_c = -h_{fe}i_b \tag{6.2-5}$$

Thus

$$A_i = \frac{i_L}{i_i} = \frac{i_L}{i_b}\frac{i_b}{i_i} = \frac{-h_{fe}}{1 + h_{ie}/R_b} = \frac{-h_{fe}}{1 + h_{fe}[(25 \times 10^{-3})/I_{EQ}R_b]} \tag{6.2-6}$$

For the current gain to approach the theoretical maximum value of h_{fe}, h_{ie}/R_b should be as small as possible, that is, $R_b \gg h_{ie}$. This result implies that for large current gain most of the signal current must flow into the base of the transistor,

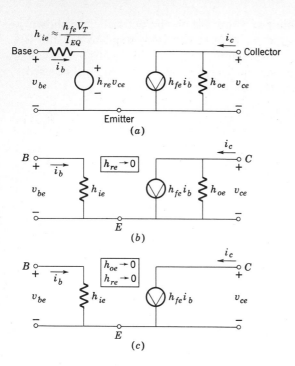

Figure 6.2-3 Small-signal equivalent circuits for the transistor in the CE configuration: (*a*) complete hybrid circuit; (*b*) circuit neglecting h_{re}; (*c*) circuit neglecting h_{re} and h_{oe}.

and a minimal amount may be lost in the bias network. In Sec. 4.1 it was found that for good stability against h_{fe} variation and temperature effects, the inequality $R_b \ll h_{fe} R_e$ should be satisfied. Thus, to meet requirements of high gain and stability simultaneously, we should design so that

$$h_{ie} = h_{fe} \frac{V_T}{I_{EQ}} \ll R_b \ll h_{fe} R_e \qquad (6.2\text{-}7)$$

If this inequality can be satisfied, the amplifier will have high current gain and good stability. Otherwise a compromise must be made in one requirement or the other.

We next calculate the input and output impedances. Looking to the right from the current source i_i, the input impedance Z_i is

$$Z_i = \frac{R_b h_{ie}}{R_b + h_{ie}} \approx h_{ie} \qquad \text{if } R_b \gg h_{ie} \qquad (6.2\text{-}8)$$

Figure 6.2-4 Complete common-emitter amplifier equivalent circuit.

This simple expression is the result of the assumption that h_{re} is negligible. The calculation of output impedance is even simpler. If h_{oe} is taken into account, then

$$Z_o = \frac{v_{ce}}{i_c}\bigg|_{i_i=0} = \frac{1}{h_{oe}} \qquad (6.2\text{-}9)$$

If we neglect h_{oe}, then $Z_o \to \infty$.

The parameters h_{re} and h_{oe} are almost never specified and are usually neglected in calculations (see Sec. 6.6 for a detailed discussion of manufacturers' specifications).

Example 6.2-1 In Fig. 6.2-5, the BJT has $h_{fe} = 50$. All bypass and coupling capacitors are assumed to have zero reactance at the signal frequency. Find

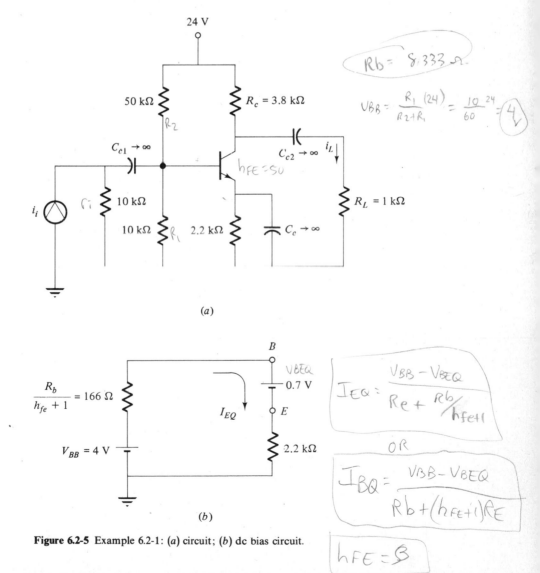

(a)

(b)

Figure 6.2-5 Example 6.2-1: (a) circuit; (b) dc bias circuit.

(a) quiescent conditions, (b) the small-signal equivalent circuit, neglecting h_{oe} and h_{re}, (c) the current gain, $A_i = i_L/i_i$, (d) the input impedance seen by the signal current source i_i, and (e) the output impedance seen by the 1-kΩ load.

SOLUTION (a) $\quad V_{BB} = \dfrac{10}{10 + 50}(24) = 4$ V

$$R_b = \dfrac{(10)(50)}{10 + 50} \text{ k}\Omega = 8.3 \text{ k}\Omega$$

A dc equivalent circuit obtained using KVL around the base-emitter loop is shown in Fig. 6.2-5b. In this circuit all components in the base circuit have been reflected into the emitter circuit by using the relation

$$I_{BQ}R_b = \dfrac{I_{EQ}R_b}{h_{fe}+1}$$

This circuit is used to calculate the Q point

$$I_{EQ} = \dfrac{4 - 0.7}{166 + 2200} \approx 1.4 \text{ mA}$$

and $\qquad V_{CEQ} \approx 24 - I_{EQ}(R_c + R_e) = 15.6$ V $\qquad Vcc - IEQ(Rc+RE)$

(b) $\quad h_{ie} \approx \dfrac{25h_{fe}}{I_{EQ}} = \dfrac{(25)(50)}{1.4} = 893 \ \Omega$

Thus $h_{ie} \ll R_b$. The resulting small-signal equivalent circuit is shown in Fig. 6.2-6.

(c) $\quad A_i = \dfrac{i_L}{i_i} = \dfrac{i_b}{i_i}\dfrac{i_L}{i_b}$

$$\dfrac{i_b}{i_i} = \dfrac{4.5 \times 10^3}{(4.5 + 0.89) \times 10^3} = 0.83$$

$$\dfrac{i_L}{i_b} = (-50)\dfrac{3.8 \times 10^3}{(3.8 + 1) \times 10^3} = -39.6$$

Thus $\qquad A_i = (0.83)(-39.6) = -33$

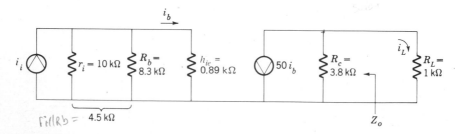

$rₑ||Rb = 4.5$ kΩ

Figure 6.2-6 Small-signal equivalent circuit for Example 6.2-1.

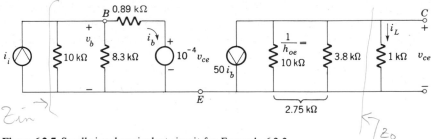

Figure 6.2-7 Small-signal equivalent circuit for Example 6.2-2.

The minus sign occurs in A_i because the positive direction for i_L is opposite to that of i_c.

(d) $Z_i = 10 \text{ k}\Omega \| 8.3 \text{ k}\Omega \| 0.89 \text{ k}\Omega \approx 740 \text{ }\Omega$

(e) $Z_o = 3.8 \text{ k}\Omega$ neglecting h_{oe} ///

Example 6.2-2 Find the current gain of the amplifier of Example 6.2-1 if $h_{re} = 10^{-4}$ and $h_{oe} = 10^{-4}$ S.

SOLUTION The equivalent circuit of the amplifier is shown in Fig. 6.2-7. Referring to the output circuit, we have

$$i_L = (-50)\frac{2.75}{2.75 + 1}i_b = -36.7i_b$$

Compare with Example 6.2-1, where $i_L/i_b = -39.6$. Note that

$$v_{ce} = -36.7 \times 10^3 i_b$$

Turning to the input circuit and applying KVL, we get

$$v_b = 890i_b - (10^{-4})(36.7 \times 10^3)i_b = (890 - 3.67)i_b \approx 890i_b$$

The effect of h_{re} is therefore negligible in this (and most) examples. The current gain is

$$A_i = \frac{i_L}{i_i} = \frac{i_L}{i_b}\frac{i_b}{v_b}\frac{v_b}{i_i}$$

where $\dfrac{i_L}{i_b} = -36.7$ $\dfrac{i_b}{v_b} = \dfrac{1}{890 \text{ }\Omega}$

and

$$i_i = v_b\left(\frac{1}{10 \text{ k}\Omega} + \frac{1}{8.3 \text{ k}\Omega}\right) + i_b \approx v_b\left(\frac{1}{10 \text{ k}\Omega} + \frac{1}{8.3 \text{ k}\Omega} + \frac{1}{0.89 \text{ k}\Omega}\right) = \frac{v_b}{740 \text{ }\Omega}$$

Thus $A_i \approx (-36.7)(\tfrac{1}{890})(740) = -31$

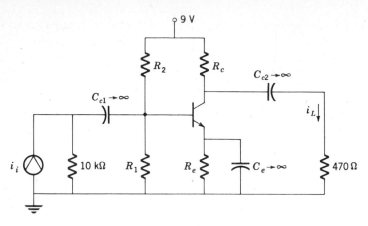

Figure 6.2-8 Circuit for Example 6.2-3.

The effect of including h_{oe} is to reduce the calculated gain from $A_i \approx -33$ to $A_i \approx -31$. This small difference is usually insignificant. ///

Example 6.2-3 A silicon *npn* transistor has $h_{fe} = 120$. Design a single-stage amplifier (Fig. 6.2-8) to achieve a small-signal current gain of 60. The load resistor is 470 Ω and is capacitively coupled to the collector. The supply voltage V_{CC} is 9 V, and the signal-source impedance is 10 kΩ. A peak load current of 0.1 mA is required.

SOLUTION The small-signal equivalent circuit is shown in Fig. 6.2-9 (it is assumed that $h_{oe} = h_{re} = 0$ and $R_1 \| R_2 = R_b$). The current gain A_i is

$$A_i = \frac{i_L}{i_b}\frac{i_b}{i_i} = -120\left(\frac{R_c}{470 + R_c}\right)\left(\frac{10^4 \| R_b}{h_{ie} + (10^4 \| R_b)}\right)$$

Since $A_i = -60$, this becomes

$$\frac{1}{2} = \left(\frac{R_c}{470 + R_c}\right)\left(\frac{10^4 \| R_b}{h_{ie} + (10^4 \| R_b)}\right)$$

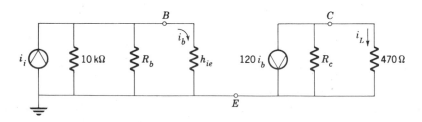

Figure 6.2-9 Equivalent circuit for Example 6.2-3.

Clearly, many combinations of R_c, R_b, and h_{ie} will satisfy this requirement. In the absence of other requirements, we *arbitrarily* choose to let each of the above factors be equal. (This simplifies calculations and usually provides a good design.) Then, to obtain a current gain greater than or equal to 60,

$$\frac{R_c}{470 + R_c} = \frac{10^4 \| R_b}{h_{ie} + (10^4 \| R_b)} \geq \sqrt{\frac{1}{2}} = 0.707$$

The first inequality yields

$$R_c \geq 1.13 \text{ k}\Omega$$

The choice of R_b and h_{ie} will determine the quiescent current and stability of the stage. Keeping this in mind, we choose R_b to be 10 kΩ. Then the second inequality leads to the relation

$$h_{ie} \leq 2.1 \text{ k}\Omega$$

(Note that if R_b were infinite, we would have $h_{ie} = 4.14$ kΩ; therefore a large change in R_b would not result in a large change in h_{ie}.)

In order to have $h_{ie} = 2.1$ kΩ, the quiescent current I_{CQ} must be [Eq. (6.2-3)]

$$I_{CQ} \approx \frac{(120)(25 \times 10^{-3})}{2100} = 1.43 \text{ mA}$$

Now R_e can be chosen to complete the design. The dc and ac load lines are shown in Fig. 6.2-10.

The emitter resistor R_e is chosen so that

$$I_{CQ} = 1.43 \text{ mA} < \frac{9}{1130 + R_e}$$

and

$$R_b \ll h_{fe} R_e - 120 R_e$$

to achieve Q-point stability. In this example Q-point stability is not critical, since a peak swing of only 0.1 mA is required and the quiescent current is 1.43 mA. Thus a value of $R_e = 1$ kΩ will satisfy both conditions. The peak swing available is approximately 1.43 mA, which more than meets the specifications.

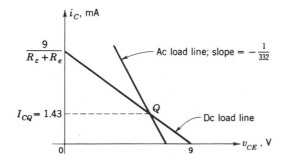

Figure 6.2-10 Load lines for Example 6.2-3.

The final values of the resistors are then

$$R_c = 1.2 \text{ k}\Omega \qquad \text{nearest standard value to } 1.13 \text{ k}\Omega$$

$$R_e = 1 \text{ k}\Omega$$

$$R_b = 10 \text{ k}\Omega \qquad \text{and} \qquad h_{ie} \ll R_b \ll h_{fe}R_e$$

To find R_1 and R_2 note that

$$I_{CQ} \approx \frac{V_{BB} - V_{BE}}{R_e + R_b/h_{fe}}$$

Thus $\qquad V_{BB} = (1.43 \times 10^{-3})(10^3 + 83.3) + 0.7 = 2.25 \text{ V}$

Knowing R_b and V_{BB}, we can find R_1 and R_2 from (2.3-1c) and (2.3-1d)

$$R_1 \approx 13 \text{ k}\Omega \qquad \text{use } 12 \text{ k}\Omega$$

$$R_2 \approx 40 \text{ k}\Omega \qquad \text{use } 39 \text{ k}\Omega$$

This completes the design. It is important to note that many designs are possible because of the loose specifications. In particular, the arbitrary division of the current-gain formula into equal factors might not be desirable if the specifications were more stringent. For example, if the gain requirement were increased from $A_i = 60$ to $A_i = 90$, R_c would have to be increased and h_{ie} decreased. In order to decrease h_{ie}, the quiescent current would have to be increased. This increase is in turn limited by the increased collector resistance, so that a compromise would have to be made. ///

6.3 THE COMMON-BASE CONFIGURATION

Has lower Z_{in}, higher Z_o than CE

The circuit of the common-base amplifier is shown in Fig. 6.3-1. This configuration does not provide current gain but does provide some voltage gain; it also has properties which make it useful at high frequencies. The hybrid small-signal equivalent circuit is shown in Fig. 6.3-1c. The hybrid equations using the notation and reference directions in the figure are (note that $i_1 = -i_e$)

$$v_{eb} = h_{ib}i_1 + h_{rb}v_{cb} = h_{ib}(-i_e) + h_{rb}v_{cb} \tag{6.3-1}$$

$$i_c = h_{fb}i_1 + h_{ob}v_{cb} = h_{fb}(-i_e) + h_{ob}v_{cb} \tag{6.3-2}$$

The input resistance h_{ib} is defined as [see (6.2-3)]

Input resistance

$$h_{ib} = \frac{v_{eb}}{i_1} = \frac{v_{eb}}{-i_e}\bigg|_{v_{cb}=0} = \frac{V_T}{I_{EQ}} \approx \frac{h_{ie}}{h_{fe} + 1} \tag{6.3-3a}$$

 Thus, if $h_{ie} = 1 \text{ k}\Omega$ and $h_{fe} = 100$, $h_{ib} = 10 \text{ }\Omega$. The input resistance of the common-base amplifier is usually significantly smaller than that of the common-emitter amplifier.

The reverse voltage gain h_{rb} is of the order of 10^{-4} and can usually be neglected.

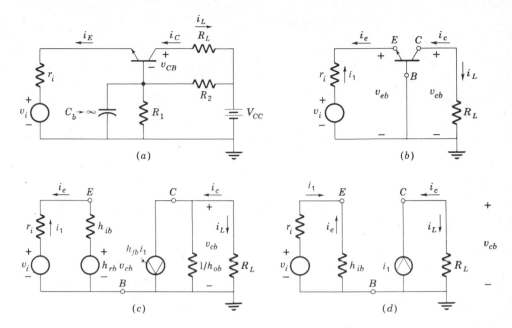

Figure 6.3-1 Common-base amplifier: (*a*) complete circuit; (*b*) ac circuit; (*c*) equivalent circuit using hybrid model; (*d*) simplified equivalent circuit.

The forward current-amplification factor h_{fb} is defined as

$$h_{fb} = \frac{i_c}{i_1}\bigg|_{v_{cb}=0} - \frac{i_c}{-i_e}\bigg|_{v_{cb}=0} = -\alpha \tag{6.3-3b}$$

Thus h_{fb} is approximately equal to -1 (note the current directions of i_c and i_e) and is equal in magnitude to the dc current-amplification factor α (see Sec. 2.1).

The output admittance h_{ob} is

$$h_{ob} = \frac{i_c}{v_{cb}}\bigg|_{i_e=i_1=0} \qquad \text{OUTPUT} \qquad \text{ADMITTANCE} \tag{6.3-3c}$$

This value is typically 1 μS, and is often neglected.

An approximate common-base equivalent circuit which neglects h_{rb} and h_{ob} and assumes h_{fb} is -1 is shown in Fig. 6.3-1*d*. Another method for obtaining the circuit of Fig. 6.3-1*c* involves rearranging the common-emitter equivalent circuit of Fig. 6.2-3*c* so that the emitter is the input terminal and the base the common terminal, as shown in Fig. 6.3-2. Now the definitions of (6.3-3) are applied to the rearranged circuit by writing KCL at the emitter terminal, with terminals *CB* shorted

$$-i_e + i_b + h_{fe}i_b = 0 \tag{6.3-4}$$

$$i_e = (1 + h_{fe})i_b = (1 + h_{fe})\frac{-v_{eb}}{h_{ie}} \tag{6.3-5}$$

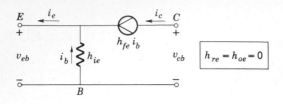

Figure 6.3-2 Rearranged CE circuit for finding h parameters of common-base configuration.

Therefore
$$-\frac{v_{eb}}{i_e}\bigg|_{v_{cb}=0} \equiv h_{ib} = \frac{h_{ie}}{1+h_{fe}} \approx \frac{h_{ie}}{h_{fe}} \qquad (6.3\text{-}6)$$

The short-circuit current gain is simply

$$h_{fb} = \frac{i_c}{-i_e}\bigg|_{v_{cb}=0} = -\frac{i_c\, i_b}{i_b\, i_e} = \frac{-h_{fe}}{h_{fe}+1} \approx -1 \qquad (6.3\text{-}7)$$

To find the output admittance of the common-base configuration, rearrange Fig. 6.2-3b as shown in Fig. 6.3-3, where h_{oe} has been included. Using (6.3-3c), we note that $i_e = 0$. Then the current through h_{oe} is $(h_{fe}+1)i_b$.

Applying KVL around the loop, we have

$$v_{cb} + (h_{fe}+1)\frac{i_b}{h_{oe}} + i_b h_{ie} = 0 \qquad (6.3\text{-}8)$$

Since $i_e = 0$, $i_b = -i_c$. Also $(h_{fe}+1)/h_{oe} \gg h_{ie}$, so that (6.3-8) becomes

$$v_{cb} \approx \frac{h_{fe}+1}{h_{oe}}\, i_c$$

and
$$h_{ob} = \frac{i_c}{v_{cb}}\bigg|_{i_e=0} = \frac{h_{oe}}{h_{fe}+1} \approx \frac{h_{oe}}{h_{fe}} \qquad (6.3\text{-}9)$$

This will be of the order of several megohms for most transistors. As in the common-emitter circuit, the reverse transmission h_{rb} and, usually, the output admittance h_{ob} are neglected for low-frequency calculations.

In order to find the CB parameters h_{ob}, h_{fb}, and h_{ib}, simply divide the corresponding CE parameter by $1 + h_{fe}$. Thus, if the CE parameters of a particular transistor are $1/h_{oe} = 10$ kΩ, $h_{fe} = 100$, and $h_{ie} = 250$ Ω, the same transistor in the CB configuration will have $1/h_{ob} = 1$ mΩ, $h_{fb} = -100/101 = -0.99$, and $h_{ib} = 2.5$ Ω. The CB stage thus has lower input impedance and higher output impedance than the CE stage.

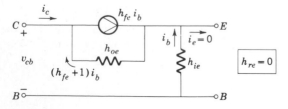

Figure 6.3-3 Common-base equivalent for finding h_{ob}.

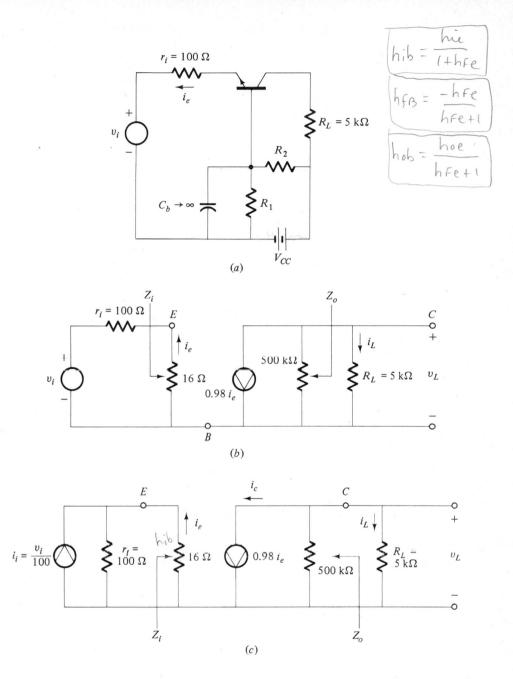

Figure 6.3-4 Circuit for Example 6.3-1: (a) complete circuit; (b) ac circuit; (c) current-source equivalent.

Example 6.3-1 (*a*) Find the CB parameters for the transistor of Example 6.2-1. Use

$$\frac{1}{h_{oe}} = 10 \text{ k}\Omega$$

(*b*) The transistor is used in a CB configuration with $r_i = 100 \ \Omega$ and $R_L = 5 \text{ k}\Omega$ (Fig. 6.3-4). Find the current gain, voltage gain, and input and output impedances.

SOLUTION (*a*) From Example 6.2-1 we have

$$h_{fe} = 50 \qquad h_{oe} \approx 10^{-4} \text{ S} \qquad h_{ie} = 0.83 \text{ k}\Omega \qquad h_{re} \approx 0$$

Using (6.3-7) and (6.3-3), we get

$$h_{fb} = \frac{-h_{fe}}{h_{fe} + 1} = \frac{-50}{51} = -0.98 \qquad h_{ib} = \frac{h_{ie}}{h_{fe} + 1} \approx \frac{830}{51} \approx 16 \ \Omega$$

$$h_{ob} = \frac{h_{oe}}{h_{fe} + 1} \qquad \frac{1}{h_{ob}} \approx 500 \text{ k}\Omega \qquad h_{rb} \approx 0$$

(*b*) The current gain (from Fig. 6.3-4*c*) is

$$A_i = \frac{i_L}{i_i} = \frac{i_e}{i_i} \frac{i_c}{i_e} \frac{i_L}{i_c} = \frac{-100}{100 + 16}(0.98)\frac{-500}{500 + 5} \approx 0.83$$

[handwritten: $A_i = \frac{i_L}{i_e} \frac{i_e}{i_i}$]

The voltage gain is

[handwritten: $A_v = \frac{v_L}{v_i} = \frac{v_L}{i_e} \frac{i_e}{v_i}$]

$$i_e = \frac{v_i}{100}\left(\frac{-100}{100 + 16}\right) = \frac{-v_i}{116}$$

$$v_L \approx -0.98 i_e(5000) \approx -5000 i_e$$

$$A_v = \frac{v_L}{v_i} \approx 43 \qquad \text{[handwritten: } (.83)\frac{5000}{100}\text{]}$$

By inspection the input and output impedances are

$$Z_i = 16 \ \Omega \qquad \text{and} \qquad Z_o = 500 \text{ k}\Omega \qquad\qquad ///$$

6.4 THE COMMON-COLLECTOR (EMITTER-FOLLOWER) CONFIGURATION

The emitter-follower configuration is characterized by a voltage gain slightly less than unity, high input impedance, and low output impedance. It is generally used as an *impedance transformer* in the input and output circuits of amplifier systems.

[handwritten: $A_v < 1 \quad Z_i = high \quad Z_o = low$]

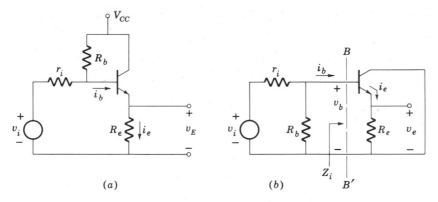

Figure 6.4-1 (a) Emitter-follower circuit; (b) ac circuit.

When placed in the input circuit, its high input impedance reduces the loading on the source. When placed in the output circuit, it serves to isolate the preceding stage of the amplifier from the load and, in addition, provides a low output impedance.

The emitter follower and its ac circuit are shown in Fig. 6.4-1. Following the procedure outlined in preceding sections, we can define a set of common-collector (CC) hybrid parameters, and draw an equivalent circuit. This, however, results in a circuit not often used for design purposes. Instead, an equivalent circuit is obtained directly from Fig. 6.4-1b, using KVL.

At terminals BB', KVL yields

$$v_b = v_{be} + i_e R_e \tag{6.4-1}$$

where, from Fig. 6.2-3c,

$$v_{be} = i_b h_{ie} \tag{6.4-2}$$

In addition, note that

$$i_e R_e = i_b[(h_{fe} + 1)R_e] \tag{6.4-3}$$

Substituting (6.4-2) and (6.4-3) into (6.4-1) gives

$$v_b = i_b h_{ie} + i_b[(h_{fe} + 1)R_e] \tag{6.4-4}$$

Equation (6.4-4) indicates that the equivalent circuit of the EF as seen looking into terminals BB' is a series combination of h_{ie} and $(1 + h_{fe})R_e$, as shown in Fig. 6.4-2.

From this circuit the voltage gain A_v is found by simple voltage division.

$$A_v = \frac{v_e}{v_i} = \left(\frac{(1 + h_{fe})R_e}{h_{ie} + (1 + h_{fe})R_e}\right)\left(\frac{R_b\|[h_{ie} + (1 + h_{fe})R_e]}{r_i + \{R_b\|[h_{ie} + (1 + h_{fe})R_e]\}}\right)$$

After some manipulation this can be put in the form

$$A_v = \frac{R_b}{r_i + R_b}\left|\frac{1}{1 + [h_{ie} + (r_i\|R_b)]/[(1 + h_{fe})R_e]}\right| \tag{6.4-5}$$

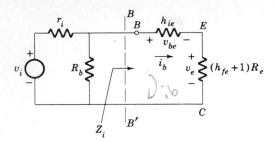

Figure 6.4-2 Equivalent circuit for emitter-follower.

Thus, if $(1 + h_{fe})R_e$ is much greater than the sum of h_{ie} and $r_i \| R_b$, as is often the case, the quantity in braces will be close to unity and the voltage gain will be determined by the $r_i - R_b$ voltage divider.

The input impedance of the emitter follower, defined as the impedance seen looking into terminals BB', is simply

$$Z_i = h_{ie} + (h_{fe} + 1)R_e \qquad (6.4-6)$$

In a similar way, the equivalent circuit looking into the emitter (output) is easily obtained by redrawing Fig. 6.4-1b as shown in Fig. 6.4-3.

Writing KVL around the emitter-base loop gives

$$v'_i = r'_i i_b + v_{be} + v_e \qquad (6.4-7)$$

where

$$v_{be} = h_{ib} i_e \qquad (6.4-8)$$

and

$$r'_i i_b = \frac{r'_i}{h_{fe} + 1} i_e \qquad (6.4-9)$$

Substituting (6.4-8) and (6.4-9) into (6.4-7), we have

$$v'_i = \frac{r'_i}{h_{fe} + 1} i_e + h_{ib} i_e + v_e \qquad (6.4-10)$$

Equation (6.4-10) yields the equivalent circuit of the EF as seen looking into terminals EE'. This circuit is shown in Fig. 6.4-4.

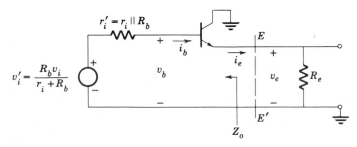

Figure 6.4-3 Emitter-follower ac circuit.

SMALL-SIGNAL LOW-FREQUENCY ANALYSIS AND DESIGN **263**

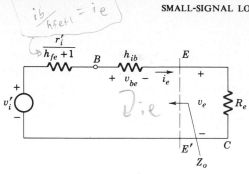

because

$\dfrac{i_b}{h_{fe}+1} = i_e$

When looking in from emitter transpose h_{ie} into h_{ib} and turn off sources.

Figure 6.4-4 Another equivalent circuit for the emitter follower.

The voltage gain A_v as calculated from this circuit is, of course, the same as that obtained from the equivalent circuit of Fig. 6.4-2.

The output impedance as seen from terminals EE' is

$$Z_o = h_{ib} + \frac{r_i'}{h_{fe}+1} \tag{6.4-11}$$

The equivalent circuits of Figs. 6.4-2 and 6.4-4 can also be obtained by replacing the transistor in Fig. 6.4-1b by the common-emitter equivalent circuit, as shown in Fig. 6.4-5.

The Thevenin output impedance at EE', with R_e removed, is

$$Z_o = \frac{v_e}{-i_e}\bigg|_{v_i=0} = \frac{v_e}{-(h_{fe}+1)i_b} = \frac{-i_b[h_{ie}+(r_i \| R_b)]}{-(h_{fe}+1)i_b}$$

$$= \frac{h_{ie}+(r_i \| R_b)}{h_{fe}+1} = h_{ib} + \frac{r_i'}{h_{fe}+1} \tag{6.4-12}$$

To obtain the Thevenin open-circuit voltage v_i' at EE', remove R_e. Then $i_e = 0$ and

$$i_b = \frac{i_e}{h_{fe}+1} = 0$$

Thus

$$v_i' = \frac{R_b}{R_b + r_i} v_i \tag{6.4-13}$$

which leads directly to the circuit of Fig. 6.4-4.

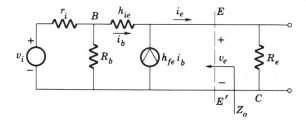

Figure 6.4-5 Emitter follower using common-emitter equivalent circuit.

Impedance reflection in the transistor Figures 6.4-2 and 6.4-4 illustrate an extremely useful small-signal property of the base-emitter circuit. For example, consider Fig. 6.4-2. Looking between terminal B (the base terminal) and ground, one sees h_{ie} in series with the emitter-to-ground impedance multiplied by $h_{fe} + 1$. All currents in the circuit are at *base-current level*. The current through the resistor $(1 + h_{fe})R_e$ is *not* the actual ac emitter current i_e but $i_e/(1 + h_{fe})$. The output voltage v_e in Fig. 6.4-2 is the same as the output voltage in the actual emitter follower of Fig. 6.4-1. Thus, when drawing an equivalent circuit, one can *reflect* the emitter circuit *through* the junction by simply multiplying the emitter-circuit impedance by $h_{fe} + 1$. Voltages throughout are preserved, while currents in the reflected impedance are *reduced* by $h_{fe} + 1$.

Consideration of Fig. 6.4-4 indicates that when reflecting in the other direction, i.e., looking into the emitter, one *divides* the base-circuit impedance by $h_{fe} + 1$.† Again, voltages are preserved, while currents in the reflected impedances are at emitter-current level, that is, $h_{fe} + 1$ times as large.

It should be pointed out that the circuit of Fig. 6.4-2 is approximate, since h_{oe} and h_{re} have been neglected. The errors introduced are negligible for most practical circuits.

As an example of the use of this property, consider the common-emitter amplifier with unbypassed emitter resistance shown in Fig. 6.4-6a. The small-signal equivalent circuit is shown in Fig. 6.4-6b. The current source can be split by employing KCL, as shown in Fig. 6.4-6c. This manipulation results in the circuit of Fig. 6.4-6d. This is then equivalent to the circuit of Fig. 6.4-6e, where the parallel combination of R_e and the $h_{fe} i_b$ source have been replaced by the *reflected* resistance $(1 + h_{fe})R_e$. The equivalence is established by noting that in both circuits $v_e = (1 + h_{fe})i_b R_e$. The current gain can then be found by inspection from the circuit of Fig. 6.4-6e

$$A_i = -h_{fe} \left(\frac{R_c}{R_c + R_L} \right) \left(\frac{r_i'}{r_i' + h_{ie} + (h_{fe} + 1)R_e} \right) \tag{6.4-14}$$

This technique is used extensively in Chap. 7, where multiple-transistor circuits are considered.

Example 6.4-1 For the emitter follower shown in Fig. 6.4-1, plot Z_i versus R_e, Z_o versus r_i, and A_v versus R_e.

SOLUTION Equations (6.4-6), (6.4-11), and (6.4-5) are the required relations. They are most easily plotted on log-log coordinates. For example, consider Z_i as a function of R_e

$$Z_i = h_{ie} + (h_{fe} + 1)R_e \tag{6.4-6}$$

† Note that $h_{ib} = h_{ie}/(h_{fe} + 1)$.

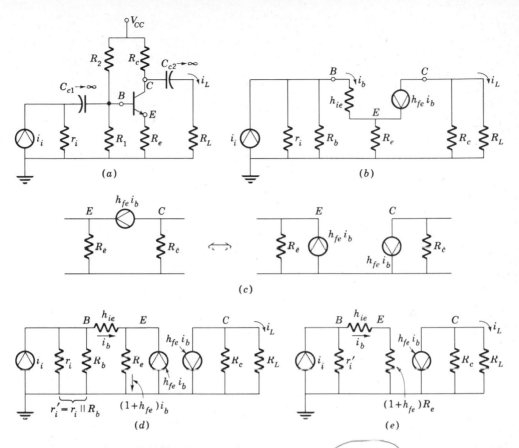

Figure 6.4-6 Impedance reflection applied to the amplifier with unbypassed emitter resistance: (a) amplifier; (b) small-signal equivalent circuit ($h_{oe} = 0$, $h_{re} = 0$); (c) splitting the current source; (d) current splitting applied to the amplifier; (f) final equivalent circuit with reflected emitter resistance.

The asymptotic values of Z_i are

$$Z_i \approx \begin{cases} h_{ie} & \text{for } h_{ie} \gg (h_{fe} + 1)R_e \\ (h_{fe} + 1)R_e & \text{for } h_{ie} \ll (h_{fe} + 1)R_e \end{cases}$$

The first of these is a constant (assuming the transistor parameters do not change if R_e is changed). The second is a straight line of slope $h_{fe} + 1$ on log-log coordinates. The two asymptotes intersect where $h_{ie} = (h_{fe} + 1)R_e$. At this point, the actual value of Z_i is $2h_{ie}$. This is shown in Fig. 6.4-7a. Equation (6.4-6) implies that $Z_i \to \infty$ as $R_e \to \infty$. However, when R_e becomes very large, the common-base output admittance must be included in the equivalent circuit. The equivalent circuit can be obtained from Fig. 6.4-2 by inserting the

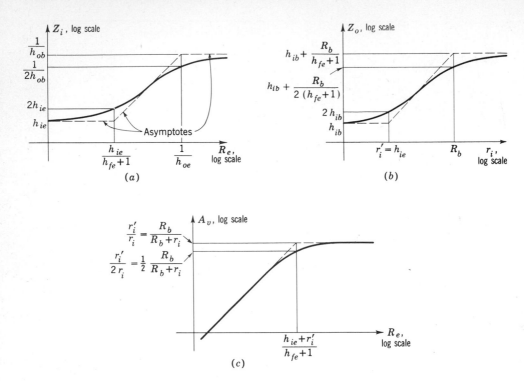

Figure 6.4-7 Variation of emitter-follower parameters: (a) $Z_i = h_{ie} + (1 + h_{fe})R_e$ [Eq. (6.4-6)];

(b) $Z_o = h_{ib} + \dfrac{r_i \| R_b}{h_{fe} + 1}$ [Eq. (6.4-11)]; (c) $A_v = \dfrac{r_i'}{r_i}\left(\dfrac{1}{1 + \dfrac{h_{ie} + r_i'}{(h_{fe} + 1)R_e}}\right)$.

resistance $1/h_{ob}$ between base and collector. The input impedance is then seen to approach $1/h_{ob}$ as R_e approaches infinity. This is shown in Fig. 6.4-7a. Z_o and A_v are plotted in a similar fashion in Fig. 6.4-7b and c. ///

Example 6.4-2 Design an emitter follower to meet the following specifications:

$A_v \geq 0.9$ for small signals $V_{im} \leq 4$ V

$r_i = 100\ \Omega$ R_L (ac-coupled) $= 50\ \Omega$

$100 \leq h_{fe} \leq 200$ $V_{CC} = 15$ V $V_{CE,\,sat} = 1$ V

The circuit is shown in Fig. 6.4-8.

SOLUTION As in previous design examples, many solutions are possible. In this example, it can be shown that the maximum value that R_e can have and still meet the specifications of a 4-V peak swing with a 1-V saturation voltage

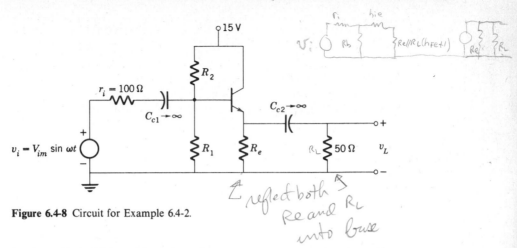

$$r_i = 100\ \Omega$$

Figure 6.4-8 Circuit for Example 6.4-2.

$$\mathcal{L}\ \text{reflect both } R_L$$
$$Re \text{ and } R_L$$
$$\text{into base}$$

is $R_e = 75\ \Omega$. Using this value of R_e, we get $I_{CQ} = 133$ mA and $V_{CEQ} = 5$ V. To use this value for R_e would require that h_{fe} be known exactly. Thus a smaller value of R_e is required to accommodate the variation of h_{fe}. To simplify the calculation, we let

$$R_e = R_L = 50\ \Omega$$

The dc and ac load lines are shown in Fig. 6.4-9. Two operating points are shown. The first, Q_1, is chosen on the basis of the minimum current which will satisfy the specifications. A peak signal-voltage swing of 4 V implies that the output voltage must be capable of swinging approximately 4 V. Since the ac

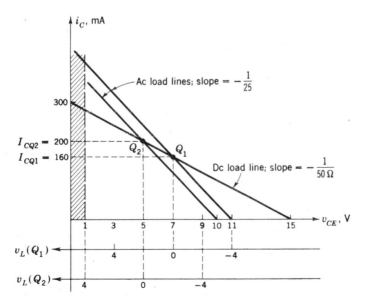

Figure 6.4-9 Load lines for Example 6.4-2.

load is 25 Ω, a peak current swing of 160 mA is required. Thus Q_1 is placed at $I_{CQ1} = 160$ mA, and the maximum signal will drive the transistor almost to cutoff.

The second operating point, Q_2, is placed to avoid the saturation voltage of 1 V. Thus $V_{CEQ2} = 5$ V and $I_{CQ2} = 200$ mA.

The base resistance R_b is chosen so that the Q point will not shift outside the range from Q_1 to Q_2 when h_{fe} varies from 100 to 200. This can be done by finding the limits on I_{CQ} as a function of h_{fe}.

The quiescent current is given by

$$I_{CQ} \approx \frac{V_{BB} - V_{BE}}{R_e + R_b/h_{fe}} = \frac{V_{BB} - 0.7}{50 + R_b/h_{fe}}$$

This will be a minimum when $h_{fe} = 100$.

Then, since the minimum allowable current is 160 mA,

$$160 \times 10^{-3} \leq \frac{V_{BB} - 0.7}{50 + R_b/100}$$

and when $h_{fe} = 200$,

$$200 \times 10^{-3} \geq \frac{V_{BB} - 0.7}{50 + R_b/200}$$

Combining these inequalities, we have

$$(0.16)\left(50 + \frac{R_b}{100}\right) \leq V_{BB} - 0.7 \leq (0.2)\left(50 + \frac{R_b}{200}\right)$$

Simplifying, we get

$$8 + (1.6 \times 10^{-3})R_b \leq 10 + 10^{-3}R_b$$

which reduces to

$$(0.6 \times 10^{-3})R_b \leq 2 \quad \text{and} \quad R_b \leq 3.3 \text{ k}\Omega$$

To achieve a gain exceeding 0.9 requires that

$$A_v \approx \left(\frac{R_b}{r_i + R_b}\right)\left(\frac{R_L \| R_e}{(R_L \| R_e) + h_{ib} + (r_i \| R_b)/h_{fe}}\right) \geq 0.9$$

Thus, with $\quad h_{ib} = (h_{ib})_{av} = \dfrac{V_T}{I_{EQ, av}} \approx \dfrac{25 \times 10^{-3}}{180 \times 10^{-3}} \approx 0.14 \ \Omega$

$$0.9 \leq \left(\frac{1}{1 + 100/R_b}\right)\left(\frac{25}{25 + 0.14 + \dfrac{100R_b}{100 + R_b} \dfrac{1}{h_{fe, \min}}}\right)$$

Note that $h_{fe, \min}$ is employed here to ensure that a gain of at least 0.9 results for any h_{fe}.

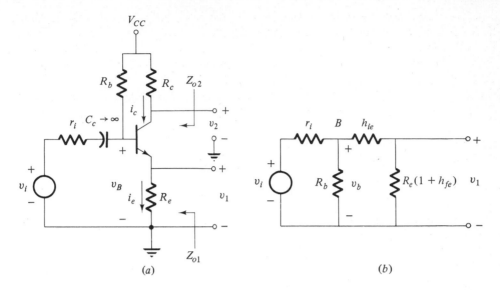

Figure 6.4-10 Phase inverter for Example 6.4-3: (a) complete circuit; (b) base-emitter equivalent.

If R_b is chosen to be 2.5 kΩ,

$$A_v = \left(\frac{1}{1 + 100/2500}\right)\left(\frac{25}{25 + 0.14 + (250{,}000/2600)(\frac{1}{100})}\right) \approx 0.92$$

and $R_b = 2.5$ kΩ satisfies the gain specification. To find V_{BB} return to the original inequality:

$$(0.16)(50 + 25) \le V_{BB} - 0.7 \le (0.2)(50 + 12.5)$$

$$12.7 \le V_{BB} \le 13.2$$

If we choose $V_{BB} = 13$ V, then, from (2.3-1), $R_2 \approx 2.9$ kΩ and $R_1 \approx 19$ kΩ. We would use standard resistors so that $R_2 = 2.7$ kΩ and $R_1 = 18$ kΩ.   ///

Example 6.4-3 The circuit in Fig. 6.4-10 is a phase inverter (phase splitter). Calculate v_2 and v_1.

SOLUTION The emitter voltage v_1 is found as if the circuit were an emitter follower, as shown in Fig. 6.4-10b. Using (6.4-5), we have

$$v_1 = v_i\left(\frac{R_b}{r_i + R_b}\right)\left(\frac{1}{1 + (h_{ie} + r_i')/[(h_{fe} + 1)R_e]}\right)$$

The emitter current i_e is

$$i_e = \frac{v_1}{R_e}$$

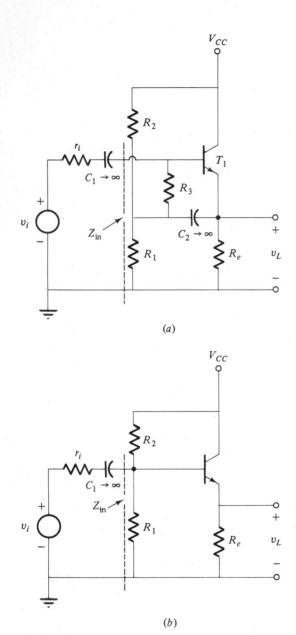

(a)

(b)

Figure 6.4-11 Emitter-follower input impedance: (a) biasing technique for high input impedance; (b) the standard bias circuit.

and $i_c = -h_{fb} i_e$, so that

$$v_2 = -R_c i_c = +R_c h_{fb} i_e = +h_{fb} \frac{R_c}{R_e} v_1$$

If $|h_{fb} R_c| = R_e$, then (since $h_{fb} = -1$) $v_1 = -v_2$. Thus a phase inverter gives two outputs that can be made equal in amplitude and are $180°$ out of phase. It is often used to provide out-of-phase input signals for the push-pull amplifier discussed in Sec. 5.3. Note that the output impedance Z_{o2} is much larger than Z_{o1} (see Prob. 6.4-8). If the external load has low resistance, the circuit may require adjustment in order to maintain the relation $v_1 = -v_2$.

///

Example 6.4-4 Show that the emitter-follower circuit in Fig. 6.4-11a has a significantly higher input impedance than the standard circuit in Fig. 6.4-11b.

SOLUTION The circuit shown in Fig. 6.4-11b employs the standard biasing technique, and its input impedance is

$$Z_{in} = R_1 \| R_2 \| [h_{ie} + (h_{fe} + 1)R_e] \qquad (6.4\text{-}6)$$

This is always less than $R_1 \| R_2$, so that the input impedance of an emitter follower biased using the standard method is limited by the parallel combination of R_1 and R_2.

In the circuit shown in Fig. 6.4-11a a technique called *bootstrapping* is used, and Z_{in} can exceed $R_1 \| R_2$ and actually approach the impedance $h_{ie} + (h_{fe} + 1)R_e$. We shall prove this statement using *impedance reflection*.

Figure 6.4-12a shows the ac equivalent of the circuit of Fig. 6.4-11a with all capacitors and voltage sources replaced by short circuits. To obtain the small-signal equivalent circuit using *reflection* it is convenient first to simplify the circuit by combining $R_1 \| R_2$ with R_e to form R'_e. In addition, since the base voltage and the voltage at R_3 are both v_b, we can remove the connection between R_3 and the base of T_1 while adding a v_b source in series with R_3, as shown in Fig. 6.4-12b. Note that all currents and voltages are the same in Fig. 6.4-12a and b.

The equivalent circuit, shown in Fig. 6.4-13, is obtained by reflecting components in the emitter circuit into the base circuit. Both emitter resistors R'_e and R_3 are increased by $h_{fe} + 1$ while their respective currents $i_{R'_e}$ and i_3 are decreased by the same factor. Thus, the voltage drop across each resistance is preserved. The voltage source v_b connected to R_3 in Fig. 6.4-12b is reflected without change since voltages remain unaltered in the reflection process.

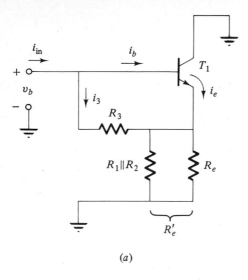

(a)

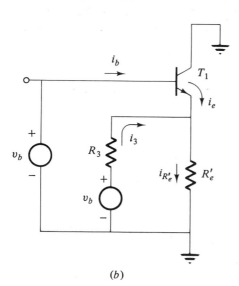

(b)

Figure 6.4-12 Emitter follower: (a) reduced small-signal circuit obtained by shorting capacitors and grounding supplies: (b) final small-signal circuit suitable for impedance reflection.

Referring to Fig. 6.4-12a, we see that the input impedance Z_{in} is

$$Z_{\text{in}} = \frac{v_b}{i_b + i_3}$$

To find Z_{in} we need only determine i_b and i_3 using Fig. 6.4-13. Since a practical circuit will have $R_3 \gg h_{ie}$, we can simplify the calculations by noting that

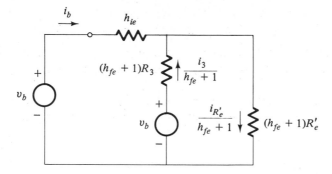

Figure 6.4-13 Equivalent circuit of the bootstrapped emitter-follower looking into the base.

$h_{ie} \ll (h_{fe} + 1)R_3$. Then the two v_b sources can be connected together, and we see that

$$i_b \approx \frac{v_b}{h_{ie} + (h_{fe} + 1)R'_e}$$

Since

$$i_3 R_3 = h_{ie} i_b$$

we have

$$i_3 = \frac{h_{ie}}{R_3} \frac{v_b}{h_{ie} + (h_{fe} + 1)R'_e}$$

The input admittance $1/Z_{in}$ is

$$\frac{1}{Z_{in}} = \frac{i_b}{v_b} + \frac{i_3}{v_b} = \frac{1 + h_{ie}/R_3}{h_{ie} + (h_{fe} + 1)R'_e}$$

Hence

$$Z_{in} = \frac{h_{ie} + (h_{fe} + 1)R'_e}{1 + h_{ie}/R_3}$$

We now see that with $R_3 \gg h_{ie}$ the input impedance of the bootstrapped circuit is significantly higher than the impedance of the standard circuit. For example, if $R_1 \| R_2 = 1$ kΩ, $R_e = 1$ kΩ, $h_{ie} = 1$ kΩ, $h_{fe} + 1 = 100$, and $R_3 = 10$ kΩ, then for the bootstrapped circuit

$$Z_{in} \approx \frac{1000 + (100)(500)}{1.1} \approx 46 \text{ kΩ}$$

while for the standard circuit with the same values

$$Z_{in} = R_1 \| R_2 \| [h_{ie} + (h_{fe} + 1)R_e] = 1000 \| 100,000 \approx 1 \text{ kΩ} \qquad ///$$

6.5 COLLECTION OF SIGNIFICANT PARAMETERS FOR THE THREE BASIC CONFIGURATIONS

The analyses of Secs. 6.2 to 6.4 yielded approximate formulas for the h parameters of the CE and CB stages and the input impedance, output impedance, and voltage gain of the EF (CC) stage. The results are summarized in Table 6.5-1.

Table 6.5-1

	CE	Configuration EF (CC)	CB
Gain	$A_i \approx -h_{fe}$	$A_v \approx 1$	$A_i \approx -h_{fb} = \dfrac{h_{fe}}{1 + h_{fe}}$
Input impedance	$h_{ie} = \dfrac{(25 \times 10^{-3})h_{fe}}{I_{EQ}}$	$Z_i = h_{ie} + (h_{fe} + 1)R_e$	$h_{ib} = \dfrac{h_{ie}}{1 + h_{fe}}$
Output impedance	$\dfrac{1}{h_{oe}} > 10^4 \ \Omega$	$Z_o \approx h_{ib} + \dfrac{r_i'}{h_{fe} + 1}$	$\dfrac{1}{h_{ob}} = \dfrac{1 + h_{fe}}{h_{oe}}$
Simplest equivalent circuit			

6.6 INTERPRETATION OF MANUFACTURERS' SPECIFICATIONS FOR LOW-POWER TRANSISTORS ($P_C < 1$ W)

In this section we discuss some common specifications given by manufacturers. To illustrate, consider the 2N3647 silicon *npn* transistor.

1. Maximum collector dissipation in free air at 25°C $P_{C, \text{max}} = 400$ mW
2. Derating factor in free air $\theta_{jc} = 0.4°\text{C/mW}$
3. Maximum junction temperature $T_{j, \text{max}} = 200°\text{C}$
4. Absolute maximum ratings at 25°C
   a. $BV_{CBO} = 40$ V
   b. $BV_{CEO} = 10$ V
   c. $BV_{EBO} = 6$ V

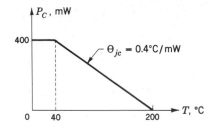

Figure 6.6-1 Derating curve for 2N3647.

d. $I_{C,\,\text{max}} = 500$ mA

e. $I_{CBO} = 25$ nA

5. Typical h parameters at 25°C

   a. $h_{fe} = 150$ (typical maximum value)

   b. $h_{oe} = 10^{-4}$ S (maximum)

   c. $h_{ie} = 4.5$ kΩ

   d. $h_{re} = 10^{-4}$

6. $C_{ob} = 4$ pF (maximum)

7. Common-base cutoff frequency $f_\alpha \geq 350$ MHz

This transistor is capable of dissipating 400 mW at room temperature using an infinite heat sink. It is derated linearly at the rate of 0.4°C/mW, as shown in Fig. 6.6-1.

The breakdown voltages differ considerably from those of the high-power transistor of Sec. 4.7. Note, for example, that the collector-base-junction breakdown voltage is only 40 V, compared with 100 V for the power transistor. In addition, I_{CBO} is 0.025 μA, an extremely small value.

Manufacturers often list typical hybrid parameters, for this transistor $h_{oe} = 10^{-4}$ S and $h_{re} = 10^{-4}$. Referring to Fig. 6.2-3, we see that h_{oe} can be neglected when this transistor is to be used as a common-emitter amplifier if

$$R_L \ll \frac{1}{h_{oe}} = 10 \text{ k}\Omega$$

In addition, h_{re} can be neglected if

$$h_{ie} i_b \gg h_{re} v_{ce} = 10^{-4} v_{ce}$$

or, using $|v_{ce}| \approx h_{fe} i_b R_L$,

$$h_{ie} i_b \gg h_{re} h_{fe} R_L i_b$$

With the typical parameters given this becomes

$$h_{ib} \gg h_{re} R_L = 10^{-4} R_L$$

Thus h_{re} can be neglected as long as h_{ib} is much larger than $10^{-4} R_L$.

The capacitance C_{ob} and the cutoff frequency are discussed in Chap. 9.

6.7 SMALL-SIGNAL EQUIVALENT CIRCUIT OF THE FET

The small-signal h-parameter equivalent circuit of the FET is shown in Fig. 6.7-1a. In a common-source FET amplifier the gate input is an open circuit at low and midfrequencies, so that h_i is infinite. Thus the h-parameter input circuit is open. In addition, at midfrequencies there is negligible feedback from output to input so that $h_r \approx 0$. Only the h-parameter output circuit is required to characterize the FET at these frequencies. The equivalent circuits at high and low frequencies are discussed in Chap. 9. The parameters needed to describe the FET are the (*forward*) *transconductance* g_m and the *drain-source output resistance* r_{ds}. These parameters are defined as follows.

Transconductance This is

$$g_m = \frac{\partial i_{DS}}{\partial v_{GS}}\bigg|_{Q\text{ point}} \tag{6.7-1}$$

The theoretical equation which describes the FET can be used to obtain an idea of the range of values of g_m. Consider the MOSFET, for which, in saturation,

$$i_{DS} = k_n(v_{GS} - V_T)^2 \tag{6.7-2}$$

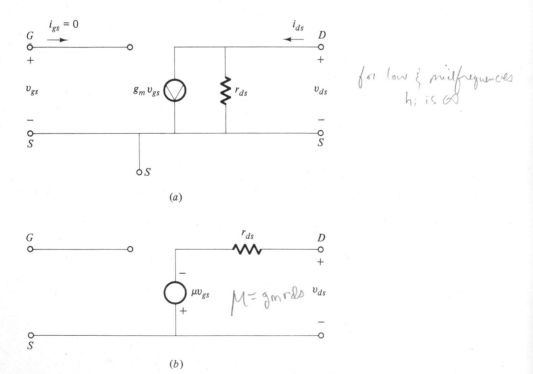

for low & midfrequencies
h_i is ∞

M = gmrds

(a)

(b)

Figure 6.7-1 Small-signal model of the FET: (a) controlled-current-source model; (b) controlled-voltage-source model.

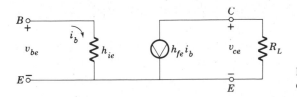

Figure 6.7-2 Transistor small-signal equivalent circuit.

Then
$$g_m = \frac{\partial i_{DS}}{\partial v_{GS}}\bigg|_{Q \text{ point}} = 2k_n(v_{GS} - V_T)\bigg|_{V_{GSQ}} = 2\sqrt{k_n I_{DSQ}} \qquad (6.7\text{-}3)$$

For example, if $k_n = 1$ mA/V² and $I_{DSQ} = 4$ mA, the transconductance $g_m = 4$ mS. This value is typical for the JFET and MOSFET. Note that the transconductance is proportional to $(I_{DSQ})^{1/2}$. Thus, if I_{DSQ} is increased by a factor of 4, g_m doubles.

The g_m of a FET is analogous to $1/h_{ib}$ in a junction transistor.† This is easily shown from Fig. 6.7-2. The output current of the junction transistor is

$$h_{fe}i_b = h_{fe}\frac{v_{be}}{h_{ie}} = \frac{1}{h_{ib}}v_{be} \quad = \quad g_m v_{be}$$

Thus the output-current source in the transistor can be, and often is (Chap. 9), replaced by $g_m v_{be}$, where $g_m - 1/h_{ib}$. As a further comparison we note that the voltage gain A_v of a common-emitter transistor amplifier is

$$A_v = \frac{v_{ce}}{v_{be}} \approx \frac{-R_L}{h_{ib}}$$

while the voltage gain of a common-source FET amplifier is, neglecting r_{ds},

$$A_v = \frac{v_{ds}}{v_{gs}} \approx -g_m R_L$$

The voltage gain of the transistor amplifier is significantly higher than the gain of the FET for the same value of R_L since

$$\left(\frac{1}{h_{ib}}\right)_{\text{transistor}} \gg (g_m)_{\text{FET}}$$

As an example, consider a transistor with $I_{CQ} = 1$ mA. Then

$$\frac{1}{h_{ib}} = 40 \text{ mS}$$

If $I_{CQ} = 10$ mA, $1/h_{ib} = 400$ mS. These values should be compared with typical values of g_m for the FET, 1 to 5 mS.

---

† It is interesting to note that the transconductance of a MOSFET is proportional to $\sqrt{I_{DQ}}$ [Eq. (6.7-3)] while the hybrid parameter $1/h_{ib}$ is proportional to I_{EQ}.

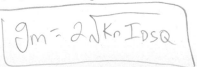

Drain-source resistance This is

$$r_{ds} = \left(\frac{\partial v_{DS}}{\partial i_{DS}}\right)_{Q\,point} \tag{6.7-4a}$$

In theory, this resistance should be infinite, since i_{DS} is not a function of drain-source voltage above pinch-off. However, the values of i_{DS} calculated from (6.7-2) represent asymptotic values not actually achieved in practice. Measured output vi characteristics do display a slight slope, as seen in the curves of Fig. 6.14-1. The range of values of r_{ds} is similar to that of the transistor output resistance $1/h_{oe}$, 20 to 500 kΩ.

The drain-source resistance is found to be, approximately, inversely proportional to the quiescent current

$$r_{ds} \propto \frac{1}{I_{DQ}} \tag{6.7-4b}$$

This is similar to the transistor, where h_{oe} is directly proportional to I_{CQ}.

Amplification factor An amplification factor μ is often defined as the product $g_m r_{ds}$. It can be calculated directly from the vi characteristics using the relation

$$\mu = -\frac{\partial v_{DS}}{\partial v_{GS}}\bigg|_{Q\,point} = g_m r_{ds} \tag{6.7-5}$$

A model which makes use of the amplification factor is shown in Fig. 6.7-1b.

6.8 THE COMMON-SOURCE VOLTAGE AMPLIFIER

The common-source voltage amplifier and its small-signal equivalent circuit are shown in Fig. 6.8-1. The input impedance seen by the source is

$$Z_i = R_3 + (R_1 \| R_2) \tag{6.8-1}$$

The output impedance seen by the load resistance R_L is

$$Z_o = R_d \| r_{ds} \tag{6.8-2}$$

and the voltage gain is

$$A_v = \frac{v_L}{v_i} = \frac{v_L}{v_{gs}} \frac{v_{gs}}{v_i} = -g_m(R_L \| Z_o) \frac{1}{1 + r_i/[R_3 + (R_1 \| R_2)]} \tag{6.8-3a}$$

Usually

$$r_i \ll R_3 + (R_1 \| R_2)$$

and if

$$R_L \ll Z_o$$

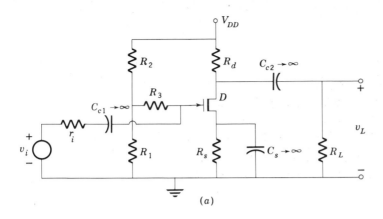

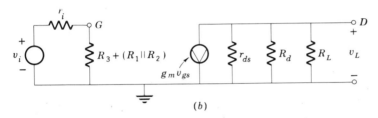

Figure 6.8-1 The common-source amplifier: (*a*) schematic and (*b*) small-signal equivalent circuit.

the voltage gain reduces to

$$A_v \approx -g_m R_L' \quad \text{if} \quad r_i \ll R_3 + R_1 \| R_2 \quad (6.8\text{-}3b)$$

Example 6.8-1 A MOSFET voltage amplifier employing feedback biasing is shown in Fig. 6.8-2*a*. Calculate the voltage gain, the input impedance, and the output impedance.

SOLUTION The equivalent circuit is shown in Fig. 6.8-2*b*. Using KCL at the drain terminal, we find

$$i = \frac{v_{gs} - v_{ds}}{R_f} = g_m v_{gs} + \frac{v_{ds}}{r_{ds} \| R_L}$$

$$= \frac{v_{gs} - v_{ds}}{10^5} = 2 \times 10^{-3} v_{gs} + \frac{v_{ds}}{6 \times 10^3}$$

Substituting $v_{gs} = v_i$ and solving this equation, we find

$$A_v = \frac{v_{ds}}{v_i} \approx -g_m(r_{ds} \| R_L) = -12$$

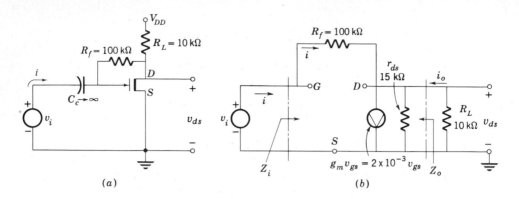

Figure 6.8-2 MOSFET amplifier for Example 6.8-1: (a) schematic; (b) small-signal equivalent circuit.

The input impedance seen by the source is

$$Z_i = \frac{v_i}{i} = \frac{v_i}{(v_i - v_{ds})/R_f} = \frac{R_f}{1 - v_{ds}/v_i} = \frac{10^5}{1 + 12} \approx 7.7 \text{ k}\Omega$$

and the output resistance seen by the 10-kΩ load is

$$Z_o = \frac{v_{ds}}{i_o}\bigg|_{v_i=0} = R_f \| r_{ds} = (100) \| (15) \approx 13 \text{ k}\Omega \qquad\qquad ///$$

6.9 THE SOURCE FOLLOWER (COMMON-DRAIN AMPLIFIER)

A source-follower circuit, employing a JFET or depletion-mode MOSFET, is shown in Fig. 6.9-1a. The biasing arrangement provides self-bias for negative gate-source-voltage operation. For this type of operation the dc-load-line equation is

$$V_{DD} = v_{DS} + i_{DS}(R_{s1} + R_{s2}) \qquad\qquad (6.9\text{-}1a)$$

and the bias voltage is, assuming zero direct current in R_1,

$$V_{GSQ} = -I_{DSQ}R_{s1} \qquad\qquad (6.9\text{-}1b)$$

Typically, V_{GSQ} will be only a few volts, while V_{DSQ} will be, roughly, one-half of V_{DD}, in order to place the operating point near the center of the load line. Therefore $R_{s1} \ll R_{s2}$.

To determine the gain and output impedance of the source follower we shall assume that R_1 is large enough to be considered infinite. (An exact analysis is postponed until Example 6.10-1.) Figure 6.9-1a can then be redrawn as in

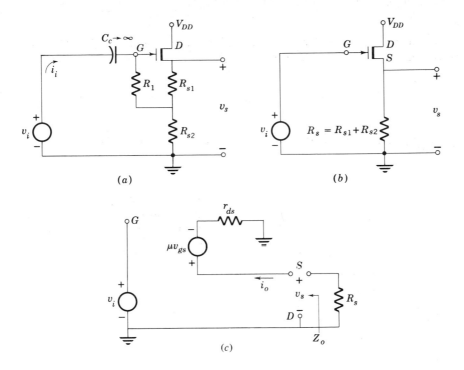

Figure 6.9-1 The source follower: (a) schematic using a JFET; (b) bias components omitted; (c) small-signal equivalent circuit.

Fig. 6.9-1b, and the resulting small-signal equivalent circuit is shown in Fig. 6.9-1c. We now determine the Thevenin equivalent circuit for this device and show that the voltage gain is nearly unity and the output resistance small.

The output impedance Z_o, as seen by the source resistance R_s, is

$$Z_o = \frac{v_s}{i_o}\bigg|_{v_i = 0} \tag{6.9-2a}$$

Then, referring to Fig. 6.9-1c, we have

$$v_s = -v_{gs} = \mu v_{gs} + i_o r_{ds} \tag{6.9-2b}$$

Using (6.9-2a) and (6.9-2b), we find that

$$Z_o = \frac{r_{ds}}{\mu + 1} \tag{6.9-2c}$$

When $\mu = g_m r_{ds} \gg 1$, the output impedance becomes

$$Z_o \approx \frac{1}{g_m} \tag{6.9-2d}$$

The open-circuit voltage gain A_v' with R_s removed is

$$A_v'\bigg|_{R_s \to \infty} = \frac{v_s}{v_g} \tag{6.9-3a}$$

The output voltage v_s is

$$v_s = \mu v_{gs} \tag{6.9-3b}$$

However,

$$v_{gs} = v_g - v_s \tag{6.9-3c}$$

Thus

$$A_v' = \frac{\mu}{\mu + 1} \tag{6.9-4a}$$

If $\mu \gg 1$, the open-circuit voltage gain A_v' becomes

$$A_v' \approx 1 \tag{6.9-4b}$$

We now determine the input impedance Z_i of the circuit. The complete output equivalent circuit with R_s connected is shown in Fig. 6.9-2. From this circuit we first calculate the gain A_v

$$A_v = \frac{v_s}{v_i} = \frac{v_s}{v_g} = \frac{A_v' R_s}{1/g_m + R_s} = \left(\frac{\mu}{\mu + 1}\right)\left(\frac{g_m R_s}{1 + g_m R_s}\right) \tag{6.9-5}$$

This will be needed shortly.

Next Z_i is defined as

$$Z_i = \frac{v_g}{i_i} \tag{6.9-6a}$$

All the input current flows in R_1. Then, since we have assumed $R_1 \gg R_{s2}$,

$$i_i R_1 \approx v_g - \frac{R_{s2}}{R_{s1} + R_{s2}} v_s \tag{6.9-6b}$$

and, solving for the ratio v_g/i_i, we get

$$Z_i \approx \frac{R_1}{1 - (v_s/v_g)[R_{s2}/(R_{s1} + R_{s2})]} \tag{6.9-6c}$$

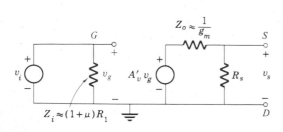

Figure 6.9-2 Thevenin equivalent circuit of the source follower.

Since R_1 is very large, the value of v_s/v_g is given in (6.9-5). Substituting (6.9-5) into (6.9-6c) and assuming $g_m R_s \gg 1$ yields

$$Z_i \approx \frac{R_1}{1 - [\mu/(\mu + 1)][R_{s2}/(R_{s1} + R_{s2})]} \tag{6.9-7a}$$

If $R_{s2} \gg R_{s1}$,

$$Z_i \approx (\mu + 1)R_1 \tag{6.9-7b}$$

The equivalent input circuit of the source follower is shown in Fig. 6.9-2. Note the similarity between the source follower and the emitter follower. Both devices are characterized by a high input impedance Z_i, a low output impedance Z_o, and a voltage gain almost equal to unity. Note also that returning R_1 to the node formed by R_{s1} and R_{s2} results in a much higher input impedance than would be obtained by returning R_1 to ground.

Example 6.9-1 Design a source follower using the JFET 2N4223 to have a Q point at $I_{DQ} = 3$ mA and $V_{DSQ} = 15$ V. The available supply voltage is 20 V. Calculate the input and output impedances and the voltage gain.

SOLUTION The circuit chosen uses self-bias, as shown in Fig. 6.9-1a. The characteristic shown in Appendix Fig. C.3-1 indicates that $V_{GSQ} \approx -1.2$ V. From (6.9-1b)

$$R_{s1} = \frac{V_{GSQ}}{I_{DQ}} = \frac{1.2}{3 \times 10^{-3}} = 400 \ \Omega$$

A standard 390-Ω resistor is used. R_{s2} is now found using (6.9-1a):

$$R_{s2} = \frac{V_{DD} - V_{DSQ}}{I_{DQ}} - R_{s1} = \frac{20 - 15}{3 \times 10^{-3}} - 390 \approx 1280 \ \Omega$$

A standard 1.2-kΩ resistor is used. The Thevenin output impedance Z_o is $1/g_m$. From Appendix Fig. C.3-1 we see that at the specified quiescent drain current (3 mA), $g_m \approx 2$ mS.

Therefore
$$Z_o \approx \frac{1}{g_m} = 500 \ \Omega$$

The resistance r_{ds} is the slope of the vi characteristic. It is found to be ≈ 83 kΩ. Thus $\mu = g_m r_{ds} \approx 166$. The voltage gain A'_v is $\mu/(\mu + 1)$. Therefore

$$A'_v \approx \frac{166}{167} \approx 1$$

and from (6.9-5)

$$A_v = \frac{v_s}{v_g} = A'_v \frac{R_s}{R_s + 1/g_m} \approx 0.77$$

The input impedance is [Eqs. (6.9-5) and (6.9-6c)]

$$Z_i \approx \frac{R_1}{1 - \dfrac{v_s}{v_g}\left(\dfrac{R_{s2}}{R_{s1} + R_{s2}}\right)} = \frac{R_1}{1 - 0.77\dfrac{1280}{390 + 1280}} \approx 2.1R_1$$

Thus if $R_1 = 100$ kΩ, $Z_i \approx 210$ kΩ.   ///

6.10 IMPEDANCE REFLECTION IN THE FET

In analyzing BJT circuits, we found that impedances and currents are reflected from the base circuit to the emitter circuit by dividing the base circuit impedances by $h_{fe} + 1$ and multiplying their associated currents by $h_{fe} + 1$. This serves to keep the voltage drop across each impedance the same while maintaining the relation which exists between the base and emitter currents, that is, $i_e = (h_{fe} + 1)i_b$. Similarly, when reflecting from the emitter circuit to the base circuit, all impedances are multiplied by $h_{fe} + 1$ and their respective currents are divided by $h_{fe} + 1$.

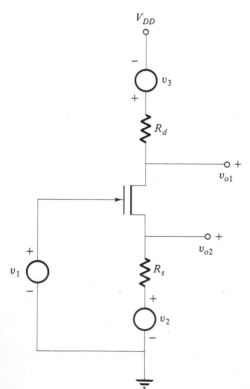

Figure 6.10-1 A FET amplifier with three independent sources (bias components omitted).

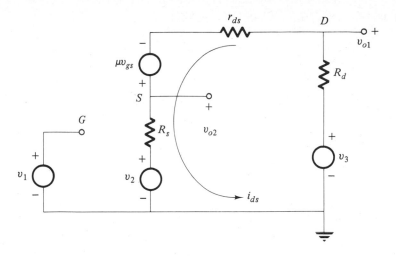

Figure 6.10-2 Equivalent circuit for Fig. 6.10-1.

In a FET the drain current and source current are equal. The rule for reflection is now that *impedances* and *voltages* located in the *drain* circuit can be reflected into the *source* circuit by dividing the impedances and the voltage sources by $\mu + 1$. Since the impedance and the voltage are divided by $\mu + 1$, the current remains the same. Similarly, impedances and voltage sources located in the source circuit can be reflected into the drain circuit by multiplying them by $\mu + 1$.

The technique of impedance reflection is illustrated using the circuit shown in Fig. 6.10-1. Here we have a FET amplifier with separate independent voltage sources in the gate, drain, and source circuits. The three sources were chosen to illustrate the effect of reflection in a variety of situations. All bias components have been omitted.

The equivalent circuit of Fig. 6.10-1 is shown in Fig. 6.10-2. From this circuit we have

$$i_{ds} = \frac{\mu v_{gs} + v_3 - v_2}{R_s + r_{ds} + R_d} \tag{6.10-1a}$$

where

$$v_{gs} = v_1 - v_2 - i_{ds} R_s \tag{6.10-1b}$$

Combining (6.10-1a) and (6.10-1b) yields

$$i_{ds} = \frac{\mu v_1 + v_3 - (\mu + 1)v_2}{(\mu + 1)R_s + r_{ds} + R_d} \tag{6.10-2}$$

The equivalent circuit seen looking between the drain and ground is found directly from (6.10-2) to be as shown in Fig. 6.10-3a. Note that no component in the drain circuit has been altered. However, the source resistor R_s has been multiplied by $\mu + 1$, the voltage v_2 in the source circuit has been multiplied by $\mu + 1$, and the gate voltage v_1 has been multiplied by μ.

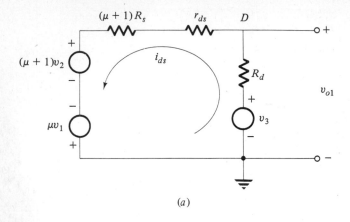

(a)

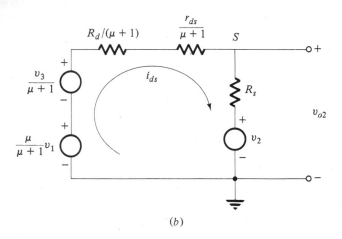

(b)

Figure 6.10-3 Impedance reflection in the FET: (a) equivalent circuit for Fig. 6.10-1 seen looking into the drain; (b) equivalent circuit for Fig. 6.10-2 seen looking into the source.

The equivalent circuit seen looking between the source and ground is found from (6.10-2) after dividing numerator and denominator by $\mu + 1$:

$$i_{ds} = \frac{[\mu/(\mu + 1)]v_1 + v_3/(\mu + 1) - v_2}{R_s + r_{ds}/(\mu + 1) + R_d/(\mu + 1)} \qquad (6.10\text{-}3)$$

The equivalent circuit is shown in Fig. 6.10-3b. Note that in this figure R_s and v_2 are unaltered since they represent the components found in the source circuit. However, the components located in the drain circuit, r_d, R_d, and v_3 are each divided by $\mu + 1$. The gate voltage v_1 is multiplied by μ to put it in the drain circuit and then divided by $\mu + 1$ to reflect it into the source.

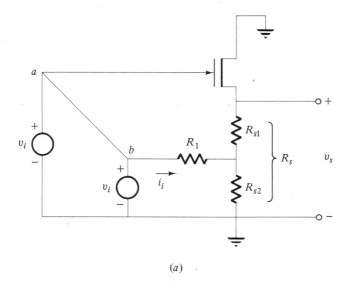

(a)

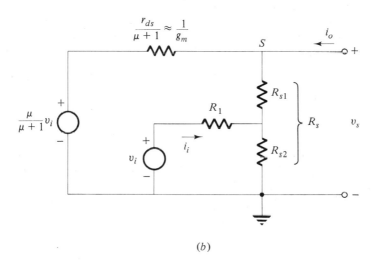

(b)

Figure 6.10-4 Source follower for Example 6.10-1: (a) Fig. 6.9-1a redrawn in simplified form; (b) output equivalent circuit obtained using reflection.

Example 6.10-1 Determine the Thevenin equivalent circuit of the source follower shown in Fig. 6.9-1a. Use the principle of reflection to reflect all components into the source circuit.

SOLUTION We begin by simplifying the circuit as shown in Fig. 6.10-4a. Since we are interested only in small signals, we have replaced the capacitor and the supply voltage by short circuits. Next we separate the gate and source circuit by removing the connection between point a and b as shown. Since points a and b are at exactly the same voltage both before and after the connection is removed, the circuit currents and voltages are not changed by this alteration.

Figure 6.10-4a is now seen to be similar to Fig. 6.10-1 with both $v_3 = 0$ and $R_d = 0$. The equivalent circuit obtained using reflection into the source is then as shown in Fig. 6.10-4b (see Fig. 6.10-3b). The gate voltage v_i is reflected into the source circuit after multiplication by $\mu/(\mu + 1)$, and r_{ds} is reflected into the source circuit after multiplication by $1/(\mu + 1)$.

It is left as a problem to show that the exact expression for the gain v_s/v_i is

$$A_v = \frac{[\mu/(\mu + 1)]g_m[R_{s1} + (R_{s2} \| R_1)] + (1/R_1)(R_{s2} \| R_1)}{1 + g_m[R_{s1} + (R_{s2} \| R_1)]} \quad (6.10\text{-}4)$$

If, however, $R_1 \gg R_{s2}$, the gain reduces to that found by straightforward circuit analysis and given by (6.9-5).

The output impedance Z_o is obtained by setting $v_i = 0$ and calculating v_s/i_o

$$Z_o = \frac{1}{g_m} \|[R_{s1} + (R_1 \| R_{s2})] \quad (6.10\text{-}5)$$

This expression reduces to $1/g_m$ if $1/g_m \ll R_s$.

Since the current into the gate is zero, the input impedance is simply $Z_1 = v_i/i_i$. This can be found directly from Fig. 6.10-4b. The result, after considerable algebra, is

$$Z_i = \frac{R_1 + [R_{s2} \| (R_{s1} + 1/g_m)]}{1 - [\mu/(\mu + 1)]g_m R_{s2}/[1 + g_m(R_{s1} + R_{s2})]} \quad (6.10\text{-}6)$$

If $R_1 \gg R_{s2}$ and $g_m(R_{s1} + R_{s2}) \gg 1$, this reduces to (6.9-7a). $\quad$ ///

6.11 THE PHASE-SPLITTING CIRCUIT

The FET can be used in a *phase-splitting* circuit as shown in Fig. 6.11-1. All biasing components have been omitted for simplicity. For the circuit of Fig. 6.11-1 to operate as a phase splitter, the two outputs v_{o1} and v_{o2} must be equal in amplitude and 180° out of phase. Since the drain current and source current are the same,

$$i_{ds} = -\frac{v_{o1}}{R_d} \quad (6.11\text{-}1a)$$

and

$$i_{ds} = \frac{v_{o2}}{R_s} \quad (6.11\text{-}1b)$$

hence

$$v_{o2} = -\frac{R_s}{R_d} v_{o1} \quad (6.11\text{-}2)$$

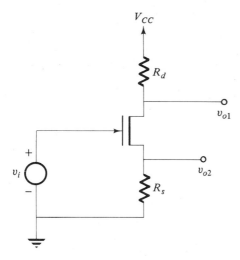

Figure 6.11-1 The phase-splitting circuit.

If $R_s = R_d$, the circuit acts as a phase splitter.

To determine the actual gain and output impedance of the phase splitter, we note that the circuit is identical to that of Fig. 6.10-1 with $v_2 = v_3 = 0$ V and $v_1 = v_i$. Hence, from the equivalent circuits of Fig. 6.10-3 we have

$$v_{o1} = -\frac{\mu R_d}{(\mu + 1)R_s + r_{ds} + R_d} v_i \qquad (6.11\text{-}3a)$$

and

$$v_{o2} = \frac{[\mu/(\mu + 1)]R_s}{R_s + r_{ds}/(\mu + 1) + R_d/(\mu + 1)} v_i \qquad (6.11\text{-}3b)$$

As expected, (6.11-3a) and (6.11-3b) have identical magnitudes if $R_d = R_s$.

The output impedance seen looking into the drain is found from Fig. 6.10-3a

$$Z_{o1} = R_d \| [r_{ds} + (\mu + 1)R_s] \qquad (6.11\text{-}4a)$$

while the output impedance seen looking into the source is (see Fig. 6.10-3b)

$$Z_{o2} = R_s \left\| \left(\frac{r_{ds}}{\mu + 1} + \frac{R_d}{\mu + 1} \right) \right. \qquad (6.11\text{-}4b)$$

For $R_d = R_s$, $(\mu + 1)R_s \gg r_{ds}$, and $\mu \gg 1$ these expressions reduce to

$$Z_{o1} \approx \frac{R_s[r_{ds} + (\mu + 1)R_s]}{r_{ds} + (\mu + 2)R_s} \approx R_s \qquad (6.11\text{-}5a)$$

and

$$Z_{o2} \approx \frac{R_s(r_{ds} + R_s)}{r_{ds} + (\mu + 2)R_s} \approx \frac{r_{ds} + R_s}{\mu} \qquad (6.11\text{-}5b)$$

Hence Z_{o2} is usually less than Z_{o1}.

6.12 THE COMMON-GATE AMPLIFIER

The common-gate amplifier is analogous to the common-base amplifier and is used primarily at high frequencies or as a switch. The schematic of such an amplifier is shown in Fig. 6.12-1a with all bias elements omitted for simplicity.

We analyze the operation of this circuit by finding an equivalent-circuit model. The input impedance is found by reflecting the resistances in the drain circuit into the source circuit. The resulting input impedance R_{sg} seen looking into the source between terminals S and G is

$$R_{sg} = \frac{v_{sg}}{i_i} = \frac{r_{ds} + R_d}{\mu + 1} \tag{6.12-1}$$

This result was obtained by noting that the resistances in the drain circuit when reflected into the source circuit are divided by $\mu + 1$. The input-circuit model is shown in Fig. 6.12-1b.

The Thevenin output equivalent circuit is found by reflecting the input voltage v_i and resistance r_i into the drain by multiplying each element by $\mu + 1$. The resulting circuit is shown in Fig. 6.12-1c. From this model we can calculate the gain of the common-gate amplifier. The result is

$$A_v = \frac{v_d}{v_i} = \frac{(\mu + 1)R_d}{R_d + r_{ds} + (\mu + 1)r_i} \tag{6.12-2}$$

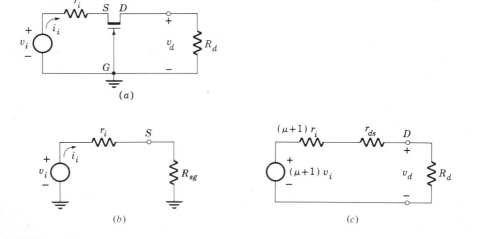

(a)

(b)

(c)

Figure 6.12-1 The common-gate amplifier. (a) schematic; (b) equivalent circuit at the input (source); (c) equivalent circuit looking into the drain.

6.13 DUAL-GATE FET

A *dual-gate* FET is shown in Fig. 6.13-1a, and its circuit symbol is given in Fig. 6.13-1b. In this type of FET the drain-source current is controlled by two gates† rather than by a single gate. Each gate is able to control the current flow independently, even to the point of cutting off the current. The device is used primarily at very high frequencies (VHF) as a multiplier, modulator, or mixer or as an amplifier characterized by very low capacitance between drain and gate (see Sec. 9.4 for a discussion of capacitance in the FET).

In this section we study the operation of the dual-gate FET as a multiplier of two signals. The *vi* characteristics of the FET are shown in Fig. 6.13-2. Figure 6.13-2a shows the variation of current i_{DS} as a function of v_{G1S} with v_{G2S} as a parameter. Note that the current ceases to flow when $v_{G1S} = V_{T1} \approx -2$ V or when $v_{G2S} = V_{T2} \approx -1$ V, showing that the FET is a depletion-mode device. Figure

---

† Of historical interest is the fact that the device is similar to the vacuum-tube *tetrode*.

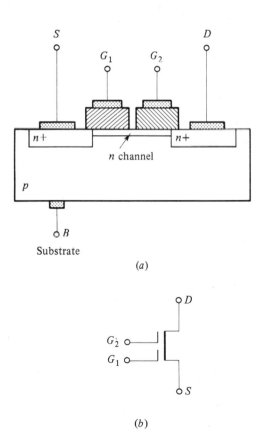

(a)

(b)

Figure 6.13-1 Dual-gate NMOS depletion-mode FET. (a) pictorial sketch; (b) circuit symbol.

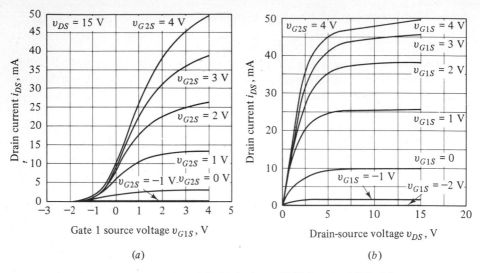

Figure 6.13-2 Vi characteristics of the MN81 dual-gate FET for $T = 25°C$: (a) i_{DS} versus v_{G_1S}; (b) i_{DS} versus v_{DS}.

6.13-2b shows the drain-source vi characteristic for $v_{G2S} = 4$ V. If the characteristics were redrawn for, say, $v_{G2S} = 2$ V, the current i_{DS} would be less than that shown when measured at the same values of v_{G1S} and v_{DS}.

To see how the dual-gate FET operates as a multiplier consider that $v_{G1S} = 0$ V, $v_{G2S} = 0.5$ V, and a small signal is applied to each gate as shown in

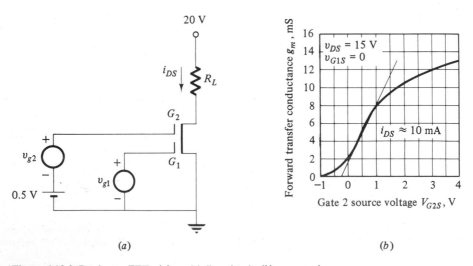

Figure 6.13-3 Dual-gate FET: (a) multiplier circuit; (b) transconductance g_m versus v_{G2S}.

Fig. 6.13-3*a*. Assuming that the FET operates above pinch-off, we can write

$$i_{DS} = I(v_{G1S}, v_{G2S}) \tag{6.13-1}$$

since above pinch-off (in saturation) the FET current i_{DS} is relatively independent of v_{DS}. Then, using a Taylor-series expansion about the quiescent voltages, we have

$$i_{DS} = I(0, 0.5) + \left[\frac{\partial I}{\partial v_{G1S}} \Big|_{\substack{v_{G1S}=0 \text{ V} \\ v_{G2S}=0.5 \text{ V}}} \right] v_{g1} + \left[\frac{\partial I}{\partial v_{G2S}} \Big|_{\substack{v_{G1S}=0 \text{ V} \\ v_{G2S}=0.5 \text{ V}}} \right] v_{g2}$$

$$+ \left[\frac{\partial^2 I}{\partial v_{G1} \, \partial v_{G2}} \Big|_{\substack{v_{G1S}=0 \text{ V} \\ v_{G2S}=0.5 \text{ V}}} \right] v_{g1} v_{g2} + \cdots \tag{6.13-2}$$

When the FET is to operate as a multiplier, we filter out the first three terms with appropriate circuitry and focus our attention on the fourth term

$$\left[\frac{\partial^2 I}{\partial v_{G1} \, \partial v_{G2}} \Big|_{Q \text{ point}} \right] v_{g1} v_{g2} = K_p v_{g1} v_{g2}$$

This term is proportional to the product of v_{g1} and v_{g2}. To determine the coefficient K_p, we write it as

$$K_p = \frac{\partial}{\partial v_{G2}} \left(\frac{\partial I}{\partial v_{G1S}} \right)_{Q \text{ point}} \tag{6.13-3a}$$

But, from (6.7-2) $g_m = \partial I / \partial v_{G1S}$; hence

$$K_p = \left(\frac{\partial g_m}{\partial v_{G2S}} \right)_{Q \text{ point}} \tag{6.13-3b}$$

Figure 6.13-3*b* shows the variation of g_m as a function of v_{G2S}. The slope of this curve at $v_{G2S} = 0.5$ V is estimated as $K_p \approx 6$ mA/V^2.

For example, if $v_{g1} = 0.5 \cos \omega_1 t$ and $v_{g2} = 0.5 \cos \omega_2 t$, i_{DS} contains the term

$$(6 \times 10^{-3})(0.5 \cos \omega_1 t)(0.5 \cos \omega_2 t) = 1.5 \cos \omega_1 t \cos \omega_2 t \quad \text{mA}$$

6.14 TYPICAL MANUFACTURERS' SPECIFICATIONS

The specifications given below are for the *n*-channel MOSFET 2N3796. This FET is a depletion-mode device that operates in the enhancement region as well as in the depletion region. It is classed as a low-power audio-frequency device.

Maximum ratings ($T_a = 25°C$) Maximum ratings are

$$V_{DS} = 25 \text{ V} \qquad P_D = 200 \text{ mW}$$

$$V_{GS} = \pm 10 \text{ V} \qquad \theta_{jc} = 1.14°C/W$$

$$I_{DS} = 20 \text{ mA} \qquad T_j = 200°C$$

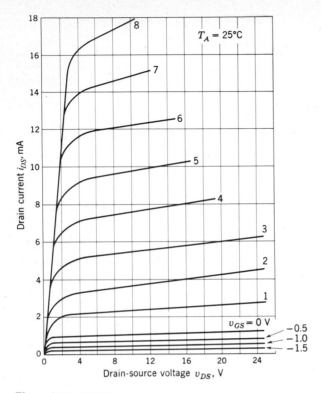

Figure 6.14-1 MOSFET output vi characteristic.

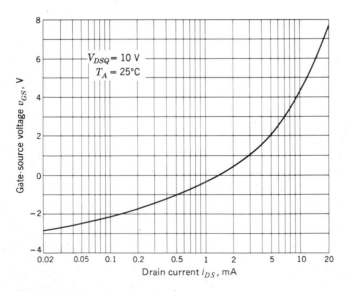

Figure 6.14-2 Transfer characterstic.

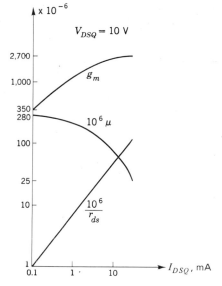

Figure 6.14-3 FET-parameter variation with drain current.

Explanation of specifications The maximum-rating specifications define the breakdown voltages, the maximum current, and the maximum power and derating characteristics. These are similar to the transistor specifications presented in Sec. 6.6 and are used in the same way.

Figure 6.14-1 shows the vi characteristic for this MOSFET. This characteristic is "typical," i.e., the actual 2N3796 FET used may differ by up to 15 percent from the typical characteristic. Thus, unlike those for the transistor, the vi characteristics can be used to set the Q point.

The transfer characteristic of Fig. 6.14-2 is sometimes provided by manufacturers. The inverse slope of this curve is the forward transconductance g_m.

Figure 6.14-3 is an extremely useful although not a necessary, characteristic. It shows that at $V_{DS} = 10$ V, $1/r_{ds}$ increases linearly with current [Eq. (6.7-4b)]. In addition, it gives the variation of g_m with drain current. Referring to (6.7-3), we see that g_m is proportional to $\sqrt{I_{DSQ}}$. Note that the product $g_m r_{ds} = \mu$ decreases with increasing drain current. This can be shown by combining (6.7-3) and (6.7-4b) to get

$$g_m r_{ds} \propto \frac{1}{\sqrt{I_{DQ}}} \qquad (6.14\text{-}1)$$

PROBLEMS

In all problems a complete equivalent circuit should accompany the solution.

6.1-1 For the amplifier shown in Fig. P6.1-1, $h_i = 2$ kΩ, $h_r = 0$, $h_f = 200$, and $1/h_o = 10$ kΩ. Find A_i; i_2/i_i and $A_v = v_2/v_i$, where $v_i = i_i r_i$.

Figure P6.1-1

6.1-2 High-frequency transistors are often specified in terms of the y parameters as defined by

$$i_1 = y_{11}v_1 + y_{12}v_2 \qquad i_2 = y_{21}v_1 + y_{22}v_2$$

(a) Draw the equivalent circuit similar to Fig. 6.1-1b using the y parameters.

(b) Give terminal definitions for the y parameters as in (6.1-4) to (6.1-7).

6.1-3 Find the relations from which one can calculate the y parameters given the h parameters.

6.1-4 Two-port parameters are often defined in terms of partial derivatives evaluated at the operating point. For the hybrid parameters (6.1-2) and (6.1-3) are written

$$v_{1T} = V(i_{1T}, v_{2T}) \qquad i_{2T} = I(i_{1T}, v_{2T})$$

where $i_{1T} = I_{1Q} + i_1$, etc. Expand v_{1T} and i_{2T} in a Taylor's series in two variables about the operating point I_{1Q}, V_{1Q}; neglect high-order terms, and find the definitions of the h parameters in terms of the partial derivatives of the functions V and I.

6.2-1 Estimate h_{oe} and h_{ie} at $I_{CQ} = 1$ and 5 mA, $V_{CEQ} = 10$ V for the 2N3904 silicon *npn* transistor whose characteristics are given in Appendix C.

6.2-2 Sketch two circuits which can be used to measure each of the h parameters. *Hint:* Refer to Fig. 6.1-1b and recall that an ac short circuit can be obtained by using a capacitor which has a very small reactance at the frequency being used.

6.2-3 For the silicon transistor in Fig. P6.2-3 $h_{fe} = 100$ and $h_{re} = h_{oe} = 0$. Find h_{ie}, A_i, Z_i, Z_o.

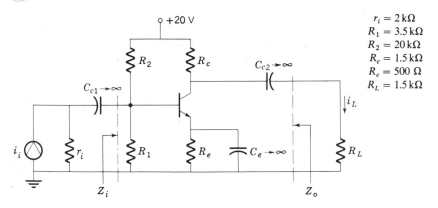

$r_i = 2\,\text{k}\Omega$
$R_1 = 3.5\,\text{k}\Omega$
$R_2 = 20\,\text{k}\Omega$
$R_c = 1.5\,\text{k}\Omega$
$R_e = 500\ \Omega$
$R_L = 1.5\,\text{k}\Omega$

Figure P6.2-3

6.2-4 For the transistor in Fig. P6.2-4, $h_{ie} = 1$ kΩ, $h_{re} = 10^{-4}$, $h_{oe} = 10\,\mu$S, and $h_{fe} = 50$.

(a) Assume that $R_b \gg h_{ie}$. Plot $A_i = i_L/i_i$ as a function of R_L.

(b) Assume that $h_{re} = h_{oe} = 0$. Plot A_i versus R_L on the same axes as in part (a). Compare the results of (a) and (b).

(c) Repeat parts (a) and (b) for Z_i.

(d) Repeat parts (a) and (b) for Z_o.

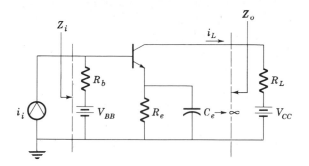

Figure P6.2-4

6.2-5 For the transistor in Fig. P6.2-5, $h_{re} = h_{oe} = 0$ and $50 < h_{fe} < 150$. Calculate the range of A_i and Z_i to be expected.

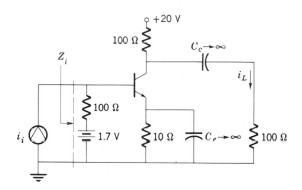

Figure P6.2-5

6.2-6 Find the voltage gain in Fig. P6.2-6 $A_v = v_c/v_i$ and Z_i. Assume that $h_{fe} = 100$ and $h_{oe} = h_{re} = 0$.

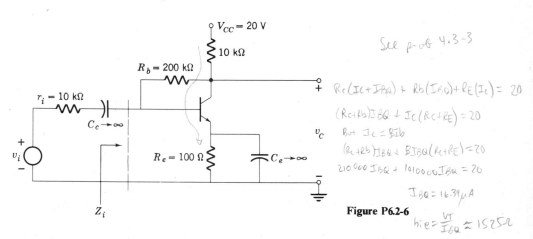

See prob 4.3-3

$R_c(I_c + I_{BQ}) + R_b(I_{BQ}) + R_E(I_c) = 20$

$(R_c + R_b)I_{BQ} + I_c(R_c + R_E) = 20$

But $I_c = \beta I_b$

$(R_c + R_b)I_{BQ} + \beta I_{BQ}(R_c + R_E) = 20$

$210\,000\,I_{BQ} + 1010000 I_{BQ} = 20$

$I_{BQ} = 16.39 \mu A$

$h_{ie} = \frac{V_T}{I_{BQ}} \approx 1525\Omega$

Figure P6.2-6

6.2-7 For the circuit of Fig. P6.2-7 find (a) $A_i = i_L/i_i$ and (b) the maximum possible symmetrical swing in i_L.

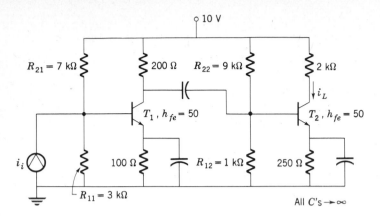

Figure P6.2-7

6.2-8 Repeat the design problem of Example 6.2-3 to achieve as high a current gain as possible.

6.3-1 (a) Find the four common-base hybrid parameters in terms of the four common-emitter hybrid parameters by substituting $i_e - i_c$ for i_b and $v_{cb} + v_{be}$ for v_{ce} into the common-emitter equations and solving for v_{be} and i_c as a function of i_e and v_{cb}.
  (b) Show that if $h_{ie} h_{oe} \ll 1$ and $h_{re} \ll 1$,

$$h_{ib} = \frac{h_{ie}}{1 + h_{fe}} \qquad h_{rb} = -\left(h_{re} - \frac{h_{oe} h_{ie}}{1 + h_{fe}}\right)$$

$$h_{fb} = \frac{-h_{fe}}{1 + h_{fe}} \qquad h_{ob} = \frac{h_{oe}}{1 + h_{fe}}$$

6.3-2 Use the Ebers-Moll equations (2.2-14) and the results of Prob. 6.1-4 to calculate h_{ib} and h_{ob}.

6.3-3 (a) Find the common-base hybrid parameters for the transistor in Prob. 6.2-4.
  (b) Use these parameters in the circuit of Fig. 6.3-4 to find A_i, A_v, Z_i, and Z_o.

6.3-4 A transistor has $h_{fe} = 10$, $h_{oe} = 0.1$ mS at 1 mA, and $h_{rb} = 0$. Design a common-base amplifier as in Fig. 6.3-4a for a maximum voltage gain if $r_i = 50\ \Omega$ and $R_L = 10$ kΩ. *what's Vcc?*

6.4-1 For the circuit of Fig. P6.4-1 find h_{ie}, A_v, Z_i, and Z_o.

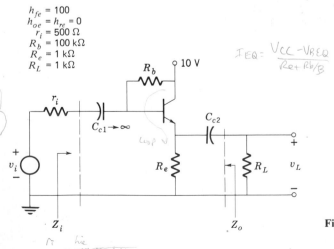

$h_{fe} = 100$
$h_{oe} = h_{re} = 0$
$r_i = 500\ \Omega$
$R_b = 100$ kΩ
$R_e = 1$ kΩ
$R_L = 1$ kΩ

IEQ = (VCC − VBEQ) / (Re + Rb/β)

IEQ = 4.65mA
hie = 537.6

Figure P6.4-1

To find Au: Vi ... hie ... [Rb+Fe][RLhfe] ... VL ... keep doing some transformation from left to right to get volt divider.

6.4-2 For the circuit of Fig. P6.4-2 plot (a) A_v versus r_i for $0 < r_i < \infty$ ($R_L = 1$ kΩ), (b) A_v versus R_L for $0 < R_L < \infty$ ($r_i = 1$ kΩ), (c) Z_i versus R_L for $0 < R_L < \infty$, and (d) Z_o versus r_i for $0 < r_i < \infty$.

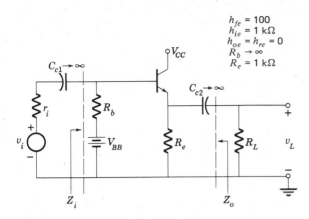

$$h_{fe} = 100$$
$$h_{ie} = 1 \text{ k}\Omega$$
$$h_{oe} = h_{re} = 0$$
$$R_b \rightarrow \infty$$
$$R_e = 1 \text{ k}\Omega$$

Figure P6.4-2

6.4-3 In Example 6.4-2 show that the maximum R_e for a 4-V peak swing is 75 Ω.

6.4-4 An *npn* silicon transistor is used in an emitter-follower circuit, with $I_{EQ} = 1$ mA, $V_{CEQ} = 5$ V. At this Q point, $h_{ib} = 42$ Ω, and $h_{fb} = -0.96$. In this circuit $r_i = 10$ kΩ, $R_b = 100$ kΩ, and $R_e = R_L = 1$ kΩ. Find A_v, Z_o, and Z_i.

6.4-5 In the circuit of Fig. 6.4-11a let $R_1 = 10$ kΩ, $R_2 = 20$ kΩ, $R_3 = 100$ kΩ, $V_{CC} = 20$ V, $R_e = 1$ kΩ, $r_i = 600$ Ω, and $h_{fe} = h_{FE} = 50$.

(a) Calculate Z_{in}.

(b) Determine the Thevenin equivalent circuit at v_L (v_T and R_T).

6.4-6 For the circuit of Fig. P6.4-6 find A_i and Z_i.

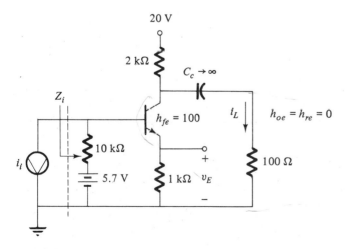

Figure P6.4-6

6.4-7 Find the Thevenin equivalent circuit at v_E in Fig. P6.4-6.

6.4-8 Calculate Z_{o1} and Z_{o2} in Fig. 6.4-10. Compare using $|h_{fb} R_c| = R_e$.

6.4-9 In the circuit of Fig. P6.4-9 find (a) R_{11}, R_{21}, R_{12}, and R_{22} so that v_L can undergo maximum swing and (b) v_L/i_i.

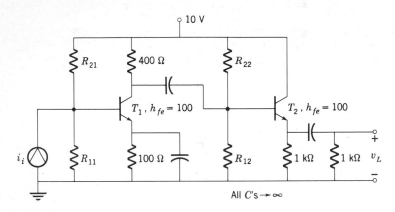

Figure P6.4-9

6.7-1 Two identical MOSFETs are connected in parallel as shown in Fig. P6.7-1. Find the equivalent MOSFET.

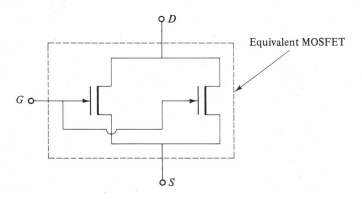

Figure P6.7-1

6.8-1 The MOSFET in Fig. P6.8-1 is biased so that $I_{DSQ} = 2$ mA.

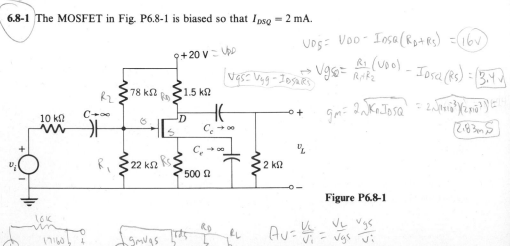

Figure P6.8-1

(a) Calculate V_{DSQ}, V_{GSQ}, and g_m assuming that $k_n = 1$ mA/V^2.

(b) If $r_{ds} = 20$ kΩ, calculate the voltage gain v_L/v_i.

6.8-2 (a) Calculate the voltage gain v_o/v_i for the circuit in Prob. 4.5-3 assuming $r_{ds} = 30$ kΩ.

(b) Find Z_i and Z_o.

3/23/81 **6.8-3** Design the MOSFET amplifier in Fig. P6.8-3 to have a gain of 10. Assume $V_{GSQ} = 3$ V, $V_{DSQ} = 4$ V, $I_{DSQ} = 5$ mA, $r_{ds} = 20$ kΩ, and $k_n = 2$ mA/V^2.

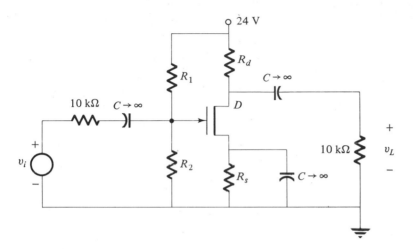

Figure P6.8-3

6.8-4 The MOSFET having the worst-case parameter variations of Fig. 4.5-4 is used in Example 6.8-1, where $V_{DD} = 20$ V, $R_L = 5$ kΩ, $R_f = 1$ MΩ, and $r_{ds} = 20$ kΩ. Estimate the variation in voltage gain v_{ds}/v_i.

6.9-1 (a) Calculate the voltage gain A'_v for the circuit in Prob. 4.5-4.

(b) Find Z_i and Z_o.

6.10-1 (a) Find the Thevenin equivalent circuit at v_o for the source follower in Prob. 4.5-4.

(b) Find Z_i.

6.10-2 A source follower is to be designed using the circuit of P6.10-2. A 2N4223 JFET is to be used (see Appendix C). The gain is to be greater than 0.8. Find R_1, R_{s1}, and R_{s2}.

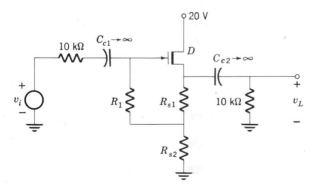

Figure P6.10-2

6.10-3 Verify Eqs. (6.10-4), (6.10-5), and (6.10-6).

6.11-1 (a) Repeat Prob. 6.8-1 with the source-resistor bypass capacitor removed. (b) Find the Thevenin resistance at v_L and (c) at the 500-Ω source resistor.

6.11-2 (a) Find the Thevenin equivalent circuit at v_L in Prob. 6.8-3 if the source-resistor bypass capacitor is removed. (Use $R_d = 3$ kΩ and $R_s = 1$ kΩ.)

(b) Find the Thevenin equivalent circuit across R_s with the source-resistor bypass capacitor removed.

6.11-3 The MOSFET having the worst-case parameter variations of Fig. 4.5-4 is to be used as a phase splitter in Fig. P6.11-3. Assume $r_{ds} = 30$ kΩ, $R_L + R_S = 10$ kΩ, and $V_{CC} = 20$ V.

(a) Determine the variations in voltage gains v_{o1}/v_i and v_{o2}/v_i.

(b) Determine the variations in the Thevenin impedances at v_{o1} and v_{o2}.

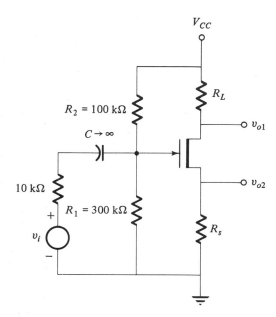

Figure P6.11-3

6.12-1 The vi characteristic of the common-gate amplifier shown in Fig. P6.12-1 is given approximately by

$$i_{DS} = (1 + v_{GS})^2 \times 10^{-4}$$

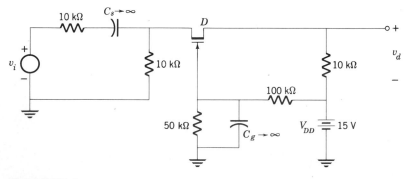

Figure P6.12-1

(a) Plot the characteristics.
(b) Find the Q point graphically.
(c) Calculate g_m.
(d) Let $r_{ds} = 10$ kΩ; calculate μ.
(e) Determine Z_i, Z_o, and the gain v_d/v_i.

6.12-2 In the circuit of Fig. P6.12-2, both FETs have identical g_m, μ, and r_{ds}. Calculate (a) i_L as a function of v_1 and v_2, (b) v_{o1}, v_{o2}, and v_{o3}, and (c) the output resistance looking into terminals $S_2 S_2'$.

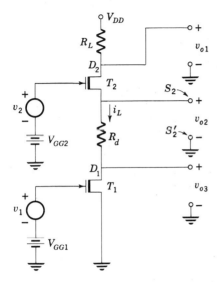

Figure P6.12-2

6.13-1 Calculate R_L in Fig. 6.13-3 to obtain the Q point: $I_{DSQ} = 10$ mA, $V_{DSQ} = 15$ V, $V_{G1SQ} = 0$ V, and $V_{G2SQ} = 0.5$ V.

MULTIPLE-TRANSISTOR CIRCUITS

Circuit designers who required an amplifier before the age of the IC often had to design the amplifier from scratch using discrete transistors, resistors, capacitors, etc. With the advent of the IC, this is no longer necessary. We now have available complete amplifiers on single chips with a large variety of characteristics. These are called op-amps (short for operational amplifiers), and engineering practice today dictates their use wherever possible because of their advantages of small size, low power consumption, and high reliability.

Most op-amps are basically the same. They consist of a difference-amplifier input stage followed by one or more high-gain amplifier stages which in turn drive some form of output stage. In this chapter we shall analyze these individual circuits and explore the fundamental properties of the overall op-amps.

7.1 THE DIFFERENCE AMPLIFIER

The difference amplifier is a versatile circuit which serves as the input stage to most op-amps and also sees use in such diverse ICs as the comparator (Sec. 15.3) and the emitter-coupled logic gate (Sec. 12.3).

The basic configuration is shown in Fig. 7.1-1. The diagram indicates that the circuit has two inputs, v_1 and v_2, and three outputs, v_{o1}, v_{o2}, and $v_{o1} - v_{o2}$. The importance of the difference amplifier lies in the fact that the outputs are proportional to the *difference* between the two input signals, as we shall show. Thus the circuit can be used to amplify the difference between the two inputs or amplify only one input simply by grounding the other. The features which distinguish the various outputs will be pointed out in the discussion which follows.

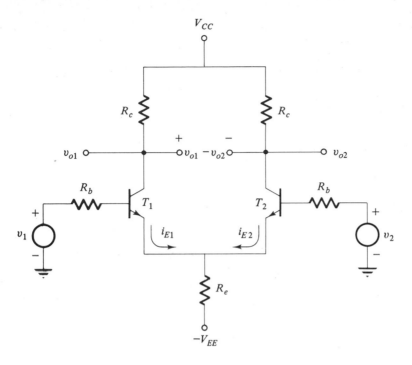

Figure 7.1-1 Basic difference amplifier.

The difference amplifier to be analyzed in this section is assumed to be fabricated on an IC chip. When this is the case, we can assume that transistors T_1 and T_2 are identical and thus that perfect symmetry exists between both halves of the circuit. In Sec. 7.4 we consider the operation of the circuit when the transistors differ slightly.

Common-mode and differential-mode signals Since the difference amplifier is most often used to amplify the difference between the two input signals, it is appropriate to express the inputs in a way which emphasizes this point, as follows. Let the difference between the input voltages be called v_d, so that

$$v_d = v_2 - v_1 \quad \text{Differential Mode} \tag{7.1-1a}$$

This is called the *differential-mode* or *difference-mode input voltage*. For completeness, we need a term for the average value of the input voltages, which we shall call v_a. It is convenient to define this voltage as

$$v_a = \frac{v_2 + v_1}{2} \quad \text{Common Mode} \tag{7.1-1b}$$

Since v_a is the *average* of the two input voltages, it is called the *common-mode* input voltage.

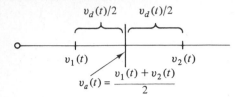

Figure 7.1-2 Decomposition of input voltages.

If we now solve (7.1-1a) and (7.1-1b) for v_2 and v_1, the result is

$$v_2 = v_a + \frac{v_d}{2} \tag{7.1-2a}$$

and

$$v_1 = v_a - \frac{v_d}{2} \tag{7.1-2b}$$

From these expressions we see that the input voltages can each be represented in terms of a *common-mode input voltage* and a *differential-mode input voltage*. A graphical interpretation of these definitions is shown in Fig. 7.1-2.

In the usual applications of the difference amplifier the differential-mode input is the desired signal and is to be amplified while the common-mode input is to be rejected and hence not amplified. The definitions above allow us to analyze the circuit directly in terms of these common- and differential-mode inputs and thus to focus on the important parameters of the difference amplifier. For test purposes we can easily design input signals which are entirely common-mode or entirely difference-mode. For example, if $v_1 = v_2$, the difference-mode input is zero and the common-mode input is simply $v_a = v_1 = v_2$. On the other hand, if $v_1 = -v_2$, the common-mode input is zero while the difference-mode input is $v_d = 2v_2 = -2v_1$.

Q-point analysis When we studied individual amplifier stages, we found that the load line (ac or dc) completely defined the path of operation of the collector circuit over the limits of variation of the input signal. This operating path remains a straight line as long as the circuit contains only resistors and sources. Now we have a different situation; we have *two* input signals. The operating path now becomes a *region* on the collector characteristics over which each transistor will operate for given maximum and minimum values of the two input signals. In this section we shall determine the boundaries of the operating region; this will lead to relations which can be used to ensure that these boundaries provide for linear operation over the expected range of variation of the input signals. The analysis is best carried through in terms of the difference-mode and common-mode inputs defined in the previous paragraph.

Usually, when we wish to determine the Q point of an amplifier, we set the input signal to zero. For the difference amplifier, it is appropriate to start the Q-point analysis by assuming that the differential-mode input is zero. This is accomplished simply by making the two inputs equal; then from (7.1-1b) we have

Q pt· analysis – tie together inputs , thus $v_d = 0$ so $v_1 = v_2 = v_a$

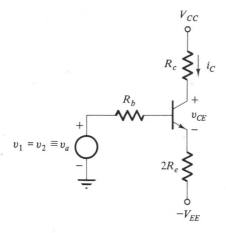

Figure 7.1-3 Equivalent circuit for either transistor T_1 or T_2 when $v_1 = v_2 = v_a$.

$v_a = v_1 = v_2$. With this assumption we begin by noting that, thanks to the circuit symmetry, we can separate the emitters, placing a resistance $2R_e$ in each emitter leg, as shown in Fig. 7.1-3. That the emitter voltage is not changed can be seen by applying KVL to the original circuit of Fig. 7.1-1 as follows:

$$v_{E1} = v_{E2} = (i_{E1} + i_{E2})R_e - V_{EE} \tag{7.1-3}$$

When $v_1 - v_2$, we shall have, again thanks to symmetry, $i_{E1} - i_{E2} - i_E$, so that (7.1-3) simplifies to

$$v_{E1} = v_{E2} = i_E(2R_e) - V_{EE} \tag{7.1-4}$$

This voltage is precisely the same as the emitter voltage found in the separated circuit of Fig. 7.1-3.

The load-line equation, which is valid when $v_a = v_1 = v_2$, is found using KVL around the collector-emitter loop in Fig. 7.1-3:

$$v_{CE} = V_{CC} - i_C R_c - i_E(2R_e) + V_{EE} \approx \underbrace{V_{CC} + V_{EE} - i_C(R_c + 2R_e)}_{V_{CE} =} \tag{7.1-5}$$

The emitter current (and therefore the collector current) is found by writing KVL around the base-emitter loop

$$\boxed{v_a = i_B R_b + v_{BE} + i_E(2R_e) - V_{EE}} \tag{7.1-6a}$$

Since $i_B = i_E/(h_{fe} + 1)$, $i_E \approx i_C$, and $v_{BE} = 0.7$ V, Eq. (7.1-6a) simplifies to

$$\boxed{i_C \approx \frac{v_a + V_{EE} - 0.7}{2R_e + R_b/(h_{FE} + 1)}} \tag{7.1-6b}$$

In obtaining (7.1-6) we neglected the effect of h_{ie} and considered v_{BE} to be constant. In this case the approximation is excellent since the impedance h_{ie} reflected into the emitter in Fig. 7.1-3 is $h_{ie}/(h_{fe} + 1)$. This is in series with $2R_e$, which is a much larger resistance in practice.

$$i_{C_Q} = \frac{v_a + V_{EE} - .7}{2R_e + \dfrac{R_b}{h_{fe}+1}}$$

$$V_{CE_Q} = V_{CC} + V_{EE} + i_C(R_c + 2R_e)$$

Q ptenations for CM load line

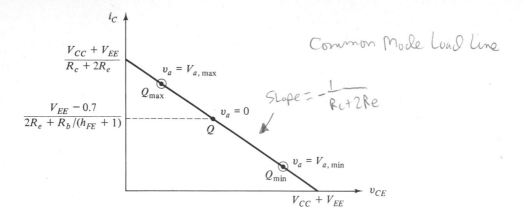

Figure 7.1-4 Common-mode load line for the circuit of Fig. 7.1-3 showing Q point when $v_a = 0$ and variation of the Q point as v_a varies from $V_{a,\,max}$ to $V_{a,\,min}$.

The load-line equation (7.1-5) is plotted in Fig. 7.1-4. Since only the common-mode input is present, we call this the *common-mode load line*. Here Q is the quiescent point obtained from (7.1-4b) by setting the common-mode input v_a to zero. The points marked Q_{max} and Q_{min} represent the operating points obtained when the common-mode input v_a varies from its most positive value $V_{a,\,max}$ to its most negative value $V_{a,\,min}$ *with the difference-mode input equal to zero.* We must note carefully that the same load line applies to each transistor since the collector current of each transistor is the same as long as the input voltages are equal, regardless of the value of v_a. Since the collector currents are the same and the circuit is symmetrical, the collector voltages will be identical and the output voltage $v_{o1} - v_{o2}$, taken between the collectors, will be *zero* regardless of the value of v_a as long as $v_1 = v_2$. The individual collector voltages v_{o1} and v_{o2} however, will vary with the variations in v_a. We shall determine the extent of this variation shortly.

The development just concluded determined the operating path when the difference-mode input is zero. We must now determine the effect of a nonzero difference-mode input. Therefore, let us set $v_2 = -v_1 = v_d/2$. For this case the common-mode input is zero and the quiescent point is the point Q shown in Fig. 7.1-4. If we refer back to the circuit of Fig. 7.1-1 and let $v_2 = v_d/2$ and $v_1 = -v_d/2$, we see that as v_2 increases, more emitter current i_{E2} flows and as v_1 decreases, the current i_{E1} decreases. If the variation of v_1 and v_2 is not excessive, the increase in i_{E2} is equal to the decrease in i_{E1} and therefore there is *no* change in the current $i_{E1} + i_{E2}$ flowing in R_e. Thus, the emitter voltage $v_{E1} = v_{E2}$ remains fixed, when a differential-mode signal is applied. However, since i_{E1} and i_{E2} are changing, v_{CE1} and v_{CE2} must also be changing such that

$$\Delta v_{CE1} = -R_c \, \Delta i_{C1}$$

or, using small-signal notation,

$$v_{ce1} = -R_c i_{c1} \qquad (7.1\text{-}7a)$$

and

$$\Delta v_{CE2} = -R_c \,\Delta i_{C2}$$

or

$$v_{ce2} = -R_c i_{c2} \qquad (7.1\text{-}7b)$$

Equations (7.1-7) are the difference-mode load-line equations of the difference amplifier, and the slope of these load lines is $-1/R_c$.

The combination of difference-mode and common-mode load lines defines the operating region for each transistor. Since each input signal will in general have both components present, we can establish the boundaries on the operating region if we know the maximum and minimum values of the signals or of their common- and difference-mode components. This is illustrated in the following example.

Example 7.1-1 In the circuit of Fig. 7.1-1, $V_{CC} = V_{EE} = 10$ V, $R_b = 0\ \Omega$, $R_e = 900\ \Omega$, $R_c = 200\ \Omega$, and the common-mode input range is -7 V $\le v_a \le$ $+7$ V. Find the maximum allowable current variation i_{C1} and i_{C2} due to the difference-mode signal so that operation is linear.

SOLUTION Using (7.1-5), we plot the common-mode load line as shown in Fig. 7.1-5. Using (7.1-6b), we must determine points Q_1 ($v_a = +7$ V) and Q_2

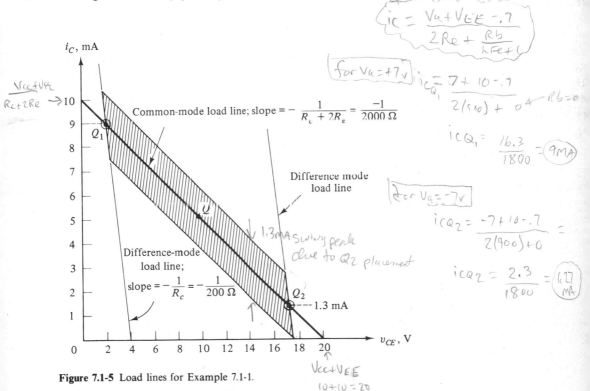

Figure 7.1-5 Load lines for Example 7.1-1.

$(v_a = -7 \text{ V})$; these represent the extreme common-mode operating points. In the worst-case condition the inputs operate at either one or the other of the common voltages $v_a = \pm 7$ V. Hence, to determine the maximum allowable current variation due to v_d we plot the difference-mode load lines (7.1-7) on Fig. 7.1-5 so as to intersect the common-mode load line at points Q_1 and Q_2. The worst case occurs at Q_2, where the current variation due to v_d must be less than 1.3 mA to avoid cutoff. If the difference-mode input signal is restricted to values which will keep the collector-current variation within ± 1.3 mA, the operating region for the two transistors is the shaded area drawn in Fig. 7.1-5. In general both transistors will operate at different points for any arbitrary pair of input voltages; however, their maximum variation must lie within the operating region. The input voltage v_d required to produce this change in current is determined in the development which follows.

$///$

Small-signal analysis Now that we have determined the conditions required to ensure linear operation we turn our attention to the small-signal operation of the difference amplifier. To this end we draw a small-signal equivalent circuit in which all components are reflected into the emitter. The result is shown in Fig. 7.1-6.

In Fig. 7.1-6a if we set $v_d = 0$, a little thought will show that as a result of the symmetry $i_{e1} = i_{e2}$. Similarly, if we set $v_a = 0$, then $i_{e1} = -i_{e2}$. We shall make use

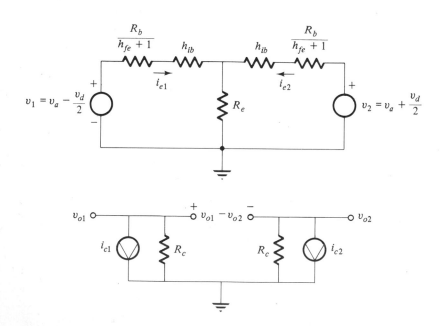

Figure 7.1-6 Difference amplifier, small-signal conditons: (a) small-signal equivalent input circuit with all components reflected into the emitter; (b) small-signal equivalent output circuit.

With v_d
set $= 0$
$v_d = 0$

With v_a
set $= 0$

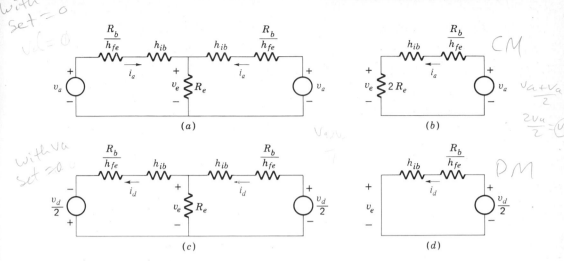

$V_a = V_e$

$V_a + V_a$ / 2

$\frac{2V_a}{2} = (V_a)$

Figure 7.1-7 Equivalent emitter circuits used to calculate the emitter current $i_{e_{2,1}} = i_a \pm i_d$: (a) circuit used to calculate common-mode emitter current i_a; (b) reduced equivalent circuit; (c) circuit used to calculate differential-mode current i_d; (d) reduced equivalent circuit.

of these facts to determine i_{e1} and i_{e2} using the principle of superposition. This will be done by first setting $v_d = 0$ and determining the average, or common-mode, emitter current $i_{e1} = i_{e2} \equiv i_a$. We then set $v_a = 0$ and determine the difference-mode emitter current $i_{e2} = -i_{e1} = i_d$.

In Fig. 7.1-7a we show the circuit of Fig. 7.1-6a when $v_d = 0$. Since the current in R_e is $2i_a$, the emitter voltage $v_e = 2R_e i_a$. As before, because of the symmetry, we can split R_e into two parallel resistors (each $2R_e$) and divide the circuit in half to obtain the reduced equivalent circuit of Fig. 7.1-7b. The emitter voltage v_a is seen to be exactly the same in this reduced circuit as in the circuit of Fig. 7.1-7a. Using Fig. 7.1-7b, we find the common-mode current i_a

$$i_a = \frac{v_a}{2R_e + h_{ib} + R_b/(h_{fe} + 1)} \quad CM \; \text{current} \quad (7.1\text{-}8)$$

Figure 7.1-7c is the equivalent circuit of Fig. 7.1-6a when $v_a = 0$. Because of the polarities of the two sources, $v_e = 0$. Therefore, this circuit can be reduced to that shown in Fig. 7.1-7d. The current i_d is

$$i_d = \frac{v_d/2}{h_{ib} + R_b/(h_{fe} + 1)} \quad DM \; \text{current} \quad (7.1\text{-}9)$$

In Example 7.1-1 this difference-mode current was restricted to be less than 1.3 mA. Knowing h_{ib}, R_b, and h_{fe}, we can find the maximum difference signal voltage v_d.

Applying superposition, we combine (7.1-8) and (7.1-9). This yields

$$i_{e1} = \frac{v_a}{2R_e + h_{ib} + R_b/(h_{fe} + 1)} - \frac{v_d/2}{h_{ib} + R_b/(h_{fe} + 1)} \tag{7.1-10a}$$

and

$$i_{e2} = \frac{v_a}{2R_e + h_{ib} + R_b/(h_{fe} + 1)} + \frac{v_d/2}{h_{ib} + R_b/(h_{fe} + 1)} \tag{7.1-10b}$$

Using the output equivalent circuit shown in Fig. 7.1-6b and assuming $i_c \approx i_e$, we can readily determine v_{o1}, v_{o2}, and $v_{o1} - v_{o2}$.

$$v_{o1} = -R_c i_{C1} = \frac{R_c/2}{h_{ib} + R_b/(h_{fe} + 1)} v_d - \frac{R_c}{2R_e + h_{ib} + R_b/(h_{fe} + 1)} v_a \tag{7.1-11a}$$

$$v_{o2} = -R_c i_{C2} = -\frac{R_c/2}{h_{ib} + R_b/(h_{fe} + 1)} v_d - \frac{R_c}{2R_e + h_{ib} + R_b/(h_{fe} + 1)} v_a \tag{7.1-11b}$$

and

$$\boxed{v_{o1} - v_{o2} = \frac{R_c}{h_{ib} + R_b/(h_{fe} + 1)} v_d} \tag{7.1-12}$$

Equation (7.1-12) shows clearly that $v_{o1} - v_{o2}$ is directly proportional to the difference-mode input voltage, $v_d = v_2 - v_1$. The outputs v_{o1} and v_{o2} are also proportional to the difference-mode voltage but contain a term proportional to the common-mode input voltage $v_a = (v_1 + v_2)/2$. In the ideal difference amplifier, the output is proportional only to v_d. Thus, if the output must be taken between one of the collectors and ground, the amplifier falls short of the ideal. The next section describes a figure of merit which provides a measure of the departure from the ideal.

7.2 COMMON-MODE REJECTION RATIO

Using (7.1-11), we can write the output voltages v_{o1} and v_{o2} as

$$v_{o1} = A_d v_d - A_a v_a \tag{7.2-1a}$$

and

$$v_{o2} = -A_d v_d - A_a v_a \tag{7.2-1b}$$

where A_d, the *difference-mode gain*, is

$$A_d = \frac{R_c/2}{h_{ib} + R_b/(h_{fe} + 1)} \tag{7.2-1c}$$

and A_a, the *common-mode gain*, is

$$A_a = \frac{R_c}{2R_e + h_{ib} + R_b/(h_{fe} + 1)} \tag{7.2-1d}$$

In an ideal differential amplifier the output voltage is proportional to v_d and does not depend on the common-mode voltage v_a. Thus, in an ideal differential

D.M.   C.M.
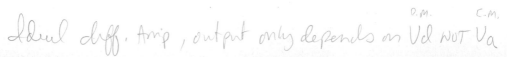
Ideal diff. Amp, output only depends on Vd NOT Va

amplifier $A_a = 0$. This condition cannot be realized in practice, since to make $A_a = 0$, R_e would have to be infinite. In order to measure the departure from the ideal, a quantity called the *common-mode rejection ratio* (CMRR) is used. It is defined as the ratio of the differential-mode gain to the common-mode gain

$$\text{CMRR} = \frac{A_d}{A_a} \tag{7.2-2a}$$

When (7.2-1c) and (7.2-1d) are used, the CMRR becomes

$$\text{CMRR} = \frac{2R_e + h_{ib} + R_b/h_{fe}}{2(h_{ib} + R_b/h_{fe})} \tag{7.2-2b}$$

If, as is often true in practice, $2R_e \gg h_{ib} + R_b/h_{fe}$, then

$$\text{CMRR} \approx \frac{R_e}{h_{ib} + R_b/h_{fe}} \qquad \text{Approximation} \tag{7.2-2c}$$
$$\text{if } 2Re \gg hib + Rb/hfe$$

To illustrate the usefulness of the CMRR consider that $v_2 = 10.5$ mV and $v_1 = 9.5$ mV. Then $v_a = 10$ mV and $v_d = 1$ mV, so that $v_a/v_d = 10$. The output voltage v_{o1} can be written, from (7.2-1a),

$$v_{o1} = A_d v_d \left(1 - \frac{A_a v_a}{A_d v_d}\right) = A_d v_d \left(1 - \frac{v_a/v_d}{\text{CMRR}}\right) \tag{7.2-3}$$

Thus, for v_{o1} be proportional to v_d the CMRR should be much greater than 10. In general, we see that the CMRR should be chosen so that

$$\text{CMRR} \gg \frac{v_a}{v_d} \tag{7.2-4}$$

if the output voltage is to be proportional to v_d.

Example 7.2-1 In Example 7.1-1 the common-mode voltage is 1 mV. (a) Find the common-mode rejection ratio. Assume

$$h_{fe1} = h_{fe2} = 100$$

(b) Find the differential-mode signal for which the differential output is at least 100 times the common-mode output.

SOLUTION (a) To calculate A_a and A_d we must first calculate $h_{ib} = V_T/I_{EQ1} = V_T/I_{EQ2}$. Referring to Fig. 7.1-3 and neglecting the current variation produced by the common-mode voltage v_a, we have (since $R_b = 0$)

remember $Rb = 0$
for Example 7.1-1

$$I_{EQ} = \frac{V_{EE} - 0.7}{2R_e} = \frac{9.3}{1800} = 5.17 \text{ mA}$$

Hence

$$h_{ib} = \frac{V_T}{I_{EQ}} = \frac{25 \times 10^{-3}}{5.17 \times 10^{-3}} \approx 4.8 \ \Omega$$

From (7.2-1d)

$$A_a = \frac{200}{(2)(900) + 4.8} \approx 0.11$$

and from (7.2-1c)

$$A_d = \frac{200}{(2)(4.8)} \approx 20.8$$

Thus
$$v_{o1} = +20.8v_d - 0.11v_a$$

The common-mode rejection ratio is

$$\text{CMRR} = \frac{A_d}{A_a} = 189 \approx 45 \text{ dB}$$

Commercially available difference amplifiers (op-amps) have CMRR ratings which are typically 80 to 100 dB.

For good differential-amplifier operation

$$v_d \gg \frac{v_a}{\text{CMRR}} = \frac{v_a}{189}$$

(b) If $v_a = 1$ mV and v_d is to be at least $100v_a$, we must have

$$v_d \geq \tfrac{100}{189} \approx 0.53 \text{ mV}$$

If v_d is less than 0.53 mV, the departure from a pure difference signal in the output will be greater than 1 percent. ///

7.3 DIFFERENCE AMPLIFIER WITH CONSTANT CURRENT SOURCE

In order to maximize the CMRR consider (7.2-2c). This equation shows that once a transistor has been chosen, our only control over the CMRR (assuming $R_b = 0$) is via the ratio R_e/h_{ib}. Since $h_{ib} = V_T/I_{EQ}$, (7.2-2c) can be written

$$\text{CMRR} = \frac{R_e}{V_T/I_{EQ} + R_b/h_{fe}} \tag{7.3-1a}$$

If R_b/h_{fe} is small, we shall have

$$\text{CMRR} < \frac{R_e I_{EQ}}{V_T} \tag{7.3-1b}$$

Thus, the CMRR can be increased only by increasing $R_e I_{EQ}$, the voltage drop across R_e. This process is limited by power dissipation in R_e, available power-supply voltage, etc.

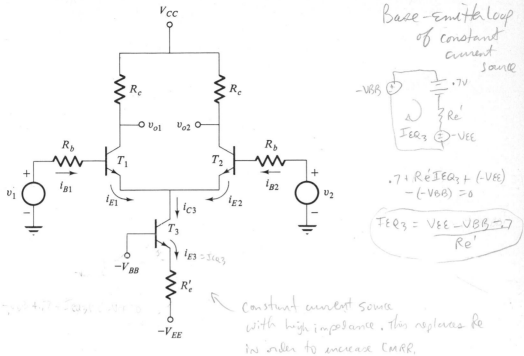

Handwritten annotations:

Base-Emitter loop of constant current source

$-V_{BB}$ $.7V$

R_e'

I_{EQ3} $-V_{EE}$

$.7 + R_e' I_{EQ3} + (-V_{EE}) - (-V_{BB}) = 0$

$I_{EQ3} = \dfrac{V_{EE} - V_{BB} - .7}{R_e'}$

Constant current source with high impedance, this replaces R_e in order to increase CMRR.

Figure 7.3-1 Difference amplifier with constant-current source.

Practical difference amplifiers often use a constant-current source in place of R_e as shown in Fig. 7.3-1. In this circuit transistor T_3 acts as the constant-current source which provides the required quiescent current and a very high impedance $1/h_{ob3}$ which is seen looking into the collector. It is this impedance which replaces R_e in (7.2-2c). For example, $1/h_{ob3}$ is typically 500 kΩ, and $I_{CQ3} \approx 0.5$ mA. In contrast, consider the basic circuit of Fig. 7.1-1 with $R_e = 500$ kΩ and $I_E = 0.5$ mA. Then V_{EE} would have to be 250 V, and the power dissipated in R_e would be 125 mW, both unreasonable figures for an IC.

Quiescent operation The quiescent current I_{CQ3} supplied by T_3 is

$B-E$ $100\,\rho$ in T_3

$$I_{CQ3} \approx \frac{V_{EE} - V_{BB} - 0.7}{R_e'} \tag{7.3-2}$$

and is constant as long as T_3 does not saturate. Referring to Fig. 2.2-7, we see that to remain in the linear region requires that

$$v_{CE} > V_T \left[2.2 + \ln \left(\frac{h_{fe}}{h_{fc}} \right) \right]$$

For a typical transistor, $h_{fe} = 100$ and $h_{fc} = 0.01$ so that $v_{CE} > 285$ mV. In order to ensure linear operation we allow a safety margin and set $V_{CEQ3} \geq 0.35$ V.

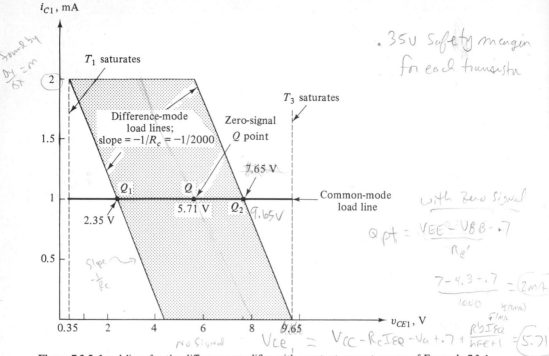

Figure 7.3-2 Load lines for the difference amplifier with constant-current source of Example 7.3-1.

To obtain the load lines that describe the circuit operation (as in Fig. 7.1-5) we first note that the common-mode load line for T_1 or T_2 is a horizontal line since with $v_1 = v_2 = v_a$, $I_{EQ1} = I_{EQ2} = \frac{1}{2}I_{CQ3}$ independent of the values of v_{CE1} and v_{CE2}. This result is also independent of the input voltages. However, even though i_{E1} ($= i_{E2}$) does not change with the common-mode input v_a, it is seen from the circuit that the collector-emitter voltage does and the change is found from the following relation, obtained using KVL:

$$v_{C1} = V_{CC} - R_c I_{EQ1} \qquad v_{E1} = v_a - R_b I_{BQ1} - 0.7$$

Then

$$v_{CE1} = v_{CE2} = v_{C1} - v_{E1} = V_{CC} - R_c I_{EQ1} - v_a + 0.7 + R_b I_{EQ1}/(h_{fe} + 1) \quad (7.3\text{-}3)$$

The positive excursion of the common-mode voltage must be restricted so that $v_{CE1} \geq 0.35$ V, and the negative excursion must not cause T_3 to saturate.

The difference-mode load line, which results when $v_2 = -v_1 = v_d/2$, has slope $-1/R_c$, as in Fig. 7.1-5. Both load lines are shown in Fig. 7.3-2, and their use in determining the maximum allowable input voltages will be illustrated in Example 7.3-1.

Example 7.3-1 In the difference amplifier shown in Fig. 7.3-1 $V_{CC} = 7$ V, $V_{EE} = 7$ V, $R_c = 2$ kΩ, $R_b = 1$ kΩ, $R'_e = 1$ kΩ, $h_{fe} + 1 = 100$, and $I_{EQ1} = I_{EQ2} =$

1 mA. Find (*a*) the required bias-supply voltage V_{BB}, (*b*) the maximum allowable difference voltage $V_{d,\,max}$, and (*c*) the maximum and minimum allowable common-mode voltage.

SOLUTION (*a*) To find V_{BB} we write KVL around the base-emitter loop of T_3

B-E loop
for T3

$$-V_{BB} = 0.7 + I_{CQ3}R'_e - V_{EE} = 0.7 + 2(1) - 7 = -4.3 \text{ V}$$

Hence $V_{BB} = 4.3$ V.

(*b*) Since the quiescent current is 1 mA and is constant, the common-mode load line is horizontal, as shown in Fig. 7.3-2. This clearly shows that the collector and hence the emitter currents can swing at most ± 1 mA before cutoff causes distortion. Since the collector resistor $R_c = 2$ kΩ, we see that the corresponding maximum change in output voltages is ± 2 V. Referring to (7.1-9), we see that when $v_a = 0$, the maximum current swing due to a difference voltage $v_2 = -v_1 = V_{d,\,max}/2$ is

$2000(1mA) = 2v$

$$I_{e1,\,max} = I_{e2,\,max} = \frac{V_{d,\,max}}{2[h_{ib} + (R_b/h_{fe} + 1)]}$$

Substituting the values $I_{e1,\,max} = I_{e2,\,max} = 1$ mA, $h_{ib} = 25$ Ω, and $R_b/(h_{fe} + 1) \approx 10$ Ω, we have

$$V_{d,\,max} = (2 \times 10^{-3})(25 + 10) = 70 \text{ mV}$$

(*c*) To determine the maximum allowable excursion of the common-mode voltage we first indicate in Fig. 7.3-2 the collector-emitter voltage v_{CE1} corresponding to the edge of saturation in both T_1 and T_3. As noted before, T_1 saturates when $v_{CE1} = 0.35$ V, and T_3 saturates when $v_{CE3} = 0.35$ V. Thus the left-hand difference-mode load-line terminates at $v_{CE1} = 0.35$ V, $i_{C1} = 2$ mA as shown. When T_3 saturates, KVL around the $T_1 - T_3$ loop indicates that

$$v_{CE1} = V_{CC} + V_{EE} - R_c I_{EQ1} - R'_e I_{EQ3} - 0.35$$
$$= 7 + 7 - 2 - 2 - 0.35 = 9.65 \text{ V}$$

This result is shown in Fig. 7.3-2 as the terminal point for the right-hand difference-mode load line. The shaded area between the two difference-mode load lines represents the linear operating region.

To obtain the maximum and minimum common-mode voltages refer again to Fig. 7.3-2, which shows that the common-mode voltage must be restricted to limit v_{CE1} to the range from 2.35 to 7.65 V. Appropriate values are found from

$$v_{CE1} = V_{CC} - R_c I_{EQ1} - v_a + 0.7 + \frac{R_b I_{EQ1}}{h_{fe} + 1} \qquad (7.3\text{-}3)$$

Substituting the values in this example, we get

$$v_{CE1} = 7 - 2 - v_a + 0.7 + 0.01$$
$$v_a = 5.71 - v_{CE1}$$

For $v_{CE1} = 2.35$ V we obtain the upper limit on v_a

$$V_{a,\,max} = 5.71 - 2.35 = 3.36\,\text{V}$$

For $v_{CE1} = 7.65$ V we obtain the lower limit

$$V_{a,\,min} = 5.71 - 7.65 = -1.94\,\text{V}$$

In conclusion, we have established that operation will lie in the shaded region of the collector characteristics of Fig. 7.3-2 as long as

$$-70\ \text{mV} < v_d < +70\ \text{mV}$$

and $\qquad\qquad -1.94\ \text{V} < v_a < +3.36\,\text{V}$ ///

Small-signal operation The small-signal equivalent circuit for the amplifier shown in Fig. 7.3-1 is identical to that shown in Fig. 7.1-6. The output voltage is given by (7.1-11). The important difference between the two-transistor amplifier shown in Fig. 7.1-1 and the three-transistor amplifier shown in Fig. 7.1-6 is the value of R_e and hence the CMRR. As noted previously, when using a constant-current transistor, $R_e \approx 1/h_{ob}$, which is much larger than the value of R_e attainable using a passive resistance. This leads to a much larger CMRR.

Input impedance To find the input impedance of the difference amplifier as seen by the input voltage sources v_1 and v_2 we must reflect all impedances into the base circuits rather than into the emitter circuits (see Fig. 7.1-6a). The result of this reflection is shown in Fig. 7.3-3, and readers should be certain they understand the difference between this circuit and that shown in Fig. 7.1-6a. The input impedance R_i is measured between the bases of T_1 and T_2, and we see that

$$R_i = 2h_{ie} \tag{7.3-4}$$

If $I_{EQ1} = I_{EQ2} = 0.1$ mA and $h_{fe} + 1 = 100$, then $h_{ie} = (h_{fe} + 1)V_T/I_{EQ} = 25$ kΩ and $R_i = 50$ kΩ. If a higher input impedance is required, one can decrease I_{EQ1} (and I_{EQ2}), replace the BJTs by FETs (Sec. 7.5), or use a Darlington (compound) amplifier (Sec. 7.6).

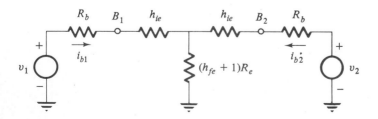

Figure 7.3-3 Small-signal equivalent circuit seen looking into the bases.

7.4 DIFFERENCE AMPLIFIER WITH EMITTER RESISTORS FOR BALANCE

The balance control When T_1 and T_2 have different characteristics, a variable resistor R_v is often connected between the emitters of T_1 and T_2 to serve as a balancing control, as shown in Fig. 7.4-1. The slider of the variable resistance is adjusted so that $I_{EQ1} = I_{EQ2}$. A small value of R_v, typically 100 Ω, is usually sufficient to compensate for large differences between h_{fe1} and h_{fe2}.

To evaluate the effect of the balance control note that KVL, applied from the arm of the control to ground, must yield the same voltage when taken around the base-emitter loop of either T_1 or T_2. Thus, with $v_1 = v_2 = 0$

$$\left(\frac{R_b}{h_{fe1}} + R_1\right)I_{EQ1} + V_{BE1} = \left(\frac{R_b}{h_{fe2}} + R_2\right)I_{EQ2} + V_{BE2} \qquad (7.4\text{-}1)$$

If $V_{BE1} = V_{BE2}$, the condition which ensures that the emitter currents of T_1 and T_2 are the same is obtained by setting $I_{EQ1} = I_{EQ2}$ in (7.4-1). This yields

$$R_2 - R_1 = R_b\left(\frac{1}{h_{fe1}} - \frac{1}{h_{fe2}}\right) \qquad (7.4\text{-}2)$$

and since

$$R_2 + R_1 = R_v = \text{const} \qquad (7.4\text{-}3)$$

balance is obtained when

$$R_2 = \frac{R_v}{2} + \frac{R_b}{2}\left(\frac{1}{h_{fe1}} - \frac{1}{h_{fe2}}\right) \qquad (7.4\text{-}4a)$$

and

$$R_1 = \frac{R_v}{2} - \frac{R_b}{2}\left(\frac{1}{h_{fe1}} - \frac{1}{h_{fe2}}\right) \qquad (7.4\text{-}4b)$$

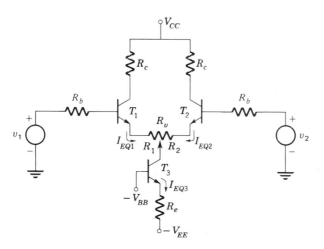

Figure 7.4-1 Difference amplifier with balance control R_v.

For example, if $h_{fe1} = 50$, $h_{fe2} = 150$, and $R_b = 1.5$ kΩ, then R_v must be at least equal to 20 Ω. For this value of R_v, we find $R_2 = 20$ Ω and $R_1 = 0$.

Inclusion of R_v results in symmetrical operation but also causes a loss in current gain. If R_1 and R_2 are adjusted as in (7.4-4) and the effective emitter resistance is very large, so that

$$(R_e)_{\text{eff}} \approx \frac{1}{h_{ob3}} \gg h_{ib1} + \frac{R_b}{h_{fe1}} + R_1 \tag{7.4-5}$$

then the differential-mode gain of (7.2-1c) becomes

$$A_d = \frac{R_c}{R_b(1/h_{fe1} + 1/h_{fe2}) + 2h_{ib} + R_v} \tag{7.4-6}$$

This should be compared with (7.2-1c), which gives the gain for an ideal circuit.

Example 7.4-1 The difference amplifier shown in Fig. 7.4-2 makes use of an IC package (shown inside the dashed lines). Find R_e and V_{EE} so that the voltage common-mode rejection ratio is 100 (40 dB). The signal sources v_1 and v_2 have internal resistances of 1 kΩ each, and the transistors have $h_{fe} = 100$. Also $V_{CC} = 10$ V and $R_c = 2$ kΩ.

SOLUTION The difference-amplifier CMRR is obtained from (7.2-2). However, because of the two 50 Ω resistors, the expressions are altered slightly. The differential-mode gain is given by (7.4-6), and the common-mode gain [see (7.2-1d)] is

$$A_a = \frac{R_c}{2R_e + h_{ib} + R_b/(h_{fe} + 1) + R_1} \tag{7.4-7}$$

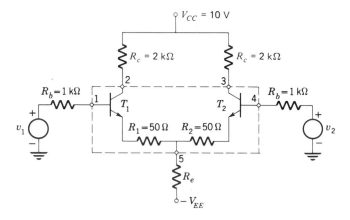

Figure 7.4-2 Circuit for Example 7.4-1.

Assuming that $2R_e \gg h_{ib} + R_b/(h_{fe} + 1) + R_1$, we find that the CMRR is

$$\text{CMRR} = \frac{A_d}{A_a} \approx \frac{2R_e}{(1000)(\frac{2}{100}) + 2h_{ib} + 100} = \frac{R_e}{60 + h_{ib}}$$

In order to obtain a CMRR = 100, we must have

$$R_e \geq (100)(60 + h_{ib})$$

If we set $I_{EQ1} = I_{EQ2} = 1$ mA, then $h_{ib} = 25\ \Omega$ and a value of $R_e = 10\ \text{k}\Omega$ satisfies the inequality. To find V_{EE} we write KVL around the base-emitter loop in Fig. 7.4-2. Letting $v_1 = v_2 = 0$, we have

$$V_{EE} = \frac{R_b}{h_{fe} + 1} I_{EQ1} + V_{BE1} + R_1 I_{EQ1} + 2R_e I_{EQ1}$$

$$= 10 \times 10^{-3} + 0.7 + 50 \times 10^{-3} + (20 \times 10^{+3}) \times 10^{-3} \approx 20.8\ \text{V} \quad ///$$

7.5 DIFFERENCE AMPLIFIER USING FETS

A very high input impedance can be realized by substituting FETs for T_1 and T_2 in Fig. 7.3-1. A difference amplifier using this technique is shown in Fig. 7.5-1.

To show that this circuit operates as a difference amplifier we reflect all components into the source circuits. The result is shown in Fig. 7.5-2. The resistor $1/h_{ob3}$ is the impedance seen looking into the collector of T_3. Figure 7.5-2 is identical in form to Fig. 7.1-6a, and therefore similar equations result, indicating that the circuit operates as a difference amplifier with a common-mode effect present.

If, for simplicity, we neglect the common-mode voltage and let $v_2 = -v_1 = v_d/2$, we find

$$i_{ds2} = -i_{ds1} = \frac{[\mu/(\mu + 1)](v_2 - v_1)}{2[r_{ds}/(\mu + 1) + R_d/(\mu + 1)]} \tag{7.5-1}$$

Therefore, since $v_{o2} = -i_{ds2} R_d$ and $v_{o1} = -i_{ds1} R_d$ (see Fig. 7.5-1),

$$v_{o2} = -v_{o1} = -\frac{\mu(v_2 - v_1)R_d}{2(r_{ds} + R_d)} \tag{7.5-2a}$$

From this relation we find the difference-mode gain

$$A_d = \frac{\mu R_d}{2(r_{ds} + R_d)} \tag{7.5-2b}$$

Impedances The input impedance of the FET difference amplifier is almost infinite since the current into the gate of a FET is often less than 1 pA.

The output impedance of the difference amplifier, as seen looking into the drain of T_2 (Fig. 7.5-1), is found by first reflecting the components R_d and r_{ds} of T_1

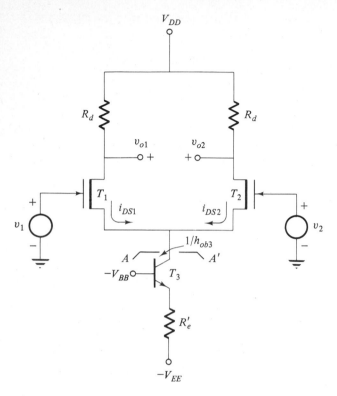

Figure 7.5-1 Difference amplifier using FETs.

into the source circuit of T_1. This manipulation then puts these components in the source circuit of T_2, as shown in Fig. 7.5-3a. Here we have omitted the collector-base impedance $1/h_{ob3}$ of T_3 to simplify the circuit.

To complete the analysis we reflect all components into the drain circuit of T_2, as shown in Fig. 7.5-3b. The output voltage v_{o2} can be obtained from this circuit, and the result is as given by (7.5-2a). The output impedance R_o is

$$R_o = R_d \| (R_d + 2r_{ds}) \tag{7.5-3}$$

Figure 7.5-2 Equivalent circuit of FET difference amplifier with all components reflected into the source circuits.

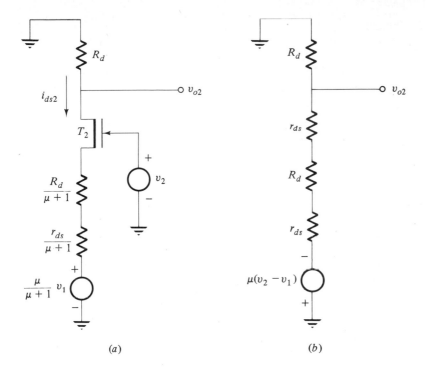

(a) $\qquad\qquad\qquad\qquad$ (b)

Figure 7.5-3 FET difference amplifier: (a) equivalent circuit with all components of T_1 reflected into its source; $1/h_{ob3}$ is neglected; (b) total equivalent circuit seen looking into the drain of T_2.

Example 7.5-1 The difference amplifier shown in Fig. 7.5-1 has the component values $R_d = 5\,\text{k}\Omega$, $r_{ds} = 100\,\text{k}\Omega$, $g_m = 5\,\text{mS}$, and $h_{ob3} = 10^{-5}\,\text{S}$. Find (a) the CMRR, (b) the output voltage $v_{o2} - v_{o1}$, and (c) the output impedance seen looking into the drain of T_2.

SOLUTION (a) Referring to (7.2-2b) and identifying $h_{ib} + R_b/(h_{fe} + 1)$ with $(r_{ds} + R_d)/(\mu + 1)$ and R_e with $1/h_{ob3}$, we have

$$\text{CMRR} = \frac{2/h_{ob3} + (r_{ds} + R_d)/(\mu + 1)}{2(r_{ds} + R_d)/(\mu + 1)} \approx \frac{\mu + 1}{h_{ob3}(r_{ds} + R_d)}$$

$$\approx \frac{500}{10^{-5} \times (105 \times 10^3)} \approx 500 = 57 \text{ dB}$$

(b) The output voltage $v_{o2} - v_{o1}$ is [see Eq. (7.5-2)]

$$v_{o2} - v_{o1} = \frac{-\mu}{1 + r_{ds}/R_d}(v_2 - v_1) = \frac{-500}{1 + \frac{100}{5}}(v_2 - v_1) \approx -25(v_2 - v_1)$$

Note that the gain

$$\frac{v_{o2} - v_{o1}}{v_2 - v_1} = 2\,\frac{v_{o2}}{v_2 - v_1} = -2\,\frac{v_{o1}}{v_2 - v_1}$$

(c) The output impedance is [Eq. (7.5-3)]

$$R_o = R_d \| (R_d + 2r_{ds}) \approx R_d = 5\ \text{k}\Omega \qquad\qquad ///$$

7.6 THE DARLINGTON AMPLIFIER

The *Darlington amplifier* (sometimes called a *compound amplifier*) is shown in Fig. 7.6-1. This configuration is used to provide increased input impedance and very high current gain, effectively $h_{fe1}h_{fe2}$. In this circuit the emitter of T_4 is connected directly to the base of T_1. Usually the two collectors are also connected together as shown, but this is not necessary.

The current gain can be found (assuming identical transistors) by noting that

$$A_i = \frac{i_o}{i_{B4}} = \frac{i_{C1} + i_{C4}}{i_{B4}} = \frac{\alpha i_{E1}}{i_{B4}} + \frac{\alpha i_{E4}}{i_{B4}} \qquad (7.6\text{-}1a)$$

Next we find

$$\frac{i_{E1}}{i_{B4}} = \frac{i_{E1}}{i_{B1}}\frac{i_{B1}}{i_{E4}}\frac{i_{E4}}{i_{B4}} = (h_{fe} + 1)(1)(h_{fe} + 1) = (h_{fe} + 1)^2 \qquad (7.6\text{-}1b)$$

Using this result in (7.6-1a), we have $\boxed{A_i = h_{fe}^2}$

$$A_i = \alpha(h_{fe} + 1)^2 + \alpha(h_{fe} + 1) = \alpha(h_{fe} + 1)(h_{fe} + 2) \approx h_{fe}^2 \qquad (7.6\text{-}1c)$$

This result would be expected from inspection of the circuit because the emitter current of T_4 becomes the base current of T_1.

Input impedance The input impedance seen looking between the base of T_4 and the emitter of T_1 is most easily found by reflecting the base-emitter impedance h_{ie1}

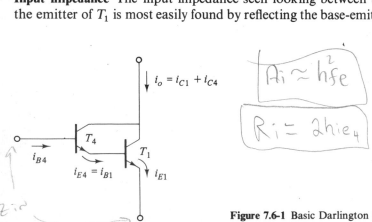

$i_o = i_{C1} + i_{C4}$

$A_i \approx h_{fe}^2$

$R_i = 2h_{ie_4}$

T_4

i_{B4}

$i_{E4} = i_{B1}$

T_1

i_{E1}

Figure 7.6-1 Basic Darlington (compound) amplifier.

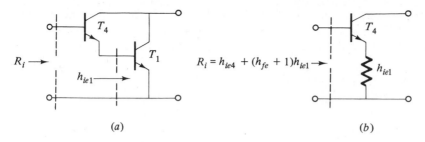

(a)   (b)

Figure 7.6-2 The use of reflection to determine the input impedance of the basic Darlington amplifier: (a) the Darlington amplifier; (b) T_1 replaced by h_{ie1}.

of T_1 from the emitter circuit of T_4 into the base circuit of T_4, as shown in Fig. 7.6-2. The result is

$$R_i = h_{ie4} + (h_{fe} + 1)h_{ie1} \qquad (7.6\text{-}2)$$

However, $h_{ie} = (h_{fe} + 1)V_T/I_{EQ}$, so that

$$R_i = \frac{(h_{fe} + 1)V_T}{I_{EQ4}} + \frac{(h_{fe} + 1)^2 V_T}{I_{EQ1}} \qquad (7.6\text{-}3)$$

where we have assumed that $h_{fe1} = h_{fe4}$. From Fig. 7.6-1 we see that

$$\frac{I_{EQ1}}{h_{fe} + 1} = I_{EQ4} \qquad (7.6\text{-}4)$$

Combining (7.6-4) with (7.6-3) yields R_i in terms of either h_{ie1} or h_{ie4}

$$R_i = 2(h_{fe} + 1)h_{ie1} = 2h_{ie4} \qquad (7.6\text{-}5)$$

Thus, we have succeeded in multiplying the input impedance of T_1 by $2(h_{fe} + 1)$.

In certain applications it is desirable to set the current in T_4 independently of T_1. This can be accomplished using the circuit of Fig. 7.6-3. Here a base-

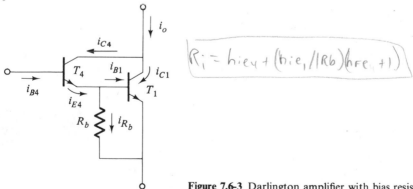

$$R_i = h_{ie4} + (h_{ie1}//R_b)(h_{FE} + 1)$$

Figure 7.6-3 Darlington amplifier with bias resistor.

biasing resistor is placed across the base-emitter terminals of T_1. The emitter current in T_4 can now be set by adjusting R_b since

$$I_{EQ4} = I_{BQ1} + \frac{0.7}{R_b} \tag{7.6-6}$$

For example, we can set $I_{EQ4} = I_{EQ1}$ or any other value greater than I_{BQ1}.

Comparing Fig. 7.6-3 with Fig. 7.6-2, we see that R_b is in parallel with h_{ie1}, so that the input impedance of this circuit is

$$R_i = h_{ie4} + (h_{fe} + 1)(R_b \| h_{ie1}) \tag{7.6-7}$$

To obtain an expression for the small-signal output current i_o we note that $i_o = i_{c1} + i_{c4}$, $i_{c1} = h_{fe} i_{b1}$, and $i_{c4} = h_{fe} i_{b4}$, where $i_{b1} = i_{e4} R_b / (R_b + h_{ie1})$. Combining these expressions and letting $i_{e4} = (h_{fe} + 1)i_{b4}$ yields

$$i_o = i_{c4} + i_{c1} = h_{fe} i_{b4} + h_{fe}(h_{fe} + 1)i_{b4} \frac{R_b}{R_b + h_{ie1}} \tag{7.6-8a}$$

The current gain is therefore

$$A_i = \frac{i_o}{i_{b4}} = h_{fe}^2 \frac{R_b}{R_b + h_{ie1}} + h_{fe}\left(1 + \frac{R_b}{R_b + h_{ie1}}\right) \tag{7.6-8b}$$

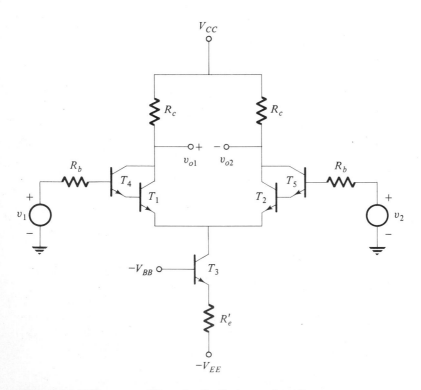

Figure 7.6-4 Difference amplifier using Darlington configuration.

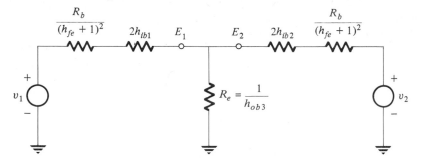

Figure 7.6-5 Equivalent circuit for Fig. 7.6-4 as seen looking into the emitters of T_1 and T_2.

The current gain given by (7.6-8b) is always less than the value obtained using the basic Darlington-amplifier configuration of Fig. 7.6-1.

Difference amplifier using the Darlington input circuit Figure 7.6-4 shows a differ-ence amplifier with Darlington input circuits. Transistors T_4 and T_5 significantly increase the differential input impedance without increasing the current flowing in T_1, T_2, and therefore T_3.

The reader should verify that the differential input impedance of this amplifier as seen between the bases of T_4 and T_5 is [refer to (7.6-5)]

$$R_i = 4(h_{fe} + 1)h_{ie1} = 4(h_{fe} + 1)^2 h_{ib1} \qquad (7.6\text{-}9)$$

Thus, if $h_{fe} + 1 = 100$ and $I_{EQ1} = 1$ mA, we find that $R_i = 1$ MΩ.

To determine the small-signal gain of the difference amplifier we reflect all components into the emitters of T_1 and T_2 as shown in Fig. 7.6-5, where R_e is the impedance seen looking into the collector of T_3. Note the similarity to Fig. 7.1-6. When we make use of this similarity, the CMRR given by (7.2-2c) becomes

$$\text{CMRR} = \frac{R_e}{2h_{ib} + R_b/(h_{fe} + 1)^2} \qquad (7.6\text{-}10)$$

Neglecting the common-mode gain, we find the output voltages

$$v_{o2} = -v_{o1} = -\frac{R_c}{2[R_b/(h_{fe} + 1)^2] + 2h_{ib}}(v_2 - v_1) \qquad (7.6\text{-}11)$$

Example 7.6-1 The Darlington differential amplifier shown in Fig. 7.6-6 is available as a self-contained integrated circuit. Determine the quiescent oper-ating conditions. Assume $h_{fe} = 100$ for all transistors.

SOLUTION Transistors T_1, T_2, and T_3 form a difference amplifier with a con-stant emitter-current supply. The quiescent current supplied by T_3 is cal-culated from the circuit of Fig. 7.6-7a. Neglecting I_{B3}, we have

$$V_{B3} = (-6)\frac{2.9}{2.9 + 1.3} = -4.14 \text{ V}$$

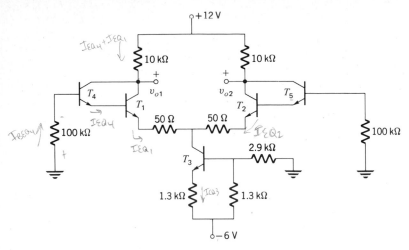

Figure 7.6-6 Darlington differential amplifier.

$$V_{B3} = \frac{2.4K \ (-6v)}{2.9K + 1.3K} \approx -4.14V$$

The emitter voltage V_{E3} is

$$V_{E3} = V_{B3} - 0.7 = -4.84 \text{ V}$$

Thus

$$I_{CQ3} \approx I_{EQ3} = \frac{6 - 4.84}{1.3 \text{ k}\Omega} \approx 0.9 \text{ mA}$$

$$\frac{-4.84 - (-6)}{1.3K} = .9mA$$

It is assumed that this current divides evenly between T_1 and T_2, so that

$$I_{EQ1} = I_{EQ2} \approx 0.45 \text{ mA}$$

To determine the quiescent operating point of T_1 and T_4 (and T_2 and T_5) consider the circuit of Fig. 7.6-7b. From the figure

$$I_{EQ4} \approx \frac{I_{EQ1}}{h_{fe}} = \frac{0.45 \text{ mA}}{100} = 4.5 \ \mu\text{A}$$

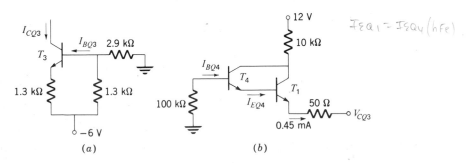

$I_{EQ1} = I_{EQ4}(h_{fe})$

Figure 7.6-7 Circuits for Example 7.6-1: (a) constant-current source for Darlington differential amplifier of Fig. 7.6-6; (b) portion of Darlington amplifier of Example 7.6-1.

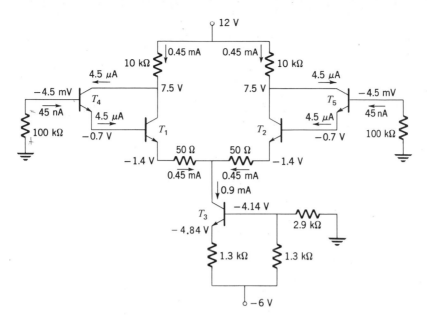

Figure 7.6-8 Compound difference amplifier with operating conditions.

and
$$I_{BQ4} \approx \frac{I_{EQ4}}{h_{fe}} = \frac{4.5 \ \mu A}{100} = 45 \ nA$$

Next the collector voltages can be found:

$$V_{CQ4} = V_{CQ1} \approx 12 - 10^4 \times [(0.45 \times 10^{-3}) + (4.5 \times 10^{-6})] \approx 7.5 \ V$$

$$V_{EQ4} \approx -10^5 \times (45 \times 10^{-9}) - 0.7 \approx -0.7 \ V$$

and
$$V_{EQ1} \approx -1.4 \ V$$

Hence
$$V_{CQ3} \approx -1.4 - (50)(0.45 \times 10^{-3}) \approx -1.4 \ V$$

The circuit with all currents and voltages is shown in Fig. 7.6-8.　　　///

7.7 THE CASCODE AMPLIFIER

The cascode amplifier shown in Fig. 7.7-1 is used in IC amplifiers as a dc level shifter when the voltage of interest consists of a small-signal ac component v_i and a fixed dc level V_i. The output voltage v_L from the level shifter is to have a dc level different from V_i; typically this final output level is to be 0 V. Resistor R_i represents the output impedance of the amplifier producing the level-shifter input voltage $v_i + V_i$.

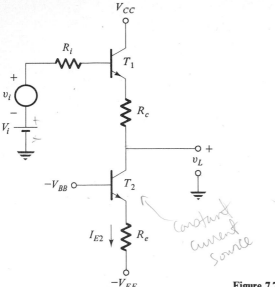

Figure 7.7-1 Cascode amplifier.

Level shifting can be accomplished using a simple resistive-voltage-divider network; i.e., we can in principle, remove T_1 and T_2 and their associated components and replace them by a single resistor connected to $-V_{EE}$, as shown in Fig. 7.7-2. Then, by appropriate adjustment of R_e and V_{EE} we can set the dc value of v_L to any desired level. For example, if the dc value of v_L is to be 0 V, we must have

$$\frac{V_{EE}}{R_e} = \frac{V_i}{R_i} \qquad (7.7\text{-}1)$$

However, this type of resistive divider results in an ac output voltage which is *less than* v_i, when V_{EE} and R_e are chosen according to (7.7-1). That is,

$$v_L = \frac{R_e}{R_e + R_i} v_i \qquad (7.7\text{-}2)$$

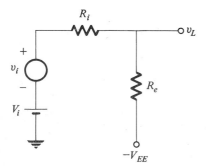

Figure 7.7-2 An elementary resistive voltage-divider network.

The cascode amplifier shown in Fig. 7.7-1 can shift the dc level without attenuating the ac signal, a feature which makes it most useful in integrated circuits.

DC analysis In the cascode amplifier of Fig. 7.7-1, T_1 is an emitter follower and T_2 acts as a constant current source. Thus, T_2 sets the dc current I_{E2} which flows through T_1 and R_c. The dc component of the output voltage is therefore

$$V_L = V_i - V_B - .7 - R_c I_{E2}$$

$$\text{DC values} \quad \boxed{V_L = V_i - \frac{R_i I_{E2}}{h_{fe} + 1} - 0.7 - R_c I_{E2}} \quad V_B \text{ usually small} \quad (7.7\text{-}3)$$

Making use of this relation, we can readily set the dc output voltage to any desired value by adjusting the voltage drop $R_c I_{E2}$. For example, if we neglect the small voltage across R_i, and if V_L is to be set to 0 V, we must have

$$R_c I_{E2} \approx V_i - 0.7 \tag{7.7-4}$$

Small-signal analysis To determine the small-signal component of the output voltage we draw the small-signal equivalent circuit of T_1 as seen looking into its emitter. The result is shown in Fig. 7.7-3 with T_2 replaced by the impedance $1/h_{ob2}$ seen looking into its collector. The small-signal component of the output voltage v_L is then

$$v_L \approx \frac{v_i}{1 + \dfrac{R_i/h_{fe} + h_{ib} + R_c}{1/h_{ob2}}} \tag{7.7-5}$$

Typically $R_i = 1\ \text{k}\Omega$, $h_{ib} = 25\ \Omega$, $R_c = 5\ \text{k}\Omega$, and $1/h_{ob2} = 100\ \text{k}\Omega$. Since $1/h_{ob2}$ is so much larger than $R_i/h_{fe} + h_{ib} + R_c$, the load voltage $v_L \approx v_i$ and negligible attenuation of the signal has resulted from the shift in dc level.

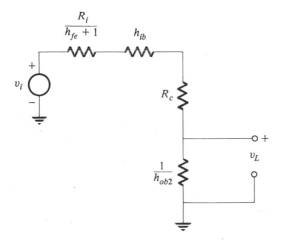

Figure 7.7-3 Equivalent circuit seen looking into the emitter of T_1.

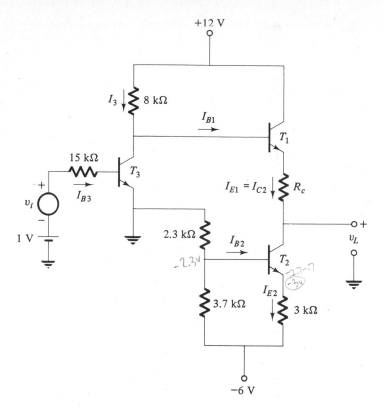

Figure 7.7-4 Cascode amplifier for Example 7.7-1.

Example 7.7-1 A cascode amplifier is shown in Fig. 7.7-4. Find the value of R_c which will set the dc component of the output voltage to 0 V. Assume $h_{fe} = 50$ for all transistors.

SOLUTION The calculation is begun by determining V_{B2} and then V_{E2} in order to calculate I_{E2}. Neglecting I_{B2}, we have

$$V_{B2} = -6\frac{2.3}{2.3 + 3.7} = -2.3 \text{ V}$$

and

$$V_{E2} = -2.3 - 0.7 = -3 \text{ V}$$

Hence

$$I_{E1} \approx I_{C2} \approx I_{E2} = \tfrac{3}{3000} = 1 \text{ mA}$$

Since the purpose of the circuit is to make $V_L = 0$ V,

$$V_{C3} = 0.7 + R_c I_{E2} = 0.7 + R_c \times 10^{-3}$$

The value of R_c can be determined once V_{C3} is found as follows. The base current I_{B3} is

$$I_{B3} = \frac{1 - 0.7}{15 \times 10^3} = 0.02 \text{ mA}$$

Hence the current $I_3 \approx I_{C3}$ is

$$I_3 \approx I_{C3} = h_{FE}I_{B3} = 50(0.02) \text{ mA} = 1 \text{ mA}$$

The collector voltage V_{C3} is therefore 4 V. Resistor R_c can now be found from the relation

$$4 = 0.7 + R_c \times 10^{-3}$$

yielding $R_c = 3.3 \text{ k}\Omega$. ///

7.8 THE OPERATIONAL AMPLIFIER

As noted previously, op-amps are characterized by differential input and very high gain, often greater than 10^5 (100 dB). A typical op-amp consists of four components, as illustrated in Fig. 7.8-1. The first is the difference amplifier, which may have a Darlington input or use FETs and may use a constant-current source. This is followed by a stage of high-gain linear amplification, often another difference amplifier. If the dc voltage present at the output of the high-gain amplifier is not 0 V when $v_1 = v_2 = 0$ V, a level-shifting circuit such as the cascode amplifier is employed. The last stage is an output amplifier, usually a complementary-symmetry or push-pull configuration.

The entire amplifier operates linearly so that

$$v_o = -A_d(v_1 - v_2) - \frac{A_a(v_1 + v_2)}{2} \tag{7.8-1}$$

as expected when using a difference amplifier. We are here assuming that A_d and A_a are positive. A typical value for A_d is 10^5 (100 dB), while a typical value for A_a is 1. Thus the CMRR is typically 10^5, or 100 dB. Referring to (7.8-1), we see that if

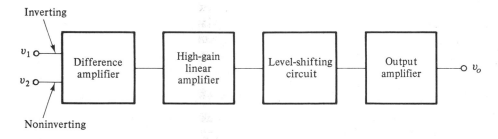

Figure 7.8-1 Typical op-amp configuration.

$v_2 = 0$, the output voltage v_o is opposite in phase to the input voltage v_1, while if $v_1 = 0$, the output voltage is in phase with v_2. It is therefore standard practice to refer to v_2 as the *noninverting* input and to v_1 as the *inverting* input.

The op-amp is usually designed so that when both input voltages are zero, the output voltage is zero over a specified range of temperature and supply-voltage variation.

In order to ensure a minimal change in output voltage due to a change in temperature, diode biasing is often used for temperature compensation. Typical output-voltage changes are of the order of 1 μV/°C. Similarly, extremely clever design techniques are employed to ensure that the effect of changes in V_{CC} and/or V_{EE} on the output voltage are minimal. The *power-supply rejection ratio* of an op-amp is typically 10^5 (100 dB); that is, a 1-V change in supply voltage causes the output voltage to change by 10 μV.

7.9 EXAMPLE OF A COMPLETE OP-AMP

A typical op-amp is a very sophisticated circuit often utilizing more than 20 transistors. In this section we analyze an IC op-amp which contains only eight transistors and yet has many of the features of more complicated op-amps. The schematic of this op-amp, which is similar to that of the National LH0005, is shown in Fig. 7.9-1.

The input stage is a Darlington difference amplifier. T_3 and T_1 constitute a Darlington configuration even though the collector of T_3 is not tied to the collector of T_1. (The essential characteristic of the Darlington is that the emitter of T_3 connect directly to the base of T_1.) The input difference amplifier does not employ a constant-current source. However, it is shown below that a high CMRR is still obtained by using both outputs v_{C1} and v_{C2} as inputs to a second difference amplifier formed by transistors T_5 and T_6. In the diagram T_5 and T_6 appear to be upside down since they are *pnp*† rather than *npn* transistors. This second stage provides high gain and a good CMRR. It also performs the required level shifting, which is accomplished by proper choice of R_4 and the number of diodes, so that $v_o = 0$ when $v_1 = v_2 = 0$. Finally the output stage is a complementary-symmetry amplifier (see Sec. 5.4), in which we have omitted the circuits used to prevent crossover distortion in order to simplify the ensuing circuit analysis.

DC analysis and design We first set $v_1 = v_2 = 0$ and determine the values of R_1 to R_4 which will ensure that $v_o = 0$ independent of temperature and supply-voltage variations. To begin, we assume that all base-emitter voltages and diode drops are equal. For simplicity, we shall call all these voltages V_D, which has the usual numerical value 0.7 V.

---

† In IC design *pnp* transistors are fabricated differently from *npn* transistors (see Sec. 16.1). As a result they have a low β, typically less than 10. Hence *pnp* transistors are rarely used as common-emitter amplifiers in ICs.

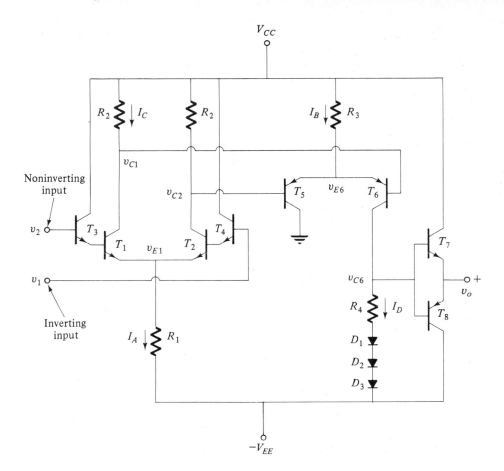

Figure 7.9 1 An IC op amp.

When $v_1 = v_2 = 0$, $v_{E1} = -2V_D$. Hence

$$I_A = \frac{V_{EE} - 2V_D}{R_1} \tag{7.9-1}$$

We now neglect the base currents I_{B1}, I_{B6}, I_{B2}, and I_{B5} in comparison with the collector currents of T_1 and T_2. Then, from the symmetry of the circuit $I_C = I_A/2$, and the voltage v_{C1} is

$$v_{C1} = V_{CC} - \frac{R_2}{2R_1}(V_{EE} - 2V_D) \tag{7.9-2}$$

This voltage is applied to the base of T_6. Since the emitter voltage of T_6 is $v_{E6} = V_D + v_{B6} = V_D + v_{C1}$, we have

$$v_{E6} = V_D + V_{CC} - \frac{R_2}{2R_1}(V_{EE} - 2V_D) \tag{7.9-3}$$

The current $I_B = (V_{CC} - v_{E6})/R_3$ is then

$$I_B = \frac{R_2}{2R_1 R_3}(V_{EE} - 2V_D) - \frac{V_D}{R_3} \tag{7.9-4}$$

Since the emitter-base voltages of T_5 and T_6 are the same, the current I_B will divide equally between transistors T_5 and T_6. Hence $I_D \approx I_B/2$. The voltage v_{C6} is therefore

$$v_{C6} = \frac{I_B}{2} R_4 + 3V_D - V_{EE}$$

$$= \frac{R_2}{2R_1} \frac{R_4}{2R_3}(V_{EE} - 2V_D) - \frac{R_4}{2R_3} V_D + 3V_D - V_{EE} \tag{7.9-5}$$

In order for v_o to be 0 V, we must have $v_{C6} = 0$ V. This condition yields

$$0 = V_{EE}\left[\frac{R_2}{2R_1}\left(\frac{R_4}{2R_3}\right) - 1\right] + V_D\left[3 - 2\left(\frac{R_2}{2R_1}\right)\frac{R_4}{2R_3} - \frac{R_4}{2R_3}\right] \tag{7.9-6}$$

In order to satisfy (7.9-6) for all values of V_{EE} and V_D we set their coefficients equal to 0. The result is

$$\left(\frac{R_2}{2R_1}\right)\left(\frac{R_4}{2R_3}\right) = 1 \tag{7.9-7a}$$

and

$$2\left(\frac{R_2}{2R_1}\right)\left(\frac{R_4}{2R_3}\right) + \frac{R_4}{2R_3} = 3 \tag{7.9-7b}$$

Combining (7.9-7a) and (7.9-7b) yields

$$\frac{R_4}{2R_3} = \frac{R_2}{2R_1} = 1 \tag{7.9-8}$$

For example, if we let $R_2 = R_4 = 10$ kΩ, then $R_1 = R_3 = 5$ kΩ.

The reader should note the full implication of prescribing R_1 to R_4 using (7.9-8). Using this relationship in design leads to an output voltage which is independent of V_{CC}, V_{EE}, and V_D and therefore independent of temperature changes, as well as supply-voltage changes. The output voltage depends only on the common-mode and difference-mode input signals.

Example 7.9-1 The op-amp shown in Fig. 7.9-1 has component values $R_1 = R_3 = 5$ kΩ and $R_2 = R_4 = 10$ kΩ. Find (a) the quiescent currents and voltages I_A, I_B, I_C, I_D, V_{C1}, V_{E1}, V_{E6}, and V_{C6} when $V_{CC} = V_{EE} = 12$ V and $V_1 = V_2 = 0$ V, (b) the maximum symmetrical peak common-mode voltage swing $V_{a, max}$ which can be accommodated if $V_{CC} = V_{EE} = 12$ V, (c) the value of V_{EE} (with $V_{CC} = 12$ V) which will maximize the allowable common-mode voltage swing.

SOLUTION (a) Since $V_1 = V_2 = 0$ V, $V_{E1} = -1.4$ V and [see (7.9-1)]

$$I_A = \frac{12 - 1.4}{5} = 2.12 \text{ mA}$$

Hence, $I_C \approx I_A/2 = 1.06$ mA and $V_{C1} = 12 - (1.06)(10) = 1.4$ V. Noting that $V_{E6} = V_{C1} + 0.7 = 2.1$ V, we find

$$I_B = \frac{12 - 2.1}{5} = 1.98 \text{ mA}$$

Since $I_D = I_B/2 = 0.99$ mA, we have

$$V_{C6} = 0.99(10) + 2.1 - 12 = 0 \text{ V}$$

as expected.

(b) The common-mode load-line equations for T_1 [see Eq. (7.1-5)] are

$$V_{CC} + V_{EE} = V_{CE1} + I_C(R_2 + 2R_1)$$

and

$$24 = V_{CE1} + I_C(20 \times 10^3) \tag{7.9-9a}$$

and [see Eq. (7.1-6b)] the Q point for different values of v_a can be found from

$$I_C = \frac{v_a + V_{EE} - 2V_D}{2R_1} = \frac{v_a + 10.6}{10^4} \tag{7.9-9b}$$

Equation (7.9-9a) is plotted in Fig. 7.9-2a for $V_{CC} = V_{FF} = 12$ V. The Q points Q, Q_1, and Q_2 are found as follows. Point Q is the Q point for which $v_a = 0$ V. Using (7.9-9b), we find that when $v_a = 0$ V, $I_C = 1.06$ mA and from the figure [or (7.9-9a)] $v_{CE1} = 2.8$ V. Q_1 is the Q point at which T_1 is at the edge of the linear region (see Sec. 7.3). This occurs when $v_{CE1} = 0.35$ V. The corresponding value of I_C is $1.06 + (2.8 - 0.35)/20 = 1.18$ mA, as shown in the figure. Substituting this result into (7.9-9b) yields $v_{a,\text{max}} = 1.2$ V. Point Q_2 is placed at the other extreme of the maximum symmetrical common-mode swing, where $v_{a,\text{min}} = -1.2$ V. At that point $I_C = 0.94$ mA and $V_{CE1} = 5.25$ V. It will be shown later that for $|v_a| < 1.2$ V the common-mode gain $A_a = v_o/v_a = 1$.

(c) Equations (7.9-9) are used to obtain the load line shown in Fig. 7.9-2b if V_{EE} is unknown. To understand why the load line is drawn as shown we must first note that as the common-mode voltage increases, currents i_{C1} and i_{C2} increase and v_{CE1} decreases until T_1 enters the saturation region, $v_{CE1} \approx 0.35$ V. The common-mode voltage when the collector-emitter voltage is 0.35 V is the largest possible common-mode voltage $V_{a,\text{max}}$. It is found by equating I_C calculated from (7.9-9a) to I_C calculated from (7.9-9b) as follows:

$$\frac{12 + V_{EE} - 0.35}{20} = \frac{V_{a,\text{max}} + V_{EE} - 1.4}{10}$$

As the common-mode voltage decreases, currents i_{C1} and i_{C2} decrease and v_{C1} increases. The increase in v_{C1} can be followed by T_6 as long as $v_{C1} \leq V_{CC} - 0.7$ (see Fig. 7.9-1). Thus, the most negative common-mode voltage allowable is that for which $v_{C1} = V_{CC} - 0.7 = 11.3$ V. Referring to

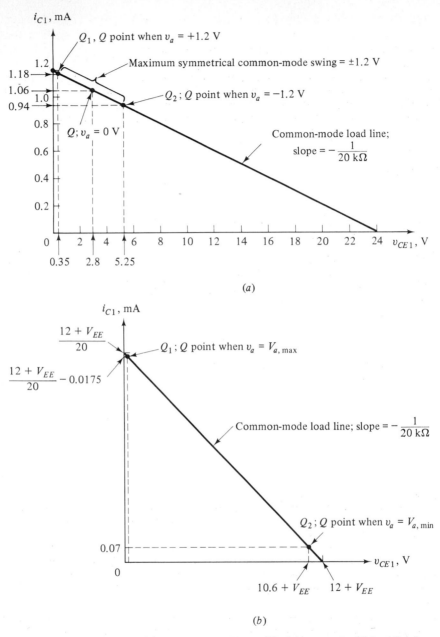

Figure 7.9-2 Example 7.9-1: (*a*) load line for T_1 in part (*b*) of the example; (*b*) load line for part (*c*) of the example.

Fig. 7.9-1, we see that when $v_{C1} = 11.3$ V, $I_C = 0.7/10 = 0.07$ mA. Using (7.9-9*a*) we find that for this case $V_{CE1} = 10.6 + V_{EE}$. This Q point, Q_2, is shown in Fig. 7.9-2*b*, and at this point the common-mode input is at the most negative value it can have without causing distortion (cutoff in T_5 and T_6 is the limiting factor here). Thus, when $I_C = 0.07$ mA, $v_a = V_{a, \min}$. In order to maximize the allowable common-mode swing V_{EE} is chosen so that the Q point when $v_a = 0$ V bisects the load line between $V_{a, \max}$ and $V_{a, \min}$. The value of I_C at this point is

$$I_C(v_a = 0) = \frac{1}{2}\left(\frac{12 + V_{EE} - 0.35}{20} + 0.07\right)$$

Combining this equation with (7.9-9b) yields

$$I_C(v_a = 0) = \frac{13 + V_{EE}}{40} = \frac{V_{EE} - 1.4}{10}$$

Solving for V_{EE}, we find

$$V_{EE} \approx 6.3 \text{ V}$$

To find $V_{a, \max}$ we can combine (7.9-9b) with I_C evaluated at $V_{a, \max}$

$$I_C(v_a = V_{a, \max}) = \frac{12 + 6.3 - 0.35}{20} = \frac{V_{a, \max} + 6.3 - 1.4}{10}$$

Solving, we find

$$V_{a, \max} \approx 4.1 \text{ V}$$

The actual maximum common-mode input range will be somewhat less than this, depending on the magnitude of the difference-mode input. The common-mode range has been considerably extended by properly choosing the negative supply voltage $-V_{EE}$. ///

Small-signal analysis The object of this analysis is to obtain an expression relating the output voltage v_o to the input voltages v_1 and v_2. As before, we shall express our result in terms of the common-mode and difference-mode components of the input rather than v_1 and v_2. This will then lead directly to the difference-mode gain and the CMRR, which are the quantities of most interest to the user of the op-amp.

The calculation will proceed through the amplifier from input to output. We begin by reflecting all components of the input difference amplifier into the emitter circuits of T_1 and T_2. The result is shown in Fig. 7.9-3. If we assume the component values $R_1 = R_3 = 5$ kΩ, $R_2 = R_4 = 10$ kΩ, and $V_{CC} = V_{EE} = 12$ V, as in Example 7.9-1, we find that $I_C \approx 1$ mA, so that $h_{ib1} = h_{ib2} \approx 25 \ \Omega$. If we also assume that each transistor has the same current amplification factor $h_{fe} = 99$, then $h_{ib3}/(h_{fe} + 1) = h_{ib4}/(h_{fe} + 1) = 25 \ \Omega$.

To find the currents i_{e1} and i_{e2} we first note the similarity between Figs. 7.9-3 and 7.1-6 and then employ (7.1-10), which describes Fig. 7.1-6. The result is

$$i_{c1} \approx i_{e1} \approx \frac{v_a}{10^4} - \frac{v_d}{2(50)} \tag{7.9-10a}$$

and

$$i_{c2} \approx i_{e2} \approx \frac{v_a}{10^4} + \frac{v_d}{2(50)} \tag{7.9-10b}$$

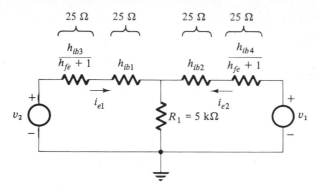

Figure 7.9-3 Small-signal equivalent circuit looking into emitters of T_1 and T_2.

The next step is to determine the collector current in T_6, since this current will determine v_{c6} via the relation

$$v_{c6} = i_{c6}(R_4 + 3r_d) \tag{7.9-10c}$$

The collector current can be found by analyzing the difference-amplifier circuit formed by T_5 and T_6 shown in Fig. 7.9-4a. In the diagram the input circuits consist of the collector resistors R_2 of T_1 and T_2 in parallel with current sources representing the collector currents i_{c1} and i_{c2}. All supply voltages have been replaced by short circuits since we are interested only in small-signal behavior. In addition, the diodes D_1, D_2, and D_3 in the collector circuit of T_6 have been replaced by their small-signal equivalent resistances $r_d = V_T/I_D \approx (25 \text{ mV})/(1 \text{ mA}) \approx 25 \ \Omega$. The small-signal equivalent circuit seen looking into the emitters of T_5 and T_6 is shown in Fig. 7.9-4b. Here the input current sources of Fig. 7.9-4a have been converted into voltage sources. In addition, the *pnp* transistors have been assumed to have $h_{fe} = 9$.

Because of the similarity between Fig. 7.9-4b and Fig. 7.1-6 we again use (7.1-10), being careful to identify the variables in the equation correctly. Observe that in Fig. 7.9-4b the input voltages are [see (7.9-10)]

$$v'_{c2} = 10^4 i_{c2} = v_a + 100v_d \tag{7.9-11a}$$

$$v'_{c1} = 10^4 i_{c1} = v_a - 100v_d \tag{7.9-11b}$$

where, as before $v_a = (v_1 + v_2)/2$ and $v_d = v_1 - v_2$. Now using (7.1-10b), we find [with i_{e6} here analogous to i_{e2} in (7.1-10b)]

$$i_{e6} \approx i_{c6} \approx \frac{v'_{c2} + v'_{c1}}{2 \times 10^4} + \frac{v'_{c2} - v'_{c1}}{2 \times 1025} \tag{7.9-12a}$$

Substituting (7.9-11) into (7.9-12a), we have

$$i_{c6} = \frac{v_a}{10^4} - \frac{200v_d}{2050} \tag{7.9-12b}$$

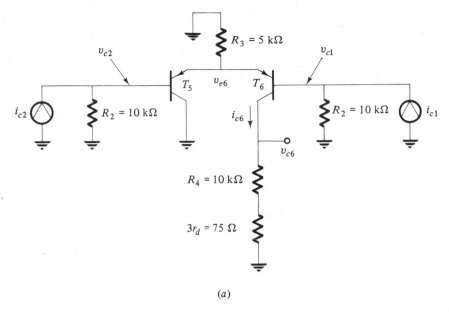

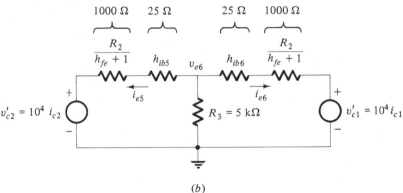

Figure 7.9-4 (a) Small-signal circuit of *pnp* difference amplifier showing inputs from collectors of T_1 and T_2; (b) small-signal equivalent circuit looking into emitters of T_5 and T_6.

The collector voltage v_{c6} and hence the output voltage v_o can now be found: from the circuit

$$v_o = v_{c6} \approx 10^4 i_{c6} = v_a - \frac{10^4 v_d}{10} \qquad (7.9\text{-}13)$$

Comparing this result with that given by (7.2-1a), we find that the magnitude of the common-mode gain is $|A_c| = 1$ and the magnitude of the difference-mode gain is $|A_d| \approx 1000$. Finally, then, the common-mode rejection ratio is

$$\text{CMRR} = 1000 = 60 \text{ dB}$$

Input impedance The input impedance of the op-amp is the impedance seen looking between the bases of T_3 and T_4 of the Darlington-difference amplifier. This impedance is given by (7.6-9),

$$R_i = 4(h_{fe} + 1)^2 h_{ib1} \tag{7.6-9}$$

If $R_1 = R_3 = 5\ k\Omega$, $R_2 = R_4 = 10\ k\Omega$, $V_{CC} = V_{EE} = 12$ V, and $h_{fe} + 1 = 100$, then $h_{ib1} = 25\ \Omega$ and $R_i = (4)(100)^2(25) = 1\ M\Omega$. The input impedance of a typical op-amp is usually between 100 kΩ and 10 MΩ, depending on its design.

Output impedance The output impedance seen looking into the output terminal is the impedance of R_4 and diodes D_1 to D_3 reflected into the emitter of T_7 or T_8 (remember that T_7 and T_8 are operating class B, so that only one of them is on at a time). Hence the output impedance is approximately

$$R_o = \frac{R_4}{h_{fe} + 1} + h_{ib7} \approx \frac{R_4}{h_{fe} + 1} \tag{7.9-14}$$

If $R_4 = 10\ k\Omega$ and $h_{fe} + 1 = 100$ for both the *npn* and *pnp* transistors,† then $R_o \approx 100\ \Omega$. The output impedance of typical op-amps lies between 50 and 200 Ω.

PROBLEMS

7.1-1 Repeat Example 7.1-1 if $R_b = 10\ k\Omega$ and $h_{fe} = 100$. Evaluate v_{o1} and v_{o2} for a quiescent common-mode voltage of 0 V.

7.1-2 Verify (7.1-11) by applying standard network-analysis techniques to the circuit of Fig. 7.1-6.

7.1-3 Find the small-signal equivalent circuit for the difference amplifier of Fig. 7.1-1 by reflecting the appropriate elements into the base circuits. Analyze the resulting circuit to verify (7.1-11).

7.1-4 Find i_L in terms of the common- and differential-mode signals in Fig. P7.1-4.

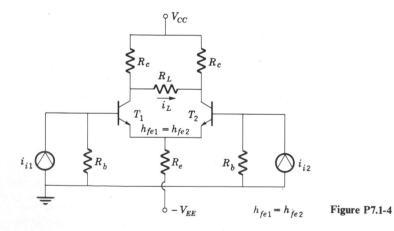

Figure P7.1-4

† Note that h_{fe} is actually quite different for the *npn* and *pnp* transistors. The value of 100 was chosen to simplify calculations.

7.1-5 Find $\Delta I_{C1}/\Delta T$ and $\Delta I_{C2}/\Delta T$ for the basic difference amplifier of Fig. 7.1-1. Assume that $I_{CBO} = 0$ and only V_{BE} varies with temperature $(\Delta V_{BE}/\Delta T = -2.0 \text{ mV/°C})$.

7.2-1 (a) Find the common-mode rejection ratio for Prob. 7.1-1.

(b) If the common-mode voltage is 2 mV, find the differential-mode signal for which the differential-mode output is at least 200 times the common-mode output.

7.2-2 (a) Design a difference amplifier as in Fig. 7.1-1 to have a CMRR of 40 dB. The voltage sources each have 1 kΩ internal resistance, and the transistors have $h_{fe} = 250$. Use two power supplies, one positive and one negative. Calculate bias currents assuming a common-mode voltage of 0 V.

(b) The common-mode voltage is 10 mV. What is the largest differential-mode signal that the amplifier can handle if the differential-mode output is to be at least 50 times the common-mode output?

7.2-3 In Fig. P7.2-3 the transistors are identical; $h_{fe} = 100$, and $h_{oe} = h_{re} = 0$.

(a) Find I_{EQ1}, and I_{EQ2} assuming that the dc common-mode voltage is negligible.

(b) Draw the small-signal equivalent circuit with all impedances reflected into the base of T_1.

(c) Find the differential-mode and common-mode gains.

(d) Determine the CMRR.

(e) If v_1 and v_2 have a common-mode dc voltage ranging from $-3 \text{ V} \leq V_1 \leq +3 \text{ V}$, find the maximum allowable current variation in i_{C1} and i_{C2} due to the difference-mode signal so that operation is linear.

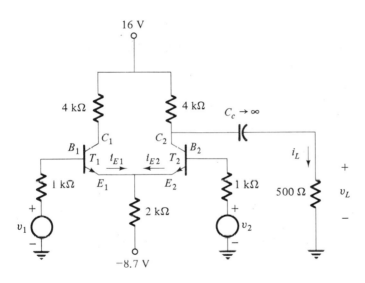

Figure P7.2-3

7.3-1 A differential amplifier is required to have a differential-mode output at least 4000 times the common-mode output. The common-mode signal is 3 mV, and the differential-mode signal is 1 mV. Design the amplifier using a constant-current source as in Fig. 7.3-1, where the voltage sources each have source resistances of 2 kΩ. Assume that the transistors are identical and each has $h_{fe} = 200$. At 1 mA these transistors have $1/h_{ob} = 1 \text{ MΩ}$ ($h_{ob} = 10^{-3} I_E \text{ S}$).

7.3-2 Repeat Example 7.3-1 if $V_{CC} = V_{EE} = 5 \text{ V}$, $R_c = 1 \text{ kΩ}$, $R_b = 3.5 \text{ kΩ}$, $R'_e = 500 \text{ Ω}$, $h_{fe} + 1 = 200$, and $I_{EQ1} = I_{EQ2} = 2 \text{ mA}$.

7.4-1 In Fig. P7.4-1 $h_{fe1} = 100$, and $h_{fe2} = 200$. If $R_c = 1.5$ kΩ, $r_i = 4$ kΩ, $R_e = 6$ kΩ, $V_{CC} = 10$ V, $V_{BB} = 5$ V, and R_x is a 100-Ω potentiometer, find (a) R_1 and R_2 and (b) the CMRR.

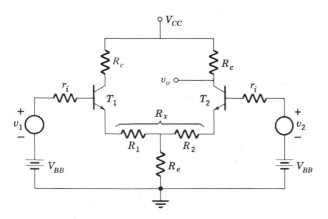

Figure P7.4-1

7.4-2 Repeat Example 7.4-1 if $V_{CC} = 5$ V and $R_c = 1.5$ kΩ.

7.5-1 In Fig. P7.5-1 $R_d = 10$ kΩ, $r_{ds} = 100$ kΩ, $g_m = 1$ mS, $R_s = 10$ kΩ, and $R_g = 1$ MΩ. Find (a) the CMRR (b) the output voltage $v_{o2} - v_{o1}$, (c) the output impedance seen looking into the drain of T_2, and (d) the input impedance at the gate of T_2.

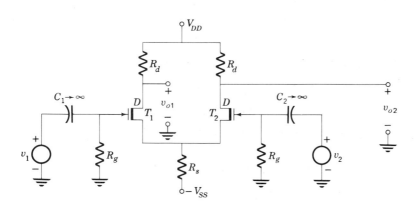

Figure P7.5-1

7.6-1 In Fig. 7.6-1 assume that $h_{fe1} \neq h_{fe4}$. Show that

(a)
$$R_i = 2h_{ie1}(1 + h_{fe4}) = 2h_{ie4}$$
(b)
$$A_i = (1 + h_{fe4})h_{fe1}$$

7.6-2 In Fig. 7.6-3 find R_i and A_i if $h_{fe1} \neq h_{fe4}$.

7.6-3 (*a*) Find the equivalent hybrid model for the compound Darlington configuration in Fig. 7.6-1 including the effects of h_{oe} and h_{re}. Refer to Fig. P7.6-3.

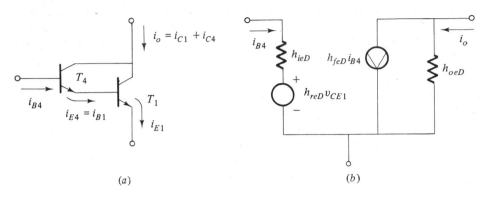

(*a*)   (*b*)

Figure P7.6-3

(*b*) Show that if $h_{re1} \ll 1$, $h_{re4} \ll 1$, $h_{ie1} h_{oe1} \ll 1$, and $h_{ie4} h_{oe4} \ll 1$, then

$$h_{ieD} \approx 2h_{ie1}(1 + h_{fe4}) = 2h_{ie4}$$
$$h_{reD} \approx \text{negligible}$$
$$h_{feD} \approx (1 + h_{fe4})h_{fe1}$$
$$h_{oeD} \approx h_{oe1} + h_{oe4}(1 + h_{fe1}) = 2h_{oe1}$$

For h_{oeD} we also need the relations that $h_{oe4} = c_1 I_{E4}$ and $h_{oe1} = c_1 I_{E1}$, where c_1 is a constant. [Recall (6.2-1*f*) and assume v_{CE} is constant.]

7.6-4 Find the hybrid equivalent (h_i, h_o, and g) circuit for the compound-transistor connection of Fig. P7.6-4. Use the simplified hybrid equivalent for each transistor but do not assume identical parameters.

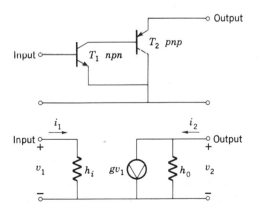

Figure P7.6-4

7.6-5 Repeat Prob. 7.6-4 for the parallel connection of Fig. P7.6-5. Assume identical transistors and comment on the possible advantages of this circuit.

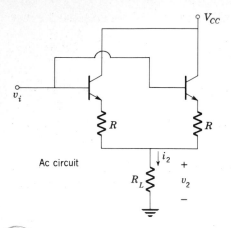

Figure P7.6-5

4/8/81

7.6-6 Find (a) quiescent conditions throughout the circuit of Fig. P7.6-6, (b) v_1/i_i and v_2/i_i, and (c) Z_i, Z_{o1}, and Z_{o2}.

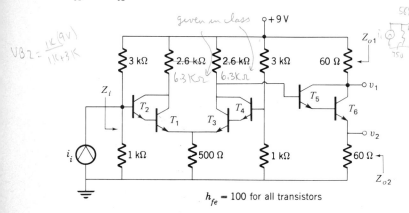

$h_{fe} = 100$ for all transistors

Figure P7.6-6

4/12/81

7.6-7 Repeat Prob. 7.6-6 for the circuit of Fig. P7.6-7.

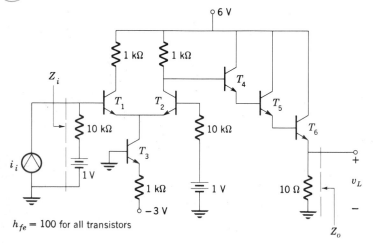

$h_{fe} = 100$ for all transistors

Figure P7.6-7

7.6-8 A Darlington configuration can be constructed using the JFET and transistor shown in Fig. P7.6-8. Resistor R_1 provides for self-biasing.

    (a) Obtain an expression for the gain v_e/v_i.

    (b) Determine Z_o.

    (c) Determine Z_i.

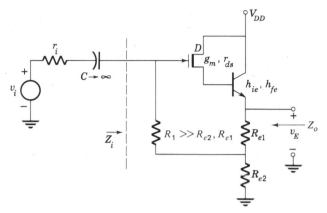

Figure P7.6-8

7.6-9 In Fig. P7.6-9 transistors T_1, T_2, and T_3 form a difference amplifier. The JFET T_4 is a Darlington amplifier used to provide a high input impedance.

    (a) If the resistance looking into the collector of T_3 is infinite, calculate Z_i.

    (b) Calculate v_L/v_i.

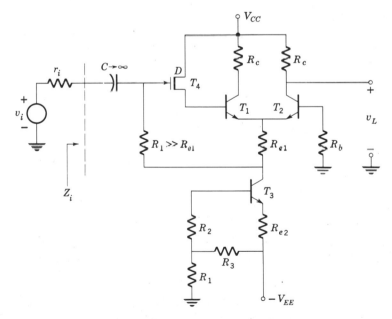

Figure P7.6-9

7.7-1 In the cascode amplifier of Fig. 7.7-1 let $R_i = 15$ kΩ, $V_{BB} = 0$, $V_{CC} = 12$ V, $V_{EE} = -6$ V, $R_e = 3.3$ kΩ, and $V_i = +6$ V. Assume $h_{FE} = h_{fe} = 200$.

    (a) Find R_c to set the dc component of the output equal to 0.

    (b) Find the small-signal gain v_L/v_i if $h_{ob} = 10$ μS at 5 mA.

7.7-2 In Fig. P7.7-2 the transistors are identical with $h_{FE} = h_{fe} = 200$ and $h_{ob} = 10\ \mu S$ at 5 mA.
    (*a*) Find R_c to set the dc component of the output equal to 0.
    (*b*) Find the small-signal gain v_L/v_i.

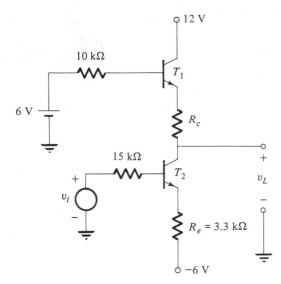

Figure P7.7-2

7.7-3 Modify the resistors R_c and R_e in Prob. 7.7-2 to obtain a dc component of 0 at v_L and unity gain for $|v_L/v_i|$. Neglect h_{oe}.

7.9-1 (*a*) In Fig. 7.9-1 use impedance reflection techniques to verify the circuit of Fig. 7.9-3.
    (*b*) Verify (7.9-10*a*) and (7.9-10*b*) by breaking Fig. 7.9-3 into common-mode and differential-mode circuits, solving each circuit, and using superposition.

7.9-2 Verify that the input impedance seen looking into the terminals between the bases of T_3 and T_4 for the op-amp in Fig. 7.9-1 is given by (7.6-9).

7.9-3 By rearranging the resistor R_4 and the diodes D_1, D_2, D_3 of Fig. 7.9-1 as shown in Fig. P7.9-3 the crossover distortion can be eliminated. Assume $V_{D1} = V_{D2} = V_{D3} = V_{BE7} = V_{BE8} = 0.7\ V$.
    (*a*) What is the quiescent value of i_L?
    (*b*) Explain the action of the circuit as I_D increases and as I_D decreases.
    (*c*) Calculate the peak value of I_D to obtain maximum i_L.
    (*d*) Determine the maximum output power and the efficiency.

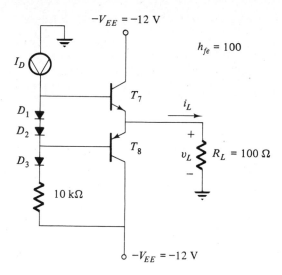

Figure P7.9-3

EIGHT

APPLICATIONS OF OPERATIONAL AMPLIFIERS

The name *operational amplifier* is applied when a very stable amplifier is used to implement a wide variety of linear and nonlinear *operations* by merely changing a few external elements such as resistors, capacitors, diodes, etc. At present a large variety of op-amps are available, many fabricated on a single chip and sold at extremely low prices, often less than $1. As a result, this device has become a primary element in analog system design. In this chapter we shall discuss several of the most important linear and nonlinear applications of op-amps.

8.1 THE LINEAR INVERTING AMPLIFIER

As shown in Sec. 7.3 and illustrated in Fig. 8.1-1a, the equivalent-circuit model of an op-amp consists of an input impedance R_i connected between the two input terminals v_1 and v_2. The output circuit consists of a controlled-voltage source $A_d v_d$ in series with an output resistance R_o connected between the output terminal and ground (one of the output terminals is always connected to ground). Also, since we have assumed the controlled-voltage source to be $A_d v_d$, we have tacitly assumed that the common-mode gain is zero. This assumption is valid for most op-amp applications. The usual circuit symbol for the op-amp is the triangle shown in Fig. 8.1-1b. It does not explicitly show any of the components of the equivalent circuit, but the inverting and noninverting input terminals are always marked, very often simply by the minus and plus signs shown.

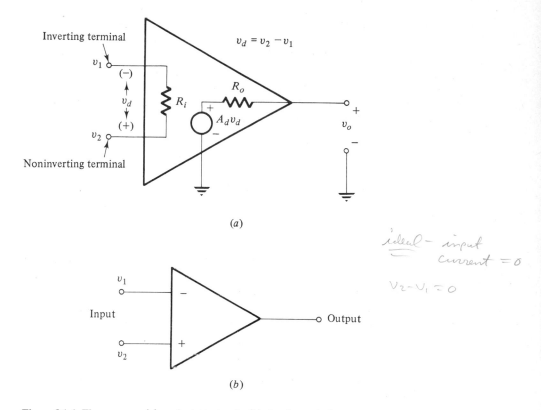

(a)

(b)

Figure 8.1-1 The op-amp: (a) equivalent circuit; (b) circuit symbol.

The voltage gain A_d of the op-amp is usually very large (typically 100,000) in comparison with the overall gain of the system in which it is employed. In fact, it is usually convenient to assume that the gain is infinite. Similarly, the input impedance R_i is much larger (typically 100 kΩ) than the external resistances in the system and is also often assumed to be infinite. The output impedance R_o on the other hand is typically 100 Ω and for many applications can be neglected. When these approximations are made, the resulting op-amp is said to be *ideal*.

A very important observation that can be made about an ideal op-amp is that the differential input voltage $v_d = v_2 - v_1 \approx 0$. The reason for this is that $v_d = v_o/A_d$ (see Fig. 8.1-1a), and if v_o is finite and A_d is infinite, v_d must be zero. In practice the output voltage of a typical amplifier is less than 10 V. If we assume that $A_d = 100,000$, the differential input voltage which produces 10 V at the output is 100 μV, an amount so small that it can usually be neglected. Thus, often $v_d \approx 0$, even in a real op-amp, and we say that *the input of an op-amp is a virtual short circuit*. This implies that $v_1 \approx v_2$, and since R_i, the impedance between v_1 and v_2, is very large, the current in R_i can usually be neglected, being of the order of $(100 \ \mu V/100 \ k\Omega) = 1$ nA.

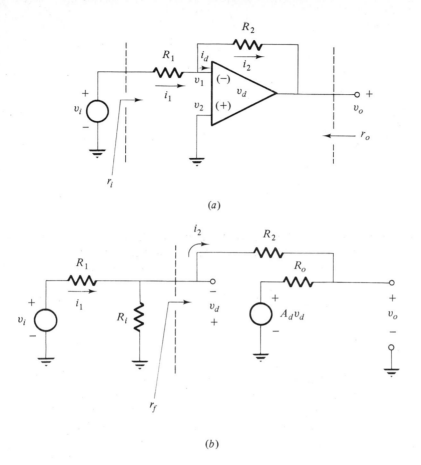

Figure 8.1-2 Linear inverting amplifier: (*a*) circuit; (*b*) equivalent circuit.

When the op-amp is to be used as a linear inverting amplifier, external re-sistors R_1 and R_2 are connected as shown in Fig. 8.1-2 and the noninverting terminal is connected to ground.

Voltage gain of the inverting amplifier If the op-amp is ideal, ($i_d = 0$) the overall gain $A_v = v_o/v_i$ of the inverter can be found by noting that $i_1 = i_2$. Hence

$$i_1 - \frac{v_i + v_d}{R_1} = i_2 = \frac{-v_d - v_o}{R_2} \qquad (8.1\text{-}1)$$

However, $v_d = v_o/A_d$, and in an ideal op-amp v_d can be assumed to be zero. Hence (8.1-1) reduces to

$$\frac{v_i}{R_1} = -\frac{v_o}{R_2} \qquad (8.1\text{-}2a)$$

Solving for A_v, we find

Inverting Amp

$$A_v = \frac{v_o}{v_i} = -\frac{R_2}{R_1} \qquad (8.1\text{-}2b)$$

The designation of this circuit as an *inverting* amplifier arises from the negative sign.

Values of R_2 and R_1 are typically such that the magnitude of the overall gain is less than 50 and $R_2 < 100$ kΩ. It is important to note that the overall gain A_v does not depend at all on the gain A_d of the op-amp but only on the ratio of the two external resistors.

Input impedance of the inverting amplifier The input impedance r_i seen looking into the inverting amplifier of Fig. 8.1-2a is

$$r_i = \frac{v_i}{i_1} \qquad (8.1\text{-}3)$$

In order to evaluate it in terms of the circuit elements we begin by writing KVL around the input loop

$$v_i = R_1 i_1 - v_d \qquad (8.1\text{-}4)$$

In an ideal op-amp $v_d = 0$ V, and we have

$$r_i \approx R_1 \qquad \text{Input Imped.} \qquad (8.1\text{-}5)$$

Instead of assuming that the op-amp is ideal, if we draw the equivalent circuit of the inverting amplifier as shown in Fig. 8.1-2b, we see that the input impedance r_i is equal to

$$r_i = R_1 + (R_i \| r_f) = \left| R_1 + R_i \right/\!\!/ \frac{R_2 + R_o}{1 + A_d} \right| \qquad (8.1\text{-}6)$$

where $r_f = -v_d/i_2$. Writing KVL around the loop which includes R_2, we have

$$-v_d = R_2 i_2 + R_o i_2 + A_d v_d \qquad (8.1\text{-}7)$$

from which we obtain

$$r_f = -\frac{v_d}{i_2} = \frac{R_2 + R_o}{1 + A_d} \qquad (8.1\text{-}8)$$

In practice $r_f \ll R_i$ and $r_f \ll R_1$, so that $r_i \approx R_1$, the value obtained for the ideal op-amp. For example, if $R_2 = 10$ kΩ, $R_o = 100$ Ω, $R_i = 100$ kΩ, and $A_d = 100{,}000$, then $r_f \approx 0.1$ Ω. Since R_1 is typically 1 kΩ and never less than about 100 Ω, the input impedance seen by the source v_i is $r_i \approx R_i$ even if the op-amp is not assumed to be ideal.

Output impedance of the inverting amplifier The output impedance of the amplifier shown in Fig. 8.1-2a is found by setting $v_i = 0$, inserting a test voltage source at the output terminal, and measuring the current drawn from the test source. Such a circuit configuration is shown in Fig. 8.1-3a.

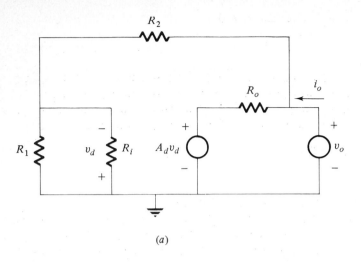

(a)

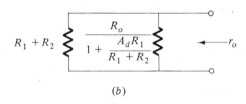

(b)

Figure 8.1-3 Output impedance: (a) equivalent circuit for calculating r_o; (b) the two parallel resistances which comprise the output resistance.

The output impedance of the amplifier is $r_o = v_o/i_o$, and the current i_o consists of two parts

$$i_o = \frac{v_o - A_d v_d}{R_o} + \frac{v_o}{R_1 + R_2} \tag{8.1-9a}$$

where we have assumed that $R_i \gg R_1$. With this assumption we have the additional relation

$$-v_d = \frac{R_1}{R_1 + R_2} v_o \tag{8.1-9b}$$

Substituting (8.1-9b) into (8.1-9a) and dividing through by v_o yields

$$\frac{1}{r_o} = \frac{i_o}{v_o} = \frac{1 + R_1 A_d/(R_1 + R_2)}{R_o} + \frac{1}{R_1 + R_2} \tag{8.1-10}$$

The output impedance r_o is seen to consist of two resistors in parallel, as shown in Fig. 8.1-3b. In most cases $R_1 + R_2 \gg R_o/[1 + A_d R_1/(R_1 + R_2)]$ so that

$$r_o \approx \frac{R_o}{1 + R_1 A_d/(R_1 + R_2)} \tag{8.1-11}$$

$r_o \approx 0$

For an ideal op-amp A_d is infinite and $r_o = 0$.

Example 8.1-1 An op-amp has $R_i = 100$ kΩ, $A_d = 100{,}000$, and $R_o = 100\ \Omega$. If $R_1 = 1$ kΩ and $R_2 = 50$ kΩ, calculate (a) A_v, (b) r_i, and (c) r_o.

SOLUTION (a) Using (8.1-2b) we have

$$A_v = -\frac{R_2}{R_1} = -50$$

(b) The input impedance is found from (8.1-6) and (8.1-8)

$$r_i = R_1 + \frac{R_2 + R_o}{1 + A_d} \approx 1000 + \frac{50{,}000}{1 + 100{,}000} \approx 1\ \text{k}\Omega$$

(c) The output impedance is found from (8.1-11)

$$r_o \approx \frac{R_o}{1 + R_1 A_d/(R_1 + R_2)} = \frac{100}{1 + (10^3 \times 10^5)/(5.1 \times 10^4)}$$

$$\approx \frac{100}{2 \times 10^3} = 0.05\ \Omega \qquad\qquad ///$$

8.2 LINEAR NONINVERTING AMPLIFIER

has higher input impedance than inverting Amp

The op-amp can also be used as a noninverting amplifier. Such a configuration, shown in Fig. 8.2-1a, results in an amplifier with an overall voltage gain greater than or equal to unity and an almost infinite input impedance.

Gain The overall gain of the noninverting amplifier is most easily determined if we assume an ideal op-amp. Then $R_o = 0$, R_i is infinite, and A_d is infinite, so that $v_d \approx 0$. The resulting equivalent circuit is shown in Fig. 8.2-1b. Using this circuit, we have

$$v_i = v_2 = v_1 \tag{8.2-1a}$$

and

$$v_1 = \frac{R_1}{R_1 + R_2} v_o \qquad \text{NONINVERTING} \tag{8.2-1b}$$

Hence

$$A_v = \frac{v_o}{v_i} = \frac{R_1 + R_2}{R_1} = 1 + \frac{R_2}{R_1} \tag{8.2-2}$$

Thus, the overall gain of the noninverting amplifier must always be greater than or equal to unity. The measured gain A_v for a real op-amp is very close to that predicted by (8.2-2) since A_d and R_i are so very large and $R_o \ll R_1 + R_2$.

The gain can be made equal to unity by replacing resistor R_2 by a short circuit and removing R_1. Under these conditions the circuit is called a *voltage follower*.

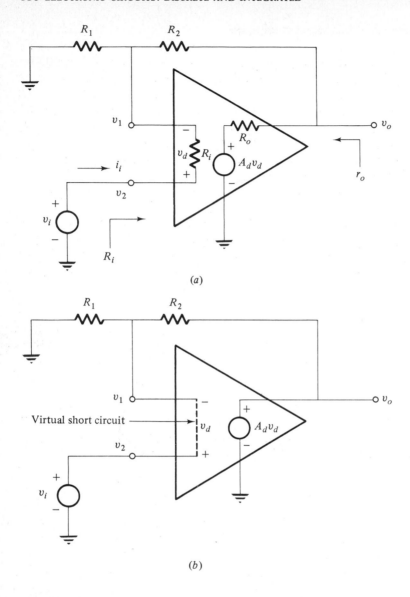

(a)

(b)

Figure 8.2-1 Linear noninverting amplifier: (a) circuit; (b) circuit with an ideal op-amp.

Input impedance The input impedance of the noninverting amplifier is $r_i = v_i/i_i$. Since $i_i = v_d/R_i$, we see that i_i must be very small ($i_i = 0$ for an ideal op-amp) and hence r_i must be very large. To determine r_i quantitatively we write

$$i_i = \frac{v_d}{R_i} \qquad\qquad (8.2\text{-}3a)$$

However, since $v_d = v_o/A_d$, (8.2-3a) becomes

$$i_i = \frac{v_o}{A_d R_i} \qquad (8.2\text{-}3b)$$

Noting that

$$v_o = \left(1 + \frac{R_2}{R_1}\right) v_i \qquad (8.2\text{-}3c)$$

we have

$$i_i = \frac{1 + R_2/R_1}{A_d R_i} v_i \qquad (8.2\text{-}3d)$$

Hence, the input impedance is

NONINVERTING INPUT Impedance

$$r_i = \frac{v_i}{i_i} = \frac{A_d R_i}{1 + R_2/R_1} \qquad (8.2\text{-}4)$$

For example, if $A_d = 10^5$, $R_i = 100 \text{ k}\Omega$, $R_2 = 10 \text{ k}\Omega$, and $R_1 = 1 \text{ k}\Omega$, then $r_i \approx 1 \text{ G}\Omega \ (= 10^9 \ \Omega)$.

Output impedance To find the output impedance r_o, we must replace the input voltage source v_i by a short circuit, apply a test voltage source at the output, and measure the current supplied by the test source. The resulting equivalent circuit is then identical to the circuit used to calculate r_o for the inverting amplifier. Hence, r_o can be found from

OUTPUT Impedance

$$\frac{1}{r_o} = \frac{1 + R_i A_d/(R_1 + R_2)}{R_o} + \frac{1}{R_1 + R_2} \qquad (8.1\text{-}10)$$

8.3 FEEDBACK

In the amplifiers described in Secs. 8.1 and 8.2, resistors R_2 and R_1 are used to feed back a portion of the output voltage to the op-amp inverting input. Any amplifier in which a portion of the output voltage or current is fed back to the input is called a *feedback amplifier*. When the feedback connection is made to the inverting input, the process is called *negative feedback*; a feedback connection to the noninverting input is called *positive feedback*. In this chapter we discuss negative feedback only; positive feedback is discussed in Chap. 10.

If we compare Figs. 8.1-2 and 8.2-1, we see that a distinction can be made in the type of negative feedback present. In Fig. 8.1-2 the current $i_d = -v_d/R_i$ is equal to the difference current $i_2 - i_1 = i_d$. If the inverting amplifier is operating properly, i_d is forced to zero by the feedback connection and i_2 is approximately equal to i_1. In Fig. 8.2-1 the voltage $v_d = v_2 - v_1$, and if the noninverting amplifier is operating properly, v_d is forced to zero by the feedback connection and v_2 is approximately equal to v_1. In each circuit, since R_i is very large, v_d and i_d are both forced to zero.

One of the benefits of the application of negative feedback is that the overall gain A_v is relatively independent of the op-amp characteristics and is a function only of R_1 and R_2. Other implications of the use of negative feedback will be discussed in Chap. 10.

In some nonlinear circuits using op-amps the feedback circuits do not operate in certain voltage ranges, in which case v_d is not always small. Such circuits require special op-amps and are discussed in a later section.

8.4 LINEAR OPERATIONS USING THE OP-AMP

In this section we describe the operation of several linear circuits which make use of the op-amp.

8.4-1 The Difference Amplifier

The difference amplifier, shown in Fig. 8.4-1, ideally produces an output voltage which is proportional to the difference between the two input voltages v_a and v_b. However, the output voltage of a practical circuit also contains a small term proportional to the common-mode voltage, $(v_a + v_b)/2$. We shall demonstrate that in the amplifier shown, the component of the output voltage which is proportional to the common-mode voltage is reduced by the CMRR while the component of the output voltage proportional to the difference-mode voltage has a gain R_2/R_1.

The output voltage of the op-amp without the external resistances is given by (7.8-1)

$$v_o = -A_d(v_2 - v_1) - A_a \frac{v_2 + v_1}{2} \tag{8.4-1}$$

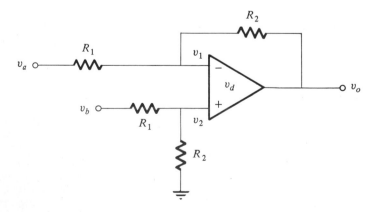

Figure 8.4-1 Difference amplifier.

From Fig. 8.4-1 we have

$$v_2 = \frac{R_2}{R_1 + R_2} v_b \tag{8.4-2a}$$

and

$$v_1 = \frac{R_2}{R_1 + R_2} v_a + \frac{R_1}{R_1 + R_2} v_o \tag{8.4-2b}$$

Substituting (8.4-2a) and (8.4-2b) into (8.4-1) yields

$$\begin{aligned}
v_o = {}& -A_d \frac{R_2}{R_1 + R_2} (v_b - v_a) + A_d \frac{R_1}{R_1 + R_2} v_o \\
& -\frac{A_a}{2} \left(\frac{R_1}{R_1 + R_2} \right) v_o - A_a \left(\frac{R_2}{R_1 + R_2} \right) \frac{v_b + v_a}{2}
\end{aligned} \tag{8.4-3}$$

Solving for v_o yields

$$v_o = \frac{\dfrac{-A_d R_2}{R_1 + R_2} (v_b - v_a) - \dfrac{A_a R_1}{R_1 + R_2} \left(\dfrac{v_b + v_a}{2} \right)}{1 - \dfrac{A_d R_1}{R_1 + R_2} + \dfrac{A_a R_1}{2(R_1 + R_2)}} \tag{8.4-4a}$$

Since $A_d \gg A_a$ and $A_d \gg 1 + R_2/R_1$, the denominator of (8.4-4a) reduces to $-A_d R_1/(R_1 + R_2)$ and (8.4-4a) becomes

$$v_o = \frac{R_2}{R_1} (v_b - v_a) + \frac{A_a}{A_d} \left(\frac{v_b + v_a}{2} \right) \tag{8.4-4b}$$

The first term of (8.4-4b) represents that part of the output voltage which is proportional to the difference voltage $v_b - v_a$. The ratio R_2/R_1 is the gain of the inverting amplifier. The second term of (8.4-4b) represents that part of the output voltage which is proportional to the common-mode signal $(v_b + v_a)/2$. However, the gain of this term is the inverse of CMRR $= A_a/A_d$; hence the common-mode signal is reduced by the CMRR. Typically, $A_d = 10^5$ while $A_a = 1$, so that the common-mode signal is attenuated by a factor of 10^5.

8.4-2 The Summing Amplifier

The amplifier shown in Fig. 8.4-2 accepts any number of inputs (only three have been shown) and provides an output voltage proportional to the linear sum of the input voltages. The equation for the output voltage can be derived by noting that the feedback causes a virtual short circuit across the op-amp input terminals, so that $v_d = i_d = 0$. Hence,

$$i_{11} + i_{12} + i_{13} = i_2 \tag{8.4-5}$$

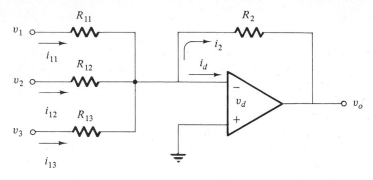

Figure 8.4-2 Summer.

Since

$$i_{11} = \frac{v_1}{R_{11}} \qquad i_{12} = \frac{v_2}{R_{12}} \qquad i_{13} = \frac{v_3}{R_{13}} \tag{8.4-6a}$$

and

$$i_2 = -\frac{v_o}{R_2} \tag{8.4-6b}$$

we find

$$v_o = -\left(\frac{R_2}{R_{11}} v_1 + \frac{R_2}{R_{12}} v_2 + \frac{R_2}{R_{13}} v_3\right) \tag{8.4-7}$$

Thus each input can be multiplied by a different scale factor and then added. The negative sign cannot be avoided in this application unless an additional amplifier is used, as illustrated in the following example.

Example 8.4-1 Design an op-amp circuit to have an output

$$v_o = 2v_1 + 5v_2$$

SOLUTION Here we use two op-amps, one to add and scale v_1 and v_2 and the other to change the sign from negative to positive. The proposed circuit consisting of two cascaded scalers is shown in Fig. 8.4-3. The resistors in the first circuit will be chosen to do the desired scaling while the second circuit will provide a gain of -1 to change the sign. The loading between the two circuits can be neglected as long as $R_3 \gg Z_{o1}$. Typical values of Z_{o1} are less than 1 Ω. In the absence of other specifications we might choose $R_2 = R_3 = R_4 = 10$ kΩ. Then with all resistance values in kilohms we have

$$v_3 = -\left(\frac{R_2}{R_{11}} v_1 + \frac{R_2}{R_{12}} v_2\right) = -\left(\frac{10}{R_{11}} v_1 + \frac{10}{R_{12}} v_2\right) \tag{8.4-8}$$

and

$$v_o = -\frac{R_4}{R_3} v_3 = -v_3 = \frac{10v_1}{R_{11}} + \frac{10v_2}{R_{12}} \tag{8.4-9}$$

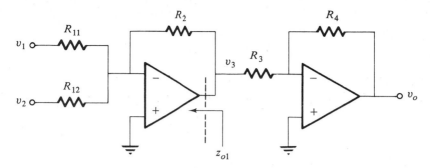

Figure 8.4-3 Circuit for Example 8.4-1.

Comparing (8.4-9) with the desired output, we see that if $R_{11} = 5 \text{ k}\Omega$ and $R_{12} = 2 \text{ k}\Omega$, the output will be

$$v_o = 2v_1 + 5v_2 \qquad\qquad ///$$

8.4-3 Integrator

The op-amp can also be used to construct active filters like the integrator shown in Fig. 8.4-4. This circuit provides an output voltage proportional to the integral of the input. Neglecting the common-mode effect and assuming that A_d is very large, so that $v_d \approx i_d = 0$, we have $i_R = i_C$, where

$$i_R = \frac{v_i}{R} = i_c = -C\frac{dv_o}{dt} \tag{8.4-10}$$

Hence
$$v_o = -\frac{1}{RC}\int^t v_i(\lambda)\,d\lambda \tag{8.4-11}$$

Equation (8.4-11) shows that the circuit does indeed provide integration of the input signal. The capacitor C is usually paralleled with a FET switch which is closed when the capacitor is to be discharged and opened when the capacitor is to

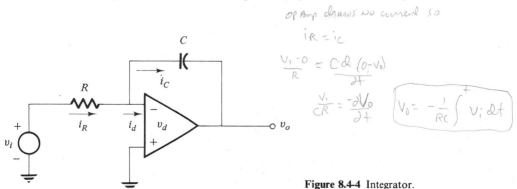

Figure 8.4-4 Integrator.

be charged, i.e., during integration. Initial voltages can be applied to the capacitor by putting the voltage in series with the switch (see Prob. 8.4-7).

Example 8.4-2 Use op-amp summers and integrators to solve the differential equation

$$\frac{d^2v_o}{dt^2} + 3\frac{dv_o}{dt} + \frac{v_o}{4} = V_{im} \cos \omega t$$

where v_o is the output voltage resulting from application of input voltage $V_{im} \cos \omega t$. We assume that all initial conditions are zero. The resulting circuit will represent the *analog-computer* simulation of the equation.

SOLUTION We begin by solving for the second derivative, to get

$$\frac{d^2v_o}{dt^2} = V_{im} \cos \omega t - 3\frac{dv_o}{dt} - \frac{v_o}{4}$$

The mathematical operations required to solve this equation include addition, integration, and scaling, all of which can be accomplished using the op-amp circuits discussed previously. Solving for the first derivative dv_o/dt by integrating d^2v_o/dt^2 and noting that an inversion results when using an op-amp circuit, we obtain

$$\frac{dv_o}{dt} = -\int^t \left(-\frac{d^2v_o}{d\lambda^2}\right) d\lambda = -\int^t \left(-V_{im} \cos \omega\lambda + 3\frac{dv_o}{d\lambda} + \frac{v_o}{4}\right) d\lambda$$

This equation can be implemented by a summer-integrator, as shown in Fig. 8.4-5a. All resistors shown are in megohms, and the capacitor is in microfarads. The output voltage v_o is obtained by integrating dv_o/dt and inverting the result as shown in Fig. 8.4-5b. The complete circuit, shown in Fig. 8.4-5c, is obtained by combining Fig. 8.4-5a and b. Readers should convince themselves that signals throughout the circuit have the proper sign to satisfy the equation. When the input signal is applied, the voltage at the output terminal is the output waveform, which can be viewed on an oscilloscope.   ///

8.4-4 All-pass Filter

The all-pass filter shown in Fig. 8.4-6 is characterized by the transfer function

$$H(j\omega) = \frac{V_o(j\omega)}{V_i(j\omega)} = -1 \underline{/ 2 \tan^{-1} \omega RC} \qquad (8.4\text{-}12)$$

From this transfer function we see that all input frequencies will appear at the output without attenuation but with a phase shift proportional to frequency.

To show how (8.4-12) is arrived at we observe that

$$V_2(j\omega) = \frac{1}{1 + j\omega RC} V_i(j\omega) \qquad (8.4\text{-}13)$$

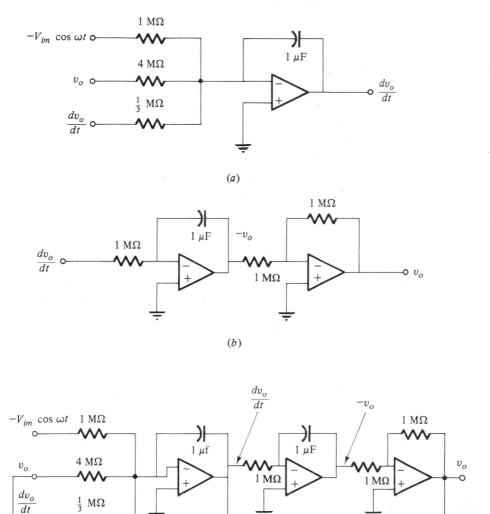

Figure 8.4-5 Analog computer solution of differential equation: (a) circuit for finding the first derivative; (b) circuit for finding the output voltage from its first derivative; (c) final combined circuit.

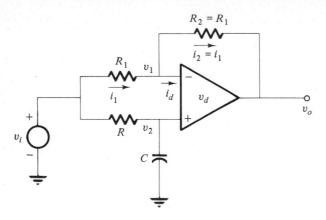

Figure 8.4-6 All-pass filter.

Assuming A_d large, so that $v_d \approx 0$ and therefore $i_d \approx 0$, we have $V_1(j\omega) = V_2(j\omega)$ and also

$$I_1(j\omega) = \frac{V_i(j\omega) - V_2(j\omega)}{R_1} = \frac{V_2(j\omega) - V_o(j\omega)}{R_1} \qquad (8.4\text{-}14)$$

Solving for $V_o(j\omega)$ and substituting (8.4-13) into (8.4-14) and setting $R_1 = R$ yields

$$V_o(j\omega) = 2V_2(j\omega) - V_i(j\omega) = \left(\frac{2}{1 + j\omega RC} - 1\right)V_i(j\omega) = \frac{1 - j\omega RC}{1 + j\omega RC} V_i(j\omega) \qquad (8.4\text{-}15)$$

Hence
$$H(j\omega) = \frac{V_o(j\omega)}{V_i(j\omega)} = -1 \big/ 2 \tan^{-1} \omega RC \qquad (8.4\text{-}12)$$

Some of the many different filters which can be constructed using op-amps will be considered in the problems.

8.5 NONLINEAR OP-AMP APPLICATIONS

In the next few sections we shall describe several nonlinear-waveform generation and shaping circuits, including the ideal rectifier, limiter, logarithmic amplifier, and multiplier. We begin the discussion by describing an op-amp circuit which acts much like an ideal diode.

8.5-1 Half-Wave Rectifier

The half-wave diode rectifier was discussed in Chap. 1, where we saw that a real diode is characterized by a cut-in voltage V_y (≈ 0.65 V for silicon), below which the current flowing through the diode is negligible, and a forward voltage V_F

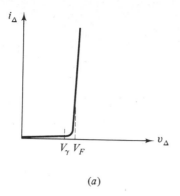

(a)

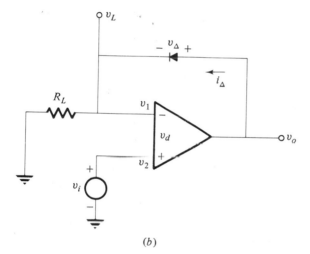

(b)

Figure 8.5-1 Half-wave rectifier: (a) typical diode characteristic; (b) half-wave rectifier using an op-amp.

(≈ 0.7 V for silicon), which represents the voltage drop across the diode when moderate currents do flow. The typical diode characteristic is shown in Fig. 8.5-1a. In the figure we have used v_Δ and i_Δ to represent the diode voltage and current in order to avoid confusion with the op-amp input voltage and current v_d and i_d.

An op-amp rectifier circuit which overcomes the effect of the cut-in voltage and which will rectify voltages in the millivolt range is shown in Fig. 8.5-1b. In this circuit the input voltage v_i is applied to the noninverting input of the op-amp, and the rectified voltage v_L is taken from the inverting input of the op-amp. The op-amp output v_o drives the diode, and resistor R_L represents the load. We next derive the input-output characteristic of this device.

From the circuit we see that when v_o is sufficiently positive to turn the diode on, the feedback loop will be closed through the conducting diode and the feedback action will cause the differential input voltage to be very small. When this is the case, we shall have $v_L \approx v_i$. However, if the output voltage v_o is less than the

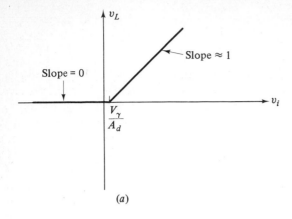

(a)

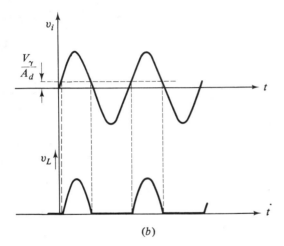

(b)

Figure 8.5-2 Half-wave rectifier: (a) input-output characteristic; (b) waveforms.

cut-in voltage of the diode, no current flows in the diode and the feedback loop is opened. The rectifier output v_L is now equal to 0 V since no current flows in R_L. Thus $v_L = 0$ when $v_o \le V_\gamma$. The input voltage required to reach this output voltage is

$$v_i = v_d = \frac{v_o}{A_d} \le \frac{V_\gamma}{A_d} \tag{8.5-1}$$

In a practical op-amp $A_d = 10^5$, and with $V_\gamma = 0.65$ V the breakpoint on the characteristic is at $v_i = 6.5\ \mu$V. The resulting input-output characteristic is shown in Fig. 8.5-2a.

On the conducting portion of the characteristic $v_i > V_\gamma / A_d$. The rectifier output v_L is now related to v_i since

$$v_i - v_L = v_d = \frac{v_o}{A_d} = \frac{v_L + 0.7}{A_d} \tag{8.5-2a}$$

Solving for v_i gives

$$v_i = v_L\left(1 + \frac{1}{A_d}\right) + \frac{0.7}{A_d} \qquad (8.5\text{-}2b)$$

In (8.5-2a) we have assumed that the diode voltage $v_\Delta = V_F = 0.7$ V. Referring to (8.5-2b), we see that for any practical op-amp rectifier, $v_L = v_i$ to within a few microvolts. Typical waveforms when v_i is sinusoidal are shown in Fig. 8.5-2b.

It has already been noted that when $v_i < V_\gamma/A_d$, the diode is an open circuit and the feedback is essentially removed. As a result $v_L = v_1 = 0$ V and v_d can be a very large negative voltage. This means that the op-amp will saturate; that is, v_o will reach its most negative voltage possible and remain at that value as v_i decreases further. As a practical matter many op-amps cannot be used in this circuit since they cannot withstand very large differential input voltages. For example, the 709 op-amp should not be used for this application since it can only take ± 5 V, while the 101 or 741 op-amps are designed to allow large variations in v_d, as much as ± 30 V, and therefore can be used.

8.5-2 Clipping Circuit

The clipping circuit shown in Fig. 8.5-3a is identical to the rectifier circuit of Fig. 8.5-1b except for the presence of a reference voltage. Thus, the clipping circuit enables one to set the level of rectification.

The operation of the circuit can be understood by noting that if the diode is operating below cut-in, the diode current is zero and therefore the voltage drop across R_L is zero. Hence $v_L = V_{ref}$ when the diode is OFF. To find the range of input voltages for which this occurs we note that we must have

$$v_\Delta = v_o - v_L < V_\gamma \qquad (8.5\text{-}3)$$

When we substitute $v_o = (v_i - v_L)A_d$, (8.5-3) can be manipulated to yield

$$v_i < \frac{V_\gamma + v_L}{A_d} + v_L \qquad (8.5\text{-}4a)$$

Since $v_L - V_{ref}$ when the diode is off, the input voltage at the breakpoint is

$$v_i \approx \frac{V_\gamma + V_{ref}}{A_d} + V_{ref} \approx V_{ref} \qquad (8.5\text{-}4b)$$

When the input voltage v_i increases above the value given in (8.5-4b), the diode turns on. Now (8.5-2) again applies, so that $v_L \approx V_i$ when $v_i > V_{ref}$. The transfer characteristic is shown in Fig. 8.5-3b and typical waveforms in Fig. 8.5-3c.

8.5-3 Full-Wave Rectifier

An op-amp full-wave rectifier is shown in Fig. 8.5-4. The voltage v_r is the input voltage v_i after half-wave rectification. This can be seen by noting that when v_i increases above 0 V, D_2 turns ON and v_o becomes negative, causing diode D_1 to

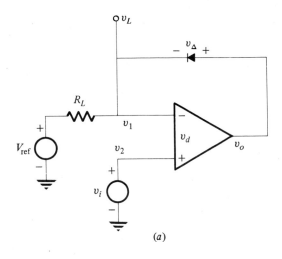

(a)

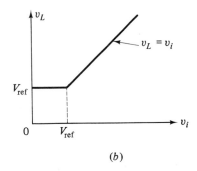

(b)

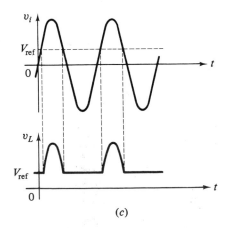

(c)

Figure 8.5-3 Op-amp clipper: (a) circuit; (b) transfer characteristic; (c) waveforms.

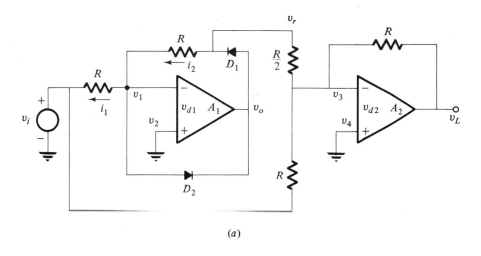

(a)

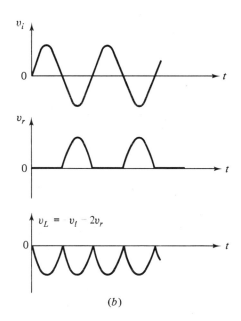

(b)

Figure 8.5-4 Full-wave rectifier: (a) circuit; (b) waveforms.

turn OFF. With D_1 OFF, the current $i_2 = 0$ and $v_r = v_1$. However, with D_2 ON, $v_o = v_1 - 0.7$. Since $v_1 = -v_d = -v_o/A_d = -(v_1 - 0.7)/A_d$, we have

$$v_r = v_1 = \frac{0.7}{A_d + 1} \approx 0 \text{ V} \qquad \text{when } v_i > 0 \qquad (8.5\text{-}5)$$

When v_i is negative, v_o is positive and diode D_2 turns OFF while diode D_1 turns ON. Now A_1 is connected as an inverting amplifier, and with $i_2 \approx i_1$ and $v_{d1} \approx 0$ we have

$$v_r = -v_i \qquad \text{when } v_i < 0 \qquad (8.5\text{-}6)$$

Combining (8.5-5) and (8.5-6), we see that the circuit of op-amp A_1 acts as a half-wave rectifier. Typical waveforms of v_i and v_r are shown in Fig. 8.5-4b.

Amplifier A_2 is a summing amplifier with an output voltage [see (8.4-6)]

$$v_L = -v_i - 2v_r \qquad (8.5\text{-}7)$$

Using the waveforms of v_i and v_r shown in Fig. 8.5-4b in (8.5-7), we arrive at the full-wave-rectified output shown in Fig. 8.5-4b.

The accuracy of the rectified output voltage depends on close matching of the resistors in the circuit. In Prob. 8.5-5 we shall investigate the effect on the output waveform of using resistors having a 10 percent tolerance.

8.5-4 Clamping Circuit

The circuit shown in Fig. 8.5-5a is used to *clamp* the minimum peaks of the input voltage waveforms to ground. This type of circuit is used extensively in television and in other systems where a pulse train may become distorted and lose its reference dc level. Figure 8.5-5b shows such a pulse train. In the original waveform each of the minimum peaks was at 0 V rather than at the varying negative voltages shown. The output of the clamp is shown in Fig. 8.5-5c. Here we see that each of the minimum peaks has been restored to 0 V and that the positive excursion ΔV of each pulse is preserved.

The operation of the clamp is readily understood if we first recall that the voltage across a capacitor cannot change instantaneously. Now assume that the capacitor C is initially uncharged and the voltage waveform shown in Fig. 8.5-5b is applied to the input. When v_i jumps from 0 to 2 V, the capacitor voltage does not change and is therefore 0 V. Hence $v_d = -2$ V, and the op-amp saturates with v_o at a large negative voltage. Diode D is then cut off, removing the feedback path. Thus $v_L = v_i = 2$ V, as shown in Fig. 8.5-5c. Note that the capacitor voltage remains unchanged at 0 V for the duration of the pulse because there is no current flowing through C.

When the input drops to -1 V, the capacitor voltage initially remains at 0 V. As a result, v_L initially drops to -1 V, and since $v_d = -v_L$, v_d initially jumps to $+1$ V. This causes the output voltage to rise toward its maximum positive value, but it is clamped to approximately $v_L + 0.7$ by diode D, which now turns ON. The capacitor voltage now changes rapidly as the diode current flows into it. When a

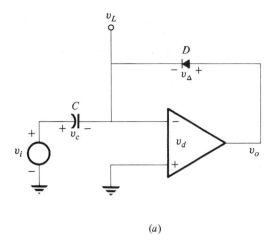

(a)

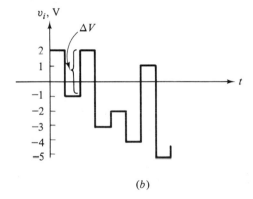

(b)

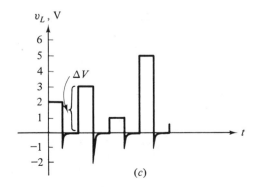

(c)

Figure 8.5-5 Clamp: (a) circuit; (b) input waveform; (c) clamped output waveform.

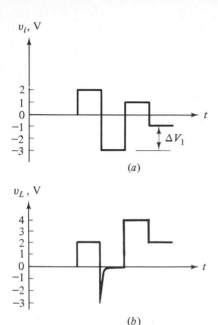

Figure 8.5-6 The difficulty in using the clamping circuit of Fig. 8.5-5a: (a) input waveform; (b) output.

steady state is reached, the capacitor current and hence the diode current are zero, as is the diode voltage v_Δ. Since the diode voltage $v_\Delta = v_o + v_d \approx v_d(A_d + 1)$, we have $v_d \approx 0$. Hence $v_L = -v_d = 0$ V. Thus, in the steady state with v_i negative the capacitor voltage v_C is equal to the input voltage v_i, and v_L is clamped to 0 V. This result is illustrated in Fig. 8.5-5c. The negative spikes result from the fact that initially the negative change in v_L is equal to the negative change in v_i because the capacitor voltage does not change instantaneously. The reader should verify that the waveform of Fig. 8.5-5c results from the input shown in Fig. 8.5-5b.

While useful for explaining the principle of operation of a clamp, the circuit of Fig. 8.5-5a will not always work. By way of example, consider the input waveform shown in Fig. 8.5-6a. When v_i jumps to 2 V, v_L also jumps to 2 V and the capacitor voltage remains at 0 V. When v_i drops to -3 V, the capacitor voltage rapidly charges to $v_C = -3$ V as v_L goes to 0 V after a brief spike (see Fig. 8.5-6b). When v_i now jumps to $+1$ V, the capacitor retains its voltage as diode D cuts off and therefore v_L jumps to 4 V. But now, when v_i falls to -1 V, $v_d = v_C - v_i = -3$ V $-$ $(-1$ V$) = -2$ V and diode D remains cut off. Since $v_L = -v_d$, v_L actually decreases to 2 V rather than clamping to 0 V.

The reason for the difficulty is that accommodating a positive voltage change of ΔV_1 requires that the capacitor discharge by ΔV_1. However, in the circuit of Fig. 8.5-5c there is no path through which the capacitor can discharge when v_d is negative. To provide such a path a resistor is always inserted from v_L to ground, as shown in Fig. 8.5-7.

The practical clamping circuit shown in Fig. 8.5-7 not only solves the problem described above but also solves another equally serious problem. The input stage

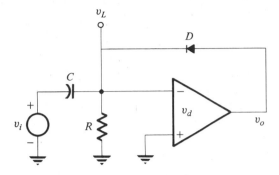

Figure 8.5-7 A practical clamping circuit.

of an op-amp is always a difference amplifier, which requires a dc return to ground so that base current can flow into the input transistors. Since with v_d negative the diode is off, and since no dc current can flow in the capacitor, resistor R must be available to provide a path to supply base current to the input transistor.

The circuit shown in Fig. 8.5-7 can be used to clamp positive peaks to ground if the direction of the diode is reversed. It can also be used to clamp a signal to a prescribed reference voltage by simply connecting the reference voltage to the noninverting input.

8.5-5 Envelope Detector

An *envelope detector*, Fig. 8.5-8a, is a device which follows the peak values of an *amplitude-modulated* sinusoidal waveform. Typical input and output waveforms are shown in Fig. 8.5-8b.

The envelope-detector circuit shown is similar to the half-wave-rectifier circuit of Fig. 8.5-1a but has an added component, capacitor C. The presence of the capacitor significantly alters the circuit operation since the voltage across the capacitor cannot change instantaneously. When the input waveform $v_i(t)$ becomes sufficiently positive, diode D turns on and the capacitor charges toward v_i, the charging current being $i_\Delta - \dfrac{v_L}{R_L}$. The diode will be ON as long as v_i increases.

Note that when v_i becomes constant and steady-state conditions are reached so that v_L is constant, the current in the capacitor is zero but $i_\Delta = v_L/R_L$. In this case since

$$v_i - v_L = v_d = \frac{v_o}{A_d} = \frac{v_L + 0.7}{A_d} \qquad (8.5\text{-}8a)$$

it can readily be shown that

$$v_L \approx v_i - \frac{0.7}{A_d} \approx v_i \qquad (8.5\text{-}8b)$$

After v_i reaches its peak, it starts to decrease. Since the voltage v_L across the capacitor cannot change instantaneously, diode D cuts off and capacitor C

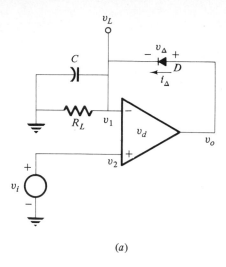

(a)

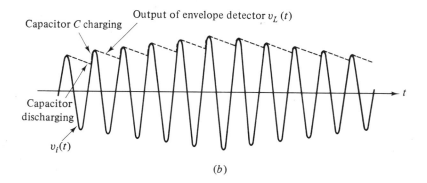

(b)

Figure 8.5-8 Envelope detector: (a) circuit; (b) waveforms.

discharges through resistor R_L. This discharge is illustrated in Fig. 8.5-8b. Note that for diode D to be cut off we must have $v_\Delta \le V_\gamma$. This happens when

$$v_i - v_L = v_d = \frac{v_o}{A_d} = \frac{v_\Delta + v_L}{A_d} \le \frac{V_\gamma + v_L}{A_d} \approx 0 \qquad (8.5\text{-}9)$$

For any practical system, we can conclude that the capacitor will charge when $v_i > v_L$ and discharge when $v_i \le v_L$.

8.5-6 Limiter

The *limiter* is a circuit which produces a constant high-level output (often a positive voltage) whenever the input signal is positive and a constant low-level output (often a negative voltage) whenever the input signal is negative.

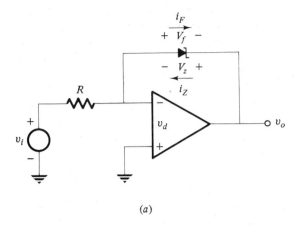

(a)

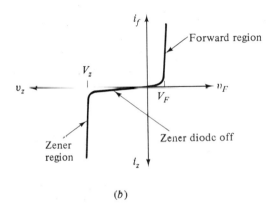

(b)

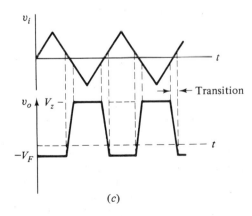

(c)

Figure 8.5-9 Op-amp limiter: (a) circuit; (b) Zener-diode characteristic; (c) waveforms.

An op-amp limiter is shown in Fig. 8.5-9a. The feedback element in this circuit is a Zener diode Z whose vi characteristic (see Sec. 1.10) is shown in Fig. 8.5-9b. When the input voltage v_i is small enough for the output voltage v_o to lie between V_Z and V_F, the Zener diode is effectively cut off. This means that the gain from v_i to v_o is essentially the gain A_d of the op-amp.

When v_i becomes sufficiently negative for v_o to go positive and cross into the Zener region $(v_o > V_Z)$, the diode turns ON and is equivalent to a battery of voltage V_Z in series with a small resistance. The feedback is now restored, so that $v_d \approx 0$, and the output voltage will be clamped to V_Z since

$$v_o = V_Z - v_d \approx V_Z \tag{8.5-10a}$$

Similarly, when v_i becomes positive enough to produce an output voltage v_o which is in the forward region, the Zener diode acts as if it were an ordinary silicon diode. Once again the diode appears to be a battery (this time its voltage is V_F) in series with a small resistance, and with the feedback again active $v_d \approx 0$ and

$$v_o \approx -V_F \tag{8.5-10b}$$

Waveforms showing typical input and output voltages are presented in Fig. 8.5-9c. Observe that the output voltage v_o switches back and forth between its two output levels with a finite transition time. During each transition the diode is cut off, and the gain v_o/v_i is

$$\frac{v_o}{v_i} \approx \frac{v_o}{-v_d} = -A_d \tag{8.5-10c}$$

Thus, the change in input voltage $|\Delta v_i|$ required to produce a change of $V_Z + V_F$ in the output voltage is

$$|\Delta v_i| \approx \frac{V_Z + V_F}{A_d} \tag{8.5-11}$$

8.6 BOOTSTRAP SWEEP GENERATOR

The *bootstrap sweep generator* shown in Fig. 8.6-1a is used to generate a linear voltage ramp. In this circuit the op-amp output is connected directly to the inverting input, as in the voltage follower of Sec. 8.2. The feedback causes a virtual short circuit across the input, so that $v_d \approx 0$ and we have $v_o = v_C$. Hence the voltage at point P is

$$v_P = V_Z + v_o = V_Z + v_C \tag{8.6-1}$$

The current flowing through resistor R is

$$i_R = \frac{v_P - v_C}{R} = \frac{V_Z}{R} = \text{const} \tag{8.6-2}$$

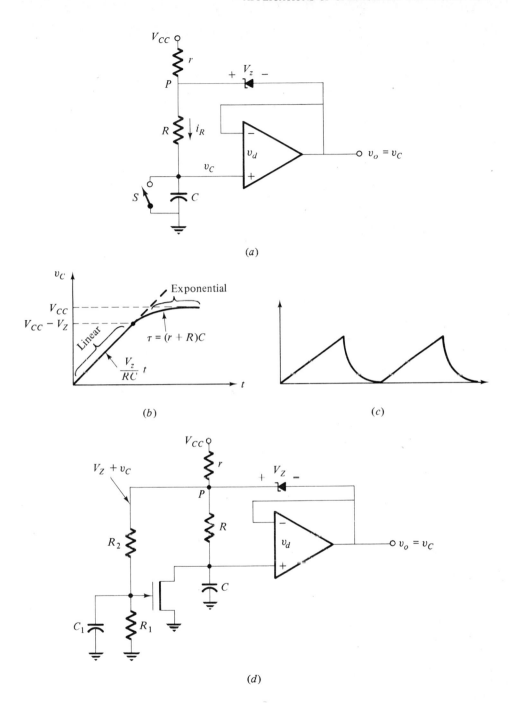

Figure 8.6-1 Bootstrap sweep generator: (a) single-sweep circuit; (b) single-sweep waveform; (c) astable waveform; (d) circuit for periodic (astable) operation.

This constant current flows through switch S when it is closed, and when the switch is opened, the current flows through the capacitor C, causing it to charge. Hence the capacitor voltage increases linearly, so that

$$v_C = \frac{1}{C} \int^t \frac{V_Z}{R} \, dt = \frac{V_Z}{RC} t \qquad (8.6\text{-}3)$$

The capacitor voltage continues to increase linearly until $v_C = V_{CC} - V_Z$. If v_C increases above this value, the Zener diode cuts off (becomes an open circuit), so that the op-amp becomes simply a voltage follower. The capacitor continues to charge toward V_{CC}, but now it follows an exponential curve with a time constant $\tau = (r + R)C$, as illustrated in Fig. 8.6-1b.

Periodically recurring linear voltages like those shown in Fig. 8.6-1c are required for many applications. In order to achieve this the switch S is replaced by a FET, as shown in Fig. 8.6-1d. Such a device is said to be operating in an *astable* mode. If the FET has a threshold voltage V_T, R_1 and R_2 are chosen so that

$$(V_Z + v_C)\frac{R_1}{R_1 + R_2} = V_T \qquad (8.6\text{-}4)$$

A value of v_C which is less than $V_{CC} - V_Z$ is chosen so that the FET will be turned ON when the capacitor voltage is still on its linear rise. When the FET turns ON, it quickly discharges C, turning the FET OFF, so that the cycle can begin again. Capacitor C_1 is employed to keep the FET on until capacitor C is completely discharged, even though the voltage at point P has decreased.

In the sweep generator shown an increase in the output causes the voltage at point P to rise, thereby producing an increase in the input, which causes the output to increase further. Thus, the circuit is called a *bootstrap circuit* since the output is being "pulled up by its bootstraps." This is an example of positive feedback. The sweep repetition frequency of the astable bootstrap sweep generator depends on the circuit components and voltages and will be considered in the problems.

8.7 LOGARITHMIC AMPLIFIER

An amplifier in which the output voltage is proportional to the logarithm of the input voltage is shown in Fig. 8.7-1. The logarithmic function is generated by transistor T, which provides the feedback.

To derive the response equation we begin by assuming that the feedback is active so that $v_d \approx 0$ V. Then

$$i_C = i_1 = \frac{v_i}{R_1} \qquad (8.7\text{-}1)$$

But using the Ebers-Moll equations for a transistor in which $v_{CB} > 0$ and $v_{BE} > 0$, we can write the collector current [see Eq. (2.2-14a)] as

$$i_C = I_o \, \epsilon^{v_{BE}/V_T} \qquad (8.7\text{-}2)$$

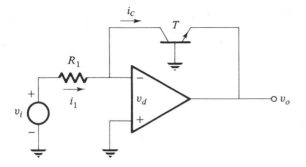

Figure 8.7-1 Logarithmic amplifier.

[Comparison of (8.7-2) with (2.2-14a) will show that we have set $I_o = \alpha_F I_{EO}$ and have assumed all other terms negligible.]

Equations (8.7-1) and (8.7-2) can be combined to yield

$$\frac{v_i}{R_1} = I_o \epsilon^{v_{BE}/V_T} \tag{8.7-3a}$$

Noting that the base voltage of the transistor is at ground (zero) potential, we have

$$v_{BE} = -v_E = -v_o \tag{8.7-3b}$$

Combining (8.7-3a) and (8.7-3b) and taking the natural logarithm of both sides, we obtain the desired response equation

$$v_o = -V_T \ln \frac{v_i}{I_o R_1} \tag{8.7-4}$$

This relation shows that v_o is proportional to the natural logarithm of v_i. The term $I_o R_1$ acts as a scale factor; that is, $\ln av_i$ is generated, where the scale factor a is set by R_1. Additional gain can be obtained by connecting v_o to a linear amplifier. Note, however, that for this log-amp to operate properly v_i must be positive. Thus, this is a unipolar device.

Temperature-compensated log-amp The accurate generation of the logarithmic function using the circuit of Fig. 8.7-1 is severely limited since both I_o and V_T are sensitive to temperature. This temperature dependence is minimized using the log-amp shown in Fig. 8.7-2a. This circuit accepts two inputs v_1 and v_2, and the output voltage is proportional to the logarithm of the ratio of these input voltages. Often v_2 is a fixed reference voltage.

In this circuit, transistors T_1 and T_2 provide feedback across the two op-amps. Hence, assuming $v_{d1} \approx 0$ V and $v_{d2} \approx 0$ V, we have

$$i_1 = \frac{v_1}{R_1} \quad \text{and} \quad i_2 = \frac{v_2}{R_2} \tag{8.7-5}$$

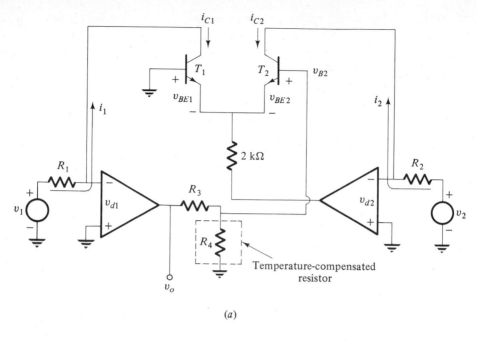

(a)

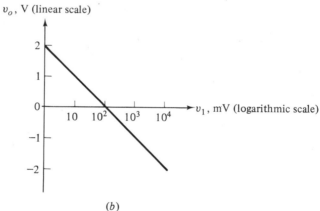

(b)

Figure 8.7-2 Temperature-compensated log-amp: (a) circuit; (b) typical response characteristic.

Collector currents i_{C1} and i_{C2} are given by (8.7-2). Thus (8.7-5) becomes

$$i_1 = i_{C1} = \frac{v_1}{R_1} = I_o \, \epsilon^{v_{BE1}/V_T} \qquad (8.7\text{-}6a)$$

and

$$i_2 = i_{C2} = \frac{v_2}{R_2} = I_o \, \epsilon^{v_{BE2}/V_T} \qquad (8.7\text{-}6b)$$

Dividing (8.7-6a) by (8.7-6b) and taking the logarithm of the result yields

$$v_{BE1} - v_{BE2} = V_T \ln \left(\frac{R_2}{R_1} \frac{v_1}{v_2} \right) \tag{8.7-7}$$

However, since $v_{E2} = v_{E1} = v_E$,

$$v_{BE1} = -v_E \tag{8.7-8a}$$

and

$$v_{BE2} = v_{B2} - v_E \tag{8.7-8b}$$

Substituting (8.7-8) into (8.7-7) yields

$$-v_{B2} = V_T \ln \left(\frac{R_2}{R_1} \frac{v_1}{v_2} \right) \tag{8.7-9}$$

Referring to Fig. 8.7-2, we see that the voltage v_{B2} is related to the output voltage v_o by

$$v_{B2} = \frac{R_4}{R_3 + R_4} v_o \tag{8.7-10}$$

Finally, combining (8.7-9) with (8.7-10), we arrive at the desired equation

$$v_o = -\left(\frac{R_3 + R_4}{R_4} V_T \right) \ln \left(\frac{R_2}{R_1} \frac{v_1}{v_2} \right) \tag{8.7-11}$$

Note that while v_o is no longer a function of I_o, it is still temperature-sensitive as a result of the term V_T. To minimize the temperature variation due to V_T, R_3 is chosen to be much larger than R_4, and R_4 is a temperature-compensated resistor designed so that

$$\frac{\Delta R_4}{\Delta T} \approx \frac{\Delta V_T}{\Delta T} \tag{8.7-12}$$

Typically $R_3 = 16$ kΩ and $R_4 = 1$ kΩ.

While v_1 and v_2 can each be a function of time, one or the other is usually a constant reference voltage. If v_o is to be positive, v_1 is the reference voltage.

The entire preceding analysis is based on T_1 and T_2 having identical characteristics. The LM194 IC provides *supermatched* transistors in which the base-emitter voltages are matched to 50 μV and the current gains h_{FE} are matched to 2 percent. Log-amps constructed using this IC have a typical useful range from $v_i = 0.1$ mV to 10 V, that is, 5 decades (100 dB).

Example 8.7-1 A log-amp using the LM194 has the parameter values

$$R_1 = R_2 = 10 \text{ k}\Omega \qquad R_3 = 16.4 \text{ k}\Omega \pm 1\%$$

$$R_4 = 1 \text{ k}\Omega \text{ (temperature-compensated)} \qquad V_T = 25 \text{ mV}$$

The input voltage v_1 varies from 1 mV to 10 V. Find the value for the reference voltage v_2 which will cause v_o to be 0 V when v_1 is 100 mV. Plot v_o versus v_1.

SOLUTION Using (8.7-11), we have

$$v_o = \frac{16.4 + 1}{1} (25 \times 10^{-3}) \ln \frac{v_1}{v_2} = -0.434 \ln \frac{v_1}{v_2}$$

Since $\log x \approx 0.434 \ln x$, this becomes

$$v_o = -\log \frac{v_1}{v_2}$$

Since $\log 1 = 0$, the reference voltage should be 100 mV in order to have $v_o = 0$ V when $v_1 = 100$ mV. Finally, then,

$$v_o = -\log \frac{v_1}{0.1} = -\log 10 v_1$$

where v_o and v_1 are in volts. This equation is plotted in Fig. 8.7-2b.   ///

8.8 FEEDBACK-REGULATED POWER SUPPLIES

In order to operate reliably, modern electronic components, particularly ICs, must be supplied with dc voltages which remain constant even though load currents or line voltages change. This requirement is met by using *regulated* power supplies. Today, regulated power supplies are available in IC form, and one such supply is often used on each printed-circuit (PC) board in a system to supply power to all the ICs on that board. This is particularly true in transistor-transistor and emitter-coupled-logic digital circuits, where sharp current pulses are drawn from the supply. The effects of these pulses are prevented from traveling between PC boards when individual supplies are used on each board.

The typical IC regulator chip contains a Zener-diode reference along with two op-amps and associated circuitry. Since high power cannot be dissipated on a chip, two external transistors are often required when high output current (greater than 1 A) is to be delivered. In this section we shall present the basic theory behind the IC voltage regulator. Simple regulators using only Zener diodes were discussed in Sec. 1.9.

A regulated supply with positive output voltage is shown in Fig. 8.8-1. The circuit is powered by the voltage V_{CC}, which comes from an unregulated power supply. The input is the reference voltage, and the output is the regulated voltage v_L. Resistor R_L represents the load.

The reference voltage is derived from a Zener diode or a regulator diode such as the LM103. Either of these devices provides a relatively constant voltage (typically 1.8 to 6.6 V) which is approximately independent of temperature. For example, the LM103 has a temperature coefficient of about -5 mV/°C when the voltage is 3.3 V and a breakdown characteristic which is 10 times sharper than that of a typical Zener diode.

The two op-amps A_1 and A_2 are chosen to provide minimum sensitivity to power supply and temperature variations. The gain of A_1 is $(R_1 + R_2)/R_1$, and the gain of A_2 including T_1 and T_2 is unity. Note that transistors T_1, T_2 and resistor r

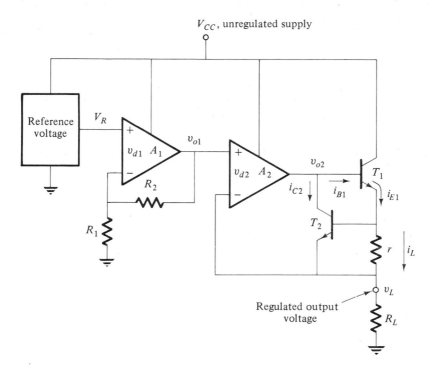

Figure 8.8-1 IC-regulated power supply.

can be considered as part of op-amp A_2. This is a consequence of the feedback connection which is from the output v_L to the inverting input of A_2. Indeed, T_1, T_2, and r are separate discrete components only when very high load current is to be supplied by T_1.

The load voltage v_L is

$$V_L = \left(1 + \frac{R_2}{R_1}\right) V_R \qquad (8.8\text{-}1)$$

Thus the load voltage can be set to any value greater than V_R (and less than V_{CC}) simply by adjusting R_2 and R_1.

Transistor T_1, as mentioned above, is connected as an emitter follower to supply current higher than that available from A_2 to the load R_L.

Transistor T_2 and resistor r provide short-circuit and overload protection for the system. For example, consider that T_1 is rated so that it can safely supply no more than 5 A, and let the power supply be designed to generate 5 V. If for some reason the load R_L is reduced to a value less than 1 Ω, i_L will exceed 5 A and T_1 may overheat and be destroyed. To avoid this difficulty we set $ri_{L,\,\text{max}} = V_\gamma = 0.65$ V. For load currents less than $i_{L,\,\text{max}}$ transistor T_2 will be OFF because its base-emitter voltage is less than 0.65 V. If, however, the load current should exceed $i_{L,\,\text{max}}$, the drop in r will cause transistor T_2 to conduct, removing current

Relative load voltage

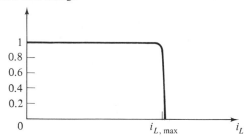

Figure 8.8-2 The constant-current characteristic obtained using T_2.

from the base of T_1. Hence the current i_{E1} tends to remain constant at the value required for $ri_{E1} = 0.65$ V. Since T_2 is barely conducting, the current i_{E2} is much less than the current i_{E1} and the load current $i_L = i_{E1} + i_{E2} \approx i_{E1}$ is kept relatively constant as the load resistor decreases. Thus, in summary, we see that the presence of T_2 means that the load voltage remains constant even though the load resistance may decrease so long as $i_L < i_{L, \max}$. When $i_L = i_{L, \max}$, further decrease in the load impedance results in the load current's remaining constant and the load voltage's decreasing. The regulating action of this voltage regulator with current limiting is plotted in Fig. 8.8-2.

Regulation The word *regulation* is used to describe the departure of a regulator from the ideal. *Line regulation*, defined as the change in output voltage for a given change in input voltage, is typically 100 μV/V. The *load regulation* of the IC supply is the change in output voltage for a change in load current. A typical value of the load regulation is 20 mV/A.

For example, consider that a 15-V unregulated supply is used to drive an IC regulator which supplies 1 A at 5 V. The unregulated supply can drop as low as 12 V, and the load current may be as high as 1.2 A. If the line regulation of the IC supply is 100 μV/V and the load regulation is 20 mV/A, the worst-case change in output voltage is

$$\Delta V_L = (15 - 12)(100) \ \mu\text{V/V} + (1.2\text{-}1)(20) \ \text{mV/A} = 0.3 \ \text{mV} + 4 \ \text{mV} = 4.3 \ \text{mV}$$

8.9 FOUR-QUADRANT ANALOG MULTIPLIER

A *four-quadrant analog multiplier* is a device in which the output voltage is directly proportional to the product of the two input voltages regardless of the polarity of the inputs. Thus

$$v_o = Kv_1v_2$$

where v_1 and v_2 can each be either positive or negative.

The basic four-quadrant multiplier is shown in Fig. 8.9-1. The circuit is similar to the Motorola MC1595. Here the two inputs v_1 and v_2 are applied to two identical difference amplifiers $T_1 - T_2$ and $T_3 - T_4$, each of which is supplied by a current source of strength I_o. To show that this circuit acts as a multiplier we

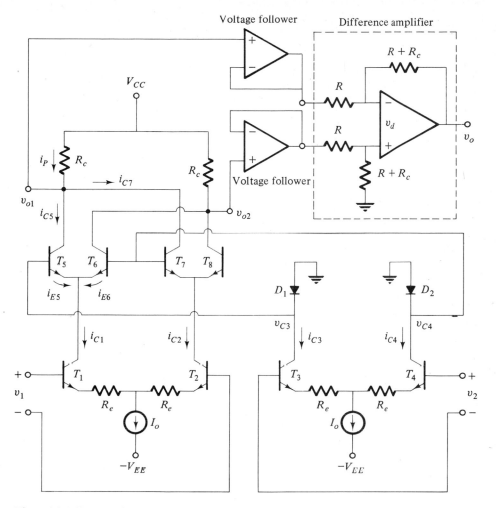

Figure 8.9-1 Four-quadrant analog multiplier.

proceed as follows. With the results of Sec. 7.1 the collector currents of $T_1 - T_4$ (see Prob. 8.9-1) can be shown to be

$$i_{C1} = \frac{I_o}{2} + \frac{v_1}{2(R_e + h_{ib})} \tag{8.9-1a}$$

$$i_{C2} = \frac{I_o}{2} - \frac{v_1}{2(R_e + h_{ib})} \tag{8.9-1b}$$

$$i_{C3} = \frac{I_o}{2} - \frac{v_2}{2(R_e + h_{ib})} \tag{8.9-1c}$$

$$i_{C4} = \frac{I_o}{2} + \frac{v_2}{2(R_e + h_{ib})} \tag{8.9-1d}$$

Also, using KCL, we have

$$i_{C1} = i_{E5} + i_{E6} \qquad (8.9\text{-}2a)$$

and

$$i_{C2} = i_{E7} + i_{E8} \qquad (8.9\text{-}2b)$$

In the Ebers-Moll equation (2.2-14b) if we let $\epsilon^{v_{BC}/V_T} \approx 0$ and neglect the -1 terms, we find i_{E5} and i_{E6} to be

$$i_{E5} \approx I_{EO}\,\epsilon^{v_{BE5}/V_T} \qquad (8.9\text{-}3a)$$

and

$$i_{E6} \approx I_{EO}\,\epsilon^{v_{BE6}/V_T} \qquad (8.9\text{-}3b)$$

Dividing these two, we obtain a relation between i_{E5} and i_{E6}

$$\frac{i_{E5}}{i_{E6}} = \epsilon^{(v_{BE5} - v_{BE6})/V_T} \qquad (8.9\text{-}4)$$

Combining (8.9-4) and (8.9-2a) yields

$$i_{E5} = \frac{i_{C1}}{1 + \epsilon^{-(v_{BE5} - v_{BE6})/V_T}} \qquad (8.9\text{-}5)$$

The base voltage v_{B5} of T_5 is the same as v_{C3}, where v_{C3} is the collector voltage of T_3. Similarly $v_{B6} = v_{C4}$, and, in addition, the emitter voltages of T_5 and T_6 are identical. Therefore using KVL, we get

$$v_{BE5} - v_{BE6} = (v_{B5} - v_{E5}) - (v_{B6} - v_{E6})$$

$$= v_{B5} - v_{B6} = v_{C3} - v_{C4} \qquad (8.9\text{-}6)$$

Thus, (8.9-5) can be written as

$$i_{E5} = \frac{i_{C1}}{1 + \epsilon^{-(v_{C3} - v_{C4})/V_T}} \qquad (8.9\text{-}7)$$

Similarly, after a bit of algebra, it can be shown that

$$i_{E7} = \frac{i_{C2}}{1 + \epsilon^{-(v_{C4} - v_{C3})/V_T}} \qquad (8.9\text{-}8)$$

Since the voltage follower has an extremely high input impedance, the current $i_P = i_{C5} + i_{C7}$; also since $i_{C5} \approx i_{E5}$ and $i_{C7} \approx i_{E7}$, we can combine (8.9-7) and (8.9-8) to obtain

$$i_P \approx i_{E5} + i_{E7} = \frac{i_{C1} + i_{C2}\,\epsilon^{-(v_{C3} - v_{C4})/V_T}}{1 + \epsilon^{-(v_{C3} - v_{C4})/V_T}} \qquad (8.9\text{-}9)$$

This current will now be shown to be directly proportional to the product of v_1 and v_2.

The current i_{C3} flows through diode D_1 and is therefore related to the voltage $v_{C3} = -v_{D1}$ by

$$i_{C3} = I_D\,\epsilon^{-v_{C3}/V_T} \qquad (8.9\text{-}10)$$

Similarly,
$$i_{C4} = I_D \epsilon^{-v_{C4}/V_T} \qquad (8.9\text{-}11)$$

To arrive at (8.9-10) and (8.9-11) we have implicitly assumed that i_{C3} and i_{C4} are much greater than the base currents in T_5, T_6, T_7, and T_8. Then

$$\frac{i_{C3}}{i_{C4}} = \epsilon^{-(v_{C3}-v_{C4})/V_T} \qquad (8.9\text{-}12)$$

Substituting (8.9-12) into (8.9-9), we find

$$i_P = \frac{i_{C1} + i_{C2}i_{C3}/i_{C4}}{1 + i_{C3}/i_{C4}} = \frac{i_{C1}i_{C4} + i_{C2}i_{C3}}{i_{C4} + i_{C3}} \qquad (8.9\text{-}13)$$

where i_{C1} to i_{C4} are given by (8.9-1). Finally, substituting (8.9-1) into (8.9-13) and simplifying yields

$$i_P = \frac{I_o^2/2 + v_1 v_2/[2(R_e + h_{ib})^2]}{I_o} \qquad (8.9\text{-}14)$$

Note that i_P contains a constant and a term directly proportional to the product $v_1 v_2$.

The output voltage v_{o1} is

$$v_{o1} = V_{CC} - i_P R_c = V_{CC} - \frac{I_o R_c}{2} - \frac{R_c}{2I_o(R_e + h_{ib})^2} v_1 v_2 \qquad (8.9\text{-}15)$$

Similarly it can be shown that

$$v_{o2} = V_{CC} - \frac{I_o R_c}{2} + \frac{R_c}{2I_o(R_e + h_{ib})^2} v_1 v_2 \qquad (8.9\text{-}16)$$

If v_{o1} and v_{o2} are the inputs to the two voltage followers, which then are inputted to the difference amplifier shown in Fig. 8.9-1, the common-mode signal $V_{CC} - I_o R_c/2$ disappears (see Sec. 8.4) and the output of the difference amplifier is

$$v_o = \frac{R_c}{I_o(R_e + h_{ib})^2} v_1 v_2 \qquad (8.9\text{-}17)$$

Some applications of the analog multiplier The circuit symbol used to represent the multiplier of Fig. 8.9-1 is shown in Fig. 8.9-2a. If v_1 and v_2 are connected together, the multiplier performs the squaring operation since

$$v_o = K v_1^2 \qquad (8.9\text{-}18)$$

Figure 8.9-2b shows how the multiplier can be used to perform the square-root operation. Since $v_d \approx 0$ and $i_i \approx 0$, we have $v_a/R + v_o/R = 0$, so that $v_a = -v_o$. But

$$v_o = v_1^2 \qquad (8.9\text{-}19)$$

Hence
$$v_1 = \sqrt{-v_a} \qquad (8.9\text{-}20)$$

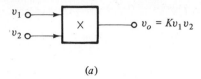

(a)

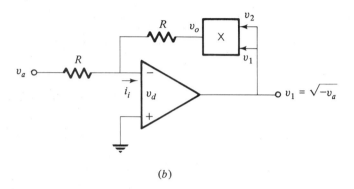

(b)

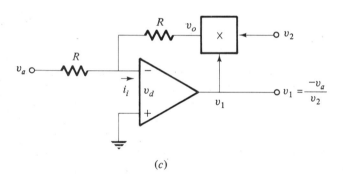

(c)

Figure 8.9-2 Analog multiplier: (a) circuit symbol; (b) square-root circuit; (c) dividing circuit.

Thus v_1 is the square root of $-v_a$, and v_a must be negative. An inverter can be included in the feedback loop so that the circuit will accommodate positive inputs.

As a final example of the use of the multiplier we consider the dividing circuit of Fig. 8.9-2c in which the input voltages are v_a and v_2. Since $v_a \approx 0$ and $i_i \approx 0$,

$$\frac{v_a}{R} + \frac{v_o}{R} = 0 \qquad \text{so } v_a = -v_o \tag{8.9-21}$$

But since $v_o = K v_1 v_2$,

$$v_1 = -\frac{v_a}{K v_2} \tag{8.9-22}$$

In practice the scale factors can be adjusted by changing the resistances.

8.10 AUTOMATIC GAIN CONTROL

A problem that arises in most communications receivers concerns the wide variation in power level of the signals received at the antenna. This variation is due to a
variety of causes. For example, in ordinary commercial AM radio, each station is
at a different distance from the receiver, and each may be transmitting at a
different power level. In a space-communications system the satellite or spaceship
transmitter may be continuously altering its position with respect to the ground
receiver. Since the received signal power decreases as the square of the distance to
the receiver, wide variations in received power level are likely to arise in many
situations.

In receiver design these variations cause serious problems which can usually
be solved by using *automatic gain control* (AGC). AGC circuits utilize feedback to
maintain a fixed signal-power level within the receiver even though the signal level
at the antenna varies widely.

Automatic gain control is achieved by using an amplifier whose gain can be
controlled by an external current or voltage. For example, the circuit shown in
Fig. 8.10-1 can be used as the gain-controlling element in an AGC system. In this
circuit v_i is the input signal of varying level, and v_o is the output signal which is to
have a relatively constant level. The circuit operates as follows. The direct current
I_{AGC} cannot flow into resistor R because of blocking capacitor C; however, it does
bias the diode into its forward region. Assuming that v_i is a small signal, the diode
resistance r_d and resistor R form a voltage divider. Hence the output voltage is

$$v_o = \frac{r_d}{R + r_d} v_i \tag{8.10-1}$$

where from (2.4-15)

$$r_d = \frac{V_T}{I_{AGC}} \tag{8.10-2}$$

When we substitute (8.10-2) into (8.10-1) and assume that $R \gg r_d$, (8.10-1)
becomes

$$v_o \approx \frac{V_T}{R I_{AGC}} v_i \tag{8.10-3a}$$

The AGC action is obtained by deriving I_{AGC} from the envelope (peak value) of v_i

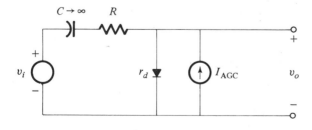

Figure 8.10-1 Basic gain-control
circuit.

which we call $V_{im}(t)$ (in a manner to be described below) so that they are in direct proportion; that is, $I_{AGC}(t) = KV_{im}(t)$. Then (8.10-3a) becomes

$$v_o \approx \frac{V_T}{RK} \frac{v_i}{V_{im}} \qquad (8.10\text{-}3b)$$

and the output is constant even though V_{im} varies. For example, if v_i is sinusoidal with an amplitude V_{im} which is time-varying, we can write

$$v_i(t) = V_{im}(t) \cos \omega_0 t \qquad (8.10\text{-}3c)$$

Hence, $v_i(t)/V_{im}(t) = \cos \omega_0 t$ and

$$v_o(t) \approx \frac{V_T}{RK} \cos \omega_0 t \qquad (8.10\text{-}3d)$$

The envelope of the output voltage is therefore

$$V_{om} = \frac{V_T}{RK} \qquad (8.10\text{-}3e)$$

and is constant, independent of the variations of the input envelope.

Commercially available AGC circuits, such as the LM170 AGC amplifier, use a controlled current source within an op-amp, as shown in Fig. 8.10-2. In this circuit T_3 acts as the constant-current source supplying current I_{AGC}, where

$$I_{AGC} = \frac{V_{AGC} - 0.7 + V_{EE}}{R_e} \qquad (8.10\text{-}4)$$

The small-signal emitter currents were found in Sec. 7.3 to be

$$i_{e1} = -i_{e2} = \frac{v_1 - v_2}{2h_{ib}} \qquad (8.10\text{-}5)$$

Hence, the output voltage v_o is

$$v_o = \frac{R_c}{h_{ib}}(v_2 - v_1) \qquad (8.10\text{-}6)$$

where

$$h_{ib} = \frac{V_T}{I_{AGC}/2} \qquad (8.10\text{-}7)$$

Substituting (8.10-7) into (8.10-6), we find the amplifier gain to be

$$\frac{v_o}{v_2 - v_1} = \frac{R_c}{2V_T} I_{AGC} \qquad (8.10\text{-}8a)$$

In contrast to the circuit of Fig. 8.10-1 we must now arrange to have I_{AGC} inversely proportional to the input voltage envelope; i.e., assuming that $v_1 = 0$,

$$I_{AGC} = \frac{K}{V_{2m}} \qquad (8.10\text{-}8b)$$

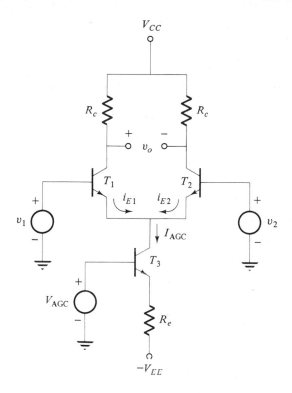

V_{CC}

Figure 8.10-2 A gain-controlled difference amplifier.

Then (8.10-8a) becomes

$$v_o = \left(\frac{R_c K}{2V_T}\right)\frac{v_2(t)}{V_{2m}(t)}$$ (8.10-8c)

and the envelope of v_o is constant even though V_{2m}, the envelope of v_2, varies. In a commercial gain-controlled op-amp the difference amplifier is followed by one or more stages of amplification.

An important parameter in a commercial AGC op-amp is the *dynamic range* of the control, i.e., the extent to which the gain can be varied. For the LM170 the gain can be varied by more than 60 dB (a factor of 1000) by adjusting the control voltage. The range of I_{AGC} in the circuit of Fig. 8.10-2 is from zero to that value which causes T_1 and T_2, or T_3 to saturate.

The circuit symbol for an AGC amplifier is shown in Fig. 8.10-3, where we show a complete AGC system, consisting of a gain-controlled op-amp A_1, such as the LM170, a linear noninverting amplifier A_2 (an inverting amplifier could also be employed), and, as the feedback element, a diode envelope- (peak-) detector circuit consisting of diode D and the RC circuit.

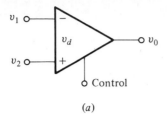

(a)

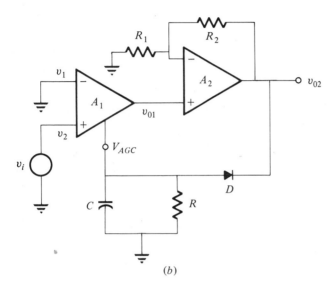

(b)

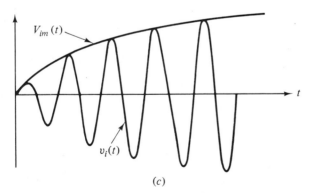

(c)

Figure 8.10-3 AGC system: (a) symbol for gain-controlled op-amp; (b) complete AGC amplifier; (c) illustrating the definition of $v_{im}(t)$.

From (8.10-8a) the output of the gain-controlled op-amp including the additional gain K_1 present in A_1 is

$$v_{o1} = \left(\frac{R_c K_1}{2V_T} I_{\text{AGC}} \right) v_i \qquad (8.10\text{-}9)$$

The output voltage v_{o2} is then

$$v_{o2} = \left(\frac{R_c}{2V_T} K_1 K_2 I_{AGC}\right) v_i \qquad (8.10\text{-}10)$$

where $K_2 = 1 + R_2/R_1$. The output of the envelope (peak) detector V_{AGC} is equal to the negative peak value of $v_{o2} = v_i$. Hence

$$V_{AGC} = -\left(\frac{R_c}{2V_T} K_1 K_2 I_{AGC}\right) V_{im}(t) \qquad (8.10\text{-}11)$$

where $V_{im}(t)$ is the envelope, or peak, value of $v_i(t)$, as illustrated in Fig. 8.10-3c.

The gain-control voltage and current are related by (8.10-4) and (8.10-11). Substituting (8.10-11) into (8.10-4) and solving for I_{AGC} yields

$$I_{AGC} = \frac{V_{EE} - 0.7}{R_e + (R_c/2V_T)K_1 K_2 V_{im}(t)} \qquad (8.10\text{-}12)$$

Substituting (8.10-12) into (8.10-10), we find

$$v_{o2} = \left(\frac{R_c}{2V_T R_e} K_1 K_2\right) \frac{V_{EE} - 0.7}{1 + (R_c/2V_T R_e)K_1 K_2 V_{im}(t)} v_i(t) \qquad (8.10\text{-}13)$$

The gain $K_1 K_2$ is usually made large enough to ensure that

$$\frac{R_c}{2V_T R_e} K_1 K_2 \gg 1 \qquad (8.10\text{-}14)$$

so that, finally,

$$v_{o2} \approx (V_{EE} - 0.7) \frac{v_i(t)}{V_{im}(t)} \qquad (8.10\text{-}15)$$

This important result indicates that v_{o2} is proportional to $v_i(t)/V_{im}(t)$. This ratio has a *constant envelope;* i.e., if $v_i(t) = A(t) \cos \omega_0 t$, where $A(t)$ changes slowly, then $V_{im}(t) = A(t)$ and $v_i(t)/V_{im}(t) = \cos \omega_0 t$, a function having a constant envelope. As a result of the peak-detector action the AGC circuit responds only to slowly varying changes in the envelope, i.e., to changes in signal power. A typical value for the RC time constant of the peak detector is 1 s.

8.11 PRACTICAL CONSIDERATIONS IN OP-AMP CIRCUITS

Maintaining symmetry in the input circuit In the op-amp circuits which we have discussed up to this point it has been assumed that perfect symmetry exists in the difference amplifiers so that when both inputs are the same, the output is zero; i.e., the common-mode rejection is perfect. In real op-amps this is not the case, and the output voltage is not exactly zero even when both input terminals are grounded.

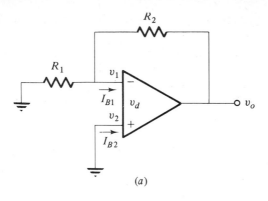

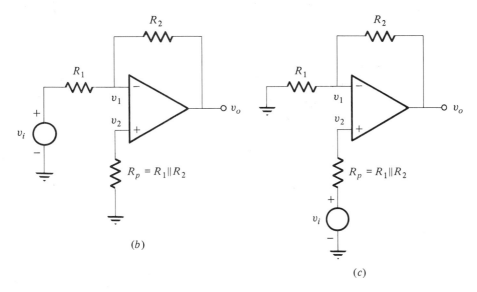

Figure 8.11-1 (a) Inverting amplifier; (b) amplifier with R_p; (c) noninverting amplifier with R_p.

This is due to a number of causes. For example, in a BJT op-amp some bias current is required to keep the input transistors turned on. These bias currents I_{B1} and I_{B2} flow into the input terminals, as shown in Fig. 8.11-1a, where we have grounded both input terminals. If we assume that v_o is small, I_{B1} flows through the parallel combination of R_1 and R_2 and then into the v_1 terminal, causing a voltage $v_1 \approx I_{B1}(R_1 \| R_2)$ to appear at input 1. Since input 2 is grounded, we have $v_2 = 0$, so that

$$v_d = v_2 - v_1 \approx -I_{B1}(R_1 \| R_2) \tag{8.11-1}$$

This differential input voltage causes an *output offset voltage* $v_o = A_d v_d$. Typical values for this offset voltage are less than 1 mV, and it can be eliminated in practice by connecting a resistor $R_p = R_1 \| R_2$ between terminal 2 and ground. When this is done, the bias currents, which are usually very nearly the same, flow through equal resistances so that there will be no differential input voltage. The inverting configuration with R_p connected is shown in Fig. 8.11-1b, and the noninverting amplifier is shown in Fig. 8.11-1c. To take into account differences in I_{B1} and I_{B2} resistor R_p can be made variable and adjusted for minimum output offset.

This compensation should be used in most op-amp circuits. For example, in the all-pass filter shown in Fig. 8.4-6, $R = R_1/2$ to ensure symmetry. In the op-amp clipper of Fig. 8.5-3 a resistor $R_p = R_L$ should be inserted in series with the input v_i. Some op-amp circuits are rather complicated, and it is not always a simple matter to find R_p. For example, consider Fig. 8.5-4. In this circuit, a resistor $R_{p1} = R/2$ should be inserted between input v_2 and ground and a resistor $R_{p2} = R/4$ connected between input v_4 and ground.

Data-Sheet Specifications

Maximum ratings (limits which, if exceeded, may permanently damage the device):

1. *Supply voltage.* Although two supplies are often required, such as ± 1.5, ± 15, or even ± 40 V, some op-amps require only a single supply, for example, $+15$ V.
2. *Power dissipation.* The maximum power the IC can dissipate without being destroyed is always specified by the manufacturer. A typical value is 0.5 W.
3. *Operating temperature.* The ambient temperature over which *military-grade* ICs can safely operate is from -55 to $+125°$C. For *commercial-grade* devices the range is from 0 to 70°C.
4. *Maximum input difference- and common-mode voltages.* If the input difference voltage v_d exceeds the amount specified by the manufacturer, excessive current flows and the op-amp may be destroyed. This will also be the case if an excessive common-mode voltage is applied to the inputs. Typical values are ± 30 V.

Electrical Characteristics (Performance Characteristics as Determined by Test)

1. *Input offset voltage.* This is the difference voltage v_d that must be applied to make the output voltage zero. Ideally $v_d = 0$ causes $v_o = 0$; however, due to the lack of perfect symmetry between transistors in the IC, some finite value of v_d is required for v_o to be zero. Typical values are 1 mV. Various nulling adjustments can be used to offset this effect.

2. *Input offset current.* This is the difference in currents entering inputs v_1 and v_2 when the output is made zero by the insertion of an offset voltage. A typical value of this current is 100 nA.

3. *Input bias current.* This is the average of the two input currents that are needed to keep the input transistors operating properly. Typical values are 300 nA.

4. *Temperature coefficient of input offset voltage.* The input offset voltage is a function of temperature and typically varies by 5 μV/°C.

5. *Large-signal voltage gain.* This is the ratio of output voltage to input voltage when the output voltage is at its maximum symmetrical, unclipped level. This value is typically greater than 100,000 and is approximately equal to A_d.

6. *Common-mode rejection ratio.* This is the ratio of the difference-mode gain to the common-mode gain A_d/A_a, typically 100 dB (100,000).

7. *Supply-voltage rejection ratio.* The lack of perfect symmetry in the circuit means that the output voltage varies with the supply voltage. The supply-voltage rejection ratio is the ratio of the change in the input offset voltage to the change in power-supply voltage producing it; i.e., suppose a particular offset voltage V_{io} at supply voltage V_{CC} is needed to make the output voltage zero. If the supply voltage is changed to $V_{CC} + \Delta V_{CC}$, the required input offset voltage becomes $V_{io} + \Delta V_{io}$. Then the supply-voltage rejection ratio is $\Delta V_{io}/\Delta V_{CC}$ and is typically -100 dB.

8. *Frequency compensation.* Frequency limitations in the op-amp are due to the finite bandwidths of the transistors and also the stray capacitances which are present in every circuit. As a result of these factors and the very high open-loop gain A_d of the op-amp, oscillations will occur unless frequency compensation is used. The *transient response* of the op-amp is also degraded by these effects. Frequency compensation of op-amps will be discussed in Chap. 10.

9. *Unity-gain bandwidth.* This is the frequency range from dc to the frequency at which the amplifier gain v_o/v_d has decreased to unity. Typical unity-gain bandwidths are approximately 1 MHz. The importance of this characteristic will be discussed in Chap. 10.

10. *Slew rate and settling time.* Referring to Fig. 8.11-2, we see that when a large input step in voltage is applied to the op-amp, the output waveform rises with a finite slope called the *slew rate*. The slewing behavior of the output is due to a nonlinear effect which occurs when a large input signal is applied to the op-amp. The transistors in the op-amp momentarily saturate or cut off until

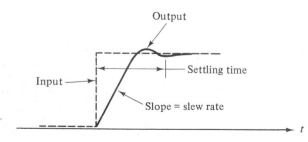

Figure 8.11-2 Transient response of an op-amp.

the external feedback elements reduce the input voltage v_d. As a result of stray and internal capacitance the output voltage takes some time to reach its final steady-state value. The time that it takes to come within a prescribed percentage of the final value is called the *settling time*. Both the slew rate and settling time are usually measured when the voltage gain is unity using the voltage-follower configuration. They depend upon whether the input step is positive or negative and whether the amplifier is inverting or noninverting.

Types of op-amps Op-amps can be classified into the five distinct categories listed in Table 8.11-1. The *general purpose* op-amps, such as the 709, 101, and 741, are the workhorses of the industry. The 709 was one of the first op-amps sold and is still widely used. The 101 is similar to the 709 but has significantly higher gain (150,000 compared with 45,000), has twice the slew rate of the 709, and is easier to frequency-compensate. The 741 is similar to the 101 but requires no external frequency compensation. It is internally compensated by the manufacturer. These general-purpose op-amps typically have a unity gain bandwidth of 1 MHz and a slew rate of 0.5 V/μs.

High-frequency, high slew-rate op-amps are available with bandwidths of 50 MHz and slew rates of 200 V/μs. In contrast, the *hybrid* op-amp 0063 has a unity gain bandwidth of 150 MHz and a slew rate of 6000 V/μs. Op-amps like these are used for high-frequency sampling and analog-to-digital and digital-to-analog conversion.

High-voltage output swings and *high output current* can be obtained using the hybrid op-amps 0004 and 0021 or the 124. The 0004 can produce an output voltage of ± 35 V into a 2-kΩ load while the 0021 can deliver in excess of 1 A at voltage levels of ± 12 V. The 124 uses a single power supply and can operate over a supply-voltage range of 5 to 32 V.

For applications requiring *low offset voltage* and *low drift with temperature* the 108 has an input offset voltage of 0.7 mV and an input-offset-voltage drift of 3 μV/°C. However, if the hybrid op-amp 0052 is selected, an input offset voltage of 0.1 mV and a drift of 2 μV/°C can be obtained.

A *programmable* op-amp such as the 4250 is arranged so that the biasing of all internal transistors can be controlled externally. Thus, these op-amps can be biased to have very low standby power dissipation, and as a result they are often called *micropower op-amps*. The main amplifier parameters governed by the bias

Table 8.11-1

Category	Representative op-amp type numbers
General purpose	709, 101, 741
High frequency, high slew rate	LH0063
High-voltage, high-power, single supply	LH0004, LH0021, LM124
Low input offset voltage and drift	LH0052, 108
Programmable	4250

control are the *open-loop gain, unity-gain bandwidth, input bias current, slew rate,* and *standby power.*

8.12 OTHER LINEAR IC AMPLIFIERS

There are many other types of *linear IC amplifiers* in addition to the op-amp and regulated voltage supply. The multiplier (which is an AM modulator) and AGC circuits discussed earlier are often considered to be *consumer IC circuits* since they are used in AM radios. Other consumer circuits include RF/IF amplifiers, AM/FM/SSB amplifier-detectors, oscillators, low-noise preamplifiers, and even power amplifiers (typically 4 W). With variety like this available, the engineer of today rarely sets out to design a system using discrete circuits.

PROBLEMS

8.1-1 In Fig. 8.1-2 let $R_1 = 1$ kΩ and $R_2 = 10$ kΩ. The op-amp is characterized by $A_d = 200,000$, $R_i = 200$ kΩ, and $R_o = 150$ Ω. Find (a) A_v, (b) r_i, and (c) r_o.

8.1-2 Repeat Prob. 8.1-1 if $R_1 = 10$ kΩ and $R_2 = 1$ MΩ.

8.1-3 (a) Show that the inverting amplifier in Fig. 8.11-1b has A_v as in (8.1-2b). Use ideal op-amp theory.

   (b) Find r_i and r_o for this case.

8.1-4 Design an inverting amplifier to have a gain of -24. The input impedance is to exceed 10 kΩ. The op-amp is characterized in Prob. 8.1-1.

8.1-5 The circuit in Fig. P8.1-5 can be used to measure the gain of the op-amp. Assume $R_i \gg R_4$ and $R_1 \gg R_4$.

   (a) Using ideal op-amp theory show that $v_o = -v_i$.
   (b) Show that the op-amp gain is given by

$$|A_d| = \frac{v_o/v_x}{R_4/R_1}$$

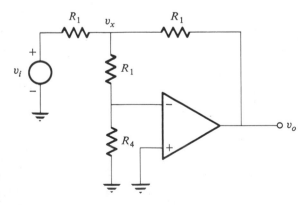

Figure P8.1-5

8.2-1 In Fig. 8.2-1 let $R_1 = 1$ kΩ and $R_2 = 8.8$ kΩ. The op-amp is described by $A_d = 200,000$, $R_i = 200$ kΩ, and $R_o = 150$ Ω. Find (a) A_v, (b) r_i, and (c) r_o.

8.2-2 Repeat Prob. 8.2-1 if $R_1 = 10$ kΩ and $R_2 = 100$ kΩ.

8.2-3 (*a*) Show that the noninverting amplifier in Fig. 8.11-1*c* has A_v as in (8.2-2).

(*b*) Find r_i and r_o for this case.

8.2-4 Design a noninverting amplifier to have a gain of $+51$ using the op-amp characterized in Prob. 8.2-1.

8.2-5 In Fig. P8.2-5 find R_1 so that $v_o = v_s = i_s \times 10 \text{ M}\Omega$.

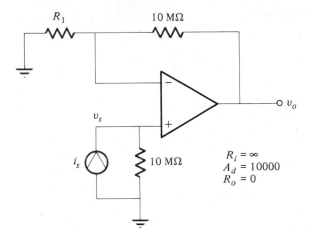

$R_i = \infty$
$A_d = 10000$
$R_o = 0$

Figure P8.2-5

8.4-1 (*a*) Design an op-amp difference amplifier to have a gain of 100. Each input should have a minimum impedance of 10 kΩ.

(*b*) Assume the op-amp has a CMRR $- 1000$ and that the maximum common-mode signal is 0.1 V. Find the differential signal for which the differential-mode output is at least 100 times the common-mode output.

8.4-2 Repeat Prob. 8.4-1 for a gain of 48.

8.4-3 Design an op-amp circuit to implement

$$v_o = 2.4v_1 - 4.6v_2 + 8.7v_3$$

8.4-4 (*a*) In Fig. 8.4-2 calculate the input impedance seen by v_1 with all other inputs shorted using ideal op-amp theory.

(*b*) Modify the result in part (*a*) if the op-amp is nonideal.

8.4-5 Find v_o in Fig. P8.4-5.

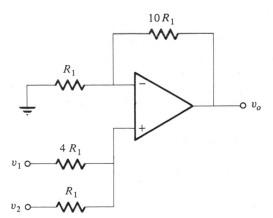

Figure P8.4-5

8.4-6 The circuit in Fig. P8.4-6 converts the voltage v_i into a current i_L. Show that if $2R_1 = 2R_2 = R_3 + R_4$ and $R_4 - R_1 = 500\ \Omega$ then $i_L = +v_i$ mA.

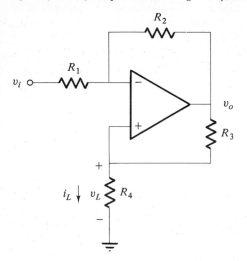

Figure P8.4-6

8.4-7 In Fig. P8.4-7 assume both T_1 and T_2 are MOSFETs and that V_N turns the switch ON. Explain the operation of the circuit and evaluate v_o.

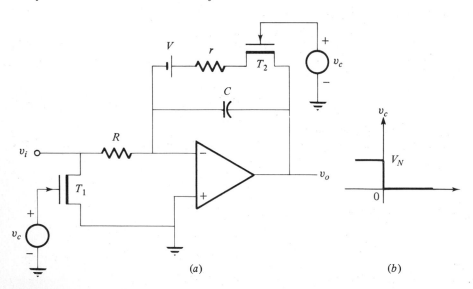

(a)

(b)

Figure P8.4-7

8.4-8 Use op-amp summers and integrators to solve the differential equations:

(a) $\dfrac{3dv_o}{dt} + 5v_o = 5 \sin 2\pi 60t$

(b) $\dfrac{d^3v_o}{dt^3} + \dfrac{3d^2v_o}{dt^2} + \dfrac{3dv_o}{dt} + v_o = 4 \cos 4t$

4/8/81

8.4-9 The circuit of Fig. P8.4-9 is a practical differentiator since it minimizes noise problems by attenuating high frequencies.

(a) Determine the voltage gain function $V_o(j\omega)/V_i(j\omega)$.

(b) If $R_1 C_1 = R_2 C_2$, to what frequencies should the input be restricted so that the circuit will function as a differentiator $[V_o(j\omega) = \text{const} \times j\omega V_i(j\omega)]$?

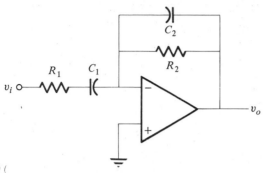

Figure P8.4-9

4/8/81

8.4-10 In Fig. P8.4-9 let C_1 be infinite.

(a) Determine the voltage-gain function $V_o(j\omega)/V_i(j\omega)$.

(b) For what range of input frequencies is the circuit a low-pass filter?

8.4-11 In Fig. P8.4-9 let C_2 be zero.

(a) Determine the voltage-gain function $V_o(j\omega)/V_i(j\omega)$.

(b) For what range of input frequencies is the circuit a high-pass filter?

8.4-12 The circuit in Fig. P8.4-12a is equivalent to the parallel RCL tuned circuit in Fig. P8.4-12b, a bandpass filter. It can be used at low frequencies where high Qs are not easily obtained (see Sec. 9.5). Assume an ideal op-amp,

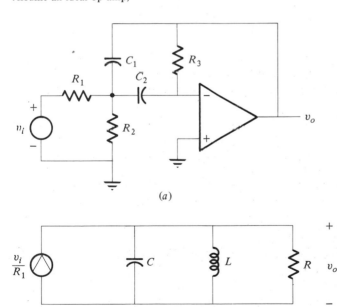

(a)

(b)

Figure P8.4-12

(a) Show that the transfer function for Fig. P8.4-12a is given by

$$\frac{V_o(j\omega)}{V_i(j\omega)} = -\frac{1/R_1}{sC_1 + \dfrac{R_1 + R_2}{R_1 R_2 R_3 C_2 s} + \dfrac{C_1 + C_2}{C_2 R_3}}$$

(b) Find the equivalent R, L, and C.

8.5-1 In Fig. 8.5-1 let $R_L = 1\ k\Omega$, $V_\gamma = 0.65\ V$, $V_F = 0.7\ V$, $A_d = 10^5$, and $v_i = 10^{-3} \sin 2\pi 1000t\ V$. Sketch v_L.

8.5-2 Repeat Prob. 8.5-1 if $v_i = 10^{-6} \sin 2\pi 1000t\ V$.

8.5-3 In Fig. 8.5-3 let $R_L = 1\ k\Omega$, $V_{ref} = 5\ V$, and $v_i = 10 \sin 2\pi 1000t\ V$. Sketch v_L.

8.5-4 Verify (8.5-6).

8.5-5 In Fig. 8.5-4 replace R between v_i and v_1 by 0.9R, R between v_1 and v_r by 1.1R, R/2 between v_r and v_s by 0.9R/2, R between v_s and v_L by 1.1R, and R between v_i and v_s by 0.9R. Sketch v_r and v_L assuming that v_i is given by the waveform in Fig. 8.5-4.

8.5-6 In Fig. 8.5-5 sketch the waveform for v_c and verify the waveform for v_L.

8.5-7 In Fig. 8.5-7 the noninverting input is connected to a reference voltage of $+2\ V$. Using v_i as in Fig. 8.5-5, sketch v_L. Assume that the capacitor is initially uncharged.

8.5-8 Apply the waveform of Fig. 8.5-6a to Fig. 8.5-7 and sketch v_L. Assume that the capacitor is initially uncharged.

8.5-9 In Fig. 8.5-8 let $R_L = 1\ k\Omega$, $C = 100\ \mu F$, and $v_i = 5 \sin 2\pi 60t\ V$. Find the ripple and the dc voltage v_L.

8.5-10 In Fig. P8.5-10 the diodes are identical and described by the curve in Fig. 8.5-1b. Let $v_i = 5 \sin \omega_0 t\ V$, $V_\gamma = 0.65\ V$, and $V_F = 0.7\ V$. $R = 1\ k\Omega$.
  (a) Sketch v_o.
  (b) Calculate $|\Delta v_i|$ if $A_d = 10^5$.

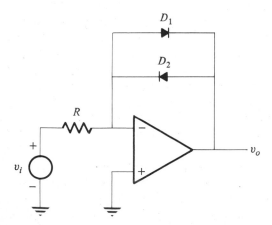

Figure P8.5-10

8.5-11 In the limiter shown in Fig. P8.5-11 the Zener diodes are identical and described by the curve in Fig. 8.5-9b. For v_i as shown in Fig. 8.5-9c sketch v_o and calculate $|\Delta v_i|$ if $A_d = 10^5$. $V_{Z1} = V_{Z2} = 5\ V$.

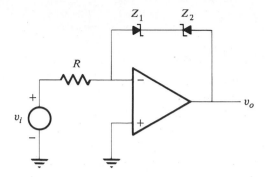

Figure P8.5-11

8.5-12 In Fig. 8.5-9 place a resistor R_f in parallel with the diode. Calculate $|\Delta v_i|$ and sketch v_o for the given v_i if $A_d = 10^5$.

8.6-1 In Fig. 8.6-1a let $C = 1\ \mu F$, $r = R = 10\ k\Omega$, $V_Z = 5$ V, and $V_{CC} = 15$ V. Assume the switch has been closed for a long time and opened at $t = 0$. Sketch $v_c(t)$ by determining the equations for $v_c(t)$ in the linear and exponential regions.

8.6-2 In Fig. 8.6-1a let $C = 0.1\ \mu F$, $r = 10\ k\Omega$, $R = 1\ M\Omega$, $V_Z - 7$ V, and $V_{CC} - 21$ V. Assume that the switch has been closed for a long time. At $t = 0$ the switch is opened for T s, then closed for 1 ms, reopened for T s, closed for 1 ms, etc.

   (a) Calculate the maximum T for $v_c(t)$ to be a linear sawtooth.
   (b) Sketch $v_c(t)$ assuming T as given in part (a).

8.6-3 (a) Design the astable bootstrap sweep generator in Fig. 8.6-1d to generate a 0.2-s sawtooth by specifying R_1 and R_2. Use the values in Prob. 8.6-2. Assume that the MOSFET has $V_T = 2$ V and that when it is ON, $R_{FET} - 500\ \Omega$. Assume further that C_1 has been properly chosen to keep the switch on for 1 ms.

   (b) Sketch $v_c(t)$ for the above design.

8.7-1 A log-amp as in Fig. 8.7-2 using the LM194 has the parameter values $R_1 = 10\ k\Omega$, $R_2 = 1\ M\Omega$, $R_3 = 16.4\ k\Omega \pm 1$ percent, $R_4 = 1\ k\Omega$ (temperature-compensated), and $v_2 = V_2 = 15$ V.

   (a) Find v_o as a function of v_1.
   (b) If v_1 varies from 1 mV to 10 V, plot v_o versus v_1.

8.7-2 Repeat Example 8.7-1, finding the value for the reference voltage v_2 which will cause v_o to be 0 V when $v_1 = 1$ V.

8.7-3 The circuit in Fig. P8.7-3 yields an output voltage which is proportional to the antilogarithm of the input voltage

$$v_o = K_1 \exp(-K_2 v_i)$$

Find K_1 and K_2 and indicate all restrictions on v_i.

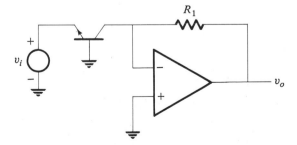

Figure P8.7-3

8.8-1 In Fig. P8.8-1 (*a*) combine the power transistor, the resistor *r*, and the op-amp into the equivalent op-amp. Use the model for the op-amp given in Fig. 8.1-1 and specify the equivalent R_i, A_d, and R_o.

(*b*) Repeat part (*a*) but include R_L in the equivalent op-amp.

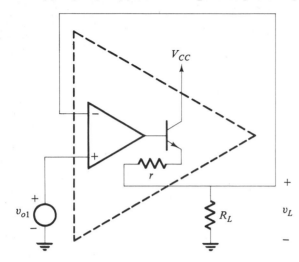

Figure P8.8-1

8.8-2 A Zener-diode–op-amp regulator is to be designed to deliver 15 V to a load current that varies between 10 and 50 mA. The supply is unregulated and varies between 20 and 25 V. Assume that an op-amp capable of delivering 50 mA and a 3.3-V Zener diode having a turn-on current of 1 mA are available. Design the regulator.

8.8-3 Repeat Prob. 8.8-2 assuming that the op-amp can supply only 5 mA but that a transistor with $h_{FE} = 100$ and a current rating of 1 A is available. What is the minimum dissipation the transistor must withstand?

8.8-4 A 20-V unregulated supply is used to drive an IC regulator which supplies 2 A at 6 V. The unregulated supply can drop as low as 15 V, and the load current may be as high as 2.3 A. If the line regulation of the IC supply is 100 μV/V and the load regulation is 25 mV/A, find the worst-case change in output voltage.

8.9-1 Verify Eqs. (8.9-1).

8.9-2 Derive (8.9-8).

8.9-3 Verify (8.9-14).

8.9-4 Modify the analog square-root circuit to handle positive inputs.

8.9-5 Design a circuit to implement $\sqrt{v_1^2 + v_2^2}$.

8.10-1 The IC amplifier shown in Fig. P8.10-1 is used as an AGC amplifier by controlling V_{AGC}. Calculate v_L/v_i as a function of V_{AGC}.

8.10-2 In Fig. 8.10-2 let $R_c = R_e = 2$ kΩ, $V_{CC} = V_{EE} = 5$ V, and $v_1 = v_2 = 0$.

(*a*) Find $V_{AGC, max}$ which causes T_1, T_2, or T_3 to saturate.

(*b*) Find $V_{AGC, min}$.

8.10-3 The gain-controlled difference amplifier in Fig. 8.10-2 is used as the controlling element in the first amplifier of Fig. 8.10-3*b*. Use the values and results in Prob. 8.10-2 and assume that $R_1 = 1$ kΩ, $R_2 = 100$ kΩ, $K_1 = 10^3$, $V_T = 25$ mV, $v_2 = 0$, and $v_1 = V_{im}(t) \sin \omega t$.

(*a*) Find the minimum $V_{im}(t)$ so that v_o is a constant amplitude. What is the amplitude of v_o?

(*b*) What is the maximum $V_{im}(t)$ that can be allowed so that v_o will not distort? What causes this distortion?

(*c*) Determine $v_o(t)$ if $V_{im}(t) = 0.5$ μV, a constant.

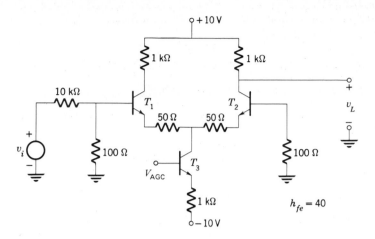

Figure P8.10-1

FREQUENCY AND SWITCHING-SPEED LIMITATIONS

A_m = gain at midfreq

In this chapter we discuss the frequency and switching characteristics of bipolar and field-effect transistors. A graph of the variation of a typical amplifier current gain or voltage gain with frequency is shown in Fig. 9.1. A_m is the maximum gain of the amplifier, which occurs in the midfrequency range, and the gain is seen to decrease at both high and low frequencies. These falloffs are due to two different types of capacitance.

At high frequencies, all transistor amplifiers have inherent limitations because of the internal capacitance of the transistors and stray, or parasitic, wiring capacitance to ground, which is always present. In the BJT, for example, capacitance exists across the collector-base and base-emitter junctions. At high frequencies the impedance of these capacitors becomes small, and consequently the gain of the amplifier is reduced. In digital circuits, where the transistor must switch rapidly from a high voltage to a low voltage or vice versa, the switching speed is limited because the voltages across these capacitances cannot change instantaneously.

At the low-frequency end of the spectrum, multitransistor amplifiers constructed with discrete transistors often employ capacitors to *couple* one transistor to the next transistor in the chain. In addition, as discussed in Sec. 2.5, emitter *bypass* capacitors are used to short-circuit the emitter resistor and thus increase the gain at high frequencies. These coupling and bypass capacitors, which are external to the transistor, cause the falloff in the low-frequency response of the amplifier because their impedance becomes large at low frequencies.

When the midfrequency range is a few octaves or more, the problem of determining the frequency response of a typical *RC*-coupled amplifier breaks down into essentially three relatively simple problems. The coupling and bypass

$X_C = \frac{1}{j\omega C}$

capacitors affect only the low-frequency response, while the transistor capacitances affect only the high-frequency response. Thus, for convenience, we divide the frequency spectrum into three regions: low, middle, and high (see Fig. 9.1). In each of these frequency bands, a different simplified equivalent circuit is used to calculate the response. In the low-frequency band, the large coupling and bypass capacitors are important, while the small transistor and stray capacitances are effectively open circuits. In the midfrequency range, the large capacitors are effectively short circuits and the small capacitances are open circuits, so that no capacitances appear in the midband equivalent circuit. At high frequencies, the large capacitors are replaced by short circuits, and the transistor and stray capacitances help determine the response.

Throughout the chapter, transfer functions are written in terms of the Laplace-transform frequency variable s (complex frequency). Thus the response to any excitation can be found using Laplace-transform techniques. The steady-state sinusoidal response, which is our main concern, is obtained simply by replacing s by $j\omega$ in the transfer function.

The boundaries separating the midfrequency range from the low- and high-frequency ranges are not well defined, and the useful operating range of an amplifier is specified in various ways, depending on the application. In some applications, the exact shape of the gain-frequency characteristic is important. For example, consider an amplifier designed for video signals. Since the *eye* is sensitive to light-intensity variations of the order of 1 dB, the response in the midfrequency range should not depart from A_m by more than about $\frac{1}{2}$ dB if high-quality pictures are to be obtained. The edges of the midfrequency region are determined by the useful frequency content of the signal. If the video signal contains important frequency components ranging from dc to 4 MHz, then no more than $\frac{1}{2}$-dB maximum variation will have to be maintained to 4 MHz. The rate at which the response can fall off above 4 MHz will depend on other factors, such as the required transient response.

For high-quality reproduction of audio signals, a frequency band from about 20 Hz to 20 kHz is required. In the usual application, the boundaries of the midfrequency range are defined as those frequencies at which the response has fallen to 3 dB below A_m. These are shown as f_L and f_h in Fig. 9-1 and are referred

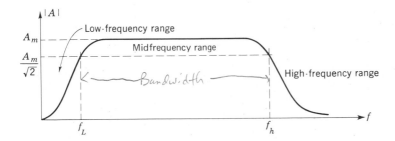

Figure 9.1 Gain versus frequency for a typical *RC*-coupled amplifier.

$3 dB freq = cutoff freq$ Bandwidth $B = f_h - f_L$

to as the 3-dB frequencies, or simply the *cutoff frequencies*. The total midfrequency range is called the *bandwidth B* (it is actually the *half-power* bandwidth), and from the figure, $B = f_h - f_L$ (when $f_h \gg f_L$, the bandwidth $B \approx f_h$). These definitions are used throughout this text.

9.1 THE LOW-FREQUENCY RESPONSE OF THE TRANSISTOR AMPLIFIER

The low-frequency response of the common-emitter amplifier is determined by the emitter bypass capacitor and the coupling capacitors. In practical amplifier circuits, the emitter bypass capacitor usually limits the low-frequency response. In the following sections we consider the effect of each capacitor separately.

9.1-1 The Emitter Bypass Capacitor — usually limits the low freq. response

Consider the single-stage amplifier of Fig. 9.1-1a. Coupling capacitors in the input and collector circuits are omitted in order to focus attention on the emitter bypass capacitor. The small-signal equivalent circuit is shown in Fig. 9.1-1b, where the

hie/hfe

$hib = \dfrac{hie}{1+hFe}$

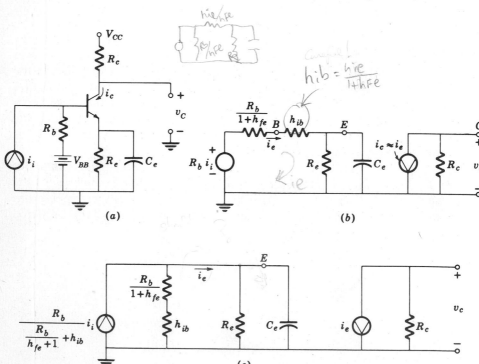

(a)   (b)

(c)

Figure 9.1-1 CE amplifier with emitter bypass capacitor: (a) a single-stage amplifier; (b) small-signal equivalent circuit valid for low frequencies; (c) reduced equivalent circuit.

base circuit has been reflected into the emitter circuit. This circuit is valid for the low-frequency and midfrequency regions.

The current-gain transfer function is obtained from the simplified equivalent circuit of Fig. 9.1-1c by routine analysis

$$A_i = \frac{i_c}{i_i} \approx \frac{i_e}{i_i} = \left(\frac{R_b}{\frac{R_b}{h_{fe}+1} + h_{ib}} \right) \left(\frac{s + 1/R_e C_e}{s + \frac{1}{[R_e\|(R_b/(h_{fe}+1)) + h_{ib})]C_e}} \right) \qquad (9.1\text{-}1a)$$

When ω is very large, the gain is at its midfrequency value A_{im}

$$A_{im} = \left| \frac{i_c}{i_i} \right|_{\omega \to \infty} \approx \frac{R_b}{R_b/(h_{fe}+1) + h_{ib}} \approx \frac{h_{fe}}{1 + h_{ie}/R_b} \qquad (9.1\text{-}1b)$$

(handwritten) $\left(\dfrac{\frac{R_b}{R_b + h_{ie}}}{h_{Fe}+1}\right)\left(\dfrac{s + \frac{1}{R_e C_e}}{s + \frac{1}{R_e\|\frac{R_b + h_{ie}}{h_{Fe}+1}} C_e}\right)$

The emitter resistor R_e is usually specified so that

$$R_e \gg \frac{R_b}{h_{fe}+1} + h_{ib} \qquad (9.1\text{-}1c)$$

in order that the zero of (9.1-1) will occur at a much lower frequency than the pole. The zero of (9.1-1a) occurs at

(handwritten) zero

$$\boxed{\omega_1 = \frac{1}{R_e C_e}} \qquad \leftarrow \text{ set Numerator } = 0 \qquad (9.1\text{-}2a)$$

and the pole at

(handwritten left margin) $W_2 = \left[\frac{r_i\|R_b + h_{ie}}{h_{Fe}+1}\|R_e\right]C_e$ pole

$$\boxed{\omega_2 = \frac{1}{\{R_e\|[R_b/(h_{fe}+1) + h_{ib}]\}C_e}} \quad \overset{\text{set}}{\text{denominator}} = 0 \qquad (9.1\text{-}2b)$$

When the condition of (9.1-1c) is valid,

$$\omega_1 \ll \omega_2 \qquad (9.1\text{-}2c)$$

(handwritten) $C_e = \{R_e\|[\frac{R_b}{h_{Fe}+1} + h_{ib}]\}\omega$

and using (9.1-1b), when $\omega > \omega_1$, (9.1-1a) becomes

$$|A_i| = A_{im}\left|\frac{j\omega + \omega_1}{j\omega + \omega_2}\right| \approx \frac{A_{im}\,\omega}{\sqrt{\omega^2 + \omega_2^2}} \qquad (9.1\text{-}3a)$$

The frequency $f = f_L$, at which the current gain is down 3 dB from its value in the midfrequency range, is found by setting

$$|A_i| = A_{im}\frac{\omega_L}{\sqrt{\omega_L^2 + \omega_2^2}} = \frac{A_{im}}{\sqrt{2}} \quad \text{(handwritten: 3 dB Value)} \qquad (9.1\text{-}3b)$$

Solving, we have

$$f_L \approx f_2 \qquad (9.1\text{-}3c)$$

If (9.1-1c) is not satisfied, the response may *never* be 3 dB down at low frequencies. This can be seen from (9.1-1a). For a 3-dB frequency to exist, we must have

$$|A_i(\omega = \omega_L)|^2 = A_{im}^2 \frac{\omega_L^2 + \omega_1^2}{\omega_L^2 + \omega_2^2} = \frac{A_{im}^2}{2}$$

Then
$$\omega_L^2 + \omega_2^2 = 2\omega_L^2 + 2\omega_1^2$$

and
$$\omega_L^2 = \omega_2^2 - 2\omega_1^2 \qquad (9.1\text{-}3d)$$

Thus, if
$$2\omega_1^2 > \omega_2^2$$

the 3-dB frequency f_L *does not exist.*

9.1-2 Asymptotic (Bode) Plots of Amplifier Transfer Functions

It is extremely helpful to be able to exhibit the frequency dependence of equations such as (9.1-1a) graphically. This is easily done using asymptotic logarithmic characteristics and logarithmic scales. To illustrate the method, we rewrite (9.1-1) in the form

$$A_i(j\omega) = \frac{i_c}{i_i} = A_{im}\frac{j\omega + \omega_1}{j\omega + \omega_2} = A_{io}\frac{1 + j\omega/\omega_1}{1 + j\omega/\omega_2} = |A_i|\,\epsilon^{j\theta} \qquad (9.1\text{-}4)$$

where
$$A_{io} = A_{im}\frac{\omega_1}{\omega_2}$$

In most practical problems, interest centers on the magnitude and phase angle of $A_i(j\omega)$. These are

$$|A_i| = A_{io}\frac{\sqrt{1 + (\omega/\omega_1)^2}}{\sqrt{1 + (\omega/\omega_2)^2}} \qquad (9.1\text{-}5)$$

$$\theta = \left(\tan^{-1}\frac{\omega}{\omega_1}\right) - \left(\tan^{-1}\frac{\omega}{\omega_2}\right) \qquad (9.1\text{-}6)$$

Let us first find A_i, in decibels:

$$|A_i|_{\text{dB}} = 20(\log A_{io}) + 20\left[\log \sqrt{1 + \left(\frac{\omega}{\omega_1}\right)^2}\right] - 20\left[\log \sqrt{1 + \left(\frac{\omega}{\omega_2}\right)^2}\right] \qquad (9.1\text{-}7)$$

The problem is now considerably simplified. We plot each of the factors of (9.1-7) and then add the individual curves graphically.

To illustrate this, consider (9.1-7). The first term, $20 \log A_{io}$, is a constant. To plot the second term, consider the asymptotic behavior at very low and very high frequencies:

At low frequencies ($\omega \to 0$):

$$20 \log \sqrt{1 + \left(\frac{\omega}{\omega_1}\right)^2} \to 20(\log 1) = 0 \text{ dB} \qquad (9.1\text{-}8a)$$

At high frequencies ($\omega \gg \omega_1$):

$$20 \log \sqrt{1 + \left(\frac{\omega}{\omega_1}\right)^2} \to 20 \log \sqrt{\left(\frac{\omega}{\omega_1}\right)^2} = 20 \log \frac{\omega}{\omega_1} \quad \text{dB} \qquad (9.1\text{-}8b)$$

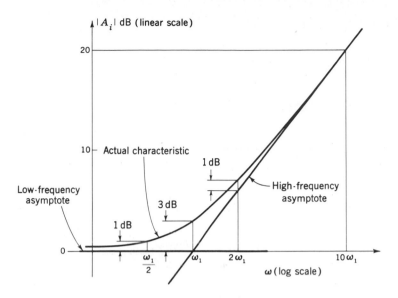

Figure 9.1-2 Asymptotic and actual frequency characteristics.

The high-frequency asymptote is a straight line when plotted against ω on a logarithmic scale, and the low-frequency asymptote is simply a constant. They intersect at $\omega = \omega_1$ because the high-frequency asymptote is 0 dB at this point. ω_1 is called the *break* or *corner frequency*. The asymptotes are sketched in Fig. 9.1-2.

The slope of the high-frequency asymptote is usually expressed in terms of octave $(2:1)$ or decade $(10:1)$ frequency ratios. Thus a frequency increase of an octave results in an increase of gain [Eq. (9.1-8b)] of

$$\Delta |A_i|_{dB} = 20 \log 2 = +6 \text{ dB} \quad OCTAVE \tag{9.1-9a}$$

A frequency increase of a decade results in a change in gain of

$$\Delta |A_i|_{dB} = 20 \log 10 = +20 \text{ dB} \quad DECADE \tag{9.1-9b}$$

When the slope and the frequency at which the asymptote goes through 0 dB are known, equations like (9.1-8b) can easily be sketched on semilog graph paper.

Now consider the third term in (9.1-7). We see that the previous discussion holds, with the exception that the break frequency is at ω_2 and the slope above ω_2 is -6 dB/octave.

Often the asymptotic curve is sufficient to supply the required information. If a more accurate curve is required, simple corrections can be applied to the asymptotic curve. The difference between the actual and asymptotic curves for a single-frequency factor of the form $1 + j\omega/\omega_1$ is usually negligible at frequencies beyond an octave on either side of the break frequency. The corrections to be applied can be found as follows. At ω_1, the asymptotic curve is at 0 dB, while the actual value is

$$20 \log \sqrt{1 + \left(\frac{\omega}{\omega_1}\right)^2} = 10 \log 2 \approx 3 \text{ dB}$$

Thus the actual curve is 3 dB above the asymptote at the break. Similar calculations at $2\omega_1$ and $\omega_1/2$ indicate that the actual curve lies about 1 dB above the asymptotic curve (note that 1 dB represents a 10 percent error). These results are illustrated in Fig. 9.1-2.

$$(1 dB = 10\% error)$$

Phase-angle variation The phase angle of the gain function can also be approximated by straight-line asymptotes. For a single-frequency factor the phase angle is $\theta = \tan^{-1}(\omega/\omega_1)$. The asymptotes often used for this are

$$\theta = \begin{cases} 0° & \omega < \omega_1/10 \\ 45\left(1 + \log \dfrac{\omega}{\omega_1}\right) & \dfrac{\omega_1}{10} < \omega < 10\omega_1 \\ 90° & \omega > 10\omega_1 \end{cases}$$

(when $\omega = \omega_1/10$, $\theta = 5.7°$, and when $\omega = 10\omega_1$, $\theta = 84.3°$). The actual and asymptotic phase curves are shown in Fig. 9.1-3. Through the use of these asymptotic characteristics, both amplitude and phase curves can be sketched quickly and accurately when the transfer function has been reduced to the factored form of (9.1-4).

Example 9.1-1 Plot the asymptotic magnitude for the gain function

$$200\left[\frac{s+10}{s+50}\right] \qquad A_i = 40\frac{1 + j\omega/10}{1 + j\omega/50}$$

$$s \to 0 = 40$$
$$s \to \infty = 200$$

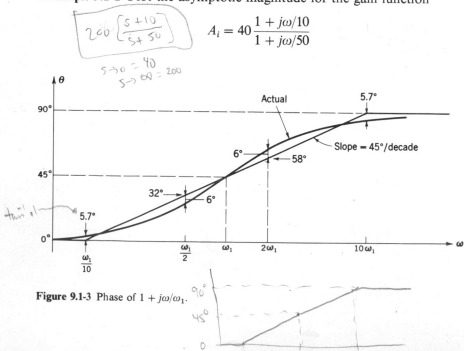

Figure 9.1-3 Phase of $1 + j\omega/\omega_1$.

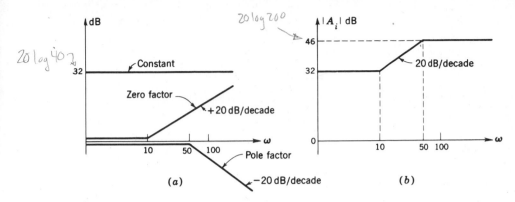

Figure 9.1-4 Asymptotic amplitude plots for Example 9.1-1: (a) sketch of individual factors; (b) asymptotic plot.

SOLUTION The constant is 20 log 40 = 32 dB. The first break occurs at the numerator corner frequency, $\omega_1 = 10$ rad/s, and the second at the denominator corner frequency, $\omega_2 = 50$ rad/s. The individual factors are shown in Fig. 9.1-4a, and their sum in Fig. 9.1-4b. The high-frequency asymptote for $|A_i|$ is a constant, which can be found graphically or from the asymptotic form

$$A_i(\omega \gg \omega_2) \approx 40\frac{j\omega/10}{j\omega/50} = (40)(5) = 200$$

or in decibels

$$|A_i(\omega \gg \omega_2)| = 20 \log 200 = 46 \text{ dB} \qquad ///$$

Example 9.1-2 Sketch the asymptotic phase curve for the gain function of Example 9.1-1.

SOLUTION The phase is

$$\theta(\omega) = \left(\tan^{-1}\frac{\omega}{10}\right) - \left(\tan^{-1}\frac{\omega}{50}\right)$$

The curves for the individual factors and their sum are shown in Fig. 9.1-5.

$$///$$

Example 9.1-3 Plot the magnitude and phase of A_i as a function of frequency for the amplifier shown in Fig. 9.1-6.

SOLUTION The equivalent circuit of Fig. 9.1-6, shown in Fig. 9.1-7a, yields the gain A_i. Thus, from (9.1-1a),

$$A_i = \frac{i_c}{i_i} = \frac{i_c}{i_e}\frac{i_e}{i_i} \approx \frac{i_e}{i_i} \approx \left(\frac{600}{8}\right)\frac{j\omega + 1/10^{-3}}{j\omega + 1/(80 \times 10^{-6})} \approx 6\frac{1 + j10^{-3}\omega}{1 + j80 \times 10^{-6}\omega}$$

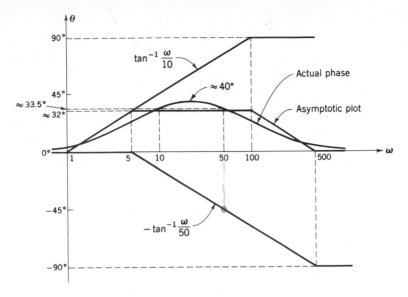

Figure 9.1-5 Asymptotic and actual phase curves for Example 9.1-2.

Hence $\qquad |A_i| = 6\dfrac{\sqrt{1 + 10^{-6}\omega^2}}{\sqrt{1 + (80^2)(10^{-12}\omega^2)}}$ ← *Magnitude*

and $\qquad \theta_{A_i} = (\tan^{-1} 10^{-3}\omega) - \tan^{-1}(80 \times 10^{-6}\omega)$ ← *angle*

These results are sketched in Fig. 9.1-7b and c. A simple way of sketching $|A_i|$ is to first plot the low-frequency and high-frequency ($\omega = 0$ and ∞) asymptotes. Mark the zero break $\omega_1 = 1/R_e C_e$ and the pole break $\omega_2 \approx 1/[(R_b/h_{fe}) + h_{ib}]C_e$. Then connect these two points with a straight line. The phase can be sketched using a similar technique. Mark the phase at 0.1 and 10 times the zero and the pole frequencies. Connect the resulting four points with straight lines.

$A_i = \dfrac{i_c}{i_i} = \dfrac{i_c}{i_e}\dfrac{i_e}{i_i} \approx \dfrac{i_e}{i_i} = \dfrac{600}{8}\left(\dfrac{j\omega + \frac{1}{10^3}}{j\omega + \frac{1}{80 \times 10^{-6}}}\right)$

$= \dfrac{6(1 + j\frac{10^{-3}}{\omega})}{1 + j\,80 \times 10^{-6}\omega}$

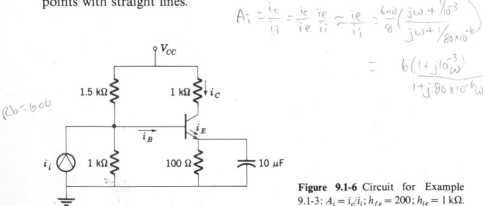

$R_b = 600$

Figure 9.1-6 Circuit for Example 9.1-3: $A_i = i_c/i_i$; $h_{fe} = 200$; $h_{ie} = 1\,k\Omega$.

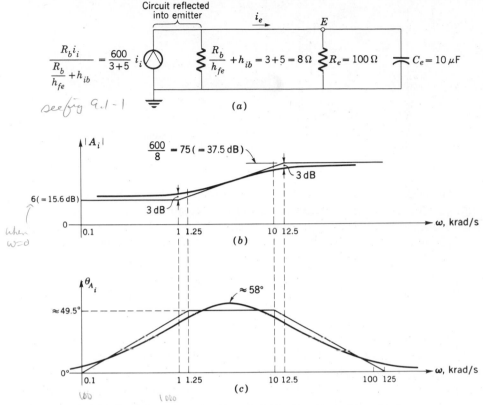

Figure 9.1-7 (a) Equivalent circuit for Example 9.1-3; (b) gain versus ω; (c) phase versus ω.

Figure 9.1-7b shows that the lower 3-dB frequency occurs at $\omega_L = 12.5 \times 10^3$ rad/s, or

$$f_L \approx 2 \text{ kHz}$$

For audio applications this is an extremely high frequency at which to end the low-frequency region. If C_e were increased to 1000 μF, f_L would be reduced to 20 Hz, which is a more reasonable value. ///

9.1-3 The Coupling Capacitor

[handwritten: Increase R_e, low freq approaches 0.]

The capacitor C_{c1} in Fig. 9.1-8 is often needed to couple the ac signal from the source to the base of the transistor, and also to block any dc so that bias conditions are not upset. Let us study the effect of C_{c1} on the circuit response, assuming that the emitter is unbypassed.†

† The response of the circuit shown in Fig. 9.1-8 is the same as the response of an amplifier circuit having a bypassed emitter resistor in the frequency region where $\omega < 1/R_e C_e$.

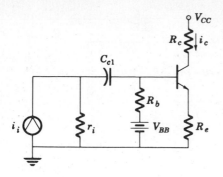

Figure 9.1-8 Common-emitter amplifier with input coupling capacitor.

The effect of C_{c1} on the low-frequency response of the amplifier of Fig. 9.1-8 is obtained by reflecting R_e into the base circuit, as shown in Fig. 9.1-9a, which is then transformed, using Thevenin's theorem, into the form shown in Fig. 9.1-9b. The current gain of the amplifier can now be obtained by inspection

$$A_i = \frac{i_c}{i_i} = \frac{i_c}{i_b} \frac{i_b}{v_b} \frac{v_b}{i_i}$$

$$= h_{fe} \left(\frac{1}{h_{ie} + (1 + h_{fe})R_e} \right) \left(\frac{r_i R_b'}{r_i + R_b' + 1/sC_{c1}} \right) \quad (9.1\text{-}10a)$$

where
$$R_b' = R_b \| [h_{ie} + (1 + h_{fe})R_e] \quad (9.1\text{-}10b)$$

This equation can now be simplified to

$$A_i \approx \left(\frac{r_i \| R_b'}{h_{ib} + R_e} \right) \left(\frac{s}{s + 1/[(r_i + R_b')C_{c1}]} \right) \quad (9.1\text{-}11)$$

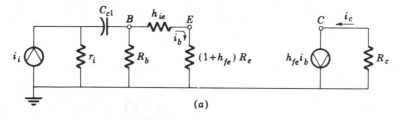

(a)

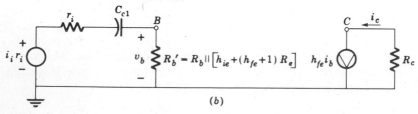

(b)

Figure 9.1-9 *(a)* Small-signal low-frequency equivalent circuit for the amplifier of Fig. 9.1-8; *(b)* reduced equivalent circuit.

$$A_i = -h_{fe} \left[\frac{r_i + \dfrac{R_b}{R_b s C_i + 1}}{r_i + \dfrac{R_b}{R_b s C_i + 1} + h_{ie} + R_e h_{fe}} \right]$$

Same

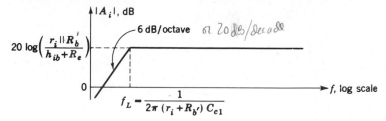

Figure 9.1-10 Bode plot for CE circuit.

The 3-dB frequency of this circuit is

$$f_L = \frac{1}{2\pi(r_i + R_b')C_{c1}}$$ (9.1-12)

3 dB freq.

A Bode plot of the current-gain magnitude is shown in Fig. 9.1-10.

Example 9.1-4 In the amplifier of Fig. 9.1-8, $h_{ie} = 1\,k\Omega$, $r_i = 10\,k\Omega$, $R_b = 1\,k\Omega$, $R_e = 100\,\Omega$, $C_{c1} = 10\,\mu F$, and $h_{fe} = 100$. (a) Find the 3-dB frequency f_L of the amplifier. (b) Assume that R_e is perfectly bypassed ($C_e \to \infty$). What is the lower 3-dB frequency f_L now?

SOLUTION (a) Using (9.1-12) and (9.1-10b), we get

$$f_L = \frac{1}{2\pi(10 \times 10^3 + \{10^3\|[10^3 + (100)(100)]\}) \times 10^{-5}} \approx 1.6 \text{ Hz}$$

(b) Again, using (9.1-12) and (9.1-10b) with $R_e = 0$, we get

$$f_L = \frac{1}{2\pi[(10 \times 10^3) + (10^3\|10^3)] \times 10^{-5}} \approx 1.6 \text{ Hz}$$

The results are the same since they are both given approximately by the equation

$$f_L \approx \frac{1}{2\pi r_i C_{c1}} \qquad \text{when } r_i \gg R_b' \qquad ///$$

9.1-4 The Base and Collector Coupling Capacitors

If the collector signal in Fig. 9.1-8 is capacitively coupled to a resistive load, the frequency response will depend on both coupling capacitors. Figure 9.1-11a shows this configuration, and Fig. 9.1-11b shows the resulting equivalent circuit. The

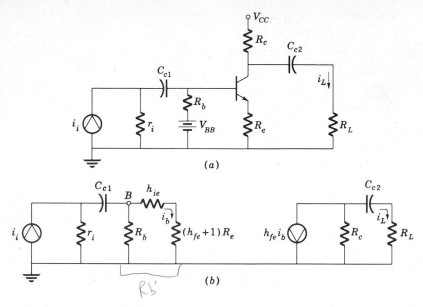

(a)

(b)

Figure 9.1-11 CE amplifier with base and collector (input and output) coupling capacitors: (a) circuit; (b) equivalent circuit $R'_b = R_b \| [h_{ie} + (h_{fe} + 1)R_e]$.

gain is

$$A_i = \frac{i_L}{i_i} = \frac{i_L}{i_b} \frac{i_b}{v_b} \frac{v_b}{i_i}$$

$$= -h_{fe} \left(\frac{R_c}{R_c + R_L + \dfrac{1}{sC_{c2}}} \right) \left(\frac{1}{h_{ie} + (1 + h_{fe})R_e} \right) \left(\frac{r_i R'_b}{r_i + R'_b + \dfrac{1}{sC_{c1}}} \right) \qquad (9.1\text{-}13a)$$

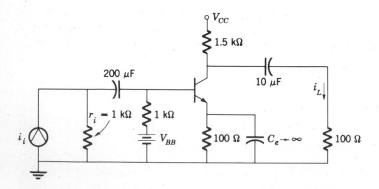

Figure 9.1-12 Amplifier for Example 9.1-5: $h_{fe} = 100$; $h_{ie} = 1$ kΩ.

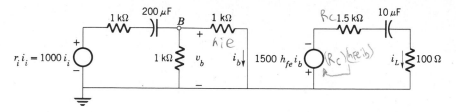

Figure 9.1-13 Equivalent circuit of amplifier shown in Fig. 9.1-12.

$$Rb' = Rb // [hie + (hfe+1)Re]$$

which can be simplified to

$$A_i \approx -\left(\frac{R_c}{R_c + R_L}\right)\left(\frac{r_i \| R_b'}{h_{ib} + R_e}\right)\left(\frac{s}{s + \dfrac{1}{(r_i + R_b')C_{c1}}}\right)\left(\frac{s}{s + \dfrac{1}{(R_c + R_L)C_{c2}}}\right) \qquad (9.1\text{-}13b)$$

Equation (9.1-13b) and Fig. 9.1-11b indicate that the two coupling circuits do not interact.

Example 9.1-5 Plot the magnitude and phase of the current gain as a function of ω for the amplifier shown in Fig. 9.1-12. The equivalent circuit of this amplifier is shown in Fig. 9.1-13.

SOLUTION The gain is

$$A_i = \frac{i_L}{i_i} = \frac{i_L}{i_b}\frac{i_b}{i_i}$$

$$= -\left(\frac{1500h_{fe}}{1600 + 1/j\omega 10^{-5}}\right)\left(\frac{1000}{1500 + 1/(j\omega 2 \times 10^{-4})}\right)\frac{1}{2} \qquad \text{why?}$$

$$= (-75 \times 10^6)\frac{(j\omega 10^{-5})(j\omega 2 \times 10^{-4})}{(1 + j\omega 16 \times 10^{-3})(1 + j\omega 30 \times 10^{-2})}$$

The magnitude of the current gain is plotted in Fig. 9.1-14. The phase angle can be roughly sketched as a function of ω, by inspection, using the technique outlined in Sec. 9.1-2. We note, for example, that

$$\theta \approx \begin{cases} -\pi & \omega \to \infty \\[4pt] 0 & \omega \to 0 \\[4pt] -\dfrac{\pi}{4} & \omega = 3.3 \text{ rad/s} \\[6pt] -\dfrac{3\pi}{4} & \omega = 62.5 \text{ rad/s} \end{cases}$$

The plot of θ versus ω is shown in Fig. 9.1-15.   ///

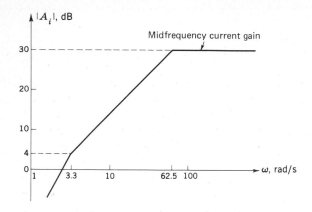

Figure **9.1-14** Gain versus frequency for Example 9.1-5.

Example 9.1-5 considered the frequency response of an amplifier with two poles widely separated from each other. To achieve this separation, C_{c1} was a very large capacitor, 200 μF. Most of the time we are interested only in the 3-dB frequency. Then we should not use a 200-μF capacitor for C_{c1} but should use a 20-μF capacitor instead, since it is smaller and less expensive. The resulting poles are now not widely separated, and the 3-dB point is *not* located at the first pole but actually occurs at a *higher* frequency. For example, consider that the two poles in (9.1-13b) coincide. Then

$$A_i = A_{im}\left(\frac{s}{s + \omega_0}\right)^2 \tag{9.1-14a}$$

where

$$A_{im} = \left(\frac{-R_c}{R_c + R_L}\right)\left(\frac{r_i \| R_b'}{h_{ib} + R_e}\right) \tag{9.1-14b}$$

and

$$\omega_0 = \frac{1}{(r_i + R_b')C_{c1}} = \frac{1}{(R_c + R_L)C_{c2}} \tag{9.1-14c}$$

Hence

$$\left|\frac{A_i}{A_{im}}\right|^2 = \frac{\omega^4}{(\omega^2 + \omega_0^2)^2} = \frac{1}{[1 + (\omega_0/\omega)^2]^2} \tag{9.1-15}$$

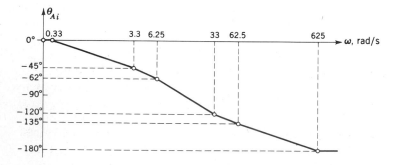

Figure **9.1-15** Phase angle versus ω for Example 9.1-5.

The 3-dB frequency ω_L is defined in (9.1-3b) as the frequency at which $|A_i/A_{im}|^2 = \frac{1}{2}$. Then

$$\left[1 + \left(\frac{\omega_0}{\omega_L}\right)^2\right]^2 = 2 \tag{9.1-16a}$$

and

$$\omega_L = \frac{\omega_0}{\sqrt{0.414}} \approx 1.55\omega_0 \tag{9.1-16b}$$

If the poles do not coincide, the quadratic equation which results when combining the two factors in (9.1-13b) yields the 3-dB point. Thus, referring to (9.1-13b), we have

$$\left|\frac{A_i}{A_{im}}\right|^2 = \frac{\omega_L^4}{(\omega_L^2 + \omega_1^2)(\omega_L^2 + \omega_2^2)} = \frac{1}{2} \tag{9.1-17a}$$

where

$$\omega_1 = \frac{1}{(r_i + R_b')C_{c1}} \tag{9.1-17b}$$

and

$$\omega_2 = \frac{1}{(R_c + R_L)C_{c2}} \tag{9.1-17c}$$

The 3-dB frequency is then found by solving

$$\omega_L^4 + (\omega_1^2 + \omega_2^2)\omega_L^2 + \omega_1^2\omega_2^2 = 2\omega_L^4 \tag{9.1-18a}$$

which yields

$$\omega_L^2 = \frac{\omega_1^2 + \omega_2^2}{2} + \frac{\sqrt{\omega_1^4 + 6\omega_1^2\omega_2^2 + \omega_2^4}}{2} \tag{9.1-18b}$$

9.1-5 Combined Effect of Bypass and Coupling Capacitors

In most applications, C_{c1}, C_{c2}, and C_e are all present, and the analysis becomes more complicated. Design information is then difficult to obtain unless simplifications are made. Our objective is to find a way to select bypass and coupling capacitors so as to achieve a specified low-frequency response. Often the design specifications will require only that the response be maintained at the midband value down to a certain frequency ω_L, at which point it may be down 3 dB. Below this break frequency the shape of the response curve is often unimportant, provided the gain continues to decrease with decreasing frequency.

When this is the case, the circuit is designed so that the emitter bypass capacitor C_e determines the specific break frequency. This is done to minimize capacitor size. Equation (9.1-2b) shows that the effective resistance, which works with C_e to yield the 3-dB frequency, is small ($\approx h_{ib} + R_b/h_{fe}$). Thus, to obtain a low 3-dB frequency, C_e will be large. If the 3-dB frequency is obtained using a different RC combination, C_e must be *still larger*, which results in increased size and cost.

✳ To obtain low 3dB frequency, Ce must be large

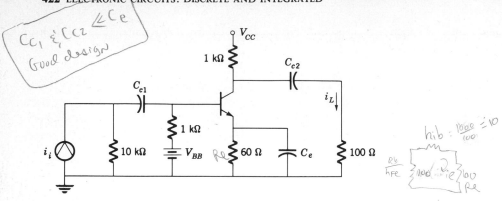

Figure 9.1-16 Circuit for Example 9.1-6: $h_{fe} = 100$; $h_{ie} = 1\ \text{k}\Omega$.

Coupling capacitors C_{c1} and C_{c2} are then chosen to yield break frequencies well below this point. When the circuit is designed in this manner, the capacitors C_{c1} and C_{c2} are usually much less than C_e.

Example 9.1-6 The amplifier shown in Fig. 9.1-16 is to have a lower 3-dB frequency at 20 Hz. Select C_{c1}, C_{c2}, and C_e to meet this specification.

SOLUTION We begin by selecting C_e to achieve the required 20-Hz 3-dB frequency. From (9.1-2b)

$$f_L = 20 = \frac{1}{2\pi C_e[60 \| (10 + 10)]} \qquad C_e \approx 530\ \mu\text{F}$$

We use a 500- or 1000-μF standard capacitor. The coupling capacitors C_{c1} and C_{c2} are selected so that the break frequencies occur well below 20 Hz. We see from Fig. 9.1-2 that to be well below the 3-dB frequency means that the break frequencies should occur more than an octave lower. As a practical matter, a decade below the 3-dB frequency represents more than adequate separation. Thus, in this example, the break frequencies due to C_{c1} and C_{c2} are each chosen to be 2 Hz. Then, from (9.1-17b),

$$C_{c1} \approx \frac{1}{2\pi(2)(10^4)} = 8\ \mu\text{F} \qquad \text{(use a 10-}\mu\text{F capacitor)}$$

and from (12.1-17c),

$$C_{c2} \approx \frac{1}{2\pi(2)(10^3)} = 80\ \mu\text{F} \qquad \text{(use a 100-}\mu\text{F capacitor)} \qquad ///$$

9.2 THE LOW-FREQUENCY RESPONSE OF THE FET AMPLIFIER

The low-frequency response of the FET amplifier is determined primarily by the source bypass capacitor because of the relative magnitudes of the circuit resistances. In this section we shall consider the effect of this capacitor by assuming that the input and output coupling capacitors are infinite. The effect of the coupling capacitors is considered in the problems.

9.2-1 The Source Bypass Capacitor

In the FET amplifier shown in Fig. 9.2-1a the coupling capacitors are assumed infinite, so that the 3-dB frequency is caused by the source bypass capacitor. It is shown in the problems that even when the coupling capacitors are considerably smaller than the bypass capacitor, it is the source bypass capacitor that determines the break frequency f_L.

The equivalent circuit of the amplifier is shown in Fig. 9.2-1b. In this figure, the source circuit has been reflected into the drain circuit. The voltage gain is found to be

$$A'_v = \frac{v_L}{v'_i} = -\frac{\mu(R_d \| R_L)}{(R_d \| R_L) + r_{ds} + (\mu + 1)R_s/(1 + sR_sC_s)} \tag{9.2-1}$$

which can be put in the form

$$A'_v = \frac{v_L}{v'_i}$$

$$= -\left[\frac{\mu(R_d \| R_L)}{(R_d \| R_L) + r_{ds}}\right]\left(\frac{s + 1/R_sC_s}{s + \dfrac{1}{C_s\{R_s[(R_d \| R_L) + r_{ds}]/[(\mu + 1)R_s + r_{ds} + (R_d \| R_L)]\}}}\right) \tag{9.2-2}$$

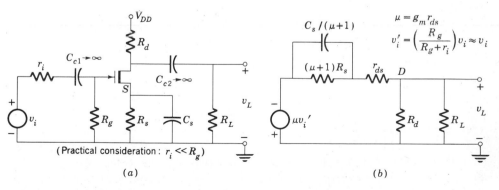

Figure 9.2-1 (a) FET amplifier circuit; (b) small-signal equivalent circuit.

We see that the form of (9.2-2) is similar to the form of (9.1-1). Substituting $\mu = g_m r_{ds}$ into (9.2-2) yields the final expression for the voltage gain:

$$A_v = \frac{v_L}{v_i} \approx \frac{v_L}{v_i'} = -g_m R_{\parallel} \left(\frac{s + 1/R_s C_s}{s + \dfrac{1}{C_s\{R_s[(R_d\|R_L) + r_{ds}]/[(\mu+1)R_s + r_{ds} + (R_d\|R_L)]\}}} \right) \tag{9.2-3}$$

where

$$R_{\parallel} = r_{ds}\|R_d\|R_L$$

To find the midband gain, let $s = j\omega \to j\infty$. Then

$$A_{vm} = \frac{v_L}{v_i}\bigg|_{\omega \to \infty} = -g_m R_{\parallel} \left(\frac{R_g}{r_i + R_g} \right) \approx -g_m R_{\parallel} \qquad r_i \ll R_g \tag{9.2-4}$$

This result can be obtained by inspection of Fig. 9.2-1, and it is the midfrequency voltage gain obtained in Sec. 3.6.

The angular frequency at which the voltage gain is 3 dB below the mid-frequency value is found as in (9.1-13b). Assuming $\omega_L \gg 1/R_s C_s$, we get

$$\omega_L = \frac{1}{C_s R_s \dfrac{r_{ds} + (R_d\|R_L)}{(\mu+1)R_s + r_{ds} + (R_d\|R_L)}} \tag{9.2-5}$$

Example 9.2-1 Find C_s to obtain a 10-Hz break frequency, when $R_s = 1$ kΩ, $r_{ds} = 10$ kΩ, $R_d = 5$ kΩ, $R_L = 100$ kΩ, and $g_m = 3$ mS. Also calculate the mid-frequency gain.

SOLUTION From (9.2-5)

$$C_s = \frac{1}{2\pi(10 \times 10^3)\dfrac{(10 \times 10^3) + (5 \times 10^3)}{(30 \times 10^3) + (10 \times 10^3) + (5 \times 10^3)}} = 47.8 \ \mu\text{F}$$

We would specify a standard 50-μF capacitor. Note that R_L does not influence the calculation since it is large relative to R_d. Also note that $(1/2\pi)R_s C_s \approx 3.3$ Hz; hence the true 3-dB frequency is 8.8 Hz [Eq. (9.1-3d)]. The mid-frequency voltage gain is, from (9.2-4),

$$A_{vm} \approx (-3 \times 10^{-3})[(10 \times 10^3)\|(5 \times 10^3)] = -10 \qquad \text{///}$$

9.3 THE TRANSISTOR AMPLIFIER AT HIGH FREQUENCIES

We have seen that the low-frequency behavior of the transistor circuit is determined by the *external* capacitors used for coupling and for emitter bypass. The upper limit on the high-frequency response of the device is limited by *internal* capacitance.

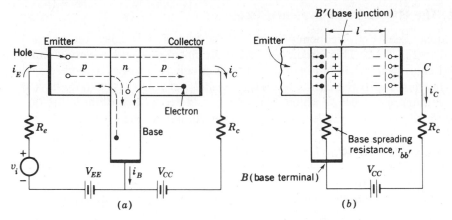

Figure 9.3-1 Collector-base capacitance: (a) *pnp* transistor; (b) collector-base region.

Consider the *pnp* transistor shown in Fig. 9.3-1a. The collector-base junction appears as a reverse-biased diode, as shown in Fig. 9.3-1b. When the collector is biased negatively, with respect to the base, the holes in the base region move into the collector region and the electrons in the collector region move into the base region. The electrons in the base move away from the base-collector junction, and the holes in the collector move away from the base-collector junction, thus forming a depletion region. The effective length *l* of the depletion region becomes larger as the reverse voltage increases. Since the electrons and holes have moved away from the junction, the base depletion region becomes positively charged and the collector depletion region becomes negatively charged (Fig. 9.3-1b). The junction therefore behaves like a capacitor whose capacitance theoretically varies inversely with V_{CB}. Actually, the collector-base junction capacitance $C_{b'c}$ is inversely proportional to either the $\frac{1}{2}$ or $\frac{1}{3}$ power of V_{CB}, depending on whether the transistor is an alloy transistor or a grown-junction transistor. This capacitance is rather small and varies from about 30 pF in low-frequency transistors to less than 1 pF in high-frequency transistors.

In addition, a diffusion capacitance exists across the base-emitter junction. The diffusion capacitance is a result of the time delay that occurs when a hole moves from emitter to collector by diffusing across the base. Consider a hole moving across the base because of a command from the signal voltage v_i. Before the hole crosses the base region, if the voltage reverses polarity, the hole will try to return to the emitter and the collector will not record this change in signal current. Thus the time required to cross the base must be small compared with the period of the signal. As the signal frequency increases, the collector current decreases because some charges are trapped in the base. Since the current decreases with increasing frequency, the effect is accounted for by a lumped capacitor $C_{b'e}$. This capacitance depends on the number of charges in the region, and thus it increases almost linearly with the quiescent emitter current I_{EQ}. $C_{b'e}$ is usually much larger than $C_{b'c}$. Typical values lie in the range of 100 to 5000 pF.

High frequency Model (handwritten)

9.3-1 The Hybrid-Pi Equivalent Circuit

The most useful high-frequency model of the transistor is called the *hybrid pi*, shown in Fig. 9.3-2. This circuit represents a refinement of the common-emitter hybrid equivalent circuit of Fig. 2.2-3a. In this circuit, the symbol B' represents the base *junction*, and B represents the base *terminal*. Between these two we have the base ohmic resistance $r_{bb'}$, which is usually considered as a constant in the range 10 to 50 Ω. This resistance is directly proportional to the base width. High-frequency transistors have smaller base widths, and thus a smaller $r_{bb'}$, than low-frequency transistors. The resistance $r_{b'e}$ is the base-emitter junction resistance ($\approx 0.025 h_{fe}/I_{EQ}$ at room temperature) and is usually much larger than $r_{bb'}$. The reader will recognize that $r_{b'e}$ is equivalent to the h_{ie} we have been using as the total base-emitter resistance up to now. Thus a more accurate expression for h_{ie} is

(rbb'—base ohmic resistance) (handwritten)

$$h_{ie} = r_{bb'} + r_{b'e} \approx r_{bb'} + \frac{0.025 h_{fe}}{I_{EQ}} \qquad T = 300 \text{ K} \qquad (9.3\text{-}1)$$

The approximation $h_{ie} \approx r_{b'e}$ is usually valid.

The output impedance $1/h_{oe}$ can often be neglected at high frequencies since it is usually much larger than the impedance of the external load R_L.

Cutoff frequency The origin of the capacitors in the equivalent circuit was discussed qualitatively at the beginning of this section. Their effect at high frequencies is often given in terms of the β cutoff frequency f_β, defined as follows. Let $v_{ce} = 0$; then the hybrid-pi model shown in Fig. 9.3-2 reduces to the form shown in Fig. 9.3-3. From this circuit, we see that the short-circuit current gain $(i_c/i_i)|_{v_{ce}=0}$ will be down 3 dB at a frequency

$$f_\beta = \frac{1}{2\pi r_{b'e}(C_{b'e} + C_{b'c})} \approx \frac{1}{2\pi r_{b'e} C_{b'e}} \qquad (9.3\text{-}2)$$

Thus f_β is the common-emitter short-circuit 3-dB frequency.

The upper-frequency limit of a transistor is sometimes defined in terms of the frequency f_T, at which the common-emitter current gain is unity. The short-circuit current gain of the ideal amplifier shown in Fig. 9.3-3 is

Short circuit Current gain (handwritten)

$$\frac{i_c}{i_i} = -\frac{h_{fe}}{1 + j\omega/\omega_\beta} \qquad (9.3\text{-}3)$$

Figure 9.3-2 Hybrid-pi model of a transistor.

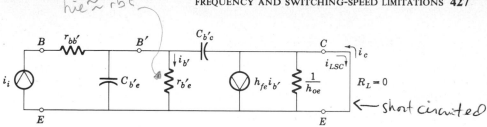

hie ≈ rb'e

Figure 9.3-3 The hybrid-pi model used to calculate f_β.

This gain becomes unity when

$f_T = (GAIN)(BW)$ product

$$f = f_T = f_\beta\sqrt{h_{fe}^2 - 1} \approx f_\beta h_{fe} \quad \text{UNITY GAIN} \quad (9.3-4)$$

The frequency f_T is often called the gain-bandwidth product of the amplifier, although, strictly speaking, this is not correct. This is considered in Sec. 9.6.

If we consider the common-base configuration at midfrequency and then include the base-emitter capacitance and the collector-base capacitance, a high-frequency common-base model similar to that shown in Fig. 9.3-3 is obtained. This model is shown in Fig. 9.3-4. The short-circuit current gain is

$$A_i = \frac{i_{sc}}{i_i}\bigg|_{v_{cb}=0} \approx \frac{h_{fb}}{1 + j\omega/h_{fe}\omega_\beta} \quad \text{circuit below} \quad (9.3-5)$$

and the 3-dB bandwidth of this amplifier is called f_α. From (9.3-5)

$$f_\alpha = h_{fe}\,f_\beta \quad (9.3-6)$$

Comparing (9.3-4) and (9.3-6), we note that $f_\alpha \approx f_T$. This result is not really correct because the circuit of Fig. 9.3-4 is not valid at f_T. Indeed, it can be shown that the α cutoff frequency is[1,2]

$$f_\alpha = (1 + \lambda)f_T \approx (1 + \lambda)h_{fe}\,f_\beta \quad (9.3-7)$$

where λ is found empirically to lie between 0.2 and 1. A typical value is 0.4.

When using the equivalent circuit of Fig. 9.3-2, it is often easier to calculate the voltage $v_{b'e}$ than the current $i_{b'}$ which flows in $r_{b'e}$. The output-current source $h_{fe}i_{b'}$ can be transformed to a voltage-controlled current source $g_m v_{b'e}$, where

circuit 9.3-2

$$h_{fe}i_{b'} = h_{fe}\frac{v_{b'e}}{r_{b'e}} = g_m v_{b'e} \quad \left(g_m = \frac{h_{fe}}{r_{b'e}}\right) \quad (9.3-8)$$

$i_{b'} = \frac{v_{b'e}}{r_{b'e}}$

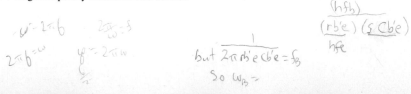

Figure 9.3-4 High-frequency common-base model.

$\frac{i_c}{i_i} = \frac{i_c}{i_e}\cdot\frac{i_e}{i_i}$

$(h_{fb})\left(\dfrac{sC_{b'e}}{sC_{b'e} + \frac{r_{b'e}}{h_{fe}}}\right)$

$\dfrac{(h_{fb})}{(r_{b'e})(sC_{b'e})}$

$\dfrac{}{h_{fe}}$

but $\overline{2\pi\,r_{b'e}\,C_{b'e}} = f_\beta$

So $\omega_\beta =$

$v = 2\pi f$

$2\pi f = \omega$

$2\pi f = \omega$

$f = 2\pi\omega$

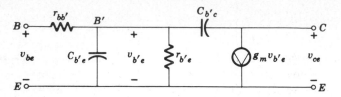

Figure 9.3-5 Hybrid-pi equivalent with voltage-controlled current source ($h_{oe} = 0$).

and

$$g_m = \frac{h_{fe}}{r_{b'e}} \approx \frac{I_{EQ}}{0.025} = 40 I_{EQ} \qquad \text{at } T = 300 \text{ K} \tag{9.3-9}$$

Note that g_m is approximately equal to $1/h_{ib}$. In order to have similar notation for the transistor and FET we use the symbol g_m.

The resulting common-emitter equivalent circuit is shown in Fig. 9.3-5.

Summary of elements of hybrid-pi equivalent circuit

$r_{bb'} \approx 10$ to $50 \ \Omega$ (smaller values for high-frequency transistors)

base ohmic resistance

$$r_{b'e} = \frac{h_{fe}}{40 I_{EQ}} \qquad g_m = \frac{I_{EQ}}{0.025} = 40 I_{EQ} \qquad h_{oe} \propto I_{EQ}$$

$\approx h_{ie}$

$$C_{b'e} \approx \frac{h_{fe}}{\omega_T r_{b'e}} = \frac{40 I_{EQ}}{\omega_T} = \frac{g_m}{\omega_T} \qquad \text{(9.3-2) and (9.3-4)}$$

$$g_m = \frac{h_{fe}}{h_{ie}}$$

$C_{ob} \rightarrow C_{b'c} \propto v_{cb}^{-p}$ where p lies between $\frac{1}{2}$ and $\frac{1}{3}$

($C_{b'c}$ is usually specified by the manufacturer as C_{ob}, the output capacitance of the common-base configuration.)

Example 9.3-1 Manufacturers' specifications for a 2N3647 silicon transistor include, at $I_{CQ} = 150$ mA, $V_{CEQ} = 1$ V

$$f_T = 350 \text{ MHz} \qquad h_{fe} \approx 150 \qquad h_{oe} \approx 0.1 \text{ mS} \qquad C_{ob} = 4 \text{ pF}$$

From these data, deduce the values of the components of the high-frequency equivalent circuit of Fig. 9.3-5 if the transistor is to be operated at $I_{CQ} = 300$ mA.

SOLUTION $\quad h_{ie} = r_{b'e} \approx \dfrac{h_{fe}}{40 I_{CQ}} = \dfrac{150}{(40)(300)(10^{-3})} \approx 12.5 \ \Omega$

Since $r_{bb'}$ is not given, we assume a value of $10 \ \Omega$.

$$h_{fe} \approx 150 \qquad g_m = (40)(0.3) = 12 \text{ S} \qquad h_{oe} \approx 0.2 \text{ mS}$$

$$C_{b'e} = \frac{g_m}{\omega_T} = \frac{12}{2\pi(350)(10^6)} = 5450 \text{ pF} \qquad C_{b'c} = C_{ob} = 4 \text{ pF} \qquad ///$$

9.3-2 High-Frequency Behavior of the Common-Emitter Amplifier; Miller Capacitance

The common-emitter amplifier represents the type of high-frequency amplifier most often used. Its high-frequency response is investigated in this section. As will be seen, this response is dominated by a single pole due to the input circuit.

A complete CE stage is shown in Fig. 9.3-6a, and its high-frequency equivalent circuit is shown in Fig. 9.3-6b. In this equivalent circuit all coupling and bypass capacitors have been assumed to be short circuits at the frequencies of interest. Wiring and other stray capacitances have been ignored, even though they sometimes inadvertently prove to be important.

For simplicity, we let R_b represent the parallel combination of r_i and R_b and also let R_L represent the parallel combination of R_c and R_L. The simplified equivalent circuit is shown in Fig. 9.3-6c.

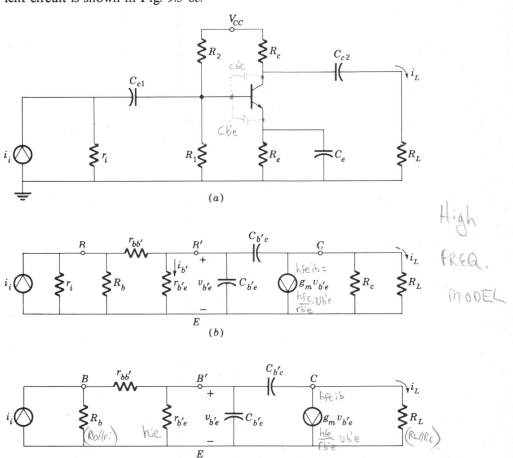

(a)

(b)

(c)

Figure 9.3-6 The common emitter amplifier at high frequencies ($h_{oe} = 0$); (a) circuit; (b) high-frequency equivalent circuit; (c) simplified equivalent circuit.

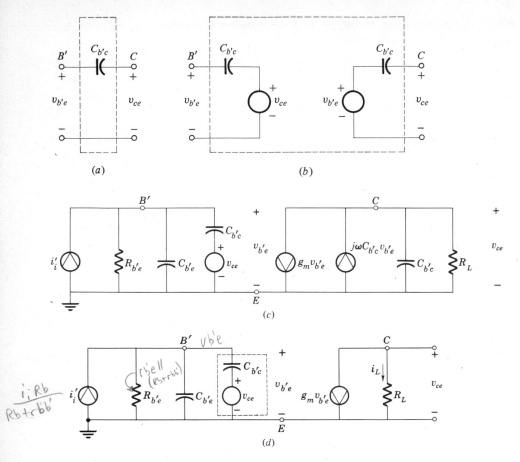

Figure 9.3-7 CE stage at high frequencies: (*a*), (*b*) four-terminal equivalent of $C_{b'c}$; (*c*) complete circuit; (*d*) simplified circuit.

For ease of analysis this circuit can be manipulated into a much more convenient form. We begin by converting $C_{b'c}$ (Fig. 9.3-7*a*) into the equivalent four-terminal network shown in Fig. 9.3-7*b*. The student can readily show that Fig. 9.3-7*a* and *b* are completely equivalent with respect to their external ports, B' and C. This operation is performed on the circuit of Fig. 9.3-6*c*, and the $v_{b'e}$ voltage source is converted into a current source to yield the circuit shown in Fig. 9.3-7*c*. In this circuit the following simplifications have also been made:

$$i_i' = \frac{i_i R_b}{R_b + r_{b'b}} \approx i_i \tag{9.3-10a}$$

and

$$R_{b'e} = r_{b'e} \| (R_b + r_{bb'}) \tag{9.3-10b}$$

In a practical application the impedance of capacitor $C_{b'c}$ is much greater than R_L, and the current source $j\omega C_{b'c} v_{b'e}$, which represents leakage current from input to output, is much less than the normal transistor current $g_m v_{b'e}$. Hence, we assume that

$$\frac{1}{\omega C_{b'c}} \gg R_L \quad \text{assumptions} \quad (9.3\text{-}11a)$$

and

$$\omega C_{b'c} \ll g_m \quad (9.3\text{-}11b)$$

for all frequencies of interest. With these assumptions, the circuit reduces to that shown in Fig. 9.3-7d.

Before continuing the analysis, let us determine the frequency range over which the inequalities of (9.3-11) are valid. If, for example, $g_m = 0.01$ S, $R_L = 1$ kΩ, and $C_{b'c} = 5$ pF, then $f \ll 32$ MHz defines the frequency range over which both inequalities are valid. Note that (9.3-11a) is almost always more restrictive than (9.3-11b) since $R_L > 1/g_m$. We shall confine the discussion to cases where (9.3-11) applies.

Input admittance Referring to the output portion of the circuit of Fig. 9.3-7d, we find

$$v_{ce} = -g_m R_L v_{b'e} \qquad vb'e = \frac{vce}{gmRL} \qquad (9.3\text{-}12)$$

Hence the current flowing through capacitor $C_{b'c}$ is

$$i_{C_{b'c}} = j\omega C_{b'c}(v_{b'e} + g_m R_L v_{b'e}) = j\omega C_{b'c}\underbrace{(1 + g_m R_L)}v_{b'e}, \qquad (9.3\text{-}13)$$

$$C_M$$

The series circuit consisting of capacitor $C_{b'c}$ and voltage v_{ce} shown in Fig. 9.3-7d can therefore be replaced by a capacitor, called the *Miller capacitor*, having the admittance

$$j\omega C_M = \frac{i_{C_{b'c}}}{v_{b'e}} = j\omega C_{b'c}(1 + g_m R_L) \qquad (9.3\text{-}14a)$$

Hence, the Miller capacitance is

$$\boxed{C_M = C_{b'c}(1 + g_m R_L')} \qquad R_L \| R_C \| \ldots \qquad (9.3\text{-}14b)$$

With this definition, the input admittance Y_{in} is

$$\boxed{Y_{\text{in}} = \frac{1}{R_{b'e}} \| j\omega C_{b'e} \| j\omega C_m} \qquad (9.3\text{-}15)$$

The final form of the high-frequency equivalent circuit including the Miller capacitor C_M is shown in Fig. 9.3-8. With this equivalent circuit the amplifier current gain A_i is

$$A_i = \frac{i_L}{i_i} = \frac{-g_m R_{b'e}}{1 + j\omega R_{b'e}(C_{b'e} + C_M)} \qquad (9.3\text{-}16)$$

$$R_{b'e} = rb'e \| [(Rb\|r_i + rbb']$$

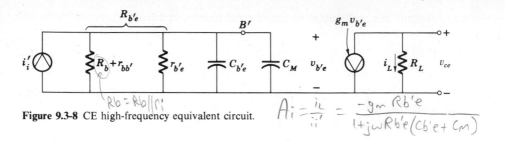

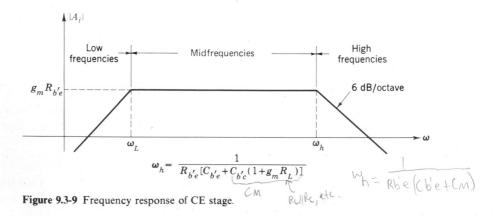

Figure 9.3-8 CE high-frequency equivalent circuit.

$$A_i = \frac{i_L}{i_i} = \frac{-g_m R_{b'e}}{1 + j\omega R_{b'e}(C_{b'e} + C_M)}$$

Equation (9.3-16) indicates that the midfrequency gain A_{im} is

$$A_{im} = -g_m R_{b'e} \qquad \text{(9.3-17a)}$$

and the upper 3-dB frequency is

$$f_h = \frac{1}{2\pi R_{b'e}(C_{b'e} + C_M)} \qquad \text{(9.3-17b)}$$

In this type of RC amplifier f_h is usually much larger than the lower 3-dB frequency f_L (see Fig. 9.1). It is therefore common practice to define the 3-dB *bandwidth* (BW) of the amplifier, which is actually $f_h - f_L$, to be simply f_h.

An asymptotic plot of the amplifier gain as a function of frequency is shown in Fig. 9.3-9.

Example 9.3-2 The amplifier shown in Fig. 9.3-6a has the following component values:

$$r_i = 10 \text{ k}\Omega \qquad C_{b'c} = 2 \text{ pF} \qquad R_b = 2 \text{ k}\Omega \qquad C_{b'e} = 200 \text{ pF}$$

$$r_{bb'} = 20 \ \Omega \qquad g_m = 0.5 \text{ S} \qquad r_{b'e} = 150 \ \Omega \qquad R_L = 200 \ \Omega \ (R_c \gg R_L)$$

Find the midband current gain and the 3-dB frequency f_h.

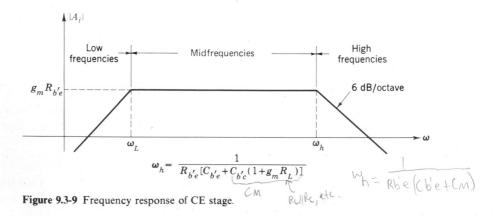

$$\omega_h = \frac{1}{R_{b'e}[C_{b'e} + C_{b'c}(1 + g_m R_L)]}$$

Figure 9.3-9 Frequency response of CE stage.

SOLUTION

approximate

Midband current gain From (9.3-17a)

$$A_{im} = -g_m R_{b'e} = -g_m[(R_b + r_{bb'})\|r_i\|r_{b'e}] \approx -(0.5)(150) = -75$$

3-dB frequency f_h From (9.3-14b) and (9.3-17b)

$$C_{b'e} + C_M = \{200 + 2[1 + (\tfrac{1}{2})(200)]\} = 400 \text{ pF}$$

$$f_h = \frac{1}{2\pi R_{b'e}(C_{b'e} + C_M)} \approx 2.6 \text{ MHz}$$

Validity of the circuit of Fig. 9.3-8 The 3-dB frequency f_h, calculated using Fig. 9.3-8, is valid when (9.3-11a) applies; i.e.,

$$f_h \ll \frac{1}{2\pi R_L C_{b'c}} \approx 400 \text{ MHz}$$

Since $f_h = 2.6 \text{ MHz} \ll 400 \text{ MHz}$, (9.3-17b) is valid and should accurately predict the upper 3-dB frequency.

Approximating f_h by f_β Sometimes f_β is used as a rough approximation to f_h. In this example,

$$f_\beta = \frac{1}{2\pi r_{b'e} C_{b'e}} = 5.3 \text{ MHz}$$

$f_\beta =$ C.E. Short circuit 3dB frequency

and the approximation is not very good. As a matter of fact, the approximation is good only when

$$C_{b'e} \gg C_M = C_{b'c}(1 + g_m R_L) \qquad \text{and} \qquad R_{b'e} \approx r_{b'e} \qquad \qquad ///$$

9.3-3 The Emitter Follower at High Frequencies

In this section we determine the frequency dependence of the emitter-follower circuit shown in Fig. 9.3-10a. In Fig. 9.3-10b the circuit is redrawn with the supply voltages replaced by short circuits and with the internal transistor capacitances $C_{b'c}$ and $C_{b'e}$ and resistance $r_{bb'}$ explicitly shown. In addition, that part of the circuit looking toward the source from the base of the transistor is replaced by its Thevenin equivalent circuit, in which the Thevenin voltage is $v_i' = [R_b/(r_i + R_b)]v_i$ and the Thevenin resistance is $r_i\|R_b$. In order to simplify the notation we define R_i as

$$R_i = (r_i\|R_b) + r_{bb'} \qquad (9.3-18)$$

The circuit of Fig. 9.3-10b is redrawn for clarity in Fig. 9.3-10c. Here we see that if $C_{b'c}$ were zero, the high-frequency response would be limited by $C_{b'e}$. At low

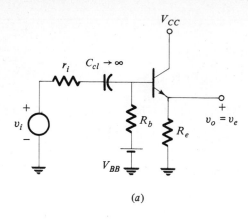

(a)

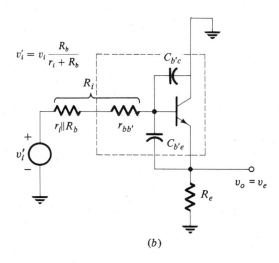

$$v_i' = v_i \frac{R_b}{r_i + R_b}$$

(b)

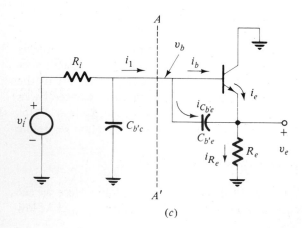

(c)

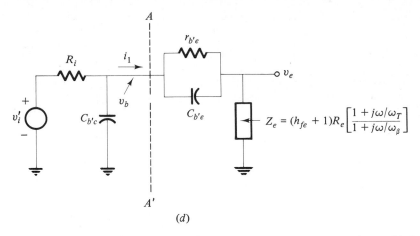

Figure 9.3-10 Emitter follower: (*a*) circuit; (*b*) circuit showing internal capacitance; (*c*) circuit redrawn for clarity; (*d*) equivalent circuit looking into the base.

frequencies the impedance of $C_{b'e}$ is much greater than the base-emitter impedance $r_{b'e}$, and therefore the current in R_e is primarily the emitter current i_e. Hence the impedance seen looking into A-A' is $r_{b'e} + (h_{fe} + 1)R_e$, and the voltage gain of the emitter follower is the midfrequency value A_{vm}, where

$$A_{vm} = \frac{(h_{fe} + 1)R_e}{(h_{fe} + 1)R_e + r_{b'e} + R_i} \tag{9.3-19a}$$

At very high frequencies the impedance of $C_{b'e}$ is much less than that of $r_{b'e}$, and the current in R_e is primarily $i_{C_{b'e}}$, the current flowing in $C_{b'e}$. Hence the impedance now seen looking into AA' is only R_e, and the gain has been substantially reduced to A_{vh}, where

$$A_{vh} = \frac{R_e}{R_e + R_i} \tag{9.3-19b}$$

At any arbitrary frequency ω, the current flowing in R_e is

$$i_{R_e} = i_e + i_{C_{b'e}} \tag{9.3-20a}$$

The current i_1 entering AA' is

$$i_1 = i_b + i_{C_{b'e}} \tag{9.3-20b}$$

where

$$r_{b'e}i_b = \frac{1}{j\omega C_{b'e}} i_{C_{b'e}} \tag{9.3-20c}$$

Substituting (9.3-20c) into (9.3-20b) and into (9.3-20a) yields

$$i_1 = (1 + j\omega r_{b'e} C_{b'e})i_b \tag{9.3-20d}$$

and

$$i_{R_e} = [(h_{fe} + 1) + j\omega r_{b'e} C_{b'e}]i_b \tag{9.3-20e}$$

Combining (9.3-20d) and (9.3-20e), we have

$$i_1 = \left(\frac{1 + j\omega/\omega_\beta}{1 + j\omega/\omega_T}\right)\left(\frac{i_{R_e}}{h_{fe} + 1}\right) \tag{9.3-20f}$$

Using (9.3-20f), we can verify that the impedance seen looking into AA' consists of the parallel combination of $r_{b'e}C_{b'e}$, which is in series with Z_e, where

$$Z_e \equiv \frac{v_e}{i_1} = \frac{v_e}{i_{R_e}}\frac{i_{R_e}}{i_1} = R_e(h_{fe} + 1)\left(\frac{1 + j\omega/\omega_T}{1 + j\omega/\omega_\beta}\right) \tag{9.3-20g}$$

as shown in Fig. 9.3-10d.

Using Fig. 9.3-10d and assuming that $C_{b'c}$ is negligibly small, we find the gain at frequency ω to be

$$A_v(\omega) = \frac{v_e}{v_{i'}} = \frac{(h_{fe} + 1)R_e(1 + j\omega/\omega_T)/(1 + j\omega/\omega_\beta)}{(h_{fe} + 1)R_e\left(\dfrac{1 + j\omega/\omega_T}{1 + j\omega/\omega_\beta}\right) + \dfrac{r_{b'e}}{1 + j\omega/\omega_\beta} + R_i} \tag{9.3-21a}$$

Simplifying yields [assuming $(h_{fe} + 1)R_e \gg r_{b'e} + R_i$]

$$A_v(\omega) \approx \frac{1 + j\omega/\omega_T}{1 + j(\omega/\omega_T)(1 + R_i/R_e)} \tag{9.3-21b}$$

The upper cutoff frequency is therefore

$$f_{h1} = \frac{f_T}{1 + R_i/R_e} \tag{9.3-21c}$$

If, however, $C_{b'c}$ is large enough for its impedance to be equal to R_i at a frequency f_{h2} much less than f_β, then at f_{h2} the impedance seen looking into AA' is $r_{b'e} + (h_{fe} + 1)R_e \gg R_i$ and the output voltage v_e at f_{h2} is decreased only because of the reduced impedance of $C_{b'c}$. In this case the voltage gain is

$$A_v(\omega) \approx \left(\frac{(h_{fe} + 1)R_e}{r_{b'e} + (h_{fe} + 1)R_e}\right)\left(\frac{1}{1 + j\omega R_i C_{b'c}}\right) \qquad \omega < \omega_\beta \tag{9.3-22a}$$

Now the upper cutoff frequency is

$$f_{h2} = \frac{1}{2\pi R_i C_{b'c}} < f_\beta \tag{9.3-22b}$$

If f_{h1} and f_{h2} are comparable in magnitude, the circuit of Fig. 9.3-10d must be analyzed without approximation in order to determine f_h. Assuming that $(h_{fe} + 1)R_e \gg R_i + r_{b'e}$, it can be shown (see Prob. 9.3-7) that

$$A_v \approx \frac{1 + j\omega/\omega_T}{1 + j\omega\left(\dfrac{1}{\omega_{h2}} + \dfrac{1 + R_i/R_e}{\omega_T}\right) + \dfrac{j\omega}{\omega_T}\dfrac{j\omega}{\omega_{h2}}} \tag{9.3-23}$$

The following example will illustrate the use of (9.3-21), (9.3-22b), and (9.3-23).

Example 9.3-3 The emitter follower of Fig. 9.3-10 has the component values

$$C_{b'e} = 1000 \text{ pF} \qquad h_{fe} = 100 \qquad C_{b'c} = 10 \text{ pF}$$
$$R_e = 1 \text{ k}\Omega \qquad r_{b'e} = 100 \text{ }\Omega \qquad R_i = 1 \text{ k}\Omega$$

Find the voltage gain A_v.

SOLUTION We first calculate f_{h1} and f_{h2}:

$$f_{h1} = \frac{f_T}{1 + R_i/R_e}$$

From (9.3-2) and (9.3-4)

$$f_T = \frac{h_{fe}}{2\pi r_{b'e}(C_{b'e} + C_{b'c})} \approx \frac{100}{2\pi(100)(1000) \times 10^{-12}} \approx 160 \text{ MHz}$$

Hence $f_{h1} = 80$ MHz

Also $$f_{h2} = \frac{1}{2\pi R_i C_{b'c}} = \frac{100}{2\pi(1000)(10 \times 10^{-12})} \approx 16 \text{ MHz}$$

Hence, in this example the upper cutoff frequency is due to the $R_i C_{b'c}$ circuit. If, however, $C_{b'c} = 2$ pF, then f_{h1} and f_{h2} are approximately equal and (9.3-23) must be employed to determine the upper cutoff frequency of the amplifier. For this example, $f_{h2} = 80$ MHz, and (9.3-23) now becomes

$$A_v \approx \frac{1 + j\omega/10^9}{1 + j\omega(2/10^9 + 2/10^9) + (j\omega)^2 2/10^{18}}$$

Solving, we find that the upper 3-dB frequency occurs at $f_h \approx 46$ MHz, which is comparable to f_{h1} and f_{h2}. The details of the solution are left to the problems. ///

9.4 THE FET AT HIGH FREQUENCIES

The FET at high frequencies can be described in terms of a hybrid-pi equivalent circuit as shown in Fig. 9.4-1. The capacitors C_{gs} and C_{gd} in a junction FET are a consequence of the back-biased gate. Figure 9.4-2 shows a JFET operating above pinch-off. The capacitance between the gate and the source and between the gate and the drain are similar to the collector-base capacitance $C_{b'c}$ since they all result from a reverse-biased pn junction (Fig. 9.3-1b). Therefore these capacitors vary, as does $C_{b'c}$:

$$C_{gs} \propto (-V_{GS})^{-1/2} \qquad V_{GS} \le 0 \qquad (9.4\text{-}1a)$$

and $$C_{gd} \propto (-V_{GD})^{-1/2} \qquad V_{GD} \le 0 \qquad (9.4\text{-}1b)$$

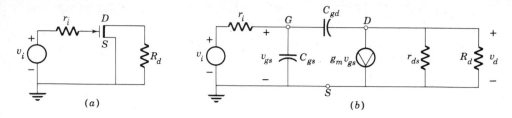

Figure 9.4-1 The FET at high frequencies: (a) circuit (bias components omitted); (b) high-frequency equivalent circuit.

Since $|V_{GD}| \gg |V_{GS}|$ in normal operation, we have

$$C_{gd} \ll C_{gs} \tag{9.4-1c}$$

Typical values of C_{gs} vary from 50 pF for a low-frequency FET to less than 5 pF for a high-frequency FET. The feedback capacitor C_{gd} is usually less than 5 pF and is often less than 0.5 pF for a high-frequency MOSFET.

9.4-1 High-Frequency Behavior of the Common-Source Amplifier; Miller Capacitance

The high-frequency equivalent circuit of the FET shown in Fig. 9.4-1b is similar to that shown in Fig. 9.3-6 for the transistor. Proceeding with a similar analysis, we find that the Miller capacitance is

$$C_M = C_{gd}[1 + g_m(r_{ds} \| R_d)] \tag{9.4-2a}$$

This result is valid [Eq. (9.3-11)] for frequencies such that

$$\omega \ll \frac{1}{C_{gd}(r_{ds} \| R_d)} \quad \text{and} \quad \omega \ll \frac{g_m}{C_{gd}} \tag{9.4-2b}$$

Using the above approximations, we obtain the high-frequency equivalent circuit shown in Fig. 9.4-3 (compare with Fig. 9.3-8).

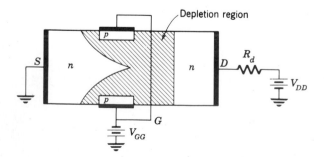

Figure 9.4-2 The JFET above pinch-off.

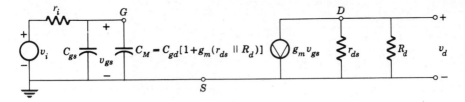

Figure 9.4-3 Common-source circuit with feedback removed.

The voltage gain of the FET amplifier is

$$A_v = \frac{v_d}{v_i} = -g_m(r_{ds} \| R_d) \frac{1}{1 + j\omega r_i(C_{gs} + C_M)} \tag{9.4-3a}$$

The 3-dB frequency is

$$f_h = \frac{1}{2\pi r_i(C_{gs} + C_M)} \tag{9.4-3b}$$

The gain-frequency plot for the FET amplifier is shown in Fig. 9.4-4.

Example 9.4-1 The FET amplifier shown in Fig. 9.4-1a has the following component values: $R_d = 10$ kΩ, $r_{ds} = 15$ kΩ, $g_m = 3$ mS, $C_{gs} = 50$ pF, and $C_{gd} = 5$ pF. Find r_i to ensure a 3-dB bandwidth of at least 100 kHz.

SOLUTION From (9.4-3b) we see that

$$r_i = \frac{1}{2\pi f_h(C_{gs} + C_M)}$$

$$\leq \frac{1}{2\pi (10^5)\{(50 \times 10^{-12}) + (5 \times 10^{-12})[1 + (3 \times 10^{-3})(6 \times 10^3)]\}} = 11 \text{ k}\Omega$$

We can therefore let $r_i = 10$ kΩ. ///

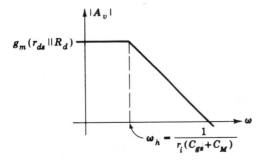

Figure 9.4-4 Gain versus frequency for the common-source amplifier.

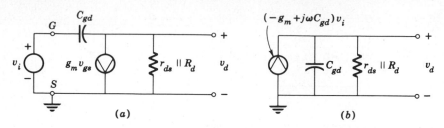

Figure 9.4-5 FET amplifier with zero input resistance ($r_i = 0$): (a) equivalent circuit; (b) simplified equivalent circuit.

Voltage gain when $r_i = 0$ Consider the FET amplifier shown in Fig. 9.4-1b, with $r_i = 0$. The equivalent circuit is shown in Fig. 9.4-5a. The voltage gain A_v is found from the simplified equivalent circuit of Fig. 9.4-5b, where we have used the fact that $v_{gs} = v_i$. Then

$$A_v = \frac{v_d}{v_i} = (-g_m + j\omega C_{gd})\frac{r_{ds}\|R_d}{1 + j\omega C_{gd}(r_{ds}\|R_d)}$$

$$= g_m(r_{ds}\|R_d)\left[\frac{-1 + j\omega C_{gd}/g_m}{1 + j\omega C_{gd}(r_{ds}\|R_d)}\right] \tag{9.4-4}$$

where

$$g_m(r_{ds}\|R_d) \gg 1 \tag{9.4-5}$$

The voltage gain versus frequency is plotted in Fig. 9.4-6.

Note that if R_d is infinite, the 3-dB bandwidth of the common-source amplifier is limited by the channel resistance r_{ds} and the gate drain capacitance C_{gd}. Hence in Example 9.4-1,

$$f_h \leq \frac{1}{2\pi C_{gd}R_{ds}} = 2 \text{ MHz}$$

9.4-2 High-Frequency Behavior of the Source Follower

The source-follower circuit is shown in Fig. 9.4-7, in which we have omitted all the bias components and have explicitly shown the internal capacitances C_{gs} and C_{gd}. The circuit is redrawn for clarity in Fig. 9.4-8a, and the equivalent circuit obtained

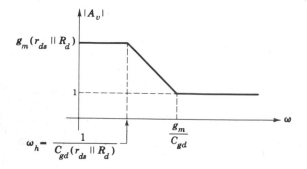

Figure 9.4-6 Voltage gain of FET amplifier of Fig. 9.4-5.

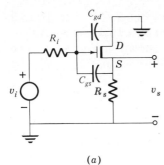

(a)

Figure 9.4-7 Source follower at high frequencies (bias components omitted).

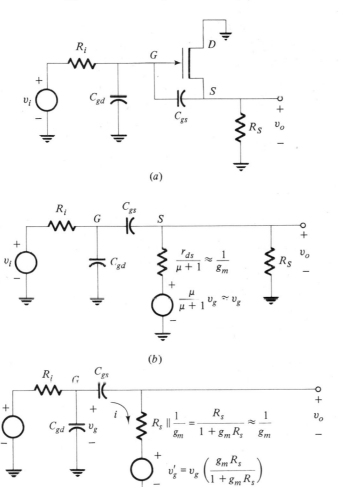

(a)

(b)

(c)

Figure 9.4-8 Source follower: (a) circuit of Fig. 9.4-7 redrawn; (b) equivalent circuit; (c) reduced equivalent circuit.

by reflection into the source (see Sec. 6.10) is shown in Fig. 9.4-8b. The reduced equivalent circuit, in which the source circuit is replaced by a Thevenin equivalent obtained by making use of the fact that $g_m R_s \gg 1$, is shown in Fig. 9.4-8c. The dependent source v'_g can be replaced by an equivalent impedance $Z_{eq} = v'_g/i$ by noting that

$$v_g - v'_g = i\left(\frac{1}{g_m} + \frac{1}{j\omega C_{gs}}\right) \tag{9.4-6a}$$

From the figure we have

$$v_g = v'_g \frac{1 + g_m R_s}{g_m R_s} \tag{9.4-6b}$$

Substituting (9.4-6b) into (9.4-6a) yields

$$v'_g\left(1 - \frac{1 + g_m R_s}{g_m R_s}\right) = i\left(\frac{1}{g_m} + \frac{1}{j\omega C_{gs}}\right) \tag{9.4-6c}$$

and therefore

$$Z_{eq} = \frac{v'_g}{i} = g_m R_s\left(\frac{1}{g_m} + \frac{1}{j\omega C_{gs}}\right) = R_s + \frac{1}{j\omega C_{gs}/g_m R_s} \tag{9.4-6d}$$

From this equation we see that the dependent source can be replaced by the series RC network shown in Fig. 9.4-9.

From Fig. 9.4-9 we see that there are two special cases. First, consider that the upper 3-dB frequency is due to C_{gs} and at that frequency the impedance due to C_{gd} is very large. In this case the voltage gain is, assuming that $g_m R_s \gg 1$,

$$A_v = \frac{v_o}{v_i} = \frac{1/g_m + R_s + g_m R_s/j\omega C_{gs}}{R_i + 1/g_m + R_s + (1/j\omega C_{gs})(1 + g_m R_s)}$$

$$\approx \frac{1 + j\omega C_{gs}/g_m}{1 + j\omega(C_{gs}/g_m)(1 + R_i/R_s)} \tag{9.4-7}$$

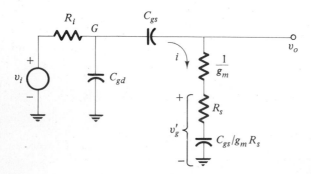

Figure 9.4-9 Reduced equivalent circuit with dependent source removed.

From this expression we see that the midfrequency gain of the source follower is unity and that the gain at very high frequencies is

$$A_v(\omega \to \infty) = \frac{1}{1 + R_i/R_s}$$

In order for a 3-dB frequency to exist we must have

$$A_v(\omega \to \infty) < 0.707 A_v(\omega_{midband})$$

Since $A_v(\omega_{midband}) \approx 1$, this requires that

$$\frac{1}{1 + R_i/R_s} < 0.707$$

which leads to the inequality

$$\frac{R_i}{R_s} > 0.414$$

If this inequality is satisfied, the upper 3-dB frequency is

$$f_{h1} = \frac{g_m R_s}{2\pi C_{gs}(R_i + R_s)} \tag{9.4-8}$$

If $R_i/R_s < 0.414$, the midband and high frequency regions are separated by less than 3-dB and there is no upper 3-dB frequency. For example, referring to (9.4-7), we see that if $R_i = 0$, $A_v \approx 1$ at all frequencies. Actually in this case, at sufficiently high frequencies stray capacitance from source to ground will short-circuit the source resistance R_s.

The second possible 3-dB frequency f_{h2} occurs if the pole is due to the $R_i C_{gd}$ circuit. If such is the case,

$$f_{h2} = \frac{1}{2\pi R_i C_{gd}} \tag{9.4-9}$$

If however f_{h1} and f_{h2} are comparable, the circuit of Fig. 9.4-9 must be analyzed without approximations in order to find the 3-dB frequency. This is left to the problems.

Example 9.4-2 The source follower shown in Fig. 9.4-7 has the following parameters: $g_m = 3$ mS, $C_{gd} = 0.2$ pF, $C_{gs} = 20$ pF, $R_i = 10$ kΩ, and $R_s = 2$ kΩ. Estimate the upper 3-dB frequency.

SOLUTION We begin by calculating f_{h1} and f_{h2} using (9.4-8) and (9.4-9). The results are

$$f_{h1} = 4 \text{ MHz} \qquad \text{and} \qquad f_{h2} = 80 \text{ MHz}$$

We conclude that the upper 3-dB frequency is approximately 4 MHz.   ///

9.5 TUNED AMPLIFIERS

In this section we discuss the amplification of signals within a narrow frequency band centered about a frequency ω_0. These tuned amplifiers are designed to *reject* all frequencies below a lower break frequency ω_L and above an upper break frequency ω_h.

Tuned circuits are used extensively in almost all communications equipment. An example with which we are all familiar is the radio receiver. When we tune a radio receiver, we are varying ω_0 while keeping the passband $\omega_h - \omega_L$ constant. The particular value of ω_0 at which we set the dial corresponds to the carrier frequency of the broadcast station whose signal we wish to receive, and $\omega_h - \omega_L$ corresponds to the bandwidth required to receive the signal information without significant distortion.

To receive AM and FM signals without significant distortion and to keep the receiver noise power as small as possible, the ideal bandpass amplifier can be shown to have an amplitude and phase response as in Fig. 9.5-1a. Consider the radio receiver. This characteristic would pass the signal from the station of interest only, rejecting completely signals emanating from stations on adjacent (and all other) channels. A typical response, achieved using the circuits to be described in this section, is shown in Fig. 9.5-1b. This characteristic represents an approximation to the ideal response obtained with a single-tuned stage. Better approximations will be considered later in the chapter.

9.5-1 Single-tuned Amplifiers

The ordinary common-emitter amplifier is converted into a tuned bandpass amplifier by including a parallel-tuned circuit as shown in Fig. 9.5-2a. All bias components have been omitted for simplicity. Let us determine the gain, center frequency, and bandwidth of this amplifier.

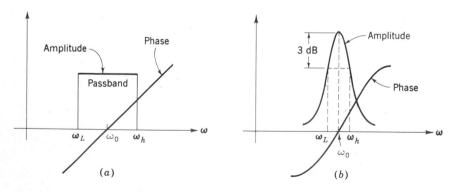

Figure 9.5-1 Response of tuned amplifiers: (*a*) ideal response; (*b*) actual response.

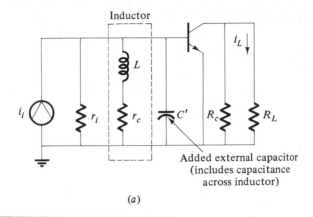

(a)

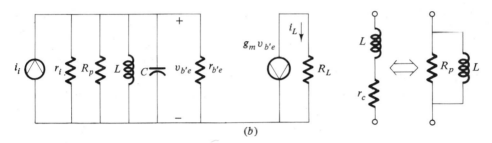

(b)

Figure 9.5-2 Single-tuned amplifier: (a) circuit; (b) equivalent circuit; (c) coil.

Before proceeding with these calculations, several practical simplifying assumptions are to be made. First, let us assume that

$$R_L \ll R_c \tag{9.5-1a}$$

and

$$r_{bb'} = 0 \tag{9.5-1b}$$

The simplified equivalent circuit of this amplifier is shown in Fig. 9.5-2b, where

$$\boxed{C = C' + C_{b'e} + (1 + g_m R_L)C_{b'c}} \tag{9.5-2}$$

C' is an added external capacitor, used to tune the circuit and/or help adjust the circuit bandwidth, and $(1 + g_m R_L)C_{b'c}$ is the Miller capacitance C_M. In the series RL circuit (Fig. 9.5-2a), which is used as a model for the actual inductor, the resistance r_c represents the losses in the coil. The parallel RL circuit of Fig. 9.5-2c is equivalent to the series circuit over the frequency band of interest if the coil losses are low, i.e., if the coil has a high Q_c

$$Q_c \equiv \frac{\omega L}{r_c} \gg 1 \tag{9.5-3a}$$

The conditions for equivalence are most easily established by equating the admittances of the two circuits shown in Fig. 9.5-2c and making use of (9.5-3a) as follows:

$$Y_c = \frac{1}{r_c + j\omega L} = \frac{r_c - j\omega L}{r_c^2 + \omega^2 L^2} \approx \frac{1}{r_c}\left(\frac{r_c}{\omega L}\right)^2 + \frac{1}{j\omega L} = \frac{1}{R_p} + \frac{1}{j\omega L} \quad (9.5\text{-}3b)$$

and

$$R_p = r_c Q_c^2 = \omega L Q_c \quad (9.5\text{-}4)$$

Note that R_p is a function of ω^2 if r_c and L are constant. Representing an inductor by a simple series RL circuit neglects the fact that every inductor has parasitic capacitance in parallel with it. In the analyses to follow, this capacitance is assumed to be part of C'.

Refer to Fig. 9.5-2b. Let

$$R \equiv r_i \| R_p \| r_{b'e} \quad (9.5\text{-}5)$$

The current gain of the amplifier is then

$$A_i = \frac{-g_m R}{1 + j(\omega RC - R/\omega L)} = \frac{-g_m R}{1 + j\omega_0 RC(\omega/\omega_0 - \omega_0/\omega)} \quad (9.5\text{-}6a)$$

where

$$\omega_0^2 = \frac{1}{LC} \quad (9.5\text{-}6b)$$

We define the Q of the tuned input circuit at the resonant frequency ω_0 to be

$$Q_i = \frac{R}{\omega_0 L} = \omega_0 RC \quad (9.5\text{-}7a)$$

The circuit analysis and the design problems discussed below assume that Q_i and Q_c are greater than 5. This is called the *high-Q* approximation. When (9.5-7a) is used, (9.5-6a) becomes

$$A_i = \frac{-g_m R}{1 + jQ_i(\omega/\omega_0 - \omega_0/\omega)} \quad (9.5\text{-}7b)$$

The gain is a maximum at $\omega = \omega_0$ and is

$$A_{im} = -g_m R \quad (9.5\text{-}8)$$

Figure 9.5-3 shows the variation of the magnitude of the gain as a function of frequency for a single-tuned amplifier. The bandwidth of the amplifier is found by setting

$$|A_i| = \frac{g_m R}{\sqrt{2}} \quad (9.5\text{-}9a)$$

in (9.5-7b) and solving the resulting equation

$$1 + Q_i^2\left(\frac{\omega}{\omega_0} - \frac{\omega_0}{\omega}\right)^2 = 2 \quad (9.5\text{-}9b)$$

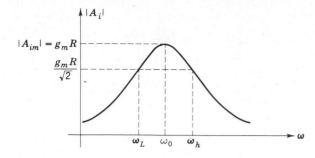

Figure 9.5-3 Gain versus frequency for single-tuned amplifier.

This equation is quadratic in ω^2 and has two positive solutions, ω_h and ω_L. The 3-dB bandwidth can be shown to be

$$\text{BW} = f_h - f_L = \frac{\omega_0}{2\pi Q_i} = \frac{1}{2\pi RC} \tag{9.5-10}$$

Example 9.5-1 Design a single-tuned amplifier to operate at a center frequency of 455 kHz with a bandwidth of 10 kHz. The transistor has the parameters $g_m = 0.04$ S, $h_{fe} = 100$, $C_{b'e} = 1000$ pF, and $C_{b'c} = 10$ pF. The bias network and the input resistance are adjusted so that $r_i = 5$ kΩ and $R_L = 500$ Ω.

SOLUTION In order to obtain a bandwidth of 10 kHz, the RC product is [Eq. (9.5-10)]

$$RC = \frac{1}{2\pi \text{BW}} = \frac{1}{2\pi 10^4}$$

where, from (9.5-5),

$$R = r_i \| R_p \| r_{b'e}$$

The input resistance is

$$r_i = 5 \text{ k}\Omega$$

and

$$r_{b'e} = \frac{h_{fe}}{g_m} = 2500 \ \Omega \qquad R_p = Q_c \omega_0 L = \frac{Q_c}{\omega_0 C}$$

Therefore

$$R = (5 \times 10^3) \| (2.5 \times 10^3) \| \frac{Q_c}{\omega_0 C}$$

and

$$C = \frac{1}{2\pi 10^4 R} = \frac{10^{-4}}{2\pi} \left[\frac{1}{5000} + \frac{1}{2500} + \frac{2\pi(455 \times 10^3)C}{Q_c} \right]$$

Solving for C yields

$$C \approx \frac{0.95 \times 10^{-8}}{1 - 45.5/Q_c}$$

The total input capacitance is

$$C = C' + C_{b'e} + (1 + g_m R_L)C_{b'c} = C' + 1200 \text{ pF}$$

Therefore

$$C' + 1200 \times 10^{-12} \approx \frac{0.95 \times 10^{-8}}{1 - 45.5/Q_c}$$

The choice of a value of Q_c to satisfy this equation is not unique. We know that Q_c must be greater than 45.5 for the capacitance to be positive. The question to be answered by the design engineer is: How large should Q_c be? If, for example, Q_c were chosen to be 45.5, C' would be infinite, C would be infinite, and $L \to 0$. This is not a practical solution! At 455 kHz, a typical range of practical values of Q_c lies between 10 and 150. Let us choose

$$Q_c = 100$$

Then

$$C' \approx 0.016 \ \mu\text{F} \qquad \text{and} \qquad C \approx 0.018 \ \mu\text{F}$$

Note that the input capacitance $C_{b'} = C_{b'e} + C_M$ is negligible. The inductance required is

$$L = \frac{1}{\omega_0^2 C} \approx 6.9 \ \mu\text{H}$$

We can now calculate R_p

$$R_p = Q_c \omega_0 L \approx 2 \text{ k}\Omega$$

Hence

$$R = r_i \| R_p \| r_{b'e} = 910 \ \Omega$$

The resulting midfrequency gain is

$$A_{im} = -g_m R = (-0.04)(910) \approx -36.4$$

If a 6.9-μH inductor with a Q_c of 100 is available, the design is complete. If the required Q_c cannot be readily obtained, a transformer may be used in order to transform the input impedances to levels which will allow the specifications to be met. This technique is discussed below. ///

The analysis-and-design problem considered above could have employed a FET instead of a transistor. When using a FET the parallel *RLC* circuit can be formed in the output circuit. Other differences between the transistor and the FET are the element values of the input impedance presented by each device.

9.5-2 Impedance Matching to Improve Gain

The circuit of Fig. 9.5-2a often yields impractical element values and low gain because of the low effective resistance in the base circuit of the transistor. Since we are dealing with parallel-tuned circuits, low resistance implies low Q_i and consequent difficulty in achieving narrow bandwidth. One method frequently used to

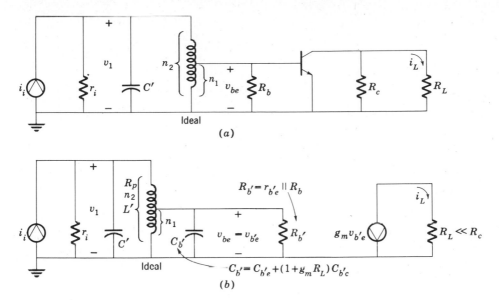

Figure 9.5-4 Tuned amplifier using an autotransformer ($r_{bb'} = 0$): (*a*) circuit (bias components omitted); (*b*) equivalent circuit.

circumvent this problem involves the use of a *tapped* inductor as an autotransformer. This serves effectively to transform the low resistance into a more reasonable value. A typical circuit is shown in Fig. 9.5-4.

The turns ratio of the ideal autotransformer is

$$a = \frac{n_1}{n_2} = \frac{v_{b'e}}{v_1} < 1 \tag{9.5-11}$$

The impedance of the transistor circuit, which consists of $R_{b'}$ in parallel with $C_{b'}$, is reflected into the input. The resulting parallel *RLC* circuit is shown in Fig. 9.5-5.

The gain, center frequency, and bandwidth of the amplifier can now be found as in Sec. 9.5-1. The results are

$$A_i = \frac{i_L}{i_i} = -ag_m R \frac{1}{1 + jQ_i(\omega/\omega_0 - \omega_0/\omega)} \tag{9.5-12}$$

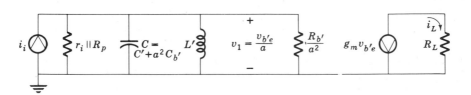

Figure 9.5-5 Tuned amplifier with reflected impedances.

where
$$Q_i = \omega_0 RC$$
$$C = C' + a^2 C_{b'} \tag{9.5-13}$$

and
$$R = r_i \| R_p \| \frac{R_{b'}}{a^2}$$

The center frequency is

$$\omega_0^2 = \frac{1}{L'C} = \frac{1}{L'(C' + a^2 C_{b'})} \tag{9.5-14}$$

If, as is often the case, $C' \gg a^2 C_{b'}$,

$$\omega_0^2 \approx \frac{1}{L'C'} \tag{9.5-15}$$

The 3-dB bandwidth and the midfrequency gain can be determined using (9.5-12):

3-dB bandwidth:
$$BW = f_h - f_L = \frac{1}{2\pi RC} \tag{9.5-16}$$

Midfrequency gain:
$$A_{im} = -a g_m R \tag{9.5-17}$$

The transformer enables us to achieve high gain and a *narrow* bandwidth.

Example 9.5-2 A single-tuned amplifier operates at $f_0 = 455$ kHz and is to have a bandwidth of 10 kHz, $L' = 6.9 \ \mu H$, $r_{b'e} = R_{b'} = 1 \ k\Omega$, $r_i = 5 \ k\Omega$, $R_p = 2 \ k\Omega$, $C_{b'e} = 1000$ pF, $g_m = 0.1$ S, $R_L = 500 \ \Omega$, and $C_{b'c} = 4$ pF. Find the required turns ratio a and the midfrequency current gain.

SOLUTION We begin by calculating C and R

$$C = C' + a^2 C_1 = C' + 1200 \times 10^{-12} a^2$$

$$\frac{1}{R} = \frac{1}{5000} + \frac{1}{2000} + \frac{a^2}{1000} = (10^{-4})(7 + 10a^2)$$

Then, from Fig. 9.5-5,

$$\omega_0^2 = 4\pi^2[(455)^2 \times 10^6] = \frac{1}{L'C} = \frac{1}{(6.9 \times 10^{-6})(C' + a^2 \, 1200 \times 10^{-12})}$$

Thus
$$C' + 12a^2 \times 10^{-10} = \frac{10^{-6}}{57.2} \approx 0.017 \times 10^{-6}$$

Since $a \leq 1$,

$$C' \approx 0.017 \ \mu F$$

Using the bandwidth relation (9.5-16) to find R, we have

$$BW = 10^4 = \frac{1}{2\pi RC} \approx \frac{1}{2\pi RC'} \approx \frac{1}{2\pi R(0.017 \times 10^{-6})}$$

Then $R \approx 930\ \Omega$, and

$$\frac{1}{R} = \frac{1}{930} = 10^{-4}(7 + 10a^2)$$

Solving, we get

$$a^2 \approx 0.4$$

and

$$a \approx 0.63$$

The midfrequency gain is

$$A_{im} = -ag_m R \approx -(0.63)(0.1)(930) = -59$$

We now see the advantage of using the autotransformer. In Example 9.5-1, $r_{b'e} = 2500\ \Omega$ and $g_m = 0.04\ \text{S}$ ($h_{fe} = 100$). The midfrequency gain was found to be 36.4 (BW = 10 kHz). In this example we let $r_{b'e} = 1\ \text{k}\Omega$ and $g_m = 0.1\ \text{S}$ ($h_{fe} = 100$). The transformer multiplied $r_{b'e}$ by $1/a^2 = 2.5$, making it look like $2500\ \Omega$. This resulted in a net current gain of $59/36.4 \approx 1.6$, with the *same* 3-dB bandwidth of 10 kHz. Note that if the transformer were not used and $r_{b'e} = 1000\ \Omega$ ($g_m = 0.1\ \text{S}$), the midfrequency gain would still be

$$A_{im} = -g_m(r_{b'e} \| R_p \| r_i) = (-0.1)(5000 \| 2000 \| 1000) \approx -59$$

However, the bandwidth would be

$$BW = \frac{1}{2\pi RC} = \frac{1}{2\pi(590)(0.017 \times 10^{-6})} \approx 16\ \text{kHz} > 10\ \text{kHz}$$

Thus we see that high gain can be achieved without the transformer but at the expense of an increased bandwidth. ///

9.5-3 Series-Resonant Circuits

At very high frequencies ($f_0 > 50\ \text{MHz}$) parallel tuning as used in Examples 9.5-1 and 9.5-2 results in very low Q circuits and hence wide bandwidths. The reason can be seen as follows. If C' were not employed, and if r_i, R_p, and R_b were infinite, the circuit Q_i would be approximately

$$Q_i \approx \omega_0 r_{b'e} C_{b'}$$

If the Miller capacitance were negligible, $C_{b'} \approx C_{b'e}$ and

$$Q_i \approx \frac{\omega_0}{\omega_\beta}$$

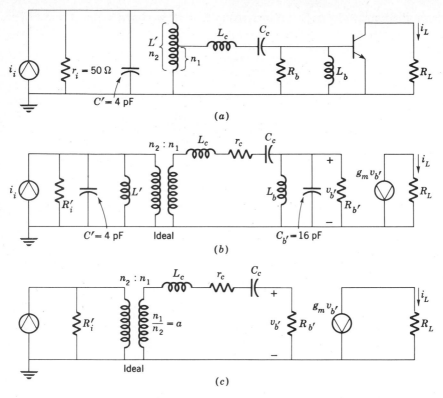

Figure 9.5-6 Amplifier using a resonant circuit: (*a*) circuit; (*b*) equivalent circuit; (*c*) equivalent circuit without low-Q tuned circuits.

which is less than unity if $\omega_0 < \omega_\beta$. We increase the Q of the circuit by adding C'. This increases the circuit capacitance but results in a reduction in the required parallel inductance. Very low values of L' are not always obtainable.

At very high frequencies a series-resonant circuit can be employed to provide a very high circuit Q with reasonable inductance values. This technique is illustrated in the following example.

Example 9.5-3 The amplifier shown in Fig. 9.5-6a is to have a 3-dB bandwidth of 2 MHz and a resonant frequency of 100 MHz [the circuit $Q_c = 10^8/(2 \times 10^6) = 50$]. The transistor employed has the parameters $r_{b'e} = 50\ \Omega$, $g_m = 0.1$ S, $C_{b'e} = 10$ pF, and $C_{b'c} = 1$ pF. The input circuit consists of a 50-Ω resistance ($r_i = 50\ \Omega$) in parallel with a 4-pF capacitor ($C' = 4$ pF). The load resistance R_L is 50 Ω. (a) Describe the circuit operation. (b) Find L', L_b, L_c, C_c, and the turns ratio a.

SOLUTION (a) This amplifier is designed so that the circuit Q is determined by the series-resonant circuit. The parallel RLC circuits at the input and the base are each designed to have a low Q. Quite often in practice the two parallel circuits are not even carefully tuned. Figure 9.5-6b shows an equivalent circuit, where

$$R_i' = r_i \| \text{ effective parallel resistance of } L'(R_p')$$

$$R_{b'} = R_b \| r_{b'e} \| \text{ effective parallel resistance of } L_b(R_p)$$

$$C_{b'} = C_{b'e} + C_M \quad \text{and} \quad \omega_0^2 = \frac{1}{L'C'} = \frac{1}{L_cC_c} = \frac{1}{L_bC_{b'}}$$

The circuit Q is determined from the simplified equivalent circuit shown in Fig. 9.5-6c. This equivalent circuit assumes that the Q's of the input and base circuits are sufficiently small to ensure that

$$\frac{1}{R_i'} \gg \omega C' - \frac{1}{\omega L'} \quad \text{and} \quad \frac{1}{R_{b'}} \gg \omega C_{b'} - \frac{1}{\omega L_b}$$

for ω between ω_L and ω_h. The circuit Q is then essentially the same as the Q of the series-resonant circuit

$$Q_c = \frac{\omega_0 L_c}{R_{b'} + r_c + a^2 R_i'}$$

(b) We begin the design procedure by finding L' and L_b. To resonate the 4-pF input capacitor C' requires that

$$L' = \frac{1}{4\pi^2 f_0^2 C'} \approx 0.65 \ \mu H$$

A Q' of 100 at 100 MHz is easily obtained. Assuming that the transformer has this Q', we find R_p'

$$R_p' = Q'(\omega_0 L') = (100)(2\pi \times 10^8)(0.65 \times 10^{-6}) \approx 41 \ k\Omega$$

Then, since $r_i = 50 \ \Omega$,

$$R_i' = r_i \| R_p' \approx 50 \ \Omega$$

Let us now consider the base circuit. To resonate $C_{b'} = 16$ pF requires

$$L_b \approx 0.17 \ \mu H$$

Assuming that $Q_b = 100$, we have

$$R_p = (100)(2\pi \times 10^8)(0.17 \times 10^{-6}) \approx 11 \ k\Omega$$

and since $r_{b'e} = 50 \ \Omega$,

$$R_{b'} = R_p \| R_b \| r_{b'e} \approx 50 \ \Omega$$

(We assume that $R_b \gg r_{b'e} = 50 \ \Omega$.) Note that the *circuit* Q's are

$$Q_i \approx \omega_0 R_i' C' \approx 0.12 \quad \text{and} \quad Q_{b'} \approx \omega_0 R_{b'} C_{b'} \approx 0.5$$

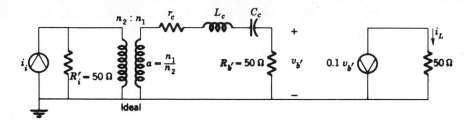

Figure 9.5-7 Simplified circuit for Example 9.5-3.

The input and base circuit Q's are each much smaller than the required circuit Q of 50. Thus, over the passband of 100 ± 1 MHz, we assume that the equivalent circuit can be represented by Fig. 9.5-7. To obtain the circuit Q_c of 50 at 100 MHz requires that

$$Q_c = 50 = \frac{1}{\omega_0 C_c(50 + r_c + 50a^2)} = \frac{\omega_0 L_c}{50 + r_c + 50a^2}$$

Note that $\omega_0 L_c / r_c$, the Q of the inductor L_c, must be greater than 50 for the overall circuit Q_c to equal 50. A Q of 250 for the inductor L_c is achievable at 100 MHz. Let us design assuming an inductance with this Q. Thus

$$\frac{\omega_0 L_c}{r_c} = 250$$

Solving for L_c yields

$$\omega_0 L_c = (1.25)(50^2)(1 + a^2)$$

Let

$$a^2 \approx 0.1$$

Then $\qquad L_c \approx 5.5 \; \mu\text{H} \qquad$ and $\qquad C_c = \dfrac{1}{4\pi^2 f_0^2 L_c} \approx 0.45 \text{ pF}$

The circuit is tuned by using a variable capacitor for C_c. $\qquad\qquad$ ///

9.5-4 The Synchronously Tuned Amplifier

In this section, we discuss the cascading of tuned amplifiers in order to achieve high gain. All amplifier stages are assumed to be identical and to be tuned to the same frequency, ω_0. This is called *synchronous tuning*, and the resulting amplifier has increased gain and a bandwidth which is narrower than the bandwidth of each of the stages.

To illustrate the effect of cascading N synchronously tuned stages, we first determine the gain and bandwidth of the single-tuned FET amplifier shown in

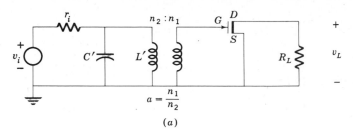

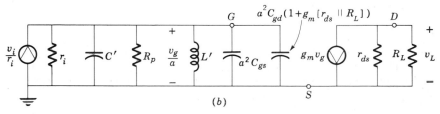

Figure 9.5-8 (*a*) Single-tuned FET amplifier circuit; (*b*) equivalent circuit.

Fig. 9.5-8*a*. The equivalent circuit is given in Fig. 9.5-8*b*, and the voltage gain is (Sec. 9.5-1)

$$A_v = \frac{-ag_m(r_{ds}\|R_L)[(r_i\|R_p)/r_i]}{1 + jQ_i(\omega/\omega_0 - \omega_0/\omega)} \tag{9.5-18a}$$

where

$$C_i = a^2\{C_{gs} + C_{gd}[1 + g_m(r_{ds}\|R_L)]\}$$

$$Q_i = \omega_0(r_i\|R_p)(C' + C_i) \tag{9.5-18b}$$

$$\omega_0^2 = \frac{1}{L(C' + C_i)} \tag{9.5-18c}$$

The center-frequency gain $(\omega = \omega_0)$ is

$$A_{vm} = -ag_m(r_{ds}\|R_L)\frac{R_p}{r_i + R_p} \tag{9.5-19a}$$

The 3-dB bandwidth is

$$BW = \frac{1}{2\pi(r_i\|R_p)(C' + C_i)} \tag{9.5-19b}$$

Let us now cascade two identical synchronously tuned FET amplifiers, as shown in Fig. 9.5-9*a*. The equivalent circuit is shown in Fig. 9.5-9*b*. In a synchronously tuned amplifier each resonant circuit is tuned to the same frequency and has the same bandwidth. Hence

$$\omega_0^2 = \frac{1}{L'(C' + C_i)} \tag{9.5-20a}$$

and

$$R \equiv r_i\|R_p = r_{ds}\|R_g\|R_p = r_{ds}\|R_L \tag{9.5-20b}$$

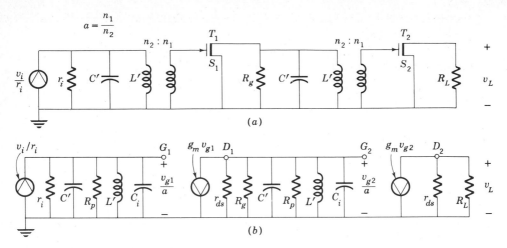

Figure 9.5-9 (a) Synchronously tuned FET amplifier circuit; (b) equivalent circuit.

The voltage gain of the amplifier is

$$A_v = \frac{v_L}{v_i} = \frac{(ag_m R)^2 (R/r_i)}{[1 + jQ_i(\omega/\omega_0 - \omega_0/\omega)]^2} \qquad (9.5\text{-}21a)$$

where

$$Q_i = \omega_0 R(C' + C_i) \qquad (9.5\text{-}21b)$$

The center-frequency gain is

$$A_{vm} = (ag_m R)^2 \frac{R_p}{r_i + R_p} \qquad (9.5\text{-}22)$$

The 3-dB bandwidth of the two-stage amplifier is obtained [Eq. (9.5-21a)] from

$$\left[1 + Q_i^2 \left(\frac{\omega}{\omega_0} - \frac{\omega_0}{\omega}\right)^2\right]^2 = 2 \qquad (9.5\text{-}23a)$$

Hence [see Eq. (9.5-10)]

$$\text{BW} = \frac{\omega_0}{2\pi Q_i} \sqrt{2^{1/2} - 1} = 0.643 \frac{f_0}{Q_i} = 0.643 \frac{1}{2\pi(C' + C_i)R} \qquad (9.5\text{-}23b)$$

The result of cascading two synchronously tuned stages is to increase the voltage gain [compare (9.5-22) and (9.5-19a)] and to decrease the bandwidth [compare (9.5-23b) and (9.5-19b)].

Equations (9.5-20) through (9.5-23) assume that the load on T_1 is resistive. This is only true at resonance. When $\omega \neq \omega_0$ these equations are only approximately correct since the load on T_1 is either slightly capacitive or inductive. However, small error is incurred so long as $Q \geq 20$.

Example **9.5-4** Let $g_m = 5$ mS, $r_i = r_{ds} = 10$ kΩ, $R_p = R_L = 100$ kΩ, $R_g = 910$ kΩ, $a = \frac{1}{2}$, $C' = 0$, $C_{gs} = 10$ pF, $C_{gd} = 0.1$ pF, and $L' = 0.25$ μH. Calculate ω_0, A_{vm}, and the bandwidth assuming a single-tuned stage (Fig. 9.5-8) or two synchronously tuned stages (Fig. 9.5-9).

SOLUTION The resonant frequency ω_0 is

$$\omega_0 = \left(\frac{1}{L'(a^2)\{C_{gs} + C_{gd}[1 + g_m(r_{ds} \| R_L)]\}} \right)^{1/2}$$

$$= \frac{1}{(0.25 \times 10^{-6})(\frac{1}{4})[(10 \times 10^{-12}) + (5 \times 10^{-12})]}^{1/2} \approx 10^9 \text{ rad/s}$$

The center-frequency gain for one stage is, from (9.5-19a),

$$(A_{vm})_1 \approx -(\tfrac{1}{2})[(5 \times 10^{-3}) \times 10^4] = -25$$

With two stages from (9.5-22),

$$(A_{vm})_2 \approx \{[(\tfrac{1}{2})(5 \times 10^{-3}) \times 10^4]\}^2 = 625$$

The 3-dB bandwidth with one stage is, from (9.5-19b) and (9.5-18b),

$$(\text{BW})_1 \approx \frac{1}{(2\pi \times 10^4)(\frac{1}{4})(15 \times 10^{-12})} \approx 4.1 \text{ MHz}$$

where

$$C_i \approx (\tfrac{1}{2})^2(10 \times 10^{-12}) + (0.1 \times 10^{-12})\{1 + [(5 \times 10^{-3}) \times 10^4]\})$$
$$\approx (\tfrac{1}{4})(15 \times 10^{-12})$$

and for two stages, from (9.5-23b),

$$(\text{BW})_2 \approx 0.643 \frac{1}{(2\pi \times 10^4)(\frac{1}{4})(15 \times 10^{-12})} \approx 2.6 \text{ MHz} \qquad ///$$

9.6 THE GAIN-BANDWIDTH PRODUCT

9.6-1 Gain-Bandwidth Product for an *RC* Amplifier

In the preliminary design of a multistage wideband amplifier it is very helpful to have some guideposts, or rules of thumb, which can be used to set up a tentative circuit. The GBW product is a figure of merit which is often used for this purpose. It is defined in terms of midband gain and upper 3-dB frequency f_h as

$$\text{GBW} = |A_{im} f_h| \qquad (9.6\text{-}1)$$

For an ideal single-stage CE amplifier $(R_L \to 0)$, the midband gain is approximately h_{fe}, and the upper 3-dB frequency is f_β. Thus

ideal

$$\text{GBW} = h_{fe} f_\beta = f_T \approx \frac{h_{fe}}{2\pi r_{b'e} C_{b'e}} = \frac{g_m}{2\pi C_{b'e}} \tag{9.6-2}$$

Manufacturers generally specify f_T, and it is used as a rough estimate of the GBW product for a given transistor. The estimate is an upper bound, the actual value being reduced because of the Miller capacitance, which was neglected in arriving at (9.6-2). To refine the estimate, we refer to (9.3-17). Then

$$\text{GBW} = g_m R_{b'e} \frac{1}{2\pi R_{b'e}(C_{b'e} + C_M)} = \frac{g_m}{2\pi(C_{b'e} + C_M)} \tag{9.6-3}$$

Comparing (9.6-3) and (9.6-2), we see that the Miller capacitance reduces the GBW product. Note that the GBW product is a function of g_m, $C_{b'e}$, $C_{b'c}$, and R_L (since C_M depends on R_L). Varying $R_{b'e}$ results in a trade-off between the gain and the bandwidth of the amplifier. Let us consider a numerical example.

Example 9.6-1 Find the gain-bandwidth product of the transistor amplifier shown in Fig. 9.6-1. All bias components have been removed for simplicity. The component values are $r_i = 1 \text{ k}\Omega$, $R_c = r_{b'e} = 100\ \Omega$, $C_{b'e} = 100$ pF, $C_{b'c} = 1$ pF, and $h_{fe} = 100$.

SOLUTION The GBW product (9.6-3) is

$$\text{GBW} = \frac{g_m}{2\pi(C_{b'e} + C_M)} = \frac{1}{2\pi(10^{-10} + 10^{-10})} = \frac{10^{10}}{4\pi} = 0.8 \text{ GHz}$$

Note that

$$f_T = \frac{g_m}{2\pi C_{b'e}} = \frac{10^{10}}{2\pi} = 1.6 \text{ GHz} \qquad ///$$

GBW of a FET amplifier The GBW product of a FET amplifier is found from (9.2-3).

use this $\longrightarrow$

$$\text{GBW}_{\text{FET}} = g_m(r_{ds} \| R_d) \frac{1}{2\pi r_i(C_{gs} + C_M)} \tag{9.6-4}$$

$g_m(r_{ds}\|R_d\|R_L)$

or G-BW = $A V_m(f_h)$

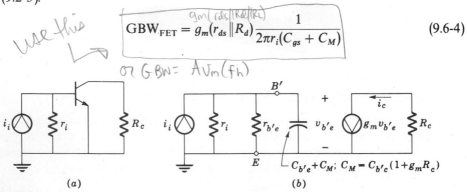

$C_{b'e} + C_M$; $C_M = C_{b'c}(1 + g_m R_c)$

(a) (b)

Figure 9.6-1 (a) Amplifier circuit for Example 9.6-1; (b) equivalent circuit.

Equation (9.6-4) is usually normalized by assuming that

$$r_i = r_{ds} \| R_d$$

(This condition could result when the FET is preceded by another FET.) The GBW then becomes

only if

if $r_i = r_{ds} \| R_d$
$$\text{GBW}_{\text{FET}} = \frac{g_m}{2\pi(C_{gs} + C_M)}$$
JFET

(9.6-5)

Example 9.6-2 Find the GBW product of a JFET amplifier having the parameters $g_m = 3$ mS, $C_{gs} = 6$ pF, $C_{gd} = 2$ pF, $r_{ds} = 70$ kΩ, and $R_d = 10$ kΩ.

SOLUTION We first determine the Miller capacitance C_M

$$C_M = C_{gd}[1 + g_m(r_{ds}\|R_d\|R_L)]$$

JFET $\quad C_M = C_{gd}[1 + g_m(r_{ds}\|R_d)] = (2 \times 10^{-12})[1 + (3)(\frac{70}{8})] \approx 54 \text{ pF}$.

Note that the Miller capacitor is not at all insignificant when using a JFET. The GBW product is, from (9.6-5),

$$\text{GBW} = \frac{g_m}{2\pi(C_{gs} + C_M)} = \frac{3 \times 10^{-3}}{2\pi(60 \times 10^{-12})} \approx 8 \text{ MHz}$$

When we compare this example with Example 9.6-1, the transistor is seen to have a much higher GBW product. The difference is due to the much larger g_m in the transistor. ///

Example 9.6-3 A MOSFET is used instead of the JFET in the preceding example. The FET parameters are $g_m = 2.5$ mS, $C_{gs} = 6$ pF, $C_{gd} = 0.6$ pF, and $r_{ds} = 60$ kΩ. $R_d = 10$ kΩ, as in Example 9.6-2. Find the GBW product.

SOLUTION The Miller capacitance is now

$$C_M = C_{gd}[1 + g_m(r_{ds}\|R_d)] = (0.6 \times 10^{-12})[1 + (2.5)(\tfrac{60}{7})] \approx 14 \text{ pF}$$

The GBW product (9.6-5) is then

MOSFET $\quad \text{GBW} = \frac{g_m}{2\pi(C_{gs} + C_M)} = \frac{2.5 \times 10^{-3}}{2\pi(20 \times 10^{-12})} \approx 20 \text{ MHz}$

Note that, while the GBW of the MOSFET is significantly higher than that of the JFET, it is still much less than the GBW of the BJT. ///

9.6-2 Gain-Bandwidth Product for a Tuned Amplifier

The GBW product of the single-tuned amplifier discussed in Sec. 9.5-1 can be found by combining (9.5-8) and (9.5-10)

$$\text{GBW} = |A_{im}| \text{BW} = \frac{g_m}{2\pi C}$$

(9.6-6)

where C is given by (9.5-2).

Comparing (9.6-6) with (9.6-3), we see that the GBW product is the same as for the RC-coupled stage and depends only on the g_m of the transistor and the total input-circuit capacitance. Thus, the addition of the high-Q tuning coil has effectively translated the frequency-response curve of the RC-coupled amplifier (for the same gain) along the frequency axis without reducing the width of the curve as measured in hertz.

9.7 THE TRANSISTOR SWITCH[3, 4]

Until now, we have considered only the linear operation of the transistor. In many electronic systems transistors are used as *controlled switches* (see Chap. 12). A digital computer, for example, will use several thousand transistor switches. The speed with which the switches operate is of paramount importance. In this section we consider the time response of a simple transistor switch.

The transistor can be made to operate as a switch by designing the associated circuit so that the transistor is either in the cutoff region or in the saturation region. When the transistor is *cut off*, no collector current flows and the switch is open. When the transistor is in saturation, maximum collector current flows and the switch is closed. The switch is controlled by the current applied to the base. Such a transistor switch is shown in Fig. 9.7-1. The response of the transistor switch to a pulse-waveform input voltage v_i is shown in Fig. 9.7-2.

Turn-on-time There is a delay between the leading edge of the input voltage pulse and the time that the collector current takes to reach 90 percent of its maximum value. This time is called the *turn-on* time t_{on}. It is divided into two time intervals; the first is called the *delay time* t_d, and the second is called the *rise time* t_r.

Delay time The delay time is the time required for the collector current to increase to $0.1 I_{C, sat}$. Another way of describing t_d is to note that it is approximately equal to the time required for the base-to-emitter (diode) voltage to increase from $-V_1$ to approximately $+0.7$ V.

Rise time The rise time is the time required for the collector current to increase from $0.1 I_{C, sat}$ to $0.9 I_{C, sat}$. During this interval the transistor is operating in the normal active region.

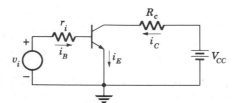

Figure 9.7-1 The transistor switch.

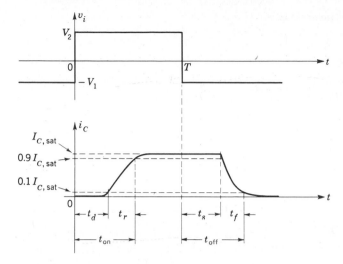

Figure 9.7-2 Response of a transistor switch to a voltage pulse.

Turn-off time The time required for the collector current to decrease from $I_{C, \text{sat}}$ to $0.1I_{C, \text{sat}}$ when v_i goes negative (Fig. 9.7-2) is called the *turn-off time* t_{off}. The turn-off time is the sum of the *storage time* t_s and the *fall time* t_f, as shown in Fig. 9.7-2.

Storage time The storage time is the elapsed time from the trailing edge of the input pulse $(t = T)$ to the point where i_C just starts to decrease toward zero.

Fall time The fall time t_f is the time it takes for the collector current to decrease from $I_{C, \text{sat}}$ to $0.1I_{C, \text{sat}}$.

Calculations of the delay time, rise time, fall time, and storage time involve many approximations and generally result in values that differ significantly from those published by manufacturers. Furthermore, most transistor switches or gates in use today are ICs containing many transistors which defy simple analysis. The reader is therefore referred to the manufacturers' catalogs for these response times.

Propagation delay time The time it takes for a transistor switch to respond to an input signal is called the *propagation delay time* t_{pd}. The propagation delay time is usually defined as the elapsed time between the midpoints of the input-signal transition voltage and the output-signal transition voltage, as shown in Fig. 9.7-3.

The propagation delay when the switch output falls from the high state to the low state is symbolized as $t_{pd, HL}$ or t_{pd-}. Similarly the propagation delay when the switch output rises from the low state to the high state is called $t_{pd, LH}$ or t_{pd+}. Usually t_{pd+} is greater than t_{pd-} because of the inevitable capacitance at the switch output. When the output falls, the capacitance is discharged through

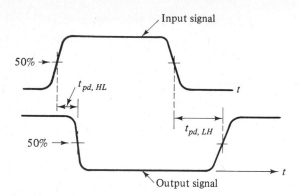

Figure 9.7-3 Propagation delay time.

the transistor (see Fig. 9.7-1), which acts like a low impedance; while when the output rises, the capacitor is charged through the load resistor, a much higher impedance, and therefore the charging time is longer. If a complementary-symmetry output stage is used, the propagation delay times are, of course, comparable.

REFERENCES

1. Motorola, Inc., Engineering Staff, "Integrated Circuits," McGraw-Hill, New York, 1965.
2. A. S. Grove, "Physics and Technology of Semiconductor Devices," Wiley, New York, 1967.
3. D. L. Schilling, and C. Belove, "Electronic Circuits: Discrete and Integrated," sec. 13.6, McGraw-Hill, New York, 1968.
4. P. E. Gray, and C. L. Searle, "Electronic Principles," Wiley, New York, 1967.

PROBLEMS

9.1-1 (a) Draw the small-signal circuit for Fig. P9.1-1.
(b) Find the transfer function $A_i = i_L(s)/i_i(s)$. *Hint:* Reflect the emitter circuit into the base.
(c) Sketch the asymptotic magnitude plot for A_i.

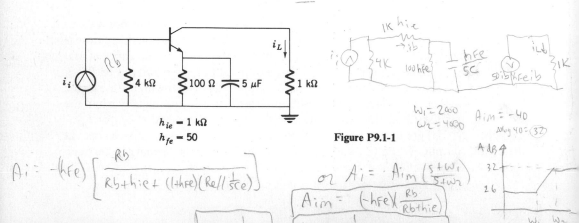

$h_{ie} = 1 \text{ k}\Omega$
$h_{fe} = 50$

Figure P9.1-1

9.1-2 (a) In Fig. P9.1-2 find R_1 and R_2 for maximum symmetrical swing (recall that for good dc stability $h_{ie} < R_b < h_{fe}R_e$). Assume $h_{fe} = 50$.

(b) Determine C_e so that the lower 3-dB frequency is at $\omega = 10$ rad/s.

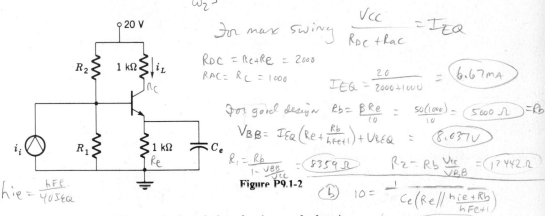

handwritten:

ω_2

for max swing $\dfrac{V_{cc}}{R_{DC}+R_{ac}} = I_{EQ}$

$R_{DC} = R_C + R_e = 2000$

$R_{AC} = R_C = 1000$

$I_{EQ} = \dfrac{20}{2000+1000} = \boxed{6.67\,mA}$

for good design $R_b = \dfrac{\beta R_e}{10} = \dfrac{50(1000)}{10} = \boxed{500\,\Omega} = R_b$

$V_{BB} = I_{EQ}\left(R_e + \dfrac{R_b}{h_{fe}+1}\right) + V_{BEQ} = \boxed{8.037\,V}$

$h_{ie} = \dfrac{h_{FE}}{40\,I_{EQ}}$

$R_1 = \dfrac{R_b}{1 - \dfrac{V_{BB}}{V_{cc}}} = \boxed{835\,\Omega}$ $R_2 = R_b\dfrac{V_{cc}}{V_{BB}} = \boxed{1244\,\Omega}$

Figure P9.1-2

(b) $10 = \dfrac{1}{C_e\left(R_e // \dfrac{h_{ie}+R_b}{h_{FE}+1}\right)}$

$\boxed{C_e = 1.07\times10^{-3}\,f}$

9.1-3 Plot the asymptotic magnitude and phase for the transfer function

$$A = 10^4\frac{(s + 10)(s + 300)(s + 4000)}{(s + 2)(s + 12)(s + 2000)}$$

Use semilog paper and plot asymptotes of the gain in decibels versus ω in radians per second and phase θ in degrees versus ω in radians per second. On the same sheet plot the actual characteristics.

9.1-4 In Fig. 9.1-8 let $r_i = 50$ kΩ, $R_b - 500$ kΩ, $R_e = 100$ Ω, and $R_c - 1$ kΩ, $h_{fe} - 100$, $h_{ie} = 1$ kΩ.

(a) Find C_{e1} so that the lower 3-dB frequency is at $f = 10$ Hz. $, 2615\mu f$

(b) Plot i_c/i_i (asymptotic plot).   p. 416 Eq. (9.1-11)

(c) Plot the phase of i_c/i_i.

9.1-5 In Fig. P9.1-5 find the transfer function v_L/i_i. Plot the asymptotic magnitude and the phase. $h_{fe} = 50$, $h_{ie} = 500$ Ω.

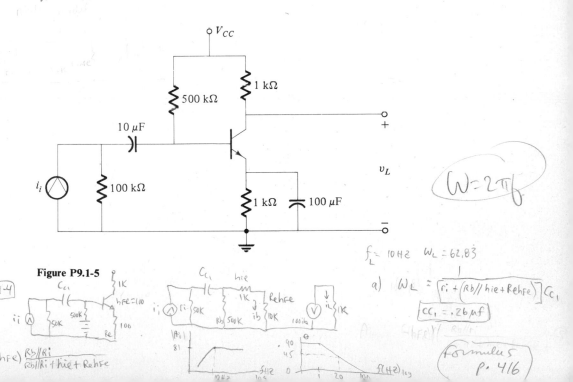

handwritten near figure:

$\boxed{\omega = 2\pi f}$

$f_L = 10$ Hz $\omega_L = 62.83$

a) $\omega_L = \dfrac{1}{[r_i + (R_b // h_{ie} + R_e h_{FE})]C_{e1}}$

$\boxed{C_{e1} = .26\,\mu f}$

Figure P9.1-5

9.1-4

$= (-h_{FE})\dfrac{R_b//R_i}{R_b//R_i + h_{ie} + R_e h_{FE}}$

9.1-6 In Fig. P9.1-6 find C_{c1} so that the lower 3-dB frequency is at $\omega = 5$ rad/s.

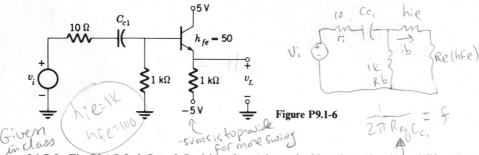

Figure P9.1-6

Handwritten annotations:
$h_{ie} = 1k$
$h_{fe} = 100$
Given in class
-5 volts is to provide for more swing

$\dfrac{1}{2\pi R_{eq} C_{c_1}} = f$

$R_{eq} = r_i + R_b // (h_{ie} + R_e h_{fe})$

9.1-7 In Fig. P9.1-7 find C_e and C_{c2} (a) so that A_i has a double pole at 10 rad/s and (b) so that the lower 3-dB frequency is at 10 rad/s.

(c) Plot the asymptotic magnitude of A_i for both (a) and (b).

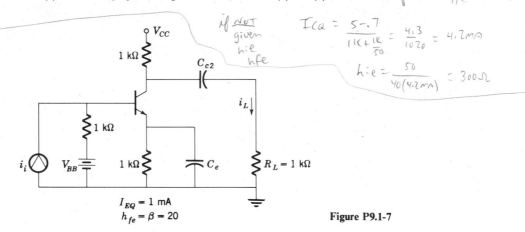

Handwritten annotations:
If NOT given h_{ie} h_{fe}

$I_{CQ} = \dfrac{5 - .7}{1K + \frac{1K}{50}} = \dfrac{4.3}{1020} = 4.2$ mA

$h_{ie} = \dfrac{50}{40(4.2\text{mA})} = 300\,\Omega$

$I_{EQ} = 1$ mA
$h_{fe} = \beta = 20$

Figure P9.1-7

9.1-8 Plot $|i_L/i_i|$ and find the lower 3-dB frequency for the circuit in Fig. P9.1-8.

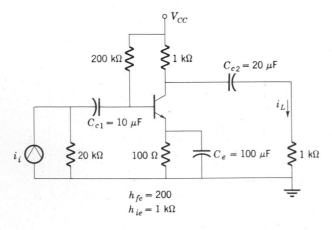

$h_{fe} = 200$
$h_{ie} = 1$ kΩ

Figure P9.1-8

9.1-9 Repeat Prob. 9.1-8 for the circuit in Fig. P9.1-9.

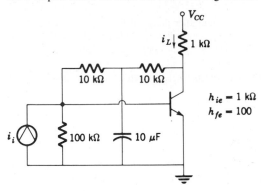

$h_{ie} = 1 \text{ k}\Omega$
$h_{fe} = 100$

Figure P9.1-9

9.1-10 Find $|i_L/i_i|$ and comment on the effect of the split emitter resistance in Fig. P9.1-10.

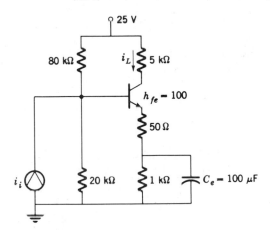

$h_{fe} = 100$

$C_e = 100 \ \mu F$

Figure P9.1-10

9.1-11 For the circuit of Fig. P9.1-8 use the technique in Sec. 9.1-5 to find new values of C_{c1}, C_{c2}, and C_e so that $f_L \leq 10$ Hz.

9.2-1 Draw the small-signal equivalent circuit for Fig. 9.2-1a and verify (9.2-1) and (9.2-2).

9.2-2 In Fig. 9.2-1a assume that $C_s \to \infty$ and $C_{c1} \to \infty$.
 (a) Show that the 3-dB radian frequency is given by

$$\omega_L = \frac{1}{C_{c2}(R_L + r_{ds} \| R_d)}$$

 (b) Use the values in Example 9.2-1 and calculate C_{c2} so that $f_L \leq 10$ Hz. What can you conclude about coupling capacitor size compared with bypass capacitor size?

9.2-3 In Fig. 9.2-1a assume that C_s is open and $C_{c1} \to \infty$.
 (a) Show that the 3-dB radian frequency is given by

$$\omega_L = \frac{1}{C_{c2}[R_L + (R_d \| r_{ds} + (\mu + 1)R_s)]}$$

 (b) Use the values in Example 9.2-1 and calculate C_{c2} so that $f_L \leq 10$ Hz. Compare with C_s in Example 9.2-1.

9.2-4 In Fig. 9.2-1a assume that $C_s \to 0$ and calculate v_L/v_i. Sketch the asymptotic magnitude assuming that the break frequency due to C_{c1} occurs first.

3/23/81 **9.2-5** Choose coupling and bypass capacitors for the amplifier designed in Prob. 6.8-3 so that the lower 3-dB frequency will be less than 5 Hz. Use the technique in Sec. 9.1-5; that is, choose C_s first and then calculate C_{c1} and C_{c2}, with C_s assumed open, to yield much lower break frequencies.

9.2-6 (a) Find $A_v = v_L/v_i$ in Fig. P9.2-6.

(b) Find A_v if the source bypass capacitor is connected across both 250-Ω resistors. Compare with part (a).

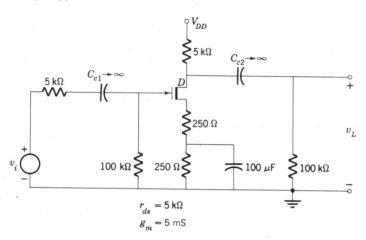

Figure P9.2-6

9.3-1 Measurements indicate that the amplifier of Fig. P9.3-1 has a midband gain i_L/i_i of 32 dB, an upper 3-dB frequency of 800 kHz, and a quiescent emitter current of 2 mA. Assuming $r_{bb'} = C_{b'c} = 0$, find h_{fe}, $r_{b'e}$, and $C_{b'e}$.

Since $C_{b'c} = 0$ $C_M = 0$!

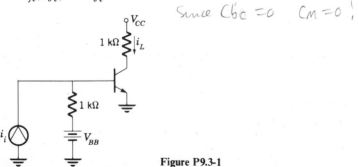

Figure P9.3-1

given in class

9.3-2 In Fig. 9.3-6 let $r_i = R_2 = 10$ kΩ, $R_1 = R_c = R_L = 1$ kΩ, $R_e = 100$ Ω, and $C_{c1} = C_e = C_{c2} = 20$ μF. The transistor is characterized by $\omega_T = 10^9$ rad/s, $h_{fe} = 100$, $C_{b'c} = 5$ pF, $r_{bb'} = 0$, and $I_{EQ} = 10$ mA. Find and plot $|i_L/i_i|$. What is the upper 3-dB frequency?

9.3-3 Repeat Prob. 9.3-2 if $C_{b'c} = 2$ pF, $h_{fe} = 20$, and $I_{EQ} = 1$ mA.

9.3-4 Design a single-stage amplifier using a transistor which has $f_T = 700$ MHz, $h_{fe} = 10$, and $C_{b'c} = 2.5$ pF. The midband gain is to be 14 dB, and the upper 3-dB frequency is to be as high as possible. The signal source has an internal resistance of 1 kΩ, and the load resistance is 50 Ω. Maximum required load current swing is ± 1 mA. Specify all resistors and find the upper 3-dB frequency.

$A_i = \dfrac{r_i // Rb}{r_i // R_L + h_{ie}} (-h_{fe}) \left(\dfrac{R_c}{R_c + R_L} \right)$, but $R_c >> R_L$

$\delta = \dfrac{r_i // Rb}{r_i // Rb + h_{ie}} (10)$ $, \delta = \dfrac{r_i Rb}{r_i + Rb}$ $\dfrac{r_i Rb}{r_i + Rb + h_{fe}}$ $\cdot \delta = \dfrac{r_i Rb}{r_i Rb + h_{fe}(r_i + Rb)}$

9.3-5 Verify (9.3-21a) and (9.3-21b).

9.3-6 Verify (9.3-22a).

9.3-7 (a) Verify (9.3-23).

    (b) Use the values of Example 9.3-3 in (9.3-23) to find the upper cutoff frequency and compare.

    (c) Verify the result when $C_{b'c} = 2$ pF.

9.3-8 For the transistor shown in Fig. P9.3-8 $\omega_T = 10^9$ rad/s, $C_{b'c} = 6$ pF, $r_{bb'} = 0$, $I_{EQ} = 1$ mA, and $h_{fe} = 20$. Find and plot the magnitude of v_L/v_i. What is the upper 3-dB frequency?

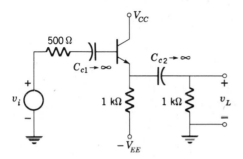

Figure P9.3-8

9.3-9 For the transistor shown in Fig. P9.3-9 $r_{bb'} = 20$ Ω, $r_{b'e} = 1$ kΩ, $C_{b'e} = 1000$ pF, $C_{b'c} = 10$ pF, and $g_m = 0.05$ S. Find and sketch the asymptotic voltage-gain characteristic for all frequencies.

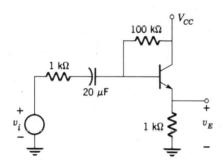

Figure P9.3-9

9.3-10 The common-base amplifier in Fig. 6.3-1b is to be analyzed.

    (a) Draw the small-signal circuit using the hybrid-pi model of Fig. 9.3-5.

    (b) Let $r_{bb'} = 0$ and show that the circuit can be redrawn as in Fig. P9.3-10.

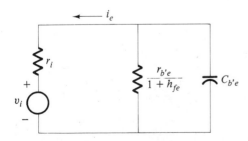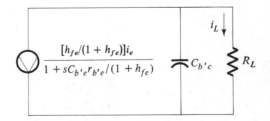

Figure P9.3-10

9.3-11 Use the results in Prob. 9.3-10 and the transistor values in Prob. 9.3-9. If $r_i = R_L = 1$ kΩ, plot the asymptotic magnitude for i_L/v_i and indicate the upper 3-dB cutoff frequency.

9.3-12 This problem determines the high-frequency response of the differential amplifier in Fig. 7.3-1. Since the CMRR is high for this configuration, we can neglect the common-mode circuit and just use the differential-mode circuit. The analysis is then exactly the same as discussed in Sec. 9.3-2. Use the hybrid-pi model of Fig. 9.3-5 with $r_{bb'} = 0$.

(a) Calculate $v_{o1}/(v_1 - v_2)$ and sketch the asymptotic magnitude indicating the 3-dB cutoff frequency.

(b) Repeat part (a) for $v_{o2}/(v_1 - v_2)$.

9.4-1 The 2N4223 JFET (see Appendix C for specifications) is used in the circuit of Fig. 9.4-1 with $R_d = 10$ kΩ.

(a) Draw the high-frequency equivalent circuit, including the Miller effect.

(b) Find the upper 3-dB frequency if $r_i = 50$ Ω.

(c) Repeat part (b) if $r_i = 10$ kΩ.

9.4-2 The 2N4223 JFET is to be used in the circuit of Fig. 9.4-1. The signal source has a 600-Ω impedance, and the load consists of a 50-kΩ resistance in parallel with 20 pF. Design the stage so that $A_{vm} \geq 20$ dB and $f_h \geq 50$ kHz.

9.4-3 (a) Show that for Fig. 9.4-9 if $g_m R_s \gg 1$

$$\frac{v_o}{v_i} = \frac{1 + sC_{gs}/g_m}{1 + s[C_{gs}(R_i + R_s)/g_m R_s + C_{gd} R_i] + s^2 C_{gs} R_i C_{gd}/g_m}$$

(b) If $R_s \gg R_i$, simplify the result in part (a) and verify (9.4-9).

(c) If $C_{gd} \to 0$, verify (9.4-7).

9.4-4 Verify the result in Example 9.4-2.

9.4-5 For the source follower in Fig. P9.4-5 sketch the asymptotic voltage gain for all frequencies.

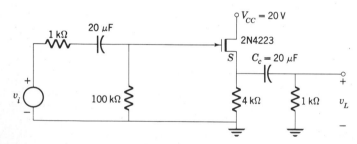

Figure P9.4-5

9.5-1 Show that the center frequency of the tuned circuit in Fig. P9.5-1 is $1/\sqrt{LC}$ and that the 3-dB bandwidth is $1/2\pi RC$, thereby verifying (9.5-10).

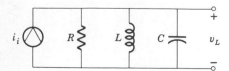

Figure P9.5-1

9.5-2 Find the resonant frequency ω_0 and the bandwidth of the tuned circuit shown in Fig. P9.5-2. Compare the results obtained with the results of Prob. 9.5-1, where

$$R = r_c \left(\frac{\omega_0 L}{r_c} \right)^2$$

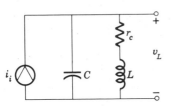

Figure P9.5-2

9.5-3 A tuned-circuit amplifier is shown in Fig. P9.5-3,
(a) Find L so that the circuit resonates at 30 MHz.
(b) What is the bandwidth of the amplifier?
(c) Calculate the current gain.

$r_{b'e} = 1\ k\Omega$
$r_{bb'} = 0$
$h_{fe} = 100$
$f_T = 500\ MHz$
$C_{b'c} = 2\ pF$

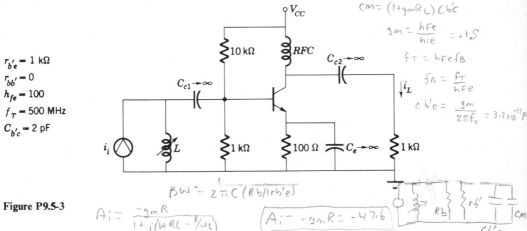

Figure P9.5-3

9.5-4 Design a single-tuned amplifier to have a current gain of 10 dB at 40 MHz. The bandwidth is to be 1 MHz. The source impedance is 1 kΩ, and the load impedance is 1 kΩ. A 10-V supply is available. Assume that the Q of the inductor used is 50.

(a) Use only one transistor. Select it from a transistor manual. Note that f_T must exceed the GBW product of the amplifier. Why?

(b) Determine the pertinent transistor parameters.

(c) Calculate L.

(d) Check the design by finding the current gain.

9.5-5 Design a bandpass filter as in Fig. P8.4-12b so that the center frequency is 160 Hz and the 3-dB bandwidth is 16 Hz. Capacitors chosen must be less than 0.1 μF.

9.5-6 The transformer in Fig. P9.5-6a is characterized by

$$E_1 = sL_1I_1 - sMI_2 \qquad \text{and} \qquad E_2 = sMI_1 - sL_2I_2$$

where $M = k\sqrt{L_1 L_2}$ is the mutual inductance between the primary L_1 and the secondary L_2 and k is the coefficient of coupling.

(a) By manipulating the above equations show that the equivalent circuit of Fig. P9.5-6b results.

(b) If the coupling is tight, so that $k = 1$, find the equivalent circuit.

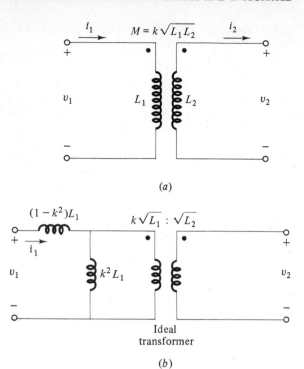

(a)

(b)

Ideal
transformer

Figure P9.5-6

9.5-7 The autotransformer shown in Fig. P9.5-7 has equations similar to those in Prob. 9.5-6.
(a) Show that the equations of the autotransformer are

$$E_1 = s(L_x + L_y + 2M)I_1 - s(M + L_y)I_2 \qquad E_2 = s(M + L_y)I_1 - sL_y I_2$$

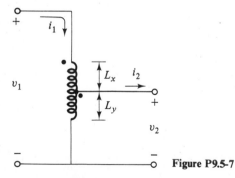

Figure P9.5-7

where M is defined as the mutual inductance between the top half of the transformer L_x and the
secondary L_y and $M = k\sqrt{L_x L_y}$.

(b) Obtain an equivalent circuit for the autotransformer similar to Fig. P9.5-6b by defining an
artificial mutual inductance M_a so that $M_a = M + L_y = k_a\sqrt{(L_x + L_y + 2M)L_2}$

(c) If the coupling is tight, so that $k = 1$, show that $k_a = 1$ and prove that the autotransformer
can be represented by the model of an inductor L' in parallel with an ideal transformer, as shown in

Fig. 9.5-6b, where

$$L' = (\sqrt{L_x} + \sqrt{L_y})^2 \quad \text{and} \quad \frac{n_2}{n_1} = \sqrt{\frac{L'}{L_y}} = \frac{\sqrt{L_x} + \sqrt{L_y}}{\sqrt{L_y}}$$

9.5-8 Verify (9.5-12).

9.5-9 In Fig. 9.5-4 $r_i = 47$ kΩ, $R_b = 10$ kΩ, $R_c = R_L = 2$ kΩ, and the autotransformer is defined by $L' = 10$ μH with $n_2/n_1 = 10$. The transistor is biased at $I_E = 5$ mA and has $r_{bb'} = 0$, $h_{fe} = 100$, $C_{b'e} = 100$ pF, and $C_{b'c} = 2$ pF.
  (a) Calculate the transfer function i_L/i_i.
  (b) Choose C' so that the circuit resonates at 10 MHz.
  (c) Determine the 3-dB bandwidth using the result in part (b).

9.5-10 In Fig. 9.5-6 let $r_i = R_L = 50$ Ω, $C' = 4$ pF, $L_c = 2$ μH, $n_1/n_2 = a = 0.1$, and $Q_{coil} = \omega_0 L/r_c = 200$. The transistor has parameters $r_{b'e} = 50$ Ω, $g_m = 0.1$ S, $C_{b'e} = 6$ pF, $C_{b'c} = 2$ pF, and $r_{bb'} = 0$.
  (a) Determine L', L_b, L_c, and C_c so that the circuit resonates at 80 MHz.
  (b) What is the bandwidth of the circuit?

9.5-11 Verify (9.5-18) and (9.5-19).

9.5-12 Verify (9.5-21).

9.5-13 Verify (9.5-23b).

9.5-14 The synchronously tuned amplifier in Fig. P9.5-14 is designed to resonate at 100 kHz and have a bandwidth of 2 kHz.
  (a) Find C_1, L_1, C_2, L_2, and R_1.
  (b) Calculate the gain v_L/i_i.

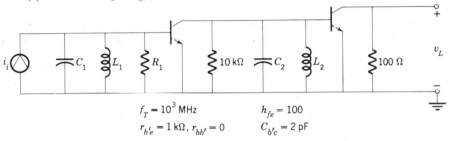

$$f_T = 10^3 \text{ MHz} \qquad h_{fe} = 100$$
$$r_{b'e} = 1 \text{ k}\Omega, \ r_{bb'} = 0 \qquad C_{b'c} = 2 \text{ pF}$$

Figure P9.5-14

9.6-1 For the transistors shown in Fig. P9.6-1, $r_{b'e} = 1$ kΩ, $C_{b'e} = 1000$ pF, $C_{b'c} = 10$ pF, and $g_m = 0.05$ S. Determine the GBW product of each configuration.

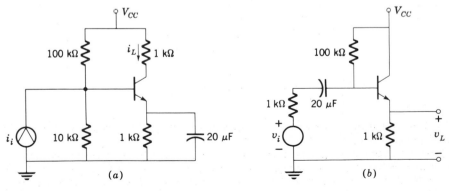

Figure P9.6-1

3/23/81

9.6-2 A one-stage FET amplifier uses a low-frequency JFET having the parameters $g_m = 3$ mS, $r_{ds} = 50$ kΩ, $R_d = 10$ kΩ, $R_g = 500$ kΩ, $C_{gs} = 30$ pF, and $C_{gd} = 5$ pF. Find the GBW product. Assume $R_L = r_i = 10$ kΩ.

9.6-3 Repeat Prob. 9.6-2 if a MOSFET having the parameters $g_m = 2.5$ mS, $C_{gs} = 30$ pF, $r_{ds} = 60$ kΩ, and $C_{gd} = 1.6$ pF is used. All other resistors are assumed to be the same.

9.6-4 Consider a cascade of n identical ideal common-emitter stages where $r_{b'b} = C_{b'c} = 0$. Each stage is characterized by a transfer function identical to (9.3-16) with $C_M = 0$ and 3-dB cutoff frequency f_h given by (9.3-17b). Find the overall 3-dB bandwidth for the cascade f_n as a function of n and tabulate for $n = 1, 2, 3, 4,$ and 5. (This is an extremely optimistic estimate since we take $C_M = 0$.)

9.6-5 An amplifier is to be designed for a midband current gain of 80 dB and a 3-dB frequency of 350 kHz. Transistors having $h_{fe} = 120$ and $f_T = 90$ MHz are to be used. Neglecting the Miller effect and assuming identical stages, find the number of stages required and the midband gain per stage.

9.6-6 A better bandwidth estimate can be obtained for the n-identical-stage cascade in Prob. 9.6-4 if the Miller effect is included in (9.3-16) and (9.3-17b). Repeat Prob. 9.6-5 including the effects of C_M. (This analysis still neglects Miller capacitor loading of the preceding stage. This case is covered in more detail in Chap. 13 of Ref. 3.)

FEEDBACK, FREQUENCY COMPENSATION
OF OP-AMPS, AND OSCILLATORS

From the discussion in Chaps. 7 and 8 it is evident that the idea of feedback is of central importance in the theory of linear and nonlinear op-amp configurations. In those chapters the output-input characteristics of the circuits were of most importance, and the effects of the feedback were largely ignored. In this chapter the feedback is the principal topic, and we shall study its effects on the stability of op-amp circuits and the design of oscillators.

10.1 BASIC CONCEPTS OF FEEDBACK

There are many different types of feedback circuits. Most of those in use today fall into two categories. Consider first the standard inverting configuration of Fig. 8.1-2, repeated in Fig. 10.1-1a. In this circuit negative feedback is obtained by connecting resistor R_f from the output to the input. The actual comparison of output and input takes place by combining current i_1 and i_2 at the amplifier input node, hence the name *current differencing*. In order to facilitate the analysis of this type of circuit the feedback resistor can be replaced by the equivalent circuit shown in Fig. 10.1-1b. If the feedback network is more complicated than the single resistor of Fig. 10.1-1a, the equivalent circuit of Fig. 10.1-1c is used since any linear circuit can be put in this form. Finally, we note that the effect of the $K_2 v_d$ source is negligible in most applications (in Fig. 10.1-1a, $K_2 v_d = v_d \approx 0$ since the op-amp input is a virtual short circuit), and so it is omitted from the final circuit of Fig. 10.1-2. This circuit is an example of a *current-differencing negative-feedback amplifier*, all of which consist conceptually of three sections:

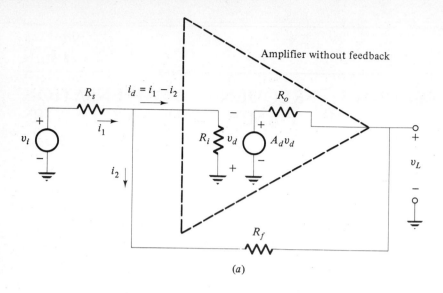

(a)

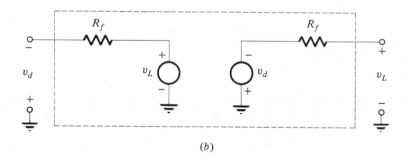

(b)

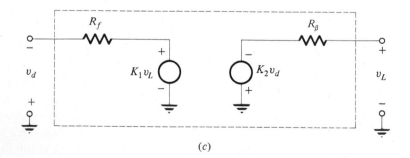

(c)

Figure 10.1-1 Feedback circuits: (a) inverting configuration; (b) feedback network equivalent circuit; (c) general equivalent circuit.

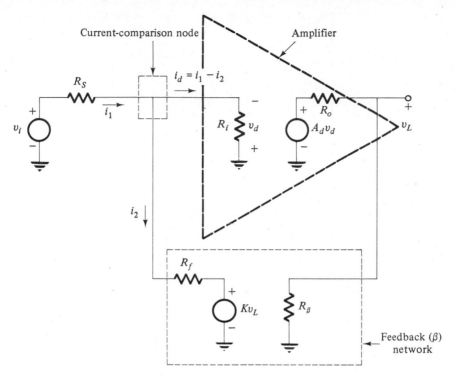

Figure 10.1-2 Current-differencing negative-feedback circuit.

1. An amplifier to which feedback is applied (usually an IC op-amp)
2. A feedback network which may contain anything from a single resistor to a circuit containing nonlinear elements
3. A differencing (or summing) circuit in which the output is compared with the input (this comparison is the essence of feedback)

The circuit of Fig. 10.1-3 differs from that of Fig. 10.1-2 in that the output of the feedback network is in series with the amplifier input resistance R_i in Fig. 10.1-3 while it is in parallel with R_i in Fig. 10.1-2. The feedback illustrated in Fig. 10.1-3 is representative of *negative feedback with voltage differencing;* this circuit is used in the noninverting amplifier shown in Fig. 8.2-1. Here the comparison of output and input takes place via the addition of voltages Kv_L and v_i in a series loop.

10.1-1 Gain of a Feedback Amplifier with Current Differencing

In order to calculate the voltage gain of the amplifier of Fig. 10.1-2 we begin by expressing the currents at the amplifier input node in the form

$$i_1 = \frac{v_i + v_d}{R_s} \qquad i_2 = -\frac{v_d + Kv_L}{R_f} \qquad (10.1\text{-}1)$$

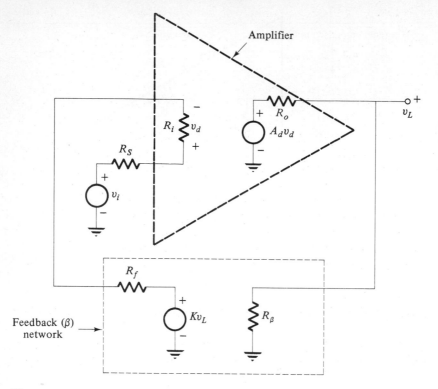

Figure 10.1-3 Voltage-differencing negative-feedback circuit.

and therefore

$$i_d = i_1 - i_2 = \frac{v_i + v_d}{R_s} + \frac{v_d + K v_L}{R_f} = -\frac{v_d}{R_i} \qquad (10.1\text{-}2)$$

Solving (10.1-2) for v_d using the assumptions $R_i \gg R_s$ or R_f and $R_\beta \gg R_o$ so that $v_L \approx A_d v_d$, we find, after some manipulation, that

$$v_L \approx -\frac{(A_d v_i / R_s)(R_s \| R_f)}{1 + (A_d K / R_f)(R_s \| R_f)} \qquad (10.1\text{-}3a)$$

When the denominator term is much greater than unity,

$$\frac{A_d K}{R_f}(R_s \| R_f) \gg 1 \qquad (10.1\text{-}3b)$$

(10.1-3a) reduces to

$$v_L \approx -\frac{R_f}{K R_s} v_i \qquad (10.1\text{-}3c)$$

which is independent of the amplifier gain A_d. For the standard inverting configuration of Fig. 10.1-1a we have $K = 1$, so that $v_L = -(R_f / R_s)v_i$, as expected [see (8.1-2b)].

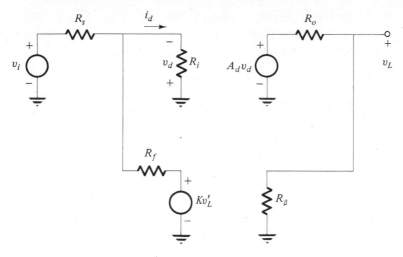

Figure 10.1-4 Circuit used to calculate the loop gain T and the gain without feedback A_o.

The term $(-A_d K/R_f)(R_s \| R_f)$ found in the denominator of (10.1-3) is called the *loop gain T* of the amplifier. This gain can be calculated directly using the circuit of Fig. 10.1-4, which differs from that of Fig. 10.1-2 in that the voltage source in the feedback network is called Kv'_L rather than Kv_L. The loop gain is defined as the gain around the circuit loop with the input voltage equal to zero

Loop Gain

$$T = \frac{v_L}{v'_L}\bigg|_{v_i=0} = \frac{v_L}{v_d}\frac{v_d}{i_d}\frac{i_d}{v'_L} \tag{10.1-4a}$$

Again assuming that $R_i \gg R_s$ or R_f and $R_\beta \gg R_o$, we have

$$T \approx A_d(-R_i)\left(\frac{K}{R_s + R_f}\right)\left(\frac{R_s}{R_s + R_i}\right)$$

$$\approx -A_d K \frac{R_s}{R_s + R_f} = -\frac{A_d K}{R_f}(R_s \| R_f) \tag{10.1-4b}$$

The numerator of (10.1-3a) is called the *open-loop gain* or the the *gain without feedback* A_o. This gain can be found using the circuit of Fig. 10.1-4 by setting $v'_L = 0$. Then

$$A_o = \frac{v_L}{v_i}\bigg|_{v_L'=0} = \frac{v_L}{v_d}\frac{v_d}{v_i} \approx A_d\frac{-R_f}{R_s + R_f} = -\frac{A_d}{R_s}(R_s \| R_f) \tag{10.1-5}$$

Using the results of (10.1-4b) and (10.1-5), we can rewrite (10.1-3) as

$$A_v = \frac{v_L}{v_i} = \frac{A_o}{1 - T} \tag{10.1-6}$$

This expression gives the overall gain A_v in terms of the gain without feedback A_o and the loop gain T for the current-difference circuit of Fig. 10.1-2.

The gain of the feedback amplifier shown in Fig. 10.1-3 can also be found using (10.1-6). To calculate A_o and T for this amplifier we first change the feedback voltage source Kv_L to Kv'_L. If we assume that $R_i \gg R_s$ and R_f and $R_\beta \gg R_o$, the loop gain is

$$T = \frac{v_L}{v'_L}\bigg|_{v_i=0} = \frac{v_L}{v_d}\frac{v_d}{v'_L} \approx A_d \frac{-KR_i}{R_i + R_f + R_s} \approx -A_d K \qquad (10.1\text{-}7)$$

The open-loop gain A_o is

$$A_o \equiv \frac{v_L}{v_i}\bigg|_{v'_L=0} = \frac{v_L}{v_d}\frac{v_d}{v_i} \approx A_d \frac{R_i}{R_i + R_s + R_f} \approx A_d \qquad (10.1\text{-}8)$$

Hence the output voltage is, from (10.1-6),

$$v_L \approx \frac{A_d}{1 + A_d K} v_i \qquad (10.1\text{-}9)$$

The reader should verify the result by direct calculation using Fig. 10.1-3. (Prob. 10.1-1.)

10.1-2 The Loop Gain T

It is seen from (10.1-6) that the loop gain regulates the "amount" of feedback present in a circuit. When $T = 0$ $(K = 0)$, there is no feedback. When $-T$ becomes very large compared with unity, the gain of the feedback amplifier approaches $A_v = v_L/v_i = -(R_f/KR_s)$ for the circuit of Fig. 10.1-2 and $A_v = 1/K$ for the circuit of Fig. 10.1-3. In both cases the gain with feedback is approximately independent of the amplifier gain A_d. As a result, the gain with feedback becomes more immune to the amplifier temperature and parameter variations as the loop gain increases. This is an important benefit of feedback. In addition, we observe that the feedback *decreases* the gain from the no-feedback value A_o to $A_o/(1 - T)$. Hence as $-T$ increases the gain with feedback is reduced by the factor $1 - T$. These ideas are considered qualitatively in the following sections.

10.1-3 Feedback Amplifiers and the Sensitivity Function

Practical feedback amplifiers are designed so that the open-loop gain A_o is extremely large (40 to 120 dB or more). The resulting closed-loop gain is then primarily a function of the feedback network. The gain is almost totally independent of variations in h_{fe}, supply voltage, temperature, etc.

Sensitivity to gain variations A quantitative measure of the effectiveness of the feedback in making A_v independent of the open-loop amplifier gain A_o is the sensitivity function $S_{A_o}^{A_v}$, defined as

$$S_{A_o}^{A_v} = \frac{dA_v/A_v}{dA_o/A_o} = \frac{A_o}{A_v}\frac{dA_v}{dA_o} \qquad (10.1\text{-}10)$$

The sensitivity of A_v with respect to variations in A_o is the ratio of the fractional (or percentage) change in A_v to the fractional (or percentage) change in A_o. Clearly, one would like to achieve zero sensitivity, because this means that changes in the open-loop gain A_o would not cause any change in the gain with feedback A_v.

We can calculate $S_{A_o}^{A_v}$ for the basic feedback amplifier of Fig. 10.1-1 by differentiating (10.1-6). This yields

$$\frac{dA_v}{dA_o} = \frac{1}{1-T} - \frac{A_o}{(1-T)^2}\frac{d(1-T)}{dA_o} \qquad (10.1\text{-}11a)$$

For most cases T is directly proportional to A_o, so that

$$A_o \frac{d(1-T)}{dA_o} = -T \qquad (10.1\text{-}11b)$$

Hence

$$\frac{dA_v}{dA_o} = \frac{1}{(1-T)^2} \qquad (10.1\text{-}11c)$$

Substituting (10.1-11) into (10.1-10) yields

$$S_{A_o}^{A_v} = \frac{1}{1-T} \qquad (10.1\text{-}12)$$

Equation (10.1-12) states that if $-T \gg 1$, a 10 percent change in the forward voltage gain A_o due to transistor replacement, temperature changes, etc., will appear as a change in overall gain A_v of approximately $10/T$ percent. Equations (10.1-12) and (10.1-6) can be used to establish preliminary design figures, as shown by the following example.

Example 10.1-1 A feedback amplifier is to be designed to have an overall gain A_v of 40 dB and a sensitivity of 5 percent to internal amplifier gain variations. Find the required loop gain and open-loop gain.

SOLUTION From (10.1-6)

$$A_v = \frac{A_o}{1-T} \approx \frac{A_o}{-T} = 100 = 40 \text{ dB}$$

From (10.1-12)

$$S_{A_o}^{A_v} \approx \frac{1}{-T} = 0.05$$

so that

$$|T| \approx 20 = 26 \text{ dB}$$

and

$$A_o \approx (20)(100) = 2000 = 66 \text{ dB}$$

Thus, in order to achieve an overall gain of 100 (40 dB), we must start with an amplifier having a gain of 2000 (66 dB). This sacrifice of gain has resulted in a considerable degree of stability. ///

10.2 FREQUENCY RESPONSE OF A FEEDBACK AMPLIFIER

In the previous section we assumed that A_o and T were independent of frequency. We also assumed that negative feedback was employed, so that T was a negative number and $1 - T$ was always a positive number. The response of a real amplifier does, of course, depend on frequency, so that while T may be negative at low frequencies, its phase will generally increase at high frequencies. Suppose that at some frequency ω_0 the magnitude of T is unity and the phase shift reaches 2π rad. Then at ω_0, $T(\omega_0) = +1$ and

$$A_v = \frac{A_o}{1 - T} \to \infty$$

What does this result mean in terms of the physical system? Qualitatively, let us see what happens as T approaches $+1$. Clearly, A_v increases. Thus, if we continually decrease v_i so as to maintain v_L at a fixed level as T approaches $+1$, we find in the limit that no signal is required to obtain an output when $T = +1$. When this happens, the amplifier is said to be *unstable*, and it may break into oscillation without any external excitation.

As will be seen, it is not necessary for T to be exactly $1/\underline{0°}$ or $1/\underline{360°}$ for instability to occur. When an amplifier is turned on, the loop gain will increase from zero to its nominal value in the time it takes for the amplifier to reach steady-state conditions. When the loop gain passes through the $1/\underline{2n\pi}$ point during this transient period, it will break into oscillation.

An alternative way to define stability is in terms of the transient response of the amplifier. Thus an amplifier is stable if its impulse response contains no modes of free vibration which persist or increase indefinitely with time.[1] In terms of the Laplace transform, an equivalent statement is that the transfer function $A_v(s)$ must have *no* poles in the right half of the $s = \sigma + j\omega$ plane, or on the imaginary axis, because such poles have $\sigma \geq 0$ and thus lead to persisting or indefinitely increasing terms in the transient response.

One of the prices we must pay for the benefits of feedback is the need to contend with the stability problem just described. In this chapter we discuss methods of predicting and avoiding high-frequency instability. In Sec. 10.7 we discuss oscillators, where instability is intentionally designed into the circuit.

10.2-1 Bandwidth and Gain-Bandwidth Product

In order to determine the frequency response of a typical feedback amplifier, we now consider the circuit shown in Fig. 10.1-2.

In Sec. 10.1 we found that the overall voltage gain for this circuit could be written in the form

$$A_v = \frac{A_o}{1 - T} \tag{10.2-1}$$

where A_o and T were assumed to be constants. We now drop this assumption so that A_o and T will, in general, be functions of the complex frequency s.

Single-pole amplifier Let us assume that the amplifier gain A_d has a single negative real pole at $s = -\omega_1$, so that

$$A_d(s) = \frac{A_{dm}}{1 + s/\omega_1} \tag{10.2-2}$$

where A_{dm} is the gain at low frequencies. If the feedback network is frequency-independent, the loop gain T, (10.1-4b) can be written as

$$T = -\left(\frac{A_{dm} K(R_s \| R_f)}{R_f}\right) \frac{1}{1 + s/\omega_1} \tag{10.2-3}$$

If we define T_m as the magnitude of the loop gain at frequencies below ω_1, then

$$T_m = \frac{A_{dm} K(R_s \| R_f)}{R_f} \tag{10.2-4a}$$

and

$$T = -\frac{T_m}{1 + s/\omega_1} \tag{10.2-4b}$$

Similarly we find the open-loop gain A_o to be (10.1-5)

$$A_o = -\frac{A_{om}}{1 + s/\omega_1} \tag{10.2-5a}$$

where A_{om}, the magnitude of the open-loop gain at low frequencies, is

$$A_{om} = \frac{A_{dm}(R_s \| R_f)}{R_s} \tag{10.2-5b}$$

Substituting (10.2-4b) and (10.2-5a) into (10.1-6) yields

$$A_v = \frac{v_L}{v_i} = \frac{-A_{om}/(1 + s/\omega_1)}{1 + T_m/(1 + s/\omega_1)}$$

$$= \frac{-A_{om}}{1 + T_m}\left(\frac{1}{1 + s/\omega_1(1 + T_m)}\right) \tag{10.2-6}$$

Equation (10.2-6) shows that the feedback amplifier has a pole at

$$s = -\omega_1(1 + T_m) \tag{10.2-7a}$$

and therefore the upper 3-dB frequency of the amplifier with feedback is

$$f_h = f_1(1 + T_m) \tag{10.2-7b}$$

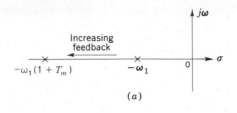

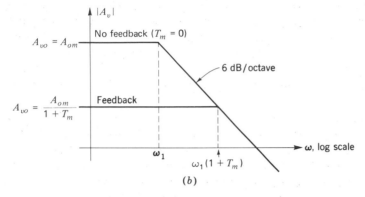

Figure 10.2-1 Characteristic of single-pole feedback amplifier: (a) locus of pole motion; (b) gain versus frequency.

Thus the 3-dB frequency is increased by the loop gain while the midfrequency gain (10.2-6) is decreased by the same amount.

The GBW product is then

$$\text{GBW} = [A_v(f = 0)]f_h = A_{om}\,f_1 \tag{10.2-8}$$

i.e., a constant, independent of the feedback.

Without feedback, the pole (10.2-7a) is at $-\omega_1$. As the amount of feedback is increased from zero, the pole moves farther into the left half-plane along the negative real axis, as shown in Fig. 10.2-1a.† This curve, called a *root locus*, shows how the poles of the overall transfer function vary in terms of the variation of a parameter (in this case, the loop gain). A graph of gain versus frequency is shown in Fig. 10.2-1b.

It is clear from (10.2-6) and Fig. 10.2-1 that no stability problem exists here because the only pole of $A_v(s)$ lies on the negative real axis so long as $-T$ is positive.

---

† Note that this is true only if T is a negative number. If T were positive, the pole would move into the right half-plane and the amplifier would become unstable.

Double-pole amplifier Now let us assume a more complicated case, an amplifier gain function with a pair of coincident real poles

$$A_d(s) = \frac{A_{dm}}{(1 + s/\omega_1)^2} \tag{10.2-9}$$

Now (10.2-3) can be written as

$$T = -\frac{A_{dm} K(R_s \| R_f)}{R_f} \left(\frac{1}{1 + s/\omega_1}\right)^2 = -\frac{T_m}{(1 + s/\omega_1)^2} \tag{10.2-10a}$$

and (10.2-5a) becomes

$$A_o = -\frac{A_{om}}{(1 + s/\omega_1)^2} \tag{10.2-10b}$$

so that (10.2-1) becomes

$$A_v(s) = \frac{-A_{vm}/(1 + s/\omega_1)^2}{1 + T_m/(1 + s/\omega_1)^2} = \frac{-A_{vm}}{1 + T_m + 2s/\omega_1 + s^2/\omega_1^2} \tag{10.2-11}$$

After some algebra this can be put in the standard form

$$A_v(s) = \frac{-A_{om}}{1 + T_m} \left(\frac{1}{1 + 2\zeta s/\omega_n + s^2/\omega_n^2}\right) \tag{10.2-12}$$

where

$$\omega_n = \omega_1 \sqrt{1 + T_m} \tag{10.2-13a}$$

and

$$\zeta = \frac{1}{\sqrt{1 + T_m}} \tag{10.2-13b}$$

The voltage gain given in (10.2-12) shows that the system is of second order, with poles at

$$s = -\omega_n(\zeta \pm \sqrt{\zeta^2 - 1}) \tag{10.2-14}$$

The poles of the overall transfer function are negative, real, and coincident when there is no feedback ($T_m = 0$ and $\zeta = 1$) but become complex as soon as $T_m > 0$. The loci of these poles for increasing T_m are simply two vertical straight lines, as shown in Fig. 10.2-2a.

The asymptotic and actual frequency responses with and without feedback are shown in Fig. 10.2-2b. Since the poles of $A_v(s)$ are in the left half-plane for all values of feedback, there is no stability problem in the strict sense. However, when the feedback is large, the frequency response exhibits a sharp peak. This leads to damped oscillations in the response of the circuit, which are usually undesirable. The magnitude of the peak depends on the damping ratio ζ and increases with increasing feedback. Normalized plots of the magnitude and phase of the frequency response are shown in Fig. 10.2-3 for various values of ζ. The graphs can be interpreted in terms of the circuit Q by noting that $Q = 1/2\zeta$ (Prob. 10.2-3).

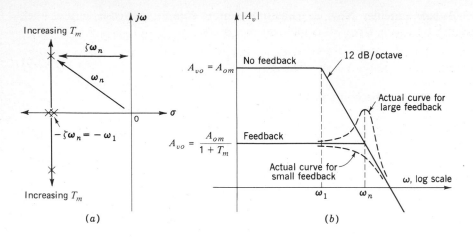

Figure 10.2-2 Characteristics of two-pole feedback amplifier: (a) root locus; (b) gain versus frequency.

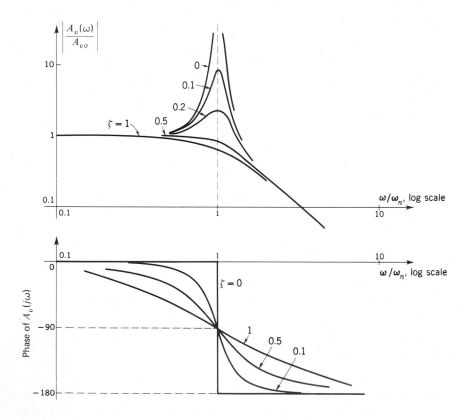

Figure 10.2-3 Response of second-order system.

The assumptions that the forward amplifier had a simple single pole or a double pole and that the feedback network was not frequency-dependent resulted in stable overall gain with feedback. In the next section we consider an example where the poles can move into the right half-plane.

10.3 THE PROBLEM OF STABILITY: A THREE-POLE AMPLIFIER

To illustrate the stability problem in an op-amp consider that the transfer function of the amplifier has three coincident poles

$$A_d(s) = \frac{A_{dm}}{(1 + s/\omega_1)^3} \tag{10.3-1}$$

so that

$$T = -\frac{T_m}{(1 + s/\omega_1)^3} \tag{10.3-2}$$

and

$$A_o = -\frac{A_{om}}{(1 + s/\omega_1)^3} \tag{10.3-3}$$

Then

$$A_v(s) = \frac{-A_{om}}{(1 + s/\omega_1)^3 + T_m} \tag{10.3-4}$$

For this case, the loci of the poles of the overall transfer function start from the triple pole at $s = -\omega_1$ when there is no feedback, $T_m = 0$. As the feedback is increased, one pole moves to the left along the negative real axis while the other two move along the lines shown in Fig. 10.3-1a. When $T_m = 8$, the complex poles lie on the $j\omega$ axis at $\omega = \sqrt{3}\,\omega_1$. Thus, for $-T_m \geq 8$, the amplifier is unstable. Referring to Fig. 10.3-1b, we see that as T_m increases, the gain with feedback A_v decreases. When T_m approaches 8, the frequency response of A_v exhibits sharp peaks because of the proximity of the complex poles to the $j\omega$ axis.

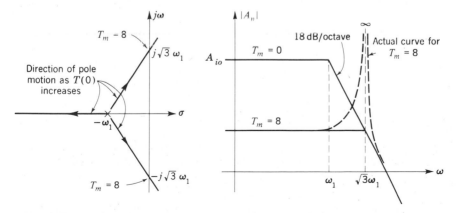

Figure 10.3-1 Characteristics of three-pole amplifier: (a) root locus; (b) gain versus frequency.

Since the factor $1 + T_m$ determines the improvement in sensitivity and the other advantages of feedback, we cannot make good use of feedback in this circuit unless we can modify it so as to realize a value greater than 8 for the magnitude of the loop gain T_m. Fortunately, it is usually a relatively easy matter to modify the basic amplifier so that larger values of T_m can be obtained with freedom from oscillations.

In the sections to follow, we present methods for analyzing feedback amplifiers with a view toward determining stability. Design methods for ensuring stability within prescribed safety margins are presented as well.

10.4 THE NYQUIST STABILITY CRITERION; BODE PLOTS

In the preceding section we indicated how the stability problem arises in feedback amplifiers when the loop gain has more than two real poles. The root-locus plots used there provided a visual picture of the effect of feedback on the poles of the overall transfer function. A complete discussion of the root-locus technique is beyond the scope of this text, and the interested reader should consult Ref. 1.

The Nyquist criterion, which we discuss in this section, forms the basis of a steady-state method of determining whether or not an amplifier is stable. We illustrate this method by considering the basic feedback equation

$$A_v(s) = \frac{A_o(s)}{1 - T(s)} \tag{10.4-1}$$

The amplifier gain $A_v(s)$ is defined as being stable if it has no poles with positive or zero real parts. Thus, to determine whether or not a given amplifier is stable, one need only determine the loop gain $T(s)$, then form the function $1 - T(s)$, factor the numerator polynomial,† and inspect the roots for positive or zero real parts. This is usually a tedious task which provides little or no design information. If we examine (10.4-1) more closely, a considerably simpler method, which also supplies design information, emerges.

In what follows, we assume that $A_v(s)$ and $T(s)$ are stable transfer functions. With this assumption, the only way for $A_v(s)$ to have poles in the right half of the s plane (RHP) is for the denominator $1 - T(s)$ to have zeros in the RHP. Thus our problem is to investigate the zeros of $1 - T(s)$. This is equivalent to finding whether there are values of s with positive or zero real parts at which $T(s) = 1$. In order to do this we plot a *Nyquist diagram* for the loop gain. The Nyquist diagram is simply a polar plot of $T(j\omega)$ for $-\infty < \omega < \infty$. A typical diagram for $T(j\omega)$ having three identical poles as in (10.3-2) is shown in Fig. 10.4-1b.

It is shown in the reference that the Nyquist diagram maps the right half of the s plane shown in Fig. 10.4-1a into the interior of the contour in the T plane

---

† The Routh-Hurwitz criterion[1] determines stability directly from the coefficients of the polynomial without the need for factoring.

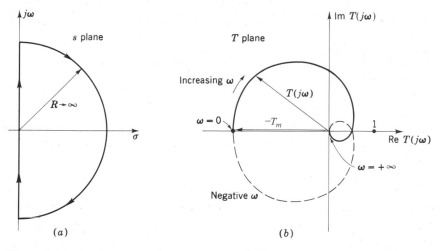

Figure 10.4-1 Nyquist diagram: (a) s-plane contour; (b) T-plane contour.

(Fig. 10.4-1b). If there are any zeros or poles of $1 - T(s)$ in the RHP, the T-plane contour will enclose the point $1\underline{/0°}$, which is called the *critical point*. The number of times that the T-plane contour encircles this critical point in a clockwise direction is equal to the number of zeros of $1 - T(s)$ with positive real parts.

The contour of Fig. 10.4-1b applies to the amplifier of Sec. 10.3. As shown, there are no encirclements of the critical point; so the amplifier is stable. However, the diagram is sketched for $T_m < 8$. If $T_m = 8$, the contour would pass through the critical point. For $T_m > 8$, the critical point would be encircled twice (once for the $0 < \omega < \infty$ range and once for $-\infty < \omega < 0$), indicating the presence of a pair of RHP zeros of $1 - T(s)$, as borne out by the root locus of Fig. 10.3-1a.

Now that we have established the idea of the critical point in the T plane, we note that it is not necessary to plot the Nyquist diagram to ascertain that the critical point is not encircled. This can be determined from Bode plots of the amplitude and phase of $T(s)$, because they contain all the information in the Nyquist diagram. Bode plots corresponding to the Nyquist diagram of Fig. 10.4-1 are shown in Fig. 10.4-2.

From a comparison of the Bode plots† and the Nyquist diagram, we see that the criterion for stability is that $T(\omega)$ cross the 0-dB (unity-gain) axis at a lower frequency than that required for the phase shift to reach 180°. If this criterion is met, the amplifier will be stable. However, we have no real knowledge of the degree of stability since in general we do not know the position of the poles of A_v, as in the root-locus analysis. We do know that when the 0-dB crossover and 180° phase frequencies are very close, the poles of the overall transfer function are very close to the $j\omega$ axis and the amplifier is on the threshold of instability.

---

† The negative sign in $T(s)$ is not included in the Bode plot; thus 180° must be added to the phase shift indicated by the curve in order to obtain the Nyquist diagram.

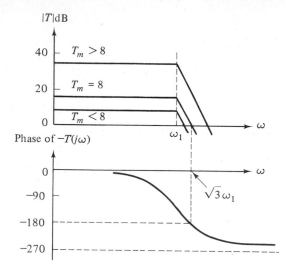

Figure 10.4-2 Bode plot for three-pole amplifier.

10.5 STABILIZING NETWORKS

In the previous sections we saw that a feedback amplifier can be unstable if its loop gain T is too large and the number of poles in the forward-gain transfer function is greater than 2. All op-amps have much more than two poles in the forward gain. When these op-amps are used with feedback, they oscillate unless properly compensated. Various compensation schemes can be employed, all having the same objective, namely, to shape the loop-gain frequency-response characteristic so that the phase shift *is less than* 180° when the magnitude of the loop gain has decreased to unity. Careful design is required in order to ensure that the amplifier will be stable, with an adequate margin of stability over the expected range of variation of transistor h_{fe}, temperature, etc.

In this section we discuss the maximum possible feedback that can be used with the various compensation techniques available.

10.5-1 No Frequency Compensation

Let us first briefly review several basic principles. To do this we consider a specific example, that of an op-amp as in Fig. 10.1-2 with open-loop gain,

$$A_o = \frac{-10^4}{\left(1 + \dfrac{s}{2\pi \times 10^6}\right)\left(1 + \dfrac{s}{2\pi 10 \times 10^6}\right)\left(1 + \dfrac{s}{2\pi 30 \times 10^6}\right)} \tag{10.5-1}$$

Bode plots for (10.5-1) are shown in Fig. 10.5-1. The amplifier will be unstable

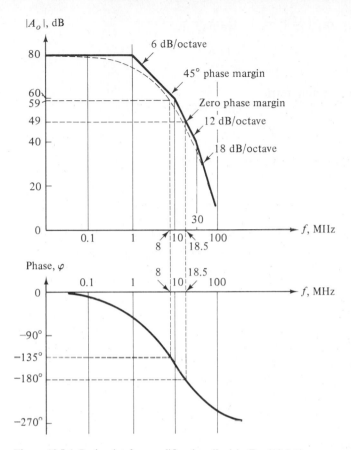

Figure 10.5-1 Bode plot for amplifier described in Eq. (10.5-1).

if used in a feedback circuit without compensation where the magnitude of the loop gain when the phase shift is 180° is greater than unity (0 dB). To ensure stability, the feedback employed must reduce the magnitude of the loop gain T to a value less than unity when the phase is 180°.

If sufficient feedback is applied to make $|T| = 1$ at 18.5 MHz (the phase of T at 18.5 MHz is 180°; see Fig. 10.5-1), the amplifier is said to be *marginally stable*, since there is no *margin of safety*. If, however, sufficient feedback is applied to make $|T| = 1$ at a frequency less than 18.5 MHz, the phase at $|T| = 1$ is less than 180°. The difference between 180° and the actual phase at $|T| = 1$ is called the *phase margin*. Designers often employ a value of 45° phase margin as a compromise between acceptable transient response and adequate stability. Thus, using this criterion in the above example, we should design the feedback network so that $|T| = 1$ at $\varphi = 135°$. From the phase curve shown in Fig. 10.5-1, we see that this occurs at approximately 8 MHz. The frequency at which $|T| = 1$ is usually referred to as the *gain crossover frequency*.

If we compare (10.1-4b) and (10.1-5), we find that

$$T(s) = \frac{A_o(s)K}{R_f/R_s} \tag{10.5-2}$$

It is convenient to define a new term

$$\beta = \frac{K}{R_f/R_s} \tag{10.5-3}$$

which represents the amount of feedback in the amplifier. Then

$$T(s) = \beta A_o(s) \tag{10.5-4a}$$

and from (10.4-1)

$$A_v(s) = \frac{A_o(s)}{1 - \beta A_o(s)} \tag{10.5-4b}$$

If $|T(s)| \gg 1$, Eq. (10.5-4b) reduces to

$$A_v(s) \approx -\frac{1}{\beta} \tag{10.5-4c}$$

In order to have $|A_v(s)| > 1$, we must have $|\beta| \leq 1$. Thus, since we almost always design so that $|A_v(s)| > 1$, it will always be true that $|T| < |A_o|$.

If the phase margin is to be 45°, we must have

$$|T(8 \text{ MHz})| = 1 = \beta|A_o(8 \text{ MHz})| \approx \beta(936) \tag{10.5-5}$$

Thus

$$\beta \approx \tfrac{1}{936} = -59 \text{ dB} \tag{10.5-6}$$

Let us now look at T. At low frequencies

$$T_m = \beta A_{om} = \frac{10^4}{936} = 10.7 = 21 \text{ dB} \tag{10.5-7}$$

If $|T|$ at low frequencies were to exceed 21 dB, the phase margin would drop below 45° and the amplifier might become unstable. A magnitude-frequency plot of T is identical with that of $|A_o|$, except that amplitude values are reduced by 59 dB (for 45° phase margin).

Summarizing, we have found that the amount of feedback necessary to achieve a 45° phase margin is

$$\beta = \frac{1}{|A_o \ (\omega \text{ corresponding to } 135°)|} \tag{10.5-8}$$

and the maximum loop gain for the amplifier of (10.5-1) is from (10.5-7)

$$|T_{max}| = \beta|A_{om}| = \frac{A_{om}}{|A_o \ (\omega \text{ corresponding to } 135°)|} \tag{10.5-9}$$

We note from the above example that at low frequencies the closed-loop gain A_v is

$$A_v = \frac{A_{om}}{1 - T_m} = \frac{A_{om}}{1 - \beta A_{om}} \approx \frac{-1}{\beta} = -936(= 59 \text{ dB}) \qquad (10.5\text{-}10)$$

Thus the loss in gain due to the feedback is 21 dB $(80 - 59 \text{ dB})$, and the maximum permissible loop gain $|T_{max}|$ is 21 dB for a phase margin of 45°. A larger loop gain reduces the phase margin below the acceptable level of 45°.

For many applications a loop gain of 21 dB is not adequate. Thus this example points up the need for compensation schemes which will allow use of larger loop gains along with sufficient phase margin. In succeeding sections we discuss several commonly used compensation techniques.

10.5-2 Simple Lag Compensation

In this section we discuss the simple lag network, which is designed to introduce an additional negative real pole in the transfer function of the open-loop amplifier gain A_o. When this network is added, the open-loop gain becomes

$$A_{o1}(s) = \frac{A_o(s)}{1 + s/\alpha} \qquad (10.5\text{-}11)$$

The pole α is adjusted so that $|T|$ drops to 0 dB at a frequency where the poles of A_o contribute negligible phase shift. Using the example of Sec. 10.5-1, shown in Fig. 10.5-1, we note that with $|T_{max}| = 21$ dB, a phase margin of 45° is achieved with no compensation. If we sketch the Bode plot for $|T_{max}| = 26$ dB, as shown in Fig. 10.5-2a, we see that the phase margin *without compensation* is reduced to zero and the crossover frequency is 18.5 MHz.

Now the question is: Can we achieve a 45° phase margin by adding a lag network to modify the forward gain as in (10.5-11)? If we set $\alpha/2\pi - 70$ kHz,† the Bode plot of $|T|$ takes the form shown in solid lines in Fig. 10.5-2a. The phase margin is now 45°, but the crossover frequency has been reduced to approximately 0.9 MHz. Thus addition of the lag network has allowed us to increase the loop gain T by 5 dB. However, the crossover frequency has been reduced by a factor of approximately 10.

The gain with feedback for the uncompensated and the compensated feedback amplifiers discussed above [Eqs. (10.5-1) and (10.5-11)] are plotted in Fig. 10.5-2b when both are adjusted to have a phase margin of approximately 45°. Note that the uncompensated feedback amplifier has a 3.6-dB peak in the response. This might be undesirable for some applications. The compensated feedback amplifier has 5 dB less low-frequency gain and a significantly narrower bandwidth than the uncompensated amplifier. However, the loop gain of the compensated amplifier is 5 dB larger, while the peak in the frequency-response curve is also 3.6 dB.

---

† This frequency was determined by a trial-and-error procedure using a digital computer.

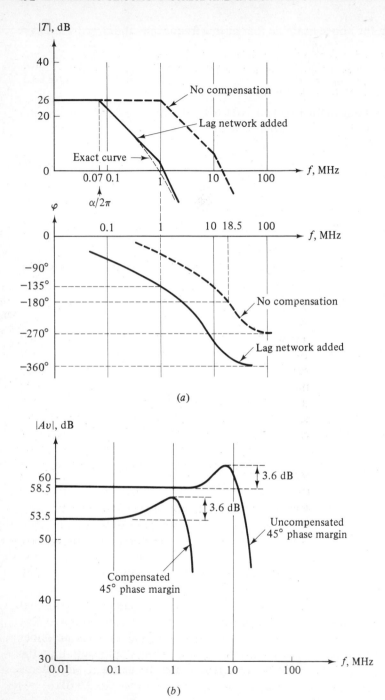

Figure 10.5-2 The effect of lag compensation: (*a*) loop gain with simple lag compensation; (*b*) amplifier frequency-response curves.

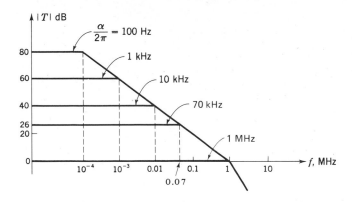

Figure 10.5-3 Loop gain for 45° phase margin with lag network $(\beta \le 1)$.

A glance at Figs. 10.5-1 and 10.5-2a indicates that, with the lag network added, we can increase the loop gain even further at low frequencies, provided that we lower the break frequency α of the lag network so that the crossover frequency occurs below 1 MHz. This ensures a minimum of 45° phase margin. The loop gain for various values of α is shown in Fig. 10.5-3. From this figure we see that the loop gain can be increased to the full amplifier gain of 80 dB and the phase margin will remain at 45°. The overall 3-dB bandwidth of the feedback amplifier would be about the same for all the loop-gain curves of Fig. 10.5-3. In a given design problem, we may not have this much flexibility because of simultaneous specifications on overall gain and bandwidth, and trade-offs may have to be made.

Stability for all T Another interesting possibility exists here. If the break frequency of the lag network is set at 100 Hz, the phase margin is 45° when $T = 80$ dB. If now T is *decreased*, the phase margin will increase. Thus, with $\alpha/2\pi = 100$ Hz, the amplifier is stable for *all* possible values of feedback. Clearly, this will result in considerable loss in bandwidth when T is much less than 80 dB, but it may be acceptable in some cases.

Design of the lag network The design of the actual lag network is easily accomplished once a satisfactory value of break frequency α has been determined. Usually a simple RC filter placed at the output of the op-amp will suffice. This is illustrated in Fig. 10.5-4 for the linear inverting amplifier. Since the resistor R is in series with the op-amp output impedance R_o, R is usually omitted and capacitor C is used alone. The break frequency of this lag filter with R included is, assuming $R + R_o \ll R_2$,

$$\frac{\alpha}{2\pi} \approx \frac{1}{2\pi(R + R_o)C} \qquad (10.5\text{-}12)$$

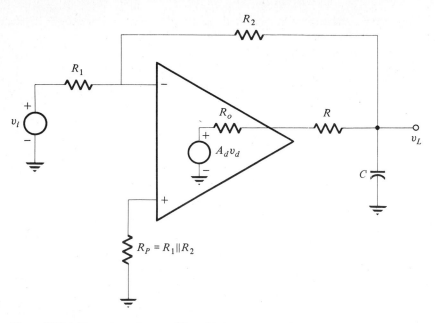

Figure 10.5-4 Linear inverting amplifier with lag compensation.

In IC amplifiers several terminals are usually provided for the connection of compensation networks. These terminals are connected to suitable high-impedance points within the amplifier. The reason for this is that at higher impedance levels smaller capacitance values will be required, and no external resistance R is needed. Examples are presented in Sec. 10.6 to illustrate this point.

10.5-3 More Complicated Lag Compensation

In the preceding section we discussed a method for modifying the loop transmission so as to achieve a 45° phase margin at crossover. The solution presented was designed to bring the loop gain down to 0 dB before the phase shift due to the basic amplifier became excessive. As a result, a considerable portion of the available bandwidth was wasted. In many cases, the specifications will call for maximum bandwidth and fixed closed-loop gain. A lag network with a pole and a zero will usually provide a much wider bandwidth than the single-pole network. The transfer function of this new lag network is

$$H(s) = \frac{1 + s/\alpha_2}{1 + s/\alpha_1} \qquad \alpha_1 < \alpha_2 \qquad (10.5\text{-}13)$$

The open-loop gain is then

$$A_{o2} = H(s)A_o \qquad (10.5\text{-}14)$$

Let us assume that A_o is characterized by three simple poles, as in (10.5-1). Then

$$A_o = \frac{-A_{om}}{(1 + s/\gamma_1)(1 + s/\gamma_2)(1 + s/\gamma_3)} \tag{10.5-15}$$

Now

$$A_{o2} = \frac{-A_{om}(1 + s/\alpha_2)}{(1 + s/\alpha_1)(1 + s/\gamma_1)(1 + s/\gamma_2)(1 + s/\gamma_3)} \tag{10.5-16}$$

From this expression we see that we can effectively cancel the smallest pole of A_o by setting the zero of the lag network at the same point; i.e.,

$$\alpha_2 = \gamma_1 \tag{10.5-17}$$

The open-loop voltage gain becomes

$$A_{o2} = \frac{-A_{om}}{(1 + s/\alpha_1)(1 + s/\gamma_2)(1 + s/\gamma_3)} \tag{10.5-18}$$

In our example [Eq. (10.5-1)]

$$A_{om} = 10^4 = 80 \text{ dB} \qquad \frac{\alpha_2}{2\pi} = \frac{\gamma_1}{2\pi} = 1 \text{ MHz}$$

$$\frac{\gamma_2}{2\pi} = 10 \text{ MHz} \qquad \frac{\gamma_3}{2\pi} = 30 \text{ MHz}$$

$$\frac{\alpha_1}{2\pi} < \frac{\alpha_2}{2\pi} \qquad T = \beta A_{o2} \tag{10.5-19}$$

The pole of $H(s)$, which occurs at α_1, is chosen so that the amplifier is stable, with a 45° phase margin independent of the value of T_m. In this example,

$$T(\omega) = \frac{-\beta A_{om}}{(1 + j\omega/\alpha_1)(1 + j\omega/\gamma_2)(1 + j\omega/\gamma_3)} \tag{10.5-20}$$

Therefore, at low frequencies $(\omega \to 0)$ $T_m = \beta A_{om}$, and at $\omega = \gamma_2$

$$-T(\gamma_2) = 1\underline{/-135°} = \frac{T_m}{(1 + j\gamma_2/\alpha_1)(1 + j)(1 + j\gamma_2/\gamma_3)} \tag{10.5-21a}$$

Now we have the inequalities

$$\alpha_1 < \alpha_2 = \gamma_1 \qquad \text{and} \qquad \gamma_1 < \gamma_2 < \gamma_3$$

Thus

$$1 + j\frac{\gamma_2}{\alpha_1} \approx j\frac{\gamma_2}{\alpha_1} \qquad \text{and} \qquad 1 + j\frac{\gamma_2}{\gamma_3} \approx 1$$

With these approximations (10.5-21a) becomes

$$-T(\gamma_2) = 1\underline{/-135°} \approx \frac{T_m}{(j\gamma_2/\alpha_1)(1 + j)} = \frac{T_m}{\sqrt{2}(\gamma_2/\alpha_1)\underline{/135°}} \tag{10.5-21b}$$

Hence
$$T_m = \sqrt{2}\frac{\gamma_2}{\alpha_1} \qquad (10.5\text{-}22a)$$

and
$$\alpha_1 = \frac{\sqrt{2}\,\gamma_2}{T_m} \qquad (10.5\text{-}22b)$$

Thus, to achieve a loop gain of 40 dB in the above example requires [see (10.5-10)] that

$$\beta = \frac{T_m}{A_{om}} = \frac{10^2}{10^4} = 10^{-2} = -40 \text{ dB} \qquad (10.5\text{-}23a)$$

and from (10.5-22b)

$$\frac{\alpha_1}{2\pi} = \frac{\sqrt{2} \times 10^7}{10^2} \approx 140 \text{ kHz} \qquad (10.5\text{-}23b)$$

These results are sketched in Fig. 10.5-5. It should be noted that the simplifications used in (10.5-21a) are valid only when $\gamma_3 \gg \gamma_2$. For the values in the example exact calculation using (10.5-21a) with $\alpha_1/2\pi = 140$ kHz yields $-T(\gamma_2) \approx 0.94\underline{/-153°}$. Thus the actual crossover frequency is slightly lower

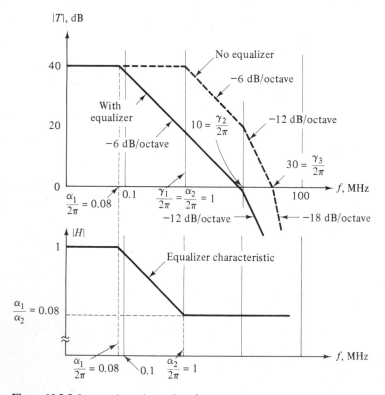

Figure 10.5-5 Loop gain and equalizer frequency characteristic with pole-zero lag network.

than γ_2, and the actual phase margin is $180° - 153° = 27°$. In order to achieve an actual phase margin of $45°$ we must choose $\alpha_1/2\pi \approx 80$ kHz. (This result was obtained using a trial-and-error computer calculation.)

Comparing Fig. 10.5-5 with Fig. 10.5-3, we see that the introduction of the zero in the lag network results in an increase in the gain-crossover frequency by a factor of 10. The bandwidth of this feedback amplifier can be shown to be approximately 12 MHz.

A network having the transfer function of (10.5-13) is shown in Fig. 10.5-6, and the design equations are given in the figure. Figure 10.5-6c shows a linear inverting amplifier with pole-zero lag compensation. Here, we have used the op-amp output impedance R_o in place of resistor R_a and have assumed that $R_2 \gg R_o$.

10.5-4 Lead Compensation

We have seen that to stabilize a feedback amplifier, the loop-gain frequency characteristic must be shaped so that the phase shift at the gain crossover frequency is removed from the critical value of $180°$ by the required phase margin. We have studied the use of the lag network to achieve this and found that the lag network could reduce the crossover frequency (and the bandwidth) without any reduction in low-frequency loop gain.

The low-frequency loop gain is often fixed by specifications on A_v and bandwidth. If a lag network is employed which results in a bandwidth that is too narrow, we must seek an alternative solution. One possibility immediately suggests itself: since stability depends only on the phase at crossover, we might try an equalizer which would introduce phase *lead* at this point. The simplest of such networks has the transfer function

$$H(s) = \frac{s + \delta_1}{s + \delta_2} \qquad \delta_2 > \delta_1 \qquad (10.5\text{-}24a)$$

Note that this transfer function is similar to (10.5-13), except that here the pole occurs at a higher frequency than the zero.

The Bode diagram for $H(s)$ is shown in Fig. 10.5-7, along with a practical network which realizes this transfer function. This should be compared with Fig. 10.5-6. Here we come across a situation which did not arise with the lag network. The lead network is seen from Fig. 10.5-7b to introduce a low-frequency attenuation.

$$H(0) = \frac{R_2}{R_1 + R_2} = \frac{\delta_1}{\delta_2} \qquad (10.5\text{-}24b)$$

This will have to be taken into account because it directly affects the low-frequency gain.

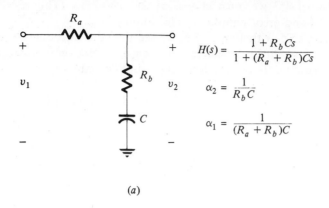

$$H(s) = \frac{1 + R_b Cs}{1 + (R_a + R_b)Cs}$$

$$\alpha_2 = \frac{1}{R_b C}$$

$$\alpha_1 = \frac{1}{(R_a + R_b)C}$$

(a)

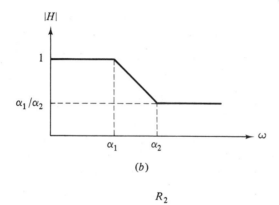

(b)

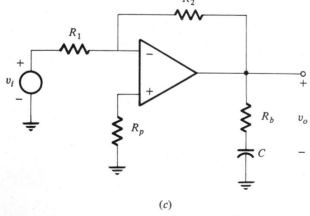

(c)

Figure 10.5-6 Pole-zero lag equalizer: (a) circuit; (b) frequency characteristic; (c) pole-zero lag compensation applied to inverting amplifier.

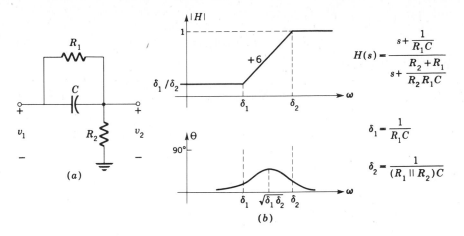

Figure 10.5-7 Lead network: (a) circuit; (b) amplitude and phase.

Let us consider how this network can be used to stabilize the amplifier of the preceding examples. The loop gain with the lead equalizer is

$$-T(s) = \beta H(0)A_{om} \frac{1 + s/\delta_1}{\left(1 + \frac{s}{\delta_2}\right)\left(1 + \frac{s}{\gamma_1}\right)\left(1 + \frac{s}{\gamma_2}\right)\left(1 + \frac{s}{\gamma_3}\right)} \qquad (10.5\text{-}25a)$$

where $A_{om} = 10^4$, $\gamma_1/2\pi = 1$ MHz, $\gamma_2/2\pi = 10$ MHz, $\gamma_3/2\pi = 30$ MHz, and $\delta_1 < \delta_2$. Now we modify the procedure employed when using the lag network, and set

$$\delta_1 = \gamma_2 \qquad (10.5\text{-}25b)$$

so that the lead-network *zero* cancels the *second-lowest* amplifier pole. We design the lead network to ensure that δ_2 is large enough to exert no significant effect at the gain crossover frequency. Then

$$T(s) = \beta H(0)A_{om} \frac{1}{\left(1 + \frac{s}{\gamma_1}\right)\left(1 + \frac{s}{\gamma_3}\right)\left(1 + \frac{s}{\delta_2}\right)} \qquad (10.5\text{-}26)$$

Equation (10.5-26) indicates that if

$$\gamma_1 < \gamma_3 \ll \delta_2 \qquad (10.5\text{-}27a)$$

then the phase shift is $-135°$ at $\omega \approx \gamma_3$, rather than at $\omega \approx \gamma_2$. Thus the gain crossover frequency has been extended by the use of a lead network.

The loop gain at the desired crossover frequency γ_3 is [Eq. (10.5-26)]

$$-T\left(\frac{\gamma_3}{2\pi}\right) = 1\underline{/-135°} \approx \frac{\beta H(0)A_{om}}{(j\gamma_3/\gamma_1)(1 + j)} = \frac{\beta H(0)A_{om}}{\sqrt{2}\,\gamma_3/\gamma_1}\underline{/-135°} \qquad (10.5\text{-}27b)$$

Since the low-frequency loop gain is

$$T_m = \beta H(0) A_{om} \qquad (10.5\text{-}27c)$$

we have

$$\sqrt{2} \, \frac{\gamma_3}{\gamma_1} = T_m \qquad (10.5\text{-}28)$$

In this example, $\gamma_1/2\pi = 1$ MHz and $\gamma_3/2\pi = 30$ MHz; hence

$$T_m = 30\sqrt{2} \approx 42.5 \approx 32 \text{ dB}$$

From (10.5-24b) and (10.5-27c)

$$\beta H(0) = \beta \frac{\delta_1}{\delta_2} = \frac{T_m}{A_{om}} = \frac{42.5}{10^4} = 42.5 \times 10^{-4} \approx -48 \text{ dB}$$

To ensure that $\delta_2 \gg \gamma_3$, we might choose

$$\frac{\delta_2}{2\pi} = 300 \text{ MHz}$$

Since [Eq. (10.5-25b)]

$$\frac{\delta_1}{2\pi} = \frac{\gamma_2}{2\pi} = 10 \text{ MHz}$$

we have

$$\beta = \frac{T_m}{A_{om}} \frac{\delta_2}{\delta_1} = (42.5)(30 \times 10^{-4}) \approx 0.13 \approx -18 \text{ dB}$$

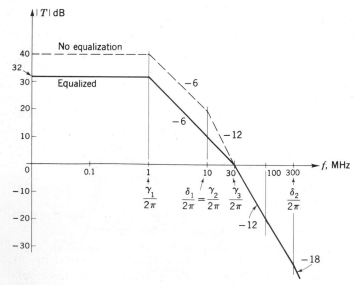

Figure 10.5-8 Loop gain with lead equalization.

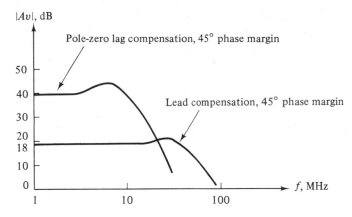

Figure 10.5-9 Overall gain of lag and lead compensated amplifiers.

The loop gain is plotted in Fig. 10.5-8 as a function of frequency. Note that a maximum low-frequency loop gain of 32 dB results. This should be compared with Fig. 10.5-5. In Fig. 10.5-5, T_m is larger, but is down 3 dB at 140 kHz rather than at 1 MHz.

The low-frequency gain with feedback of the lead-compensated amplifier is, for this example,

$$|A_{vm}| = \frac{H(0)A_{om}}{1 + T_m} = \frac{H(0)A_{om}}{1 + \beta H(0)A_{om}} \approx \frac{1}{\beta} \approx 18 \text{ dB}$$

The low-frequency gain with feedback of the lag-compensated amplifier was 40 dB. The bandwidth of the lead-compensated amplifier can be shown to be approximately 50 MHz, compared with the 12-MHz bandwidth obtained using lag compensation when both have 45° phase margin. Figure 10.5-9 shows the overall-gain–versus–frequency curves for these amplifiers. These examples illustrate the trade-off that the designer can make between loop gain T and the bandwidth of the feedback amplifier.

An important disadvantage of lead compensation is that if the phase near the gain crossover frequency increases rapidly with frequency, it may not be possible to obtain sufficient phase lead to produce effective stabilization. Combination lead-lag equalizers are often used in this situation. A second disadvantage of lead compensation is the low-frequency loss introduced in the lead network, which results in a decreased loop gain.

10.6 FREQUENCY COMPENSATION OF OP-AMPS

As mentioned above, real op-amps always have more than two poles and therefore will almost always oscillate in a feedback configuration unless frequency compensation is employed. The 741 op-amp is internally compensated by the

Open-loop gain $|A_o|$, dB

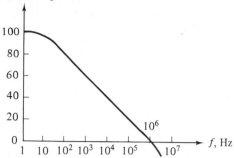

Figure 10.6-1 Open-loop gain response curve for the 741 op-amp.

addition of the simple lag network. However, this type of compensation, when applied by the manufacturer, results in a very narrow-band open-loop frequency response.

The open-loop gain of the 741 is shown in Fig. 10.6-1 as a function of frequency. It is seen from this figure that a single pole exists at 10 Hz. Additional poles occur at frequencies exceeding 1 MHz, at which point the amplifier gain A_o is equal to unity. Since $\beta \leq 1$, the loop gain $|T| \leq 1$ at 1 MHz. Hence the phase margin is a minimum of approximately 45°, and the amplifier is stable for all values of loop gain.

10.6-1 Externally Applied Compensation

If an op-amp is to be used to amplify high frequencies, external compensation must be employed. In such cases the designer must refer to the manufacturer's specifications like those shown in Fig. 10.6-2 for the LH0024 high-slew-rate op-amp.

Figure 10.6-2a shows the open-loop gain variation with frequency. Here we see that the open-loop gain is 0 dB at 70 MHz. Figure 10.6-2b shows the op-amp with pin connections and compensation capacitors C_1, C_2, and C_3. The values given in Fig. 10.6-2c are required to stabilize the amplifier at the closed-loop gains shown. When the amplifier is designed to have a gain of 0 dB, the amplifier bandwidth is 50 MHz. Higher gains produce proportionally less bandwidth.

Some manufacturers present the voltage-gain response curves ($|A_v|$ versus ω) as well as the open-loop-gain response curve. For example, Fig. 10.6-3 shows a variety of voltage-gain curves for the 1539 op-amp and the compensation component values required for each of them.

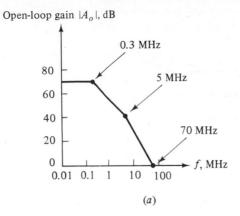

(a)

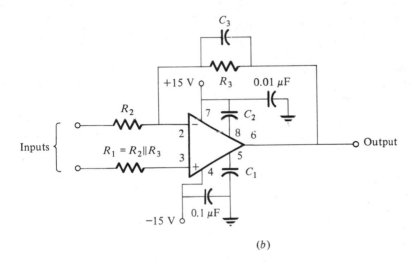

(b)

Closed-loop gain $\lvert Av \rvert$, dB	C_1 pF	C_2 pF	C_3 pF
40	0	0	0
26	0	0	0
20	0	20	1
0	30	30	3

(c)

Figure 10.6-2 LH0024 high-slew-rate op-amp: (a) open-loop frequency response; (b) circuit showing pin connections; (c) values of compensation elements.

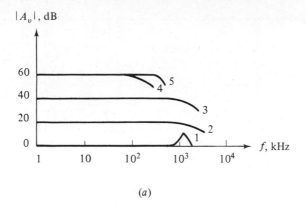

(a)

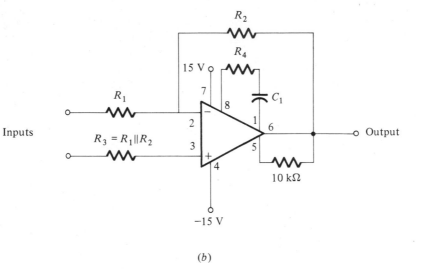

(b)

Curve	R_1, kΩ	R_2, kΩ	R_4, kΩ	C_1, pF
1	10	10	0.39	2200
2	1	10	1	2200
3	1	100	10	2200
4	1	1000	30	1000
5	1	1000	0	10

(c)

Figure 10.6-3 1439/1539 op-amp: (a) voltage gain versus frequency; (b) circuit showing pin connections; (c) compensation component values.

10.7 SINUSOIDAL OSCILLATORS

In preceding sections we saw that when the loop gain $T = 1/\underline{0°}$, the feedback amplifier became unstable.† To prevent this from occurring we employed lead- and lag-compensation networks. Now consider the possibility of producing an unstable amplifier such that $T = 1/\underline{0°}$ at a single frequency ω_0, the magnitude of T being less than unity at all other frequencies. Then the feedback amplifier will be unstable at only one frequency ω_0. Physically, this means that an output is possible with no input present at the one frequency ω_0. Thus the output must be sinusoidal, and such a device is a sinusoidal oscillator.

10.7-1 The Phase-Shift Oscillator

One of the simplest oscillators to design and construct at low frequencies is the phase-shift oscillator shown in Fig. 10.7-1. To determine the conditions for oscillation we must calculate $T(\omega)$ and set it equal to $1/\underline{0°}$.

Before analyzing this circuit, let us attempt to anticipate the results to be obtained. The transistor will provide a phase shift of 180°. Thus, if T is to be equal to $1/\underline{0°}$, an additional 180° must be provided by the three RC circuits [the third $R = R' + (h_{ie} \| R_b)$]. If each RC circuit could act independently, it could be adjusted to provide a 60° phase shift at ω_0. Under these conditions the transfer function for each section would be

$$\frac{s}{s+\alpha} = \frac{\omega}{\sqrt{\alpha^2 + \omega^2}} \exp j \left(\frac{\pi}{2} - \tan^{-1} \frac{\omega}{\alpha} \right) \qquad \alpha = \frac{1}{RC} \qquad (10.7\text{-}1)$$

† If $T > 0$, the feedback is called *positive*.

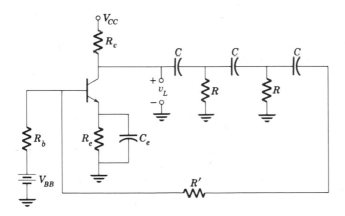

Figure 10.7-1 Phase-shift oscillator.

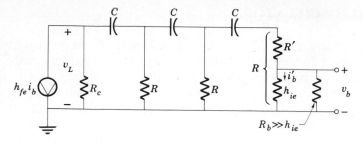

Figure 10.7-2 Equivalent circuit of phase-shift oscillator.

To obtain a 60° phase shift at ω_0

$$\frac{\pi}{2} - \left(\tan^{-1} \frac{\omega_0}{\alpha} \right) = \frac{\pi}{3} \tag{10.7-2}$$

or

$$\frac{\pi}{6} = \tan^{-1} \frac{\omega_0}{\alpha} \tag{10.7-3}$$

and

$$\omega_0 = \frac{\alpha}{\sqrt{3}} = \frac{1}{\sqrt{3}\,RC} \tag{10.7-4}$$

This is the condition for oscillation, assuming independent RC sections. The loop gain T is adjusted to unity at ω_0 by adjusting the attenuation in the circuit. This oscillator works well as long as $\omega_0 \ll \omega_\beta$, since near ω_β, the transistor input appears capacitive.

In the circuit of Fig. 10.7-1, the RC circuits are not independent, and we must proceed with an analysis of the equivalent circuit shown in Fig. 10.7-2.

The loop gain for this circuit is most conveniently defined in terms of currents as

$$T = \frac{i_b}{i_b'} \tag{10.7-5}$$

Solving the circuit of Fig. 10.7-2 yields

$$-h_{fe}i_b \approx i_b'\left(3 + \frac{4}{sRC} + \frac{1}{s^2R^2C^2} + \frac{R}{R_c} + \frac{6}{sR_cC} + \frac{5}{s^2C^2RR_c} + \frac{1}{s^3R^2R_cC^3}\right) \tag{10.7-6}$$

Since $i_b = i_b'$ for $T = 1\underline{/0°}$ at ω_0, we can equate real and imaginary parts on both sides of this equation to determine the conditions for oscillation. From the imaginary terms,

$$\frac{1}{\omega_0^2 RR_cC^2} = 4 + 6\frac{R}{R_c}$$

Thus

$$\omega_0 = \frac{1}{RC}\frac{1}{\sqrt{6 + 4R_c/R}} \tag{10.7-7}$$

Setting the real parts equal on both sides and using (10.7-7), we get

$$-h_{fe} = 3 + \frac{R}{R_c} - \frac{4 + 6R/R_c}{R/R_c} - 5\left(4 + 6\frac{R}{R_c}\right) \tag{10.7-8}$$

Solving for R/R_c in terms of h_{fe} yields

$$\frac{R}{R_c} = \frac{h_{fe} - 23}{58} + \sqrt{\left(\frac{h_{fe} - 23}{58}\right)^2 - \frac{4}{29}} \tag{10.7-9}$$

Hence, given h_{fe} and ω_0, we can determine R/R_c from (10.7-9) and then RC from (10.7-7).

In order to have the term under the square root positive, we must have

$$h_{fe} > 23 + 21.6 = 44.6 \tag{10.7-10}$$

If h_{fe} is less than this value, the circuit will not oscillate since $T < 1\underline{/\,0°}$. At this value of h_{fe}, from (10.7-9),

$$\frac{R}{R_c} = 0.375$$

If this ratio is increased, the oscillator waveform will be distorted. Thus, when constructing this type of oscillator, h_{fe} is not known exactly, and R_c is usually a variable resistor, which is adjusted to eliminate the distortion.

10.7-2 The Wien-Bridge Oscillator

An IC Wien-bridge oscillator is shown in Fig. 10.7-3a. The operation of the oscillator can be explained using Fig. 10.7-3b. The loop gain is (Prob. 10.7-3)

$$T = \frac{v_L}{v_L'} = \left[\frac{R/(1 + sRC)}{R/(1 + sRC) + R + 1/sC} - \frac{R_i}{R_i + R_f}\right]A_d$$

$$= \left[\frac{s/\omega_0}{s/\omega_0 + (1 + s/\omega_0)^2} - \frac{R_i}{R_i + R_f}\right]A_d \tag{10.7-11a}$$

where
$$\omega_0 = \frac{1}{RC} \tag{10.7-11b}$$

The condition for oscillation is $T = 1\underline{/\,0}$. Solving (10.7-11b), with $T = 1\underline{/\,0°}$ and $s = j\omega$, yields

$$1 - \left(\frac{\omega}{\omega_0}\right)^2 + j\frac{\omega}{\omega_0}\left[3 - \frac{A_d(R_i + R_f)}{R_i(1 + A_d) + R_f}\right] = 0 \tag{10.7-12a}$$

Equating real and imaginary parts to zero, we find that

$$\omega = \omega_0 = \frac{1}{RC} \tag{10.7-12b}$$

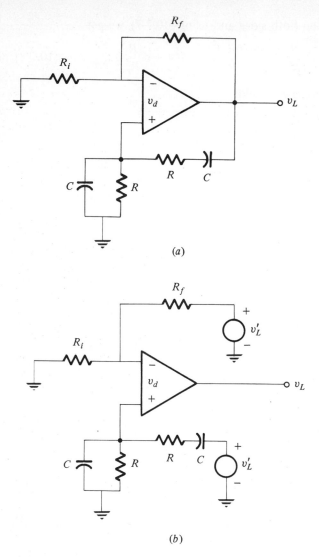

Figure 10.7-3 Wien-bridge oscillator: (a) circuit; (b) circuit redrawn to clarify loop gain.

and
$$\frac{R_f}{R_i} = \frac{2A_d + 3}{A_d - 3} \approx 2 \tag{10.7-13}$$

10.7-3 The Tuned-Circuit Oscillator

A simple tuned-circuit oscillator is shown in Fig. 10.7-4. The operation of the circuit can be explained by assuming that a small collector current flows at the frequency $\omega_0 \approx 1/\sqrt{LC}$. Then a voltage at the frequency ω_0 will appear at

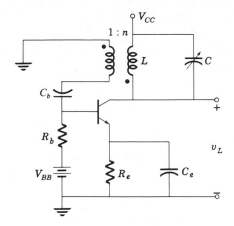

Figure 10.7-4 Tuned-circuit oscillator.

the collector, part of which is fed back to the base through the transformer. The polarity of the transformer is adjusted so that the feedback is positive, and the base current (also at ω_0) therefore tends to increase. Eventually, of course, the nonlinearities of the transistor limit the current swing.

Now, to put this on a quantitative basis, let us study the equivalent circuit shown in Fig. 10.7-5. Figure 10.7-5a shows the basic equivalent circuit, where $T = v_b/v_b'$. Note that the voltage nv_b is reversed because of the phase reversal in the transformer. It is this phase reversal which provides the extra 180° phase shift needed for positive feedback.

Figure 10.7-5b is the final equivalent circuit with the base-circuit impedance reflected through the transformer into the collector. The loop gain T is now

$$T = \frac{v_b}{v_b'} = \frac{g_m/n}{1/n^2 h_{ie} + 1/j\omega L + j\omega[C + (C_{in}/n^2)]} \tag{10.7-14}$$

The criterion for oscillation is that T must be $1 \underline{/0^\circ}$. This leads to the relations

$$\omega_0^2 = \frac{1}{L[C + (C_{in}/n^2)]} \tag{10.7-15}$$

and
$$nh_{fe} = 1 \tag{10.7-16}$$

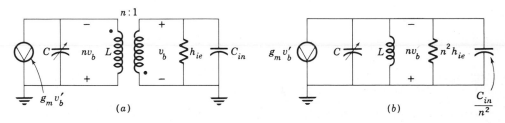

Figure 10.7-5 Small-signal equivalent circuit of tuned-circuit oscillator ($R_b \gg r_{b'e}$): (a) transformer included; (b) transformer eliminated.

The frequency of oscillation is dependent on C_{in}. Since

$$C_{in} = C_{b'e} + C_M$$

the frequency ω_0 depends on the quiescent point. Note also that

$$n = \frac{1}{h_{fe}}$$

Since h_{fe} is very large, n can be small. If nh_{fe} exceeds unity, the amplitude of oscillation will be limited by the transistor nonlinearities. However, the tuned collector circuit tends to filter out the resulting harmonics present in the collector current. The collector voltage is therefore almost sinusoidal.

10.7-4 The Colpitts Oscillator

A Colpitts oscillator, which is used at radio frequencies, is shown in Fig. 10.7-6a. The equivalent circuit is shown in Fig. 10.7-6b, where

$$C_2 = C_2' + C_{b'e} + C_M \qquad (10.7\text{-}17a)$$

and

$$R = R_b \| r_{b'e} \qquad (10.7\text{-}17b)$$

The loop gain T is

$$T = \frac{v_b}{v_b'} = 1 \underline{/\,0°} = \frac{-g_m R}{sRC_1 + (1 + sRC_2)(1 + s^2LC_1)} \qquad (10.7\text{-}18)$$

Solving (10.7-18), we obtain

$$\omega^2 = \omega_0^2 = \frac{1}{L[C_1 C_2/(C_1 + C_2)]} \qquad (10.7\text{-}19a)$$

and

$$\omega_0^2 LC_1 = (1 + g_m R) \qquad (10.7\text{-}19b)$$

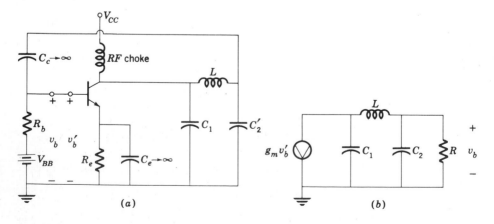

Figure 10.7-6 (a) The Colpitts oscillator circuit; (b) small-signal equivalent circuit.

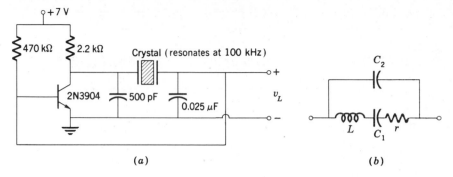

Figure 10.7-7 (a) A Colpitts crystal oscillator; (b) equivalent circuit of crystal.

Dividing (10.7-19a) by (10.7-19b) yields

$$1 + g_m R = 1 + \frac{C_1}{C_2} \tag{10.7-20a}$$

If $R \approx r_{b'e}$, this yields

$$h_{fe} \approx \frac{C_1}{C_2} \tag{10.7-20b}$$

Thus the conditions for oscillation are strongly dependent on the h_{fe} of the transistor.

The inductor L in the feedback network of the Colpitts oscillator can be replaced by a piezoelectric crystal, as shown in Fig. 10.7-7a. The equivalent circuit of the crystal is shown in Fig. 10.7-7b. A typical crystal will have an extremely high Q (several thousand) and thus tends to stabilize the oscillator and prevent frequency variation with transistor replacement. The analysis of the circuit of Fig. 10.7-7a is left as an exercise for the reader.

10.7-5 The Hartley Oscillator

The final oscillator circuit to be considered here is the Hartley oscillator, shown in Fig. 10.7-8a. This oscillator employs the common-base configuration and is generally used at very high frequencies. Positive feedback is achieved by returning a portion of the output to the input through a transformer. In practice, an auto-transformer is often used, rather than the transformer shown. The loop gain of unity is obtained by adjusting the turns ratio. R_e is used as a bias resistance and is much greater than h_{ib}.

The loop gain T can be found from the equivalent circuit shown in either Fig. 10.7-8b or c.

$$T = 1 \underline{/0°} = \frac{(n_2/n_1)h_{fb}(L/C)}{L/C + (n_2/n_1)^2 h_{ib}(j\omega L + 1/j\omega C)} \tag{10.7-21}$$

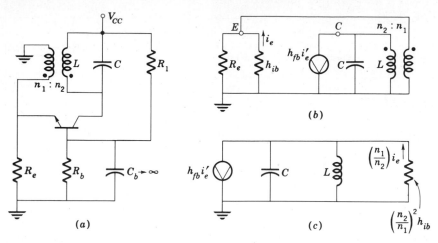

Figure 10.7-8 Hartley oscillator: (a) circuit; (b) equivalent circuit; (c) transformer eliminated $(R_e \gg h_{ib})$.

The conditions for oscillation, which are obtained by solving (10.7-21), are

$$\omega_0^2 = \frac{1}{LC} \tag{10.7-22a}$$

and

$$\frac{n_1}{n_2} = h_{fb} \tag{10.7-22b}$$

Thus, the turns ratio is adjusted to make up the loss in current gain of the common-base configuration. The circuit shown in Fig. 10.7-9 is a Hartley oscillator which operates in the common-collector mode instead of the common-base mode. The emitter voltage is stepped up by the autotransformer and returned to the base. The analysis of this circuit is left as an exercise (Prob. 10.7-9).

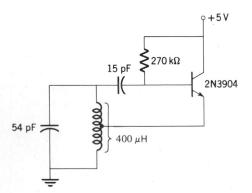

Figure 10.7-9 A common-collector Hartley oscillator.

REFERENCE

1. J. D'Azzo, and C. Houpis, "Linear Control System Analysis and Design," McGraw-Hill, New York, 1975.

PROBLEMS

10.1-1 By direct calculation use Fig. 10.1-3 to verify (10.1-9).

10.1-2 (a) Calculate the input impedance in Fig. 10.1-2, where

$$Z_i = \frac{v_i}{i_1}$$

(b) Calculate the output impedance in Fig. 10.1-2 by applying an external generator v_o, as in Fig. 8.1-3, and determining

$$Z_o = \frac{v_o}{i_o}\bigg|_{v_i = 0}$$

Hint: Refer to Sec. 8.1 and note that, there, $K = 1$.

10.1-3 Repeat Prob. 10.1-2 for Fig. 10.1-3.

10.1-4 By writing node equations prove that Fig. 10.1-1b is equivalent to the resistor R_f which is connected between v_L and $-v_d$ in Fig. 10.1-1a.

10.1-5 In this problem we show that ideal op-amp theory is equivalent to the feedback amplifier having a loop gain whose magnitude $\gg 1$. In Fig. P10.1-5a we have combined the practical inverting and noninverting op-amps of Fig. 8.11-1b and c ($R_p = R_1 \| R_2$ for symmetry) along with the equivalent op-amp circuit.

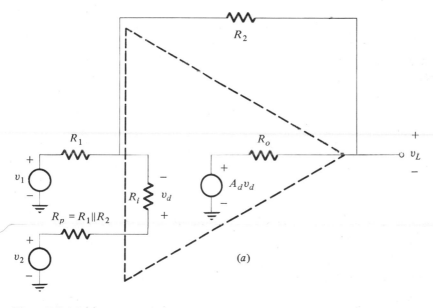

(a)

Figure P10.1-5 (a)

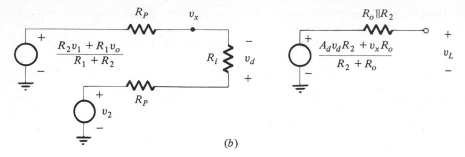

(b)

Figure P10.1-5 (b)

(a) Use the results of Prob. 10.1-4 and show that the amplifier is equivalent to Fig. P10.1-5b. Note that since $|A_d v_d R_2| \gg |v_x R_o|$, the output circuit becomes a voltage source $A_d v_d$ in series with a resistor $R_o \| R_2$.

(b) Find v_L as a function of v_1 and v_2. Define the loop gain.

(c) Show that when the loop gain has a magnitude $\gg 1$, the results can be obtained more easily using ideal op-amp theory.

(d) Assume a load R_L is connected across v_L to ground. Find v_L and define the loop gain. Under what conditions will the load cause a result different from that of ideal op-amp theory, assuming, of course, that the amplifier remains linear?

10.1-6 Repeat Example 10.1-1 if the overall gain is to be 36 dB with a sensitivity of 10 percent to internal amplifier-gain variations.

10.1-7 This problem illustrates the effect of feedback on internal disturbances such as noise or power-supply drift. In Fig. 10.1-2 replace the voltage source $A_d v_d$ by the two voltage sources in Fig. P10.1-7. The source $A_1 v_n$ is an equivalent noise source due to internally generated noise in the amplifier, and A_1 is usually some fraction of A_d. (For power-supply drift $A_1 = 1$.) Find v_L assuming a high loop gain and show how feedback reduces the disturbance.

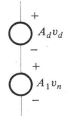

Figure P10.1-7

10.2-1 In Fig. 10.1-2 let $R_o \ll R_\beta$, $R_i \gg R_s$, and $K = 1$ and replace A_d by $A_d(s)$ of (10.2-2).

(a) Solve for v_L, verifying (10.2-3) and (10.2-5a).

(b) Verify (10.2-7a).

10.2-2 Repeat Prob. 10.2-1 for $A_d(s)$ as given in (10.2-9).

10.2-3 The transfer function of (10.2-11) can be synthesized using an RLC circuit and ideal current amplifier.

(a) Find the circuit.

(b) Show that the Q of the circuit is $1/2\zeta$.

10.2-4 Low-frequency instability can occur in a multistage ac-coupled feedback amplifier. Consider a two-stage amplifier with two coupling capacitors adjusted so that in Fig. 10.1-2 we can replace A_d by

$$A_d(s) = \frac{A_{dm} s^2}{(s + \omega_L)^2}$$

Solve for v_L and comment on the stability. Assume $R_o \ll R_\beta$ and $R_i \gg R_s$.

10.2-5 In Fig. 10.1-2 let $R_o \ll R_\beta$, $R_i \gg R_s$ and place a stray capacitance C_L across v_L to ground. Find $A_d(s)$ and determine v_L/v_i. What is the loop gain? What is the open-loop gain?

10.2-6 In Fig. 10.1-2 let $R_o \ll R_\beta$, $R_i \gg R_s$, $R_s = R_f$, and $K = 1$ and replace A_d by

$$A_d(s) = \frac{164C}{[1 + s/(50 \times 10^3)][1 + s/(120 \times 10^3)]}$$

(a) Find v_L/v_i, indicating the open-loop gain and the loop gain.
(b) If $C = 1$ and 100, find ζ and ω_m.
(c) Sketch the asymptotic gain for T for each value of C.
(d) Sketch the asymptotic gain for $|v_L/v_i|$ for each value of C. Estimate the 3-dB bandwidth in each case.

10.4-1 In Fig. 10.1-2 $R_o \ll R_\beta$, $R_i \gg R_s$, $K = 1$, $R_s = 10$ kΩ, and $R_f = 50$ kΩ. The amplifier gain is

$$A_d(s) = \frac{24 \times 10^3}{(1 + s/10^5)[1 + s/(2 \times 10^5)]^2}$$

Find and sketch the magnitude and phase of T. Is the amplifier stable?

10.4-2 The amplifier gain in Prob. 10.4-1 is

$$A_d(s) = (19.2 \times 10^3)\left(\frac{s/50}{1 + s/50}\right)^2 \frac{s/100}{1 + s/100}$$

Find and sketch the magnitude and phase of T. Is the amplifier stable?

10.4-3 Sketch the Nyquist diagrams for Probs. 10.4-1 and 10.4-2 and compare.

10.4-4 In the feedback amplifier of Prob. 10.4-1, find the value of R_f for which the phase margin is between 45 and 60°. Use $R_s = 10$ kΩ.

10.4-5 Repeat Prob. 10.4-1 if

$$A_d(s) = (19.2 \times 10^3)\left(\frac{s/0.1}{1 + s/0.1}\right)\left(\frac{s/0.5}{1 + s/0.5}\right)\left(\frac{s/5000}{1 + s/5000}\right)$$

10.5-1 The linear inverting amplifier of Fig. 10.1-1 is to be stabilized using pole-zero lag compensation, as in Fig. 10.5-6. The two circuits are equivalent (see Prob. 10.1-5) since $R_i \gg R_p$, $R_a = R_o \ll R_2$, and $|A_d v_d R_2| \gg v_x R_o$. Design for a phase margin of approximately 45° using R_s, R_f, and $A_d(s)$ as given in Prob. 10.4-1. If $R_o = 100$ Ω, determine C and R_b.

10.5-2 Find and plot A_v for Prob. 10.5-1 and estimate the overall bandwidth.

10.5-3 Verify Fig. 10.5-2b.

10.5-4 Plot $|A_v|$ versus frequency for the compensated feedback amplifier of Sec. 10.5-3. Verify that the 3-dB bandwidth is approximately 12 MHz.

10.5-5 Plot $|A_v|$ as a function of frequency for the lead-compensated amplifier of Sec. 10.5-4. Show that the 3-dB bandwidth is approximately 50 MHz.

10.5-6 The op-amp in Fig. P10.5-6 is used to implement the lead-compensated amplifier of Sec. 10.5-4.

(a) Show that if $R_2 \ll R_f$ and $R_i \gg R_p$, the gain of the op-amp is given by (10.1-3a), where A_d is replaced by

$$A_d(s) = \frac{A_d(0)\left(\dfrac{R_2}{R_1 + R_2 + R_o}\right)\left|\dfrac{1 + sR_1C_1}{1 + s[R_1\|(R_2 + R_o)]C_1}\right|}{\left(1 + \dfrac{s}{2\pi \times 10^6}\right)\left[1 + \dfrac{s}{2\pi(10 \times 10^6)}\right]\left[1 + \dfrac{s}{2\pi(30 \times 10^6)}\right]}$$

(b) If $R_1 = 1$ kΩ and $R_f = 100$ kΩ, find the value of $A_d(0)$ required to obtain the open-loop gain of $A_o(0) = -10^4$ with no compensation.

(c) If $R_o = 100$ Ω, determine R_1, R_2, and C_1 in the lead network.

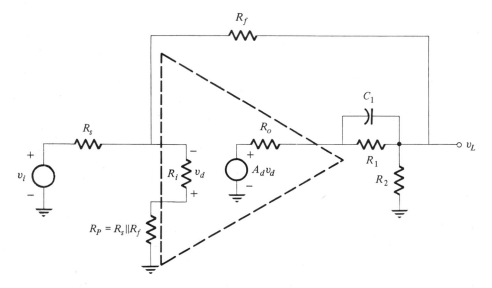

Figure P10.5-6

10.5-7 In the integrator circuit of Fig. 8.4-4, the parameters are $R = 1$ MΩ and $C = 1$ μF. The op-amp is nonideal, having $R_i = 10$ MΩ, $R_o = 100$ Ω, and $A_d = 10^4$. Find and plot v_L when v_i is a unit step function. Compare with the output of an ideal integrator. Up to what point does the actual output depart from the ideal by less than 1 percent? 10 percent?

10.5-8 The lag-compensated amplifier of Fig. 10.5-6c has $R_1 = 2$ kΩ, $R_2 = 10$ kΩ, $R_i = 100$ kΩ, $R_o = 100$ Ω, and

$$A_d(s) = \frac{10^4}{[1 + s/(2\pi \times 10^6)]\{1 + s/[2\pi(4 \times 10^6)]\}}$$

(a) Design for a phase margin of about 45°. Make reasonable approximations.

(b) Find R_b and C.

10.5-9 The op-amp in Fig. 10.1-2 has $R_o \ll R_\beta$, $R_i \gg R_s$, $K = 1$, $R_s = 10$ kΩ, and $R_f = 50$ kΩ. The amplifier gain has a double pole and is given by

$$A_d(s) = \frac{24 \times 10^3}{[1 + s/(2 \times 10^5)]^2}$$

Stabilize the amplifier by connecting a capacitor C_x across the input of the op-amp ($-v_d$ to ground). Design for a phase margin of approximately 45°. Calculate C_x.

10.5-10 Find and plot A_v for Prob. 10.5-9 and estimate the overall bandwidth.

10.7-1 Verify (10.7-6) and (10.7-9).

10.7-2 In Fig. 10.7-1 let $V_{CC} = 9$ V, $V_{BB} = 5$ V, $R_e = R = 1$ kΩ, $C = 0.0068$ μF, $R_c = 420$ Ω, and $R_b = 900$ Ω. Find (a) R' if $h_{fe} = 150$ and (b) f_o.

10.7-3 (a) Verify (10.7-11a).
    (b) In Fig. 10.7-3 let $A_d = 10^4$, $R_i = 10$ kΩ, $R = 10$ kΩ, and $C = 0.001$ μF. Find f_o and R_f.

10.7-4 (a) Verify (10.7-14).
    (b) In Fig. 10.7-5 let $L = 400$ μH, $C = 50$ pF, $C_M = 100$ pF, $I_E = 1$ mA, $h_{fe} = 200$, and $R_b = 50$ kΩ. Find f_o and n.

10.7-5 (a) In Fig. 10.7-4 include the effect of R_b and find T.
    (b) Find f_o and n.

10.7-6 (a) Verify (10.7-18).
    (b) In Fig. 10.7-6 let $I_E = 1$ mA, $h_{fe} = 200$, $R_b = 5$ kΩ, $L = 4$ μH, $C_2 = 100$ pF. Find f_o and C_1.

10.7-7 Find the condition under which the circuit shown in Fig. 10.7-7a will oscillate.

10.7-8 (a) Verify (10.7-21).
    (b) In Fig. 10.7-8 let $L = 400$ μH, $C = 54$ pF, $R_e = 1$ kΩ, $I_E = 1$ mA, and $h_{fe} = 100$. Find f_o and n_1/n_2.

10.7-9 In Fig. 10.7-9 let $L = 400$ μH, $C_1 = 54$ pF, $C_2 = 15$ pF, $R_b = 270$ kΩ, $I_E = 1$ mA, and $h_{fe} = 150$.
    (a) Assume that the coupling is tight (see Prob. 9.5-7) so that the autotransformer can be represented as in Fig. 9.5-6 and R_b can be treated as an open circuit. Show that

$$\omega_0 = \frac{1}{\sqrt{LC_1}} \quad \text{and} \quad n = \frac{n_2}{n_1} \geq h_{fe} + 1$$

    (b) Find f_o and n.

LOGIC FUNCTIONS AND BOOLEAN ALGEBRA

INTRODUCTION

In previous chapters we studied diodes, BJTs, and FETs primarily from the *analog*, or continuous-signal, point of view. In this chapter we begin to examine the use of these same devices in *digital*, or discrete, systems. The digital computer is perhaps the best-known of such systems, all of which exist today because of the reliability of ON-OFF types of operations, compared with analog operations. Within the computer, transistors of one type or another are used to perform these operations; they operate like switches, being either ON or OFF. Thus the complicated functions performed by the computer can be thought of as being accomplished by interconnecting large numbers of switches called *logic gates*. These interconnections are designed to implement the laws of *logic* and are manipulated so that the desired calculations are performed. The mathematical technique used for logical analysis of systems containing large numbers of ON-OFF elements is called *boolean algebra*. It is characterized by a set of rules similar to those of ordinary algebra but based on only two possible states, ON-OFF (or TRUE-FALSE), called *logic states*. In this chapter we introduce boolean algebra and show how the operations in this algebra are implemented with various types of logic gates.

11.1 LOGIC FUNCTIONS

11.1-1 The NOT Function

There are three basic logic functions that are performed by the components of a digital computer, the NOT, AND, and OR functions. The one we shall consider first is perhaps the simplest of the three, the NOT function. To illustrate this, consider a light bulb. If the bulb is lit, we signify its state by the letter L. If it is NOT lit, its state is NOT L, symbolically $\bar{L}$. The logic variable L is a *binary* (two-valued) variable.

The bulb can be in only one of two states, lit or NOT lit. Thus L and $\bar{L}$ are said to be *complementary*. Since there are only two possible states of the logic variable L, we often represent them with the symbols **1** and **0**,† corresponding to ON and OFF (or TRUE and FALSE); i.e.

$$L = \begin{cases} \mathbf{1} & \text{if light is lit (ON, TRUE)} \\ \mathbf{0} & \text{if light is NOT lit (OFF, FALSE)} \end{cases} \tag{11.1-1}$$

These symbols do not have any numerical significance at this point. The significance of the NOT function in the system where there are only two symbols, **1** and **0** is that a state which is NOT represented by the symbol **1** is represented by the symbol **0**.

A device which implements the logic NOT function is called a NOT (or *inverter*) gate. In logic diagrams, different symbols are used to distinguish between the various types of gates. The symbols usually used for the NOT gate are shown in Fig. 11.1-1. If the switch is connected to the battery, the input to the gate is $S = \mathbf{1}$ and the output of the gate will then be $L = \mathbf{0}$. Since $L = \mathbf{0}$, the bulb does NOT

† For clarity we shall use boldface **0** and **1** to distinguish logic symbols from scalars.

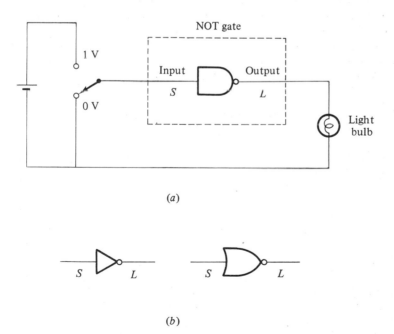

(a)

(b)

Figure 11.1-1 The NOT gate: (a) circuit which inverts switch output; (b) alternate NOT-gate symbols.

light. If the switch is moved to the 0-V side, we have $S = 0$, $L = 1$, and the bulb will light.

The logic equation which describes this operation is

$$L = \bar{S} \qquad (11.1\text{-}2)$$

In the next chapter we shall show that this function is readily implemented by a BJT or FET single-stage amplifier similar to those discussed in Chaps. 2 and 3.

11.1-2 The Truth Table

Every logic problem can be phrased in terms of the three logic functions NOT, AND, and OR. The more complicated the problem, the larger the number of logic variables and gates becomes, until we have difficulty in stating the problem precisely in words, i.e., taking into account all the possibilities. Fortunately, organized procedures have been developed by means of which we can tabulate all the possibilities that can arise. Using these procedures, we can obtain the correct symbolic statement of the problem (logic equation) in a systematic way.

The most basic procedure employed, and also one of the best, is the *truth table*. A truth table makes use of the fact that if we have N two-state variables, there are 2^N different ways of combining them. These 2^N possible combinations are set forth in the truth table. For the NOT function of (11.1-2) which describes the circuit of Fig. 11.1-1a we have the following two possibilities:

$$S = \begin{cases} 1 & \text{switch contact to upper terminal, light OFF} \\ 0 & \text{switch contact to lower terminal, light ON} \end{cases}$$

This can be written in tabular form as

Switch	Light
Up	Off
Down	On

If we represent the switch being up by a **1** and the switch being down by a **0** and the light being on by a **1** and being off by a **0**, we have the truth table (Table 11.1-1) for the NOT function.†

As the number of variables increases, so does the complexity of the truth table. Its usefulness will become more apparent when we consider more complicated logic functions.

---

† Boldface for binary digits is not used in truth tables, or Karnaugh maps (Sec. 11.5), where the binary nature of the digits is clear.

Table 11.1-1 Truth table for the NOT function

S	$L = \bar{S}$
1	0
0	1

11.1-3 The AND Function

In a typical problem we may encounter many binary logic variables. The logic AND function relates two or more such logic variables in the following way. For example, we might require that if two separate switches S_1 AND S_2 are closed, light bulb L will go on. This is written symbolically as

$$L = S_1 \cdot S_2 \qquad (11.1\text{-}3)$$

where the dot is read AND. Frequently, the dot is omitted, and the AND operation is written simply $Z = AB$. The reader should note that the two forms we have used for writing the AND operation are identical to ordinary algebraic notation for multiplication. This is not meant to imply that they are the same. However, there are some similarities, which will be pointed out as we go along.

A circuit which implements the AND function with switches is shown in Fig. 11.1-2a. From the circuit, it is clear that the light cannot go on unless switches S_1 AND S_2 are *both* closed.

The standard circuit symbol for an AND gate is shown in Fig. 11.1-2b. When switch S_1 is connected to its associated battery, input logic variable A is considered to be **1**; otherwise it is **0**. A similar statement holds for S_2 and input B. The gate is designed so that when both A AND B are **1**, the output L will be **1** and the light will go on.

In order to set down the truth table for the AND function, we note that since there are two variables A and B, there are four possible combinations:

A	B
0	0
0	1
1	0
1	1

From the previous description of the AND function, we see that the combinations in the first three rows of the table all lead to an output $L = 0$, while the

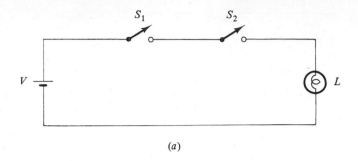

(a)

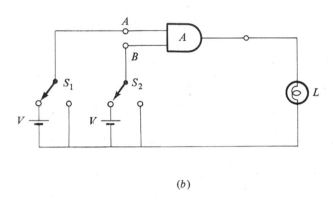

(b)

Figure 11.1-2 The AND function: (a) switching-circuit realization; (b) AND-gate circuit.

fourth row leads to an output $L = 1$. Thus the truth table is as shown in Table 11.1-2. In logic-equation form, the four lines of the truth table are written:

$$0 \cdot 0 = 0$$
$$0 \cdot 1 = 0$$
$$1 \cdot 0 = 0$$
$$1 \cdot 1 = 1$$

The AND function can relate more than two inputs. For example, we might require that if events A AND B AND C are each equal to 1, then X is equal to 1. This is written symbolically as

$$X = A \cdot B \cdot C \tag{11.1-4}$$

Commercial AND gates are often designed for multiple inputs.

Table 11.1-2 Truth table for the AND function

A	B	L
0	0	0
0	1	0
1	0	0
1	1	1

11.1-4 The OR Function

The logic OR function relates two or more logic variables, like the AND function. However, in the case of the OR function we would require that if *either* or both of two separate switches S_1 OR S_2 are closed, light bulb L will go on. This is written

$$L = S_1 + S_2 \qquad (11.1\text{-}5)$$

Here the plus sign is read OR.

A switching circuit which represents the OR function is shown in Fig. 11.1-3a. Clearly the bulb will light if either S_1 OR S_2 OR both are closed. The standard circuit symbol for an OR gate is shown in Fig. 11.1-3b. For the gate, the logic equation relating output L to inputs A and B is

$$L = A + B \qquad (11.1\text{-}6)$$

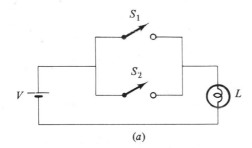

(a)

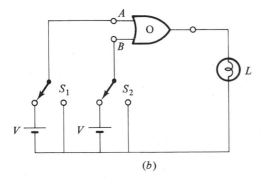

(b)

Figure 11.1-3 The OR function: (*a*) switching circuit; (*b*) OR-gate circuit.

Table 11.1-3 Truth table for the OR function

A	B	L
0	0	0
0	1	1
1	0	1
1	1	1

When either switch 1 or switch 2 or both are connected to their associated batteries, input logic variables A or B or both are considered to be **1** and the output L will be **1**.

The truth table for the OR function is given in Table 11.1-3.

In logic-equation form we have the four possibilities

$$0 + 0 = 0$$

$$0 + 1 = 1$$

$$1 + 0 = 1$$

$$1 + 1 = 1$$

As with the AND function, the OR function can relate more than two logic variables. For example, consider

$$L = A + B + C + D \tag{11.1-7}$$

Here $L = 1$ as long as one or more of the four variables are **1**; $L = 0$ only when all four variables are **0**.

11.1-5 Combinations of the Basic Logic Functions

In a practical circuit, different combinations of gates will appear, and we often have to analyze the circuit in order to determine the logic equation which governs its operation. An example follows.

Example 11.1-1 Consider the circuits shown in Fig. 11.1-4. Find the relation between the output X and the indicated inputs.

SOLUTION

Figure 11.1-4a We begin by noting that in terms of the inputs to the AND gate we have

$$X = B \cdot C \tag{11.1-8}$$

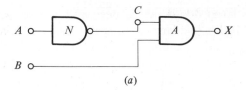

(a)

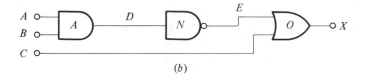

(b)

Figure 11.1-4 Logic circuits for Example 11.1-1.

Noting that the NOT-gate output variable C is related to input A by

$$C = \bar{A} \tag{11.1-9}$$

We substitute this into (11.1-8) and have the final result

$$X = B \cdot \bar{A} \tag{11.1-10}$$

Figure 11.1-4b In this circuit, we define additional variables D and E as shown. The relations covering the individual gates are

$$X = E + C \tag{11.1-11}$$
$$E = \bar{D} \tag{11.1-12}$$
$$D = A \cdot B \tag{11.1-13}$$

Substituting (11.1-13) into (11.1-12), we have

$$E = \bar{D} = \overline{A \cdot B} \tag{11.1-14}$$

Note that the overbar covers both A and B; that is, we must first find A AND B and then complement it. This is substituted into (11.1-11) to yield the final result

$$X = \overline{A \cdot B} + C \qquad /\!/\!/ \tag{11.1-15}$$

In a typical logic design problem we are given a statement in words which we must "translate" into logic equations. Circuits containing AND, OR, and NOT gates are then synthesized to realize the logic functions in the equations. This process is illustrated in the example which follows.

Example 11.1-2 A farmer has a large dog which is part wolf, a goat, and several heads of cabbage. In addition, the farmer owns two barns, a north barn and a south barn. The farmer, dog, cabbages, and goat are all in the south barn. The farmer has chores to perform in both barns. However, if the dog is left with the goat when the farmer is absent, he will bite the goat, and if the goat is left alone with the cabbages, he will eat the cabbages. To avoid either disaster, the farmer asks us to build a small portable computer having four switches, representing the farmer, dog, goat, and cabbage. If a switch is connected to a battery, the character represented by the switch is in the south barn; if the switch is connected to ground, the character is in the north barn. The output of the computer goes to a lamp, which lights if any combination of switches will result in a disaster. Thus the farmer can go about his chores, using the computer to tell him what he must take with him from one barn to another in order to avoid a disaster. How do we build this computer?

SOLUTION To design the computer we must state very precisely what it is we wish to do. We want the lamp to light if any of the following four possibilities occurs:

1. The farmer is in the north barn AND the dog AND the goat are in the south barn, OR if
2. The farmer is in the north barn AND the goat AND the cabbages are in the south barn, OR if
3. The farmer is in the south barn AND the dog AND the goat are in the north barn, OR if
4. The farmer is in the south barn AND the goat AND the cabbages are in the north barn.

We now represent symbolically the farmer's being in the south barn by the letter F. Hence, if the farmer is NOT in the south barn, i.e., if he is in the north barn, he is represented by $\bar{F}$. Similarly, we have

$$D = \text{dog in the south barn}$$

$$\bar{D} = \text{dog in the north barn}$$

$$G = \text{goat in the south barn}$$

$$\bar{G} = \text{goat in the north barn}$$

$$C = \text{cabbages in the south barn}$$

$$\bar{C} = \text{cabbages in the north barn}$$

We can now write the symbolic logic statement which combines all the possibilities leading to a disaster:

$$L = \bar{F} \cdot D \cdot G + \bar{F} \cdot G \cdot C + F \cdot \bar{D} \cdot \bar{G} + F \cdot \bar{G} \cdot \bar{C} \qquad (11.1\text{-}16)$$

where L indicates that the light is ON.

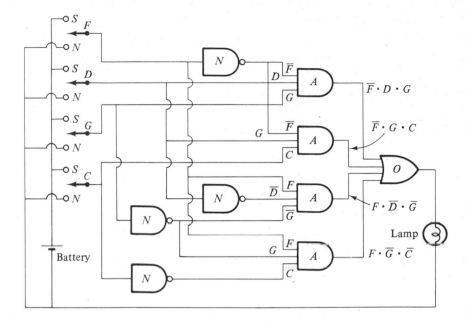

Figure 11.1-5 A special-purpose computer to solve the dog-goat-cabbage problem.

Each term in the equation represents one of the four possible combinations for which we want the lamp to light, and we see that each term will require a three-input AND gate. The outputs of the four AND gates will then be the inputs to a four-input OR gate. In addition, the complement of each of the variables is required, so that we shall need four inverters. A circuit which realizes (11.1-16) is shown in Fig. 11.1-5. The circuit contains a switch for each of the characters F, D, G, and C, a battery, lamp, and the gates mentioned above. The student should verify that the lamp will light only if any one of the four terms in (11.1-16) is true. ///

11.2 BOOLEAN ALGEBRA

Equation (11.1-16) and the resulting circuit of Fig. 11.1-5 are not unique. Using the rules of boolean algebra, it is possible to manipulate the expression for L into many different forms that are equivalent to the original. Each form will, in general, lead to a circuit that looks different but does exactly the same job as the original. The advantages to the designer are numerous: as we shall show later, the designer is free to manipulate the logic equation so that the resulting circuit will contain only one type of gate or to try several designs in order to find the one which uses the fewest gates.

In this section, we set down and illustrate some of the rules of boolean algebra which are most appropriate for the study of electronic logic circuits.

11.2-1 Boolean Theorems

We first present the most important boolean theorems in Table 11.2-1. They are presented in pairs, i.e., with both AND and OR versions of each.

Many of the theorems are identical to ordinary algebra, e.g., Theorems 1, 2, and 3a. Note that the last of these, the *distributive law* Theorem 3a, allows us to multiply through just as we do in ordinary algebra. We can also use it in reverse to *factor* expressions like $XY + XZ = X(Y + Z)$.

The proof of some of the theorems is not obvious and requires explanation. Consider Theorem 3b, for example. This can be proved using the truth table or by algebraic manipulation. We illustrate the algebraic method by expanding the right-hand side. With each step the applicable theorem will be listed.

$$(A + B) \cdot (A + C) = AA + AB + AC + BC \qquad \text{Theorem } 3a$$
$$= A + A(B + C) + BC \qquad \text{Theorems } 3a \text{ and } 4b$$
$$= A(1 + B + C) + BC \qquad \text{factoring and Theorem } 3a$$
$$= A + BC \qquad \text{Theorem } 7c$$

Thus, the theorem is proved.

Table 11.2-1 Boolean algebra theorems

No.	Theorem	Name
1a	$A + B = B + A$	Commutative law
1b	$A \cdot B = B \cdot A$	
2a	$(A + B) + C = A + (B + C)$	Associative law
2b	$(A \cdot B) \cdot C = A \cdot (B \cdot C)$	
3a	$A \cdot (B + C) = A \cdot B + A \cdot C$	Distributive law
3b	$A + (B \cdot C) = (A + B) \cdot (A + C)$	
4a	$A + A = A$	Identity law
4b	$A \cdot A = A$	
5a	$\bar{\bar{A}} = \bar{A}$	Negation
5b	$\bar{\bar{A}} = A$	
6a	$A + A \cdot B = A$	Redundancy
6b	$A \cdot (A + B) = A$	
7a	$0 + A = A$	
7b	$1 \cdot A = A$	
7c	$1 + A = 1$	
7d	$0 \cdot A = 0$	
8a	$\bar{A} + A = 1$	
8b	$\bar{A} \cdot A = 0$	
9a	$A + \bar{A} \cdot B = A + B$	
9b	$A \cdot (\bar{A} + B) = A \cdot B$	
10a	$\overline{A + B} = \bar{A} \cdot \bar{B}$	De Morgan's laws
10b	$\overline{A \cdot B} = \bar{A} + \bar{B}$	

Table 11.2-2 Proof of Theorem 6a

A	B	$A \cdot B$	$A + A \cdot B$
0	0	0	0
0	1	0	0
1	0	0	1
1	1	1	1

Table 11.2-3 Proof of Theorem 9a

A	B	$A + B$	$\bar{A}$	$\bar{A}B$	$A + \bar{A}B$
0	0	0	1	0	0
0	1	1	1	1	1
1	0	1	0	0	1
1	1	1	0	0	1

Next consider Theorem 6a, which states that $A = A + A \cdot B$. This does not resemble anything in our ordinary algebra, but it is easily proved with Theorem 3a since $A + AB = A(1 + B) = B$. In order to illustrate another method of proof we construct the truth table (Table 11.2-2). A column for $A \cdot B$ is added to provide an intermediate step. Since the first and last columns are identical for all possible combinations of A and B, the theorem is proved. This kind of proof is called a proof by *perfect induction*.

Theorems 7 and 8 provide a number of useful relations between arbitrary logic variables and the fixed values **0** and **1**.

Theorems 9a and b are not at all obvious. They can be verified most easily with a truth table, given in Table 11.2-3 for Theorem 9a. The third and sixth columns are identical for all possible combinations, thereby proving the theorem. Proof of Theorem 9b is left as an exercise for the reader.

Our final example will be to prove the first of De Morgan's laws (Theorem 10a) using a truth table (Table 11.2-4). Separate columns are provided for intermediate steps in the comparison. The fourth and seventh columns are identical for all possible combinations, proving that $\overline{A + B} = \overline{A} \cdot \overline{B}$.

The reader will note that each of the theorems is given in two different forms, called *duals*. In general, the dual of a theorem is obtained by interchanging AND and OR operations and also interchanging 1s and 0s if they are present.

The boolean theorems are used to manipulate logic equations like (11.1-16) into different forms which are completely equivalent to each other. Each form then leads to a different circuit each of which realizes exactly the same relation between input and output. The designer can then choose among the various forms to satisfy the requirements of the problem at hand. Typical practical requirements

Table 11.2-4 Proof of Theorem 10a

A	B	$A + B$	$\overline{A + B}$	$\bar{A}$	$\bar{B}$	$\bar{A} \cdot \bar{B}$
0	0	0	1	1	1	1
0	1	1	0	1	0	0
1	0	1	0	0	1	0
1	1	1	0	0	0	0

involve choosing the circuit using the minimum number of gates or only one type of gate.

The manipulation of logic equations is illustrated in the following examples.

Example 11.2-1 Simplify the logic equation

$$L = \bar{X}Y + XY + \bar{X}\bar{Y}$$

SOLUTION The steps are carried out in detail, with the applicable theorem noted in each case:

$$
\begin{aligned}
L &= Y(\bar{X} + X) + \bar{X}\bar{Y} & &\text{Theorem } 3a \\
&= Y(1) + \bar{X}\bar{Y} = Y + \bar{X}\bar{Y} & &\text{Theorem } 8a \\
&= Y + \bar{X} & &\text{Theorem } 9a
\end{aligned}
$$

The reduced equation is considerably simpler than the original equation.

///

Example 11.2-2 Frequently, alternate forms of an expression result in better circuit implementation. Find an alternate form for

$$L = \overline{X + YZ}$$

SOLUTION Here De Morgan's laws are used in the following way. Let $YZ = A$. Then, by Theorem 10a, we have

$$L = \overline{X + A} = \bar{X}\bar{A} = \bar{X}(\overline{YZ})$$

Now, using Theorem 10b, we have $\overline{YZ} = \bar{Y} + \bar{Z}$, so that, finally,

$$L = \bar{X}(\bar{Y} + \bar{Z}) \qquad\qquad ///$$

Example 11.2-3 If $L = \bar{X}Y + X\bar{Y}$, find an expression for $\bar{L}$.

SOLUTION

$$
\begin{aligned}
\bar{L} &= \overline{\bar{X}Y + X\bar{Y}} \\
&= (\overline{\bar{X}Y})(\overline{X\bar{Y}}) & &\text{Theorem } 10a \\
&= (\bar{\bar{X}} + \bar{Y})(\bar{X} + \bar{\bar{Y}}) & &\text{Theorem } 10b \\
&= (X + \bar{Y})(\bar{X} + Y) & &\text{Theorem } 5b \\
&= X\bar{X} + XY + \bar{Y}\bar{X} + \bar{Y}Y & &\text{Theorem } 3a \\
&= XY + \bar{Y}\bar{X} & &\text{Theorem } 8b \qquad ///
\end{aligned}
$$

11.3 THE NAND AND NOR FUNCTIONS

Electronic switches (gates) do not readily perform the OR and AND logic operations, but most commercially available gates do perform the combined operations AND-NOT (NAND) and OR-NOT (NOR).

Table 11.3-1 Truth table for the NAND operation

A	B	Output L	A · B
0	0	1	0
0	1	1	0
1	0	1	0
1	1	0	1

11.3-1 The NAND Function

The NAND function provides a logic **0** output only when all inputs are logic **1**. The truth table is given in Table 11.3-1. From the table we see that the output column is the complement of the fourth column, where the AND function AB is listed. Thus, for the NAND function we have

$$L = \overline{AB} \tag{11.3-1}$$

The word NAND is clearly a contraction of NOT-AND. The NAND operation is illustrated in Fig. 11.3-1a, as is the standard symbol for the NAND gate. The standard symbol is the same as for the AND gate; the small circle at the output represents the NOT operation.

When the NAND operation is applied to more than two variables, it is written

$$L = \overline{ABCD \cdots} \tag{11.3-2}$$

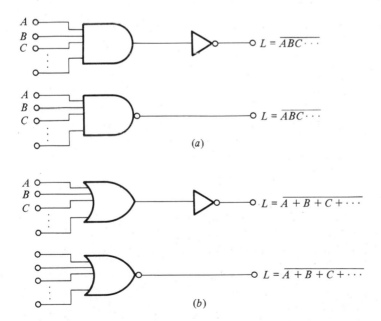

(a)

(b)

Figure 11.3-1 Circuit symbols: (a) NAND gate; (b) NOR gate.

Table 11.3-2 Truth table for the NOR operation

A	B	Output L	$A + B$
0	0	1	0
0	1	0	1
1	0	0	1
1	1	0	1

The NAND operation is commutative; i.e.,

$$L = \overline{ABC} = \overline{BAC} = \cdots \tag{11.3-3}$$

However, it is not associative, since (see Prob. 11.3-1)

$$\overline{(\overline{AB})C} \neq \overline{A(\overline{BC})} \tag{11.3-4}$$

11.3-2 The NOR Function

As we might expect, there is an operation dual to the NAND operation called the NOR operation, for which the output is logic **0** when any one or more inputs are logic **1**. The truth table is given in Table 11.3-2. From the table we see that the output column is the complement of the fourth column, where the OR function $A + B$ is listed. Thus, for the NOR function

$$L = \overline{A + B} \tag{11.3-5}$$

The word NOR is a contraction for NOT-OR, and the symbol for a NOR gate is shown in Fig. 11.3-1b.

As with the NAND gate, the symbol is the same as for the OR gate with a small circle at the output which represents the NOT operation.

Like the NAND operation, the NOR operation is commutative; i.e.,

$$L = \overline{A + B + C + \cdots} = \overline{B + A + C + \cdots} \tag{11.3-6}$$

and again it is not associative, since (see Prob. 11.3-2)

$$\overline{\overline{A + B} + C} \neq \overline{A + \overline{B + C}} \tag{11.3-7}$$

When more than two variables are ORed, the meaning is as given in (11.3-6).

In later sections we shall discuss synthesis methods which lead to circuits involving only NAND and/or NOR gates.

11.3-3 The EXCLUSIVE-OR Function

A function which arises frequently is given by the truth table in Table 11.3-3. It is called the EXCLUSIVE-OR operation and is written symbolically as

$$L = A \oplus B \tag{11.3-8}$$

Table 11.3-3 Truth table for the EXCLUSIVE-OR function

A	B	Output L
0	0	0
0	1	1
1	0	1
1	1	0

In words, the output L is logic 1 if input A or input B is logic 1 *exclusively*, that is, when they are not 1 simultaneously.

We can express the EXCLUSIVE-OR in terms of the previously defined NOT, AND, and OR functions by the following line of reasoning based on the truth table. We first form all possible combinations that lead to $L = 1$. They are derived from the second and third lines of the truth table, i.e., lines for which $L = 1$. The conditions from the second line are $A = 0$ AND $B = 1$, which will yield 1 when written in logic form as $\bar{A}B$, while those from the third line are $A = 1$ AND

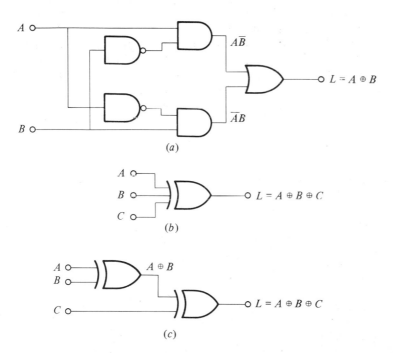

Figure 11.3-2 (*a*) A circuit which realizes the EXCLUSIVE-OR operation; (*b*) symbol for EXCLUSIVE-OR gate; (*c*) arrangement which accommodates more than two inputs.

$B = 0$, which yield **1** when written as $A\bar{B}$. Note that either the second line OR third line leads to the desired $L = 1$, so that we can write the desired relation in logic-equation form as

$$L = \bar{A}B + A\bar{B} \tag{11.3-9}$$

The reader should construct a truth table from this equation to verify that it does indeed meet the specifications for the EXCLUSIVE-OR function.

The EXCLUSIVE-OR function is both commutative and associative, so that the notation

$$L = A \oplus B \oplus C \oplus D \cdots \tag{11.3-10}$$

can be used without parentheses to indicate grouping.

One possible circuit which realizes the EXCLUSIVE-OR operation using NOT, AND, and OR gates is shown in Fig. 11.3-2a. The standard symbol for the EXCLUSIVE-OR gate is shown in Fig. 11.3-2b. In practice, EXCLUSIVE-OR gates with more than two inputs are not available, and so the arrangement of Fig. 11.3-2c is used to accommodate additional inputs.

11.3.4 Expressions with all NAND or all NOR Operations

In practice it is often convenient to design logic circuits using only one type of gate, i.e., NAND or NOR. This is particularly true when using ICs. By proper use of De Morgan's theorems all possible logic functions can be manipulated into forms which are easily synthesized using only one type of gate. Let us begin by considering only NAND logic. The AND operation is obtained by using a NAND gate followed by an inverter, as shown in Fig. 11.3-3. For the OR operation we use De Morgan's theorem as follows

$$L = \overline{\bar{A} \cdot \bar{B} \cdot \bar{C}} = \bar{\bar{A}} + \bar{\bar{B}} + \bar{\bar{C}} = A + B + C \tag{11.3-11}$$

The corresponding circuit is shown in Fig. 11.3-3b.

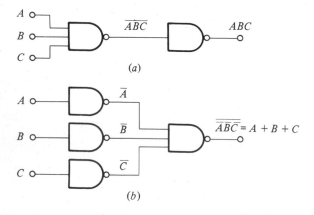

Figure 11.3-3 NAND-only logic: (a) the AND operation; (b) the OR operation.

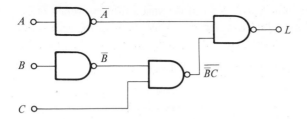

Figure 11.3-4 NAND-only implementation for $L = A + \bar{B}C$.

In an analogous fashion the NOR gate can be used to synthesize both AND and OR functions (see Prob. 11.3-3).

To illustrate this technique for a more complicated expression, consider expressing in terms of only NAND operations

$$L = A + \bar{B}C \tag{11.3-12}$$

Using De Morgan's theorems, we have

$$\bar{L} = \overline{A + \bar{B}C} = (\bar{A})(\overline{\bar{B}C})$$

Then

$$\bar{\bar{L}} = L = \overline{(\bar{A})(\overline{\bar{B}C})} \tag{11.3-13}$$

The corresponding circuit, using only NAND gates, is shown in Fig. 11.3-4.

Since these problems can become quite involved, systematic procedures are useful. One such procedure is discussed in the next section.

11.4 STANDARD FORMS FOR LOGIC FUNCTIONS

11.4-1 Sum of Products   *written from truth table "1's"*

In the first standard form we shall consider, the *sum-of-products* form, the logic function is written as a simple sum of terms. For example, consider the function

$$L = (\bar{W} + XY)(X + YZ) \tag{11.4-1}$$

In order to express this as a sum of products we expand the expression using Theorem 3a:

$$L = (\bar{W} + XY)X + (\bar{W} + XY)YZ = \bar{W}X + XXY + \bar{W}YZ + XYYZ$$

$$= \bar{W}X + XY + \bar{W}YZ + XYZ \tag{11.4-2}$$

This is the desired sum of products.

Sometimes De Morgan's laws must be used. For example, consider

$$L = (\overline{AB} + C)\bar{D} \tag{11.4-3}$$

Using Theorem 3a, we get

$$L = (\overline{AB})\bar{D} + C\bar{D} \tag{11.4-4}$$

This is not yet in the desired sum-of-products form because of the presence of the term $\overline{AB}$, where the combination AB is complemented. We use De Morgan's law $\overline{AB} = \bar{A} + \bar{B}$ to get

$$L = (\bar{A} + \bar{B})\bar{D} + C\bar{D} = \bar{A}\bar{D} + \bar{B}\bar{D} + C\bar{D} \tag{11.4-5}$$

This is the desired form.

The foregoing examples indicate that it is always possible to manipulate a logic expression into the form of a simple sum of terms in which each term is a product of some combination of the variables. Some of the variables may be complemented. The same variable will never appear twice in any term because if it does, we eliminate the repetition using $AA = A$ or $\bar{A}\bar{A} = \bar{A}$. If the form $A\bar{A}$ appears in any term, that term is identically zero and can be dropped.

Note that not every term in (11.4-2) and (11.4-5) involves all the variables. A further standardization leads to an expression in which *all* variables appear in each term, either complemented or uncomplemented. This is often called the *expanded sum-of-products form*. It arises naturally when the logic equation is derived from a truth table, as illustrated in the example which follows.

Example 11.4-1 Find the logic equation for L as described by the truth table

Row	X	Y	Z	L
1	0	0	0	0
2	0	0	1	0
3	0	1	0	1
4	0	1	1	1
5	1	0	0	1
6	1	0	1	0
7	1	1	0	1
8	1	1	1	1

SOLUTION The rows of the truth table have been numbered for convenience. We see that $L = 1$ for the conditions of rows 3, 4, 5, 7, and 8. Consider row 3. From this row we see that $L = 1$ if $X = 0$ AND $Y = 1$ AND $Z = 0$. These three conditions can be combined into one expression

$$\bar{X}Y\bar{Z} = 1 \tag{11.4-6}$$

The reader should check that this expression is **1** only when $X = 0$, $Y = 1$, and $Z = 0$. For the conditions of row 4 we have

$$\bar{X}YZ = 1 \tag{11.4-7}$$

Note that either row 3 OR row 4 leads to $L = 1$; thus, considering only these two rows, we can write

$$L = \bar{X}Y\bar{Z} + \bar{X}YZ$$

where the two terms on the right represent the third and fourth rows of the truth table.

Rows 5, 7, and 8 lead to the terms $X\bar{Y}\bar{Z}$, $XY\bar{Z}$, and XYZ, respectively. If any one of these is **1**, L will be **1**. Thus they are simply ORed with the expression for rows 3 and 4. The final logic equation is

$$L = \bar{X}Y\bar{Z} + \bar{X}YZ + X\bar{Y}\bar{Z} + XY\bar{Z} + XYZ \qquad (11.4\text{-}8)$$

This is the expanded sum-of-products form. Each individual term is called a *minterm*.

The expression given by (11.4-8) can be written directly from the truth table by noting that the variables within each row having $L = 1$ are connected by the AND function (any variable appearing as a **1** is left unchanged, and each variable appearing as **0** in the table is complemented) and the terms for each row are connected by the OR function.

The expanded sum-of-products form is often more complex than necessary. For example, the expression for L in (11.4-8) can be reduced to the much simpler form (see Prob. 11.4-3)

$$L = Y + X\bar{Z} \qquad (11.4\text{-}9)$$

The expanded sum-of-products form is useful for synthesis methods to be discussed in a later section. ///

Written from "0's rows. Take opposite of each term and put + sign between them. AND" the terms from each row.

11.4-2 Product of Sums

The alternative *product-of-sums* form consists of a product of terms in which each term consists of a *sum* of all or part of the variables. It can be arrived at in a number of ways. We illustrate the algebraic method by considering (11.4-1). In order to convert the products XY and YZ into sums of individual variables we use Theorem 3b. This yields

$$L = (\bar{W} + XY)(X + YZ) = (\bar{W} + X)(\bar{W} + Y)(X + Y)(X + Z) \qquad (11.4\text{-}10)$$

Each factor of the product contains only two of the four variables.

The *expanded product-of-sums form* has all variables appearing in each factor. This form can be obtained directly from the truth table in a manner analogous to that used for the sum-of-products form. Refer to the truth table in Example 11.4-1. We now consider rows 1, 2, and 6, for which $L = 0$. For each of these rows a sum of terms is formed (if a variable has the value **1**, it is complemented, while if the variable has the value **0**, it is left unchanged). Thus for row 1 we have the sum $X + Y + Z$, that is, when $X = 0$ and $Y = 0$ and $Z = 0$ simultaneously, $L = 0$. Similarly for row 2 the sum is $X + Y + \bar{Z}$, and for row 6 it is $\bar{X} + Y + \bar{Z}$. When any one of these sums is **0**, that is, each term in the sum is **0**, we must have $L = 0$. Hence, the three terms must be ANDed together, leading to the logic equation in expanded product-of-sums form

$$L = (X + Y + Z)(X + Y + \bar{Z})(\bar{X} + Y + \bar{Z}) \qquad (11.4\text{-}11)$$

Each of the three terms is called a *maxterm*. The reader should verify that the values of X, Y, and Z in each row of the truth table lead to the proper value of L, either **0** or **1**.

To summarize this section, we note that the standard sum-of-products form indicates by inspection the combinations of variable values which lead to $L = 1$ while the standard product-of-sums form shows those combinations which lead to $L = 0$. The sum-of-products form is expressed in terms of *minterms*, only one of which must be **1** in order to have $L = 1$. The product-of-sums form is expressed in terms of *maxterms*, only one of which must be **0** in order to have $L = 0$.

Equations (11.4-8) and (11.4-11) are different forms for the same logic function L which must be consistent. This consistency is easily verified by substituting values of X, Y, and Z from each row of the truth table into the two forms. Note that we have three variables, so that there are $2^3 = 8$ possible combinations. From the truth table we see that five of these lead to $L = 1$, and thus we have five minterms in (11.4-8). The other three combinations lead to $L = 0$ and appear as the maxterms in (11.4-11).

11.4-3 Synthesis Using Standard Expressions

Consider the sum-of-products expression

$$L = \bar{A}\bar{D} + \bar{B}\bar{D} + C\bar{D} \tag{11.4-5}$$

The logic circuit which results is shown in Fig. 11.4-1. We see that it consists of a number of AND gates equal to the number of terms followed by a single OR gate. This type of circuit is often referred to as a two-level AND-OR circuit because the inputs first go through the AND gates (called the first level) and then through the OR gate at the second level. Clearly, all sum-of-products equations lead to similar two-level AND-OR structures.

If we begin with a product-of-sums expression, the resulting circuit will be a two-level OR-AND structure, where the first-level gates are OR gates and the second-level gate is an AND gate.

The two-level gates described above may not always provide the most efficient or economical realization of the given expression but they are easily designed and have smaller propagation delay times than higher-level structures.

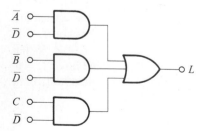

Figure 11.4-1 Logic circuit for $L = \bar{A}\bar{D} + \bar{B}\bar{D} + C\bar{D}$.

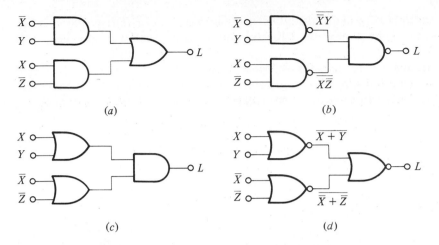

Figure 11.4-2 NAND-only or NOR-only structures: (*a*) AND-OR circuit; (*b*) corresponding circuit with NAND only; (*c*) OR-AND circuit; (*d*) corresponding circuit with NOR only.

11.4-4 Synthesis Using Only NAND or NOR gates[1,2]

In Sec. 11.3 we illustrated a procedure for logic-circuit design using only NAND or NOR gates. In this section we shall show that conversion of two-level AND-OR or OR-AND circuits into NAND only or NOR only is an extremely simple step. Consider, for example, the sum-of-products expression

$$L = XY + XZ \tag{11.4-12}$$

This is synthesized in two-level form as shown in Fig. 11.4-2*a*. Next we apply De Morgan's theorem to (11.4-12) to get

$$\bar{L} = \overline{XY + X\bar{Z}} = (\overline{XY})(\overline{X\bar{Z}}) \tag{11.4-13a}$$

and therefore

$$L = \bar{\bar{L}} = \overline{(\overline{XY})(\overline{X\bar{Z}})} \tag{11.4-13b}$$

Equation 11.4-13*b* is synthesized using only NAND gates in Fig. 11.4-2*b*. The resulting NAND-only circuit is exactly the same as the AND-OR circuit except that every gate is a NAND gate.

The product-of-sums expression for (11.4-12) is

$$L = (X + Y)(\bar{X} + \bar{Z}) \tag{11.4-14}$$

This is synthesized in OR-AND form in Fig. 11.4-2*c*.

Using De Morgan's theorem, we can manipulate this as follows:

$$\bar{L} = \overline{(X + Y)(\bar{X} + \bar{Z})} = \overline{(X + Y)} + \overline{(\bar{X} + \bar{Z})} \tag{11.4-15a}$$

Hence

$$L = \bar{\bar{L}} = \overline{\overline{(X + Y)} + \overline{(\bar{X} + \bar{Z})}} \tag{11.4-15b}$$

This is synthesized using only NOR gates in Fig. 11.4-2d. The resulting NOR circuit is exactly the same as the OR-AND circuit except that every gate is a NOR gate.

The design procedure can now be set down.

To design a NAND-only circuit convert the desired logic expression to sum-of-products form then draw the corresponding two-level AND-OR circuit and change all gates to NAND.

To design a NOR-only circuit convert the desired logic expression to product-of-sums form then draw the corresponding two-level OR-AND circuit and change all gates to NOR.

11.5 KARNAUGH MAPS

As we found in the last section, logic equations come in all shapes and sizes, and their algebraic simplification is not always easy. Considerable experience is required before facility can be gained in the simplification process. The *Karnaugh map* provides a graphical technique for reducing logic equations to a minimal form. It may be used for any number of variables, but we shall not go beyond four.

The Karnaugh map is an array of cells which contains all the information in the truth table arranged in a way that allows a quick visual simplification of the logic equation according to some very simple rules. The simplest map involves two variables and is shown in Fig. 11.5-1a. The map contains four cells, one for each of the possible combinations of the variables, and thus one for each row of the truth table. In Fig. 11.5-1b we have shown the variable combinations in the individual cells of the Karnaugh map and the logic-equation terms which would result from a **1** in each row of the truth table. It is seen that each row of the truth table corresponds to exactly one cell of the Karnaugh map. In use, we place a **1** in each cell of the map for which the corresponding truth-table row would be **1** and **0** in each cell corresponding to a row in the truth table which is equal to **0**. Usually the zeros are not explicitly written, and an empty cell is assumed to contain a zero. With these designations a map can be read in exactly the same way as a truth table in order to write the logic equation, as shown in Fig. 11.5-1c. If a cell contains a **1**, the term in the equation corresponding to that cell will contain all the variables ANDed together; if the row or column of the map in which the **1** appears is headed by a **0** in the variable, the complement of that variable appears in the term; otherwise the variable appears uncomplemented. For example, in Fig. 11.5-1c a **1** appears in the cell formed by the intersection of the column $A = 0$ AND the row $B = 1$. Hence that cell is represented by the term $\bar{A} \cdot B$. The individual terms, one for each cell containing a **1**, are ORed together to form the final logic equation.

The usefulness of the Karnaugh map lies in the fact that adjacent cells can be grouped visually to eliminate redundant variables. For example, consider the map in Fig. 11.5-2a. The combination of 1s shown would lead us to write the logic equation for L_1 as

$$L_1 = A\bar{B} + AB \tag{11.5-1}$$

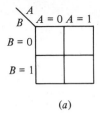

(a)

Truth table

A	B	Term in logic equation equal to 1
0	0	$\overline{A}\overline{B}$
0	1	$\overline{A}B$
1	0	$A\overline{B}$
1	1	AB

(b)

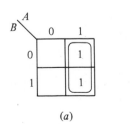

K map for $L = \overline{A}B + A\overline{B} + AB$

Truth table

A	B	L
0	0	0
0	1	1
1	0	1
1	1	1

(c)

Figure 11.5-1 Karnaugh maps: (a) two-variable map; (b) correspondence with truth table; (c) example.

This can be simplified as

$$L_1 = A(\overline{B} + B) = A \tag{11.5-2}$$

Thus the *grouping* of the two cells, as shown, immediately leads to the simplified expression $L_1 = A$, which contains only one variable.

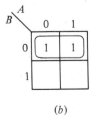

(a)   (b)

Figure 11.5-2 Karnaugh maps: (a) $L_1 = A$; (b) $L_2 = \overline{B}$.

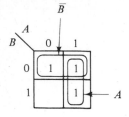

Figure 11.5-3 Karnaugh map for $L_3 = \bar{B} + BA$.

For the map in Fig. 11.5-2*b* we read immediately that $L_2 = \bar{B}$. In order to prove that this simple expression is indeed correct we proceed as before by writing the complete expression

$$L_2 = \bar{A}\bar{B} + A\bar{B} = \bar{B}(\bar{A} + A) = \bar{B} \qquad (11.5\text{-}3)$$

For a more interesting case, consider the logic equation

$$L_3 = \bar{B} + BA \qquad (11.5\text{-}4)$$

Let us see whether we can use a Karnaugh map to simplify this expression. The map is shown in Fig. 11.5-3. Both a horizontal and a vertical grouping are possible, as shown. The horizontal grouping shows that $L_3 = 1$ when $B = 0$, that is, when $\bar{B} = 1$, and the vertical grouping indicates that $L_3 = 1$ when $A = 1$; hence

$$L_3 = \bar{B} + A \qquad (11.5\text{-}5)$$

This is readily checked as being a statement of Theorem 9*a*.

A set of rules for simplification using a two-variable map can now be listed:

1. A group of two adjacent cells combines to yield a single variable.
2. A single cell which cannot be combined represents a two-variable term.

It is permissible for groups to overlap because of the fact that, in boolean algebra, $A + A = A$.

11.5-1 Three-Variable Maps

When the map involves three variables, each cell represents the logic product of all three variables, as shown in Fig. 11.5-4*a*. The nonbinary ordering of the AB axis is required so that only one variable at a time changes between adjacent cells, a condition which makes the visual simplification by grouping of individual cells possible. It is important to note that the map is continuous in the sense shown in Fig. 11.5-4*b*, where we have surrounded the primary map by two auxiliary maps

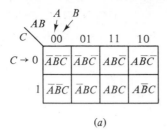

(a)

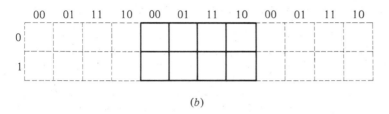

(b)

Figure 11.5-4 Three-variable Karnaugh maps: (a) primary map; (b) primary map (solid lines) with auxiliary maps (dashed lines).

each of which contains the same array of **1**s and **0**s as the primary map. In practice we seldom bother to draw the auxiliary maps, but we must always remember that they are present.

The rules for simplification of the three-variable map are as follows:

1. A group of four adjacent cells (in-line or square) combines to yield a single variable.
2. A group of two adjacent cells combines to yield a two-variable term.
3. A single cell which cannot be combined represents a three-variable term.

The use of these rules is illustrated by the maps shown in Fig. 11.5-5. Figure 11.5-5a and b illustrates the first rule; in Fig. 11.5-5b the continuous nature of the map appears. We note that in all four cells where the **1** appears $B = 0$ and that in no other cell is $B = 0$. Note also that in these four cells A and C take on both values **0** and **1**; hence $L = 1$ when $B = 0$ or $L = \bar{B}$. Rules 2 and 3 are illustrated in Fig. 11.5-5c.

11.5-2 Four-Variable Maps

A four-variable map has 16 cells, as shown in Fig. 11.5-6a. The grouping rules are as follows:

1. Eight adjacent cells yield a single variable.
2. Four adjacent cells yield a two-variable term.

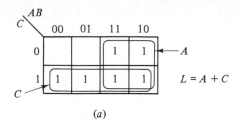

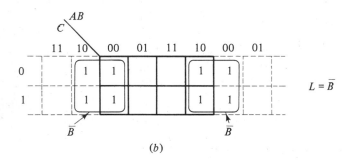

$L = A + C$

(a)

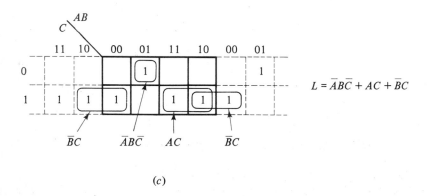

$L = \bar{B}$

(b)

$L = \bar{A}B\bar{C} + AC + \bar{B}C$

(c)

Figure 11.5-5 Three-variable maps: (a) four-cell grouping; (b) map continuity; (c) two-cell groupings and single cell.

3. Two adjacent cells yield a three-variable term.
4. Individual cells represent four-variable terms.

The four-variable map is continuous from the left edge to the right edge, as was the three-variable map, and is also continuous from the top edge to the bottom edge, as illustrated in Fig. 11.5-6b. Use of the four-variable map is illustrated in the following examples.

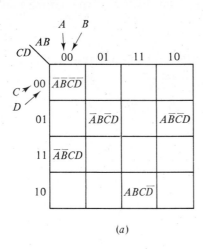

(a)

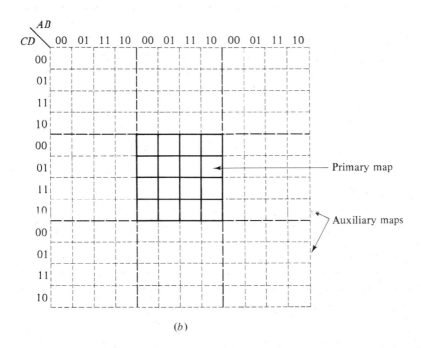

(b)

Figure 11.5-6 Four-variable maps: (a) primary map; (b) continuity with auxiliary maps.

Example 11.5-1 Write the logic equation for the Karnaugh map shown in Fig. 11.5-7a.

SOLUTION The grouping which results in the minimum number of terms is shown in Fig. 11.5-7b. The resulting logic expression is

$$L = \bar{A}\bar{B}\bar{C}\bar{D} + A\bar{B}D + AB\bar{D} + CD \qquad\qquad ///$$

Example 11.5-2 Use the Karnaugh map to simplify the expression

$$L = \bar{A}\bar{B}C\bar{D} + A\bar{B}\bar{C}\bar{D} + \bar{A}BC + \bar{A}C\bar{D} + \bar{A}\bar{C}\bar{D}$$

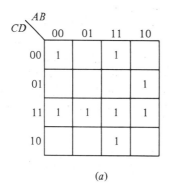

(a)

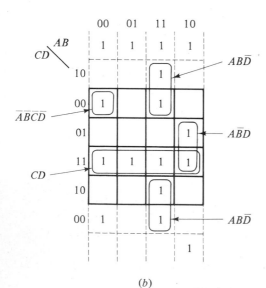

(b)

Figure 11.5-7 Karnaugh map for Example 11.5-1: (a) map; (b) groupings.

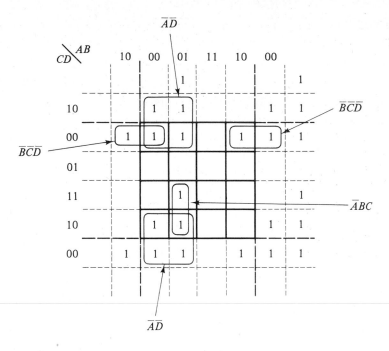

Figure 11.5-8 Karnaugh map for Example 11.5-2.

SOLUTION The map is shown in Fig. 11.5-8, and the indicated grouping leads to the simplified expression

$$L = \bar{A}\bar{D} + \bar{A}BC + \bar{B}\bar{C}\bar{D} \qquad ///$$

Often a map will lead to several possible minimum groupings, all of which lead to expressions which appear about the same. When this happens, the designer will have to base his choice on some other criteria, and often there are several "optimum" solutions.

11.6 DESIGN EXAMPLE: A VOTING MACHINE

Specifications for logic circuits, i.e., the statement of the problem, are often given in word statements. These statements must then be translated either into a truth table or logic equation from which the desired circuit can be designed using the principles set forth in previous sections. As an example of this process consider the following problem.

In a certain corporation the board members own all the stock, which is distributed as follows:

A owns 45 percent
B owns 30 percent
C owns 15 percent
D owns 10 percent

Each member has a percentage vote equal to his holdings, and a total vote greater than 50 percent is required to pass a motion.

We have been asked to design an electronic voting system for the corporation. In the board room each member is to have a switch with which to indicate a YES or NO vote. A lamp is to light if the total vote cast is more than 50 percent, indicating that the measure being voted on is passed.

In order to design such a voting system we proceed as follows:

1. Write the truth table
2. Convert the truth table into a Karnaugh map
3. Draw the logic circuit using (1) NAND gates only and (2) NOR gates only.

Truth table A NO vote is signified as **0** (open switch) and a YES vote as **1** (closed switch). We have four inputs, A, B, C, D, and one output L which is **1** (ON) when the vote is more than 50 percent. The truth table is shown in Table 11.6-1, and the percentage YES vote for each row is included along with each member's percentage.

Table 11.6-1 Truth table showing when vote exceeds 50%

Row	45% A	30% B	15% C	10% D	L	%
0	0	0	0	0	0	0
1	0	0	0	1	0	10
2	0	0	1	0	0	15
3	0	0	1	1	0	25
4	0	1	0	0	0	30
5	0	1	0	1	0	40
6	0	1	1	0	0	45
7	0	1	1	1	1	55
8	1	0	0	0	0	45
9	1	0	0	1	1	55
10	1	0	1	0	1	60
11	1	0	1	1	1	70
12	1	1	0	0	1	75
13	1	1	0	1	1	85
14	1	1	1	0	1	90
15	1	1	1	1	1	100

Combinations leading to a vote greater than 50 percent can be set down by inspection using common sense, but the truth table provides a well-organized procedure for this process and gives us a means of handling problems too complex to be solved by inspection. From the truth table we see that there are eight combinations of the four input variables that lead to $L = 1$ (the motion's being passed). Any one of these combinations is sufficient to pass a motion. Thus, if the first one of these combinations, OR the second, OR the third, etc., is realized, the lamp will light. Our logic equation for L will thus be a sum of products containing eight terms, each one corresponding to one of the lines of the truth table for which $L = 1$.

Next consider the first combination, which leads to $L = 1$, that is, $A = 0$ AND $B = 1$ AND $C = 1$ AND $D = 1$. This combination will yield the logic value 1 when written as $\bar{A}BCD$. The second combination is $A = 1$ AND $B = 0$ AND $C = 0$ AND $D = 1$, which is written $A\bar{B}\bar{C}D$, and so on for all eight combinations. The complete equation is

$$L = \bar{A}BCD + A\bar{B}\bar{C}D + A\bar{B}C\bar{D} + A\bar{B}CD + AB\bar{C}\bar{D}$$

$$+ AB\bar{C}D + ABC\bar{D} + ABCD \tag{11.6-1}$$

Karnaugh map Instead of attempting to simplify this expression algebraically we transfer the information from the truth table to the Karnaugh map shown in Fig. 11.6-1.

Simplified equation The logic expression derived from the map is

$$L = BCD + AB + AC + AD = BCD + A(B + C + D) \tag{11.6-2}$$

This agrees with commonsense reasoning when one considers the percentages. Either B, C, and D all vote YES, OR A and any one other vote YES in order to pass a measure.

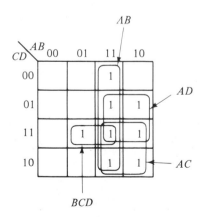

Figure 11.6-1 Karnaugh map for voting-machine example.

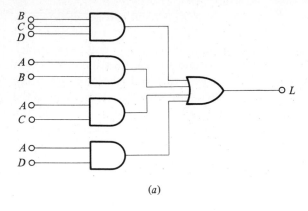

(a)

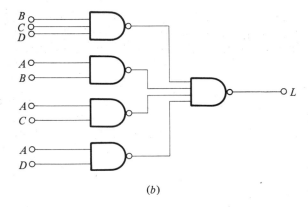

(b)

Figure 11.6-2 Voting-machine circuit: (a) AND-OR; (b) NAND-only.

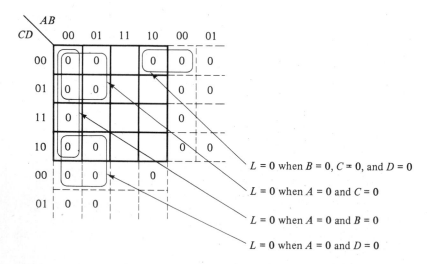

Figure 11.6-3 Karnaugh map from which the product-of-sums expression is obtained.

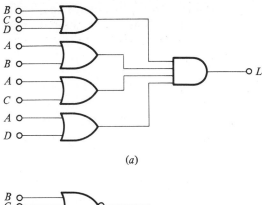

(a)

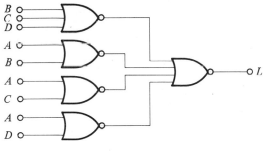

(b)

Figure 11.6-4 Voting-machine circuit: (a) OR-AND; (b) NOR-only.

Logic circuit The logic circuit is shown using two-level AND-OR logic in Fig. 11.6-2a and using NAND gates in Fig. 11.6-2b. In order to obtain the logic circuit using NOR gates only we return to the Karnaugh map of Fig. 11.6-1, which is redrawn in Fig. 11.6-3 to show the cells where $L = 0$. Two auxiliary maps are drawn to aid in the combining of the cells. This results in four terms for which $L = 0$: $B = 0$, $C = 0$, and $D = 0$; $A = 0$ and $C = 0$; $A = 0$ and $B = 0$; and $A = 0$ and $D = 0$. Thus expressing L as a product of sums yields

$$L = (B + C + D)(A + C)(A + B)(A + D) \qquad (11.6\text{-}3)$$

The circuit describing (11.6-3) is shown in Fig. 11.6-4a using a two-level OR-AND configuration and in Fig. 11.6-4b using NOR gates only.

11.7 THE BINARY NUMBER SYSTEM

In the binary number system just two digits are used, **0** and **1**. Because these two digits can be identified directly with the ON and OFF states of the logic variables discussed earlier in this chapter, the binary system becomes the basic number system of the digital computer.

One way to introduce the concept of a number system is to explain the underlying principles in terms of the decimal system, with which we are all familiar. If we take the decimal number 634.72 and break it down into its fundamental components, we write

$$N = 634.72 = (6 \times 10^2) + (3 \times 10^1) + (4 \times 10^0) + (7 \times 10^{-1}) + (2 \times 10^{-2})$$
$$(11.7\text{-}1)$$

When the number is written this way, we see the significance of the position occupied by each digit. We also see that the number which specifies this position relative to the decimal point is a power of 10. Thus 10 is the fundamental unit of the decimal system, called the *base*. The system requires 10 symbols, which we know as the digits 0, 1, 2, ..., 8, 9.

In the binary system, the base is 2 and the symbols are **0** and **1** These digits are thus the coefficients of powers of 2. For example, consider the binary number **1011.01**. This is expanded as

1011.01

$$
\begin{aligned}
&= (1 \times 2^3) + (0 \times 2^2) + (1 \times 2^1) + (1 \times 2^0) + (0 \times 2^{-1}) + (1 \times 2^{-2}) \\
&= \quad 8 \quad + \quad 0 \quad + \quad 2 \quad + \quad 1 \quad + \quad 0 \quad + \quad \tfrac{1}{4} \\
&= 11.25 \text{ (decimal)} \tag{11.7-2}
\end{aligned}
$$

In the binary system the separation between positive and negative exponents is called the *binary point*. If it is necessary to indicate the base of a number explicitly, we use a subscript. For example, from (11.7-2) we have $1011.01_2 = 11.25_{10}$.

Several methods exist for converting from decimal to binary and vice versa and will be explored in the problems. Table 11.7-1 gives the equivalent decimal and binary numbers for numbers containing five binary digits (bits). These numbers range from **00000** = 0 to **11111** = 31. In general the largest decimal number N that can be represented by a binary number containing B bits is

$$N = 2^B - 1 \tag{11.7-3}$$

Thus, in the above example where $B = 5$, $N = 2^5 - 1 = 31$.

Table 11.7-1 Conversion table for decimal to binary

Decimal	Binary	Decimal	Binary	Decimal	Binary	Decimal	Binary
0	00000	8	01000	16	10000	24	11000
1	00001	9	01001	17	10001	25	11001
2	00010	10	01010	18	10010	26	11010
3	00011	11	01011	19	10011	27	11011
4	00100	12	01100	20	10100	28	11100
5	00101	13	01101	21	10101	29	11101
6	00110	14	01110	22	10110	30	11110
7	00111	15	01111	23	10111	31	11111

11.7-1 Octal and Hexadecimal Numbers

Systems with base 8 (octal) and base 16 (hexadecimal) are used in many digital computers because they are easily related to the binary system. In octal the eight required digits are 0 to 7, and a typical number is

$$241_8 = 2 \times 8^2 + 4 \times 8^1 + 1 \times 8^0 = 128 + 32 + 1 = 161_{10}$$

In hexadecimal, 16 symbols are required, $0, 1, 2, \ldots, 7, 8, 9, A, B, C, D, E, F$. For example,

$$C1A.F_{16} = 12 \times 16^2 + 1 \times 16^1 + 10 \times 16^0 + 15 \times 16^{-1}$$
$$= 3072 + 16 + 10 + 0.9375 = 3098.9375_{10}$$

Conversions between these systems and the binary system are quite simple. For example, to convert from binary to octal we separate the binary number into groups of three bits which are converted directly to octal:

011	**101**	**110**	.	**001**	← original binary number
= 3	5	6	.	1	← octal equivalent

For the reverse procedure, we simply write a three-place binary number for each octal digit:

	6	7	1	.	3
=	**110**	**111**	**001**	.	**011**

Table 11.7-2 Binary-octal-hexadecimal equivalents

Hexadecimal	Octal	Binary	Hexadecimal	Octal	Binary
0	0	0	10	20	10000
1	1	1	11	21	10001
2	2	10	12	22	10010
3	3	11	13	23	10011
4	4	100	14	24	10100
5	5	101	15	25	10101
6	6	110	16	26	10110
7	7	111	17	27	10111
8	10	1000	18	30	11000
9	11	1001	19	31	11001
A	12	1010	1A	32	11010
B	13	1011	1B	33	11011
C	14	1100	1C	34	11100
D	15	1101	1D	35	11101
E	16	1110	1E	36	11110
F	17	1111	1F	37	11111

To convert from hexadecimal to binary and vice versa the same procedure is used except that each binary groups contains 4 bits. Examples are

$$0110\ 1111\ 0101_2 = 6F5_{16}$$

$$A2C_{16} = 1010\ 0010\ 1100_2$$

Table 11.7-2 shows some binary-octal-hexadecimal equivalents.

REFERENCES

1. F. J. Hill and G. R. Peterson, "Switching Theory and Logical Design," Wiley, New York, 1974.
2. H. Taub and D. L. Schilling, "Digital Integrated Electronics," sec. 3.25, McGraw-Hill, New York, 1977.

PROBLEMS

11.1-1† John has decided to go to the movies if Alice will go with him and if he can use the family car. However, Alice has decided to go to the beach if it is not raining and if the temperature is above 80°F. John's father has made plans to use the car to visit friends if it rains or if the temperature is above 80°F. Under what conditions will John go to the movies? Construct a special-purpose computer using NOT, AND, and OR gates, switches, battery, and a light bulb to help you solve the problem. The bulb should light if John goes to the movies.

11.1-2† Buses leave the terminal every hour on the hour unless there are fewer than 10 passengers or if the driver is late. If there are fewer than 10 passengers, the bus will wait 10 min or until the number of passengers increases to 10. If the bus leaves on time, it can travel at 60 mi/h. If the bus leaves late, or if it rains, it can travel at only 30 mi/h. Under what conditions can the bus travel at 60 mi/h?

11.1-3‡ Seven switches operate a lamp in the following way. If switches 1, 3, 5, and 7 are closed and switch 2 is open, or if switches 2, 4, and 6 are closed and switch 3 is open, or if all seven switches are closed, the lamp will light. Show, using NOT, AND, and OR gates, how the switches must be connected.

11.1-4 Figure P11.1-4 shows the waveform at the input terminals of a two-input AND gate. Sketch the output waveform.

---

† Adapted from C. Belove, H. Schachter, and D. L. Schilling, "Digital and Analog Systems Circuits and Devices," p. 58. Copyright 1973 by McGraw-Hill, Inc. Used by permission of McGraw-Hill Book Company.

‡ Adapted from H. Taub and D. Schilling, "Digital Integrated Electronics," p. 585. Copyright 1977 by McGraw-Hill, Inc. Used by permission of McGraw-Hill Book Company.

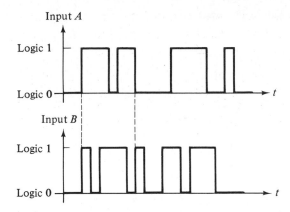

Input A

Input B

Figure P11.1-4

11.1-5 Repeat Prob. 11.1-4 if the gate is an OR gate.

11.2-1 Prove Theorem 3*b* using a truth table.

11.2-2 Prove Theorem 6*a* using (*a*) algebraic manipulation and (*b*) a truth table.

11.2-3 Prove Theorem 9*b* using (*a*) algebraic manipulation and (*b*) a truth table.

11.2-4 Simplify the following expressions as far as possible:
(*a*) $\bar{X}YZ + \bar{X}\bar{Y}$
(*b*) $\bar{X}Z + \bar{Y}Z$
(*c*) $(A + \bar{B})(\bar{A}\bar{B}\bar{C})$
(*d*) $X(\bar{Y} + \bar{Z}) + XY$
(*e*) $XY + XYZ + XY\bar{Z} + \bar{X}YZ$

11.2-5 Complement and simplify the following expressions (see Example 11.2-3):
(*a*) $\bar{X}Y$
(*b*) $(\bar{X} + Y)(\bar{S}\bar{T})$
(*c*) $(X + \bar{X}Y)(U + \bar{V})$
(*d*) $XYZ + \bar{X}\bar{Y}$
(*e*) $\bar{A}\bar{B} + AB$

11.2-6 Evaluate the following expressions for $A = 1, B = 0, C = 0, D = 1, E = 1$:
(*a*) $(\bar{A}\bar{B} + AB) + (\bar{B} + C) + D\bar{E}$
(*b*) $A\bar{B} + \bar{A}B$
(*c*) $A\bar{B}\bar{C}DE + \bar{A}(D + E)$

11.2-7 Prepare a truth table for each of the following expressions:
(*a*) $X\bar{Y} + XY$
(*b*) $\bar{X}YZ + \bar{X}\bar{Y}$
(*c*) $AB(\bar{A}\bar{B}\bar{C} + \bar{B}C)$
(*d*) $\bar{A}\bar{B}C + AB\bar{C} + \bar{A}BC$

11.3-1 Prove that the NAND operation is not associative, i.e., that

$$\overline{(\overline{AB})C} \neq \overline{A(\overline{BC})}$$

11.3-2 Prove that the NOR operation is not associative, i.e., that

$$\overline{\overline{A + B} + C} \neq \overline{A + \overline{B + C}}$$

_{4/24/81}
11.3-3 Show how NOR gates can be used to achieve both AND and OR functions.

_{4/24/81}
11.3-4 Convert the EXCLUSIVE-OR function $L = \bar{X}Y + X\bar{Y}$ into all-NAND form and show the corresponding circuit.

11.3-5 Repeat Prob. 11.3-4 for all-NOR.

11.3-6 Manipulate the following expressions into all-NAND form:

(a) $AB + \bar{A}C + A\bar{C}$

(b) $(AC + \bar{B})(\bar{B} + C)$

11.3-7 Repeat Prob. 11.3-6 for all-NOR.

11.4-1 Convert the following into sum-of-products form:

(a) $(A + B)(\bar{B} + C)(\bar{A} + C)$

(b) $(X + Y\bar{Z})(\bar{X}\bar{Y} + \bar{X}Y)$

(c) $(\bar{X} + Z)(XY + \bar{X}Z)$

(d) $(\overline{XY} + A)(\overline{XZ} + Y)$

11.4-2 Convert the expressions of Prob. 11.4-1 into product-of-sums form.

11.4-3 Prove that Eq. (11.4-8) reduces to Eq. (11.4-9).

11.4-4 For the truth table below, find the logic equations for L_1 and L_2 in expanded sum-of-products form.

Inputs			Outputs		Inputs			Outputs	
A	B	C	L_1	L_2	A	B	C	L_1	L_2
0	0	0	1	1	1	0	0	1	1
0	0	1	1	0	1	0	1	0	0
0	1	0	0	1	1	1	0	0	1
0	1	1	1	0	1	1	1	1	0

11.4-5 Simplify L_1 and L_2 in Prob. 11.4-4 as far as possible using algebraic manipulation.

11.4-6 For the truth table of Prob. 11.4-4, find L_1 and L_2 in expanded product-of-sums form.

11.4-7 Simplify L_1 and L_2 in Prob. 11.4-6 as far as possible using algebraic manipulation.

11.4-8 Synthesize two-level AND-OR circuits for L_1 and L_2 from Prob. 11.4-4.

11.4-9 Convert the circuit of Prob. 11.4-8 into all-NAND form. Prove the equivalence of the two circuits using algebraic manipulation.

11.4-10 Synthesize two-level OR-AND circuits for L_1 and L_2 from Prob. 11.4-6.

11.4-11 Convert the circuits of Prob. 11.4-10 to all-NOR form. Prove the equivalence of the two circuits using algebraic manipulation.

11.4-12 For the logic circuit of Fig. P11.4-12:

(a) Derive an expression for the output.

(b) Convert the expression found in part (a) to product-of-sums form.

11.4-13 In the circuit of Fig. P11.4-13 the input waveforms are shown. Sketch the waveforms at points 1 to 5.

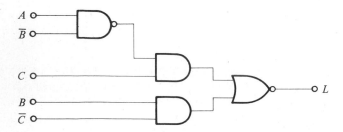

Figure P11.4-12

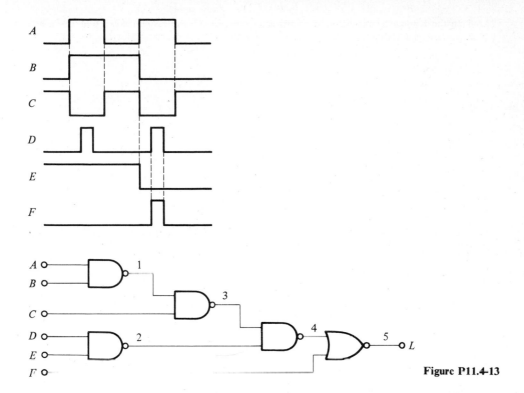

Figure P11.4-13

11.5-1 Find the simplest expression for L in the following Karnaugh maps:

(a)

C \ AB	00	01	11	10
0	1	1		1
1		1	1	

(b)

Z \ XY	00	01	11	10
0	1	1		
1	1	1	1	1

11.5-2 Repeat Prob. 11.5-1 for the following maps:

(a)

CD \ AB	00	01	11	10
00				
01	1	1	1	1
11	1	1		
10	1	1		

(b)

CD \ AB	00	01	11	10
00	1	1		
01		1	1	
11			1	
10	1	1		

11.5-3 In the Karnaugh map below, d means *don't care*; i.e., we can assign *either* a **0** or **1** to a box which contains a d. Using these *don't-care* terms, find the simplest expression for L.

CD \ AB	00	01	11	10
00			d	1
01	1		d	1
11	1		d	d
10	1		d	d

11.5-4 Plot the following expressions on a Karnaugh map. From the map obtain a simpler expression if possible.

(a) $XYZ + X\bar{Y}Z + \bar{X}YZ + \bar{X}\bar{Y}Z$

(b) $AB\bar{C} + A\bar{B}C + A\bar{B}\bar{C} + ABC$

(c) $\bar{A}\bar{B}\bar{C}\bar{D} + A\bar{B}\bar{C}D + \bar{A}\bar{B}CD + ABC\bar{D} + A\bar{B}C\bar{D} + \bar{A}\bar{B}\bar{C}D$

11.5-5 Given that $L = Z + \bar{X}\bar{Y}$. Plot $\bar{L}$ on a Karnaugh map. From the map prove that $\bar{L} = Y\bar{Z} + X\bar{Z}$.

11.5-6 Using maps, simplify the following expressions:

(a) $\bar{A}\bar{B}C + A\bar{B}\bar{C} + \underbrace{ABC + A\bar{B}C + \bar{A}\bar{B}C}$

don't care terms

(b) $ABCD + \bar{A}\bar{B}\bar{C}D + \bar{A}BCD + \underbrace{A\bar{B}CD + \bar{A}\bar{B}CD + ABC\bar{D}}$

don't care terms

11.6-1 An investor proposed the following technique to make money in the stock market:

1. If the dividends paid on a stock exceed those paid on a bond, buy the stock.
2. If the dividends paid on a bond exceed those paid on a stock, buy the bond unless the growth rate of the stock is at least 25 percent per year for the past 5 years, in which case the stock should be purchased.

The investor required a special-purpose computer to tell her what to buy. The computer requires three switches; for higher dividend in stock, higher dividend in bond, and 25 percent growth rate, and two lamps, one to light if a stock is selected and the other to light if a bond is selected. Design the computer using all-NAND logic.

11.6-2 Design a logic circuit which will activate a relay whenever either only one or all of three switches are closed.

11.6-3 Design a logic circuit which has four inputs and provides an output of **1** whenever the four inputs are not all the same.

11.6-4 A lamp is to be controlled independently by switches in two different locations. Design a suitable logic circuit.

11.6-5 Design a logic circuit which will give a **1** output when *any two* of the three inputs A, B, and C are **1**.

11.7-1 To convert from decimal to binary the *successive division-by-two* process can be used, in which the decimal number is divided by 2, and then each quotient divided by 2 until a 0 results. The remainder generated by each division forms the binary number, with the first remainder becoming the least significant bit in the binary number. The process can be organized as shown below.

work this way ←————

0	1	2	4	9	19	÷ 2
	1	0	0	1	1	Remainder

Read answer this way ————→

Using this method, convert the following to binary (a) 67, (b) 942, (c) 631.

11.7-2 Convert the following binary numbers to decimal: (a) **11.0011**, (b) **1100101.1**, (c) **11110.001**.

11.7-3 Convert the following binary numbers to octal: (a) **111001010.101**, (b) **11000100101.10001**, (c) **00111.1111**.

11.7-4 Convert the binary numbers in Prob. 11.7-3 to hexadecimal.

11.7-5 The scheme given in Prob. 11.7-1 can be used to convert from decimal to any base by simply changing the divisor to the new base. Use this method to convert the following decimal numbers to base 8 (octal): (a) 37, (b) 694, (c) 3642.

11.7-6 Convert the decimal numbers in Prob. 11.7-5 to base 16 (hexadecimal).

TWELVE

LOGIC GATES

INTRODUCTION

In the last chapter we considered the logical aspect of digital systems. In an actual system design, the first step is to derive the logic equations to be solved by the system and from them to set down a system design using logic blocks. The next part of the process involves going from the logic-block formulation to an actual circuit realization which will perform the desired functions. The basic building block for the digital system is the logic gate, which will be discussed in this chapter. Logic circuits have evolved into *families*, each of which has its own advantages and disadvantages. Usually a system is built with circuits from a single logic family, and the logic gates in this family will be used to implement all the required logical operations. Sometimes more than one family will be used in a system, and it is then necessary to take account of the fact that the output of one family may not be compatible with the input of the other, so that special *interfacing* circuits may be required.

The choice of family to be used in a particular application depends on such factors as speed, cost, noise immunity, power dissipation, availability of different logic functions, and others. Examples of systems in which low power consumption is a necessity would be space satellite applications and electronic digital wristwatches. On the other hand, a large scientific computer would probably be designed for the highest speed possible in order to minimize calculation time.

The advent of the *integrated circuit* (IC), in which many discrete components (diodes, transistors, and resistors) are fabricated at the same time on one chip of silicon, has led to many different types of logic circuits in IC form. These ICs are inexpensive and very reliable.

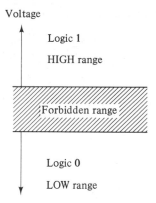

Voltage

Logic 1

HIGH range

Forbidden range

Logic 0

LOW range

Figure 12.1 Logic convention for positive logic.

In all the circuits to be studied, logic variables will be represented by gate output and input voltages, which must be identified as either logic **1** or logic **0**. All the gates operate between two voltage levels, one of which will be HIGH (logic **1**) and the other LOW (logic **0**). These logic levels, corresponding to *positive* logic, are illustrated in Fig. 12.1. In the diagram we have shown a *forbidden* region, which separates the two logic ranges, and our analysis of the various gate families will establish firm boundaries for this region.

In the following discussion we shall consider only the most popular logic families, complementary-symmetry MOS (CMOS), transistor-transistor logic (TTL), and emitter-coupled logic (ECL), along with those characteristics which are of importance in the design of digital systems. The CMOS logic family is the slowest of the three logic families but also dissipates significantly less power than the medium-speed family, TTL, or the high-speed family, ECL. TTL is the most often employed logic family, being significantly faster than CMOS but slower than ECL. The power dissipated in a TTL gate is usually less than in the corresponding ECL gate.

12.1 THE INVERTER (NOT GATE)

The *inverter* (NOT gate) is available in each of the logic families. As its name implies, this gate has an output which is the *inverse* of the input; i.e., if the input is in the HIGH (logic **1**) range, the output will be in the LOW (logic **0**) range, and vice versa. The bipolar transistor, connected in the common-emitter configuration, can serve as a rudimentary inverter, as shown in Fig. 12.1-1a. The output voltage V_O is seen to be the inverse of the input voltage V_I. In this section we discuss the inverter primarily to illustrate the important characteristics common to all families of gates. The inverter symbol is shown in Fig. 12.1-1b, and the input-output characteristic describing its operation is shown in Fig. 12.1-1c.

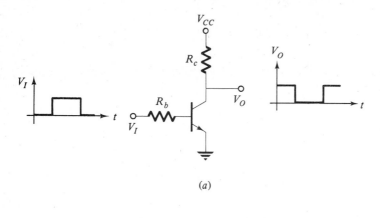

(a)

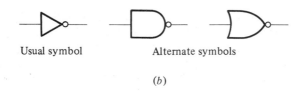

Usual symbol   Alternate symbols

(b)

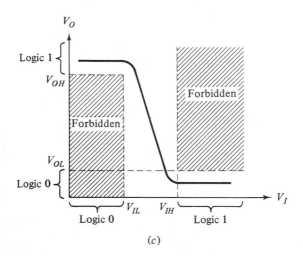

(c)

Figure 12.1-1 The inverter: (a) rudimentary transistor inverter; (b) circuit symbols; (c) input-output characteristic.

Inverter specifications typically include the following maximum and minimum voltage levels:

V_{IH} = minimum gate input voltage which will reliably be recognized as logic **1** (HIGH)

V_{IL} = maximum gate input voltage which will reliably be recognized as logic **0** (LOW)

V_{OH} = minimum voltage at gate output when output is at logic **1** (HIGH)

V_{OL} = maximum voltage at gate output when output is at logic **0** (LOW)

The manufacturer of the gate guarantees that when the input signal V_I is less than V_{IL} (logic **0**), the output V_O will be greater than V_{OH} (logic **1**) under worst-case conditions. He also guarantees that when the input voltage exceeds V_{IH} (logic **1**), the output will be less than V_{OL} (logic **0**) under worst-case conditions. Thus, in effect, the manufacturer guarantees that the gate will not operate in the shaded regions in Fig. 12.1-1c.

The voltages V_{IL}, V_{IH}, V_{OL}, and V_{OH} are specified for input and output current levels not to exceed I_{IL}, I_{IH}, I_{OL}, and I_{OH}, respectively. If, for example, the inverter output is HIGH, V_O will be greater than V_{OH} provided that the output current delivered by the inverter is less than I_{OH}. If the inverter drives a load which causes the output current to exceed I_{OH}, the output voltage may decrease below V_{OH} because of the voltage drop across the output resistance R_o. This is illustrated graphically in Fig. 12.1-2.

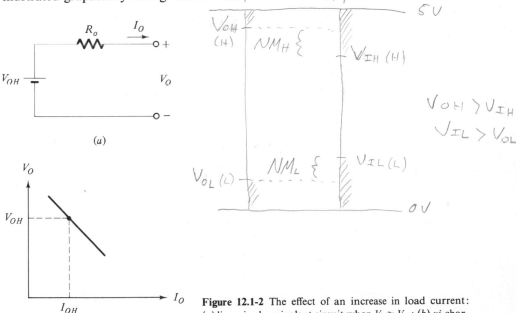

Figure 12.1-2 The effect of an increase in load current: (a) linearized equivalent circuit when $V_o \approx V_{oh}$; (b) *vi* characteristic showing decrease of gate output voltage with increasing load current.

12.1-1 Noise Margin

In practice, noise is always present in a physical system. It may be internally generated or picked up from signals on the power-supply lines or by radiation from other nearby signals, fluorescent lights, etc. If a noise pulse of sufficient amplitude appeared at the input terminal of a gate, it might cause the circuit to switch from one logic state to the other, which, in turn, would cause a false logic signal to appear at the output. In order to give the designer an idea of the amount of noise that can be tolerated we define LOW-state and HIGH-state *noise margins.* The noise margin is illustrated graphically in Fig. 12.1-3 for a circuit containing two identical gates.

For the driven gate to reliably recognize the output V_O of the driver gate as logic **0** when it is LOW we must have $V_{OL} < V_{IL}$; consequently, the LOW-state noise margin NM_L is defined as the difference between the LOW-level input threshold V_{IL} and the LOW-level output voltage V_{OL} of the preceding gate. Thus

LOW STATE NOISE MARGIN
$$MAXIMUM \qquad maximum$$
$$NM_L = V_{IL} - V_{OL} \tag{12.1-1}$$

For the HIGH state, for the driven gate to reliably recognize the output V_O of the driver gate as logic **1** when it is HIGH we must have $V_{OH} \geq V_{IH}$. The HIGH-level noise margin NM_H is then defined as

High STATE NOISE MARGIN
$$NM_H = V_{OH} - V_{IH} \tag{12.1-2}$$

The LOW-state noise margin represents that positive noise-voltage amplitude

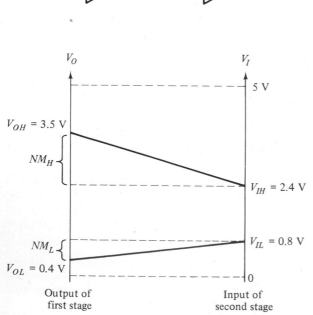

Figure 12.1-3 Diagram illustrating noise margin (values are for TTL).

which, when added to the driver-gate output V_O, may cause the input voltage V_I of the next state to exceed the threshold V_{IL} and thus cause false triggering. Similarly, the HIGH-state noise margin represents that negative noise-voltage amplitude which, when added to V_O, may cause the input voltage V_I of the next stage to fall below the threshold V_{IH} and thus cause false triggering.

It should be clear that the higher the noise margins the less susceptible the system to false triggering due to noise and thus the more reliable the design.

Example 12.1-1 A TTL inverter has the parameters $V_{IL} = 0.8$ V, $V_{IH} = 2.4$ V, $V_{OL} = 0.4$ V, and $V_{OH} = 3.5$ V. A CMOS inverter has the parameters $V_{IL} = 1.5$ V, $V_{IH} = 3.5$ V, $V_{OL} = 0.01$ V, and $V_{OH} = 4.99$ V. Calculate the noise margin when two TTL inverters are cascaded and when two CMOS inverters are cascaded. Compare the results.

SOLUTION Using (12.1-1) and (12.1-2), we find for TTL

$$NM_L = V_{IL} - V_{OL} = 0.8 - 0.4 = 0.4 \text{ V}$$

and
$$NM_H = V_{OH} - V_{IH} = 3.5 - 2.4 = 1.1 \text{ V}$$

CMOS have higher Noise margin than TTL

For CMOS we find

$$NM_L = V_{IL} - V_{OL} = 1.5 - 0.11 = 1.49 \text{ V}$$

and
$$NM_H = V_{OH} - V_{IH} = 4.99 - 3.5 - 1.49 \text{ V}$$

Thus CMOS gates have a substantially higher noise margin than TTL gates. It can similarly be shown (Prob. 12.1-1) that ECL gates have a lower noise margin than TTL gates. Since CMOS has such a large noise margin, it is often used in industrial applications where noise levels are high.   ///

12.1-2 Fan-out

In practice, the output of one transistor inverter gate is often connected to the input of one or more gates, as shown in Fig. 12.1-4. In the diagram T_O is the driving gate and T_1 to T_N are the load gates, which we have drawn as NAND gates although NOR gates could have been employed. The number of load gates N is called the *fan-out*, and the upper limit on N is determined by the tolerable values of V_{IL}, I_{IL}, V_{IH}, I_{IH}, V_{OL}, I_{OL}, V_{OH}, and I_{OH}.

For example, consider that the driving inverter has its output in the HIGH state. As the fan-out increases, the current that must be supplied by the driving gate increases. Referring to Fig. 12.1-2, we see that as the output current of the inverter increases, the output voltage decreases until if the fan-out is increased beyond the value of N recommended by the manufacturer, the output current will exceed I_{OH} and the output voltage may fall below V_{OH}, thereby decreasing the noise margin below the value set by the manufacturer. As the noise margin decreases, the logic state of the driven gates becomes more sensitive to outside interference.

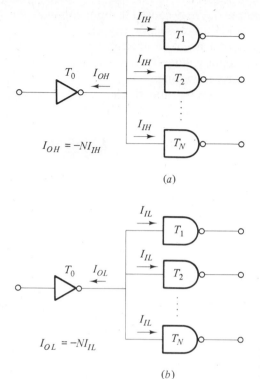

(a)

(b)

Figure 12.1-4 Fan-out: (a) when the inverter output is HIGH; (b) when the inverter output is LOW.

12.2 TRANSISTOR-TRANSISTOR LOGIC (TTL)

The basic circuit of the TTL gate is shown in Fig. 12.2-1 and is seen to consist of a multiemitter input transistor T_0 and an output transistor T_1.

The input transistor T_0 contains two to eight emitters and is fabricated as shown in Fig. 12.2-2a. In practice, it is found that this dual emitter transistor behaves very much like two transistors having their bases and their collectors connected as in Fig. 12.2-2b; however, the multiemitter transistor requires significantly less chip area ("real estate") than the parallel connection of transistors.

In order to determine the logic associated with this gate (Fig. 12.2-1) consider first that if input V_A is LOW ($V_A = 0.2$ V), the base-emitter junction of T_0 will be forward-biased and current will flow from V_{CC} through R, then through the base-emitter junction, and finally out of input terminal A. The base voltage of transistor T_0 is then

$$V_{BO} = V_{BE} + V_A = 0.7 + 0.2 = 0.9 \text{ V}$$

We now show that when $V_{BO} = 0.9$ V, T_1 is OFF. If T_1 were ON, the base-collector junction of T_0 would be forward-biased, so that base current could be

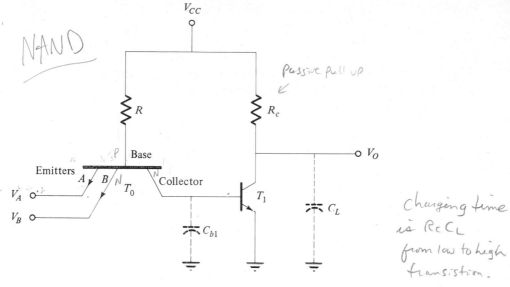

NAND

Passive pull up

Changing time is $R_C C_L$ from low to high transistion.

Figure 12.2-1 Basic TTL gate.

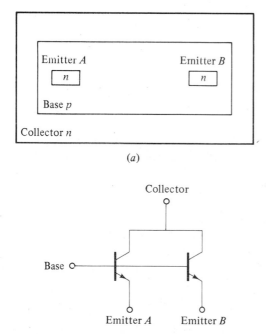

(a)

(b)

Figure 12.2-2 Multiemitter transistors: (a) construction; (b) equivalent parallel transistors.

supplied to T_1. For this condition to hold we would require a base voltage $V_{BO} = V_{BE1} + V_{BCO} = 0.7 + 0.7 = 1.4$ V. The actual voltage $V_{BO} = 0.9$ V is thus not sufficient to turn both the base-collector junction of T_0 and the base-emitter junction of T_1 ON. Since T_0 is saturated due to its large base current $(I_{CO}/I_{BO} \approx 0)$, its collector-emitter voltage is approximately 0.2 V (see Sec. 2.2-1). The base voltage of T_1 is therefore equal to the sum of the collector-emitter voltage of T_0 and the emitter voltage V_A:

$$V_{B1} = V_{CEO, \text{sat}} + V_A = 0.2 + 0.2 = 0.4 \text{ V}$$

Thus T_1 is OFF and $V_O \approx V_{CC}$; that is, V_O is HIGH. This result is true as long as any of the inputs is low.

Now consider that all inputs are HIGH (V_{CC}). The base-emitter junctions of T_0 will now all be reverse-biased, so that no current will flow into emitter A or B. However, the base-collector pn junction of T_0 is now forward-biased, as will be the base-emitter junction of T_1. Current will flow from V_{CC}, through R, and then into the base of T_1. This current is sufficient to saturate T_1, so that the output voltage V_O will be LOW (0.2 V).

The logic performed by this gate will be recognized as NAND, i.e.,

$$V_O = \overline{V_A V_B} \tag{12.2-1}$$

Consider the operation of the multiemitter transistor in the TTL circuit of Fig. 12.2-1. When the inputs all drop from HIGH to LOW, transistor T_1 must come out of saturation. This is accomplished by removing the charge stored in the base of T_1. The base-emitter junction of T_0 becomes forward-biased, and the charge stored in the base of T_1 flows into the collector of T_0. The removal of the stored charge is rapid since $\Delta Q/\Delta t = I_{CO}$ and the collector current I_{CO} in T_0 is large.

12.2-1 Output Stages; The Active Pull-up

In the previous section, we saw that an advantage of the TTL gate of Fig. 12.2-1 is the quick removal of charge from the base of the output transistor when the output switches from LOW to HIGH. With this problem taken care of, the principal remaining limitation on speed is the capacitive load on the output transistor. This is shown on dashed lines as C_L in Fig. 12.2-1. It consists of the capacitances of any driven gates, the output capacitance of T_1, and wiring capacitance. Using the collector resistor R_c as shown in Fig. 12.2-1 (R_c is called a *passive pull-up*), we have the situation that when the output switches from LOW to HIGH, capacitor C_L charges exponentially through R_c from $V_{CE, \text{sat}} = 0.2$ V to $V_{CC} = 5$ V. The charging time constant is $R_c C_L$, and this may be large enough to introduce prohibitively long propagation delays. For example, if $C_L = 5$ pF and $R_c = 1$ kΩ, the output changes from $V_{OL}(0.2$ V$)$ to $V_{OH}(3.3$ V$)$ in the time $T = 5.2$ ns (see Prob. 12.2-1).

The delay can be shortened by decreasing R_c, but this increases the power dissipation when the output is LOW. A better solution is to effectively reduce R_c

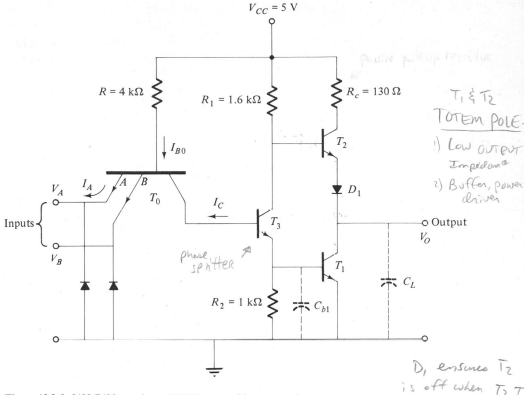

Figure annotations (handwritten):
- Passive pull-up resistor
- T_1 & T_2 TOTEM POLE —
  1) Low output Impedance
  2) Buffer, power driver
- D_1 ensures T_2 is off when T_3, T_1 on
- phase splitter

Figure 12.2-3 5400/7400 two-input NAND gate with totem-pole output stage.

by replacing it by an active resistance, called an *active pull-up* The circuit of a 5400/7400 two-input NAND gate which accomplishes this is shown in Fig. 12.2-3. Operation of this circuit is as follows.

Transistor T_0 is the multiemitter transistor input stage. Transistor T_3 acts as a *phase splitter*, since any change in its emitter voltage is out of phase (opposite) to the corresponding change in collector voltage. Transistors T_2 and T_1 form a totem-pole amplifier, which in addition to having low output resistance acts as a power driver, or buffer output stage.

Consider that both inputs are HIGH, so that the output is LOW (0.2 V). For this state T_3 and T_1 are both saturated. Then, to determine the state of T_2 we note that the collector voltage of T_3 is

$$V_{C3} = V_{CE3,\,sat} + V_{BE1} = 0.2 + 0.7 = 0.9 \text{ V}$$

Since the base of T_2 is tied to the collector of T_3, we have $V_{B2} = 0.9$ V. The voltage from the base of T_2 to the collector of T_1 is

$$V_{BE2} + V_{D1} = V_{C3} - V_{CE1} = 0.9 - 0.2 = 0.7 \text{ V}$$

This is not enough to turn both T_2 and D_1 ON. Diode D_1 provides the offset voltage required to ensure that T_2 is OFF when T_3 and T_1 are ON.

Now consider that the output is LOW, and we have a sudden change of state due to one input, going LOW. T_3 will be turned OFF, which will cause T_1 to turn OFF because V_{BE1} goes to zero after the charge stored in the base of T_1 is removed through resistor R_2. (This is analogous to the stored charge residing on capacitor C_{b1} and discharging through R_2.) The voltage across C_L cannot change instantaneously and thus will remain close to 0.2 V for a short time. During this time, T_2 turns ON. With T_2 ON, its base voltage is

$$V_{B2} = V_{BE2} + V_{D1} + V_O = 0.7 + 0.7 + 0.2 = 1.6 \text{ V}.$$

The base current in T_2 is

$$I_{B2} = \frac{V_{CC} - V_{B2}}{1.6} = \frac{5 - 1.6}{1.6} = 2.13 \text{ mA}$$

The collector current is

$$I_{C2} = \frac{V_{CC} - V_{CE2,\,sat} - V_{D1} - V_O}{0.13} = \frac{5 - 0.2 - 0.7 - 0.2}{0.13} = 30 \text{ mA}$$

The ratio of these two is $I_{C2}/I_{B2} = 30/2.13 = 14.1$. Since $\beta_2 \approx 50$, T_2 will be in saturation. Now T_2 supplies the current to charge C_L. For this reason T_2 is called a *source*. With T_2 in saturation the output voltage rises almost exponentially with the time constant $(130 + R_{S2} + R_{D1})C_L$, where R_{S2} is the saturation resistance of T_2 and R_{D1} is the forward resistance of D_1. Typical values are $R_{S2} \approx 10 \ \Omega$ and $R_{D1} < 10 \ \Omega$; thus the charging time constant is very small, approximately 6.7 times smaller than for the passive pull-up. In very high speed TTL gates $R_c \approx 55 \ \Omega$, which reduces the time constant by an additional factor of 2. As V_O rises, the current in T_2 decreases and eventually T_2 comes out of saturation. Finally V_O reaches a steady state when T_2 arrives at the edge of saturation. Its value is

$$V_O = V_{CC} - R_1 I_{B2} - V_{BE2} - V_{D1} \approx 5 - 0.7 - 0.7 = 3.6 \text{ V}$$

where we have neglected $R_1 I_{B2} = R_1 I_{E2}/\beta$, a voltage drop which is usually negligible.

The function of R_c in this circuit is to limit the transient current spike in R_c which occurs when V_O changes state from LOW to HIGH. Note that when V_O was low T_2 was cut off and so the current in R_c was zero. When V_O is HIGH, T_2 is ON, but the current flowing is only large enough to drive a typical output load of about 0.5 mA. However, it takes approximately 0.9 ns for V_O to change state, and during that time the current attains a peak value of 30 mA. Since current spikes of this magnitude cause extremely undesirable voltage spikes on the V_{CC} power-supply lines, bypass capacitors are added from the V_{CC} terminal of each IC to ground. A rule of thumb is to use 0.02 μF per IC package, and the bypass capacitors should be placed as closely as possible to the V_{CC} pin connection. It is also good

[handwritten marginal note:] T_2 ?

[handwritten note at bottom:] ✗ Rule of thumb is to use a .02 µf capacitor from Vcc to ground to reduce spikes in Vcc due to switching of transistor

practice to use an IC regulated voltage supply on each printed-circuit (PC) board rather than to distribute the voltage from one main supply.

The diodes connected between the input terminals and ground have no effect on dc conditions but do minimize undesired negative noise transients by clamping the input to -0.7 V.

12.2-2 Wired-AND Logic

If the outputs of two or more transistor gates are tied together, as in Fig. 12.2-4a, an additional logic function is created. This function has been given several names; collector logic, wired AND, implied AND, or dot AND. The scheme often saves logic gates compared with other methods. The circuit operation is as follows. Consider that the output X of gate 1 is LOW. Then regardless of the state of the output Y of gate 2, the output V_O will be LOW. Only when both outputs X and Y are simultaneously HIGH will V_O be HIGH. This is recognized as the AND function; that is, $V_O = X \cdot Y$, and hence the name wired AND. The output is

$$V_O = X \cdot Y = (\overline{AB})(\overline{CD}) \tag{12.2-2}$$

If a wired-AND circuit is not used, the circuit of Fig. 12.2-4b is required to realize (12.2-2) using only NAND gates. This saving of two gates is not always possible using the wired-AND connection, and each case must be considered on its own merits, various approaches being tried in order to achieve the most economical design.

It is not good practice to interconnect TTL gates with active pull-up in the wired-AND configuration because if one gate output is HIGH while the other gate output is LOW, the gate with the output HIGH will dissipate a large amount of power (see Prob. 12.2-4).

TTL logic gates are available with *open-collector* outputs, which can be wired-AND connected. The circuit of a two-input NAND gate with open-collector output, type 5401/7401, is shown in Fig. 12.2-4c. This should be compared with Fig. 12.2-3, which shows the same circuit with an active pull-up and totem-pole output. In practice, all the open-collector outputs are tied together and share one common pull-up resistor R_L, which is shown in dashed lines in Fig. 12.2-4c. The size of the pull-up resistor depends on the number of open collectors, fan-out, required noise margin, etc., and is considered in the example which follows.

> **Example 12.2-1: Determination of pull-up resistor** In the circuit of Fig. 12.2-4c, four 5401/7401 NAND gates with open-collector outputs are to drive a fan-out of five 5400/7400 load gates. Determine a suitable value for the pull-up resistor R_L.
>
> SOLUTION We first determine a maximum resistor value when the output V_O is HIGH so that we have available sufficient load current for the load gates and OFF current for the driving gates. Next, a minimum resistor value is determined when V_O is LOW such that current through this resistor and

You can WIRE-AND Open Collector TTL with (use only 1 pullup)

Do NOT WIRE-AND TTL with ACTIVE PULLUP (TOTEM POLE)

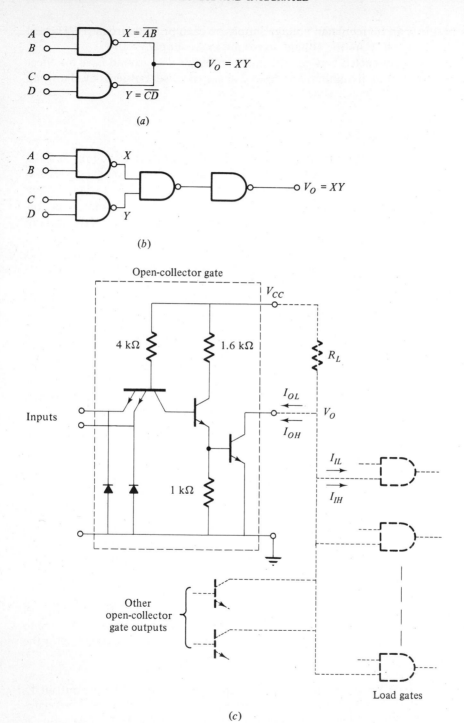

Figure 12.2-4 Wired-AND logic: (*a*) gate outputs tied together; (*b*) all-NAND circuit; (*c*) 5401/7401 two-input NAND gate with open-collector output.

currents from the load gates will not cause the output voltage V_O to rise above $V_{OL, \, max}$ even if only one driving gate is sinking all the currents.

Determination of $R_{L, \, max}$ When V_O is HIGH (driving gates OFF), the drop across R_L must be less than

$$V_{RL, \, max} = V_{CC} - V_{OH, \, min}$$

where $V_{CC} = 5$ V and $V_{OH, \, min} = 2.4$ V for 5400/7400 gates. The total current through R_L is the sum of the load gate currents I_{IH} and the OFF currents of each driving gate I_{OH}. Thus   *You.c. gutes wire anded driving 5 other gate inputs*

$$I_{RL} = 5I_{IH} + 4I_{OH}$$

For 5400/7400 gates $I_{OH} = 0.25$ mA and $I_{IH} = 40$ μA, both flowing in the directions shown in Fig. 12.2-4c. Then

$$R_{L, \, max} = \frac{V_{RL}}{I_{RL}} = \frac{5 - 2.4}{(5)(0.04) + (4)(0.25)} = 2.17 \text{ k}\Omega$$

Determination of $R_{L, \, min}$ Here V_O is LOW and the load gates are all ON. The worst case occurs when only one of the driving transistors is ON and the current through R_L must be limited so that the recommended maximum I_{OL} for this ON transistor (16 mA for the 5401/7401) will not be exceeded. The LOW-state currents I_{IL} (-1.6 mA) flowing into the load gates contribute to I_{OL}, so that taking account of the directions in Fig. 12.2-4c and neglecting the OFF currents of the three OFF driving transistors, we have

$$I_{RL} = I_{OL} + 5I_{IL}$$

Also $V_{RL, \, min} = V_{CC} - V_{OL, \, max}$ where $V_{OL, \, max} = 0.4$ V

Then $R_{L, \, min} = \dfrac{V_{RL}}{I_{RL}} = \dfrac{5 - 0.4}{16 + (5)(-1.6)} = 0.575 \text{ k}\Omega$

Thus R_L must lie between 575 Ω and 2.17 kΩ.   ///

An additional use for open-collector gates is to drive external loads such as relays, indicator lamps, or other types of logic. Some manufactures provide gates with open-collector high-voltage outputs especially for this type of application. An example is the 5406/7406 inverter, which has a minimum output breakdown voltage rating of 30 V.

12.2-3 TTL Transfer Characteristics

Typical input-output transfer characteristics for the TTL gate shown in Fig. 12.2-3 are plotted in Fig. 12.2.5. The shape of the curve comes about mainly because of the characteristics of the active pull-up and totem-pole output. Consider first that input V_B is at 5 V and that input V_A is less than the input voltage represented by point *a* on the characteristic, that is, $V_A < 0.7$ V. For this case,

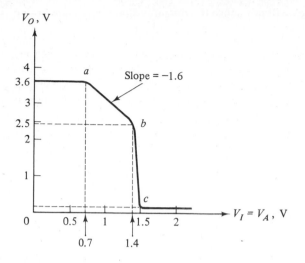

Figure 12.2-5 TTL transfer characteristic.

transistors T_3 and T_1 are OFF, while T_2 is conducting so that the output is HIGH. The output voltage is approximately (we neglect the small drop in R_1)

$$V_{OH} \approx V_{CC} - V_{BE2} - V_{D1} = 5 - 0.7 - 0.7 = 3.6 \text{ V} \qquad (12.2\text{-}3)$$

For this HIGH range, current flows out of emitter A of T_0, the amount of current being determined primarily by V_{CC} and R. In our example (see Fig. 12.2-3) with $V_A = 0.2$ V

$$I_A = \frac{V_{CC} - V_{BE0} - V_A}{R} = \frac{5 - 0.7 - 0.2}{4 \times 10^3} \approx 1 \text{ mA}$$

The gate output connected to emitter A must be capable of *sinking* this much current.

At point a the transition region between HIGH output and LOW output begins, and we see that there are two separate line segments having different slopes in this region. These arise in the following way. When $V_A < 0.7$ V, $I_C = 0$ and transistor T_0 is saturated and therefore $V_{CE0} \approx 0.2$ V. Thus, the base voltage of T_3 is $V_{B3} \approx V_A + 0.2$ V. When V_A rises to about 0.5 V, $V_{B3} \approx 0.7$ V and current begins to flow in T_3. T_3 will now be operating in its linear region, with a gain from base to collector which is $A_v \approx -R_1/R_2$ (see Prob. 12.2-7). T_1 is still OFF, and emitter-follower T_2 is ON. Thus, the output-input curve follows the gain characteristic of T_3 and decreases at a slope of $R_1/R_2 = 1.6$ V out per volt in until point b (Fig. 12.2-5) is reached. At point b the input voltage $V_A \approx 1.4$ V, so that base current flows in T_1 and T_1 turns ON. As this happens, the collector of T_3 is falling, thereby cutting T_2 OFF. As the collector current of T_2 (which is also the collector current of T_1) decreases simultaneously with the increasing base current of T_1, T_1 is driven rapidly into saturation. At point c, $V_A \approx 1.5$ V, T_2 turns OFF, and the output is LOW, i.e., at the saturation voltage of T_1 (≈ 0.2 V). In addition, T_0 is now operating in the inverse mode since the collector of T_0 is clamped to 1.4 V while the emitter is at a higher positive voltage (1.5 V).

12.2-4 Schottky-clamped TTL

The transistors in the TTL gate shown in Fig. 12.2-3 switch from saturation to cutoff as the gate changes states. Such gates are characterized by relatively long time delays due primarily to the time required to remove a transistor from saturation. When speed is important, the Schottky transistor (Sec. 2.2-2) is employed.

A typical two-input NAND gate using Schottky transistors is shown in Fig. 12.2-6. The output circuit has been modified compared with the 54/74 series to give a symmetrical transfer characteristic. These gates are called the 54S/74S series, where the S denotes Schottky. A series 54LS/74LS is also available, in which the gate resistors are larger and the gates therefore dissipate significantly less power than in the S series. However, the LS gates are also slower.

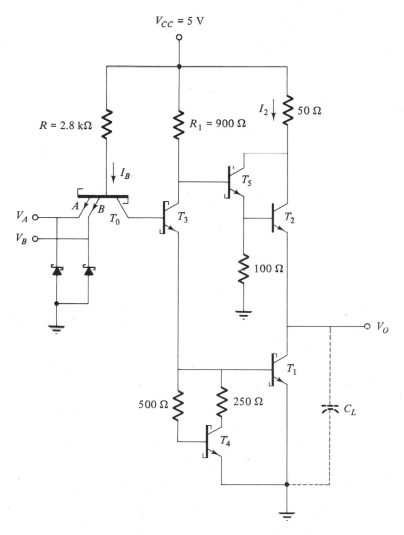

Figure 12.2-6 Schottky TTL NAND gate (54S/74S).

The operation of the Schottky TTL NAND gate is as follows. Assume for simplicity that $V_B = 5$ V and the gate output changes in response to changes in the input voltage V_A. As V_A increases, the base current I_B flows out of emitter A until T_3 and T_1 turn on *simultaneously*. Remember that in Fig. 12.2-2 T_3 turned on first (at point a). This was possible because a resistor was used to connect the emitter of T_3 to ground. In the present circuit the emitter current of T_3 must flow into the base of T_1 and into the base and collector of T_4. Assuming the cut-in voltage of a transistor to be 0.65 V, we see that T_3 and T_1 both turn on when $V_{B3} \approx 1.3$ V. The 500-Ω resistor in series with the base of T_4 permits T_1 to turn on before T_4.

Since T_0 is a Schottky transistor, the collector-emitter voltage V_{CE0} when T_3 and T_1 turn on is $V_{CE0} \approx 0.35$ V. Hence, with $V_{B3} = 1.3$ V,

$$V_A = V_{B3} + V_{CE0} = 1.3 + 0.35 = 1.65 \text{ V} \qquad (12.2\text{-}4)$$

When $V_A < 1.65$ V, T_1 and T_3 are off and T_2 and T_5, which form a compound amplifier (Sec. 7.6), are on. Hence $V_O \approx 3.6$ V, as before (again we neglect the voltage drop in resistor R_1). As V_A increases above 1.65 V, the base current in T_3 increases, as do the collector and emitter currents of T_3. The collector voltage of T_3 falls, reducing the emitter current in T_2 and thereby also reducing the collector current in T_1. Simultaneously the base current in T_1 rises so that T_1 saturates and V_O is clamped to 0.35 V in the LOW state. The resulting characteristic is shown in Fig. 12.2-7.

Current spikes also occur in the Schottky TTL gate. For, if $V_O = 0.35$ V and V_A is suddenly decreased, base current stops flowing into T_3 and the collector of T_3 rises suddenly. The collector-emitter voltage of T_2 is then

$$V_{CE2} = V_{CE5} + V_{BE2} = 0.35 + 0.7 = 1.05 \text{ V} \qquad (12.2\text{-}5)$$

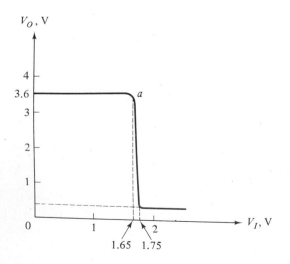

Figure 12.2-7 Schottky TTL output-input characteristic.

With $V_{CE1} = 0.35$ V, the current in the 50-Ω collector resistor jumps to

$$I_2 = \frac{V_{CC} - V_{CE2} - V_{CE1}}{50} = \frac{5 - 1.05 - 0.35}{50} = 72 \text{ mA} \qquad (12.2\text{-}6)$$

This is more than twice the 30-mA spike found for the regular TTL gate.

It is interesting to note that T_2 is not a Schottky transistor since T_2 is prevented from saturating by T_5.

12.2-5 Interpreting Manufacturers' Data Sheets

The data sheet for a Texas Instrument QUAD two-input NAND gate is shown in Fig. 12.2-8. This IC comes in various 14-pin packages. Each of the four NAND gates requires three external connections, using a total of 12 pins, the remaining two being used for V_{CC} and ground.

The data sheet contains circuit diagrams of the gates covered as well as pin connections for the different package types. Also, there are tables of recommended operating conditions and electrical characteristics. Operation of the IC outside of the recommended range may cause immediate failure or gradual deterioration of the device. The table of electrical characteristics gives measured values obtained under the test conditions specified in the table and carefully documented in other literature available from the manufacturer.[1] These measured values are usually worst-case values at maximum fan-out.

The recommended power supply voltage V_{CC} for the military grade 5400 is 4.5 V minimum, 5.5 V maximum, with a 5-V nominal value, i.e., a 10 percent variation about the nominal. For the 7400, which is the commercial version of the gate, the variation is reduced to 5 percent. The maximum fan-out for these gates driving similar gates is given by the manufacturer as 10. It must be emphasized that this figure holds *only* when the load consists of other 54/74 series gates. Any other type of load may change the fan-out considerably. The operating temperature range extends from -55 to $+125°C$ for the 5400, but for the 7400 it is restricted to 0 to 70°C. The electrical characteristics are summarized below.

Input voltage V_{IH} The first parameter listed is the logic **1** input voltage required at both input terminals to ensure a logical **0** at the output. The worst-case value for this parameter is 2.0 V minimum, measured at the lower limit of V_{CC}. That this is the worst case can be seen by considering the circuit of Fig. 12.2-3. If the output is LOW, then from Fig. 12.2-5 we have typically $V_A \approx 1.5$ V. For any input voltage greater than this calculated value we would expect the output to be LOW. The manufacturer guarantees 2.0 V as the worst-case minimum value, which is considerably more conservative than our calculated value of 1.5 V. Similarly we see from Fig. 12.2-5 that for the output voltage to exceed 2.5 V, we must have $V_I < 1.4$ V. The manufacturer specifies $V_{IL} = 0.8$ V measured at the maximum permissible value of V_{CC}, again more conservative than our calculated figures.

schematic (each gate)

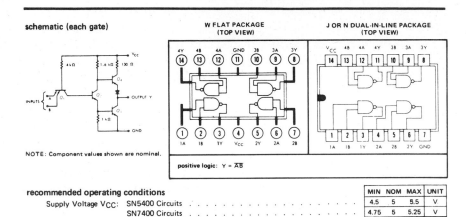

NOTE: Component values shown are nominal.

W FLAT PACKAGE (TOP VIEW)

J OR N DUAL-IN-LINE PACKAGE (TOP VIEW)

positive logic: Y = $\overline{AB}$

recommended operating conditions

	MIN	NOM	MAX	UNIT
Supply Voltage V_{CC}: SN5400 Circuits	4.5	5	5.5	V
SN7400 Circuits	4.75	5	5.25	V
Normalized Fan-Out From Each Output, N			10	
Operating Free-Air Temperature Range, T_A: SN5400 Circuits	−55	25	125	°C
SN7400 Circuits	0	25	70	°C

electrical characteristics over recommended operating free-air temperature (unless otherwise noted)

	PARAMETER	TEST FIGURE	TEST CONDITIONS[†]		MIN	TYP[‡]	MAX	UNIT
V_{IH}	Logical 1 input voltage required at both input terminals to ensure logical 0 level at output	1			2			V
V_{IL}	Logical 0 input voltage required at either input terminal to ensure logical 1 level at output	2					0.8	V
V_{OH}	Logical 1 output voltage	2	V_{CC} = MIN, V_{in} = 0.8 V, I_{load} = −400 µA		2.4	3.3		V
V_{OL}	Logical 0 output voltage	1	V_{CC} = MIN, V_{in} = 2 V, I_{sink} = 16 mA			0.22	0.4	V
I_{IL}	Logical 0 level input current (each input)	3	V_{CC} = MAX, V_{in} = 0.4 V				−1.6	mA
I_{IH}	Logical 1 level input current (each input)	4	V_{CC} = MAX, V_{in} = 2.4 V				40	µA
			V_{CC} = MAX, V_{in} = 5.5 V				1	mA
I_{OS}	Short-circuit output current[§]	5	V_{CC} = MAX	SN5400	−20		−55	mA
				SN7400	−18		−55	
I_{CCL}	Logical 0 level supply current	6	V_{CC} = MAX, V_{in} = 5 V			12	22	mA
I_{CCH}	Logical 1 level supply current	6	V_{CC} = MAX, V_{in} = 0			4	8	mA

switching characteristics, V_{CC} = 5 V, T_A = 25°C, N = 10

	PARAMETER	TEST FIGURE	TEST CONDITIONS		MIN	TYP	MAX	UNIT
t_{PHL}	Propagation delay time to logical 0 level	65	C_L = 15 pF,	R_L = 400 Ω		7	15	ns
t_{PLH}	Propagation delay time to logical 1 level	65	C_L = 15 pF,	R_L = 400 Ω		11	22	ns

[†] For conditions shown as MIN or MAX, use the appropriate value specified under recommended operating conditions for the applicable device type.
[‡] All typical values are at V_{CC} = 5 V, T_A = 25°C.
[§] Not more than one output should be shorted at a time.

Figure 12.2-8 5400/7400 data sheet.

Output voltage This parameter specifies the range of values we can expect at the output when the input voltage meets the requirements of the previous paragraph. The HIGH output voltage is specified as $V_{OH} = 2.4$ V minimum, 3.3 V typical, measured when $V_I = 0.8$ V $= V_{IL}$, $V_{CC} = 4.5$ V (the minimum value), and $I_{LOAD} = -400$ μA (the negative sign indicates that the current is flowing out of the gate). An approximate value for this was calculated in Sec. 12.2-3 and found to be 3.6 V for $V_{CC} = 5$ V. Thus for $V_{CC} = 4.5$ V we would have $V_{OH} = 3.1$ V, which agrees well with the typical value given. The LOW output voltage $V_{OL} = 0.22$ V typical, 0.4 V maximum, measured when $V_{CC} = 4.5$ V, $V_I = 2$ V $= V_{IH}$, and the fan-out is maximum, so that the gate is sinking 16 mA. We previously assumed that the LOW output voltage would be the $V_{CE, \text{sat}} \approx 0.2$ V of transistor T_1 in Fig. 12.2-3.

The information in the previous paragraphs is summarized graphically in the voltage transfer characteristic shown in Fig. 12.2-9. Although these curves are not provided in the data sheets, they are often available in the manufacturers' literature.[1] An alternate graphical representation is shown in Fig. 12.2-10.

Input current Any device used to drive these gates must be capable of both sourcing and sinking current. Thus it is important that we know the currents required at the input to the gate for both logic levels. This is especially true when the driving device is something other than a 54/74 gate. The input-current parameters provide this information. It should be noted that conventionally current flowing *toward* a device terminal is designated as positive and current flowing away as negative. The data sheet specifies a logic **0** level input current for each input of -1.6 mA when V_{CC} is at its maximum and $V_I = V_{IL} = 0.4$ V. Since this is *negative*, it flows out of the input terminal. We can estimate this current by referring to Fig. 12.2-3 and noting that this input current flows from V_{CC} through R and the forward-biased base-emitter junction of T_0 to the input terminal. Thus

$$I_{IL} = \frac{V_{CC} - V_{BE0} - V_{IL}}{R} = \frac{5.5 - 0.7 - 0.4}{4} = 1.1 \text{ mA} \qquad (12.2\text{-}7)$$

The manufacturer guarantees that this current will always be less than 1.6 mA.

When the input is at the logic **1** level, the manufacturer guarantees that the driving device will have to supply no more than 40 μA per input when the input voltage is $V_I = V_{OH, \text{min}} = 2.4$ V. Referring to Fig. 12.2-3, we see that when $V_I = 2.4$ V, transistor T_0 is operating in the inverse mode since the collector of T_0 is approximately 1.4 V. Thus, the base current I_{B0} is

$$I_{B0} = \frac{V_{CC} - V_{BE0} - V_{B3}}{R} = \frac{5 - 0.7 - 1.4}{4 \times 10^3} \approx 0.7 \text{ mA} \qquad (12.2\text{-}8)$$

The emitter current I_A is now flowing into the emitter and therefore looks like collector current with an inverse current gain h_{FC}. The manufacturer, stating that $|I_A| < 40$ μA, guarantees that

$$h_{FC} < \frac{|I_A|}{I_{B0}} = \frac{0.04}{0.7} \approx 0.06 \qquad (12.2\text{-}9)$$

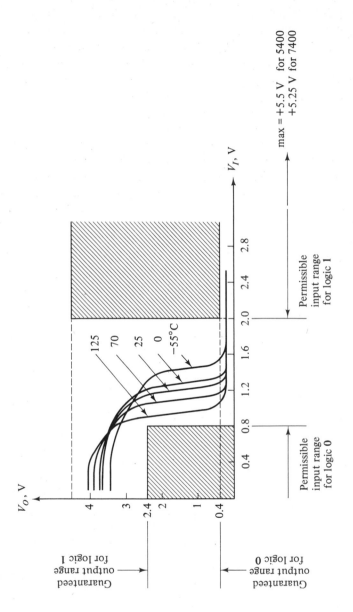

Figure 12.2-9 Voltage transfer characteristic for typical 54/74 TTL NAND gate measured for a typical gate at $V_{CC} = 5$ V and fan-out = 10. The shaded areas define regions prohibited by the manufacturers' guarantees.

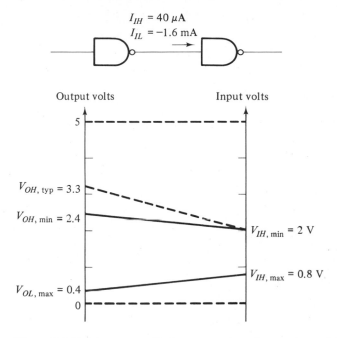

Figure 12.2-10 Alternate graphical representation of output-input characteristics of 5400/7400 gates.

The other value specified, $I_{IH} = 1$ mA when $V_I = 5.5$ V, is a value that will ensure that the emitter-base junction will not be damaged due to excessive power dissipation. It is good practice to connect any unused emitter to the 5-V supply through a 2-kΩ resistor. This resistor serves to prevent noise pickup from causing the gate to change state.

Short-circuit output current This parameter, I_{os}, is the logic 1 output current when the output is shorted to ground. When this test is made by the manufacturer, it checks the value of the current-limiting resistor R_c and proper operation of transistor T_2 and diode D_1 (Fig. 12.2-3). In the test all inputs are grounded, and maximum V_{CC} is used. Limits are given in the data sheet, negative values indicating that the current flows out of the gate. Too large a value of I_{os} would damage the device, and too small a value would degrade HIGH output switching times since load capacitance could not be charged rapidly enough to meet switching-time specifications. Note that this is the maximum current in the current spike. If this experiment is performed on more than one of the gates in the package simultaneously, the heat developed by the extraordinary power dissipation may damage the gates permanently.

Supply current This parameter is extremely important to the design engineer, who has to estimate system power-supply requirements. The data sheet lists logic-0-level supply current I_{CCL} and logic-1-level supply current I_{CCH}. Both are measured with maximum V_{CC} applied, $V_I = 0$ V for the HIGH level and $V_I = 5$ V for the LOW level. The maximum values specified apply to all four of the gates on the chip; therefore in order to get the per gate supply current we must divide by 4. Thus the low-level supply current is 5.5 mA maximum per gate, and the high level is 2.0 mA maximum per gate.

We can estimate these currents for the gate of Fig. 12.2-3 by noting that when the input is HIGH, the supply provides current for the base of T_0 and the collector of T_3. Since T_2 is cut off, there will be no supply current through R_c. The base voltage for T_0 when the inputs are HIGH is $V_{BC0} + V_{BE3} + V_{BE1} = 2.1$ V. Thus

$$I_{B0} = \frac{5 - 2.1}{4} = 0.725 \text{ mA}$$

The collector current of T_3, which is ON, is

$$I_{C3} = \frac{V_{CC} - V_{CE3, \text{ sat}} - V_{BE1, \text{ sat}}}{R_1}$$

$$= \frac{5 - 0.2 - 0.7}{1.6} = 2.56 \text{ mA}$$

Then
$$I_{CCL} = I_{B0} + I_{C3} = 3.28 \text{ mA}$$

This compares favorably with the manufacturer's typical value of 3.0 mA.

For the other logic state, the only significant current, since T_3 and T_1 are cut off, is the base current of T_0. Thus

$$I_{CCH} = I_{B3} = \frac{V_{CC} - V_{BE0, \text{ sat}} - V_I}{R_1}$$

The stated conditions are $V_{CC, \text{max}} = 5.5$ V and $V_I = 0$ V; thus

$$I_{CCH} = \frac{5.5 - 0.7 - 0}{4} = 1.2 \text{ mA}$$

This lies between the specified typical and maximum values.

The supply currents specified in the data sheet are for static dc conditions only. When a gate switches from one logic state to the other, a transient occurs in the supply current. These transient currents serve to charge internal and load capacitance. As a result the average supply current increases with load capacitance and with increasing switching frequency.[1]

Propagation delay times *Propagation delay time* is an important parameter of logic gates. Because of the finite switching speed of the transistors used and the inevitable circuit capacitance, there is a delay from the time a signal is applied to the input of a gate until the corresponding signal appears at the output. This is shown using idealized trapezoidal waveforms in Fig. 12.2-11 (see Sec. 9.7). The switching times specified on the data sheet are the propagation delay time t_{PHL} when the output goes from a HIGH level to a LOW level and t_{PLH} when the

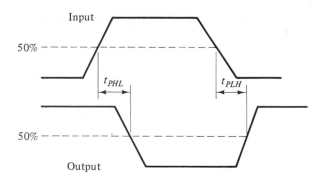

Figure 12.2-11 Waveforms for measurement of propagation delay times.

output goes from a LOW level to a HIGH level. The delay times are measured from the 50 percent point on each waveform, as shown on the figure. They are, of course, measured with the gate loaded by the equivalent of ten 54/74 series TTL gates, i.e., maximum fan-out. Note that both nominal and maximum values are given on the data sheet.

Sometimes a propagation delay time t_{PD} is defined as the average of the turn-on delay time t_{PHL} and the turnoff delay time t_{PLH}

$$t_{PD} = \frac{t_{PHL} + t_{PLH}}{2}$$

For the 5400/7400 gates t_{PD} would have a nominal value of 9 ns and a maximum value of 18.5 ns.

Noise margin Although not listed explicitly on the data sheets, noise margins are a very important characteristic of a logic-gate family. For the 54/74 series which we have been discussing the LOW-level noise margin [see (12.1-1)] is

$$NM_L = V_{IL} - V_{OL} = 0.8 - 0.4 = 0.4 \text{ V}$$

The HIGH-level noise margin [see (12.1-2)] is

$$NM_H = V_{OH} - V_{IH} = 2.4 - 2.0 = 0.4 \text{ V}$$

These are *guaranteed* dc noise margins. However, these gates typically exhibit noise margins in excess of 1.0 V. A change of state will typically take place when the input passes through a threshold voltage of about 1.35 V. The HIGH output is typically 3.3 V, and the low output is typically 0.2 V. Thus, in the HIGH state, the output can typically take 1.95 V of negative-going noise before causing the driven gate to change state falsely. Similarly 1.15 V of positive-going noise can typically be tolerated in the LOW state.

Schottky TTL characteristics A typical Schottky TTL NAND-gate data sheet is shown in Fig. 12.2-12. Note the similarity between the Schottky and standard TTL characteristic of Fig. 12.2-8. The major differences are $V_{OL} = 0.35$ V rather

FAIRCHILD SUPER HIGH SPEED TTL/SSI • 9S00 (54S00/74S00)

QUAD 2-INPUT NAND GATE

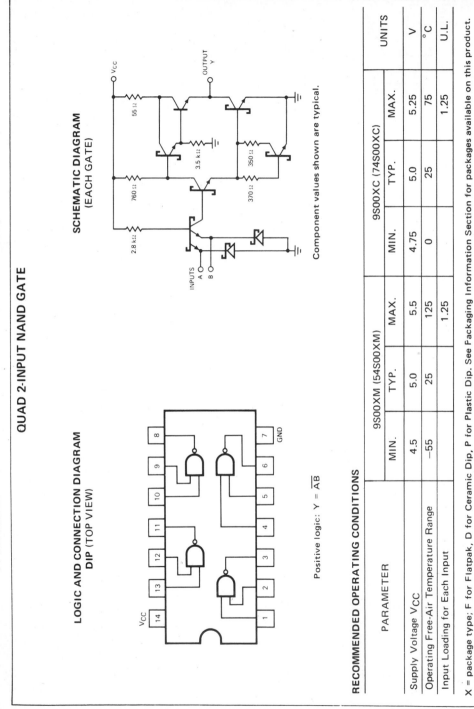

LOGIC AND CONNECTION DIAGRAM
DIP (TOP VIEW)

Positive logic: $Y = \overline{AB}$

SCHEMATIC DIAGRAM
(EACH GATE)

Component values shown are typical.

RECOMMENDED OPERATING CONDITIONS

PARAMETER	9S00XM (54S00XM)			9S00XC (74S00XC)			UNITS
	MIN.	TYP.	MAX.	MIN.	TYP.	MAX.	
Supply Voltage VCC	4.5	5.0	5.5	4.75	5.0	5.25	V
Operating Free-Air Temperature Range	−55	25	125	0	25	75	°C
Input Loading for Each Input			1.25			1.25	U.L.

X = package type; F for Flatpak, D for Ceramic Dip, P for Plastic Dip. See Packaging Information Section for packages available on this product.

DC CHARACTERISTICS OVER OPERATING TEMPERATURE RANGE (Unless Otherwise Noted)

SYMBOL	PARAMETER		LIMITS MIN.	LIMITS TYP. (Note 2)	LIMITS MAX.	UNITS	TEST CONDITIONS (Note 1)
V_{IH}	Input HIGH Voltage		2.0			V	Guaranteed Input HIGH Voltage
V_{IL}	Input LOW Voltage				0.8	V	Guaranteed Input LOW Voltage
V_{CD}	Input Clamp Diode Voltage			-0.65	-1.2	V	V_{CC} = MIN., I_{IN} = -18 mA
V_{OH}	Output HIGH Voltage	XM	2.5	3.4		V	V_{CC} = MIN., I_{OH} = -1.0 mA, V_{IN} = 0.8 V
		XC	2.7	3.4			
V_{OL}	Output LOW Voltage			0.35	0.5	V	V_{CC} = MIN., I_{OL} = 20 mA, V_{IN} = 2.0 V
I_{IH}	Input HIGH Current			1.0	50	µA	V_{CC} = MAX., V_{IN} = 2.7 V Each Input
					1.0	mA	V_{CC} = MAX., V_{IN} = 5.5 V
I_{IL}	Input LOW Current			-1.4	-2.0	mA	V_{CC} = MAX., V_{IN} = 0.5 V Each Input
I_{OS}	Output Short Circuit Current (Note 3)		-40	-65	-100	mA	V_{CC} = MAX., V_{OUT} = 0 V
I_{CCH}	Supply Current HIGH			10.8	16.0	mA	V_{CC} = MAX., V_{IN} = 0 V
I_{CCL}	Supply Current LOW			25.2	36.0	mA	V_{CC} = MAX., Inputs Open

AC CHARACTERISTICS: T_A = 25° C

SYMBOL	PARAMETER	LIMITS MIN.	LIMITS TYP.	LIMITS MAX.	UNITS	TEST CONDITIONS	TEST FIGURES
t_{PLH}	Turn Off Delay Input to Output	2.0	3.0	4.5	ns	V_{CC} = 5.0 V	A
t_{PHL}	Turn On Delay Input to Output	2.0	3.0	5.0	ns	C_L = 15 pF	

NOTES:
1. For conditions shown as MIN. or MAX., use the appropriate value specified under recommended operating conditions for the applicable device type.
2. Typical limits are at V_{CC} = 5.0 V, 25° C.
3. Not more than one output should be shorted at a time.

Figure 12.2-12 Schottky TTL data sheet. (*Fairchild Semiconductor.*)

than 0.22 V and $I_{OS} = 100$ mA rather than 55 mA. In addition the propagation delay is much reduced and is now typically 3 ns rather than 9 ns.

For low-power Schottky (54LS/74LS) the propagation delay is typically 10 ns, and the power requirements are less than half those of a standard TTL. Since the cost of these two types is about the same, low-power Schottky TTL is rapidly replacing the standard type in new designs.

12.3 EMITTER-COUPLED LOGIC (ECL)

In TTL logic operation depended on transistors being driven into or toward saturation. In addition, relatively large voltage swings were necessary in order to switch states reliably. This led to reliable operation, low power dissipation, and good noise immunity. However, operation into and out of saturation and the switching of large voltages increases the propagation delay time considerably. In order to decrease propagation delay time dramatically, a nonsaturating logic family, *emitter-coupled logic* (ECL), was introduced. This family has a switching voltage range of $V_{OH} - V_{OL} = 0.8$ V rather than 3 V as needed in TTL but dissipates more power than TTL.

The basic gate The basic component of the ECL gate is the *voltage-comparator circuit* shown in Fig. 12.3-1, also called a *current switch* or *difference amplifier*. The operation of the circuit is as follows. The input signal is applied to the base of T_1

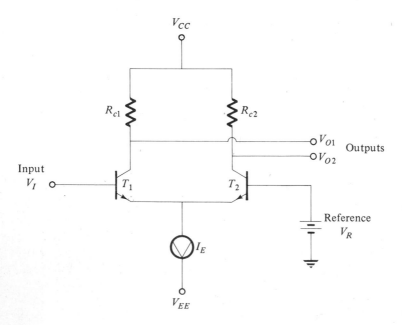

Figure 12.3-1 Basic component of ECL gate.

and a constant reference signal to the base of T_2. Current I_E is supplied by a constant-current source. Consider, for convenience, that the reference voltage is 0 V and that the input signal can be either $+0.4$ or -0.4 V, corresponding to HIGH or LOW logic levels. When V_I is -0.4 V, transistor T_1 will be cut off. Current I_E will then flow through T_2, the voltage on the emitters will be -0.7 V, and the base-emitter junction voltage of T_1 will be

$$V_{BE1} = V_I - V_E = -0.4 - (-0.7) = +0.3 \text{ V} \qquad (12.3\text{-}1)$$

Thus T_1 is OFF, and its collector voltage will be V_{CC}. Note carefully at this point that since T_1 is OFF, all of the current I_E flows through T_2. Thus the collector voltage of T_2 will be, assuming $\beta \gg 1$,

$$V_{O2} \approx V_{CC} - I_E R_{c2} \qquad (12.3\text{-}2)$$

By properly selecting V_{CC} and R_{c2} we can ensure that T_2 is not in saturation.

Now when V_I switches to $+0.4$ V, T_1 turns on, the emitter voltage changes to -0.3 V, and

$$V_{BE2} = V_R - V_{E2} = 0 - (-0.3) = +0.3 \text{ V}$$

Thus T_2 is OFF, and all the current I_E must flow through T_1. The collector voltage of T_2 will be V_{CC}, and the collector voltage of T_1 will be $V_{CC} - I_E R_{c1}$. Again, by appropriately selecting V_{CC} and R_{c1} we can ensure that T_1 will be ON but not into saturation.

To summarize the preceding discussion, we see that when the input voltage is switched from HIGH to LOW, the current I_E is switched from T_1 to T_2. This switching is accomplished rapidly, and the transistors are not saturated in the process. Because of this current switching, ECL is often called *current-mode logic* (CML).

If the two collector resistors are equal, the voltages at the two collectors will be identical in amplitude and *complementary;* i.e., when one is HIGH, the other is LOW. The output at the collector of T_2 follows the input voltage and is sometimes called the *in-phase output*, while the voltage at the collector of T_1 is opposite to the input voltage and is called the *out-of-phase output*.

12.3-1 A Commercial ECL Gate, Motorola Semiconductor MECL 10,000

In practice, the input transistor T_1 of Fig. 12.3-1 is replaced by a number of transistors in parallel, as shown in Fig. 12.3-2, in order to provide multiple inputs, and the current source I_E is replaced by resistor R_e, connected to a large negative voltage $V_{EE} = -5.2$ V. Transistors T_3 and T_4 are emitter followers, which provide level shifting and a low output impedance for driving transmission lines. Internal emitter resistors are not used because the transmission lines which this circuit is designed to drive provide the required load, and an internal resistor would waste power. The 50-kΩ input resistors serve to drain off input-transistor leakage current.

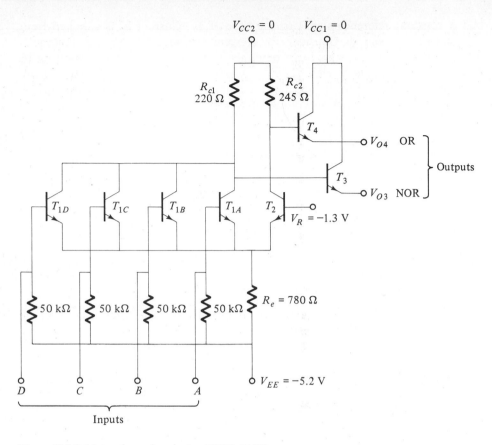

Figure 12.3-2 Motorola semiconductor MECL 10,000.

The reference voltage is provided by an internal temperature and voltage-compensated bias driver, which is not shown in the diagram. The bias driver sets the reference V_R at the midpoint of the logic swing independently of the temperature. With the V_{EE} supply set at -5.2 V, the reference voltage turns out to be approximately -1.3 V, as indicated. The use of a negative supply for V_{EE} with both V_{CC1} and V_{CC2} at ground potential reduces the effects of noise voltages which may be coupled into the power-supply connections. This aspect will be discussed in Sec. 12.3-5.

As with the other logic families we have discussed, ECL gates are commercially available in several types which differ in speed and power dissipation. The units with larger resistance values dissipate less power and operate at lower speeds. The higher-speed units have smaller resistance values and dissipate more power.

The gate shown in Fig. 12.3-2 is a Motorola Semiconductor MECL 10,000 unit, which features propagation delay times from 1.5 to 2 ns. As noted on the circuit diagram, both OR and NOR outputs are available from the emitter followers. This can be seen by considering that input A, for example, is a HIGH-level voltage. Then T_{1A} is ON and T_2 will be OFF. Since the collector of T_2 is high, there is a HIGH level at the base and the emitter of T_4. The same is true if there is a HIGH-level signal at either input B or C or D. The logic expression is then

$$V_{O4} = A + B + C + D \qquad (12.3\text{-}3)$$

When A or B or C or D is high, so that the collector of T_2 is HIGH, the collector of T_1 is LOW, so that the base and emitter of T_3 are LOW, leading to

$$V_{O3} = \overline{A + B + C + D} \qquad (12.3\text{-}4)$$

When all inputs change to LOW level, T_1 will be OFF and T_2 will be ON. Now there will be a LOW level at the emitter of T_4 and a HIGH level at the emitter of T_3, as required. This gate would be designated a four-input OR-NOR gate.

12.3-2 Calculation of Current and Voltage Levels

In this section we calculate some pertinent voltage levels and discuss the transfer characteristic. To determine the output logic levels we note that when the input voltages are at logic **0**, transistors T_1 are OFF and T_2 is ON. The voltage at the common emitter point is

$$V_E = V_R + V_{BE2} \qquad (12.3\text{-}5)$$

For the transistors used in ECL logic it is reasonable to use the value $V_{BE} = 0.8$ V in the active region. Thus

$$V_E = -1.3 - 0.8 = -2.1 \text{ V}$$

The emitter current of T_2 is

$$I_{E2} = \frac{V_E - V_{EE}}{R_e} = \frac{-2.1 + 5.2}{0.78} \approx +4 \text{ mA} \qquad (12.3\text{-}6)$$

Now, ignoring the small base current in T_4, we find the collector voltage

$$V_{C2} \approx V_{CC} - I_{E2}R_{c2} = 0 - (4)(0.245) \approx -1 \text{ V} \qquad (12.3\text{-}7)$$

The OR output is then

$$V_{OL} = V_{O4} = V_{C2} - V_{BE4} = -1 - 0.8 = -1.8 \text{ V} \qquad (12.3\text{-}8)$$

If the base current of T_3 is neglected, the base of T_3 is at 0 V, so that the NOR output is

$$V_{OH} = V_{O3} = -0.8 \text{ V} \qquad (12.3\text{-}9)$$

If now one or more of the input voltages changes to logic **1** level ($V_I = -0.8$ V), T_2 will turn OFF and T_1 will turn ON. The emitter voltage will be [see Eq. (12.3-5)]

$$V_E = V_I + V_{EB1} = -0.8 - 0.8 = -1.6 \text{ V} \qquad (12.3\text{-}10)$$

The emitter current becomes

$$I_{E1} = \frac{V_E - V_{EE}}{R_e} = \frac{-1.6 + 5.2}{0.78} \approx 4.6 \text{ mA} \tag{12.3-11}$$

and

$$V_{C1} \approx V_{CC} - I_{E1} R_{c1} = 0 - (4.6)(0.220) \approx -1.0 \text{ V} \tag{12.3-12}$$

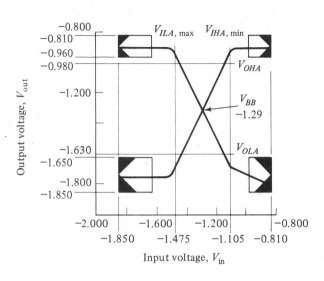

(a)

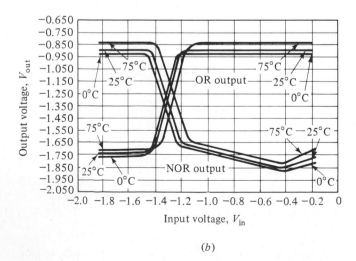

(b)

Figure 12.3-3 MECL 10,000 characteristics: (a) normal operating range (25°C); (b) transfer characteristics ($V_{EE} = -5.2$ V). (*Motorola, Inc.*)

The NOR output for T_3 will be

$$V_{OL} = V_{O3} = V_{C1} - V_{BE3} = -1.0 - 0.8 = -1.8 \text{ V} \qquad (12.3\text{-}13)$$

Since $V_{C2} = 0$ V, the OR output from T_4 will be

$$V_{OH} = -0.8 \text{ V} \qquad (12.3\text{-}14)$$

These calculations are summarized in the transfer characteristics shown in Fig. 12.3-3a. The HIGH-level output voltages for both OR and NOR are between -0.81 and -0.96 V rather than -0.8 V, as predicted by Eq. (12.3-9). The difference is accounted for by the voltage drop in R_{c1} or R_{c2} when T_1 or T_2, respectively, is off.

From the characteristics we see that the logic swing is quite symmetrical about the reference voltage $V_R = -1.3$ V.

The reader might assume that the transfer characteristics for OR and NOR should be *mirror images* because of the symmetry of the circuit. This is true except in the region where the NOR output is LOW (see Fig. 12.3-3b). The reason for the difference in the characteristics is that as the input, say V_A, goes positive, the collector current of T_{1A} continues to increase until saturation is reached when the input voltage is about -0.4 V. For more positive input voltages, the base-collector junction of T_1 is forward-biased, and hence the collector of T_{1A} and the output voltage V_{O3} rise as the input voltage rises. This effect occurs well outside the normal operating region, as can be seen by comparing Fig. 12.3-3a and b. The HIGH-level OR output depends on the collector voltage of T_2, which is independent of the input voltage except in the transition region where both transistors are conducting.

12.3-3 Voltage-Reference Circuit

The circuit which supplies the reference voltage V_R is on the same IC chip as the comparator (T_1 and T_2) and the emitter followers (T_3 and T_4). A schematic is shown in Fig. 12.3-4. The diodes and the base-emitter junction of T_5 provide temperature compensation by maintaining a V_R level in the midpoint of the transition region despite changes in temperature.

If we assume that the diode voltages and the base emitter voltage of T_5 are 0.8 V, the base voltage of T_5 is

$$V_{B5} = \frac{800}{5800} (5.2 - 1.6) \approx -0.5 \text{ V}$$

Then

$$V_R = V_{B5} - V_{BE5} = -0.5 - 0.8 = -1.3 \text{ V}$$

12.3-4 Wired-OR Connection

In Sec. 12.2 we found that the outputs of several TTL gates could be connected together directly, thereby providing an additional logic function, the wired-AND operation. The same can be done with ECL gates, providing an additional OR

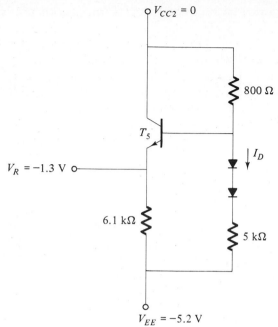

Figure 12.3-4 Reference voltage supply.

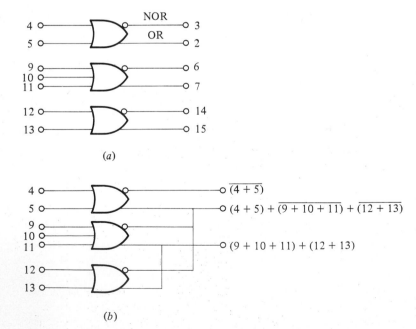

Figure 12.3-5 ECL gates: (a) MC10105 OR-NOR gate (numbers refer to IC socket pins); (b) example of wired-OR-NOR.

operation (see Prob. 12.3-2). Additional flexibility is available when both OR and NOR outputs are available. The logic diagram of a Motorola Semiconductor MC10105 triple two-, three-, two-input OR/NOR gate is shown in Fig. 12.3-5a. From the circuit diagram of the typical 10,000 series gate we see that the outputs are open emitters. In order to take advantage of the additional logic function it is only necessary to connect the appropriate output terminals together to the proper transmission line or load resistance. In the example using the MC10105 gate shown in Fig. 12.3-5b several additional OR operations are obtained without using additional gates.

12.3-5 Power-Supply Noise Reduction

The reader will have noted that ECL uses a negative supply voltage in contrast to TTL and CMOS (Sec. 12.4), which both use positive supply voltages. The advantage of this scheme is peculiar to ECL and may be seen qualitatively by referring to Fig. 12.3-6a. Here we have included voltage sources V_{n1} and V_{n2}, which represent the effect of noise voltages or transient changes on the power-supply lines. Assuming that the gate is from the 10,000 series, the -5.2 V supply is used for the differential amplifier current switch consisting of T_1 and T_2. Consider that a positive supply is used and T_2 is OFF. Then any noise on the supply is transmitted through R_{c2} to the base of T_4 and then to the output through the emitter follower. When a negative supply is used with T_2 ON and T_1 OFF, T_2 acts as a grounded base amplifier and the noise voltage present at the emitter of T_2 is greatly attenuated (see Prob. 12.3-6).

The -2-V supply which feeds the emitter-follower transmission line drivers may also be subject to noise pickup. The approximate equivalent circuit for the noise as viewed from the terminals of the noise generator V_{n2} is shown in Fig. 12.3-6b. From this circuit we see that for typical values of h_{ib4} (1 to 5 Ω) the noise voltage will be attenuated by a factor of 10 or more depending on the emitter current.

12.3-6 Interconnecting Gates Using Transmission Lines

The design and construction of high-speed ECL circuits involves several considerations that differ from the design and construction of TTL circuits: (1) time delays through the wires connecting two or more gates now become important, and (2) waveforms distort as a result of reflections on wires that have not been properly terminated. For example, a propagation delay of 2 ns, which is the propagation delay of an ECL 10,000 gate, is equivalent to 1 ft of interconnect wiring.

When ECL 10,000 is used, connecting wires up to 8 in long can be employed without significant overshoot, but if ECL III is used, only a 1-in interconnecting wire can be employed. Even in these cases, a *pull-down* resistor must be used since

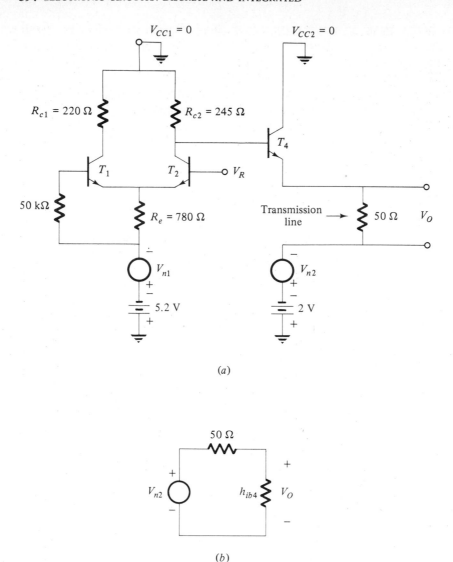

(a)

(b)

Figure 12.3-6 Power supply noise considerations: (a) circuit including noise sources; (b) equivalent circuit.

the output emitters are not terminated. Two possible terminating resistor connections[2] are shown in Fig. 12.3-7. When longer lead lengths are needed, transmission lines[3] must be used. Figure 12.3-8 shows the interconnection of gates using a 50-Ω transmission line. In Fig. 12.3-8a the 50 Ω terminating resistor is connected directly to -2 V, while in Fig. 12.3-8b the standard -5.2-V supply is used with two attenuating resistors, R_1 and R_2, where $R_1 \| R_2 = 50 \ \Omega$.

When connecting ECL gates that are on different circuit boards or over

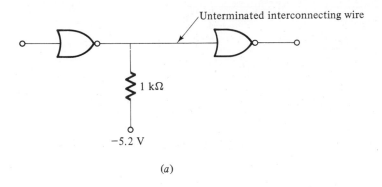

(a)

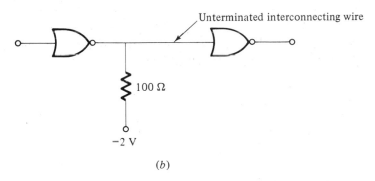

(b)

Figure 12.3-7 Terminations when using ordinary wire: (a) -5.2-V supply; (b) -2-V supply.

extremely long distances such that there is no high-frequency common ground connection it becomes necessary to use an ECL receiver or a driver-receiver combination, as shown in Fig. 12.3-9a. The driver is a high current OR-NOR gate which drives two 50-Ω coaxial cables each terminated in its characteristic impedance. The receiver is a difference amplifier with a differential-mode gain $A_d = 7$ and a common-mode gain $A_a = 0.16$. Thus, the CMRR of the receiver is

$$\text{CMRR} = \frac{A_d}{A_a} = \frac{7}{0.16} \approx 44 = 33 \text{ dB}$$

Figure 12.3-9b illustrates the circuit operation when a receiver is not employed. Since the voltage at the "ground" connection of gate G_1 is not the same as the voltage at the "ground" of G_2,† there may be a noise voltage V_n between them, as shown in the figure. Then the signal received by gate G_2 is $V_s + V_n$. As is often the case, V_n is much larger than V_s, and the output of G_2 changes state as a result of variations in V_n.

---

† Even though two points may be connected by a wire, the inductance of the wire presents a high impedance at the frequencies of 0.1 to 4 GHz that are common with ECL.

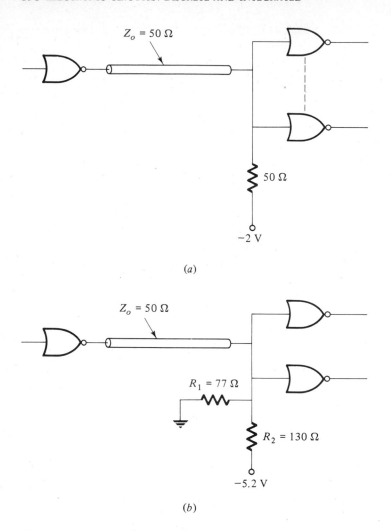

Figure 12.3-8 Transmission-line interconnections: (a) -2-V supply; (b) -5.2-V supply.

Figure 12.3-9c shows how the problem is alleviated by using a difference-amplifier receiver. If for the moment we assume that the common-mode gain $A_a = 0$, then

$$V_o = A_d[(V_s + V_n) - (\bar{V}_s + V_n)] = A_d(V_s - \bar{V}_s) = 2A_d V_s$$

and the noise voltage has been eliminated. In practice, however [see Eq. (7.8-1)],

$$V_o = A_d[(V_s + V_n) - (\bar{V}_s + V_n)] + A_a\left[\frac{(V_s + V_n) + (\bar{V}_s + V_n)}{2}\right]$$

$$V_o = A_d(2V_s) + A_a(V_n) = 2A_d V_s\left(1 + \frac{V_n/2V_s}{\text{CMRR}}\right)$$

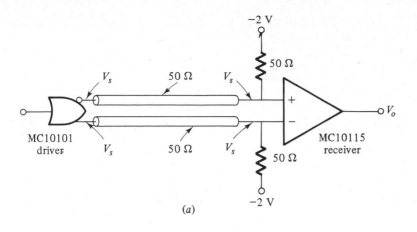

(a)

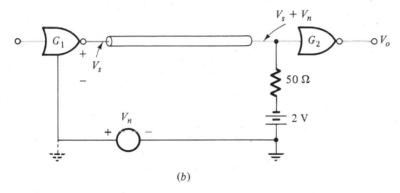

(b)

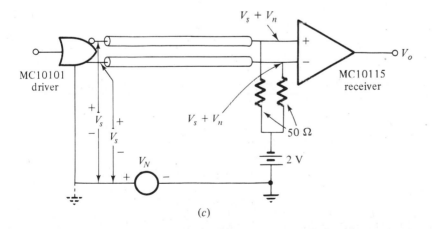

(c)

Figure 12.3-9 Techniques for reducing effects of physical separation in ECL circuits: (a) Driver-receiver combination; (b) effective noise source between grounds; (c) circuit showing effect of noise when driver-receiver combination is used.

For the 10115 receiver the CMRR = 44, and if $V_n/2V_s \ll 44$, the gate will change state only when V_s changes state.

12.3-7 Manufacturer's Specifications

The data sheet for the Motorola Semiconductor MC10102 quad two-input NOR gate is shown in Fig. 12.3-10. Because it is in a 16-pin package, only one OR output is available at pin 9. All the outputs for the MECL 10,000 are unterminated, and external *pull-down* resistors must be supplied. In this section we discuss some of the specifications on the data sheet.

Power-supply drain current The MC10102 consists of four NOR gates with emitter-follower output stages, as shown in Fig. 12.3-2 and one reference-voltage supply, as shown in Fig. 12.3-4. The power-supply drain current I_E specified on the data sheet refers to the current furnished by the V_{EE} supply. Both the V_{CC1} terminal and the V_{CC2} terminal are grounded using the shortest possible wires.

To calculate the average current drain I_E we need the supply current when the inputs are HIGH and then when the inputs are LOW. These currents were calculated earlier in this section, and we found, in Eq. (12.3-6), that when the input was LOW, the emitter current was

$$I_{E2} \approx 4 \text{ mA}$$

From Eq. (12.3-11) we have for the input HIGH

$$I_{E1} \approx 4.6 \text{ mA}$$

The average is then one-half the sum of these two currents

$$I_{AV} = \frac{4 + 4.6}{2} = 4.3 \text{ mA}$$

To find the reference supply current, consider the circuit of Fig. 12.3-4. The reference supply current is the emitter current of T_5 plus the current through the series diodes. Since we know that the voltage at the emitter of T_5 is -1.3 V, we have

$$I_{E5} = \frac{V_R - V_{EE}}{R_{e5}} = \frac{-1.3 - (-5.2)}{6.1} \approx 0.64 \text{ mA}$$

To find the diode current we assume that the base current of T_5 is negligible, so that

$$I_D = \frac{V_{CC2} - (V_{EE} + 2V_D)}{800 + 5000} = \frac{0 - (-5.2 + 1.6)}{5.8 \times 10^3} = 0.62 \text{ mA}$$

Finally, the current requirement of the reference supply is

$$I_R = 0.64 + 0.62 = 1.26 \text{ mA}$$

L SUFFIX
CERAMIC PACKAGE
CASE 620

ELECTRICAL CHARACTERISTICS

Each MECL 10,000 series has been designed to meet the dc specifications shown in the test table, after thermal equilibrium has been established. The circuit is in a test socket or mounted on a printed circuit board and transverse air flow greater than 500 linear fpm is maintained. Outputs are terminated through a 50-ohm resistor to −2.0 volts. Test procedures are shown for only one gate. The other gates are tested in the same manner.

MC10102L Test Limits

Characteristic	Symbol	Pin Under Test	−30°C Min	−30°C Max	+25°C Min	+25°C Typ	+25°C Max	+85°C Min	+85°C Max	Unit
Power Supply Drain Current	I_E	8				20	26			mAdc
Input Current	I_{inH}	12					265			μAdc
	I_{inL}	12			0.5					μAdc
Logic "1" Output Voltage	V_{OH}	9	−1.060	−0.890	−0.960		−0.810	−0.890	−0.700	Vdc
		9	−1.060	−0.890	−0.960		−0.810	−0.890	−0.700	
		15	−1.060	−0.890	−0.960		−0.810	−0.890	−0.700	
		15	−1.060	−0.890	−0.960		−0.810	−0.890	−0.700	
Logic "0" Output Voltage	V_{OL}	9	−1.890	−1.675	−1.850		−1.650	−1.825	−1.615	Vdc
		9	−1.890	−1.675	−1.850		−1.650	−1.825	−1.615	
		15	−1.890	−1.675	−1.850		−1.650	−1.825	−1.615	
		15	−1.890	−1.675	−1.850		−1.650	−1.825	−1.615	
Logic "1" Threshold Voltage	V_{OHA}	9	−1.080		−0.980			−0.910		Vdc
		9	−1.080		−0.980			−0.910		
		15	−1.080		−0.980			−0.910		
		15	−1.080		−0.980			−0.910		
Logic "0" Threshold Voltage	V_{OLA}	9		−1.655			−1.630		−1.595	Vdc
		9		−1.655			−1.630		−1.595	
		15		−1.655			−1.630		−1.595	
		15		−1.655			−1.630		−1.595	
Switching Times (50-ohm load) Propagation Delay	t_{12+15-}	15	1.0	3.1	1.0	2.0	2.9	1.0	3.3	ns
	t_{12-15+}	15								
	t_{12+9+}	9								
	t_{12-9-}	9								
Rise Time (20 to 80%)	t_{15+}	15	1.1	3.6	1.1		3.3	1.1	3.7	
	t_{9+}	9								
Fall Time (20 to 80%)	t_{15-}	15								
	t_{9-}	9								

TEST VOLTAGE VALUES (Volts)

@ Test Temperature	V_{IH} max	V_{IL} min	V_{IHA} min	V_{ILA} max	V_{EE}
−30°C	−0.890	−1.890	−1.205	−1.500	−5.2
+25°C	−0.810	−1.850	−1.105	−1.475	−5.2
+85°C	−0.700	−1.825	−1.035	−1.440	−5.2

TEST VOLTAGE APPLIED TO PINS LISTED BELOW:

	V_{IH} max	V_{IL} min	V_{IHA} min	V_{ILA} max	V_{EE}	(Vcc) Gnd
					8	1,16
	12				8	1,16
		12			8	1,16
	12 13				8	1,16
	12 13				8	1,16
			12 13	12 13	8	1,16
			12 13	12 13	8	1,16
	Pulse In 12			Pulse Out 15 9 9 9	−3.2 V	+2.0 V
				15 9 15 9	8	1,16

Figure 12.3-10 Data sheet for MC10102 quad two-input NOR gate. (*Motorola, Inc.*)

The total current drain from the V_{EE} supply encompasses four current switches and one reference supply; thus

$$I_{V_{EE}} = (4)(4.3) + 1.26 \approx 18.5 \text{ mA}$$

This compares well with the typical value of 20 mA specified on the data sheet.

Logic levels The output voltage logic levels specified on the data sheet are best interpreted in terms of the transfer curves of Fig. 12.3-11a, which clearly show the specification test points. The test data can be divided into two groups. In the first group $V_{IH, \text{max}}$ and $V_{IL, \text{min}}$ are the input test voltages. The corresponding outputs are V_{OH} and V_{OL}, for which minimum and maximum values are given in the data sheet of Fig. 12.3-10 and shown as $V_{OH, \text{max}}$, $V_{OH, \text{min}}$, $V_{OL, \text{max}}$, and $V_{OL, \text{min}}$ on the transfer curve of Fig. 12.3-11a. For example, from the data sheet for $+25°C$, $V_{EE} = -5.2$ V, we have the following data: for $V_{IH, \text{max}} = -0.810$ V and $V_{IL, \text{min}} = -1.850$ V applied to pins 12 and 13 the output ranges will be (measured at pins 9 and 15)

$$V_{OH, \text{min}} = -0.960 \text{ V} \qquad V_{OH, \text{max}} = -0.810 \text{ V}$$

$$V_{OL, \text{min}} = -1.850 \text{ V} \qquad V_{OL, \text{max}} = -1.650 \text{ V}$$

The second group relates to the switching thresholds, i.e., maximum and minimum values for which performance is specified. These data are distinguished by an A in the subscript. They are obtained by applying a test voltage $V_{ILA, \text{max}} = -1.475$ V and measuring the OR output to see that it is above the $V_{OHA, \text{min}} = -0.980$ V level and the NOR output to see that it is below the $V_{OLA, \text{max}} = -1.630$ V level. Similar checks are made using a test voltage $V_{IHA, \text{min}} = -1.105$ V.

These specifications ensure that:

1. The switching threshold ($\approx V_{BB}$) falls within the darkest rectangle (Fig. 12.3-11a); i.e., switching does not begin outside this rectangle.
2. Quiescent logic levels fall in the lightest shaded ranges.
3. Guaranteed noise immunity is met.

Figure 12.3-11b presents the manufacturer's input and output voltage levels as derived from the data sheet for one ECL gate driving another ECL gate.

Noise margin The guaranteed noise margins depend on the parameters which carry the A subscript, defined as follows. For the LOW state

$$NM_L = V_{ILA, \text{max}} - V_{OLA, \text{max}} = -1.475 - (-1.630) = 155 \text{ mV}$$

For the HIGH state

$$NM_H = V_{OHA, \text{min}} - V_{IHA, \text{min}} = -0.980 - (-1.105) = 125 \text{ mV}$$

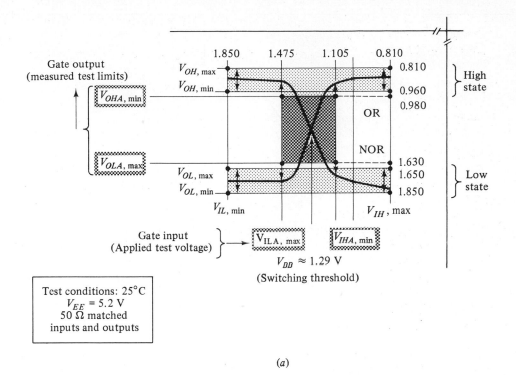

(a)

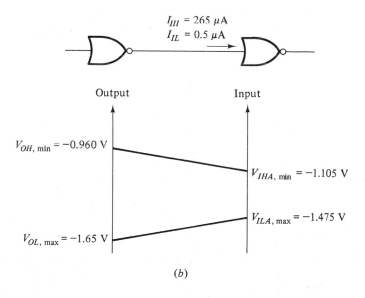

(b)

Figure 12.3-11 ECL: (a) MECL transfer curves (10,000 series) and specification test points; (b) alternate representation.

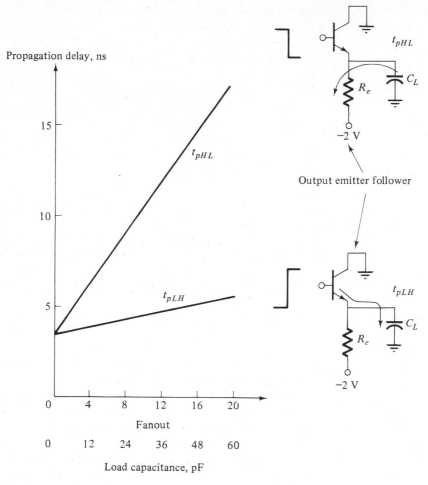

Figure 12.3-12 Propagation delay versus loading for ECL gates ($T = 25°C$).

These noise margins are absolute worst-case conditions. The smaller of the two is 125 mV, which constitutes the guaranteed margin against signal undershoot and power or thermal disturbances. Typical noise margins are usually better than the guaranteed value by about 75 mV. Comparing with Sec. 12.2-5, we see that ECL gates are much more susceptible to noise than TTL, which has guaranteed margins of 0.4 V.

Fan-out[2] The differential input to ECL circuits offers a number of advantages. Its ability to reject common-mode input signals offers immunity against noise injected on the V_{EE} supply line, and its relatively high input impedance makes it possible for any ECL gate to drive a large number of load gates without causing deterioration of the guaranteed noise margin. The manufacturer specifies a dc loading factor (the number of gate inputs of the same family that can be driven by a gate output) of 90 for MECL 10,000. It is further stated that best performance at fan-outs greater than 10 will occur with the use of transmission lines. However, since each driven gate introduces a capacitance (typically 3 pF), increasing the fan-out increases the propagation delay of the gate. Curves of propagation delay versus fan-out are shown in Fig. 12.3-12. t_{pLH} is the delay when the output switches from LOW to HIGH, so that the load capacitance charges through h_{ib}. When the output switches from HIGH to LOW, the output emitter follower is cut off for a short time, so that the load capacitance must discharge through the emitter resistor or transmission line. As a result, t_{pHL} is much larger than t_{pLH}.

12.4 CMOS LOGIC

*Slow but use little power, high Z_{in}
High Noise Immunity*

MOS devices see considerable use in logic circuits. They are considerably slower than TTL and ECL gates but because of the simplicity of their geometry and very small physical size, they can be packed quite densely on a silicon chip. This leads to *large-scale integration* (LSI), in which thousands of MOSFETS are contained in one circuit which occupies a fraction of a square inch of area. These circuits are used in applications where speed is not the primary factor. Another advantage of MOS logic is that since the inputs are insulated gates, the dc loading is minimal, leading to high fan-out capability. Also, the dc power dissipation is extremely small because of the complementary nature of the circuit. This is a distinct advantage in many applications, e.g., the pocket electronic calculator.

 Three types of MOS logic gates are possible, PMOS, NMOS, and CMOS. Because of their faster switching speeds and extremely low dc power consumption, CMOS circuits are now the most favored type of discrete IC MOS logic.

 In this section we discuss CMOS gates and their characteristics.

12.4-1 Basic CMOS Gates

For the CMOS family both NAND and NOR gates are available. The circuit of a two-input NAND gate is shown in Fig. 12.4-1, and we see that it consists of two series-connected *n*-channel driver transistors and two parallel-connected *p*-channel load transistors. The operation of the circuit is as follows. Consider that both inputs are LOW. Then both *p*-channel devices will be conducting (their channel resistances will be relatively low, of the order of 500 Ω). The *n*-channel transistors will be cut off since the LOW input level will be below their threshold voltage (their channel resistance will therefore be relatively high). The voltage-divider equivalent circuit shown in Fig. 12.4-1*b* then leads to the output voltage V_O being approximately V_{SS},† a HIGH level. Even if only one of the inputs is LOW,

---

† Since the supply voltage is connected to the *source* of the *p*-channel FET, we call it V_{SS} so that the subscript identifies the terminal to which the supply is connected.

NAND

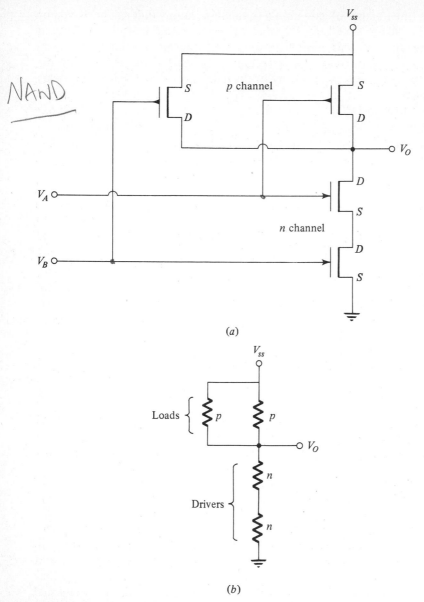

(a)

(b)

Figure 12.4-1 CMOS NAND gate: (*a*) circuit; (*b*) voltage-divider equivalent circuit.

the corresponding *n*-channel FET will be off and the voltage-divider equivalent indicates that the output will remain HIGH. Only when both inputs are HIGH will both *p*-channel transistors be OFF (their channel resistance large) and both *n*-channel transistors ON (their channel resistance low) so that the output voltage is approximately 0 V, corresponding to a LOW level.

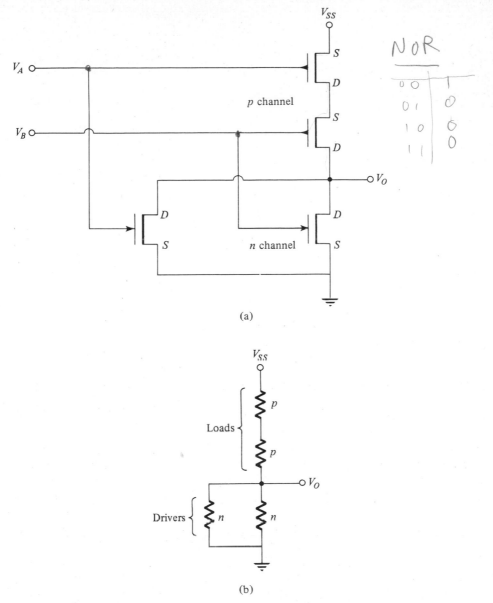

NOR

0	0	1
0	1	0
1	0	0
1	1	0

Figure 12.4-2 CMOS NOR gate: (a) circuit; (b) voltage-divider equivalent circuit.

A NOR gate is shown in Fig. 12.4-2. Here the load transistors are in series, while the drivers are in parallel. When both inputs are low, both p-channel devices are conducting (low resistance), while both n-channel devices are cut off (high resistance). The voltage divider action then leads to a HIGH output (approximately V_{SS}). With either or both inputs HIGH, one or both of the series p-channel

CD4001M/CD4001C

absolute maximum ratings

Voltage at Any Pin (Note 1)	$V_{SS} - 0.3V$ to $V_{DD} + 0.3V$
Operating Temperature Range	
CD4001M	−55°C to +125°C
CD4001C	−40°C to +85°C
Storage Temperature Range	−65°C to +150°C
Package Dissipation	500 mW
Operating V_{DD} Range	$V_{SS} + 3.0V$ to $V_{SS} + 15V$
Lead Temperature (Soldering, 10 seconds)	300°C

dc electrical characteristics CD4001M

PARAMETER	CONDITIONS	LIMITS									UNITS
		−55°C			25°C			125°C			
		MIN	TYP	MAX	MIN	TYP	MAX	MIN	TYP	MAX	
Quiescent Device Current (I_L)	$V_{DD} = 5V$			0.05		0.001	0.05			3	μA
	$V_{DD} = 10V$			0.1		0.001	0.1			6	μA
Quiescent Device Dissipation/Package (P_D)	$V_{DD} = 5V$			0.25		0.005	0.25			15	μW
	$V_{DD} = 10V$			1		0.01	1			60	μW
Output Voltage Low Level (V_{OL})	$V_{DD} = 5V$, $V_I = V_{DD}$, $I_O = 0A$			0.01		0	0.01			0.05	V
	$V_{DD} = 10V$, $V_I = V_{DD}$, $I_O = 0A$			0.01		0	0.01			0.05	V
Output Voltage High Level (V_{OH})	$V_{DD} = 5V$, $V_I = V_{SS}$, $I_O = 0A$	4.99			4.99	5		4.95			V
	$V_{DD} = 10V$, $V_I = V_{SS}$, $I_O = 0A$	9.99			9.99	10		9.95			V
Noise Immunity (V_{NL}) (All Inputs)	$V_{DD} = 5V$, $V_O = 3.6V$, $I_O = 0A$	1.5			1.5	2.25		1.4			V
	$V_{DD} = 10V$, $V_O = 7.2V$, $I_O = 0A$	3			3	4.5		2.9			V
Noise Immunity (V_{NH}) (All Inputs)	$V_{DD} = 5V$, $V_O = 0.95V$, $I_O = 0A$	1.4			1.5	2.25		1.5			V
	$V_{DD} = 10V$, $V_O = 2.9V$, $I_O = 0A$	2.9			3	4.5		3			V
Output Drive Current N-Channel (I_{DN})	$V_{DD} = 5V$, $V_O = 0.4V$, $V_I = V_{DD}$	0.5			0.40	1		0.28			mA
	$V_{DD} = 10V$, $V_O = 0.5V$, $V_I = V_{DD}$	1.1			0.9	2.5		0.65			mA
Output Drive Current P-Channel (I_{DP})	$V_{DD} = 5V$, $V_O = 2.5V$, $V_I = V_{SS}$	−0.62			−0.5	−2		−0.35			mA
	$V_{DD} = 10V$, $V_O = 9.5V$, $V_I = V_{SS}$	−0.62			−0.5	−1		−0.35			mA
Input Current (I_I)						10					pA

CD4001M/CD4001C

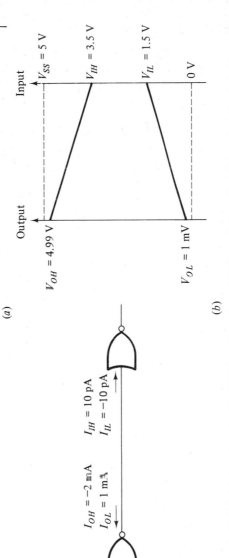

ac electrical characteristics CD4001M

$T_A = 25°C$ and $C_L = 15$ pF anc input rise and fall times = 20 ns. Typical temperature coefficient for all values of $V_{DD} = 0.3\%/°C$.

PARAMETER	CONDITIONS	MIN	TYP	MAX	UNITS
Propagation Delay Time High to Low Level (t_{PHL})	$V_{DD} = 5V$		35	65	ns
	$V_{DD} = 10V$		25	40	ns
Propagation Delay Time _ow to High Level (t_{PLH})	$V_{DD} = 5V$		35	65	ns
	$V_{DD} = 10V$		25	40	ns
Transition Time High to _ow Level (t_{THL})	$V_{DD} = 5V$		65	125	ns
	$V_{DD} = 10V$		35	70	ns
Transition Time Low to High Level (t_{TLH})	$V_{DD} = 5V$		65	175	ns
	$V_{DD} = 10V$		35	75	ns
Input Capacitance (C_I)	Any Input		5		pF

(a)

$V_{SS} = 5$ V

Input

$V_{IH} = 3.5$ V

$V_{IL} = 1.5$ V

0 V

Output

$V_{OH} = 4.99$ V

$V_{OL} = 1$ mV

$I_{OH} = -2$ mA $I_{IH} = 10$ pA
$I_{OL} = 1$ mA $I_{IL} = -10$ pA

(b)

Figure 12.4-3 CMOS; (a) data sheet for CD4001M quad two-input NOR gate; (b) graphical illustration of typical characteristics. (*National Semiconductor, Inc.*)

devices will turn off (high resistance), and one or both of the parallel n-channel devices will turn on (low resistance). For all these combinations, the voltage-divider action leads to a LOW output.

Both the NOR gate and the NAND gate can accommodate more than two inputs. This is accomplished by adding, for each input, an additional series-parallel pair of MOSFETs.

12.4-2 Manufacturer's Specifications

Some of the characteristics of the National Semiconductor CD4001M quad two-input NOR gate are shown in Fig. 12.4-3a. From the dc characteristics we see that at 25°C, V_{OL} is less than 10 mV and V_{OH} is within 1 mV of V_{SS}. The noise immunity (noise margin) is the same for both HIGH and LOW levels and is 30 to 45 percent of the supply voltage at 25°C. These characteristics are shown graphically in Fig. 12.4-3b for $V_{SS} = 5$ V.

The quiescent dc power dissipation at $V_{SS} = 5$ V and $T = 25$°C is typically 5 nW, an extremely small figure. This dissipation increases linearly with frequency and also depends on the amount of capacitive load as shown in Fig. 12.4-4.[4] The input current is typically 10 pA, an amount small enough to be ignored in many cases.

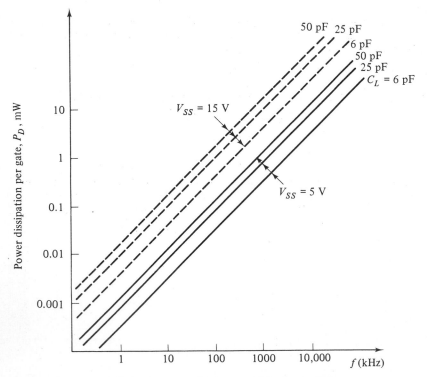

Figure 12.4-4 Power dissipation versus frequency and load capacitance.

From the ac characteristics we see that the average propagation delay with $V_{SS} = 5$ V is typically 35 ns. The transition time given in this data sheet represents the rise and fall times of the output waveform.

In a later section we shall compare the important characteristics of the most popular logic families.

12.5 INTERFACING

Situations often arise where we have portions of a digital system which require high-speed logic gates and other portions which can tolerate much slower speeds. For such cases, it is often advantageous to use more than one logic family. We can then take advantage, for example, of the high speed of Schottky TTL and the high packing density of CMOS in the same system. When this is done, we must take careful note of the fact that the various logic families operate at different voltage and current levels. Thus care must be taken to ensure that the conditions at the interface will lead to proper operation on both sides. Manufacturers provide a large variety of IC's designed to effect proper interfacing.

TTL-CMOS interfacing Consider interfacing TTL and CMOS gates. Let the power-supply requirements be 5 V for TTL and 5 V for CMOS. Figure 12.5-1 shows a diagram which includes a scale of input and output voltages. Typical values for TTL series 74LS and CMOS series 4000 are given in Table 12.5-1.

There are two cases to consider, depending on which type of gate is the driver and which is the load. In Fig. 12.5-2 we show a TTL gate driving N CMOS gates. If this configuration is to operate properly, conditions at the TTL output must be

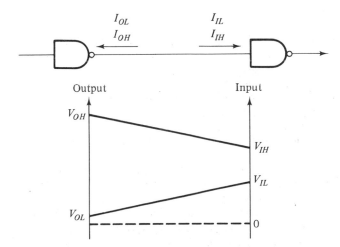

Figure 12.5-1 Voltage levels for interfacing TTL and CMOS.

Table 12.5-1 Typical values for interfacing TTL and CMOS

	TTL	CMOS
V_{IH}	2 V	3.5 V
I_{IH}	20 μA	10 pA
V_{IL}	0.8 V	1.5 V
I_{IL}	−0.36 mA	−10 pA
V_{OH}	2.7 V min	4.99 V
	3.4 V typ	
I_{OH}	−400 μA	−2 mA
V_{OL}	0.5 V	0.01 V
I_{OL}	8 mA	0.4 mA min
		1 mA typical

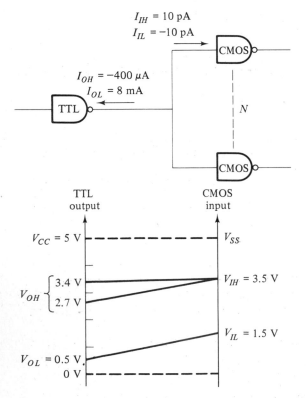

Figure 12.5-2 A TTL gate driving N CMOS gates.

such as to supply the requirements at the input to the CMOS gates. These requirements are

	TTL		CMOS	
	$-I_{OH}$	$\geq$	NI_{IH}	(12.5-1)
	I_{OL}	$\geq$	$-NI_{IL}$	(12.5-2)
	V_{OL}	$\leq$	V_{IL}	(12.5-3)
	V_{OH}	$\geq$	V_{IH}	(12.5-4)

From Table 12.5-1 we see that because the input current to the CMOS is so
small, Eqs. (12.5-1) and (12.5-2) will be easily satisfied for any reasonable value of
the fan-out N. Equation (12.5-3) is also readily satisfied. Only (12.5-4) causes
difficulty since the HIGH-level output of the TTL varies from 2.7 V minimum to
3.4 V typical at full-load current while the HIGH-level input to the CMOS gate
must be greater than 3.5 V for reliable operation. Since the V_{IH} of the CMOS gate
cannot be lowered, the solution is to raise the V_{OH} of the TTL gate above 3.5 V.
One way to accomplish this is to add a resistor R (typically 2 to 6 kΩ) to the TTL
output, as shown in Fig. 12.5-3, so that when the output goes HIGH, $V_O \approx 5$ V.
Another solution is to use an open-collector output TTL gate if one is available.

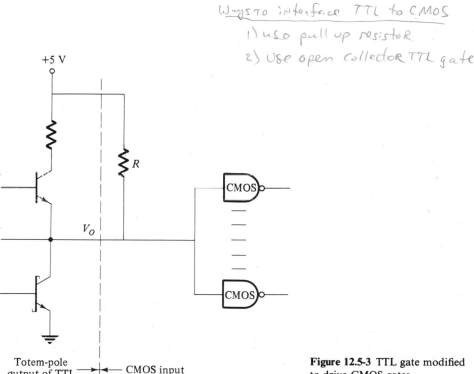

Totem-pole
output of TTL — CMOS input

Figure 12.5-3 TTL gate modified
to drive CMOS gates.

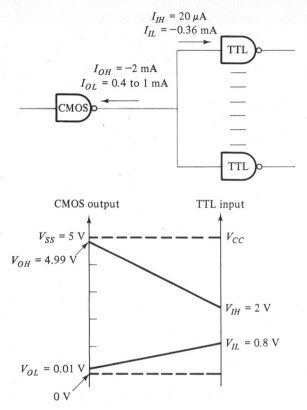

Figure 12.5-4 CMOS-to-TTL circuit.

When the situation is reversed, as shown in Fig. 12.5-4, where we have CMOS driving TTL, the conditions which must be satisfied are

	CMOS		TTL	
	$-I_{OH}$	$\geq$	NI_{IH}	(12.5-5)
	I_{OL}	$\geq$	$-NI_{IL}$	(12.5-6)
	V_{OL}	$\leq$	V_{IL}	(12.5-7)
	V_{OH}	$\geq$	V_{IH}	(12.5-8)

Of these, only (12.5-6) leads to problems since the LOW-level output current of the CMOS gate varies from 0.4 mA minimum to 1 mA typical while the TTL gate requires a LOW-level input current of 0.36 mA per gate. Thus the CMOS can drive one TTL gate in the LOW state without a safety factor. The solution here is to use a CMOS buffer between the two gates. Various types of buffers are available; e.g., the National Semiconductor CD 4010 noninverting hex buffer can supply up to 4 mA of output drive current at 25°C with $V_{DD} = 5$ V or 10 mA with $V_{DD} = 10$ V.

ECL-TTL interfacing When interfacing TTL with ECL, translator chips, e.g., the 10124 TTL-to-ECL translator or the 10125 ECL-to-TTL translator, are used. Typical propagation delays in such chips are 4 ns.

12.6 COMPARISON OF LOGIC FAMILIES

At the time of writing (1978) the most popular logic families are TTL, ECL, and CMOS. Table 12.6-1 lists the most important characteristics of each of these families.

In design, the intended application will determine the speed (propagation delay time) required. The choice of logic family is made on this basis, and once the type of logic is determined, other characteristics follow. From the table we see that ECL is by far the fastest logic. However, there is considerable sacrifice in power dissipation and noise immunity compared with the other types. ECL III has the smaller propagation delay of the two ECL series, but this is obtained at the expense of high power dissipation. CMOS is the slowest of the three but claims the advantages of extremely small power dissipation and high noise immunity.

The TTL family is the most popular at present. It exhibits good logic swing and noise immunity and is quite fast, especially in the Schottky versions listed in the table. The 54S/74S series has the highest speed and is the closest rival to ECL in this regard. The 54LS/74LS (low-powered Schottky) features relatively high speed along with significantly lower power dissipation. Because of the popularity of TTL, several hundred different circuit configurations are available commercially.

Figure 12.6-1*a* shows a graph of propagation delay versus power dissipation

Table 12.6-1 Comparison of logic families

	TTL		ECL		CMOS
	Series 54S/74S	Series 54LS/74LS	Series 10,000	Series III	Series 4000
V_{OL}, V	0.5	0.35	-1.75	-1.75	0
V_{OH}, V	3.5	3.4	-0.9	-0.9	10
Logic swing, V	3	3	0.85	0.85	10
NM_L, V, typical	1.15	1.15	0.21	0.2	4.5
NM_H, V, typical	1.95	1.95	0.21	0.2	4.5
Fan-out	10	10	90	70	> 50
Supply volts, typical	5	5	-5.2	-5.2	10
Quiescent power dissipation per gate	19 mW	2 mW	25 mW	60 mW	10 nW
Propagation delay time, ns	2	8	1.5	0.7	25

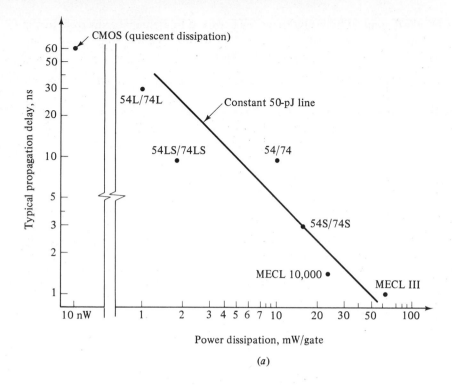

(a)

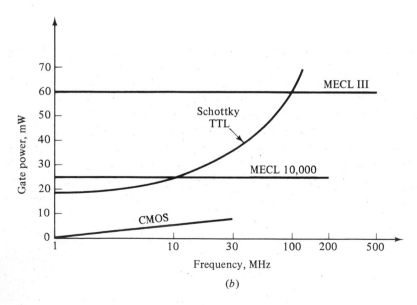

(b)

Figure 12.6-1 Logic-family comparisons: (a) speed-power characteristics; (b) power dissipation versus frequency.

for CMOS, TTL, and ECL. These two properties are related approximately by the equation

$$t_{pd} P_d \approx 50 \text{ pJ} \qquad (12.6\text{-}1)$$

and we see that in order to decrease propagation delay we must increase the power dissipation.

Figure 12.6-1b compares the power dissipated in a gate versus frequency for CMOS, TTL, and ECL gates. Note that above 20 MHz TTL dissipates more power than ECL 10,000 and above 100 MHz dissipates more power than ECL III.

REFERENCES

1. R. L. Morris and J. R. Miller, (eds.), "Designing with TTL Integrated Circuits," McGraw-Hill, New York, 1971.
2. W. R. Blood, Jr., "MECL System Design Handbook," 2 ed., Motorola Semiconductor Products, Inc., December 1972.
3. H. Taub and D. L. Schilling, "Digital Integrated Electronics," McGraw-Hill, New York, 1977.
4. "McMos Handbook," Motorola Semiconductor Products, Inc., 1974.

PROBLEMS

12.1-1 The 10,000 series ECL gates are characterized by the following typical threshold voltages: $V_{IH} = -1.105$ V, $V_{OH} = -0.960$ V, $V_{IL} = -1.475$ V, $V_{OL} = -1.65$ V. Find the HIGH- and LOW-state noise margins.

12.1-2 In the transistor inverter of Fig. 12.1-1a, $V_{CC} = 5$ V, $V_{CE, \text{sat}} = 0.2$ V, $V_{BE, \text{sat}} = 0.7$ V, and $\beta = 50$. Find values for R_c and R_b so that the transistor is at the edge of saturation with $I_{c, \text{sat}} = 10$ mA and $V_{IH} = 2$ V.

12.1-3 In the transistor inverter of Fig. 12.1-1a, a *base pull-down* resistor R_2 is connected from the base to a negative supply V_{BB}, as shown in Fig. P12.1-3. Values in the circuit are $R_b = 5$ kΩ, $R_2 = 20$ kΩ,

4/28/8|

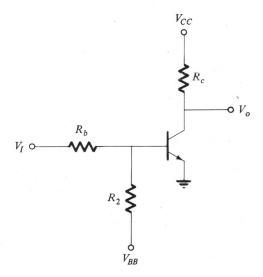

Figure P12.1-3

$R_c = 1.5$ kΩ, $V_{CC} = 5$ V, $V_{BB} = -5$ V, $\beta_{\min} = 20$, $V_{CE,\,\text{sat}} = 0.2$ V, $V_{BE,\,\text{sat}} = 0.7$ V, and $V_{BE,\,\text{on}} = 0.5$ V.

(a) Find the output when the input is $+5$ V. Is the transistor saturated?

(b) Find V_{BE} and the output when the input is 0 V.

(c) Find the noise margins and draw an output-input characteristic as in Fig. 12.1-3.

12.1-4 In the transistor inverter of Prob. 12.1-2 find the fan-out so that V_{OH} does not fall below 2 V.

12.2-1 In the TTL circuit of Fig. 12.2-1 $V_{CE,\,\text{sat}} = 0.2$ V, $V_{CC} = 5$ V, $C_L = 5$ pF, and $R_c = 1$ kΩ. Prove that the output charges from V_{OL} (0.2 V) to V_{OH} (3.3 V) in 5.2 ns.

12.2-2 In the circuit of Fig. 12.2-1 $V_{CC} = 5$ V, $\beta = 50$ for all transistors, collector saturation voltages are 0.2 V, and all pn-junction ON voltages are 0.7 V. Find suitable values of R and R_c so that $I_{c,\,\text{sat}} = 20$ mA and $V_{IH} = 2.4$ V.

12.2-3 In the TTL NAND gate shown in Fig. 12.2-3, assume that T_1 and T_2 saturate at the same time. Find the resulting current in R_c and the terminal voltages of T_2. Is T_3 saturated?

12.2-4 Assume that two TTL gates with active pull-up, as in Fig. 12.2-3, are inadvertently connected in the wired-AND configuration. Calculate the appropriate currents in the output circuits when one output is HIGH while the other is LOW. Assume a single TTL gate load and show that the gate with the HIGH output dissipates a large amount of power.

12.2-5 A wired-AND connection of open-collector TTL gates (see Fig. 12.2-4c) is shown in Fig. P12.2-5. Write the logic expression for the output and use De Morgan's theorems to convert it into OR form.

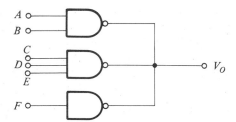

Figure P12.2-5

12.2-6 In the circuit of Fig. 12.2-4c three Schottky open-collector gates are to drive a fan-out of eight similar gates. Using Manufacturer's data sheet for the 54LS/74LS03, determine a suitable value for the pull-up resistor R_L. These specification sheets are not provided in the text.

12.2-7 For the TTL transfer characteristic of Fig. 12.2-5 prove that the slope of segment ab is -1.6. The TTL circuit is shown in Fig. 12.2-3.

12.2-8 Manufacturer's specifications for a TTL gate include a guarantee at the output of 12-mA sink current with $V_{OL} \le 0.4$ V and 6-mA source current with $V_{OH} \ge 2.4$ V. At the input $I_{IH} = 100$ μA when $V_I = 2.4$ V and $I_{IL} = -0.8$ mA when $V_I = 0.4$ V. Find the fan-out for both LOW- and HIGH-state conditions.

12.3-1 The circuit of a MECL III gate is identical to the 10,000 series gate shown in Fig. 12.3-2 except for the following component values: $R_{c1} = 290$ Ω, $R_{c2} = 300$ Ω, $R_e = 1.18$ kΩ, and $V_R = -1.175$ V. Find V_{OL} and V_{OH} for both OR and NOR outputs.

12.3-2 Show that when two or more emitter-follower outputs, as in the ECL 10,000 gates, are connected to a common load, the OR operation results; i.e., in Fig. P12.3-2 $V_O = V_{B3}$ OR V_{B4}.

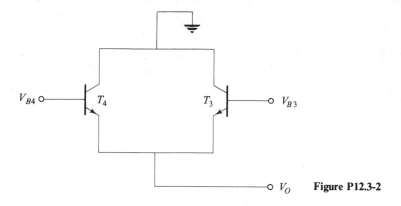

Figure P12.3-2

12.3-3 The logic diagram of the 10105 OR/NOR gate is shown in Fig. 12.3-5a. Using the wired OR-NOR technique, show how to obtain the following logic operations using only one 10105 gate:

$$L_1 = A + B + \overline{F + G}$$

$$L_2 = \overline{C + D} + \overline{A + B} + F + G$$

$$L_3 = \overline{F + G}$$

12.3-4 In Fig. 12.3-6a assume that $V_{n1} = 1$ mV and $V_{n2} = 0$. Find the noise component of the output for T_2 OFF and T_1 ON.

12.3-5 Repeat Prob. 12.3-4 for T_1 OFF and T_2 ON.

12.3-6 Repeat Prob. 12.3-4 and 12.3-5 with $V_{n1} = 0$ and $V_{n2} = 1$ mV.

12.3-7 Determine the dc fan-out for the ECL gate of Fig. 12.3-2 assuming that the OR output drives N similar gates. All transistors have $h_{fe} = 40$, and the driven gates must recognize an input voltage of -0.8 V as logic 1.

12.4-1 Draw the schematic of a four-input CMOS NOR gate.

12.4-2 Draw the schematic of a four-input CMOS NAND gate.

12.4-3 Design a two-input CMOS AND gate using only FETs.

12.5-1 A TTL gate is to drive a fan-out of 10 CMOS gates (data for both gates given in Table 12.5-1). Determine a suitable value for the pull-up resistor R in Fig. 12.5-3.

12.5-2 A CMOS gate is to drive a fan-out of six TTL gates (data for both gates in Table 12.5-1). What must be the characteristics of a buffer connected between the gates?

12.5-3 When interfacing CMOS to ECL, the positive voltages of the CMOS must be translated into the negative voltages of the ECL. Design a circuit to accomplish this translation. Specify the fan-out of your circuit.

12.5-4 Repeat Prob. 12.5-3 for an ECL-to-CMOS translator.

12.5-5 Repeat Prob. 12.5-3 for TTL to ECL.

12.5-6 Repeat Prob. 12.5-3 for ECL to TTL.

THIRTEEN

FLIP-FLOPS

INTRODUCTION

In Chap. 11, we considered the analysis of combinational networks composed of logic gates. With this capability we can specify a circuit which will provide any function of a number of input signals providing these signals are all available at the same time. The resulting circuits provide output signals upon application of the input signals.

Combinational circuitry alone is not sufficient to implement the many requirements of a system such as a digital computer. For as signals pass through such a system, they must often be held at a certain point long enough for other signals to arrive or for various operations to be performed. In order to satisfy these requirements we need a *memory element*, in which digital information can be stored temporarily. When such elements are added to combinational circuits, we call the result a *sequential circuit*.

The basic memory element is the *flip-flop*. This is a circuit, often constructed from logic gates, which is capable of storing 1 bit of information, and it can store that bit as long as required, keeping it available for use by other circuits. It will change the stored information only upon application of proper control signals. Flip-flops are connected together in various configurations to form, among other things, *registers*, which store and manipulate multibit data, and *counters*, which count the number of bits applied to their input terminals.

The nature and complexity of the operations performed by a digital system require that some means be provided to synchronize the many operations which are performed. This is the function of the *master clock*, which provides a train of

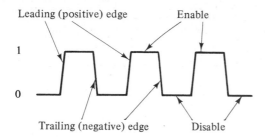

Leading (positive) edge   Enable

1

0

Trailing (negative) edge   Disable   **Figure 13.1** Clock waveform.

carefully regulated pulses. The flip-flops are usually arranged so that they change state only upon application of a clock pulse. How they change, if at all, depends on their inputs before the clock pulse arrives. Thus, it is often convenient to think of a sequential system as one which solves a sequence of combinational problems, each of which lasts for one clock period.

A clocked system as just described is called *synchronous*. The alternative, in which combinational operations trigger other operations as they occur, is called *asynchronous* operation. This is often unsatisfactory for digital systems of any complexity because of the difficulty of design and maintenance.

At this point we introduce some definitions which are important in connection with the clock. An idealized clock waveform is shown in Fig. 13.1. The important characteristics of this waveform are as follows:

1. The transition from logic **0** to logic **1**, called the *leading* or *positive* edge of the pulse.
2. The transition from logic **1** to logic **0**, called the *trailing* or *negative* edge of the pulse. The distinction between these two is important because later we shall find that the triggering of our flip-flops takes place at one or the other of these edges.
3. The time during which the clock is at logic **1** has been labeled *enable* and the time during which it is at logic **0** has been labeled *disable*.

In dealing with flip-flops we often select one input terminal of an input gate to the flip-flop and use it to either *enable* or *disable* the flip-flop. If the gate is AND or NAND, this input will take control when it is at logic **0** and the gate will be totally *disabled* since the output cannot respond to any of the other inputs; i.e., the output is the same whether the other inputs are **0** or **1** or any combination of them. When the input is at logic **1**, the gate is *enabled* since the output can respond to the other gate inputs.

If the gate is an OR or NOR gate, the *disable* signal is a logic **1**, for with a logic **1** on any input the output of the gate is the same regardless of the state of any of the other inputs. If the enable-disable input is at logic **0**, the *enable* condition, the output of the gate will be determined by the other inputs. The reader should be sure that these ideas are clearly understood at this point because they arise often in the systems we are going to study.

In this chapter we consider the most important flip-flop types. First we shall study their characteristics purely from the point of view that they manipulate logic 0s and 1s; then we shall discuss their implementation in the various logic families.

13.1 THE *RS* FLIP-FLOP

The flip-flop (or bistable multivibrator) is a circuit capable of storing 1 bit and is the basic building block of sequential circuits. A flip-flop constructed from NOR gates is shown in Fig. 13.1-1a. Only one input to each NOR gate is used, the other input being connected to logic 0 (ground in our example) so that it will not affect the operation of the gate. Thus, the NOR gates are acting as inverters. The notation Q and $\bar{Q}$ for the outputs will be justified shortly.

The circuit indicates that the two inverters are in cascade, with the output connected directly back to the input. This is positive feedback (see Chap. 10), which results in unstable operation in the sense that each output is always driven either to saturation or cutoff and remains in that state as long as power is applied. These are called *stable* states even though they are the result of unstable operation. If we consider saturation as the LOW state (logic 0) and cutoff as the HIGH state (logic 1), the two states are:

1. $Q = 1$ ($\bar{Q} = 0$), called the SET state, in which a 1 is stored
2. $Q = 0$ ($\bar{Q} = 1$), called the RESET or CLEAR, state in which a 0 is stored

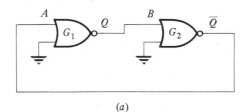

(a)

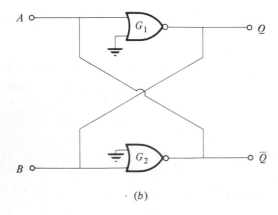

(b)

Figure 13.1-1 NOR-gate flip-flop: (a) Circuit drawn to emphasize positive feedback; (b) conventional representation.

That these two possibilities are consistent can be seen by considering the conventional representation of the circuit in Fig. 13.1-1b. If we start out with $Q = 1$, then the feedback connection causes $B = 1$, which leads via NOR gate 2 to $\bar{Q} = 0$. The other feedback connection causes $A = 0$, which leads via NOR gate 1 to $Q = 1$, the condition we assumed at the start. For the other state we start with $Q = 0$, and as we go around the loop we easily find that $\bar{Q} = 1$ and $Q = 0$, so that again consistent conditions are obtained and we have a *bistable* circuit.

The flip-flop shown in Fig. 13.1-1 is of no practical use because it has no input terminals permitting the flip-flop to change states conveniently. In order to change the state of the flip-flop we employ the two previously grounded NOR-gate inputs, as shown in Fig. 13.1-2a. The two input terminals are called the SET S and RESET R inputs. The circuit in this form is called an *RS flip flop*, and the implications of the words SET and RESET will become clear shortly.

The signals applied to the S and R terminals are *control inputs* (often called *data inputs*) and permit us to store 1 binary digit (*bit*). Determining the output states corresponding to all possible combinations of these inputs will lead to the truth table for the flip-flop.

Consider that $S = R = 0$, as in the first row of the truth table shown in Fig. 13.1-2b. For these particular inputs we have precisely the conditions which held when we discussed the circuit of Fig. 13.1-1. We found at that time that the circuit had two stable states, $Q = 1$ ($\bar{Q} = 0$) or $Q = 0$ ($\bar{Q} = 1$), and as long as $S = R = 0$, the NOR gates are not affected by the control inputs and the state of the RS flip-flop will not change. This is indicated in the truth table (Fig. 13.1-2b), where the notation is to be interpreted as follows:

$$Q_n = \text{value of } Q \text{ } before \text{ } S = R = 0 \text{ condition was imposed}$$

$$Q_{n+1} = \text{value of } Q \text{ } after \text{ } S = R = 0 \text{ condition was imposed}$$

The subscripts n and $n + 1$ are used because there will in general be numerous times during which the control inputs will change. When they both go to logic **0**, the output will remain in the state it had before the change. With regard to the output terminals, it is usually convenient to think of the Q terminal as the main output even though both Q and $\bar{Q}$ signals are available. When the Q output is HIGH (logic **1**), we say that the flip-flop is SET (a **1** is stored), and when the Q output is LOW (logic **0**), we say that it is RESET or CLEARED (a **0** is stored).

Now assume that $S = 0$ and $R = 1$. From the circuit we see that the output of the upper NOR gate will be

$$Q = \overline{R + \bar{Q}} = \overline{1 + \bar{Q}} = \bar{1} = 0 \tag{13.1-1}$$

Since the two inputs to the lower NOR gate are **0**, that is, $Q = 0$ and $S = 0$, its output is

$$\bar{Q} = \overline{S + Q} = \overline{0 + 0} = 1 \tag{13.1-2}$$

Thus the only possible state for this combination of control inputs is $Q = 0$ ($\bar{Q} = 1$). As noted above, this state is called the *reset* (*or clear*) *state*. If now the R

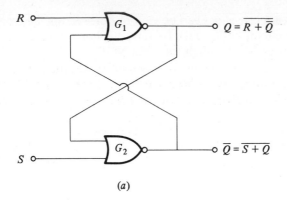

$$Q = \overline{R + \overline{Q}}$$

$$\overline{Q} = \overline{S + Q}$$

(a)

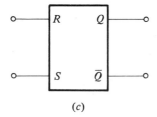

S	R	Q_{n+1}
0	0	Q_n
0	1	0
1	0	1
1	1	Not allowed

Reset state ⟶ 0   1   0

Set state ⟶ 1   0   1

(b)

R	Q
S	$\overline{Q}$

(c)

Figure 13.1-2 *RS* flip-flop: (*a*)NOR circuit; (*b*) truth table; (*c*) circuit symbol.

input changes to **0**, so that $S = R = 0$, the inputs to the upper NOR gate will be $R = 0$ and $\overline{Q} = 1$, so that the state which existed before the change ($Q = 0$) remains. The flip-flop has *remembered* what its output state was before the change of control inputs took place, and in fact we can state that the flip-flop was in the reset state ($S = 0, R = 1$) before the change. It is this ability to *remember a previous state* that sets the flip-flop apart from the combinational circuits discussed in the previous chapter.

In order to achieve the *set state* (row 3 of the truth table) we must have $S = 1$ and $R = 0$. Then since $S = 1$, the output of the lower NOR gate will be

$$\overline{Q} = \overline{S + Q} = \overline{1 + Q} = 0 \tag{13.1-3}$$

and the output of the upper NOR gate will be

$$Q = \overline{R + \bar{Q}} = \overline{0 + 0} = 1 \qquad (13.1\text{-}4)$$

Finally, if $S = R = 1$, we are led to conclude that both outputs are **0**, which contradicts our assumption that the two outputs are complementary. *This condition is therefore not allowed.* We shall see later that another type of flip-flop, the JK unit, avoids this ambiguity.

To summarize, we have shown that the RS flip-flop constructed from cross-coupled NOR gates obeys the truth table in Fig. 13.1-2b. It is capable of storing a **1** or a **0** on its output terminal under the control of the R and S inputs. The circuit symbol for this flip-flop is shown in Fig. 13.1-2c.

RS flip-flop constructed using NAND gates By using two inverters and two NAND gates, as shown in Fig. 13.1-3a, it is possible to achieve the same truth

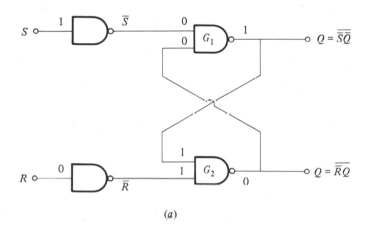

(a)

S	R	$\overline{S}$	$\overline{R}$	Q_{n+1}
1	1	0	0	Not allowed
1	0	0	1	1
0	1	1	0	0
0	0	1	1	Q_n

(b)

Figure 13.1-3 RS flip-flop using NAND gates: (a) circuits and logic levels when $S = 1$ and $R = 0$; (b) truth table.

table as for the NOR-gate flip-flop of Fig. 13.1-2. Consider first the action of the circuit as viewed from the $\bar{S}$ and $\bar{R}$ terminals. For $\bar{S} = \bar{R} = 0$ we have

$$Q = \overline{\bar{S}\bar{Q}} = \overline{0 \cdot \bar{Q}} = 1 \qquad (13.1\text{-}5)$$

$$\bar{Q} = \overline{\bar{R}Q} = \overline{0 \cdot Q} = 1 \qquad (13.1\text{-}6)$$

As in the NOR-gate flip-flop, this combination is not allowed. When $\bar{S} = 0$ and $\bar{R} = 1$, we find the logic values shown on the circuit of Fig. 13.1-3a, which indicate that the flip-flop is SET $(Q = 1)$. In a similar manner, the other rows of the truth table of Fig. 13.1-3b can be verified. If S and R columns are added to the table, it becomes identical with that shown in Fig. 13.1-2b and thus the NAND-gate flip-flop with inverters is identical in external behavior to the NOR-gate flip-flop; both are represented by the truth table of Fig. 13.1-2b and circuit symbol of Fig. 13.1-2c.

Debouncing a switch An interesting application of the RS flip-flop involves "debouncing" the output of a hand-operated push-button switch. In the typical application using an RS flip-flop constructed with NOR gates, as shown in Fig. 13.1-4a, the switch is normally set at position A and is used to apply a

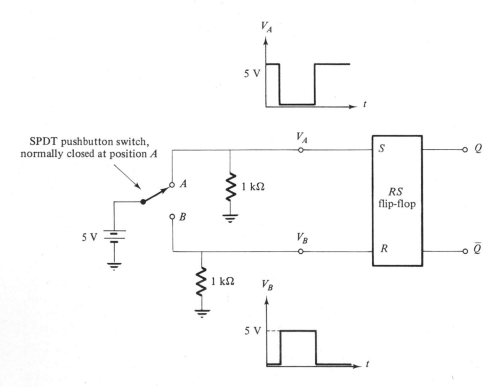

Figure 13.1-4a

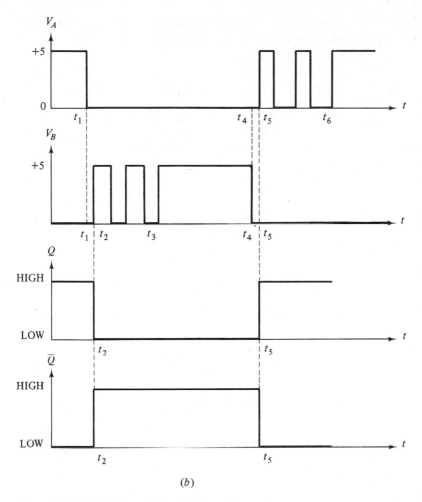

(b)

Figure 13.1-4 Switch debouncing: (a) circuit; (b) waveforms.

5-V (logic **1**) pulse to terminal B when the push button is depressed. The pulse duration is equal to the length of time the operator keeps the button depressed. Once the button is released, an internal spring restores the contact to position A. The problem arises because the switch contact bounces when it arrives at position B at the beginning of the pulse and then again at position A at the end of the pulse. This leads to undesired pulses which could cause false triggering in a logic system. Figure 13.1-4b shows typical waveforms which might occur during a switching cycle. Details of the process are as follows:

1. Before time t_1 the switch contact is at position A, so that $V_A = 5$ V and $V_B = 0$ V. Thus $S = 1$ and $R = 0$. From the truth table of Fig. 13.1-2b the flip-flop is SET, and so $Q = 1$, $\bar{Q} = 0$, as shown.

2. At time t_1 the operator pushes the button down, and the switch contact leaves A and starts moving to B. As soon as the contact leaves A, the voltage V_A drops to 0 V until time t_2, when the contact reaches B, causing V_B to rise to 5 V. The time interval between t_1 and t_2 may amount to a few milliseconds. From t_1 to t_2, S and R are both at logic **0**, and so the flip-flop will maintain the SET state that it was in before t_1.

3. At t_2, R goes to logic **1** with S at logic **0**, and so the flip-flop RESETS. This is shown in the Q and $\bar{Q}$ waveforms, Q going LOW and $\bar{Q}$ going HIGH.

4. Between t_2 and t_3 the contact bounces several times from terminal B. In a good switch the contact will not bounce all the way back to A. During the time that the switch is bouncing we have either $R = 1$, $S = 0$ or $R = S = 0$, so that the output remains in the RESET state no matter how many times the contact bounces.

5. At t_4 the operator releases the push button and the contact leaves B, arriving at A at t_5.

6. Between t_4 and t_5 $R = S = 0$, so that the flip-flop remains in the RESET state.

7. At t_5 we have $S = 1$, $R = 0$, so that the flip-flop returns to the SET state, as shown in the Q and $\bar{Q}$ waveforms.

8. Between t_5 and t_6 we have several contact bounces at A, and in this interval either $S = 1$, $R = 0$ or $S = R = 0$, and so the SET state is maintained.

The waveforms of Q and $\bar{Q}$ indicate that we have the desired debounced waveform available at the output of the flip-flop with either a HIGH-LOW-HIGH or a LOW-HIGH-LOW transition.

For the flip-flop to change from the set to the reset state the push-button switch must remain at position B for a time greater than the propagation delay time of the two gates which form the flip-flop. This can be seen by considering Fig. 13.1-2. When R goes to the **1** state, Q goes to **0** after the delay present in gate G_1 and then $\bar{Q}$ goes to **1** after an additional propagation delay in gate G_2. It is only after $\bar{Q} = 1$ that R can return to **0** without the flip-flop's again changing state since $\bar{Q} = 1$ *disables* gate G_1.

13.1-1 The Clocked RS Flip-Flop

In a digital system which involves many gates and flip-flops, it is impossible to guarantee that the S and R control signals will arrive at the precise times required for the desired logic operations. False logic commands may be generated under certain conditions because one signal arrives before or after another. This difficulty can usually be circumvented by allowing the flip-flop to change state only in synchronism with an external *clock*. In this way the output waveforms are synchronized with the clock and do not depend on the time of arrival of the S and R signals.

We illustrate the concept of external clock synchronization with the circuit of Fig. 13.1-5a, which shows an RS flip-flop with two additional AND gates at the input. The clock signal which accomplishes the synchronization is inputted to

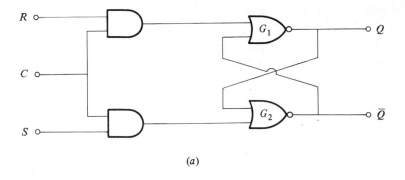

(a)

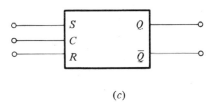

S_n	R_n	Q_{n+1}
0	0	Q_n
0	1	0
1	0	1
1	1	Not allowed

(b)

(c)

Figure 13.1-5 Clocked *RS* flip-flop (latch): (*a*) circuit; (*b*) truth table; (*c*) circuit symbol.

each AND gate while the *S* and *R* control signals constitute the other inputs. Thus, the inputs to the NOR gate flip-flop can now be activated only when the clock is HIGH. In this way the *S* and *R* inputs will determine the final state of the flip-flop, but the time at which the flip-flop may change state is determined by the clock signal. The control signals can arrive at different times without affecting the state of the output. For this system to operate properly the *S* and *R* inputs must arrive while the clock is in the LOW state, during which time the AND gates are *disabled*. The truth table for this circuit is given in Fig. 13.1-5*b*, and it is seen to be identical to the truth table for the *RS* flip-flop shown in Fig. 13.1-2. The clocked *RS* flip-flop is referred to in manufacturers' literature as a *latch*.

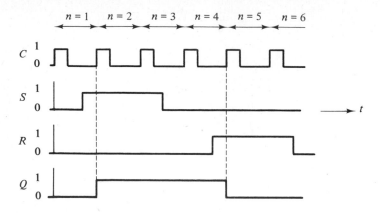

Figure 13.1-6 Waveforms in the clocked RS flip-flop.

Figure 13.1-6 shows a series of waveforms to illustrate all possible state transitions of the RS flip-flop. Before the first clock pulse both S and R are 0 and the flip-flop is reset, so that $Q = 0$. According to the truth table, the result of $R = S = 0$ is to produce no change. Thus, the reset condition holds during the interval $n = 1$.

In the interval between pulses 1 and 2 we find that $S = 1$ and R is unchanged at 0. This leads to the set state when pulse 2 is applied, and we have $Q = 1$. The output remains HIGH through the third pulse because S and R do not change.

During the $n = 3$ interval the S input drops to 0, so that $S = R = 0$, and as a result we have no change in Q in the $n = 4$ interval. Between the fourth and fifth pulses R becomes 1, S remaining at 0, so that the flip-flop is reset when pulse 5 is applied and remains in that state through the $n = 6$ interval. This completes all possible combinations of the inputs with the exception of $R = S = 1$, which is not allowed. If this condition is inadvertently applied, the output will generally be unpredictable.

Asynchronous (direct) inputs In the foregoing discussion we have always assumed a certain initial state for the flip-flop and we have brought about a change of this state only by application of the clock pulse. In practice it is often desirable to have the means whereby we can *preset* the flip-flop so that $Q = 1$ or *clear* the flip-flop so that $Q = 0$ independently of the inputs S and R and the clock C. Figure 13.1-7 illustrates a circuit in which the preset (Pr) and clear (Cl) signals completely override all other inputs. When Pr $=$ Cl $= 0$, they have no effect on the OR gates to which they are connected and the state of the flip-flop is determined by the $C, S,$ and R inputs. When Pr $= 1$ and Cl $= 0$, the output of the upper OR gate is 1 and that of the lower OR gate is 0 independent of the logic levels of all other signals. Thus the Pr terminal presets the flip-flop to $Q = 1$ independent of $C, S,$ and R. In a similar manner when Pr $= 0$, Cl $= 1$ the flip-flop is reset to $Q = 0$ (see Prob. 13.1-6) independent of $C, S,$ and R.

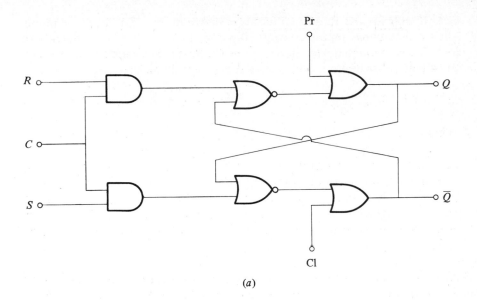

(a)

Pr	Cl	Q_{n+1}
0	0	Q_n
0	1	0
1	0	1
1	1	Not allowed

(b)

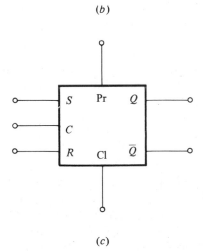

(c)

Figure 13.1-7 Asynchronous inputs: (a) circuits; (b) truth table; (c) circuit symbol.

A truth table for the Pr and Cl inputs is shown in Fig. 13.1-7*b*. These inputs are called *asynchronous* inputs since the flip-flop changes state independently of the clock. They are sometimes referred to as the *direct-set* S_d and the *direct-reset* R_d inputs. Note that the Pr and Cl inputs cannot simultaneously be in the 1 state.

13.2 THE *RS* MASTER-SLAVE FLIP-FLOP

In many flip-flop applications it is desirable to determine the states of the S and R inputs, then decouple the inputs from the flip-flop so that subsequent changes in S and R are not noted, and finally, after the inputs are decoupled, to change the flip-flop outputs Q and $\bar{Q}$. In practice there are two ways to accomplish this sequence of operations, *master-slave* and *edge triggering*. In this section we discuss the master-slave flip-flop, shown in Fig. 13.2-1. The edge-triggered flip-flop is described in Sec. 13.4.

The master-slave flip-flop employs two flip-flops, FF_1 and FF_2, operated 180° out of phase with one another. Thus, when the clock is HIGH, C_1 is HIGH and the state of the S and R inputs results in Q_1 and $\bar{Q}_1$ of the *master* flip-flop FF_1 changing state when necessary. However, during this time clock C_2 is LOW, and FF_2 does not change state. When the clock pulse C goes LOW, C_1 is LOW and FF_1 is decoupled from the S and R inputs. Now clock C_2 is HIGH, and the *slave* flip-flop FF_2 responds to its S_2 and R_2 inputs. Hence, the outputs Q and $\bar{Q}$ of the master-slave flip-flop change state only after the clock pulse C has returned to the 0 state.

The clock waveforms and their effect on the circuit operation are shown in Fig. 13.2-2, and typical waveforms are shown in Fig. 13.2-3. Observe carefully that Q changes only when the C_1 clock pulse returns to the LOW state.

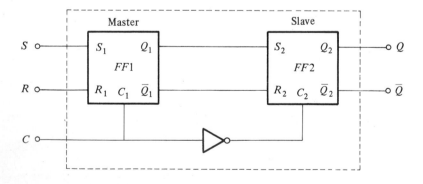

Figure 13.2-1 Master-slave flip-flop.

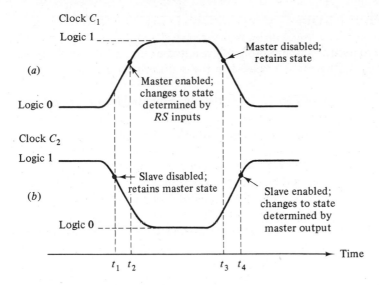

Figure 13.2-2 Clock pulse waveforms illustrating events in masterslave flip-flop triggering: (*a*) clock pulse applied to master; (*b*) inverted clock pulse applied to slave.

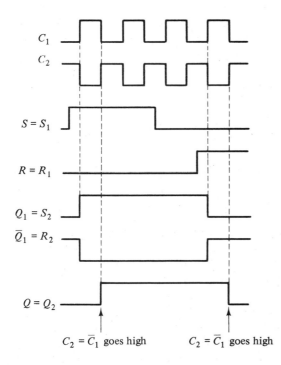

Figure 13.2-3 Waveforms illustrating possible states of the *RS* master-slave flip-flop.

13.3 THE *JK* FLIP-FLOP

In the *RS* flip-flop we found that an ambiguity resulted when both *S* and *R* inputs were at logic **1**. This ambiguity is avoided in the *JK* flip-flop discussed in this section.

The clocked *RS* flip-flop of Fig. 13.2-1 is readily converted into *JK* operation by adding an additional terminal to each of the input AND gates of the master flip-flop and providing a feedback connection from these terminals to the outputs, as shown in Fig. 13.3-1a. The *S* input (S_1 of the master flip-flop) is now called the *J* input, and the *R* input (R_1 of the master flip-flop) is called the *K* input. Before the feedback connection was added, the *S* and *R* inputs "steered" the clock waveform to one or the other of the input AND gates. If the *S* input was high, that gate was enabled and the clock pulse was steered through the gate to set the flip-flop. The addition of the feedback allows the output state of the flip-flop to participate in steering the clock pulse to one gate or the other.

The truth table for the *JK* flip-flop is shown in Fig. 13.3-1b. The first three

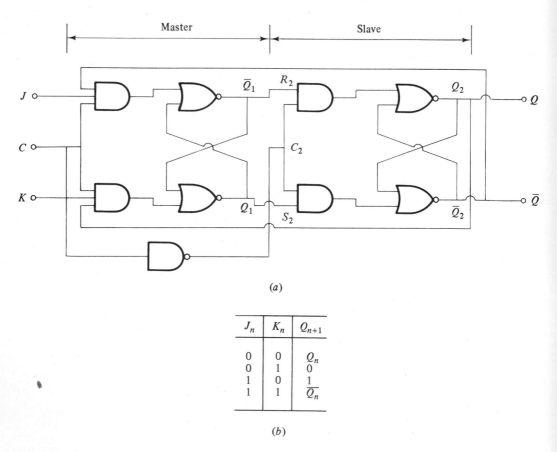

(a)

J_n	K_n	Q_{n+1}
0	0	Q_n
0	1	0
1	0	1
1	1	$\bar{Q}_n$

(b)

Figure 13.3-1 *JK* master-slave flip-flop: (a) circuit; (b) truth table.

rows are identical to those for the clocked RS truth table. The circuit operation is as follows:

1. Consider the first row. When $J_n = K_n = 0$, both input AND gates are disabled, so that a clock pulse will not affect the state of the flip-flop and we have $Q_{n+1} = Q_n$.
2. For the second row we have two possibilities. First $J_n = 0$, $K_n = 1$, and $Q_n = 0$; that is, the flip-flop is in the reset state. Here the upper AND gate is disabled because $J_n = 0$, and the lower AND gate is disabled as a result of the feedback connection because $Q_n = 0$. Thus, the state of the flip-flop will not change upon application of a clock pulse. If now $J_n = 0$, $K_n = 1$, and $Q_n = 1$ so that the flip-flop is in the set state, the upper AND gate is disabled since $J_n = 0$ but the lower AND is enabled since $Q_n = 1$ and $K_n = 1$ and a clock pulse will cause the flip-flop to shift to the reset state.
3. For the third row, similar arguments show that application of a clock pulse will shift the flip-flop to the set state if it is not already there.
4. Finally, for the fourth row we have $J_n = K_n = 1$. Thus, the output state of Q and $\bar{Q}$ will determine which of the AND gates is enabled. If $Q_n = 1$, the lower AND gate is enabled as a result of the feedback connection and a clock pulse will reset the flip-flop to $Q_{n+1} = 0$. If $Q_n = 0$ ($\bar{Q}_n = 1$), the upper AND gate is enabled and a clock pulse will set the JK flip-flop to $Q_{n+1} = 1$. Each of these cases leads to $Q_{n+1} = \bar{Q}_n$. To describe this operation, wherein each time the clock pulse arrives the state of the flip-flop changes, we say that the flip-flop *toggles*. This is analogous to the action of an ordinary toggle switch, and it is this toggling when $J_n = K_n = 1$ that differentiates the JK flip-flop from the RS flip-flop.

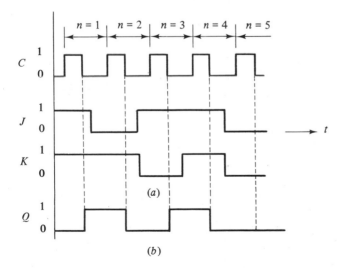

Figure 13.3-2 Waveforms for Example 13.2-1: (*a*) input; (*b*) output.

Example 13.3-1 The C, J, and K waveforms shown in Fig. 13.3-2a are applied to a master-slave JK flip-flop. Plot the Q output waveform assuming that the initial state is $Q = 0$.

SOLUTION The Q waveform is shown in Fig. 13.3-2b. Observe that since this is a master-slave flip-flop, Q changes only when the clock pulse returns to the LOW state. Specific actions are as follows:

Clock pulse 1: Both J and K are **1** before this pulse, so that Q toggles, going to **1** on the negative edge of the pulse.
Clock pulse 2: $J = 0$ and $K = 1$ before the pulse. Since this is the *reset* condition, Q goes to **0**.
Clock pulse 3: $J = 1$ and $K = 0$; this is the *set* condition, and so Q goes to **1**.
Clock pulse 4: $J = 1$ and $K = 1$ and so the flip-flop toggles, Q going to **0**.
Clock pulse 5: Here $J = K = 0$. Since this is the *no-change* condition, Q remains at **0**. ///

13.4 EDGE-TRIGGERED *JK* FLIP-FLOP

An edge-triggered flip-flop is one in which the inputs can change when the clock is in the LOW state or in the HIGH state without causing a change in the output Q. The output changes only during the transient operation of the flip-flop brought about by the *trailing edge* of the clock.

The 54S114 Schottky edge-triggered flip-flop is shown in Fig. 13.4-1a. When the clock is LOW, gates G_1 and G_2 are disabled and changes in the J and K inputs cannot affect the output of the flip-flop. To see what takes place when the clock is HIGH let $Q_n = 1$ ($\bar{Q}_n = 0$). Then since $C = 1$ and $Q_n = 1$, the output of $G_3 = 1$; hence $\bar{Q}_n$ remains **0** independent of the J or K input. Also, since $\bar{Q}_n = 0$, the outputs of gates G_5 and G_6 are LOW and $Q_n = 1$ independent of J or K.

We now show that the flip-flop changes state on the trailing edge of the clock pulse. This is the edge which disables gates G_1 and G_2. We assume that $J = K = 1$, so that we are in the toggling mode, and let $Q_n = 1$ ($\bar{Q}_n = 0$).

Refer to Fig. 13.4-1a and the waveforms of Fig. 13.4-1b and let the clock go from **0** to **1**. Then G_1 is enabled, and its output falls after a time delay t_{pd1}. Simultaneously G_3 is enabled and its output rises after a delay t_{pd3}. It is seen that the delay encountered in G_1 far exceeds the delay in G_3. The flip-flop is purposely constructed so that the delays in G_1 and G_2 are greater than the delays in any of the other gates. For the purpose of clearly illustrating the operation we have arbitrarily assumed an 8 : 1 ratio in propagation delay. After G_1 goes to **0**, the output of G_4 becomes **0**. The outputs Q and $\bar{Q}$ do not change.

Now let the clock pulse return to **0**, thereby disabling gates G_1 and G_2. After a delay t_{pd1}, G_1 rises to the **1** state. However, after a much shorter delay t_{pd3}, G_3 falls to the **0** state. Now the outputs of both G_3 and G_4 are **0**, and therefore the

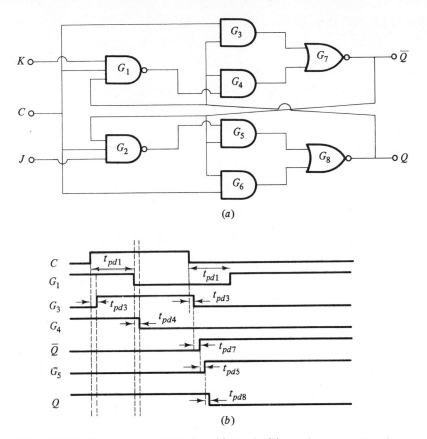

(a)

(b)

Figure 13.4-1 JK edge-triggered flip-flop: (a) circuit; (b) waveforms, $J = K = 1$.

output of G_7, which is $\bar{Q}$, rises to the **1** state. With $\bar{Q}$ high, the output of gate G_5 goes high, and then Q falls. It is left for the problems to show that

$$t_{pdi} \geq 4t_{pdj} \qquad \begin{aligned} i &= 1, 2; \\ j &= 3, 4, 5, 6, 7, 8 \end{aligned}$$

in order for the edge-triggered flip-flop to operate properly.

13.5 THE D (DELAY) FLIP-FLOP

In a synchronous digital system it is often necessary to have a signal delayed by exactly one cycle of the clock, and a special flip-flop is available for this operation called the D (delay) flip-flop. The logic diagram of a type 7474 D-type flip-flop using NAND gates is shown in Fig. 13.5-1a and the truth table in Fig. 13.5-1b. Its function is to transfer the data on the D line to the Q output at the next clock pulse. In the edge-triggered flip-flop shown the change occurs when the clock

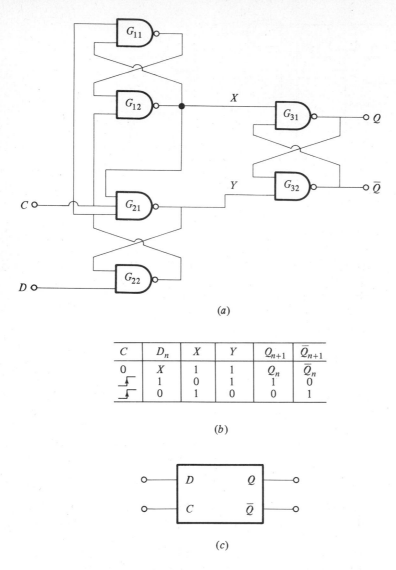

(a)

C	D_n	X	Y	Q_{n+1}	$\bar{Q}_{n+1}$
0	X	1	1	Q_n	$\bar{Q}_n$
⌐_	1	0	1	1	0
⌐_	0	1	0	0	1

(b)

(c)

Figure 13.5-1 D-type flip-flop: (a) logic diagram (preset and clear circuits have been omitted); (b) truth table; (c) logic symbol.

pulse goes HIGH. Operation of the flip-flop is best explained by considering each line of the truth table:

1. $C = 0$ (clock LOW); the outputs of gates G_{21} and G_{12} are both HIGH, that is, $X = Y = 1$, so that gates G_{31} and G_{32} are enabled. For this condition either state $Q = 0$, $\bar{Q} = 1$ or $Q = 1$, $\bar{Q} = 0$ is valid, and so Q will remain at the value it had previously; that is, $Q_{n+1} = Q_n$. The D input has no effect on the output.

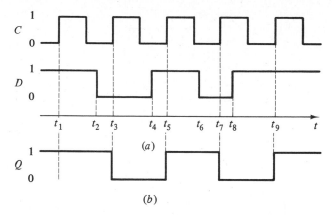

Figure 13.5-2 Waveforms for Example 13.5-2: (*a*) input; (*b*) output.

2. $D_n = 1$, C going HIGH; here Y will remain **1**, as it was while C was LOW. The inputs to gate G_{12} go HIGH, so that $X = 0$, and Q_{n+1} goes to **1** and $\bar{Q}_{n+1}$ to **0**. As long as C is HIGH, changes in D produce no change in X, Y, or Q. When C goes LOW, Y remains HIGH and X goes HIGH so that the G_{31}-G_{32} flip-flop remains in the previous state, that is, $Q_{n+1} = Q_n = 1$.

3. $D_n = 0$, C going HIGH. Here Y goes to **0** (all three inputs to gate G_{21} are **1**) while X remains at **1**. Thus, $Q_{n+1} = 0$. While $C = 1$, changes in D will not affect X, Y, or Q. When C goes LOW, X remains HIGH while Y goes HIGH so that the output remains in its previous state; that is, $Q_{n+1} = Q_n = 0$.

These steps show that the data on the D input when the clock is LOW will be transferred to the Q output when the clock goes HIGH, i.e., on the positive edge of the clock pulse. They also show that while the clock is HIGH, changes in D have no effect on the output Q (see Prob. 13.5-1).

Example 13.5-1 Figure 13.5-2*a* shows typical waveforms for the clock and the D input. Sketch the Q output waveform assuming the flip-flop is initially SET.

SOLUTION The resulting waveform is shown in Fig. 13.5-2*b*. Note that the Q output assumes the same state as the D input but that the output is synchronized to the positive edge of the clock pulse. Thus at t_1, $D = Q = 1$. At t_3, Q goes to **0**; at t_4, D goes to **1** and Q follows at t_5. At t_6, D goes to **0**, Q following at t_7. At t_8, D goes to **1** while the clock is still at **1**; however, the change in Q to **1** does not occur until the positive clock edge at t_9.　　　///

In Fig. 13.5-3 we show the logic diagram for a D latch. The *latch* is a clocked flip-flop which is neither master-slave nor edge-triggered. The D latch is a much less complicated device than the edge-triggered D flip-flop shown in Fig. 13.5-1 and therefore less costly. However, while the D flip-flop can be used as a storage cell in a shift register (Sec. 14.1), the D latch cannot.

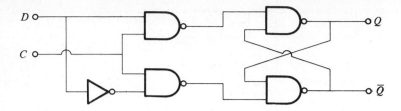

Figure 13.5-3 *D* latch.

13.6 COMMERCIALLY AVAILABLE FLIP-FLOPS

In this section we consider briefly the types of flip-flops available in the different logic families. As noted previously, most designs at the time of this writing use either Schottky TTL, ECL, or CMOS. We tabulate the most important characteristics of each family for comparison.

13.6-1 TTL

The popular TTL family has four major types at present:

1. Standard 54/74
2. High-speed 54/74H
3. Schottky 54/74S
4. Low-powered Schottky 54/74LS

The different flip-flop types are available in some or all of these varieties. For example, the 1976 Signetics catalog lists 17 integrated circuit flip-flops; 15 are *JK* units and 2 are *D* type. A few of the flip-flops are listed below:

1. 54/7470 *JK* Positive edge-triggered. Inputs are J_1, J_2, $\bar{J}$, K_1, K_2, $\bar{K}$, preset, clear, and clock. Available in standard 54/74 only. One unit per 14-pin package.
2. 54/7472 *JK* Master-Slave. Inputs J_1, J_2, J_3, K_1, K_2, K_3, preset, clear, and clock. Available in standard 54/74, 54/74H, and 54/74LS. One unit per 14-pin package.
3. 54/7474 Dual *D* type positive edge-triggered. Inputs *D*, preset, clear, and clock. Available in all four types. Two completely separate units per 14-pin package.

Manufacturer's specifications The data sheet for a TTL *JK* master-slave flip-flop is shown in Fig. 13.6-1. It includes a complete schematic of the unit, a truth table, and typical clock waveform with a description of the operation at each significant point on the clock pulse. The schematic is quite complicated, containing 22 transistors, 8 of them with multiple emitters, and 11 diodes.

J-K MASTER-SLAVE FLIP-FLOP | S5472 N7472

S5472—A,F,W • N7472—A,F

DIGITAL 54/74 TTL SERIES

DESCRIPTION

These J-K flip-flops are based on the master-slave principle and each has AND gate inputs for entry into the master section which are controlled by the clock pulse. The clock pulse also regulates the state of the coupling transistors which connect the master and slave sections. The sequence of operation is as follows:

1. Isolate slave from master

2. Enter information from AND gate inputs to master

3. Disable AND gate inputs

4. Transfer information from master to slave.

TRUTH TABLE

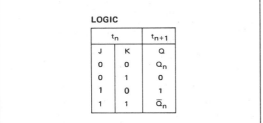

LOGIC

t_n		t_{n+1}
J	K	Q
0	0	Q_n
0	1	0
1	0	1
1	1	$\bar{Q}_n$

NOTES:

1. $J = J1 \cdot J2 \cdot J3$

2. $K = K1 \cdot K2 \cdot K3$

3. t_n = Bit time before clock pulse.

4. t_{n+1} = Bit time after clock pulse.

5. NC = No Internal Connection.

PIN CONFIGURATIONS

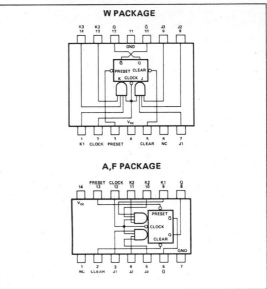

W PACKAGE

A,F PACKAGE

CLOCK WAVEFORM

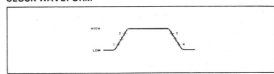

SCHEMATIC DIAGRAM

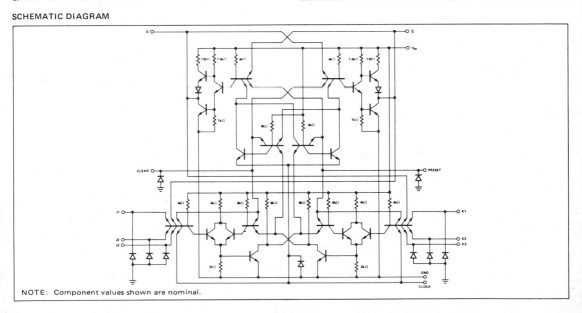

NOTE: Component values shown are nominal.

RECOMMENDED OPERATING CONDITIONS

		MIN	NOM	MAX	UNIT
Supply Voltage V_{CC}:	S5472 Circuits	4.5	5	5.5	V
	N7472 Circuits	4.75	5	5.25	V
Operating Free-Air Temperature Range, T_A:	S5472 Circuits	-55	25	125	°C
	N7472 Circuits	0	25	70	°C
Normalized Fan-Out From each Output, N				10	
Width of Clock Pulse, $t_{p(clock)}$		20			ns
Width of Preset Pulse, $t_{p(preset)}$		25			ns
Width of Clear Pulse, $t_{p(clear)}$		25			ns
Input Setup Time, t_{setup}		$\geqslant t_{p(clock)}$			
Input Hold Time, t_{hold}		0			

ELECTRICAL CHARACTERISTICS (over recommended operating free-air temperature range unless otherwise noted)

PARAMETER		TEST CONDITIONS*			MIN	TYP**	MAX	UNIT
$V_{in(1)}$	Input voltage required to ensure logical 1 at any input terminal	V_{CC} = MIN			2			V
$V_{in(0)}$	Input voltage required to ensure logical 0 at any input terminal	V_{CC} = MIN					0.8	V
$V_{out(1)}$	Logical 1 output voltage	V_{CC} = MIN,	I_{load} = −400μA		2.4	3.5		V
$V_{out(0)}$	Logical 0 output voltage	V_{CC} = MIN,	I_{sink} = 16mA			0.22	0.4	V
$I_{in(0)}$	Logical 0 level input current at J1, J2, J3, K1, K2, or K3	V_{CC} = MAX,	V_{in} = 0.4V				−1.6	mA
$I_{in(0)}$	Logical 0 level input current at preset, clear, or clock	V_{CC} = MAX,	V_{in} = 0.4V				−3.2	mA
$I_{in(1)}$	Logical 1 level input current at J1, J2, J3, K1, K2, or K3	V_{CC} = MAX, V_{CC} = MAX,	V_{in} = 2.4V V_{in} = 5.5V				40 1	μA mA
$I_{in(1)}$	Logical 1 level input current at preset, clear, or clock	V_{CC} = MAX, V_{CC} = MAX,	V_{in} = 2.4V V_{in} = 5.5V				80 1	μA mA
I_{OS}	Short circuit output current†	V_{CC} = MAX,	V_{in} = 0	S5472 N7472	−20 −18		−57 −57	mA
I_{CC}	Supply current	V_{CC} = MAX,	V_{in} = 5V			10	20	mA

SWITCHING CHARACTERISTICS, V_{CC} = 5V, T_A = 25°C, N = 10

PARAMETER		TEST CONDITIONS		MIN	TYP	MAX	UNIT
f_{clock}	Maximum clock frequency	C_L = 15pF,	R_L = 400Ω	15	20		MHz
t_{pd1}	Propagation delay time to logical 1 level from clear or preset to output	C_L = 15pF,	R_L = 400Ω		16	25	ns
t_{pd0}	Propagation delay time to logical 0 level from clear or preset to output	C_L = 15pF,	R_L = 400Ω		25	40	ns
t_{pd1}	Propagation delay time to logical 1 level from clock to output	C_L = 15pF,	R_L = 400Ω	10	16	25	ns
t_{pd0}	Propagation delay time to logical 0 level from clock to output	C_L = 15pF,	R_L = 400Ω	10	25	40	ns

* For conditions shown as MIN or MAX, use the appropriate value specified under recommended operating conditions for the applicable device type.
** All typical values are at V_{CC} = 5V, T_A = 25°C.   † Not more than one output should be shorted at a time.

Figure 13.6-1 *JK* master-slave flip-flop data sheet. (*Signetics, Inc.*)

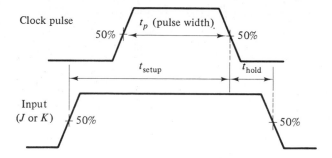

Figure 13.6-2 Waveforms illustrating pulse timing intervals.

In the table of recommended operating conditions, we have some new terms which require definition, all of which are illustrated by the waveforms shown in Fig. 13.6-2.

1. Pulse width t_p. This is the time between the 50 percent points of the pulse, and minimum values are 20 ns for the clock pulse and 25 ns for preset or clear pulses.
2. Input setup time t_{setup}. The time interval during which the J or K input must be stable *before* the triggering edge of the clock pulse in order for the J or K level to be reliably clocked into the flip-flop. It will be recalled that for the JK master-slave unit the triggering edge is the negative edge. The recommended condition here is that t_{setup} be at least as much as the clock pulse width, that is, 20 ns. All measurements are made as shown to the appropriate 50 percent points.
3. Input hold time t_{hold}. This is the time interval during which the data inputs must be constant *after* the triggering edge of the clock pulse has dropped below the 50 percent point to ensure proper flip-flop operation. For most TTL flip-flops $t_{hold} = 0$.

The data sheet also includes electrical and switching characteristics. The various parameters associated with these characteristics have been defined in Sec. 12.3-5. It is interesting to note that the propagation delay time for the 7472 flip-flop is approximately twice the delay time for the 7400 NAND gate. Thus, as far as signal transmission time is concerned, the flip-flop is equivalent to two gates in cascade.

13.6-2 ECL

The ECL family is the fastest of the currently available logic types. The high speed is not obtained without cost, however, since high power dissipation comes with it.

Currently popular ECL types are the 10,000 series and MECL III (Motorola). For the 10,000 series, toggle speeds can exceed 200 MHz, while for MECL III the figure is 500 MHz. Type D ECL flip-flops are available which operate at

4 GHz. Power dissipation for the 10,000 series is approximately 140 mW per flip-flop, while MECL III requires 220 mW.

Typical flip-flops available in the 10,000 series include

1. MC10231 dual type-D master-slave
2. MC10135 dual JK master-slave
3. MC10176 hex type-D master-slave (common clock)

For MECL III the selection includes

1. MC1666 dual clocked RS.
2. MC1668 dual clocked latch. This unit is a combination clocked latch and RS flip-flop. When the clock is LOW, the RS inputs control the output state. When the clock is HIGH, the output follows the D input.
3. MC1670 type-D master-slave.

13.6-3 CMOS

The CMOS family is the slowest logic family but has the advantage of very low dissipation. The maximum clock frequency for the CD4013 dual type-D flip-flop is listed as 10 MHz, while the quiescent device dissipation per package is typically 50 nW. There are 10 different flip-flops listed in the RCA catalog including

1. CD4013 dual D with set-reset
2. CD4027 dual JK master-slave

CMOS flip-flops are not constructed in the same manner as those of the other families but do make use of the FET switches described in Sec. 3.8. The circuit of a D-type master-slave unit is shown in Fig. 13.6-3. The switches (labeled TG for *transmission gate*) are controlled by the clock, a HIGH clock signal causing the

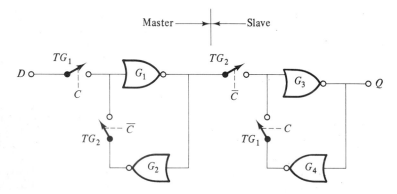

Figure 13.6-3 CMOS D-type flip-flop.

Table 13.7-1

	Schottky TTL	Low-power Schottky TTL	10,000 series ECL	III ECL	CMOS
Maximum toggle frequency, MHz	125	45	250	500	10
Power dissipation, mW	150	50	140	220	10

switch to be open and a LOW clock signal causing the switch to be closed. Note that TG_1 is controlled by the clock signal C while TG_2 is controlled by $\bar{C}$. The two NOR gates G_1 and G_2 form the master flip-flop when the switch TG_2 is closed, and gates G_3 and G_4 form the slave flip-flop when TG_1 is closed.

Operation of the circuit is as follows. Consider that the clock is HIGH so that TG_1 is open and TG_2 closed. The master flip-flop will have at its output the data from the previous cycle, which it feeds to the slave through TG_2. Now when the clock goes LOW, TG_2 will open, TG_1 will close, and the slave will act as a flip-flop. The data received from the master will appear at the output. Note that if both TG_1 and TG_2 were to be closed at the same time, there would be a direct path from input to output. In order to avoid this, the clocking signals applied to the transmission gates are arranged so that TG_2 opens before TG_1 closes.

13.7 COMPARISON OF FLIP-FLOPS OF VARIOUS FAMILIES

In Table 13.7-1 we compare the flip-flop characteristics of ECL, TTL, and CMOS. As we found in a similar comparison of logic gates in Chap. 12, ECL is the fastest but dissipates considerable power while CMOS is the slowest and dissipates relatively little power. Note that Schottky TTL dissipates as much power as ECL at very high toggle frequencies.

PROBLEMS

13.1-1 The waveforms in Fig. P13.1-1 are applied to the RS flip-flop of Fig. 13.1-2a. Sketch the resulting Q waveform assuming that Q is initially LOW.

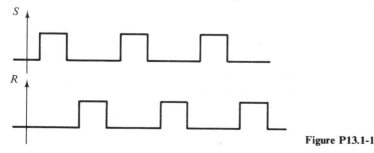

Figure P13.1-1

13.1-2 Repeat Prob. 13.1-1 for the waveforms in Fig. P13.1-2.

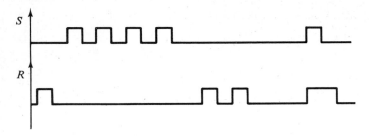

Figure P13.1-2

13.1-3 The circuit of Fig. P13.1-3 is a flip-flop using cross-coupled transistor inverters for which $V_{CE, \text{sat}} = 0$ and $V_{BE, \text{sat}} = 0.7$ V. Assuming that T_1 is ON and T_2 OFF, find all currents and voltages in the circuit and prove that the assumption leads to consistent conditions.

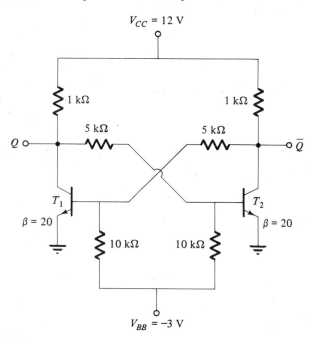

Figure P13.1-3

13.1-4 Devise a circuit which will trigger the flip-flop of Fig. P13.1-3, that is, cause it to switch state so that T_1 is OFF and T_2 is ON.

13.1-5 The waveforms in Fig. P13.1-5 are applied to the clocked RS flip-flop of Fig. 13.1-5a. Sketch the resulting Q waveform assuming that Q is initially HIGH.

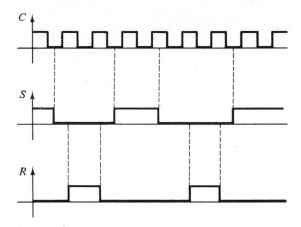

Figure P13.1-5

13.1-6 The waveforms in Fig. P13.1-6 are applied to the *RS* flip-flop with asynchronous inputs shown in Fig. 13.1-7. Sketch the *Q* output waveform assuming *Q* initially LOW.

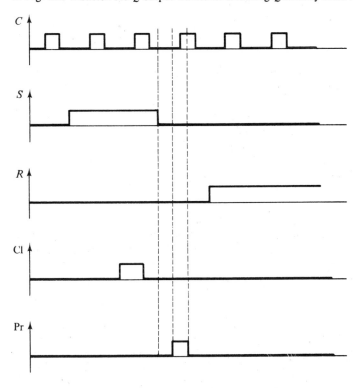

Figure P13.1-6

13.2-1 The waveforms of Fig. P13.1-5 are applied to the *RS* master-slave flip-flop of Fig. 13.2-1. Sketch the resulting *Q* waveform assuming that *Q* is initially LOW.

13.2-2 Repeat Prob. 13.2-1 for the waveforms in Fig. P13.1-6. Assume that the Pr and Cl action is the same as in the non-master-slave flip-flop.

13.3-1 Repeat Example 13.3-1 assuming that all gates have a propagation delay time t_{pd} per gate which is 5 percent of the width of one clock pulse. Plot C, C_2, J, K, Q_1, $\bar{Q}_1$, Q_2, and $\bar{Q}_2$ taking into account the propagation delay.

13.3-2 The waveforms of Fig. P13.3-2 are applied to the JK master-slave flip-flop of Fig. 13.3-1. Sketch the Q output waveform assuming Q LOW initially.

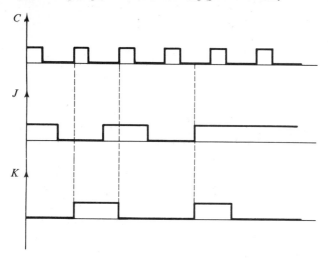

Figure P13.3-2

13.3-3 The JK master-slave flip-flop with $J = K = 1$ is called a T (toggle) type flip-flop. Sketch the Q waveform if the clock input is a 1-MHz square wave.

13.3-4 Find a pair of J and K waveforms which will produce the Q waveform shown in Fig. P13.3-4. Assume a master-slave flip-flop.

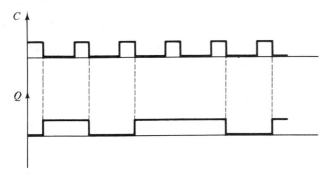

Figure P13.3-4

13.4-1 The waveforms of Fig. P13.3-2 are applied to the edge-triggered JK flip-flop of Fig. 13.4-1a. Assuming that the delay in gates 1 and 2 is 50 percent of the width of one clock pulse and is 5 times the delay in gates 3 to 8, plot the Q output waveform. Assume Q LOW initially.

13.4-2 In accordance with the discussion in Sec. 13.4, show that we must have $t_{pd1,\,2} \geq 4t_{pd3\ \text{to}\ 8}$.

13.5-1 The waveforms of Fig. P13.5-1 are applied to the D flip-flop shown in Fig. 13.5-1a. Assuming Q LOW initially, plot the Q waveform.

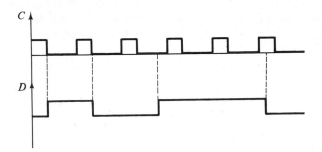

Figure P13.5-1

13.6-1 In the circuit of Fig. P13.6-1 the *JK* TTL flip-flops are master-slave.

(*a*) Sketch Q_1 and Q_2 assuming negligible time delays.

(*b*) Assuming an input setup time t_{setup} of 20 ns, a hold time $t_{hold} = 0$, and a propagation delay t_{pd} from clock to output of 40 ns for each flip-flop, determine the maximum clock frequency for reliable operation.

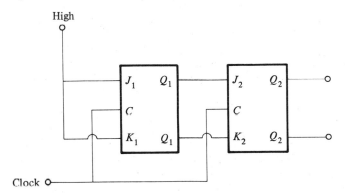

Figure P13.6-1

13.6-2 Repeat part (*b*) of Prob. 13.6-1 for an ECL flip-flop for which $t_{setup} = 2.5$ ns, $t_{hold} = 1.5$ ns, and t_{pd} (clock to output) = 3 ns.

13.6-3 Repeat part (*b*) of Prob. 13.6-1 for a CMOS flip-flop for which $t_{setup} = 70$ ns, $t_{hold} = 0$, and t_{pd} (clock to output) = 175 ns.

FOURTEEN

REGISTERS, COUNTERS, AND ARITHMETIC CIRCUITS

INTRODUCTION

In this chapter we consider some of the more important circuits used in digital systems. These include the register, a device which is used for storage and transfer of data; the counter, which does just what its name implies; and arithmetic circuits, chiefly those which add binary numbers. Emphasis will be placed on devices which are currently available in IC form.

14.1 SHIFT REGISTERS

Registers are devices which are used to store and/or shift data entered from external sources. They are constructed by connecting a number of flip-flops in cascade. As we found in Chap. 13, a single flip-flop can store 1 bit of data, thus an n-bit register will require n flip-flops. In a digital system such registers are generally used for the temporary storage of data.

There are two types of data with which we are concerned, serial and parallel. Serial data consist of a time sequence of binary digits which are transferred on a single line; parallel-data transfer means that all bits in a given group are transferred simultaneously on separate lines. The shift function of a register allows the stored data to be moved serially from stage to stage or into or out of the register. Types of data movement include serial shift right or left and parallel shift into and out of the register. In addition, registers are used for conversion of data from serial to parallel and vice versa.

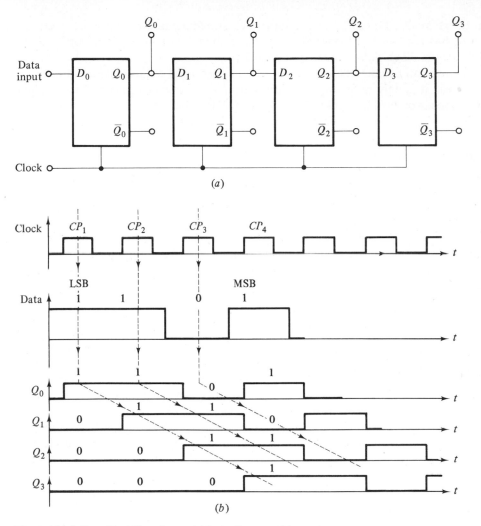

Figure 14.1-1 Four-bit shift registers: (*a*) logic diagram, (*b*) waveforms.

14.1-1 Serial-in Shift Registers

A 4-bit shift register using type-D flip-flops is shown in Fig. 14.1-1a. In order to illustrate its operation we consider the data word **1011** as the input. We assume that the register is initially cleared; i.e., at $t = 0$, Q_0 to Q_3 are all **0**. The serial input waveform containing the data word is shown in Fig. 14.1-1b. Recall (Sec. 13.5) that the D-type flip-flop transfers data to the output on the positive edge of the clock pulse. Thus, on the positive edge of clock pulse 1 (CP_1) the first bit appears at the Q_0 terminal. When the second clock pulse is applied, the bit at Q_0 is transferred to Q_1 while the next data bit is transferred to Q_0. Next CP_3 causes the bit at Q_1 to appear at Q_2, the bit at Q_0 to appear at Q_1, and the next data bit appears at Q_0. Operation continues in this manner until at CP_4 the complete word has been shifted from left to right into the register. If the data word is to be stored, the clock pulses must be stopped at this point.

Data can be taken from this register in either serial or parallel form. For serial removal, it is only necessary to apply four additional clock pulses; then the data will appear on the Q_3 terminal in serial form. After the last bit has been shifted out, the register will be clear until new data are inserted.

The clock frequency need not be the same while information is being shifted out as it was when the information was fed in, so that the register can act as a frequency changer.

In order to read the data in parallel form it is only necessary first to enter the data serially, as described above. Once the data are stored, each bit appears on a separate output line, Q_0 to Q_3 in Fig. 14.1-1a, so that each bit is available simultaneously, as required for parallel output.

14.1-2 Parallel-in Shift Registers

In modern digital systems, where speed is important, parallel-data transfer is often employed. The reason for this can be seen by considering for example, a 16-bit shift register. If the clock period for the system is 1 μs, it takes 16 clock pulses, or 16 μs, to transfer a word in or out serially. This same word can be transferred in 1 μs if parallel-data transfer is used. The cost of this faster operating rate is high because more signal lines are required and thus more logic gating.

The 54/7494 4-bit shift register shown in Fig. 14.1-2 is an example of a shift register whose flip-flops have asynchronous *preset* and *clear* capability. This unit has synchronous serial or asynchronous parallel-load capability and a clocked serial output. It is cleared (all flip-flop outputs LOW) by applying a HIGH voltage level to the *clear* input while the internal presets (labeled *A*, *B*, *C*, and *D*) are inactive (HIGH). This operation is independent of the clock input level.

For parallel loading, all stages are cleared (the clear input is HIGH). The NAND gates feeding the *preset* terminals are initially disabled by applying a LOW preset-enable (*PE*) voltage. Next the data word is applied to the parallel data inputs P_1, P_2, P_3, and P_4, and then a HIGH voltage level is applied to the PE terminal, enabling the NAND gates. This transfers the data word to the internal preset terminals *A*, *B*, *C*, and *D*, and hence to the flip-flop outputs, independent of the clock. The PE level now goes LOW, disabling the NAND gates, and application of four clock pulses will cause the data word to appear in serial form at the output terminal Q_3.

In the 54/7494 information is transferred to the outputs on the positive-going edge of the clock pulse. Thus the proper information must be available at the *RS* input of each flip-flop before the leading edge of the clock waveform. The clock pulse should be low when asynchronously presetting the register. When the clock pulse is HIGH and during the clock transitions the clear input should be LOW and the preset-enable pulse must also be inactive (LOW).

Parallel data inputs

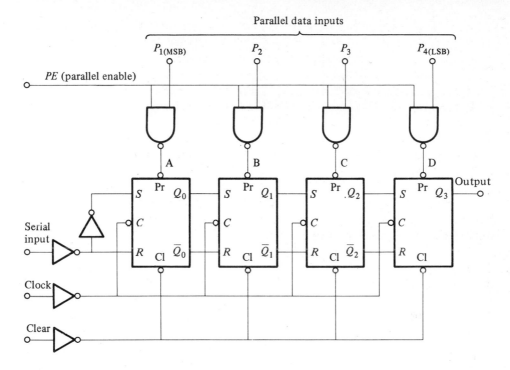

Figure 14.1-2 Simplified 54/7494 4-bit shift register.

14.1-3 Universal Shift Register

In Fig. 14.1-3*a* we show the logic diagram of the Fairchild 9300 shift register. This unit is available in regular and Schottky TTL and can function in both serial and parallel synchronous modes with shift-left, shift-right capability. It contains four clocked master-slave flip-flops with D inputs. The logic gates at the D inputs provide the versatility required for the many functions performed by this unit. For example, with the *parallel enable* $\overline{PE}$ line LOW, the AND gates G_0 to G_3 are enabled, thereby connecting the asynchronous preset inputs P_0 to P_3 to the D inputs of the four stages. These inputs then determine the outputs of each of the D flip-flops. While the $\overline{PE}$ line is LOW, the register is essentially four separate D flip-flops with a common clock because the AND gates A_1 to A_3, which connect Q_0 to D_1, Q_1 to D_2, and Q_2 to D_3, are disabled.

When the $\overline{PE}$ signal is HIGH, AND gates A_1 to A_3 are enabled and the outputs Q_0, Q_1, and Q_2 are connected to the inputs D_1, D_2, and D_3, respectively, thus forming a 4-bit shift register.

When the $\overline{PE}$ is HIGH, the gates A_J and A_K are enabled and the input D_0 is

$$D_0(n + 1) = J\overline{Q}_0(n) + \overline{K}Q_0(n) \qquad (14.1\text{-}1)$$

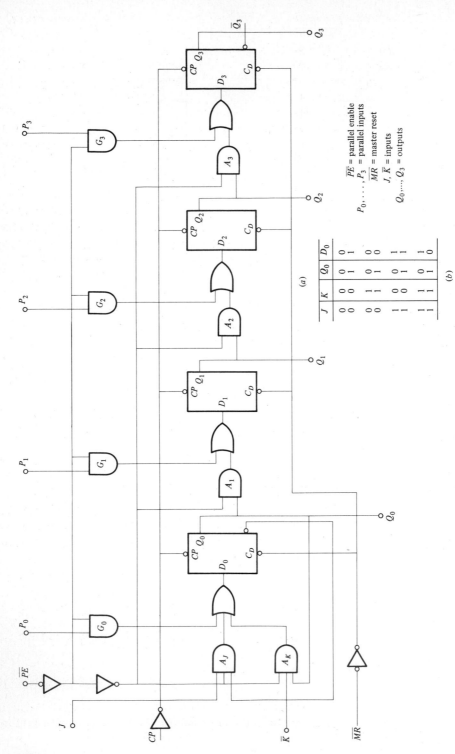

Figure 14.1-3 Fairchild 9300/74195 4-bit shift register: (*a*) circuit; (*b*) truth table.

Here $Q_0(n)$ is the current value of Q_0, and $D_0(n + 1)$ is the value that $Q_0(n + 1)$ will have after the clock pulse. The truth table for (14.1-1) is shown in Fig. 14.1-3b.

For this device the outputs change state after the LOW to HIGH clock transition. The action of the $\overline{MR}$ (master reset) input is to set all stages to zero when it is low, regardless of all other inputs.

As an example of the waveforms encountered with this register, consider that it is to be used for parallel-to-serial conversion in the shift-right mode. In this case we set $J = \overline{K} = 0$, thereby setting the outputs of A_J and A_K to 0. This allows P_0 as well as P_1 to P_3 to be shifted through the D flip-flops.

Let the input word be $P_0 P_1 P_2 P_3 = 1011$, as shown in Fig. 14.1-4. Since we are shifting to the right, the leftmost flip-flop inputs should be arranged so that the

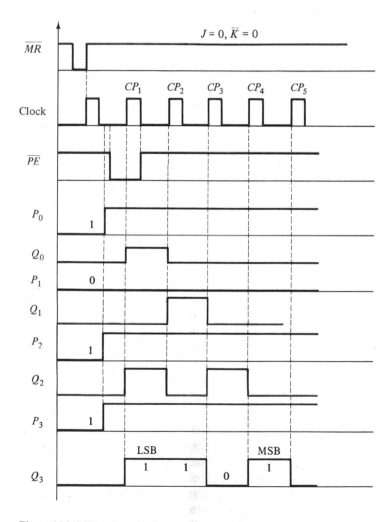

Figure 14.1-4 Waveform in the parallel to serial converter.

Q_0 output is **0** after the first shift. This will be accomplished since $J = \bar{K} = 0$. The time sequence of events is as follows:

1. The $\overline{MR}$ (master reset) signal goes LOW, setting all flip-flops to zero.
2. The parallel inputs P_0, P_1, P_2, and P_3 are applied to their respective terminals.
3. The $\overline{PE}$ signal goes LOW for one clock pulse, causing the parallel inputs to be loaded into their respective flip-flops.
4. Each clock pulse causes the data to shift one stage to the right on the LOW-to-HIGH clock transition, producing the Q_3 output waveform shown.
5. The fifth clock pulse returns the output to zero.

To operate the 9300 as a standard shift-right serial-in, serial-out register we connect the J and $\bar{K}$ terminals together. This new terminal becomes the input-data terminal. We then set $\overline{PE} = 1$ to disable gates G_0 to G_3 and enable gates A_J, A_K and A_1 to A_3.

To operate the 9300 as a shift-left serial-in serial-out register we connect Q_3 to P_2, Q_2 to P_1, and Q_1 to P_0. The input data are applied to P_3 and $\overline{PE} = 0$, thereby disabling gates A_J, A_K, A_1, A_2, and A_3 and enabling gates G_0 to G_3. When the clock pulse occurs, the data bit on P_3 is transferred to Q_3. Since Q_3 is connected to P_2 and hence to D_2, the output of Q_3 is shifted to Q_2. Similarly the data bit on Q_2 is shifted to Q_1 and the output level of Q_1 is shifted to Q_0.

The reader is urged to refer to the appropriate manufacturers' literature for many other modes of operation.[1-3]

14.1-4 First-In–First-Out Shift Register

The *first-in-first-out* (FIFO) shift register is shown in Fig. 14.1-5. It consists of a standard shift-right register with parallel input capability; it is the use of these parallel inputs that distinguishes the FIFO from the ordinary shift register.

The input data bit D is passed through gate G_D when the input clock (CL_i) is high. The D flip-flop which accepts this bit of data is always chosen to be the first empty register. For example, if the entire register is clear, steering is arranged, using gates G_0 to G_7, so that the first bit of data will go into FF_0. If before the clock out (CL_o) pulse arrives a second bit of data appears, it is stored in FF_1. Now suppose that data are stored in FF_0 to FF_4 when the first CL_o pulse arrives. Then the stored data are shifted out (the first bit of data to arrive is the first to leave). If now another bit of data arrives, it is stored in the first empty register FF_4.

As a result of this steering process, data can enter the register at one rate and leave at another rate. If the register is completely empty, the CL_o pulse is prevented from shifting the register, and if the register is full and another bit of data arrives, the shift register shifts to the right and the new bit of data is stored in FF_7. The circuitry which accomplishes these two operations is not shown.

The steering is accomplished by a circuit which operates similar to an up-down counter (Prob. 14.2-17). The counter counts up whenever the input clock

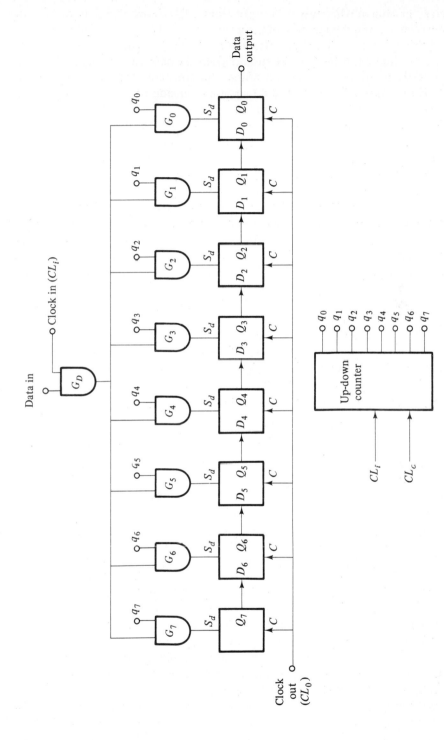

Figure 14.1-5 Simplified diagram of a first-in–first-out shift register (FIFO).

CL_i arrives, and down whenever the output clock CL_o arrives. The eight outputs of the counter determine the steering of the input data. Only one of the eight outputs is HIGH at any time. Thus, if q_2 is HIGH, it means that two more CL_i pulses than CL_o pulses arrived and therefore we should insert the next bit of data in FF_2. As a practical consideration, it must be noted that the counter will not operate properly if the edges of the CL_i and CL_o pulses coincide.

14.2 COUNTERS

Applications of digital electronic counters are numerous. They include control circuits in large digital computers, industrial applications such as counting the number of revolutions a motor makes in a given time, and the frequency division required to produce the hour, minute, and second outputs in a digital wristwatch.

Many different types of electronic counters are available; they are all either of the asynchronous or synchronous type and are usually constructed using JK flip-flops.

Before we begin a detailed discussion of counters it will be well to define some terminology. Digital counters, of course, generate 0s and 1s. However, they do not always count in what we would consider the normal numerical binary sequence, **0, 1, 10, 11, 100, 101**, ..., but often employ some arbitrary sequence which must be decoded. For example, a counter which starts at **0** and has a maximum count of 3 might count according to the sequence

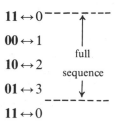

If the actual decimal count is required, a *decoding* circuit must be included. In the above example, the counter goes through four states before it recycles, and this number represents the *modulus* of the counter; often the counter is referred to as a mod-4 counter. Since each flip-flop has two possible states, a counter with N flip-flops has a maximum of 2^N states, each state being a different combination of the 0s and 1s stored in the individual flip-flops. Thus, the maximum modulus of an N-flip-flop counter is 2^N. Often less than the maximum number of states is used, as, for example, in a four-flip-flop counter used to count to 10 (a *decade* or *mod-10* counter). Here only 10 out of 16 possible states are used.

In the following sections we shall discuss the characteristics of a number of different counters.

14.2-1 The Ripple Counter

A three-stage ripple counter using JK flip-flops is shown in Fig. 14.2-1a. It uses the maximum count capability of the three stages and would thus be classified as mod-8. The flip-flops are connected to toggle and change state on the negative-going transition of the waveform applied to the clock input. The input signal whose pulses are to be counted is applied to the clock input of the first flip-flop, FF_0, and the output of each flip-flop is connected directly to the clock input of the next. The waveforms at the input and three output terminals are shown in Fig. 14.2-1b. The input is shown for convenience as being periodic even though it may actually be completely random. Note that each output is at **0** before we start counting. This is done in practice by utilizing the *clear* input to each flip-flop (not shown on the diagram). The output Q_0 is obtained by noting that FF_0 changes state only when the input goes from HIGH to LOW and similarly Q_1 changes state only when Q_0 changes from HIGH to LOW, and so on. The count is

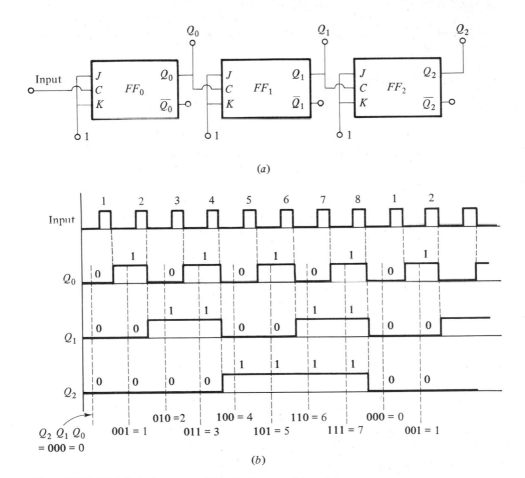

Figure 14.2-1 Mod-8 ripple counter: (a) logic diagram; (b) waveforms.

obtained by observing the sequence $Q_2 Q_1 Q_0$ which starts off at **000** = 0. Some of the counts are shown on the diagram under the Q_2 waveform, and we see that the count cycles from 0 to 7, after which it returns to its initial state. The actual decimal count can be obtained by assigning Q_2 a weight of $2^2 = 4$, Q_1 a weight of $2^1 = 2$, and Q_0 a weight of $2^0 = 1$. Thus, the formula for the count C is

$$C = Q_2 \times 2^2 + Q_1 \times 2^1 + Q_0 \times 2^0 \qquad (14.2\text{-}1)$$

If it is required to obtain the decimal output for display or other reason, a binary-to-decimal decoder like that shown in Fig. 14.2-2 may be used. The circuit consists of eight four-input AND gates, one for each of the possible outputs. One input of each AND gate is a *strobe* signal, and the other three come from the flip-flops of the counter; the output of each AND gate can be used to operate a printer or other display device. Circuit operation is easily understood by assuming the strobe signal HIGH, so that the gates are enabled, and considering, for example, gate 0. The output of this gate is

$$C_0 = \overline{Q_2 Q_1 Q_0} \qquad (14.2\text{-}2a)$$

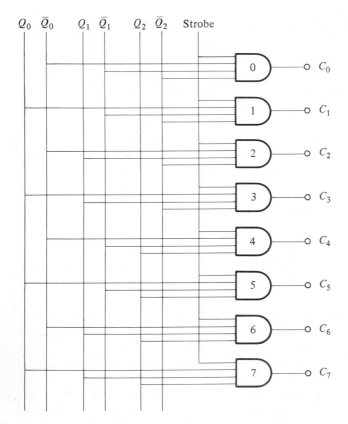

Figure 14.2-2 Binary-to-decimal decoder.

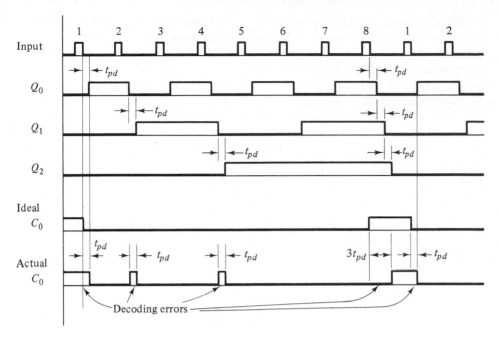

Figure 14.2-3 Waveforms illustrating the effect of propagation delay.

When the count is **0**, so that $Q_2 = Q_1 = Q_0 = 0$, we find that $C_0 = 1$. It is easily verified that all other gate outputs are **0**. As another example consider gate 5. Here

$$C_5 = Q_2 \overline{Q_1} Q_0 \qquad (14.2\text{-}2b)$$

When the count is 5, we have $Q_2 Q_1 Q_0 = 101$, so that $C_5 = 1$. Again it is easily verified that all other gate outputs are **0**. Each of the other outputs is readily shown to be **1** when the count is at the corresponding value.

The strobe input is necessary because of decoding errors which result when finite propagation delays are taken into account.[4] The waveforms of Fig. 14.2-1 are repeated in Fig. 14.2-3 with propagation delays included. The last two waveforms in the figure are the *ideal* decoder output at the C_0 gate and the *actual* C_0 output when the strobe is set continually to **1**. The *decoding error* is the difference between these two. Besides the fact that the actual C_0 pulse is reduced in width by the cumulative effect of the propagation delay we see that undesired additional pulses have been added to C_0. To prevent these pulses from causing false counts the strobe signal is arranged to be LOW so that all AND gates are disabled except for a short interval when the strobe goes HIGH. During this interval we read the decoder output. The strobe signal is carefully designed to cover a short interval which avoids the undesired pulses.

It should now be clear why the counter shown in Fig. 14.2-1 is called a ripple counter. Refer to the input after the seventh pulse. Outputs Q_0, Q_1, and Q_2 are high. When the eighth input pulse is applied, Q_0 goes from HIGH to LOW; this causes Q_1 to go from HIGH to LOW, which causes Q_2 to go from HIGH to LOW. Thus, the trailing edge of the eighth pulse causes a transition in each of the succeeding flip-flops, the effect *rippling* through the counter. It is precisely the delay caused by this ripple which results in a limitation on the maximum frequency of the input signal.

Consider that in an N-stage ripple counter the $(2^N - 1)$th input pulse has just returned to the LOW state and each flip-flop $Q_0, \ldots, Q_{N-1}$ is in the HIGH state. After the next input pulse occurs, the output Q_0 goes LOW after a time t_{pd}, Q_1 goes LOW after an additional time t_{pd}, etc. If we let T_i denote the period of the input waveform (assumed periodic for this discussion) and T_s the time interval required for strobing, then, in order to be able to read the counter output we must have

$$T_i \geq N t_{pd} + T_s \qquad (14.2\text{-}3a)$$

where N is the number of flip-flops, each having a delay t_{pd}. Since $f_i = 1/T_i$, the maximum input frequency is

$$f_i \leq \frac{1}{N t_{pd} + T_s} \qquad (14.2\text{-}3b)$$

Ripple counters are not restricted to the type of counting described above. They are also available in *countdown* form (Prob. 14.2-16), in which a number is stored in the counter and as pulses are applied, the value stored in the counter decreases. Usually, logic is included to indicate when the count reaches zero. Such a counter would be used in the clock which counts the seconds remaining before a missile is fired.

The counter shown in Fig. 14.2-1 is a mod-2^3 counter. If N flip-flops were used, we would obtain a mod-2^N counter. A few simple changes can be made to convert the mod-2^N counter into a counter which counts to any non-mod-2^N number such as 11, etc. (see Probs. 14.2-6 to 14.2-8).

Frequency division A glance at the waveform of Fig. 14.2-1 indicates that the Q_0 output waveform has a frequency exactly equal to one-half that of the input, the Q_1 output frequency is one-quarter of the input frequency, and the Q_2 output frequency is one-eighth of the input frequency. Thus, the circuit acts as a *frequency divider*, or *frequency scaler*. For an N-flip-flop circuit the input frequency is divided by 2^N in steps of 2.

The decoding error mentioned in connection with the use of the circuit as a counter does not affect its operation as a frequency divider. The upper limit on the input frequency of the frequency divider is the same as the maximum rate at which the first flip-flop in the cascade can toggle. Since the second and subsequent flip-flops toggle at frequencies one-half or less that of the first, they do not affect the maximum input frequency that can be accommodated.

14.2-2 Synchronous Counters

The main problem with ripple counters is the cumulative flip-flop delay as the count progresses down the line. This is eliminated in the synchronous counter, in which all flip-flops are controlled by a common clock. The circuit of a 4-bit synchronous parallel counter is shown in Fig. 14.2-4. The first two flip-flops each have one J and K input; however, FF_2 requires two J and two K inputs connected to an AND gate; FF_3 requires three, and so on.

Table 14.2-1 shows all 16 states of the counter. From it we observe the following:

1. FF_0 must change state (toggle) with each clock pulse. This is easily accomplished by connecting J_0 and K_0 to a HIGH level, as shown in Fig. 14.2-4.
2. FF_1 must change state whenever $Q_0 = 1$. This is achieved by connecting J_1 and K_1 directly to Q_0.
3. FF_2 changes state only when $Q_0 = Q_1 = 1$. Thus, Q_0 and Q_1 are connected through AND gates to J_2 and K_2 as shown.
4. FF_3 changes state only when $Q_0 = Q_1 = Q_2 = 1$. This requires three-input AND gates connecting Q_0, Q_1, and Q_2 to J_3 and K_3.
5. Each following stage (not shown in the figure) requires an additional input to the AND gate.

Table 14.2-1 State table for the counter in Fig. 14.2-4

State	Q_3	Q_2	Q_1	Q_0
0	0	0	0	0
1	0	0	0	1
2	0	0	1	0
3	0	0	1	1
4	0	1	0	0
5	0	1	0	1
6	0	1	1	0
7	0	1	1	1
8	1	0	0	0
9	1	0	0	1
10	1	0	1	0
11	1	0	1	1
12	1	1	0	0
13	1	1	0	1
14	1	1	1	0
15	1	1	1	1
0	0	0	0	0
1	0	0	0	1
2	0	0	1	0
⋮	⋮	⋮	⋮	⋮

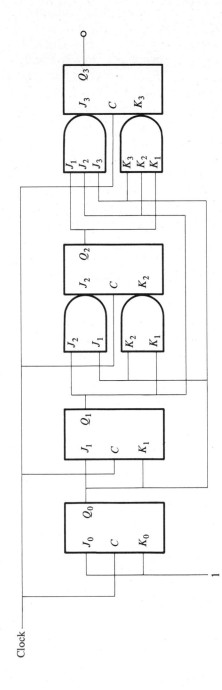

Figure 14.2-4 Circuit of synchronous parallel counter.

The waveforms and individual flip-flop outputs are similar to those shown in Fig. 14.2-1. A disadvantage of this type of counter is that the source of clock pulses must be capable of supplying all the flip-flops in the counter. Also the fan-out required of each flip-flop increases by 1 for each stage added to the counter. Since the propagation delay of a flip-flop increases with load, this limits the speed attainable with the counter. From the circuit we see that FF_3 has the largest fan-out and thus the maximum propagation delay time. This time is often specified by manufacturers as a function of fan-out. The overall repetition rate is limited only by the delay of one flip-flop plus the delay of one AND gate and the strobe time. Thus

$$f_{max} \leq \frac{1}{t_{pd}(FF_3) + t_{pd}(\text{AND}) + t_s} \tag{14.2-4}$$

14.2-3 A Synchronous Nonbinary Counter: Mod-3

The counters considered up to this point have utilized all the possible states available, that is, 2^N. Examples included mod-8 $(N = 3)$ and mod-16 $(N = 4)$ counters. In this section we consider a mod-3 counter, which has only three states. In general, for a mod-m counter, we must use N flip-flops, where N is chosen so that it is the smallest number for which the maximum modulus 2^N is greater than m. Thus, a three-stage counter $(N = 3)$ would be used for $m = 5$ to 8, for $N = 4$ we accommodate $m = 9$ to 16, and so on. For a mod-3 counter, clearly only two flip-flops are required. Since two flip-flops are capable of generating four different states, we must omit one possible state. For convenience, we specify that the three states of the counter are to be as shown in Table 14.2-2. From the table, we see that the **11** state is omitted, and for convenience we let the decimal count be the same as the counter state.

The two-flip-flop circuit shown in Fig. 14.2-5a implements the state table for the mod-3 counter. The waveforms are shown in Fig. 14.2-5b. To see how the

Table 14.2-2 State table

Counter state	Q_1 (2^1)	Q_0 (2^0)
0	0	0
1	0	1
2	1	0
0	0	0
1	0	1
2	1	0
0	0	0
⋮	⋮	⋮

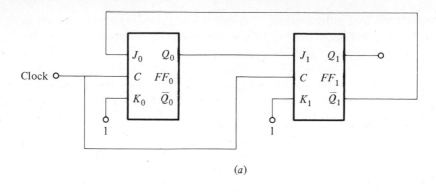

(a)

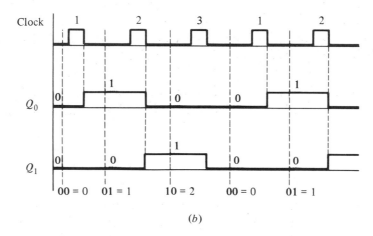

(b)

Figure 14.2-5 Mod-3 counter: (a) circuit; (b) waveforms.

counter operates, assume that FF_0 and FF_1 are initially reset so that $Q_1 Q_0 = 00$. Then $\bar{Q}_1 = 1$ and $J_0 K_0 = 11$ so that FF_0 is set by the first clock pulse. The count registered by the counter is now $Q_1 Q_0 = 01$.

Since $J_0 K_0 = 11$, the next clock pulse will reset FF_0, making $Q_0 = 0$ once more. Simultaneously, since $Q_0 = 1$ before the second clock pulse arrived, FF_1 is set by the second clock pulse and Q_1 becomes 1. There is no ambiguity in this process since each flip-flop is either master-slave or edge-triggered.

With $Q_1 Q_0 = 10$ (the count is now 2), $J_0 = 0$ while $K_0 = 1$, so that the third clock pulse leaves FF_0 reset. However, since $Q_0 = J_1 = 0$ and $K_1 = 1$, clock pulse 3 resets FF_1. The counter state, after the third clock pulse, is then $Q_1 Q_0 = 00$, which was the initial state. Hence, the process repeats with the next clock pulse.

14.2-4 Shift-Register Counters

The *ring counter* shown in Fig. 14.2-6 is similar to the 4-bit shift register shown in Fig. 14.1-1, the only difference being that the output Q_3 of the ring counter is

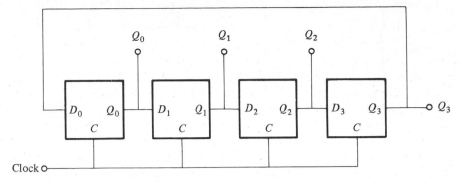

Figure 14.2-6 Ring counter.

connected to D_0. In normal operation the counter is initially preset, so that $Q_0 = 1$ while $Q_1 = Q_2 = Q_3 = 0$. Then each clock pulse shifts the **1**, first to Q_1, then to Q_2, then to Q_3, then back to Q_0 and so on around the ring.

The ring counter has numerous applications, among them *multiplexing*. A time-division multiplexor is shown in Fig. 14.2-7. In this circuit $d_0, d_1, d_2,$ and d_3 are four independent data sources and $Q_0, \ldots, Q_3$ are the outputs of the ring counter. When $Q_0 = 1$, the output $v_o = d_0$, when $Q_1 = 1$, $v_o = d_1$, etc. Thus, v_o is the time-division-multiplexed data sequence

$$v_o = d_0 d_1 d_2 d_3 d_0 d_1 \cdots$$

Other counters or frequency dividers can readily be implemented using the configuration shown in Fig. 14.2-8. For example, consider designing a mod-3 counter as in Sec. 14.2-3. A two-stage shift register is required, and the state table shown in Table 14.2-3 can be used for the synthesis. Note that this state table

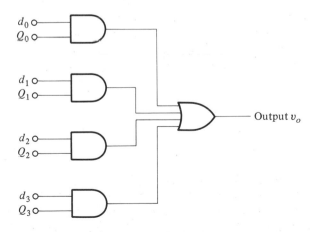

$$v_o = d_0 d_1 d_2 d_3 d_0 d_1 d_2 d_3 \cdots$$

Figure 14.2-7 Time-division multi-plexing.

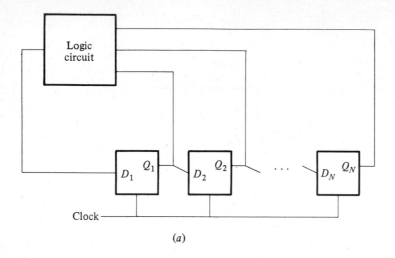

(a)

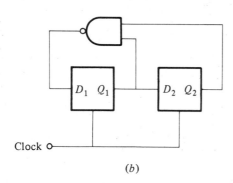

(b)

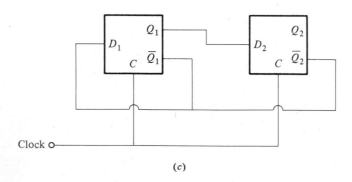

(c)

Figure 14.2-8 Shift-register counters: (a) general form; (b) mod-3 counter; (c) mod-3 ECL counter using wired-or connection.

Table 14.2-3

State count, k	Q_1	Q_2
1	0	1
2	1	0
3	1	1
1	0	1
$\vdots$	$\vdots$	$\vdots$

Table 14.2-4

k	$D_1(k)$	$Q_1(k)$	$Q_2(k)$
1	1	0	1
2	1	1	0
3	0	1	1
1	1	0	1
$\vdots$	$\vdots$	$\vdots$	$\vdots$

differs from the one shown in Table 14.2-2, primarily because

$$Q_2(k + 1) = Q_1(k) \tag{14.2-5}$$

where k is the state count. That is, the output Q_1 of FF_1, after the k clock pulse will be the same as the output Q_2 of FF_2, after the $(k + 1)$th clock pulse.

The state table can be used to form a truth table for the logic circuit of Fig. 14.2-8a. The truth table can then be used to find the logic equation which relates D_1 to Q_1 and Q_2. The input and output of the D flip-flop obey the relation

$$D_1(k) = Q_1(k + 1) \tag{14.2-6}$$

From this equation the column for $D_1(k)$ in the truth table of Table 14.2-4 can be set down directly from the column for $Q_1(k)$. From the truth table we see that $D_1 = 0$ when $Q_1 = 1$ AND $Q_2 = 1$. This can be written

$$D_1 = \overline{Q_1 Q_2} = \overline{Q_1} + \overline{Q_2} \tag{14.2-7}$$

One implementation of (14.2-7) is shown in Fig. 14.2-8b, where a NAND gate is employed. Another implementation using ECL flip-flops is shown in Fig. 14.2-8c (see Prob. 14.2-18).

Other shift-register counters are discussed in the problems.

14.2-5 Commercially Available IC Counters

Asynchronous counters A large variety of ripple counters is available. In the TTL family, for example, the current catalog lists a number of decade counters and 4-bit binary counters in both single and dual configurations, i.e., two counters in a single package. These are available in regular and Schottky TTL. Typical count frequencies range from 3 MHz for low-power TTL to 32 MHz for Schottky TTL. For example, the 74S196 decade counter is capable of clock rates up to 100 MHz.

The CMOS family also features counters of many different types. For example, the MC14024 is a seven-stage ripple counter composed of seven toggle flip-flops with a common reset. Triggering takes place on the negative edge of the clock pulse, and clock rates to 8 MHz are typical. Twelve- and fourteen-bit counters are also available in CMOS.

There is a smaller selection of ECL counters than in TTL or CMOS. The MECL III MC1654 is a 4-bit ripple counter which can toggle at rates up to

325 MHz. However, 750 mW is dissipated. In this unit triggering takes place on the positive-going edge of the clock pulse. Individual set (preset) and common reset (clear) inputs are provided which override the clock, providing asynchronous set or clear. Outputs are available from all four stages, as well as complementary outputs from the first and last stages.

Synchronous counters Synchronous 4-bit counters capable of binary and decade operation are available in all logic families. Some feature count-up or count-down capability, and various types of preset and clear are available. For example, in TTL, the 74192 (decade) and the 74193 (binary) counters can both be connected to count up or down at speeds up to 40 MHz. Similar counters are available in CMOS, e.g., the 4510 (decade) and 4516 (binary), and in ECL, e.g., the 10136 (binary) and 10137 (decade). The Motorola catalog lists the MECL III MC1696 as a divide-by-10 counter capable of 1-GHz operation.

As an example of versatility, the Fairchild 9305 *variable-modulo* counter can be connected to count with mod 2, 4, 5, 6, 7, 8, 10, 12, 14, or 16 without additional external logic. With some additional logic it can count with mod 11, 13, or 15.

14.3 ARITHMETIC CIRCUITS

At the heart of the digital computer lies the arithmetic unit, which performs the operations of addition, subtraction, multiplication, and division. In this section we consider techniques for performing some of these operations with binary numbers along with circuits which implement them.

14.3-1 Addition of Two Binary Digits; The Half Adder

The rules for the addition of two binary digits are

$$
\begin{array}{llll}
\text{addend} \quad 0 & 0 & 1 & 1 \\
\text{augend} \ +0 & +1 & +0 & +1 \\
\hline
\underset{\text{carry}}{\underbrace{00}} \leftarrow \text{sum} & \underset{\text{carry}}{\underbrace{01}} \leftarrow \text{sum} & \underset{\text{carry}}{\underbrace{01}} \leftarrow \text{sum} & \underset{\text{carry}}{\underbrace{10}} \leftarrow \text{sum}
\end{array}
$$

Here the 0s and 1s are *numerical values*. However, we shall also assume that they represent their respective *logic values*.

As in decimal addition, binary addition generates a sum S and a carry C. For the first three additions shown, the carry is **0**; only for the last rule is it **1**. These additions are readily tabulated in the truth table (Table 14.3-1), from which we find the logic equation for the carry to be

$$C = AB \tag{14.3-1}$$

while for the sum we have

$$S = \bar{A}B + A\bar{B} \tag{14.3-2}$$

Table 14.3-1 Truth table for a half adder

A	B	S	C
0	0	0	0
0	1	1	0
1	0	1	0
1	1	0	1

From (11.3-8) we see that S is given by the EXCLUSIVE-OR operation, so that we can also write

$$S = A \oplus B \tag{14.3-3}$$

A circuit with inputs A and B and outputs S and C which obey these equations is called a *half adder*. One possible circuit using AND, OR, and NOT gates is shown in Fig. 14.3-1a, and another using an EXCLUSIVE-OR gate is shown in

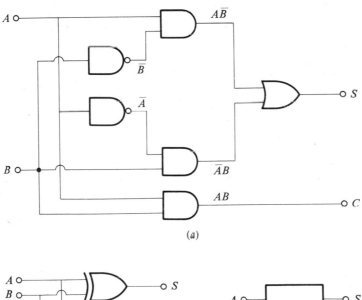

(a)

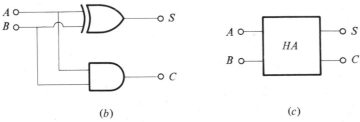

(b)   (c)

Figure 14.3-1 The half-adder: (a) circuit with AND, OR, and NOT gates; (b) circuit with EXCLUSIVE-OR gate; (c) block diagram symbol.

Fig. 14.3-1*b*. The half adder is represented by the symbol shown in Fig. 14.3-1*c*. It is quite restricted in use because it cannot accept a carry from a previous addition. This is overcome in the full adder, described next.

14.3-2 Addition of More than Two Bits; The Full Adder

When more than two binary digits are involved, the addition is complicated by the carries which are generated. Consider, for example, finding the sum of **00101** and **00111**. This can be broken down as follows, keeping the carries separate as we add from right to left:

$$
\begin{array}{rl}
\text{A} & \textbf{00101} \\
\text{B} & \textbf{00111} \\
\hline
\text{Sum 1} & \textbf{00010} \\
\text{Carry 1} & \textbf{01010} \\
\\
\text{Sum 2} & \textbf{01000} \\
\text{Carry 2} & \textbf{00100} \\
\hline
\text{Final Sum 3} & \textbf{01100}
\end{array}
$$

This example shows the fundamental process involved in addition. When A and B are added, sum 1 and carry 1 are generated simultaneously. Then sum 1 and carry 1 are added. The result is sum 2 and carry 2, which are generated at the same time. Then sum 2 and carry 2 are added, yielding the final sum 3. The process ends when there are no further carries.

When we add two numbers mentally, the carries are taken care of automatically; however, any circuit designed to add multibit numbers must have provisions for accepting carries from an addition to the right and passing carries on to

Table 14.3-2 Truth table for adding A_n, and B_n and a carry C_{n-1}

A_n	B_n	C_{n-1}	S_n	C_n
0	0	0	0	0
0	1	0	1	0
1	0	0	1	0
1	1	0	0	1
0	0	1	1	0
0	1	1	0	1
1	0	1	0	1
1	1	1	1	1

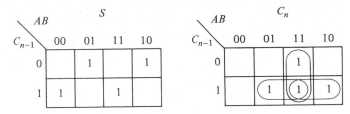

Figure 14.3-2 Karnaugh maps for S and C_n.

the left. The truth table for adding two bits and a carry from the right (C_{n-1}) is shown in Table 14.3-2, from which we see that the sum $S_n = 1$ if an odd number of the input variables A_n, B_n, and C_{n-1} are **1** and the sum $S_n = 0$ when an even number of the variables are **1**. Thus

$$S_n = \bar{A}_n \bar{B}_n C_{n-1} + \bar{A}_n B_n \bar{C}_{n-1} + A_n B_n C_{n-1} + A_n \bar{B}_n \bar{C}_{n-1} \qquad (14.3\text{-}4)$$

This result can also be obtained from the Karnaugh map representing the truth table (Fig. 14.3-2). Equation (14.3-4) can be expressed in terms of the EXCLUSIVE-OR operation by first factoring S_n to get

$$S_n = C_{n-1}(\bar{A}_n \bar{B}_n + A_n B_n) + \bar{C}_{n-1}(\bar{A}_n B_n + A_n \bar{B}_n) \qquad (14.3\text{-}5a)$$

Then since $\bar{A}_n \bar{B}_n + A_n B_n = \overline{\bar{A}_n B_n + A_n \bar{B}_n} - \overline{A_n \oplus B_n}$, we have

$$S_n = C_{n-1}(\overline{A_n \oplus B_n}) + \bar{C}_{n-1}(A_n \oplus B_n) \qquad (14.3\text{-}5b)$$

and finally
$$S_n = C_{n-1} \oplus (A_n \oplus B_n) \qquad (14.3\text{-}5c)$$

For the carry bit C_n, the truth table indicates that $C_n = 1$ when two or three of the input variables are **1**. Thus

$$C_n = A_n B_n + A_n C_{n-1} + B_n C_{n-1} = A_n B_n + C_{n-1}(A_n + B_n) \qquad (14.3\text{-}6)$$

This equation can also be obtained from the Karnaugh map shown in Fig. 14.3-2.

The logic equations for S_n and C_n can be implemented in a number of ways. One popular method is to use two half adders as a *full adder*, as shown in Fig. 14.3-3a. That this combination satisfies (14.3-5) and (14.3-6) can be seen by noting first that the sum output of HA_1 is the EXCLUSIVE-OR function of its two inputs $A_n \oplus B_n$, which then becomes input A_2 to HA_2. The other input to HA_2 is $B_2 = C_{n-1}$ so that its sum output is $C_{n-1} \oplus (A_n \oplus B_n)$, as in (14.3-5c). To verify the expression for C_n we note that the carry output of each half adder is the AND function of its two inputs. In Fig. 14.3-3, C_n is

$$C_n = A_n B_n + A_2 B_2 = A_n B_n + (A_n \oplus B_n)C_{n-1} = A_n B_n + (A_n + B_n)C_{n-1} \qquad (14.3\text{-}7)$$

since, when $A_n = B_n = 1$, $A_n B_n = 1$ independently of the second term. The circuit symbol for the full adder is shown in Fig. 14.3-3b.

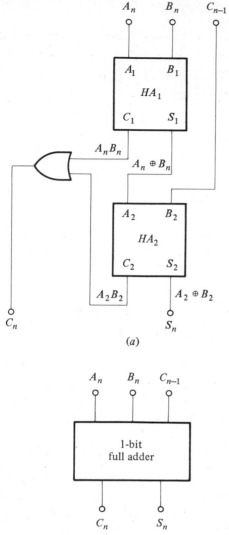

Figure 14.3-3 The full adder: (a) combination of two half adders; (b) circuit symbol.

An alternate technique for implementing the 1-bit full adder is to generate (14.3-4) and (14.3-6) directly. Equation (14.3-4) requires four three-input AND gates followed by a four-input OR gate, and therefore the time required to achieve the sum S_n is equal to the propagation delay of two gates, or $2t_{pd}$. Equation (14.3-6) requires three two-input AND gates followed by a three-input OR gate. Thus, the carry term C_n is also obtained in a time $2t_{pd}$.

Parallel addition The addition of multibit numbers can be accomplished by using several full adders connected in tandem, as shown in Fig. 14.3-4. In this figure we

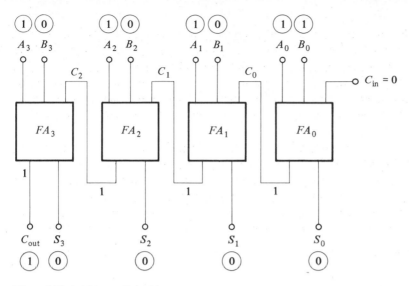

Figure 14.3-4 4-bit parallel adder.

see the addition of two 4-bit numbers. This type of adder is called a *ripple-carry adder*. Not shown in the figure are the three 4-bit *storage registers*, for the addend A_n, the augend B_n, and the sum S_n. These registers may consist of D type flip-flops, one for each bit, under control of a common clock. The input carry to the first full adder FA_0 is set to zero since there is no carry input.

In the example shown in Fig. 14.3-4 we add the numbers $A = \mathbf{1111}$ and $B = \mathbf{0001}$. The numbers in the circles represent the binary values throughout the circuit. This example points up one of the factors which limit the speed of ripple-carry parallel adders, i.e., the time required for carries to ripple through the circuit. In the example, A and B are both presented to the chain of adders at the same time. If we assume that the propagation delay time $t_A = 2t_{pd}$ is the same for each adder, then the carry from FA_0 appears at FA_1 after the time t_A, the carry from FA_1 to FA_2 after $2t_A$, and so on; the final C_{out} appears after $4t_A$. If we add two N-bit numbers, the addition will take a time Nt_A. This limitation can be overcome by using a special *look-ahead-carry circuit*, described in the next section.

14.3-3 Look-ahead-carry Adders

Commercially available adders such as the TTL 7483 and the CMOS 14008 4-bit full adders have *fast look-ahead-carry circuits* included on the chip. In such circuits the carry is derived directly from gates, whose inputs are the original bits of the augend and addend. This process requires many more gates than the ripple-carry circuit shown in Fig. 14.3-4 but has a considerable advantage in speed over that circuit.

For example, to add two 4-bit numbers, we could construct a combinational logic circuit which took the nine input variables C_{in}, A_0, B_0, A_1, B_1, A_2, B_2,

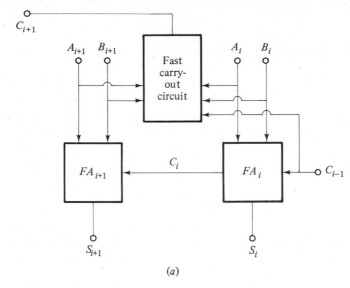

(a)

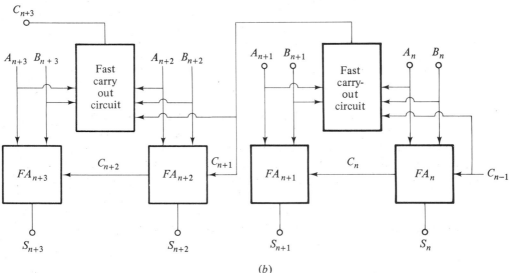

(b)

Figure 14.3-5 Carry circuits: (a) 2-bit adder with a fast carry output; (b) 4-bit adder with fast carry output.

A_3, and B_3 and using a large number of AND gates followed by a single OR gate generate S_0, S_1, S_2, S_3, and C_3 in the time t_A. Such a scheme is not practical due to the large number of gates required (Prob. 14.3-18).

One technique, which typifies the circuits used to speed addition, is shown in Fig. 14.3-5a. Here we see two 1-bit adders FA_n and FA_{n+1}. The inputs C_{n-1}, A_n, B_n, A_{n+1}, and B_{n+1} are inputs to a combinational logic circuit which produces C_{n+1} directly. (The actual circuit is found in Prob. 14.3-15.) Thus C_{n+1} occurs at a

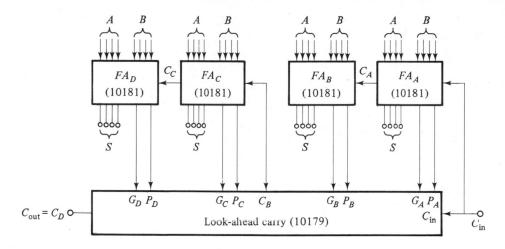

Figure 14.3-6 Addition using look-ahead techniques.

time t_A after the five input variables are applied. Note that the sum S_{n+1} still takes a time $2t_A$ to occur since C_n is applied in the ordinary way. The advantage of this procedure is gained in the 4-bit adder shown in Fig. 14.3-5b. Here the sum S_{n+2} takes $2t_A$ since the carry C_{n+1} is produced after the time t_A and S_{n+3} occurs after the time $3t_A$, representing a saving of t_A compared with the ripple adder shown in Fig. 14.3-4.

In addition to speeding the sum and the carry outputs within a 4-bit adder, similar fast carry-out circuits are available on IC chips to speed addition when adding two numbers having 16 or more bits. A 16-bit adder using a *look-ahead-carry* IC is shown in Fig. 14.3-6. The four 4-bit adders FA_A to FA_D are shown as ECL adders although similar circuits exist for TTL and CMOS. Each full adder has four A inputs, four B inputs, and a carry input. The outputs of each adder are the four sum outputs, a G and a P output, and a carry output, which is not always used. The G (generate) and P (propagate) outputs are used by the look-ahead-carry circuit to form carries C_B and C_D.

To see how the G and P outputs are formed we refer to the block diagram of the 4-bit adder shown in Fig. 14.3-5b. An output carry C_{n+3} is equal to **1** if $C_{n-1} = 1$ and (see Prob. 14.3-16) if

$$(A_n + B_n)(A_{n+1} + B_{n+1})(A_{n+2} + B_{n+2})(A_{n+3} + B_{n+3}) = 1 \quad (14.3\text{-}8a)$$

We say that the carry input C_{n-1} can be *propagated* through the adder if

$$P = (A_n + B_n)(A_{n+1} + B_{n+1})(A_{n+2} + B_{n+2})(A_{n+3} + B_{n+3}) = 1 \quad (14.3\text{-}8b)$$

This propagate P term is formed in the 10181 4-bit adder and is an input to the 10179 look-ahead carry circuit.

The carry output C_{n+3} can also be *generated* even when $C_{n-1} = 0$. For example, if $C_{n-1} = 0$ but $A_n = B_n = 1$, we have generated a carry $C_n = 1$. Then

$C_{n+3} = 1$ if $(A_{n+1} + B_{n+1})(A_{n+2} + B_{n+2})(A_{n+3} + B_{n+3}) = 1$. We can show (Prob. 14.3-17) that $C_{n+3} = 1$ if $C_{n-1} = 0$ and

$$G = (A_n \cdot B_n)(A_{n+1} + B_{n+1})(A_{n+2} + B_{n+2})(A_{n+3} + B_{n+3}) +$$
$$(A_{n+1} \cdot B_{n+1})(A_{n+2} + B_{n+2})(A_{n+3} + B_{n+3}) +$$
$$(A_{n+2} \cdot B_{n+2})(A_{n+3} + B_{n+3}) + (A_{n+3} \cdot B_{n+3}) = 1 \quad (14.3\text{-}9)$$

Equation (14.3-9) shows the ways of *generating* an output carry internally and is the *carry-generate*, or G, term shown in Fig. 14.3-6. The G term is formed within each adder.

The carry-out term can either be generated or propagated. Referring to Fig. 14.3-6, we see that

$$C_A = G_A + P_A C_{in} \tag{14.3-10a}$$

and
$$C_B = G_B + P_B C_A \tag{14.3-10b}$$

Combining (14.3-10a) and (14.3-10b) yields

$$C_B = G_B + P_B G_A + P_A P_B C_{in} \tag{14.3-11}$$

Equation (14.3-11) is formed in the look-ahead carry chip, which also forms C_D in the same way:

$$C_D = G_D + P_D G_C + P_C P_D C_B \tag{14.3-12}$$

The advantage of using the look-ahead-carry circuit is shown in Table 14.3-1, which compares the total add times of 10181 fast adders when look-ahead is and is not used. Note that significant improvement occurs when adding numbers represented by more than 24 bits.

14.3-4 Addition of a Sequence of Numbers: Accumulation

In many digital systems it is necessary to determine the sum of a sequence of numbers:

$$S(k) = X(k) + X(k-1) + \cdots + X(1) + X(0) \tag{14.3-13}$$

Table 14.3-1 Add time using the 10181 and 10179 ECL

Number of bits to be added	Total add time, ns	Add time if 10179 chips are not used, ns
4		8 (1-10181)
8		11 (2-10181)
16	16 (1-10179, 4-10181)	17
24	17 (1-10179, 6-10181)	23
32	19 (2-10179, 8-10181)	30
64	25 (4-10179, 16-10181)	54

Equation (14.3-13) states that S evaluated at time kT (the T is understood to be present and for simplicity is omitted) is equal to the sum of the sequence of numbers $X(k)$, $X(k-1)$, ..., $X(1)$, $X(0)$. This sum can be found by noting that

$$S(0) = X(0) \tag{14.3-14a}$$

$$S(1) = S(0) + X(1) \tag{14.3-14b}$$

$$S(2) = S(1) + X(2) \tag{14.3-14c}$$

$$\cdots\cdots\cdots\cdots\cdots\cdots\cdots$$

$$S(k) = S(k-1) + X(k) \tag{14.3-14d}$$

Equation (14.3-14d) is implemented in the accumulator shown schematically in Fig. 14.3-7a. In the circuit the sequence of X's is stored in a shift register. At the time shown, the output of the register is $X(k)$, and the output of the adder is $S(k)$. The value of S at time $k-1$ is obtained using the D flip-flop SR_1.

The actual circuit of the accumulator is more complicated than shown in Fig. 14.3-7a since $X(k)$ is an N-bit word, as are $S(k)$ and $S(k-1)$. Thus the adder is an N-bit adder, and the SR consists of $N1$-bit-shift registers, as shown in Fig. 14.3-7b.

14.3-5 Subtraction

Up to this point we have discussed only addition of positive numbers. In practice either the addend or augend or both may be negative, and when the signs of the numbers are taken into account, subtraction becomes simply a form of addition and virtually the same circuitry can be used to accomplish both operations. By a process called *complementation* the subtraction is readily converted into addition. As an illustration using the decimal system consider performing the subtraction $423 - 239$ in the following way:

Step 1: $\quad 423 - 239 = [423 + (1000 - 239)] - 1000$

Step 2: $\quad\quad\quad\quad = (423 + \quad\quad 761) \quad\quad - 1000$

Step 3: $\quad\quad\quad\quad = 1184 - \quad\quad\quad\quad\quad 1000$

Step 4: $\quad\quad\quad\quad = 184 \tag{14.3-15}$

The number $1000 - 239 = 761$ is called the *tens complement* of 239, and it is found by subtracting 239 from the appropriate power of 10. The technique appears roundabout but can be implemented to use only addition after the complement is found. The answer can be obtained from step 3 of (14.3-15) by simply erasing the most significant digit from the sum S, that is, the digit on the extreme left: $\boxed{1}\,184$. This is the procedure when the minuend is greater than the subtrahend and both are positive, leading to a positive answer. If the answer is negative, as

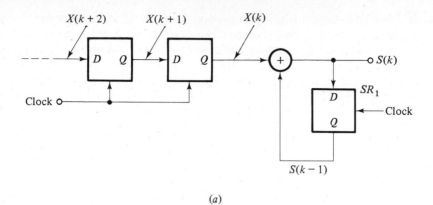

(a)

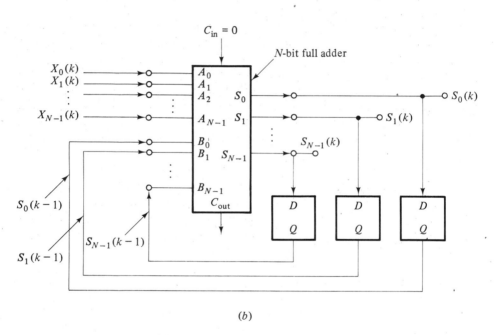

(b)

Figure 14.3-7 Accumulators: (a) schematic representation; (b) circuit of an accumulator to add N-bit words (clock pulses not shown.)

would be the case for $239 - 423$, we proceed as follows:

Step 1: $239 - 423 = [239 + (1000 - 423)] - 1000$

Step 2: $= (239 + 577) - 1000$

Step 3: $= 816 - 1000$

Step 4: $= -184$ (14.3-16)

Here $577 (= 1000 - 423)$ is the tens complement of 423. We recognize that the answer will be negative by inspection of Step 3 where we find that sum S is 0816, a number less than 1000. The magnitude of the answer is 184 where 184 is the tens complement of 816.

To summarize, if S exceeds 1000, the answer is S after removal of the most significant digit. If S is less than 1000, the answer is negative and the magnitude is the tens complement of S. Clearly the number $1000 = 10^3$ is the result of the example chosen; if the number chosen consisted of four digits, we would use $10^4 = 10,000$.

While the above method may appear far more complicated than the standard method of subtraction it has considerable advantages when the binary system is to be used since complementing a number is easily implemented with simple logic circuits.

Twos complement The twos complement of a binary number N is defined as

$$C_2(N) = 2^n - N \tag{14.3-17}$$

where n is the bit capacity of the register and $2^n \geq N$. Fortunately, the subtraction process indicated in (14.3-17) does not have to be used because there is a simple rule which allows for determination of the twos complement by inspection. The rule is: *Copy the number from right to left up to and including the first bit which is a 1. All further bits are complemented.* Some examples are given below:

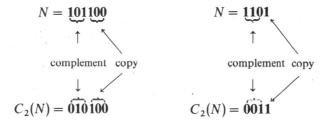

This type of complementing is accomplished in the logic circuit shown in Fig. 14.3-8a. In order to explain the operation of the circuit, we make use of the waveform shown in Fig. 14.3-8b. The input data to be complemented are in serial form and synchronized to the clock. Initially, the flip-flop is in the RESET state with $Q = 0$, so that the output of the EXCLUSIVE-NOR gate G_1 is LOW. Thus the data bits will pass through G_1 until after the first **1** bit occurs. (Note that the LSB is at the left end of the graph of input vs. time because it must occur *earliest* in time and be the first bit presented to the circuit.) When the first bit which is **1** occurs, the J input to the flip-flop goes HIGH, causing the flip-flop to SET ($Q = 1$) after the next clock pulse. This, in turn, causes the output of the EXCLUSIVE-NOR gate G_1 to be the complement of the input data. A pulse must be applied to the clear (Cl) terminal of the flip-flop immediately after the last bit in order to reset the flip-flop so that $Q = 0$ at the start of the next word.

Another method for obtaining the twos complement will be explored in Prob. 14.3-7.

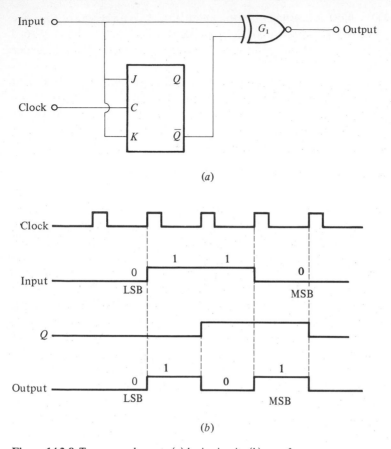

Figure 14.3-8 Twos complement: (*a*) logic circuit; (*b*) waveforms.

14.3-6 Signed Numbers

Some means must be provided to tell whether a number in a register is positive or negative. One procedure, called *sign-and-magnitude notation*, utilizes the leftmost bit as a *sign bit;* it is **0** for positive numbers and **1** for negative numbers. With this method the 16 possible combinations in a four-place register would have the numerical significance shown in Fig. 14.3-9*a*. Note that there are two representations for the number 0, **0000** and **1000**, and the largest numerical magnitude that can be represented is 7.

Generally an association of register position using the twos complement is used. Here the register position assigned to a negative number is the twos complement of the corresponding positive number. For example, consider representing the number −6 in a 4-bit register. We take the representation for +6, which is **0110**, and find its twos complement to be **1010**. Thus **1010** is the representation for −6. The 16 possible combinations in a four-place register are shown in

Register	0000	0001	0010	...	0111	1000	1001	1010	...	1111
Decimal	0	1	2	...	7	0	−1	−2	...	−7

(a)

Register	0000	0001	0010	...	0111	1000	1001	1010	...	1111
Decimal	0	1	2	...	7	−8	−7	−6	...	−1

(b)

Figure 14.3-9 Negative-number representation: (a) sign and magnitude; (b) twos complement.

Fig. 14.3-9b. Comparing with the sign-magnitude representation, we see that the twos-complement system has only one zero and can handle negative numbers up to −8 but has a different ordering. However, since all negative numbers have a 1 as the leftmost digit, they are immediately identifiable. For example, if the register reads **1100**, we know that the number is negative because the leftmost bit is 1 and we find the twos complement to be **0100**, so that its magnitude is 4. Thus the number represented by **1100** is −4.

Other schemes are possible since we can associate each of the 16 register combinations with any number. However, the twos-complement representation has been found to be extremely useful for carrying out addition and subtraction.

14.3-7 Addition and Subtraction Using Twos-Complement Notation

Addition of two signed numbers is illustrated in the following examples:

1. Both numbers positive

$$
\begin{array}{rr}
+7 & \mathbf{00111} \\
+ \quad +5 & \mathbf{00101} \\
\hline
+12 & \mathbf{01100}
\end{array}
$$

2. Both numbers negative; each is therefore written using twos-complement notation:

$$
\begin{array}{rr}
-7 & \mathbf{11001} \\
+ \quad -5 & \mathbf{11011} \\
\hline
-12 & \mathbf{110100}
\end{array}
$$

sign bit

answer → **10100**

The **1** in the leftmost place will not appear in the 5-bit sum register. The sum is negative since the sign bit is **1** and its magnitude is the twos complement of the number in the sum register, that is, **01100** = 12.

3. Larger number positive, smaller negative

$$+7 \qquad \mathbf{00111}$$

$$+ \quad -5 \qquad \mathbf{11011} \qquad \qquad \nearrow \text{sign bit}$$

$$+2 \qquad \mathbf{100010} \qquad \text{answer} \quad \mathbf{\dot{0}0010}$$

Again the leftmost **1** will not appear in the 5-bit register, and the result is positive since the sign bit is **0**.

4. Larger number negative, smaller number positive

$$-7 \qquad \mathbf{11001}$$

$$+ \quad +5 \qquad \mathbf{+00101}$$

$$-2 \qquad \nearrow\mathbf{11110}$$

$$\text{sign bit}$$

Since the sign bit is **1**, the result is negative and we take the twos complement to find its magnitude. This is **00010** = 2, and so the result represents -2.

Overflow Using twos-complement arithmetic a 5-bit register can accommodate numbers from -16 to $+15$ only. If we add two numbers which have a larger sum, an *overflow error* occurs:

$$+9 \qquad \mathbf{01001}$$

$$+ \quad +8 \qquad \mathbf{+01000}$$

$$+17 \qquad \mathbf{10001}(= -15)$$

This result would be interpreted as a negative number of magnitude $\mathbf{1111} = 15$, so that the difference between the true answer and the register reading is $17 - (-15) = 32$, a significant amount. Similar errors occur when the overflow is in the negative direction. (It is shown in Prob. 14.3-10 that when an N-bit adder overflows, the error is 2^N.) In order to avoid overflow error, the bit capacity of the adder and registers must be sufficient to accommodate the largest sums expected.

Subtraction In order to subtract a number A from B we first take the twos complement of A and add that result to B. Thus, subtraction is equivalent to addition.

14.3-8 Scaling

The situation often arises where a variable must be multiplied by a constant, a process called *scaling*. If the constant is a power of 2, scaling is readily accomplished using shift registers.

On paper, multiplication of a binary number by a power of 2, say 2^n, is easily accomplished by moving the binary point n places to the right. Dividing by 2^n involves moving the binary point n places to the left. For example, consider the number $5.25 = \textbf{0101.01}$, which is to be multiplied and divided by the scale factor $a = 2^2$. The result of the multiplication is $\textbf{010101.0} = 21$, and the result of the division is $\textbf{01.0101} = 1.3125$.

For binary numbers in registers it is more convenient to shift the digits so that the physical location of the binary point remains fixed. Shifting the digits *left* by n places then multiplies the number by 2^n, while a shift *right* divides by 2^n. Along with the shift, zeros must be placed into any vacant register positions if the number is positive.

Errors occur if one or more digits are lost because of the shift. If the shift is to the right, bits of low significance may be shifted out of the register. If the LSB is lost, the error (called *quantization error*) is equal to one-half the numerical value of the rightmost bit position of the register.

When the shift is to the left, care must be taken to avoid shifting the most significant **1** in the number into the register storing the sign bit. If the most significant **1** were to shift into the sign-bit position, the resulting *overflow* error would be unacceptable since it affects both magnitude and sign.

When negative numbers in twos-complement form are used, scaling by 2^n can again be accomplished by moving the binary point or shifting the digits. For example, consider the number -5.25 multiplied or divided by $a = 2^2$. The twos complement of 5.25 is readily shown to be **11010.1100**. Note that adding 1s to the left of the number or 0s to the right of the number does not alter its numerical value. Thus $\textbf{1111010.1100000} = -5.25$. Now

$$(-5.25)(2^2) = \textbf{11010 11,00} = \textbf{1101011.00} = -21 \qquad (14.3\text{-}18)$$

For division by 2^2 we have

$$(-5.25)(2^{-2}) = \textbf{110.10,1100} = \textbf{110.101100} = -1.3125$$

When scaling by 2^n it is possible to avoid using shift registers and actually to *hard-wire* the scaling circuit. A circuit wired to produce $A = 2^2 B$ using twos complements where 5-bit numbers are used and $B = 3 = \textbf{00011}$ is shown in Fig. 14.3-10a. Several points are to be noted in connection with the circuit. First, the sign bit is not shifted since a sign change is not possible as a result of scaling. Second, B_3 and B_2 are lost in the scaling process because of the two-digit shift to the left. Thus the range of values of B that can be handled without an overflow error in this simplified circuit is $B = +3$ to -4 (**00011** to **11100**). Finally, the ground connection on A_1 and A_0 comes about because zeros are shifted into the two rightmost positions regardless of whether B is a positive or a negative number.

In Fig. 14.3-10b we show a scaling circuit where $A = 2^{-2} B$. Here the vacant positions A_3 and A_2 are connected to the sign bit A_4 because if A is positive, then $A_4 = A_3 = A_2 = \textbf{0}$, while if A is negative, $A_4 = A_3 = A_2 = \textbf{1}$, as shown in the

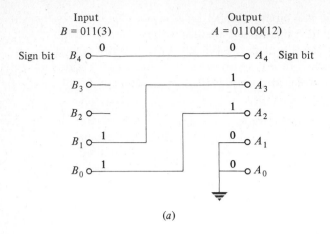

(a)

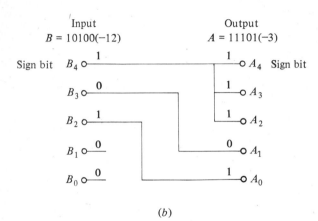

(b)

Figure 14.3-10 Scaling with numbers in twos-complement form: (a) $A = 2^2B$, where $B = 3$; (b) $A = 2^{-2}B$, where $B = -12$.

example on the diagram. The range of B that can be accommodated without overflow error is $+15$ to -16 (**01111** to **10000**). However, B_1 and B_0 are lost, so that if either of them is **1**, there will be a quantization error.

14.4 DIGITAL FILTERS

As discussed in Chap. 8, active filters, which operate on analog signals, utilize op-amp adders, subtractors, scalers, and integrators to simulate and solve the differential equations which describe physical systems. Digital filters perform the same function in the digital domain but instead of operating on continuous signals they operate purely on numbers. A block diagram of a generalized digital filter system is shown in Fig. 14.4-1a. The system operates as follows. At time $t = n \, \Delta t$ the analog input signal is sampled by the sample-hold circuit (see Chap. 15). A

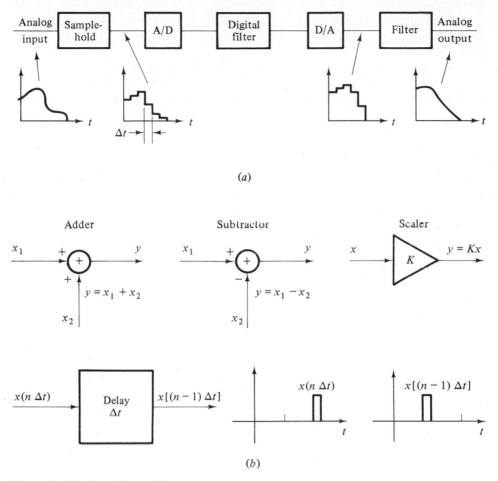

(a)

(b)

Figure 14.4-1 Digital filters: (a) block diagram of generalized system and typical waveforms; (b) digital components.

constant value equal to the sample value results, and the input signal is converted into a stepwise waveform, as shown. The total range of voltages in this signal is then divided into equal intervals (a process called *quantizing*), and in the A/D (analog-to-digital) converter (Chap. 15) any amplitude within a given quantization interval is converted into a specific number in the form of a digital word.[5] Thus quantization is an approximation which is required when digitizing a signal so that a finite number of digital numbers will result. The digital word at the A/D converter output may, of course, be *coded* in any desired form, and the twos-complement system is often used. The required operations are then carried out in digital form in the digital filter, which is essentially a special-purpose digital computer. After the signal is filtered, a D/A (digital-analog) converter (Chap. 15) converts the numbers at the digital filter output into the analog values in the waveform shown. An analog filter follows the D/A converter and converts the stepwise waveform into a continuous output signal thereby eliminating undesired frequency components.

In this section we introduce the digital filter. Since digital signals are discrete, they can be considered simply as sequences of numbers. The basic arithmetic operations required of digital networks which process these signals are those of addition (or subtraction) and scaling; a delay element which delays any given pulse exactly one period Δt is also required.

Digital adders (and subtractors) were discussed earlier in this chapter and are readily available. The circuit symbol we shall use to represent a digital adder (or subtractor) is shown in Fig. 14.4-1b. Digital scalers will be represented as shown, and the delay elements are simply single-input single-output type-D flip-flops, as discussed in Sec. 14.1 and shown schematically in Fig. 14.4-1b.

These three elements are interconnected as desired to form digital networks. A basic network using one of each type of element is shown in Fig. 14.4-2a. The input to the network is the sequence of numbers $x(n\,\Delta t)$, and the output is the sequence $y(n\,\Delta t)$. The present value of the output $y(n\,\Delta t)$ is delayed by Δt, at which time it becomes $y[(n-1)\,\Delta t]$. This is then multiplied by the constant K and added to the present value of the input, $x(n\,\Delta t)$. Thus the output can be written

$$y(n\,\Delta t) = Ky[(n-1)\,\Delta t] + x(n\,\Delta t) \qquad (14.4\text{-}1)$$

This is a first-order linear *difference* equation and should be thought of as an algorithm, i.e., a systematic, step-by-step process for solving a particular digital signal-processing problem. The digital network of Fig. 14.4-2, called a *first-order digital filter*, is the implementation of the algorithm which provides the solution. Some typical responses are found in the following example.

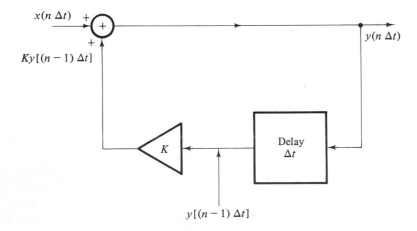

Figure 14.4-2 First-order digital filter.

Example 14.4-1† Find the output of the network of Fig. 14.4-2 if the input is a single pulse of amplitude A at $t = 0$ and if $y(-\Delta t) = 0$.

SOLUTION The input pulse at $t = 0$ indicates that

$$x(n\,\Delta t) = \begin{cases} A & n = 0 \\ 0 & n \neq 0 \end{cases}$$

† Adapted from C. Belove, H. H. Schachter, and D. L. Schilling, "Digital and Analog Systems, Circuits and Devices," secs. 3.4 and 4.3. McGraw-Hill, New York.

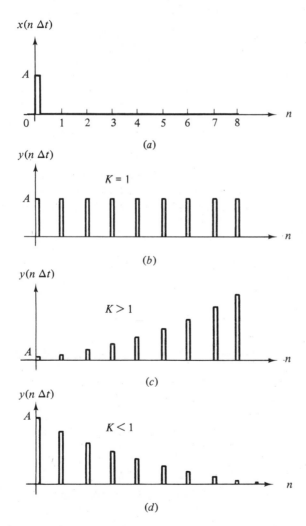

Figure 14.4-3 Digital-network response: (a) input; (b) output for $K = 1$; (c) output for $K > 1$; (d) output for $K < 1$.

This is shown in Fig. 14.4-3a. We write (14.4-1) for increasing values of n as follows:

$n = 0$: $\qquad\qquad\qquad y(0) = Ky(-\Delta t) + x(0)$

$n = 1$: $\qquad\qquad\qquad y(\Delta t) = Ky(0) + x(\Delta t)$

$n = 2$: $\qquad\qquad\qquad y(2\,\Delta t) = Ky(\Delta t) + x(2\,\Delta t)$

$n = 3$: $\qquad\qquad\qquad y(3\,\Delta t) = Ky(2\,\Delta t) + x(3\,\Delta t)$ $\qquad\qquad$ (14.4-2)

In the first equation, $y(-\Delta t) = 0$ and $x(0) = A$ are the given initial conditions; thus $y(0) = A$. Using this in the second equation with $x(\Delta t) = 0$, we find $y(\Delta t) = AK$. Continuing in this way, we obtain the sequence

n	0	1	2	3	$\cdots$
$y(n\,\Delta t)$	A	AK	AK^2	AK^3	$\cdots$

By induction, we have the general solution

$$y(n\,\Delta t) = AK^n$$

This solution is shown in Fig. 14.4-3 for various values of K. Note that if $K > 1$, the output of the network increases without bound. This is an *unstable* condition and is generally undesirable. However, if $K < 1$, the output decays. This is a *stable* condition; consequently most digital systems are designed with $K < 1$. When $K = 1$ the digital filter is identical to the accumulator shown in Fig. 14.3-7, and (14.4-1) becomes identical to (14.3-14d). $\qquad$ ///

14.4-1 Sinusoidal Response†

In this section we determine the response of the linear digital filter to a sampled sinusoidal signal. This response consists of two parts, the transient solution and the steady-state solution, which has the same form as the input sinusoid. The transient response of the filter has already been considered. In this section we restrict our discussion to the steady-state response.

The first-order digital filter Consider the first-order digital filter shown in Fig. 14.4-2. This filter is characterized by the difference equation

$$y(n\,\Delta t) = Ky[(n-1)\,\Delta t] + x(n\,\Delta t) \qquad (14.4\text{-}1)$$

where $x(n\,\Delta t)$ is the amplitude of the input signal at the sample time $n\,\Delta t$.

---

† Adapted from C. Belove, H. H. Schachter, and D. L. Schilling, "Digital and Analog Systems, Circuits and Devices," secs. 3.4 and 4.3. McGraw-Hill, New York.

To find the response of this filter to the sinusoidal input $x(t) = X_m \cos \omega t$, we first find the response to $x_1(t) = X_m e^{j\omega t}$. The desired input is then the real part of $x_1(t)$. The filter response to $x_1(t)$ will then be $y_1(t) = Y_m e^{j\omega t}$, where Y_m is a complex number. Hence the response to $x(t)$ is $y(t)$, which is the real part of $y_1(t)$.

The sampled input $x_1(n\,\Delta t)$ is $X_m e^{j\omega n\,\Delta t}$. Hence, the response $y_1(n\,\Delta t) = Y_m e^{j\omega n\,\Delta t}$ and $y_1[(n-1)\,\Delta t] = Y_m e^{j\omega(n-1)\,\Delta t}$. Substituting into (14.4-1) yields

$$Y_m e^{j\omega n\,\Delta t} = KY_m e^{j\omega(n-1)\,\Delta t} + X_m e^{j\omega n\,\Delta t} \tag{14.4-3}$$

We first cancel the term $e^{j\omega n\,\Delta t}$ from both sides of the equation. Then, the ratio Y_m/X_m is the transfer function $H(f)$ of the filter:

$$H(f) = \frac{Y_m}{X_m} = \frac{1}{1 - Ke^{-j\omega\,\Delta t}} \tag{14.4-4}$$

The transfer function $H(f)$ is a complex function of frequency having a magnitude $|H(f)|$ and a phase $\varphi(f)$:

$$|H(f)| = \left| \frac{1}{1 - Ke^{-j\omega\,\Delta t}} \right| = \left| \frac{1}{1 - K\cos\omega\,\Delta t + jK\sin\omega\,\Delta t} \right|$$

$$= \frac{1}{(1 + K^2 - 2K\cos\omega\,\Delta t)^{1/2}} \tag{14.4-5}$$

and
$$\varphi(f) = -\tan^{-1}\left(\frac{K\sin\omega\,\Delta t}{1 - K\cos\omega\,\Delta t} \right) \tag{14.4-6}$$

Note that the magnitude and the phase of the transfer function $H(f)$ are functions of K and Δt, as well as the applied frequency f.

The steady-state response

$$v(t) = \text{Re}\,(Y_m e^{j\omega n\,\Delta t}) \tag{14.4-7}$$

to the sampled input $x(n\,\Delta t) = X_m \cos \omega n\,\Delta t$ is then

$$y(n\,\Delta t) = \text{Re}\,[X_m|H(f)|e^{j\varphi(f)}e^{j\omega n\,\Delta t}]$$

$$= \frac{X_m}{(1 + K^2 - 2K\cos\omega\,\Delta t)^{1/2}} \cos\left(\omega n\,\Delta t - \tan^{-1}\frac{K\sin\omega\,\Delta t}{1 - K\cos\omega\,\Delta t}\right) \tag{14.4-8}$$

It is useful to plot $|H(f)|$ as a function of frequency in order to determine the frequency band passed by the filter. Since the sampling frequency is $f_s = 1/\Delta t$, we plot $|H(f)|$ vs. f/f_s:

$$|H(f)| = \frac{1}{(1 + K^2 - 2K \cos 2\pi f/f_s)^{1/2}} \tag{14.4-9a}$$

The magnitude of the transfer function $|H(f)|$ given in Eq. (14.4-9a) is a periodic function of frequency. It has a maximum value when

$$\cos 2\pi \frac{f}{f_s} = 1 \tag{14.4-9b}$$

which occurs when

$$\frac{f}{f_s} = 0, 1, 2, \ldots \tag{14.4-9c}$$

This maximum value is

$$|H(f)|_{max} = \frac{1}{(1 + K^2 - 2K)^{1/2}} = \frac{1}{1 - K} \tag{14.4-10}$$

The minimum value of $|H(f)|$ occurs when

$$\cos 2\pi \frac{f}{f_s} = -1 \tag{14.4-11a}$$

which results whenever

$$\frac{f}{f_s} = \frac{1}{2}, \frac{3}{2}, \frac{5}{2}, \ldots \tag{14.4-11b}$$

This minimum value is

$$|H(f)|_{min} = \frac{1}{(1 + K^2 + 2K)^{1/2}} = \frac{1}{1 + K} \tag{14.4-12}$$

Comparing Eqs. (14.4-10) and (14.4-12), we see that the difference between $|H(f)|_{max}$ and $|H(f)|_{min}$ increases as K approaches unity. Some typical values are given in Table 14.4-1.

Table 14.4-1

K	$|H|_{max}$	$|H|_{min}$
0.99	100	0.502
0.8	5	0.55
0.5	2	0.67
0.1	1.1	0.9

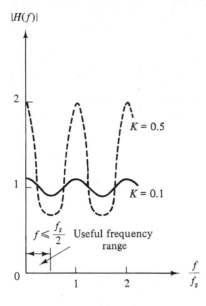

Figure 14.4-4 Frequency response of first-order digital filter.

Note that $|H|_{\max} > 1$, thereby providing amplification from input to output. This is a characteristic of digital filters.

The variation of $|H(f)|$ vs. frequency is plotted in Fig. 14.4-4 for $K = 0.5$ and 0.1. The filter appears to be low pass for frequencies less than one-half the sampling frequency. However, unlike its analog counterpart (an RC low-pass filter), the digital filter is periodic, and when $f = f_s$, the signal is actually enhanced. To avoid this problem and ensure that the digital low-pass filter is indeed always low pass, we set the sampling frequency f_s to be greater than or equal to twice the maximum applied signal frequency (Nyquist sampling rate)[5]

$$f_s \geq 2f_{\max} \tag{14.4-13}$$

Then the transfer function, as plotted in Fig. 14.4-4 is only of practical interest up to the frequency $f_{\max} < f_s/2$.

14.4-2 Circuit Realization of the First-Order Digital Filter

Operations within the digital filter can be carried out using the adder, D flip-flop, and scaler. A typical circuit is shown in Fig. 14.4-5. This circuit implements (14.4-1) with the constant $K = \frac{1}{2}$. The adder is as described in Sec. 14.3-3. For simplicity only five bit positions have been shown, and it is assumed that the A/D unit provides twos-complement numbers for negative values of $x(n \, \Delta t)$. The 1-bit delay is provided by a D-type flip-flop in each line. The scaler is of the hard-wired variety shown in Fig. 14.3-10. Binary values throughout the circuit are shown for

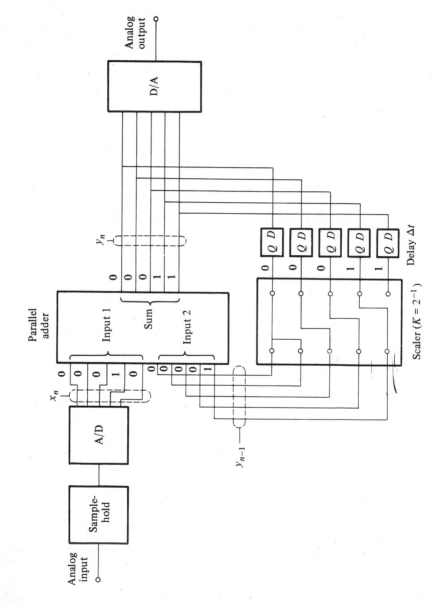

Figure 14.4-5 First-order digital filter with parallel processing $[x_n = x(n\,\Delta t),\, y_n = y(n\,\Delta t)]$.

$x(n\,\Delta t) = 2$, $y(n-1)\,\Delta t = 2$, which lead to the output value $y(n\,\Delta t) = 3$. It should be pointed out that the filter must be capable of performing the addition within the sampling time Δt.

REFERENCES

1. Fairchild Semiconductor, "The TTL Applications Handbook," Mountain View, Calif., 1973.
2. RCA, "COS/MOS Integrated Circuits Manual," Somerville, N.J., 1972.
3. Signetics "Applications; Digital, Linear, MOS," Sunnyvale, Calif., 1974.
4. H. Taub and D. L. Schilling, "Digital Integrated Electronics," p. 328, McGraw-Hill, New York, 1977.
5. H. Taub and D. L. Schilling, "Principles of Communication Systems," McGraw-Hill, New York, 1972.

PROBLEMS

14.1-1 The data word **101011** is stored in a 6-bit shift register as in Fig. 14.1-1. The data word **011010** is fed in without clearing the previously stored word. Plot the clock, data, and Q_0 to Q_5 waveforms.

14.1-2 Using four D-type flip-flops, design a 4-bit register which can be used for parallel-in parallel-out data transfer. Show the four input and four output waveforms when the data word **1101** is transferred in at clock pulse 1 and out at clock pulse 5.

14.1-3 Figure P14.1-3 shows clock and data input waveforms for the 4 bit shift register of Fig. 14.1-1a. Assuming that the register contains all 1s initially, plot the Q_0 to Q_3 waveforms.

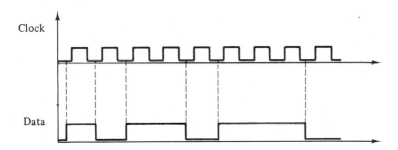

Figure P14.1-3

14.1-4 Repeat Prob. 14.1-3 if the flip-flops in the register are of the JK master-slave type.

14.1-5 In a *recirculating* shift register, the binary information circulates through the register as clock pulses are applied. The register of Fig. 14.1-1 can be converted into a circulatory register by connecting Q_3 to the data-input terminal. Assume that the register starts out with $Q_0 = 1, Q_1 = 1, Q_2 = 0, Q_3 = 1$. List the states that each register cycles through as eight clock pulses are applied.

14.1-6 Figure P14.1-6a shows a bidirectional shift register, i.e., one in which the data can be shifted left or right.

    (a) Explain its operation.

    (b) Show the Q_0 to Q_3 waveforms if the register starts out with $Q_0 = Q_1 = Q_3 = 1, Q_2 = 0$ and the *clock* and *right/left* waveforms are as shown in Fig. P14.1-6b. The *serial-data-in* line is LOW.

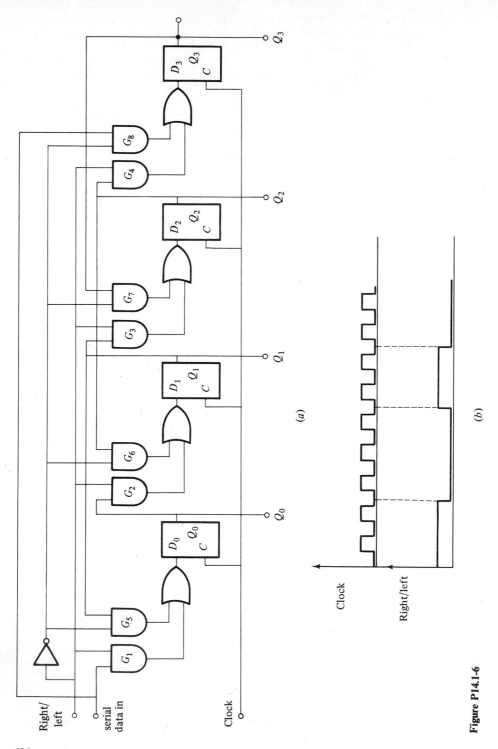

Figure P14.1-6

14.2-1 How many flip-flops are required to count to (a) 7, (b) 29, (c) 65, (d) 121?

14.2-2 What is the maximum modulus for a counter which contains (a) 3, (b) 5, (c) 7, (d) 10 flip-flops?

14.2-3 In the mod-8 ripple counter of Fig. 14.2-1 the clock frequency is 10 MHz, and the clock pulse width is 25 ns. Each flip-flop has a propagation delay of 5 ns. Sketch the Q_0, Q_1, and Q_2 waveforms taking this delay into account. What is the total delay from the trailing edge of a clock pulse until a corresponding change can occur in Q_2?

14.2-4 The maximum operating frequency for a ripple counter is to be 5 MHz. The flip-flops to be used have a propagation delay of 15 ns and a strobing time of 60 ns. Find the maximum number of stages the counter can have.

14.2-5 The flip-flops of Prob. 14.2-4 are used with a strobing time of 50 ns in an eight-stage ripple counter. What is the maximum frequency which can be applied to the counter and have the output reliable?

14.2-6 Figure P14.2-6 shows the mod-8 ripple counter of Fig. 14.2-1 modified to count mod-6 by clearing the counter when the count of 6 occurs. Operation is as follows: (1) The NAND gate output does not affect the flip-flops until it goes LOW, at which time all flip-flops are cleared. (2) The NAND gate output can go LOW only when $Q_1 = Q_2 = 1$. This occurs when the counter goes from 101 to 110. Thus the count sequence is 000, 001, 010, 011, 100, 101 (temporary 110), 000, 001, ... The temporary state lasts only a short time until the flip-flops clear, at which point the NAND gate output goes HIGH again. Thus the counter has six different states.

(a) Sketch the clock, Q_0, Q_1, and Q_2 waveforms. Be sure to include a narrow pulse at the appropriate points when the temporary state occurs.

(b) If the clock frequency is 100 kHz, what is the frequency of Q_2?

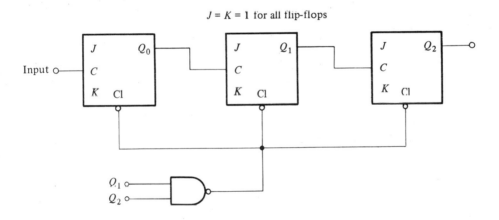

$J = K = 1$ for all flip-flops

Figure P14.2-6

14.2-7 Using the technique of Prob. 14.2-6, design a mod-7 counter and draw the timing diagram.

14.2-8 Repeat Prob. 14.2-7 for mod 10.

14.2-9 A five-stage ripple counter is driven by an 8-MHz clock. Find the output frequency and duty cycle if the clock is a square wave with (a) a 50 percent duty cycle and (b) a 20 percent duty cycle.

14.2-10 In the synchronous parallel counter of Fig. 14.2-4, the propagation delay of the flip-flops for a fan-out of 3 is specified as 24 ns. The AND gate has $t_{pd} = 4$ ns, and the strobe time is 20 ns. Find f_{max}.

14.2-11 Show that the counter in Fig. P14.2-11 is mod 5 by constructing a state table. Assume that all flip-flops are initially cleared.

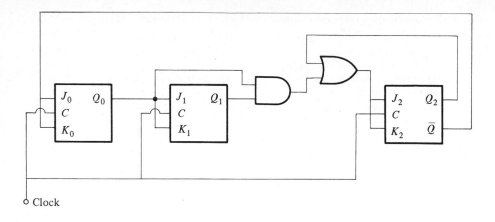

Figure P14.2-11

14.2-12 A five-stage ring counter has the initial state **10100**. Sketch the Q_0 to Q_4 waveforms through five clock pulses.

14.2-13 Design a shift-register counter using the state-table technique of Sec. 14.2-4 to count mod 5.

14.2-14 Design a counter such that the counter sequence is **00, 10, 01, 11**. This is a nonbinary mod-4 counter. *Hint:* Use two JK flip-flops.

14.2-15 A 10-MHz clock signal is to be changed. The available components are single JK flip-flops, mod-5, and decade counters. Design circuits using the available components to produce frequencies of (*a*) 2.5 MHz, (*b*) 100 kHz, (*c*) 500 kHz, (*d*) 40 kHz, and (*e*) 20 kHz.

14.2-16 Design a four-stage counter that counts from **1111** to **0000**, that is, **1111, 1110, 1101, 1100**, and so forth.

14.2-17 Explain the operation of the up-down counter shown in Fig. P14.2-17. *Hint:* Set $m = 1$ with the count at **0000**.

    (*a*) Apply four clock pulses. Sketch your result.
    (*b*) Now set $m = 0$ and apply an additional three clock pulses. Sketch your result.
    (*c*) Which way do we count when $M = 1$? $M = 0$?

14.2-18 Explain the operation of the counter shown in Fig. 14.2-8*c*.

14.3-1 Design a half adder using (*a*) all-NAND logic and (*b*) all-NOR logic.

14.3-2 Perform the following binary additions, keeping the carries separate:
    (*a*) **1110011 + 011001**
    (*b*) **1010111 + 101011**
    (*c*) **10001 + 11111**
    (*d*) **11100011 + 0011011**

14.3-3 Design a full-adder using (*a*) EXCLUSIVE-OR, AND, and OR gates and (*b*) all-NAND logic.

14.3-4 Verify that the combination of two half adders shown in Fig. 14.3-3 does operate as a full adder.

14.3-5 In the ripple carry adder of Fig. 14.3-4 the numbers **1101** and **0011** are added.
    (*a*) Show the binary values throughout the circuit.
    (*b*) Each full adder has a propagation delay of 8 ns. Find the time delay from the time that the numbers are applied to the adder to the time the complete sum appears at the output.

14.3-6 Find the twos complement of **111010** using the simplified method of Sec. 14.3-5. Verify your answer using (14.3-17).

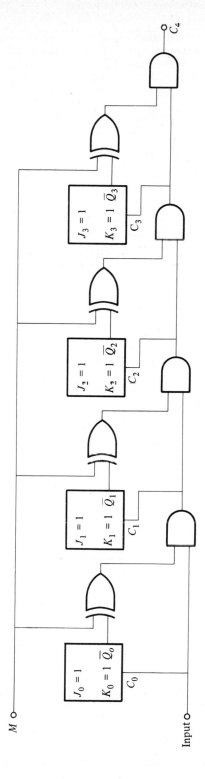

Figure P14.2-17

14.3-7 The ones complement of a binary number is easily found by simply complementing each bit of the number. The twos complement can then be found by adding **1** to the ones complement. Using these ideas, the parallel adder of Fig. 14.3-4 can be used to perform twos-complement subtraction by making $C_{in} = 1$ and adding $A_3 A_2 A_1 A_0$ to $\bar{B}_3 \bar{B}_2 \bar{B}_1 \bar{B}_0$.

    (a) Prove that $C_1(N) = C_2(N) - 1$ and that $C_1(N)$ can be obtained by simply inverting each bit of N.

    (b) Show binary values throughout the circuit of Fig. 14.3-4 for the twos-complement subtraction $D = 3 - 5$.

14.3-8 Use twos-complement subtraction to find (a) **111001 − 001101** and (b) **00110 − 10101**. Specify the minimum register length for each case so that there is no overflow.

14.3-9 The numbers 22, − 130, 82, −4 are to be stored in separate 10-bit registers using twos complements. Show the register contents for each case.

14.3-10 Show that when an N-bit adder overflows, the error is 2^N.

14.3-11 Prove that if overflow occurs when adding two positive numbers, the sign bit of the sum will be **1**; if both numbers are negative, the sign bit will be **0**.

14.3-12 Using the result of Prob. 14.3-11, design a logic circuit for the adder of Fig. 14.3-4 (with A_3 and B_3 as sign bits) which will give a **1** output when there is overflow.

14.3-13 Design a hard-wired scaling circuit for which $A = 2^3 B$ and both numbers are 12 bits long. Find the decimal range of B which can be accommodated without error.

14.3-14 Repeat Prob. 14.3-14 with $A = 2^{-3}B$.

14.3-15 Design a combinational logic circuit to produce C_{n+1} directly in the circuit of Fig. 14.3-5a.

14.3-16 In connection with Fig. 14.3-5b and Eq. (14.3-8a), prove that $C_{n+3} = 1$ if $C_{n-1} = 1$ and Eq. (14.3-8a) = 1.

14.3-17 Prove the assertion made in connection with Eq. (14.3-9).

14.3-18 Refer to Sec. 14.3-3. Consider constructing a combinational logic circuit which takes the nine input variables $C_{in}, A_0, B_0, A_1, B_1, A_2, B_2, A_3$, and B_3 and using a two-level AND/OR circuit generates S_0, S_1, S_2, S_3, and C_3. How many gates are required? *Hint:* Derive the truth table.

14.4-1 Find the output of the digital network of Fig. 14.4-2 if the input is as shown in Fig. P14.4-1 and $K = \frac{1}{2}$ and $y(-1) = 0$.

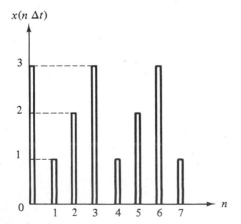

Figure P14.4-1

14.4-2 Repeat Prob. 14.4-1 if the circuit of Fig. P14.4-2 is used with $y(-1) = 0$. Compare results.

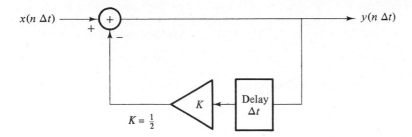

Figure P14.4-2

14.4-3 The circuit shown in Fig. P14.4-3 is analogous to a differentiator. Show that this is so by finding the response to a constant-amplitude pulse train.

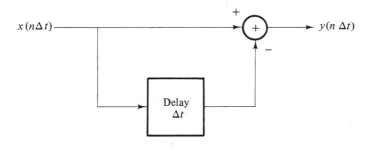

Figure P14.4-3

14.4-4 A first-order low-pass digital filter has $K = 0.5$ and the sampling frequency is 1 kHz.
    (a) What is the maximum signal frequency that can be accommodated safely?
    (b) Find the output if the input is

$$x(t) = 2 \cos (2\pi \, 100t) + 3 \cos (2\pi \, 400t)$$

14.4-5 The first-order filter of Prob. 14.4-4 is used to filter signals. Compare the output amplitudes of signals at 10, 100, and 500 Hz. Assume the input signals have identical amplitudes.

14.4-6 A high-pass first-order digital filter can be described by the difference equation

$$y(n \, \Delta t) = Ky[(n - 1) \, \Delta t] + x(n \, \Delta t) - x[(n - 1) \, \Delta t]$$

where $x(t)$ is the input signal and $y(t)$ the filter output.
    (a) Synthesize a digital circuit which will produce this difference equation.
    (b) Determine $|H(f)|$.
    (c) Plot $|H(f)|$ versus f/f_s for $K = 0$, 0.5, and 0.9.

14.4-7 The circuit for a second-order digital filter is shown in Fig. P14.4-7.
    (a) Find the difference equation which describes this filter.
    (b) Find the response when $k_1 = 0.5$, $k_2 = -0.5$, $x(0) = 1$, and $x(n \, \Delta t) = 0$ for $n \neq 0$; that is, the input is a single pulse occurring at $t = 0$. The initial conditions are $y(-\Delta t) = y(-2 \, \Delta t) = 0$.

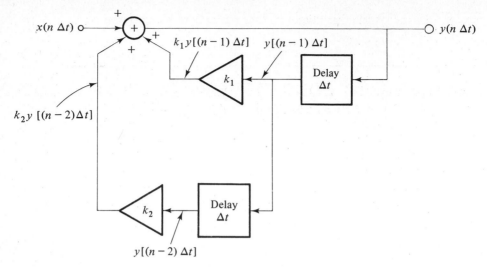

Figure P14.4-7

14.4-8 Find the response of the digital filter of Fig. P14.4-7 to a constant-amplitude pulse train if $k_1 = +1.2$ and $k_2 = -0.4$.

14.4-9 Find the response of the digital filter of Fig. P14.4-7 to a constant-amplitude pulse train if $k_1 = +1.5$ and $k_2 = -1.5$.

14.4-10 Design the circuit realization of a first-order digital filter as in (14.4-1) with $K = 2^{-2}$. Use eight bit positions and assume twos-complement representation for negative numbers.

FIFTEEN

SAMPLE-HOLD CIRCUITS, DIGITAL-TO-ANALOG AND ANALOG-TO-DIGITAL CONVERTERS, AND TIMING CIRCUITS

15.1 SAMPLE-AND-HOLD CIRCUITS

Sample-and-hold circuits are used to *sample* an analog signal at a particular instant of time and *hold* the value of the sample as long as required. The sampling instants and hold duration are determined by a logic control signal, and the hold interval depends on the application in which the circuit is being used. For example, in the digital filters discussed in Sec. 14.4 the samples must be held long enough for the analog-to-digital conversion to take place.

Most sample-hold circuits utilize a capacitor to hold the sample voltage. An electronically controlled switch provides a means for rapidly charging the capacitor to the sample voltage and then removing the input so that the capacitor can retain the desired voltage. Such a circuit is shown in Fig. 15.1-1a, with v_A the analog source and R_g its internal impedance. Idealized waveforms are shown in Fig. 15.1-1b. The switch is closed while the control logic waveform v_C is HIGH and, on the assumption that the $R_g C$ time constant is very small, the output voltage will very closely follow the input voltage and will be equal to it at the instant that the control logic goes LOW, opening the switch. During the HOLD interval, while the control signal is low, the switch is open and capacitor C will hold the last input value. Ideally the output will remain constant at that value throughout the HOLD interval.

In practice, the electronic switches and capacitors are not perfect, and various departures from the ideal occur. Among the important specifications given by

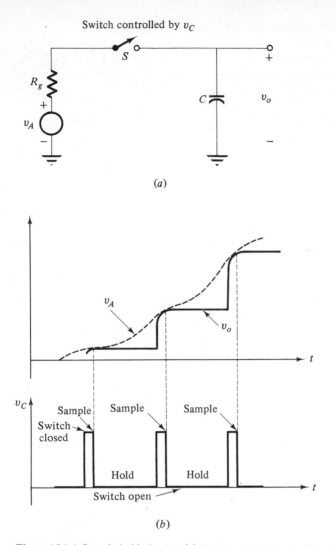

Figure 15.1-1 Sample-hold circuits: (a) simple switch circuit; (b) waveforms.

manufacturers are *aperture time* and *acquisition time.* Aperture time can be understood by referring to Fig. 15.1-2, from which we see that the aperture time is the maximum delay between the time that the control logic tells the switch to open and the time that it actually does open. The aperture time requirement of a system essentially determines the type of switch to be used. If it is of the order of milliseconds, S can be a relay. With FET switches aperture times are typically 50 to 100 ns, while for very fast diode switches the aperture time is much less than 1 ns. As a result of the aperture time there is an uncertainty in the sampling rate which may degrade system performance. Usually one selects a switch with an aperture time which is much less than the reciprocal of the sampling rate.

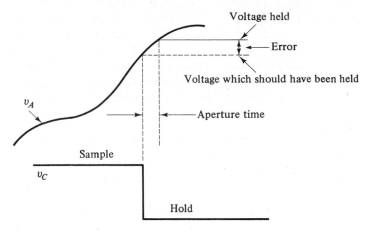

Figure 15.1-2 Aperture time and its effect.

Since the input signal is changing during the sampling interval, it takes a finite amount of time before the output is identical to the input. The *acquisition time* (Fig. 15.1-3) is the shortest time after a SAMPLE command has been given that a HOLD command can be given and result in an output voltage which approximates the input voltage with the necessary accuracy. The worst case occurs when the input is a step function which has an amplitude equal to the maximum peak-to-peak voltage swing of the circuit. In the circuit of Fig. 15.1-1a the speed with which the output can follow such an input depends on the characteristics of the signal source v_A. Considering the effect of the source impedance R_g, v_o will be an exponential with time constant $R_g C$, and in order for v_o to be within 0.01 percent of the input the time required is approximately $9R_g C$. In addition the

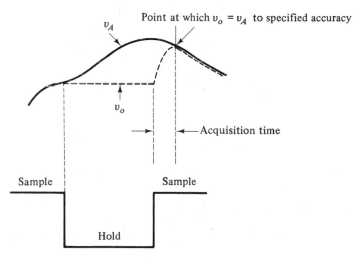

Figure 15.1-3 Acquisition time.

signal source must be capable of supplying the charging current required by capacitor C. Usually the analog input is buffered from the switch by an op-amp voltage follower (see Sec. 8.2) to ensure that R_g is very small.

The *settling time* is also usually specified by the manufacturer. This is the time from the opening of the switch (HOLD) to the point when the output has settled to its final value, within a specified percentage (usually 0.01 percent of full scale). If the sample-hold circuit is followed by an A/D converter, conversion should not begin until the signal has settled or the wrong voltage may be converted.

Sometimes the *output decay* rate, or *droop*, is specified. This represents the voltage change across the capacitor during the HOLD time and is inversely proportional to the capacitance since $dv_o/dt = I/C$, where I is the capacitor leakage current. The leakage current can arise as the result of bias current in an op-amp, leakage current through the switch, or internal leakage in the capacitor.

15.1-1 A Practical Circuit

Practical sample-hold circuits make extensive use of op-amps in order to achieve a low driving-circuit impedance and a high load impedance across the holding capacitor. Low-frequency sample-hold circuits utilize FET switches rather than BJTs because of the linearity and lack of offset of their transfer characteristics in the vicinity of the origin where the switching action takes place. If extreme speed is required, diode bridges (see Sec. 1.6) are used for switching.

An inverting sample-hold circuit is shown in Fig. 15.1-4a. It operates as follows. When switch S is closed (SAMPLE), the circuit acts as a conventional op-amp RC filter. If a step of amplitude V_A is applied to the input, the output will be (see Prob. 15.1-4)

$$v_o(t) = -V_A \frac{R_2}{R_1} (1 - \epsilon^{-t/R_2 C}) \tag{15.1-1}$$

Clearly, the $R_2 C$ time constant must be much shorter than the SAMPLE interval so that the output can track the input.

When the switch is opened, the voltage V_A will be HELD on the capacitor. As noted previously, the capacitor cannot hold this voltage indefinitely due to the input-bias-current requirement of the op-amp and internal leakage current in the capacitor and switch. To minimize this droop effect, the largest value of C consistent with acquisition time requirements should be used. The op-amp should have the smallest possible input bias current so that an op-amp with a FET input stage would be appropriate. In addition, the capacitor should be of the high-quality low-leakage type.

In Fig. 15.1-4b we show the inverting sample-hold circuit with a FET switch. T_1, the switching transistor, is a p-channel depletion FET. In this type of transistor, current flows when the source-to-gate voltage is zero and decreases as the source-to-gate voltage becomes more negative. A typical threshold voltage is $V_T = V_{SG} = -4$ V. Let us consider how this operates in the circuit of Fig. 15.1-4b.

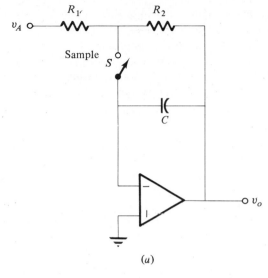

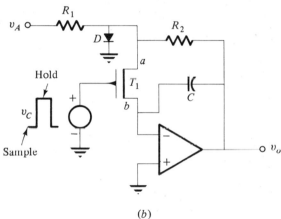

Figure 15.1-4 Inverting sample-hold circuit: (a) circuit with switch; (b) use of FET switch.

Consider first that the control voltage v_C is $+5$ V, that is, we are in the HOLD interval, where the switch is to be open, and let the analog voltage input v_A become positive. For this case, any current through T_1 must flow from a to b, making a effectively the source and b the drain. The presence of diode D causes the voltage at a to be clamped to 0.7 V, so that V_{SG} will never exceed $0.7 - 5 = -4.3$ V regardless of the value of V_A. Thus T_1 will remain OFF (an open switch), and if the threshold voltage V_T is -4 V, the source-gate voltage is 0.3 V more negative than needed to turn T_1 OFF. We then have a *safety margin* of 0.3 V.

If the analog input v_A goes negative, the roles of terminals a and b are reversed, b becoming the source. With the op-amp input terminal at a virtual ground, T_1 stays OFF because the 5 V on the gate provides a 1-V margin over the threshold voltage of -4 V.

In the SAMPLE interval, v_C goes to zero, and T_1 is turned ON (the switch is closed). The circuit then operates essentially like the simplified circuit in Fig. 15.1-4a. When T_1 is ON, diode D is connected across the FET switch. Since the drop across the FET is small, the voltage across the diode will be small and much less than the 0.7 V required for conduction; thus the diode has no effect during the SAMPLE interval. It is interesting to note that in this circuit the FET operation is bilateral, in the sense that source and drain are interchanged as the polarity of the input is changed.

15.2 DIGITAL-TO-ANALOG CONVERTERS

A large number of physical devices generate output signals which are of the analog, or continuous, variety. Examples include temperature, pressure, and flow transducers. In today's technology, signal processing is often accomplished using digital methods, and the processed signal is then often converted back to analog form. Many different types of data converters which are the interfaces between analog devices and digital systems have been developed. They are used in a wide variety of applications, including automatic process control, measurement and testing, data telemetry, and voice and video communicators. In this section, we consider digital-to-analog (D/A) converters, which produce an analog output from a given digital input.

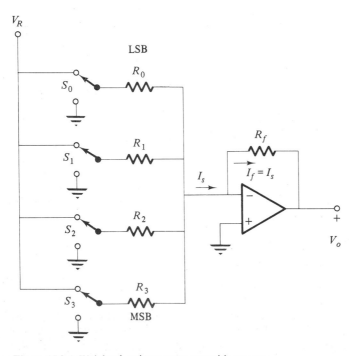

Figure 15.2-1 Weighted-resistor converter with op-amp.

15.2-1 The Weighted-Resistor D/A Converter

In Fig. 15.2-1 we show a passive-resistance network which converts a 4-bit parallel digital word $A_3 A_2 A_1 A_0$ to an analog voltage which is proportional to the binary number represented by the digital word. If the digital word is in serial form, a shift register (Sec. 14.1) can be used to convert it to parallel form. A 4-bit word is used for illustrative purposes only; the extension to more than 4 bits is easily made.

The logic voltages which represent the individual bits A_3, A_2, A_1, and A_0 are not applied directly to the converter but are used to operate electronic switches S_3, S_2, S_1, and S_0, respectively. When any of the A's are **1**, the corresponding switch is connected to a reference voltage V_R; when an A is **0**, the switch is connected to ground. The resistors in the network are weighted so that successive resistors are related by a factor of 2 and individual resistors are inversely proportional to the numerical significance of the appropriate binary digit. Thus for this 4-bit converter we have

LSB:
$$R_0 = \frac{R}{2^0} = R$$

$$R_1 = \frac{R}{2^1} = \frac{R}{2}$$

$$R_2 = \frac{R}{2^2} = \frac{R}{4} \tag{15.2-1}$$

MSB:
$$R_3 = \frac{R}{2^3} = \frac{R}{8}$$

where R is an arbitrary resistance which can be chosen to set the impedance level of the network.

In order to find the relation between the analog output voltage V_o at the output of the op-amp and the digital input we note that the op-amp input is a virtual short circuit. Hence the current I_s is

$$I_s = V_R \left(\frac{A_3}{R_3} + \frac{A_2}{R_2} + \frac{A_1}{R_1} + \frac{A_0}{R_0} \right) \tag{15.2-2a}$$

When (15.2-1) is used, this becomes

$$I_s = \frac{V_R}{R} (2^3 A_3 + 2^2 A_2 + 2^1 A_1 + 2^0 A_0) \tag{15.2-2b}$$

where $A_i = \mathbf{1}$ if switch S_i is connected to V_R and $A_i = \mathbf{0}$ if S_i is grounded. Equation (15.2-2b) shows clearly that the numerical value of the short-circuit current is directly proportional to the binary number $A_3 A_2 A_1 A_0$. For example, if the input is $A_3 A_2 A_1 A_0 = \mathbf{1111}$, $I_s = 15V_R/R$, while if $A_3 A_2 A_1 A_0 = \mathbf{0110}$, $I_s = 6V_R/R$, etc.

The output voltage V_o is

$$V_o = -I_f R_f = -I_s R_f = -\frac{R_f V_R}{R}(2^3 A_3 + 2^2 A_2 + 2^1 A_1 + 2^0 A_0) \quad (15.2\text{-}3)$$

Thus the output voltage is directly proportional to the numerical value of the binary input.

Some comments on this circuit are in order at this point. The first concerns accuracy and stability, both of which depend on the resistance ratios being powers of 2 and on their ability to track each other when the temperature changes. Since all the resistors have different values, it is difficult to obtain identical tracking characteristics. Also, since succeeding resistors differ by a factor of 2, the ratio of the largest resistor to the smallest resistor is 2^{n-1}, where n is the number of bits in the digital word. Thus in a 10-bit converter in which R_0 is to be 1 kΩ, the LSB resistor R_9 must be $2^{10} \times 1$ kΩ $= 1024$ kΩ. If the actual R_9 differs from the theoretical value of 1024 kΩ by 1 kΩ, that is, an accuracy of approximately 0.1 percent, the error voltage will be as large as the voltage produced by the least significant bit A_0. In this case the D/A converter will be capable of accurately converting only 9 rather than 10 bits. Because of this problem, the circuit is used primarily in low-resolution applications. A circuit which is not subject to this source of error is discussed in the next section.

The circuit shown in Fig. 15.2-1 will provide output voltages which are negative, as the output ranges from $V_o = 0$ when $A_3 A_2 A_1 A_0 = \mathbf{0000}$ to $V_o = -15 V_R R_f / R$ when $A_3 A_2 A_1 A_0 = \mathbf{1111}$. This range can be offset to any desired value by removing the ground connections from the lower switch terminals and connecting them to an appropriate negative voltage.

15.2-2 The R-2R Ladder Converter

Figure 15.2-2a shows a *resistive-ladder* D/A converter which does not require a wide range of resistance values. In fact, it requires only two values of resistance, R and $2R$. The resistance ladder is available in a single package in the circuit form shown in Fig. 15.2-2b. This circuit has the interesting property that the resistance seen looking into any of the terminals A, B, S_0, S_1, S_2, or S_3, with the remainder of these terminals grounded, is $3R$ (see Prob. 15.2-1). For ease of explanation the circuit shown in Fig. 15.2-2a accommodates a 4-bit parallel digital word, as in the weighted-resistance converter of Fig. 15.2-1. Extension to more than 4 bits is easily accomplished by adding additional switches and sections to the ladder.

To explain the converter operation assume that in Fig. 15.2-2a all switches are at ground except for S_0; the resulting resistive circuit is redrawn for clarity in Fig. 15.2-3a. The property which makes this circuit useful as a D/A converter is illustrated by the successive Thevenin conversions shown in Fig. 15.2-3b to d. In Fig. 15.2-3b we have replaced everything to the left of node 3 by a Thevenin equivalent in which the Thevenin voltage is $V_R/2$ and the Thevenin resistance is $2R \| 2R = R$. In Fig. 15.2-3c everything to the left of node 2 has been replaced by a

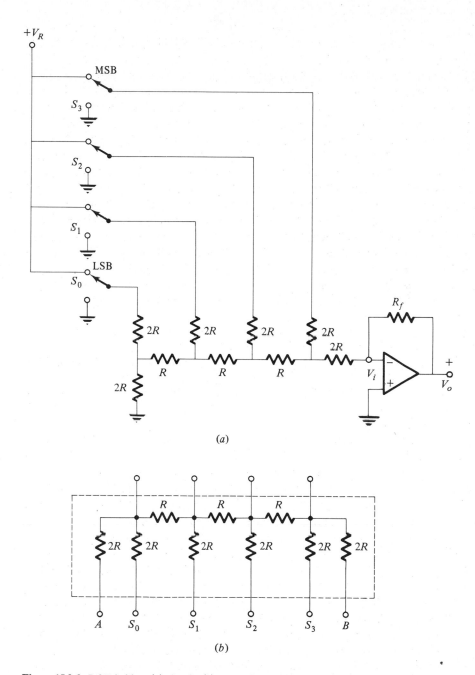

(a)

(b)

Figure 15.2-2 R-$2R$ ladder: (a) circuit; (b) network available as a package.

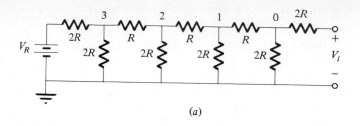

(a)

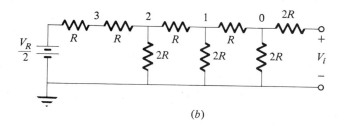

(b)

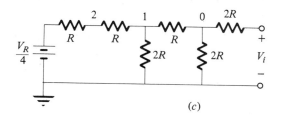

(c)

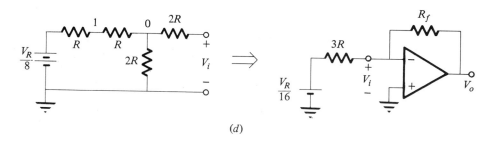

(d)

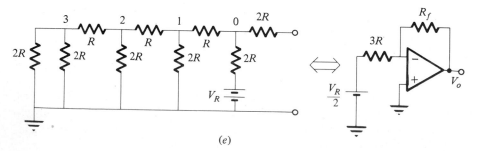

(e)

Figure 15.2-3 R-$2R$ ladder: (a) circuit with LSB = 1; (b) first Thevenin conversion; (c) second conversion; (d) third conversion; (e) circuit with MSB = 1.

710

Thevenin equivalent with the same resistance as before, that is, R, and a voltage $V_R/4$. Finally, in Fig. 15.2-3d everything to the left of node 1 is replaced by a Thevenin equivalent with resistance R and Thevenin voltage $V_R/8$. The output V_i, which is the input to the op-amp, is found from this circuit to consist of the voltage $V_R/16$ in series with a resistor of value $3R$.

If we let S_1 be connected to V_R and connect S_0 to ground, we find that the voltage in Fig. 15.2-3d is now $V_R/8$ rather than $V_R/16$.

As a further illustration in Fig. 15.2-4e we show the circuit when the switches are all at ground except the MSB switch S_3. Here the input to the op-amp consists of a voltage of value $V_R/2$ in series with a resistor $3R$.

It must be remembered that the switch S_i is connected to V_R when $A_i = 1$ and the switch arm is grounded when $A_i = 0$. Using the notation that $S_i = 1$ when $A_i = 1$ and $S_i = 0$ when $A_i = 0$, by superposition we can show the output of the circuit of Fig. 15.2-2 to be

$$V_o = -\frac{R_f V_R}{3R}\left(\frac{S_3}{2^1} + \frac{S_2}{2^2} + \frac{S_1}{2^3} + \frac{S_0}{2^4}\right) \tag{15.2-4a}$$

which yields

$$V_o = \frac{-R_f V_R}{48R}(2^3 S_3 + 2^2 S_2 + 2^1 S_1 + 2^0 S_0) \tag{15.2-4b}$$

Resolution and accuracy An important specification of a D/A converter is the *resolution* of which it is capable. Resolution is defined as the smallest increment in voltage that can be discerned by the circuit and depends primarily on the number of bits in the digital word. In our example, with a 4-bit word, the LSB has a weight of $\frac{1}{16}$ [see (15.2-4a)]. This means that the smallest increment in V_o is one-sixteenth the reference voltage V_R. For illustrative purposes, let us assume that $V_R = 16$ V. and $R_f - 3R$. Since the LSB has a weight of $\frac{1}{16}$, a change of one unit will result in a change of 1 V in the output, and we see that the output changes in steps of 1 V for each unit change in the value of the digital word input. If the switches were connected to a 4-bit counter which counts up from 0 to 15 (**0000** to **1111**), the output of the converter would be the staircase waveform shown in Fig. 15.2-4. The smallest voltage increment in this example is 1 V, which is the *voltage resolution* of the converter. The *percent resolution* is $(\frac{1}{16})(100) = 6.25$ percent. If this resolution is not sufficient, we must use a converter with more bits in the input word. For example, the LSB for a 10-bit converter has a weight of $\frac{1}{1024}$, and the percent resolution would be $(\frac{1}{1024})(100) \approx 0.1$ percent. If this converter has $V_R = 16$ V, the voltage resolution is approximately 16 mV.

Another specification of D/A converters is the *accuracy of conversion*, which depends on the difference between the actual analog output voltage and the theoretical output. This is a function of the accuracy of the precision resistors used in the ladder and the precision of the reference voltage source. As a practical matter a converter should have an accuracy better than $\pm\frac{1}{2}$LSB for accuracy and resolution to be compatible.

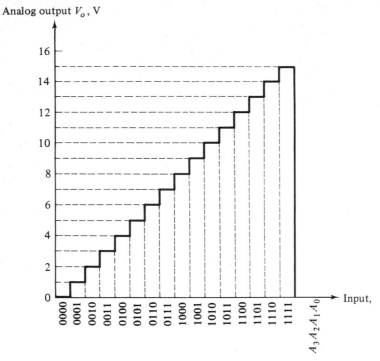

Figure 15.2-4 Output of 4-bit D/A converter with $V_R = 16$ V when connected to a 4-bit counter.

15.2-3 Switches Used in D/A Converters

The switches used in D/A converters are constructed using either BJTs or FETs, and they tend to fall into two classes, voltage-driven or current-driven. The voltage-driven converter utilizes BJTs or FETs which are switched from ON to OFF. These circuits are useful for relatively low-speed low-resolution applications because of the inherent inaccuracies in the system and the relatively slow speed obtained when switching a FET or BJT. In the current-driven converter, switching is accomplished using ECL current switches, which do not saturate but are driven from the active region to cutoff. This type of converter is capable of much faster operation than the voltage-driven type.

A voltage-driven converter The circuit of a basic 4-bit D/A converter using FET inverting switches is shown in Fig. 15.2-5. It operates as follows. Since each FET switch closes when the input control is high (logic **1**), each input is connected to the corresponding $\bar{Q}$ output of the register which holds the input word $A_3 A_2 A_1 A_0$. This is indicated on the diagram by labeling the gate inputs; $\overline{2^0}, \overline{2^1}, \overline{2^2}$, and $\overline{2^3}$. Consider that the $\overline{2^1}$ signal $\bar{A}_1$ is LOW. Then switch S_1 will be OFF, and the ladder arm $2R$ will be connected to the reference voltage V_R through the drain resistor R_d. If $R_d \ll 2R$, the resulting circuit is identical to that shown in Fig. 15.2-2. When $\bar{A}_1$ is HIGH, switch S_1 is on and the $2R$ ladder arm is connected

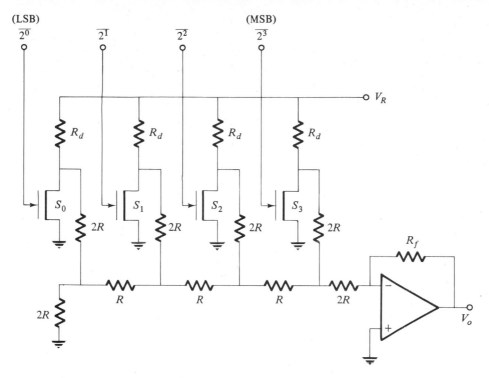

Figure 15.2-5 D/A converter with FET inverter switches.

to ground through the FET. The *voltage* V_R or ground is therefore connected to each ladder arm, hence the classification *voltage-driven converter*.

The accuracy of this circuit is limited by the tolerance of the resistors, accuracy and stability of the power supply, differences in the impedance of the individual switches, and the ratio of R_d to $2R$. It can be improved by various means but because of its inherent limitations this circuit is not used for high-performance converters.

A current-driven converter Many of the problems encountered with the voltage-driven converter can be solved by using a method in which currents are switched into and out of the ladder network. The voltage-driven ladder of Fig. 15.2-2a is converted into current drive by first replacing the voltage source V_R by individual sources of voltage V_R in series with each of the arm resistors of value $2R$. These Thevenin circuits are then replaced by equivalent Norton circuits, consisting of current sources $I = V_R/2R$ in parallel with resistors of value $2R$, as shown in Fig. 15.2-6a. In the converter circuit of Fig. 15.2-6b each switch is connected to a resistance of value $2R/3$ rather than ground. This is done so that the current I flows through the same resistance for either switch position. [The reader can verify (Prob. 15.2-13) that the resistance seen at the other switch terminal is indeed

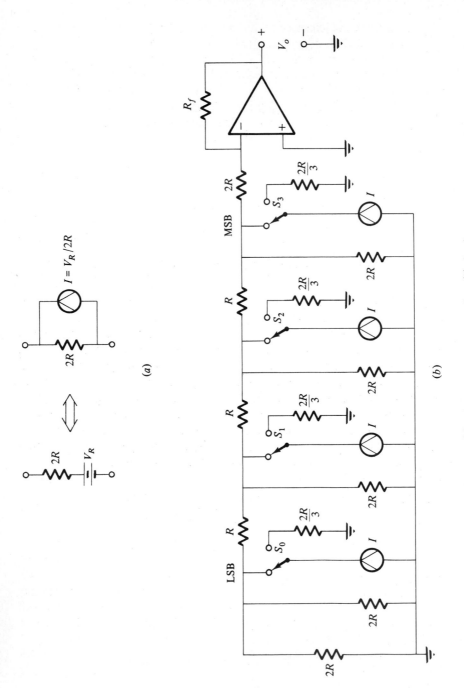

Figure 15.2-6 Basic current-driven D/A converter: (*a*) voltage-to-current conversion; (*b*) circuit.

$R \| 2R = 2R/3$.] The circuit of Fig. 15.2-6b is exactly equivalent to that of Fig. 15.2-2 if $2RI = V_R$.

The current-source-switch combinations shown in Fig. 15.2-6 can be implemented using standard ECL difference-amplifier circuits (Sec. 12.4). Such a circuit for 4-bit conversion is shown in Fig. 15.2-7. It makes use of the packaged resistance ladder of Fig. 15.2-2b, a quad ECL switch circuit, a special bias supply, and an op-amp. The circuit operation is as follows. The circuit is designed to be used in an ECL system in which logic 1 is typically about -0.8 V and logic 0 is about -1.6 V. The voltage on the base of the right-hand transistor of each pair is set about halfway between these two values by the reference bias supply, i.e., to about -1.2 V. From the bias-supply circuit we see that

$$V_B = V_{EB}(T_B) + V_{D1} + V_{D2} + V_Z - V_{EE} \qquad (15.2\text{-}5)$$

Assuming that all *pn* junction drops are equal yields

$$V_B = V_D + V_Z - V_{EE} \qquad (15.2\text{-}6)$$

If we choose $V_Z = 6.2$,† then $-V_{EE} \approx -1.2 - 0.8 - 6.2 = -8.2$ V.

Now consider any switch, say S_0. When A_0 is at logic 0 (-1.6 V), the left-hand transistor will be OFF, the right-hand transistor will be ON, and current will be drawn from node 0 of the ladder. The magnitude of the current is [see (15.2-6)]

$$I = \frac{V_{Re}}{R_e} = \frac{(V_B - V_{BE}) - (-V_{EE})}{R_e} = \frac{V_Z}{R_e} \qquad (15.2\text{-}7)$$

Thus I is a constant independent of the switch or V_{EE}.

When A_0 switches to logic 1 (-0.8 V), the left-hand transistor will turn ON and the right-hand transistor will turn OFF, drawing no current from the ladder. From the foregoing description we see that the switch action is virtually identical to that of the circuit of Fig. 15.2-6. The ECL type switches provide high-speed operation along with a considerable degree of temperature independence[1] and high accuracy.

The reference voltage V_R in the circuit of Fig. 15.2-7 is included in case bipolar operation is required. This is discussed in the next section.

Output offset In the circuit of Fig. 15.2-2 let the reference voltage be 16 V and the feedback resistor $R_f = 48R$. Then, when the input is **0000**, the output is 0 V; when the input is **1111**, the output is -15 V [see (15.2-4b)] and there is a 1-V step size as the binary value of the input changes. The important thing to note here is that the output voltage *range* extends from 0 to -15 V. For this case the output is called *unipolar* because it is all negative (it could be all positive). In some cases, it is desirable to *offset* the output swing so that the output is *bipolar*, i.e., swings from negative to positive voltages. In Fig. 15.2-2, if we had chosen $V_R = 8$ V and instead

---

† When $V_Z = 6.2$ V, the temperature coefficient of the Zener diode is approximately 0 (see Sec. 1.10).

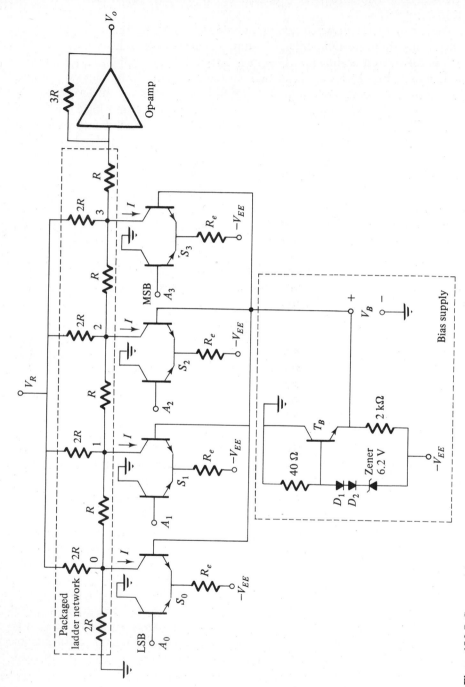

Figure 15.2-7 Current-driven D/A converter.

of grounding the other switch terminals had connected them to -8 V, the output voltage would have been bipolar. The correspondence between the analog voltage and the binary value of the digital input can be changed as desired by adjusting the offset. It is also possible to arrange the circuit so that negative numbers in twos-complement form can be converted; this will be considered in the problems.

In Fig. 15.2-7 bipolar operation is obtained by using a positive value of V_R. In Prob. 15.2-15 it is shown that if $V_R \leq 0$, the output is unipolar and always positive. If $V_R > 0$, the output is bipolar.

15.2-4 Manufacturers' Specifications

The important parameters *resolution* and *accuracy* were discussed in Sec. 15.2-2. Some additional specifications follow.

Linearity In Fig. 15.2-4 the actual analog voltages at the output of a D/A converter for a given digital input will not fall exactly on the steps of the staircase waveform because of inaccuracies in the resistors, etc. Linearity is a specification of the maximum deviation of the measured output from a straight line which extends over the full range of the waveform. It may be expressed as a percentage of the full-scale voltage or as a fraction of the voltage equivalent of the LSB and should be less than $\frac{1}{2}$LSB.

Settling time This is the elapsed time between the application of an ideal input pulse and the time at which the output voltage has settled to or approached its final value within a specified limit of accuracy. Typically, the settling time specification describes how soon after the input pulse the output can be relied upon as accurate to within $\frac{1}{2}$LSB.

Temperature sensitivity For a given fixed digital input, the analog output varies with temperature because reference-voltage supplies and resistors are temperature-sensitive. The temperature sensitivity of the offset voltage and the bias current of the op-amp also affects the output voltage. Typical sensitivities range from ± 50 to ± 1.5 ppm/°C† for high-quality converters.

15.3 ANALOG-TO-DIGITAL CONVERTERS

In many applications it is required to convert an analog signal into a digital form suitable for processing by a digital system. A number of methods are available for this type of conversion, and in this section we shall discuss two common techniques.

A *quantization error* is inherent in all A/D converters. To illustrate this consider the block diagram and graph of Fig. 15.3-1. The diagram in Fig. 15.3-1a

---

† ppm is an abbreviation for *parts per million*.

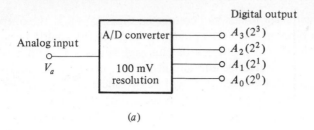

(a)

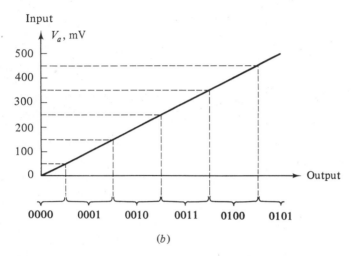

(b)

Figure 15.3-1 A/D converters: (a) symbol; (b) quantization error.

shows a 4-bit A/D converter which has a resolution of $(1 \text{ count})/(100 \text{ mV})$. Figure 15.3-1b shows the analog input and the digital output which results for the different *ranges* of the input. We see, for example, that all analog voltages between 50 and 150 mV produce the same digital output, **0001**; between 150 and 250 mV the output is **0010**, etc. Thus we have one count at the output for each 100 mV at the input. Now, if the digital output is **0011** and is the input to a D/A converter, the output of the D/A converter is assigned the analog value of 300 mV. However, we know that the original analog input voltage was between 250 and 350 mV. Thus the quantization error in this case can be specified as ± 50 mV, which is $\pm\frac{1}{2}$LSB. This is an ideal result; in practice the error might be somewhat larger than $\pm\frac{1}{2}$LSB because of inherent inaccuracies in reference voltages, etc.

15.3-1 Analog Comparators

All the techniques used to convert analog voltages into digital sequences require the use of an analog *comparator*. In Fig. 15.3-2a we show the logic symbol for an analog comparator. The inputs are two analog voltages V_a and V_b, and the output is a binary voltage. The circuit compares the two inputs so that if $V_a > V_b$, the output is a HIGH level signal (logic **1**). If, on the other hand, $V_a < V_b$, the output is

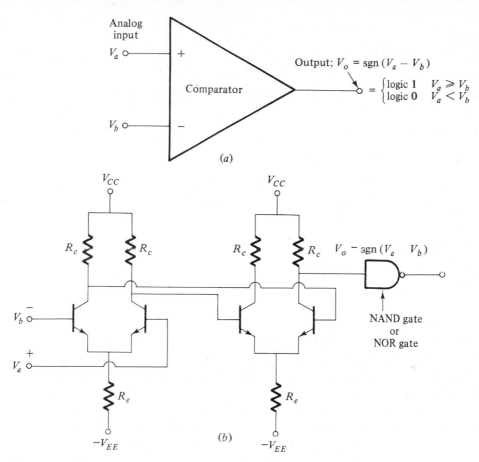

Figure 15.3-2 Analog comparator: (*a*) symbol; (*b*) schematic.

a LOW level signal (logic **0**). Thus

$$V_o \equiv \text{sgn}(V_a - V_b) = \begin{cases} 1 & V_a \geq V_b \\ 0 & V_a < V_b \end{cases} \qquad (15.3\text{-}1)$$

Figure 15.3-2*b* shows an analog comparator which consists of an op-amp comprising two difference amplifiers in cascade, followed by a logic gate. If the comparator output voltages are intended for TTL, the logic gate is often a TTL NAND gate connected as an inverter; an ECL NOR gate is used if ECL output levels are desired.

15.3-2 Counter-controlled A/D Converter

One of the simplest methods of A/D conversion is shown in Fig. 15.3-3. It uses three main elements, a counter, a D/A converter, and an analog comparator. For simplicity most of the control logic has been omitted from the diagram.

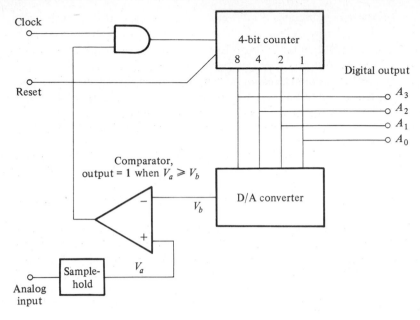

Figure 15.3-3 Counter-controlled A/D converter.

The converter operates as follows. At the beginning of a cycle the counter is reset to zero. This produces a D/A output voltage $V_b = 0$ which is applied to one input of the comparator. The analog input is fed through a sample-hold circuit (Sec. 15.1), whose output V_a is applied to the other input of the comparator. As long as the analog signal V_a is greater than V_b, the comparator output will be **1** and the AND gate will be enabled, allowing clock pulses to enter the counter. The counter will then count up, starting at zero. With each count the D/A output V_b will increase by one voltage step, as shown in Fig. 15.3-4. This count will continue

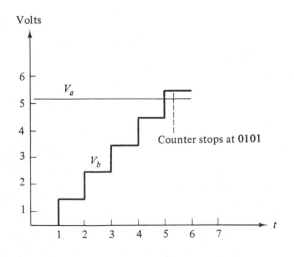

Figure 15.3-4 Waveforms in counter-controlled A/D converter.

until the staircase waveform exceeds the value of the analog signal V_a. At this time the comparator output drops to zero, disabling the AND gate, thereby stopping the counter. The output is then read from the output terminals of the counter. In Fig. 15.3-4 the output would be **0101**, corresponding to 5 V.

This type of counter is comparatively slow since as many as $2^N - 1$ clock periods (15 for a 4-bit converter) may be required for conversion. The converter to be described in the next section is much faster. The conversion time can be reduced if an up-down counter is used and the converter arranged to count up when $V_b < V_a$ and count down when $V_b > V_a$. In this case the comparator operates the up-down mode control of the counter.

In an actual circuit realization of this converter, additional logic is required to control the hold time of the sample-hold circuit and provide synchronization of the clock, reset, and hold signals.

15.3-3 Successive-Approximation D/A Converter

In the counter-controlled converter, the analog signal is compared first with 0 V, then with 1 V, then with 2 V, and so on, until the unknown voltage is found. This is a slow process. It can be speeded up considerably by making the comparisons in the following way. Assume that we wish to find an integral value of a voltage which we know lies between 0 and 16 V. We begin by having a comparator ask: Is the unknown voltage 8 V or more? If yes, we know that the unknown lies between 8 and 16 V; if no, it lies between 0 and 8 V. Assume that the answer is yes. We next compare the unknown to 12 V and we have the comparator ask: Is the unknown voltage greater than 12 V? If yes, the unknown lies between 12 and 16 V; if no, then it lies between 8 and 12 V. Continuing to bisect the interval in this way, we arrive at the correct answer very quickly. We can get an idea of the number of steps involved by assuming that the unknown voltage is $V_a - 10$ V. Then the process just described can be set forth as shown in Table 15.3-1 The digital answer is **1010**, which is the binary notation for 10, and we have found our result in just four comparisons. It is readily shown that in general an N-bit conversion will require N comparisons. This is considerably less than the 2^N required by the counter-controlled converter.

Table 15.3-1 Steps in successive approximation when $V_a = 10$ V

Step	Comparison	Answer	Digital answer (yes = **1**, no = **0**)
1	$V_a \geq 8$?	Yes	1 MSB
2	$V_a \geq 12$?	No	0
3	$V_a \geq 10$?	Yes	1
4	$V_a \geq 11$?	No	0 LSB

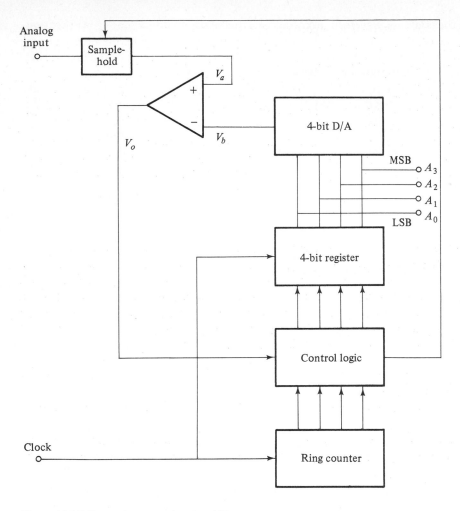

Figure 15.3-5 Successive-approximation A/D converter.

The simplified block diagram of a 4-bit successive-approximation D/A converter is shown in Fig. 15.3-5. The converter operates by successively dividing in half the voltage range in which the comparator has placed the analog voltage being converted. Its main components are the 4-bit register, which has independent set and reset capability for each stage, the 4-bit D/A converter, and the analog comparator. The ring counter and control logic are required to synchronize operation to the system clock. The ring counter (Sec. 14.2) provides timing waveforms to control the operation of the converter.

For simplicity, let us assume a unipolar input from 0 to 15 V, the output being the binary equivalent of the input voltage. We also assume that the HOLD cycle of the sample-hold circuit is timed so that V_a is constant over the conversion cycle.

Table 15.3-2 Successive approximation when $V_a = 10$ V

Step	V_b	Comparison	Answer	$A_3 A_2 A_1 A_0$
1	8	$V_a \geq 8$ V?	Yes	1000
2	12	$V_a \geq 12$ V?	No	1100
3	10	$V_a \geq 10$ V?	Yes	1010
4	11	$V_a \geq 11$ V?	No	1011
	10	Read output		1010

The steps in the conversion cycle are then as follows, each step taking one clock period:

1. The D/A unit, 4-bit register, and ring counter are all reset by the first pulse from the ring counter, so that the MSB bit in the D/A converter is $A_3 = 1$ and all others are **0**. Thus the V_b output of the D/A is 8 V. This is compared with V_a, and if $V_a \geq V_b$ ($V_a \geq 8$ V), the MSB flip-flop in the register is left at **1**; otherwise it is reset to **0**.
2. The second pulse from the ring counter sets $A_2 = 1$, A_1 and A_0 remaining at **0**, and A_3 being either **0** or **1** depending on the outcome of step 1. If $A_3 = 1$, then $V_b = 12$ V; if $A_3 = 0$, then $V_b = 4$ V. Let us assume $V_b = 12$ V. This is compared with V_a, and if $V_a \geq 12$ V, the A_2 flip-flop in the register is left at **1**; otherwise it is reset to **0**.
3. Same as step 2 but the A_1 flip-flop is either reset to **0** or left at **1**; the A_2 and A_3 flip-flops retain their states from step 2.
4. Same as step 3 but the A_0 flip-flop is used, the A_1, A_2, and A_3 flip-flops retaining their states from step 4. The desired number is now in the counter and is read out. Table 15.3-2 illustrates the process for $V_a = 10$ V.

The actual conversion is seen to take four clock periods. Thus, for example, a 10-bit converter with a 10-MHz clock will have a conversion time of approximately $10 \times 10^{-7} = 1$ μs.

15.3-4 Manufacturers' Specifications

Some specifications provided by manufacturers are listed below.

1. *Input signal.* This is the maximum allowable analog input voltage range and may be unipolar, that is, 0 to 10 V, or bipolar, that is, ± 5, ± 10 V, etc.
2. *Conversion time.* This depends on the type of converter. Ultrafast parallel converters have conversion times in the 10- to 60-ns range; successive approximation converters vary from 1 to 100 μs.
3. *Output format.* A variety of formats are available, including unipolar binary, offset binary, ones and twos complement, and various standard codes. Output circuits are often designed to interface directly to TTL, ECL, or CMOS.

4. *Accuracy.* The accuracy includes errors from both the analog and digital parts of the system. The digital error is due to the quantization discussed in Sec. 15.3, and the resulting *quantization error* is usually $\pm\frac{1}{2}$LSB. The main source of analog error is the comparator. Other sources are the reference voltage supply, ladder resistors, etc. The required accuracy and the number of bits should be compatible. For example, consider a 10-bit converter with an analog input range from 0 to + 10 V. The quantization error is $(1/2^{10}) \times 10$ V ≈ 10 mV. If we assume the analog error to be approximately equal to 10 mV, the overall error is 20 mV referred to the input. In this case the system operates like a 9-bit A/D converter which is free of analog error, since a 9-bit converter has a quantization error of $1/2^9 \times 10$ V ≈ 20 mV.

15.4 TIMING CIRCUITS

In many of the circuits described previously, *clock* pulse trains are required in order to synchronize operations throughout the system. Pulse trains are generated by timing circuits called *astable multivibrators*. Often a single pulse of predetermined length is required. Such pulses are generated by *monostable multivibrators*. These two timing circuits will be described in this section.

15.4-1 Monostable Multivibrator

The *monostable multivibrator*, or *one-shot*, is basically a flip-flop which has *one* stable state. It can be constructed using NOR gates, as shown in Fig. 15.4-1*a*. This circuit should be compared with the basic NOR gate flip-flop shown in Fig. 13.1-1 and can be readily constructed using CMOS or ECL. The operation of a CMOS monostable multivibrator can be explained using the waveforms shown in Fig. 15.4-1*b*. A quantitative discussion of the ECL monostable multivibrator is given in Prob. 15.4-2.

Initially, V_{o2} is LOW due to V_R, so that V_{o1} is HIGH. To begin the monostable operation an input pulse V_{i1} is applied. Since the input to gate G_1 is now HIGH, the output V_{o1} will become LOW after the propagation delay t_{pd}. Since the voltage across capacitor C cannot change instantaneously, the input V_{i2} to gate G_2, instantly falls. After a delay t_{pd} the output V_{o2} of G_2, rises, thereby keeping the output of G_1 low even after the input V_{i1} has returned to the **0** state. From the waveforms we see that the minimum input pulse width W is

$$W_{\min} > 2t_{pd} \tag{15.4-1}$$

As time increases, capacitor C charges toward V_R. When C has charged to the point that V_{i2} reaches the threshold voltage V_t of G_2, then G_2 turns ON and V_{o2} falls (after a delay t_{pd}), turning G_1 off. After an additional delay t_{pd}, V_{o1} goes high. Instantly V_{i2} goes high, and capacitor C must now discharge until a steady state is reached.

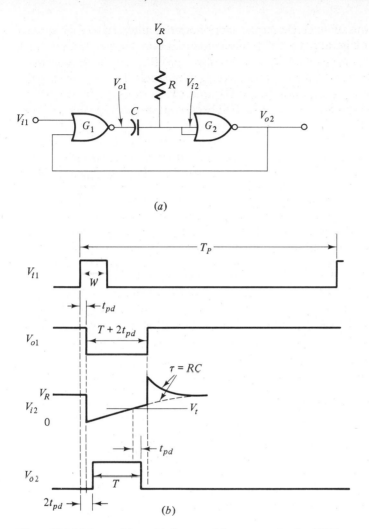

Figure 15.4-1 Monostable multivibrators: (*a*) constructed using NOR gates; (*b*) waveforms.

The circuit now remains in the steady state with $V_{o2} = 0$ V until the next input pulse arrives. It is shown in Prob. 15.4-1 that the pulse duration T is

$$T = RC \ln \frac{V_R}{V_R - V_t} \tag{15.4-2}$$

To ensure proper monostable operation V_{o2} must remain HIGH until after V_{i1} becomes LOW. Otherwise gate G_1 will not respond when V_{o2} becomes LOW. Thus, the maximum input pulse width is

$$W_{\max} < T + 2t_{pd} \tag{15.4-3}$$

Furthermore, the next input pulse should not be permitted to arrive until a steady state is reached. If we assume that this occurs after a time equal to $3RC$, we must have

$$T_p > T + 3t_{pd} + 3RC \qquad (15.4\text{-}4)$$

15.4-2 Astable Multivibrator

The *astable multivibrator* used to generate a clock pulse train is basically a flip-flop with no stable states. An astable multivibrator formed using NOR gates is shown in Fig. 15.4-2a. The waveforms shown in Fig. 15.4-2b assume CMOS operation, but ECL gates can also be employed (see Prob. 15.4-5).

The operation of the clock is most easily explained by starting with voltage V_{o1} rising from 0 V to V_{SS}. After a delay t_{pd}, V_{o2} falls, and instantaneously V_{i1} falls by V_{SS}. At this point $V_{o1} = V_{SS}$. Let us assume that V_{i1} was initially at the threshold voltage so that after falling by V_{SS}, $V_{i1} = V_t - V_{SS}$. This assumption is

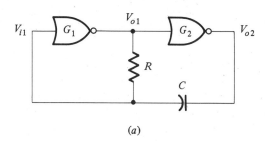

(a)

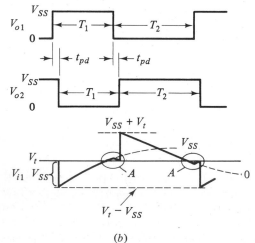

(b)

Figure 15.4-2 Astable multivibrator: (a) constructed using NOR gates; (b) waveforms.

verified below. Now the capacitor must charge in an attempt to make $V_{i1} = V_{SS}$ so that the current in R will become zero. The charging equation is (see Prob. 15.4-6)

$$V_{i1} = (V_t - V_{SS}) + (2V_{SS} - V_t)(1 - \epsilon^{-t/RC}) \tag{15.4-5}$$

The charging cycle is terminated when $V_{i1} = V_t$, that is, after a time

$$T_1 = RC \ln \frac{2V_{SS} - V_t}{V_{SS} - V_t} \tag{15.4-6}$$

At this time gate G_1 turns on, and after a time t_{pd}, V_{o1} falls to 0 V. After an additional delay t_{pd}, V_{o2} rises to V_R. The capacitor C now jumps from V_t to $V_t + V_{SS}$, and since $V_{o1} = 0$ V, the capacitor discharges, bringing V_{i1} toward 0 V. The equation of this discharge is

$$V_{i1} = (V_{SS} + V_T)\epsilon^{-t/RC} \tag{15.4-7}$$

When $V_{i1} = V_t$, that is, when $t = T_2$, where

$$T_2 = RC \ln \frac{V_{SS} + V_t}{V_t} \tag{15.4-8}$$

gate G_2 goes off and after a time t_{pd}, V_{o1} rises to V_{SS}. After a further delay of t_{pd}, V_{o2} falls to 0 V and V_{i1} falls from V_t to $V_t - V_{SS}$, thereby verifying our earlier assumption.

This charge-discharge operation continues indefinitely, and the period of oscillation is therefore

$$T_p = T_1 + T_2 \tag{15.4-9}$$

Note that $T_2 \geq T_1$ and T_2 is equal to T_1 only if $2V_t = V_{SS}$.

It is interesting to consider the behavior of V_{i1} during the time interval $2t_{pd}$ after it reaches threshold. These regions are circled and marked A in Fig. 15.4-2b. For a time t_{pd} following threshold V_{o1} does not change, and so the capacitor continues to charge or discharge as the case may be. However, in the interval t_{pd} to $2t_{pd}$, V_{o1} has changed states, but V_{o2} has not yet done so. In this region the capacitor will reverse its charging direction, as shown in the figure.

15.4-3 The 555 IC Timer

A general-purpose TTL *timer* is available in IC form with either one or two individual units in a single package. This IC, the 555 timer, can be wired to operate as a monostable or astable multi and to perform many other functions, a few of which are explored in the problems.

The basic circuit of the 555 timer is shown in Fig. 15.4-3a, and its connection as a monostable multi is shown in Fig. 15.4-3b. The timer consists of two comparators, an RS flip-flop, and a transistor. When connected as a one-shot, capacitor C_d is initially discharged. The operation begins when a negative trigger pulse is applied. The output of comparator C_2 now goes HIGH, setting the RS flip-flop.

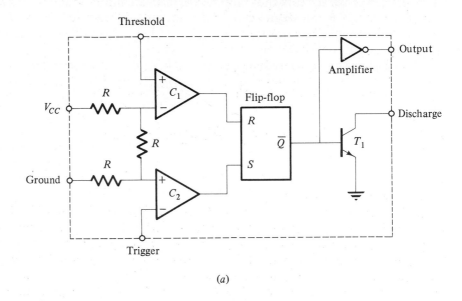

(a)

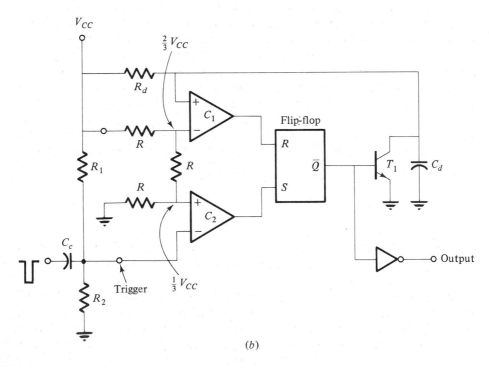

(b)

Figure 15.4-3 555 timer: (a) block diagram; (b) 555 timer connected as a monostable multivibrator.

The flip-flop output $\bar{Q}$ goes LOW, turning T_1 OFF. Capacitor C_d now begins to charge through resistor R_d toward V_{CC}. When the voltage across C_d reaches $2V_{CC}/3$, the output of comparator C_1 goes HIGH, resetting the flip-flop, so that $\bar{Q}$ goes HIGH, turning T_1 ON and thereby discharging C_d.

The result of the above operation is a negative output pulse whose duration is determined by the time required for C_d to charge from 0 V to $2V_{CC}/3$. Since

$$V_{Cd} = V_{CC}(1 - \epsilon^{-t/R_d C_d}) \tag{15.4-10}$$

the pulse width is

$$T = RC \ln 3 = 1.1RC \tag{15.4-11}$$

Many other applications of the 555 timer can be found in the manufacturers' literature.[4] Specification sheets that include instructions for connecting the 555 as an astable multivibrator will be found in Appendix C, Fig. C4.7. Its operation is investigated in Prob. 15.4-11.

REFERENCES

1. H. Taub and D. L. Schilling, "Digital Integrated Electronics," McGraw-Hill, New York, 1977.
2. David F. Hoesschele, Jr., "Analog-to-Digital/Digital-to-Analog Conversion Techniques," Wiley, New York, 1968.
3. Astable and Monostable Oscillators, *RCA Appl. Note* ICAN 6267.
4. Semiconductor Data Library, vol. 6, series B, p. 8–43, Motorola, Inc., 1975.

PROBLEMS

15.1-1 In connection with the discussion of *acquisition time* prove that $9R_s C$ s is required for v_o to be within 0.01 percent of the input.

15.1-2 A sample-hold circuit has a holding capacitor of 50 pF, and the leakage current in the HOLD mode is 1 nA. If the HOLD interval is 50 μs and the HOLD voltage is 1 V, find the percentage *droop*.

15.1-3 In a certain sample-hold circuit the holding capacitor is 100 pF, and the equivalent leakage resistance in the HOLD mode is 15 GΩ. Estimate the percentage *droop* if the hold interval is 100 μs.

15.1-4 Verify (15.1-1).

15.1-5 In the circuit of Fig. 15.1-4a, $R_1 = R_2 = 15$ kΩ, and the SAMPLE interval is 50 ns. Find C so that the output tracks the input to 0.1 percent.

15.1-6 In the circuit of Fig. 15.1-4a, $C = 500$ pF, $R_1 = R_2 = 15$ kΩ, and the input bias current of the op-amp is 300 nA. Estimate the percentage droop if the HOLD interval is 1 ms and the HOLD voltage is 1 V.

15.2-1 Verify that the resistance network of Fig. 15.2-2b has a resistance $3R$ looking into any of the terminals A, B, S_0, S_1, S_2, or S_3 with the remaining terminals grounded.

15.2-2 Verify (15.2-4a).

15.2-3 A 6-bit D/A converter has a unipolar current output. When the digital input is **110100**, the output current is 5 mA. Find the output current when the digital input is **111100**.

15.2-4 A 12-bit D/A converter has a step size of 10 mV. Find the full-scale output voltage and the percent resolution.

15.2-5 Find the resolution of the D/A converter of Prob. 15.2-3. Express your answer in terms of current and percentage.

15.2-6 Show that the percent resolution can be expressed as

$$\%R = \frac{1}{2^N}$$

where N is the number of input bits.

15.2-7 In the circuit of Fig. 15.2-1 $R_o = 8$ kΩ, $R_f = 1$ kΩ, and $V_R = 5$ V. Make a table showing the output voltage for all possible digital inputs.

15.2-8 The circuit of Fig. 15.2-1 is converted into 8-bit operation by adding four resistor-switch combinations. The reference $V_R = 10$ V, $R_o = 160$ kΩ, and $R_f = 10$ kΩ. Find the output corresponding to the input **11100110**.

15.2-9 In the D/A converter of Prob. 15.2-8, the step size is to be changed to 0.3 V. Determine a new value for R_f.

15.2-10 An 8-bit D/A converter has a full-scale output of 20 V. Find the output voltage when the input is **11011011**.

15.2-11 A D/A converter is to have a full-scale output of 10 V and a resolution less than 40 mV. How many bits are required?

15.2-12 An 8-bit D/A converter has an *accuracy* of 0.2 percent of full scale, and its full-scale output is 10 mA. What is the maximum possible error? If resolution error is included, would a 10-μA output for zero input be within the specified range of accuracy?

15.2-13 In the circuit of Fig. 15.2-6b show that the resistance seen at either switch terminal is $2R/3$.

15.2-14 In Fig. 15.2-7, find the base-emitter voltage of the OFF transistor in any transistor pair.

15.2-15 In Fig. 15.2-7, show that if $V_R \leq 0$ the output is unipolar and always positive, while if $V_R > 0$ the output is bipolar.

15.2-16 A 3-bit D/A converter is to be arranged so that it will accept numbers from -4 to $+3$ with the negative numbers in twos-complement form. The analog output is to vary from 0 V to $+7$ V, as shown in the table. This can be done by applying the digital input to the D/A converter with the MSB complemented so that digital input **100** produces 0 V output as in the table. Show how the converter of Fig. 15.2-1 can be modified to accomplish this translation.

Decimal value of input	Twos-complement representation	Offset binary format when **100** $\equiv$ 0 V	Analog output voltage, V
+3	011	111	7
+2	010	110	6
+1	001	101	5
0	000	100	4
−1	111	011	3
−2	110	010	2
−3	101	001	1
−4	100	000	0

15.3-1 An 8-bit A/D converter has a full-scale input range of 10 V. Find the resolution and the quantization error.

15.3-2 An analog comparator as shown in Fig. 15.3-2 has a linear gain $A_v = V_o/(V_a - V_b) = 5000$. The output V_o is $+5$ or 0 V. If $V_a = 3.274$ V, what value of V_b is required to cause the output to switch states?

15.3-3 An 8-bit counter-controlled A/D converter has a resolution of 40 mV and a clock frequency of 2 MHz. Find (a) the digital output if $V_a = 6$ V and (b) the maximum conversion time.

15.3-4 A certain 12-bit A/D converter has a full-scale output of 5 V and accuracy of 0.05 percent of full scale. Find the quantization error and the total possible error in volts.

15.3-5 For the successive-approximation A/D converter, show that an N-bit conversion will require N comparisons.

15.3-6 A 6-bit successive approximation A/D converter has a resolution of 0.05 V per step. If the analog input is 2.2 V, construct a table like Table 15.3-2 to find the final register reading.

15.3-7 Design a 3-bit successive approximation A/D converter using three D flip-flops for the ring counter, three two-input AND gates for the control logic circuit, and three JK flip-flops for the register. Draw a timing diagram for your converter, showing the clock, ring-counter waveforms, register-output waveforms, and comparator-output waveforms for an input corresponding to a digital output **010**.

15.4-1 For the monostable multivibrator of Fig. 15.4-1 show that the pulse duration is given by (15.4-2) assuming that the LOW output is 0 V.

15.4-2 Consider that the monostable multivibrator of Fig. 15.4-1 is constructed from ECL gates, for which -1.6 V is the LOW output and -0.8 V is the HIGH output. Sketch the waveforms corresponding to Fig. 15.4-1b.

15.4-3 A monostable multivibrator is constructed using TTL NAND gates as shown in Fig. P15.4-3.

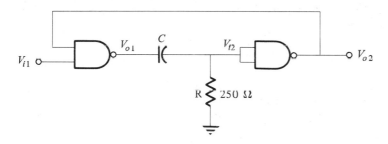

Figure P15.4-3

For the gates assume that $V_{OL} = 0$ V, $V_{OH} = 3.6$ V, and $V_{IL} = V_{IH} = 1.5$ V.
    (a) Sketch the waveforms of the trigger, V_{o1}, V_{i2}, and V_{o2}. State any assumptions made.
    (b) Derive an equation similar to (15.4-2) for the output pulse width.
    (c) Find the minimum input pulse width.

15.4-4 A one-shot constructed with discrete transistors is shown in Fig. P15.4-4. In the stable state T_2 is ON, and T_1 is OFF. The short trigger pulse turns T_1 ON.
    (a) Describe the circuit operation after T_1 turns ON.
    (b) Show that after the trigger is applied, the voltage at the base of T_2 is

$$V_{B2} = V_{CC} + (V_o - V_{CC})\epsilon^{-t/R_{b2}C}$$

where $V_o = -(V_{CC} - 0.7)$
    (c) Show that the pulse width is the time required for T_2 to reach a turn-on voltage $V_{B2} = 0.5$ V and

$$T = -R_{b2}C \ln \frac{V_{b2} - V_{CC}}{V_o - V_{CC}} \approx 0.7 R_{b2} C$$

    (d) Find the pulse width if $V_{CC} = 12$ V, $R_{b2} = 6.8$ kΩ, and $C = 0.002$ μF.

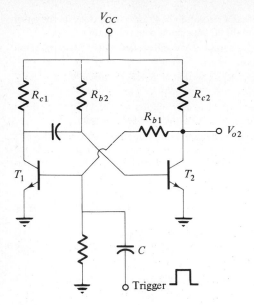

Figure P15.4-4

15.4-5 Sketch the waveforms shown in Fig. 15.4-2b if ECL NOR gates are used.

15.4-6 Verify Eqs. (15.4-5) and (15.4-6).

15.4-7 Figure P15.4-7 shows an astable multivibrator using discrete transistors.

(a) Describe its operation qualitatively.

(b) Using the techniques of Prob. 15.4-4, show that

$$T_{ON1} = T_{OFF2} \approx 0.7 R_{b2} C_2 \qquad T_{OFF1} = T_{ON2} \approx 0.7 R_{b1} C_1$$

(c) Show that the frequency of the output is

$$f \approx \frac{1.4}{R_{b1} C_1 + R_{b2} C_2}$$

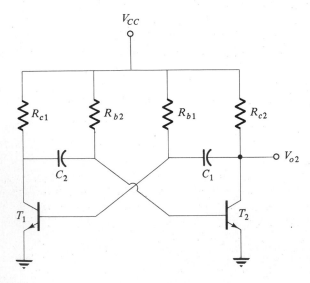

Figure P15.4-7

15.4-8 In the circuit of Fig. P15.4-7, $R_{b2} = 10$ kΩ, $C_1 = C_2 = 0.01$ μF, and $R_{b1} = 20$ kΩ. Find the frequency and duty cycle of the output. Sketch the waveform.

15.4-9 Figure P15.4-9a shows an astable multivibrator constructed with comparator. The comparator transfer characteristic is shown in Fig. P15.4-9b.

(a) Describe the operation of the multivibrator qualitatively, including a sketch of waveforms at the output and both inputs of the comparator.

(b) Show that

$$T = 2R_f C \ln \left(\frac{R_1 + 2R_2}{R_1} \right)$$

(c) Find the frequency if $R_f = 10$ kΩ, $R_1 = 6.8$ kΩ, $R_2 = 3.3$ kΩ, and $C = 0.1$ μF.

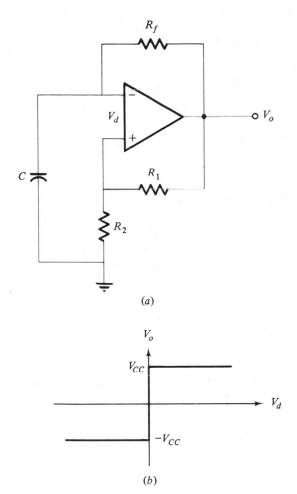

(a)

(b)

Figure P15.4-9

15.4-10 The astable multivibrator of Fig. P15.4-9 is to generate a pulse train with a period of 20 μs and a 50 percent duty cycle. If $R_1 = R_2 = 4.7$ kΩ and $R_f = 2$ kΩ, find C.

15.4-11 Figure P15.4-11 shows the connections required to have the 555 timer act as an astable multivibrator.

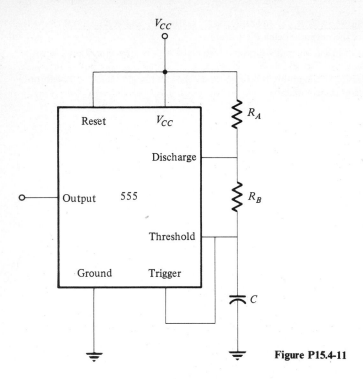

Figure P15.4-11

(a) Describe its operation qualitatively.

(b) The output is HIGH for a time $t_1 = 0.7(R_A + R_B)C$ and LOW for a time $t_2 = 0.7R_B C$, so that the frequency of oscillation is $f \approx 1.4/(R_A + 2R_B)C$. Find f and the duty cycle if $R_A = 1.5$ kΩ, $R_B = 6.8$ kΩ, and $C = 0.01$ μF.

SIXTEEN

INTEGRATED CIRCUITS

INTRODUCTION

During the 1960s, techniques were developed for the construction of integrated circuits (ICs) in which a whole amplifier or electronic system is fabricated on one small piece of silicon. These ICs occupy orders of magnitude less volume than the discrete circuits used previously and are much more reliable. Figure 16-1a shows a single conventional transistor, while Fig. 16-1b shows a nine-transistor IC amplifier. The transistor and the IC amplifier are each contained in a TO-5 package, a standard case used to package transistors and ICs. Most ICs today are contained in a *dual-in-line* package (DIP) (Fig. 16-1c) or a *flat pack* (Fig. 16-1d).

Two types of ICs are currently being manufactured, the *monolithic* and *hybrid* circuits. The monolithic circuit is fully integrated, i.e., the complete amplifier is made at one time, using diffusion techniques. In the hybrid circuit, separate microminiaturized components, constructed using both diffusion and thin-film techniques, are connected either by deposition techniques, to be described below, or by wire bonds. Thus the hybrid IC has the form of a discrete circuit which is packaged in a single case. Several fabrication techniques are used to form the monolithic and hybrid IC. In this chapter only the basic fabrication process is discussed. More detail can be found in the References.

A major result of this technology is a tremendous saving of space. Consider, for example, a digital computer. In 1955, using vacuum tubes, the space it occupied would have been equivalent to several large rooms. In 1965, using discrete transistors, one room might suffice. Today, the same computer constructed using

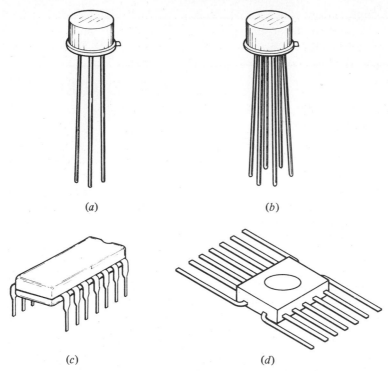

Figure 16-1 Comparison of a single transistor and a complete IC amplifier: (*a*) single transistor; (*b*) IC; (*c*) dual-in-line package; (*d*) flat pack.

ICs fits on a desk top. The advent of the microcomputer, a " computer-on-a-chip," further illustrates the amount of miniaturization that has taken place.

A second advantage of ICs lies in their increased reliability, due partly to the fact that all interconnections, along with the transistors and other elements, are made in the initial manufacturing process. The use of highly refined manufacturing and testing techniques which are economically feasible in the mass production of ICs also contributes to increased reliability.

A third advantage of the IC amplifier is its extended frequency response (Chap. 9). Since there are no lengthy wire connections, and since the size of each element is decreased, the frequency response of the integrated circuit is extremely good.

With each new technology, the engineer is faced with new and different problems. The design of transistor circuits differed greatly from that of vacuum-tube circuits. For example, the first transistor radios did not operate properly in very hot or cold temperatures, as on a beach or on ski slopes, because temperature compensation was not well understood. Because of this fact automobile radios until 1960 used vacuum tubes in the output power amplifier. Similar problems are arising now in the use of ICs, which are basically low-power amplifiers. Care must be taken in the design of their biasing networks, and feedback must be used with compensating networks (Chap. 10) in order to avoid oscillation.

The underlying causes of these problems can best be understood from a discussion of the fabrication techniques employed in the manufacture of ICs. This chapter deals primarily with these techniques. The application of ICs, biasing, and related problems have been considered in earlier chapters.

16.1 AN INTRODUCTION TO THE FABRICATION OF AN IC TRANSISTOR

The fabrication process begins with a p-type silicon crystal wafer which is cut and polished until it is 0.005 to 0.007 in thick. This p-type silicon is called the *substrate* material. The silicon crystal wafer is then placed in an oven having an oxygen atmosphere, and the temperature is raised to 1200°C. The result of this process is oxidation of the silicon, which results in a silicon dioxide (SiO_2) layer on top of the wafer, as shown in Fig. 16.1-1. The wafer is then coated with a photosensitive emulsion, as shown in Fig. 16.1-2. A prescribed mask is then placed over the wafer, as illustrated in the figure. The masked wafer is next exposed to light. Those portions of the emulsion not exposed to the light are removed, using a solvent, and the corresponding portions of the silicon dioxide coating are also removed. The result is shown in Fig. 16.1-3.

The wafer is then placed in a high-temperature oven and exposed to an n-type dopant such as arsenic, which diffuses into the substrate. The heavy concentration of the doping material results in a heavily doped n region called an n^+ region. The result of this process is shown in Fig. 16.1-4. An n-type layer is now formed above

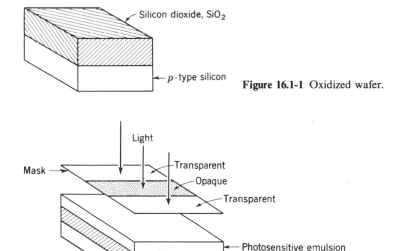

Figure 16.1-1 Oxidized wafer.

Figure 16.1-2 Photolithographic isolation masking.

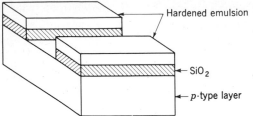

Hardened emulsion

← SiO₂

← *p*-type layer **Figure 16.1-3** The wafer ready for the first diffusion.

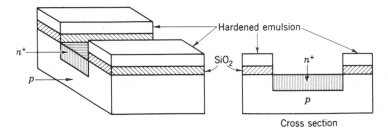

Figure 16.1-4 The first diffusion.

the n^+ layer by passing a gas containing an *n*-type impurity over the n^+ region. The concentration of the *n*-type dopant is less than in the n^+ region. The *n*-type impurity deposits a *layer* on top of the n^+ region. This is called an *epitaxial layer* and will eventually become the collector.

To form the base of the transistor, a *p*-type diffusion is applied as shown in Fig. 16.1-5. An additional *n-type diffusion* (the emitter) ends the diffusing process. A silicon dioxide protective covering is now applied, except at the terminals of the emitter, base, and collector, where aluminum metallic bonds are made. The final result is shown in Fig. 16.1-6.

Each *wafer*, or *slice*, of substrate material is a disk with a diameter of about 1 in. This is subdivided into squares or rectangles, called *dice* or *chips*, of the order of 0.050 to 0.100 in. Each *die*, or *chip*, becomes one complete circuit. Thus each *wafer* may contain as many as 300 or 400 complete circuits. In practice, the masks are cut so that all the elements of the desired circuit are formed at the same time.

The interconnection of the emitter, base, and collector terminals determines whether the element becomes a transistor, diode, resistor, or capacitor. These are

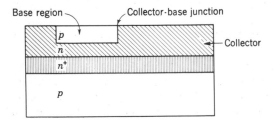

Figure 16.1-5 Base diffusion.

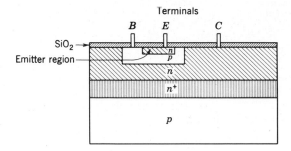

Figure 16.1-6 Emitter diffusion, protective covering, and location of terminals.

described individually in succeeding sections. Since the construction of all transistors of a complete amplifier system occurs simultaneously, the transistors can be made almost identical. Thus, matching integrated transistors on one wafer is accomplished automatically. A problem that does arise, however, is the accuracy of the doping required to achieve a specified h_{fe}. Thus, while all transistors on one wafer will have nearly the same h_{fe}, the value of h_{fe} still may vary by 3 : 1 from wafer to wafer. Another problem is that resistors cannot be accurately specified (10 percent tolerances are typical) although *ratios* between resistors can be kept to within 3 percent.

In the sections that follow, the connection and interconnection of the finished elements (shown in Fig. 16.1-6) are discussed, and some of the resulting problems analyzed.

16.2 THE EQUIVALENT CIRCUIT OF THE INTEGRATED TRANSISTOR

Let us now consider the equivalent circuit of the transistor of Fig. 16.1-6. To do this we represent the circuit as an ideal transistor with the undesirable (parasitic) elements attached externally.

The first "external" element considered is the series collector resistance r_{sc}, shown in Fig. 16.2-1. This resistor can be calculated from the voltage drop v_{ab} and

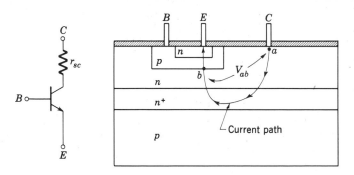

Figure 16.2-1 The integrated transistor.

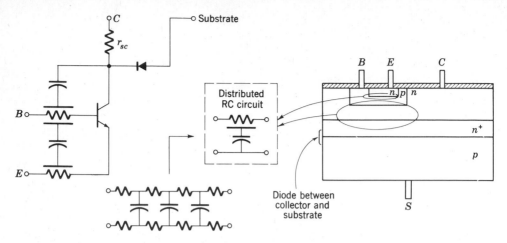

Figure 16.2-2 Approximate equivalent circuit of the integrated transistor.

the collector current. The reason for the insertion of the n^+ material can now be explained. This material shunts the collector n material and has a low resistance. The resulting r_{sc} is then typically less than 1 Ω. The importance of r_{sc} can be seen by considering a transistor in saturation. Let the ideal transistor have a saturation voltage of 0.1 V at a current of 100 mA. Then if r_{sc} is 1 Ω, the apparent saturation voltage is 0.2 V, an increase of 100 percent. Without the n^+ region, r_{sc} is typically 6 to 15 Ω, which would result in an apparent saturation voltage of 0.7 to 1.6 V.

The second and third external elements are *RC transmission lines* formed by the distributed *RC* circuits along the collector-base and base-emitter junctions. This is shown in Fig. 16.2-2.

A most important external element is the *pn* junction diode, shown in Fig. 16.2-2, which is formed by the *p*-type substrate and *n*-type collector regions. If this diode is forward-biased under any operating conditions, the collector will essentially be short-circuited to the substrate. Thus most integrated circuits have

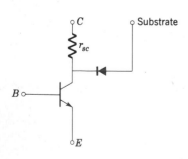

Figure 16.2-3 Integrated transistor showing the collector-subtrate diode.

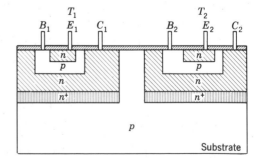

Figure 16.2-4 Two transistors diffused on one chip.

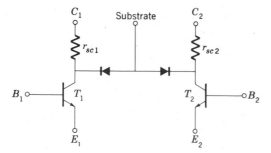

Figure 16.2-5 Equivalent circuit of the two-transistor chip.

the substrate connected to the most negative part of the circuit in order to ensure that the diode is reverse-biased.

Since integrated amplifiers are capable of operating at very high frequencies (above 1 GHz), and since most amplifiers operate at frequencies well below this value, the midfrequency equivalent circuit of the transistor is generally used. A practical equivalent circuit for the transistor of Fig. 16.2-2 is shown in Fig. 16.2-3. The transistor shown has the same characteristics as an ordinary transistor.

Using the equivalent circuit of Fig. 16.2-3, let us now determine the equivalent circuit of two adjacent transistors on the same chip, as shown in Fig. 16.2-4. From this figure it is apparent that the two transistors are isolated from each other by the p-type substrate. However, the collectors of T_1 and T_2 and the substrate form two pn junction diodes. These diodes are placed back to back and therefore present a high impedance. The resulting equivalent circuit is shown in Fig. 16.2-5. As long as the substrate is connected to the most negative point of the circuit, both diodes are reverse-biased and will not affect normal circuit operation.

16.3 THE INTEGRATED DIODE

To form a diode, the terminals of a transistor are connected so as to obtain diode action. There are five such possible combinations, as shown in Fig. 16.3-1. Each of the five configurations has its own advantages and disadvantages. They are not

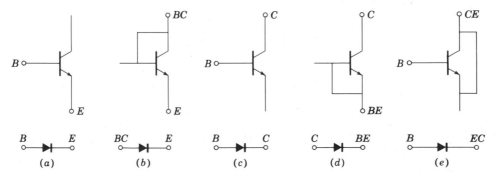

Figure 16.3-1 The five ways in which a transistor can be connected as a diode.

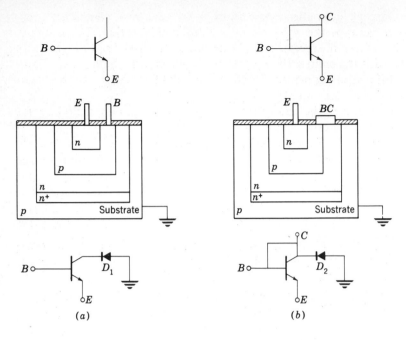

Figure 16.3-2 The integrated diode.

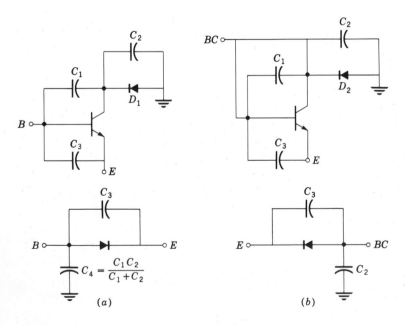

Figure 16.3-3 Equivalent circuits for integrated diode.

discussed here, except for connections *a* and *b*, which are most commonly used. Connection *a* is perhaps the most obvious diode configuration. The problem associated with it is that the reverse breakdown voltage is small (5 to 7 V) because it is associated with the emitter-base junction. Connection *b*, the most common diode configuration, also has a 5- to 7-V reverse breakdown voltage but is able to pass much higher currents.

Let us take a look at the equivalent circuits of these diodes, including the effect of the substrate material, as shown in Fig. 16.3-2.

The diodes D_1 and D_2 formed by the *p* substrate and the collector are reverse-biased. This provides isolation between the various elements (for example, see Fig. 16.2-5 or 16.7-2). Thus, since a reverse-biased diode is effectively a capacitor, D_1 and D_2 behave as capacitors. Since collector-base and base-emitter capacitance are always present, the equivalent circuits of the diodes formed by connections *a* and *b* are as shown in Fig. 16.3-3. Values for these capacitances are usually of the order of 0.5 to 3 pF. Discrete diodes always have a capacitance C_3 present. The integrated diode has an extra capacitance (C_4 or C_2).

16.4 THE INTEGRATED CAPACITOR

Two types of capacitors are used in ICs. The first to be discussed is the junction capacitor, which is formed from an integrated transistor (Fig. 16.1-6) by using the collector and base terminals. Two disadvantages of this type of capacitor are that its capacitance varies with collector-to-base voltage and that the maximum capacitance is small.

The second type to be discussed is the thin-film capacitor. This type is capable of yielding larger capacitance values than the junction capacitor, and its capacitance is not a function of the terminal potential. However, its construction is far more expensive, and it uses a large area on the chip.

16.4-1 The Junction Capacitor

The junction capacitor is formed from the collector-base terminals of an ordinary transistor. Thus the circuit of Fig. 16.2-2, after connecting the substrate to ground, becomes the circuit of Fig. 16.4-1. At low frequencies the distributed *RC* circuit

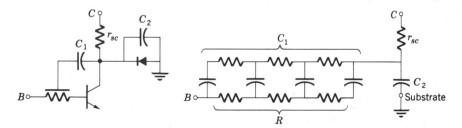

Figure 16.4-1 The junction capacitor and its equivalent circuit.

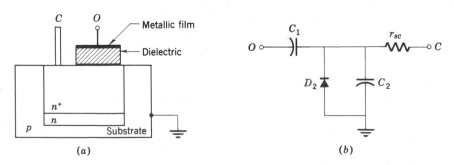

Figure 16.4-2 Simplified lumped equivalent circuit of the junction capacitor.

appears capacitive since the impedance C_1 of the distributed capacitor is considerably higher than that of the distributed resistance R. The resulting equivalent circuit is shown in Fig. 16.4-2.

Consideration of the physics of the junction yields the result that the effective capacitance of a reverse-biased diode is inversely proportional either to the square root or to the $\frac{1}{3}$ power of the diode voltage. Since C_1 and C_2 represent capacitors formed by reverse-biased diodes, C_1 is inversely proportional to the voltage V_{CB} and C_2 is inversely proportional to V_{CS}. To make the "capacitor" of Fig. 16.4-2 approach the ideal $C_1 \gg C_2$, the voltage V_{CS} is made as large as possible and the voltage V_{CB} is made as small as possible. Typical ratios of C_1 to C_2 vary between $1:1$ and $7:1$.

The junction capacitor of Fig. 16.4-2 is further degraded by the series collector resistor r_{sc}. This resistance is minimized by the use of the n^+ region shown in Fig. 16.2-2.

16.4-2 The Thin-Film Capacitor

The thin-film capacitor is constructed by first diffusing an n region and then an n^+ region. A thin layer of dielectric material (usually silicon dioxide) is applied to cover the n^+ material. A metallic film (aluminum) which acts as the second terminal of the capacitor is then applied to the dielectric, as shown in Fig. 16.4-3a. The equivalent circuit of this capacitor is shown in Fig. 16.4-3b.

There are two major advantages to using the thin-film capacitor. First, C_1 is fixed and is not a function of collector voltage. This is extremely important, since variations in voltage cause variations in the *junction* capacitance which may result

Figure 16.4-3 Thin-film capacitor and equivalent circuit.

in possible frequency modulation of the signal. Second, the thin-film capacitance is nonpolar, that is, V_{CO} can be positive or negative. Capacitance C_2, due to the substrate material, still varies with variations of the voltage across it.

These capacitors have a capacitance per unit area which varies between 0.2 and 0.5 pF/mil^2, depending on the process (and the dielectric used, in the case of thin-film capacitors). Thus, to make a large capacitor requires a significant amount of area, and hence is costly. At present maximum capacitance values range from 500 to 5000 pF, again depending on the process employed. This is considerably less than the 20- or 100-μF capacitors used as bypass and coupling capacitors in ordinary transistor circuits. The problems associated with small bypass and coupling capacitors when considering low- and high-frequency circuits are discussed in Chap. 9. It is also of interest to note that although adjacent capacitors can be made almost identical on one wafer, their capacitance may vary by ± 20 percent from wafer to wafer.

16.5 THE INTEGRATED RESISTOR

Resistors, as well as capacitors, can be constructed either directly from the basic transistor element or using thin-film techniques. Let us first consider the junction resistor formed from the transistor.

16.5-1 The Junction Resistor

The junction resistor is formed by stopping the diffusion process after depositing the p-type material which forms the base of the transistor, as shown in Fig. 16.1-5. The silicon dioxide and aluminum contacts are then applied as shown in Fig. 16.5-1a. The resistance of the junction resistor is determined by the resistivity ρ, length L, and area A of the p-type base material, as shown in Fig. 16.5-1b. An approximate expression for the resistance, assuming uniform fields, is

$$R_{AB} = \rho \frac{L}{A} \tag{16.5-1}$$

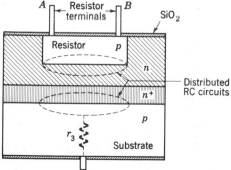
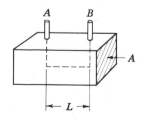

Figure 16.5-1 The junction resistor.

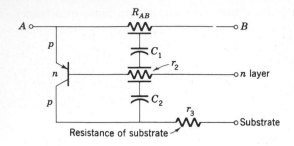

Figure 16.5-2 Equivalent circuit of junction resistor.

Figure 16.5-2 shows the complete equivalent circuit of the resistance element. We see from Fig. 16.5-1a that the *pnp* materials form a transistor. The collector of this transistor is the *p*-type substrate, and the emitter is the *p*-type material to be used as the resistor. (The substrate is always connected to the lowest potential point in the circuit; thus it becomes the collector in the *pnp* configuration.)

Figure 16.5-2 shows that a distributed *RC* circuit exists between the emitter and the base and between the base and the collector. This can be seen from Fig. 16.5-1a. Resistor r_3 is the series collector resistance shown diagrammatically in Fig. 16.5-1a.

The h_{fe} of the transistor formed is small, since the base *n* region was intended to be a collector and therefore is relatively large. In Sec. 2.1, it was pointed out that the base current of a well-designed transistor is small compared with the collector current because the base region is narrow. In this case the transistor formed in Fig. 16.5-2 has a wide base region, and hence the base current is large, thus resulting in a low h_{fe}.

Typical values for a 4-kΩ resistor are

$$h_{fe} \approx 0.5 \text{ to } 5 \qquad r_3 \approx 1 \text{ Ω} \qquad R_{AB} = 4 \text{ kΩ}$$
$$C_1 \approx 5 \text{ pF} \qquad r_2 \approx 50 \text{ Ω} \qquad C_2 \approx 15 \text{ pF}$$

At low frequencies ($f \ll 1/2\pi R_{AB}C_1 = 8$ MHz for this example) capacitors C_1 and C_2 can be neglected. The result is an effective resistance R_{AB}. It is to be noted that the *n* region must be connected to the most positive potential in the circuit to reverse-bias the *pn* emitter-base diode. If this were not done, current would flow in this diode and the resistance R_{AB} would be effectively short-circuited.

Values of resistance typically range from 100 Ω to 30 kΩ. Absolute variations are of the order of ±10 percent, while variations between resistances on the same chip can be less than ±3 percent. Thus resistance *ratios* can be maintained more accurately in production than resistance *values*. For this reason, it is good practice to design ICs so that performance depends on resistance ratios rather than absolute values. Resistance is also a function of temperature because of the properties of the semiconductor material.

16.5-2 The Thin-Film Resistor

The thin-film resistor is constructed by depositing a resistive material such as nitrided tantalum, nichrome, or tin oxide over the silicon dioxide covering the p-type substrate, as shown in Fig. 16.5-3a. The resistance obtained using this process is given by the standard resistance equation (16.5-1). When this relation is applied to the configuration of Fig. 16.5-3b, the resistance between terminals A and B is

$$R_{AB} = \frac{\rho L}{dW} \tag{16.5-2}$$

If $L = W$, we have a *unit square* of the material and the resistance per unit square is

$$R = \frac{\rho}{d} \; \Omega/\text{sq} \tag{16.5-3}$$

Note that the dimension of ρ is in ohm-centimeters and the dimension of d is in centimeters; thus the square is *dimensionless*. R is called the *sheet resistance* of the material, and the overall resistance is

$$R_{AB} = (\Omega/\text{sq})(\text{number of squares}) = R\frac{L}{W} \tag{16.5-4}$$

Some typical values are

$$R - \begin{cases} 50 \; \Omega/\text{sq} & \text{nitrided tantalum} \\ 400 \; \Omega/\text{sq} & \text{nichrome} \\ 1000 \; \Omega/\text{sq} & \text{tin oxide} \end{cases}$$

If a 5-kΩ resistor is to be designed using nitrided tantalum, the length-to-width ratio L/W is

$$\frac{R_{AB}}{R} = \frac{L}{W} = \frac{5000}{50} = 100 \; \text{sq} \tag{16.5-5}$$

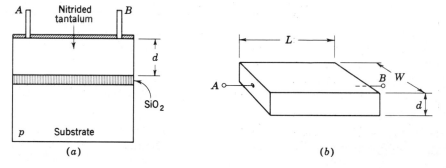

Figure 16.5-3 Thin-film resistor.

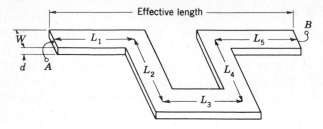

Figure 16.5-4 Zigzagging a thin-film resistor to reduce the overall length.

If a width of 5 mils is selected, the length will be 500 mils, or $\frac{1}{2}$ in, a very long distance on the IC scale. In order to shorten the overall length, the resistor is often constructed in a zigzag pattern, as shown in Fig. 16.5-4.

If

$$L_1 = L_2 = L_3 = L_4 = L_5 \qquad \text{and} \qquad W = 5 \text{ mils}$$

then the effective length required is only about 300 mils. Additional zigzagging will reduce the length further.

Another technique for reducing the length is to reduce the width W. However, power-dissipation limitations preclude decreasing W indefinitely. For example, nitrided tantalum on silicon dioxide has a maximum power rating of 10 W/in² of film. Therefore, assuming a dc voltage V impressed across the resistor, we have

$$\frac{P}{A} = 10 = \frac{V^2/R_{AB}}{LW} \tag{16.5-6}$$

If V is 10 V and $R_{AB} = 5$ kΩ,

$$LW = \tfrac{1}{500} \text{ in}^2 \tag{16.5-7}$$

Combining (16.5-5) and (16.5-7) yields

$$W \approx 4.5 \text{ mils} \tag{16.5-8a}$$

and

$$L \approx 450 \text{ mils} \tag{16.5-8b}$$

in order not to exceed the power rating of the material.

The foregoing analysis assumed that the thin-film resistor was ideal. However, from Fig. 16.5-3a we see that a capacitor is formed between the resistive and substrate materials, the dielectric being the silicon dioxide. Thus the thin-film resistor is actually a distributed RC circuit, as shown in Fig. 16.5-5. The equivalent circuit reduces to a simple resistance for frequencies less than about 1 GHz.

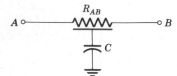

Figure 16.5-5 Equivalent circuit of the thin-film resistor.

16.6 THE INTEGRATED INDUCTOR

The integrated inductor has not been perfected at the time of this writing. Very small inductors, not much larger than 5 μH, can be built with a Q between 30 and 50. Practical amplifier systems still use external inductors attached to the IC. These inductors require a great deal of space, so that RC or direct-coupled amplifiers are preferred.

16.7 DESIGN OF A SIMPLE IC

Consider designing a simple monolithic IC amplifier as shown in Fig. 16.7-1, i.e., an IC not using thin-film techniques. The first step is to lay out the circuit so that no wires cross, as shown in Fig. 16.7-2a. The cross-sectional view is shown in Fig. 16.7-2b. After the required masks are designed and constructed, the manufacturing process outlined in Sec. 16.1 follows.

Figure 16.7-3a shows a photomicrograph of an integrated differential-amplifier circuit. It is interesting to note that the required diffusion masks can be generated directly by a digital computer. The reader should try to identify the individual transistors, diode, resistors, and capacitors by comparing the photomicrograph with the circuit diagram.

16.8 LARGE-SCALE INTEGRATION

An IC chip containing an op-amp, gate, flip-flop, or other circuits with fewer than 20 transistors is called a *small-scale* integrated circuit (SSI). In contrast to SSI, IC chips are available that contain tens of thousands of transistors. A typical example is an NMOS memory chip that has 16,000 bits of memory in which each memory cell consists of more than four MOSFETS. Such ICs are called *large-scale integrated* circuits (LSI), and many different types are available today off the shelf.

In order to accommodate customers whose requirements cannot be met by stock ICs, several companies have begun to produce LSI circuits to order. The customer presents a circuit diagram of his system to the manufacturer, who then designs the masks from which the IC chip containing the complete system is manufactured. This type of IC is called *custom LSI* and is very expensive, approximately $100,000 for the design and tooling at this writing. However, after the initial outlay each chip costs approximately $20. Hence, custom LSI is only used for large production runs.

An alternative to custom LSI is semicustom LSI. Here, the manufacturer has available several different LSI chips. The layout of one such chip for linear circuit applications is shown in Fig. 16.8-1. It measures 70 by 70 mils and contains 200 components, including 68 transistors. Chips are also available for digital applications. The manufacturer furnishes detailed instructions from which the engineer can design his system. The design requires the interconnection of the transistors

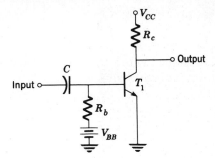

Figure 16.7-1 An IC amplifier.

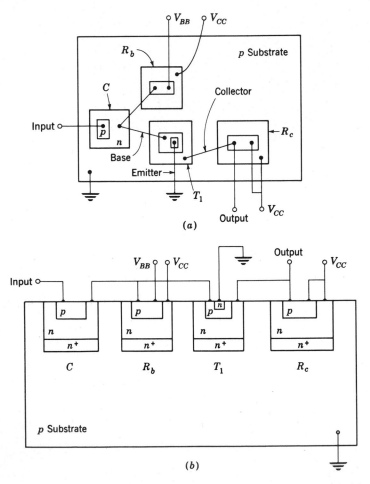

(a)

(b)

Figure 16.7-2 Realization of IC amplifier of Fig. 16.7-1 (not to scale): (a) top view; (b) distorted cross-sectional view drawn to avoid overlap of R_b and T_1.

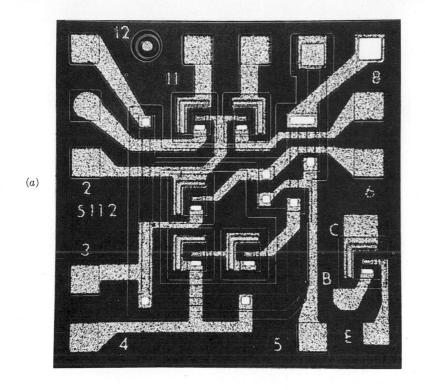

(a)

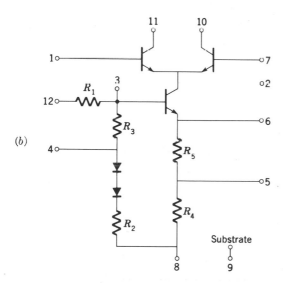

(b)

Figure 16.7-3 IC differential amplifier: (a) photomicrograph of the RCA 3005 differential amplifier (RCA); (b) circuit diagram.

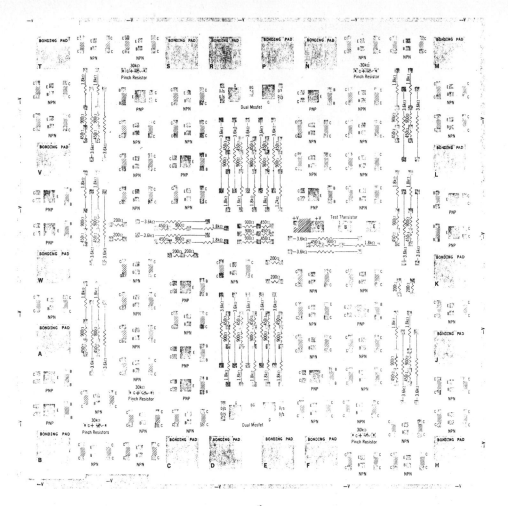

Figure 16.8-1 A semicustom LSI circuit for linear designs. Monochip MO-E. Linear, bipolar; 200 components; 70 × 70 mils. 18 pins max; 20 volts max. 46 small NPN transistors; 18 dual PNP transistors; 4 MOSFETs; 8 resistors 200 Ω; 28 resistors 450 Ω; 28 resistors 900 Ω; 25 resistors 1.8 kΩ; 26 resistors 3.6 kΩ; 5 resistors 30 kΩ (pinch).

and resistors which are available on the chip without having any wires cross. The number of components is not extraordinarily high even though the chip is equivalent to 10 SSI circuits. At this time semicustom LSI requires an initial investment of only $3000 per design. If the complexity of the design is such that one of the available chips can be used and quantities of about 5000 chips per year are required, then semicustom LSI is the best approach available today.

When designing with LSI care must be taken not to dissipate an excessive amount of power. As the number of transistors on a given size chip is increased, the power dissipation of each transistor must be decreased in order to avoid overheating. For this reason most LSI circuits today employ MOS devices. In the next section we discuss an alternate technology called integrated injection logic (I^2L), in which BJTs are used.

16.9 INTEGRATED INJECTION LOGIC (I^2L)

When the speeds attainable with MOS are not sufficient, BJTs can be used in a new LSI techniques called *integrated injection logic* (I^2L) or *merged transistor logic* (MTL). In this method regions from several transistor are merged into one region, and as a result it is possible to construct on a given chip area a larger number of gates than is possible with TTL, for example.

To illustrate how I^2L serves to reduce the real estate used in a system design we consider as an example the NOR gate shown in Fig. 16.9-1*a*. In this circuit transistors T_1 and T_2 constitute the NOR gate and T_3 and T_4 are the driving transistors. To show that T_1 and T_2 indeed form a NOR gate we note that if either V_1 OR V_2 is in the **1** state, the corresponding transistor (T_1 or T_2) will saturate and the output V_o will be in the **0** state; otherwise V_o is in the **1** state. The physical layout of this NOR gate without the driving transistors is shown in Fig. 16.9-1*b*.

Figure 16.9-2*a* shows the schematic of a NOR gate which behaves like the circuit of Fig. 16.9-1*a*. However, in Fig. 16.9-2*a* resistors R_1 and R_2 have been replaced by the *pnp* transistor T_I. This transistor is called the *injector* because it injects currents into the bases of T_1 and T_2. For example, if T_3 is OFF, V_1 is high and current flows from collector C_1 of T_I into the base of T_1, causing T_1 to saturate so that V_o is in the **0** state. On the other hand if T_3 is ON, V_1 is low and current flows out of collector C_1 of T_I into the collector of T_3. If T_3 and T_4 are both ON (saturated), no current flows into the bases of T_1 and T_2; hence T_1 and T_2 are cut off, and V_o is therefore in the **1** state. Thus the circuit of Fig. 16.9-2*a* is indeed a NOR gate since V_o is in the **0** state if V_1 OR V_2 is in the **1** state.

The physical layout of the I^2L circuit is shown in Fig. 16.9-2*b*. Observe that no *p* substrate is used in I^2L. The *n* material serves as the emitters of T_1 and T_2 as well as the base of T_I. The two *lateral pnp* transistors share a common emitter and a common base; hence they form a single two-collector *pnp* transistor, which we have called T_I. Notice the difference in geometry between the *lateral* transistor T_I and the *vertical npn* transistors T_1 and T_2. Typically, vertical transistors have much higher current gains than lateral transistors. However, the *npn* transistor in I^2L has a small current gain since the emitter is large and the collector small (we can think of the *npn* transistor as being operated in the inverse mode in this application).

Comparing Fig. 16.9-1*b* and 16.9-2*b*, we see that the I^2L circuit (as a result of merging the emitters of T_1 and T_2 and the base of T_I, the base of T_1 with one collector of T_I, and the base of T_2 with the other collector of T_I) results in a much less complex structure. This is, of course, also due to the fact that no resistors are required in I^2L, the injector transistor being used to supply the required current to

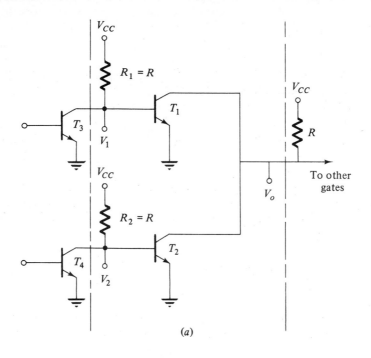

(a)

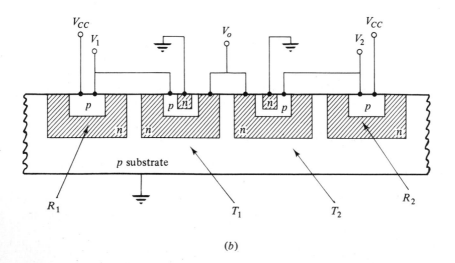

(b)

Figure 16.9-1 Transistor NOR gate: (a) schematic; (b) physical structure of the NOR gate.

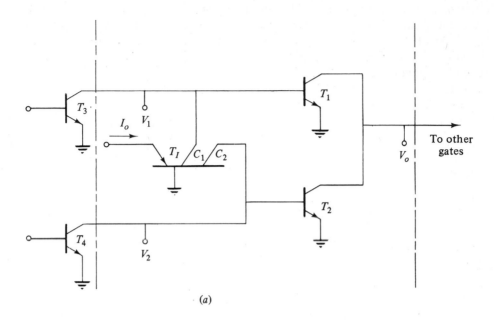

(a)

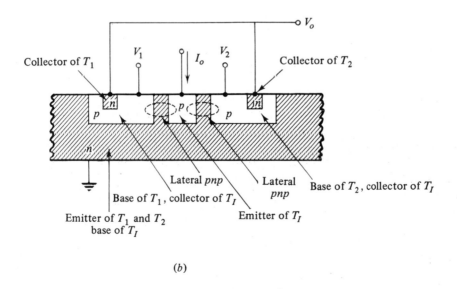

(b)

Figure 16.9-2 I^2L: (a) NOR-gate schematic; (b) physical structure of NOR gate.

the bases of the driven transistors. Although we have shown T_I to be a two-collector transistor injecting current into two bases, T_I can actually be designed to have many collectors, thereby driving many transistors. The current I_o is obtained by externally connecting the emitter of T_I to a resistor which is connected to the V_{CC} supply. The manufacturer specifies the amount of current required for a given speed of operation.

Transistors T_1 and T_2 are shown as single-collector transistors in Fig. 16.9-2. In most practical I²L circuits the driven transistors have multiple collectors.

As a result of the increased efficiency obtained with I²L, these circuits are today finding wide application in calculators and microprocessors.

REFERENCE

1. J. Glaser and J. Subak-Sharpe, "Integrated Circuit Engineering," Addison-Wesley, Reading, Mass., 1978.

PROBLEMS

16.2-1 Fig. P16.2-1 is a model of an IC transistor at low frequencies.
  (a) Explain how r_b, r_e, and r_{sc} arise.
  (b) Should S be connected to the most positive or most negative voltage?

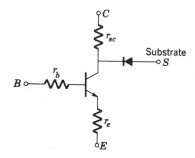

Figure P16.2-1

16.2-2 The IC transistor shown in Fig. P16.2-1 is connected to the circuit shown in Fig. P16.2-2.
  (a) If $r_b = r_{sc} = r_e = 10\ \Omega$, draw an approximate small-signal equivalent circuit for the amplifier.
  (b) Calculate the current gain A_i.

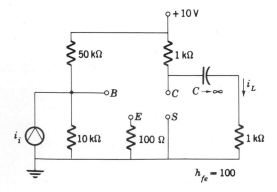

Figure P16.2-2

16.3-1 A 10-V step is applied to the diode circuit of Fig. P16.3-1. The IC diode models shown in Fig. 16.3-3 are used. Sketch the response. Compare the results obtained using each model.

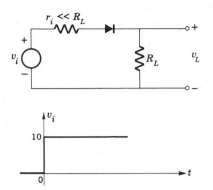

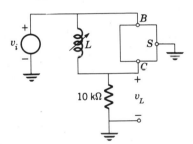

Figure P16.3-1

16.4-1 A junction capacitor having the model shown in Fig. 16.4-2 is used to tune an inductance to 100 Mrad/s in Fig. P16.4-1. $C_1 = 500$ pF, $C_2 = 70$ pF, and $r_{sc} = 2$ Ω.
   (a) Determine the voltage v_L when $v_i = \cos \omega t$, where $\omega = \omega_0 = 100$ Mrad/s.
   (b) Calculate L.

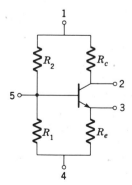

Figure P16.4-1

16.4-2 Plot v_L in Fig. P16.4-1 as ω varies from zero to infinity.

16.5-1 Show that at low frequencies the circuit of Fig. 16.5-2 is a resistor. *Hint:* Neglect capacitances, and draw the model for the transistor. Is S connected to a positive or negative voltage? Why?

16.5-2 Design a 4-kΩ thin-film resistor using nitrided tantalum. Assume a maximum power rating of 10 W/in^2 of film. The maximum applied voltage is 10 V. Design for *minimum area*.

16.7-1 Design a layout for the IC amplifier shown in Fig. P16.7-1. Show a top view. Your layout should avoid overlapping connections.

Figure P16.7-1

16.7-2 Design a layout for the IC Darlington amplifier of Fig. P16.7-2.

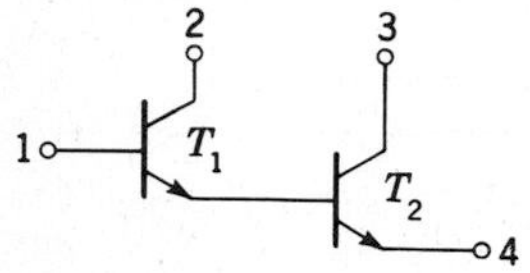

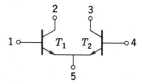

Figure P16.7-2

16.7-3 Design a layout for the IC difference amplifier of Fig. P16.7-3.

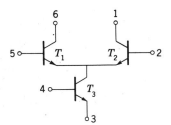

Figure P16.7-3

16.7-4 Design a layout for the IC difference amplifier with a constant-current source shown in Fig. P16.7-4.

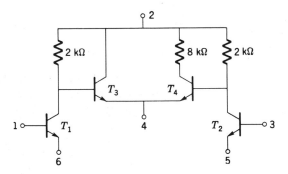

Figure P16.7-4

16.7-5 Design a layout for the amplifier shown in Fig. P16.7-5.

Figure P16.7-5

16.7-6 In the photomicrograph of the IC differential amplifier shown in Fig. 16.7-3a, identify the various resistors and transistors in the circuit of Fig. 16.7-3b.

16.9-1 Draw the schematic for a three-input I^2L NOR gate using a multicollector injector transistor. Design an approximate layout showing the merging of terminals where appropriate.

GAIN EXPRESSED IN LOGARITHMIC UNITS; THE DECIBEL

It is often convenient to express the magnitude of the gains of individual stages in logarithmic units. When this is done, the overall gain is found by the simple addition of the individual stage gains. This logarithmic unit also simplifies plotting of frequency response (Chap. 9). For most purposes, the most convenient unit is the decibel (dB), which, when used to express the power gain of an amplifier, is defined as

$$A_p = 10 \left(\log \frac{P_2}{P_1} \right) \qquad \text{dB} \tag{A.1}$$

where log means $\log_{10}$.

This definition is used in two different ways: (1) to express a power ratio in logarithmic units and (2) to express power level with respect to a fixed reference power. A common reference is 1 mW, for which the units are abbreviated dBm. Thus, for a power level in decibels with respect to 1 mW, we have

$$A_p = 10 \left(\log \frac{P_2}{10^{-3}} \right) = 10 \log (P_2 \times 10^3) \qquad \text{dBm} \qquad P_2 \text{ in W} \tag{A.2}$$

The decibel was originally defined in terms of power gain as in (A.1). Because of its usefulness, it is generally applied directly to voltage and current gain in the following way: Referring to (A.1), if P_2 and P_1 are dissipated in identical resistances R, then

$$P_2 = \frac{V_2^2}{R} = I_2^2 R \qquad \text{and} \qquad P_1 = \frac{V_1^2}{R} = I_1^2 R$$

Substituting in (A.1) gives

$$A_v = 10\left(\log \frac{V_2^2}{V_1^2}\right) = 20\left(\log \frac{V_2}{V_1}\right) \text{ dB} \tag{A.3}$$

or

$$A_i = 10\left(\log \frac{I_2^2}{I_1^2}\right) = 20\left(\log \frac{I_2}{I_1}\right) \text{ dB} \tag{A.4}$$

Thus voltage and current gains in decibels are equal to power gain only if the resistances in which the powers are dissipated are equal. However, by convention, (A.3) and (A.4) are used to express the voltage and current gains in decibels independent of the resistance level.

Example A.1 Measurements on a certain amplifier yield $R_{in} = 1\ k\Omega$; $R_L = 100\ \Omega$. When the input voltage is 1 mV peak, the load voltage is 10 V peak. Find, in decibels, the voltage, current, and power gains and the output power level in decibels above 1 mW.

SOLUTION From (A.3)

$$A_v = 20\left(\log \frac{V_2}{V_1}\right) = 20\left(\log \frac{10}{10^{-3}}\right) = 80 \text{ dB}$$

To find A_i, note that $I_{in} = 1\ \mu A$ and $I_2 = 0.1$ A so that

$$A_i = 20\left(\log \frac{I_2}{I_1}\right) = 20\left(\log \frac{10^{-1}}{10^{-6}}\right) = 100 \text{ dB}$$

The powers are

$$P_1 = \frac{10^{-6}}{(2)(10^3)} = \frac{1}{2} \times 10^{-9} \text{ W} \qquad P_2 = \frac{10^2}{(2)(100)} = \frac{1}{2} \text{ W}$$

and so

$$A_p = 10\left(\log \frac{\frac{1}{2}}{\frac{1}{2} \times 10^{-9}}\right) = 90 \text{ dB}$$

Relative to 1 mW, the output power level is

$$P_2 = 10\left(\log \frac{\frac{1}{2}}{10^{-3}}\right) = 10(3 \log 10 - \log 2) = 27 \text{ dBm} \qquad ///$$

STANDARD VALUES OF RESISTANCE
AND CAPACITANCE

The following lists of standard components are included for use in conjunction with the design problems. These lists are typical and, especially in the case of capacitors, are subject to some variation from one manufacturer to another.

B.1 RESISTORS

Carbon resistors of 10 percent tolerance are available in power ratings of $\frac{1}{4}, \frac{1}{2}, 1$, and 2 W in the following range:

$$
\left.
\begin{array}{lll}
2.7 & 5.6 & 12 \\
3.3 & 6.8 & 15 \\
3.9 & 8.2 & 18 \\
4.7 & 10 & 22
\end{array}
\right\} \quad \text{all} \times 10^n, \text{ where } n = 0, 1, 2, 3, 4, 5, 6
$$

B.2 CAPACITORS

Typical ranges of capacitor values available from one manufacturer are shown in Table B.2-1.

Table B.2-1

Ceramic-disk capacitors, 10% tolerance, pF

3.3	30	200	560	2200
5	39	220	600	2500
6	47	240	680	2700
6.8	50	250	750	3000
7.5	51	270	800	3300
8	56	300	820	3900
10	68	330	910	4000
12	75	350	1000	4300
15	82	360	1200	4700
18	91	390	1300	5000
20	100	400	1500	5600
22	120	470	1600	6800
24	130	500	1800	7500
25	150	510	2000	8200
27	180			

Tantalum capacitors, 10% tolerance, μF

0.0047	0.010	0.022	
0.0056	0.012	0.027	all $\times 10^n$, where $n = 0, 1, 2, 3, 4, 5$ (to 330 μF)
0.0068	0.015	0.033	
0.0082	0.018	0.039	

Electrolytic capacitors for bypass application, μF

250	2000
500	3000
1000	4000
1500	5000

DEVICE CHARACTERISTICS

C.1 ZENER-DIODE SPECIFICATIONS

Figure C.1-1 shows the variation of Zener test voltage V_{ZT} with Zener resistance r_{ZT} for several typical power levels P_Z (see Sec. 1.10). Note that the test current I_{ZT} is measured at 25 percent of the maximum power rating,

$$I_{ZT} = \frac{1}{4} \frac{P_Z}{V_{ZT}}$$

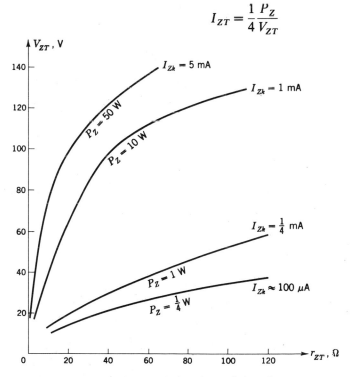

Figure C.1-1 Zener-diode specifications ($I_{ZT} = \frac{1}{4}P_Z/V_{ZT}$).

In addition, it should be noted that the current at the knee of the characteristic I_{Zk} is approximately constant for a specified P_Z and is independent of V_{ZT}.

C.2 TRANSISTOR CHARACTERISTICS

MAXIMUM RATINGS

Characteristic	Symbol	Rating	Unit
Collector-Base Voltage	V_{CB}	60	Vdc
Collector-Emitter Voltage	V_{CEO}	40	Vdc
Emitter-Base Voltage	V_{EB}	6	Vdc
Collector Current	I_C	200	mAdc
Total Device Dissipation @ $T_A = 60^\circ C$	P_D	210	mW
Total Device Dissipation @ $T_A = 25^\circ C$ Derate above $25^\circ C$	P_D	310 2.81	mW mW/$^\circ$C
Thermal Resistance, Junction to Ambient	θ_{JA}	0.357	$^\circ$C/mW
Junction Operating Temperature	T_J	135	$^\circ$C
Storage Temperature Range	T_{stg}	-55 to +135	$^\circ$C

Figure C.2-1 Characteristics of the 2N3903 and 2N3904 *npn* silicon transistors, designed for general-purpose switching and amplifier applications and for complementary circuitry with types 2N3905 and 2N3906. (*Motorola, Inc.*)

ELECTRICAL CHARACTERISTICS ($T_A = 25°C$ unless otherwise noted)

Characteristic	Fig. No.	Symbol	Min	Max	Unit
OFF CHARACTERISTICS					
Collector-Base Breakdown Voltage ($I_C = 10$ μAdc, $I_E = 0$)		BV_{CBO}	60	-	Vdc
Collector-Emitter Breakdown Voltage* ($I_C = 1.0$ mAdc, $I_B = 0$)		BV_{CEO}^*	40	-	Vdc
Emitter-Base Breakdown Voltage ($I_E = 10$ μAdc, $I_C = 0$)		BV_{EBO}	6.0	-	Vdc
Collector Cutoff Current ($V_{CE} = 30$ Vdc, $V_{EB(off)} = 3.0$ Vdc)		I_{CEX}	-	50	nAdc
Base Cutoff Current ($V_{CE} = 30$ Vdc, $V_{EB(off)} = 3.0$ Vdc)		I_{BL}	-	50	nAdc
ON CHARACTERISTICS					
DC Current Gain* ($I_C = 0.1$ mAdc, $V_{CE} = 1.0$ Vdc) 2N3903 / 2N3904	15	h_{FE}^*	20 / 40	- / -	–
($I_C = 1.0$ mAdc, $V_{CE} = 1.0$ Vdc) 2N3903 / 2N3904			35 / 70	- / -	
($I_C = 10$ mAdc, $V_{CE} = 1.0$ Vdc) 2N3903 / 2N3904			50 / 100	150 / 300	
($I_C = 50$ mAdc, $V_{CE} = 1.0$ Vdc) 2N3903 / 2N3904			30 / 60	- / -	
($I_C = 100$ mAdc, $V_{CE} = 1.0$ Vdc) 2N3903 / 2N3904			15 / 30	- / -	
Collector-Emitter Saturation Voltage* ($I_C = 10$ mAdc, $I_B = 1.0$ mAdc) ($I_C = 50$ mAdc, $I_B = 5.0$ mAdc)	16, 17	$V_{CE(sat)}^*$	- / -	0.2 / 0.3	Vdc
Base-Emitter Saturation Voltage* ($I_C = 10$ mAdc, $I_B = 1.0$ mAdc) ($I_C = 50$ mAdc, $I_B = 5.0$ mAdc)	17	$V_{BE(sat)}^*$	0.65 / -	0.85 / 0.95	Vdc

Figure C.2-1 (*Continued*)

SMALL-SIGNAL CHARACTERISTICS

Characteristic	Type	Fig	Symbol	Min	Max	Unit
Current-Gain–Bandwidth Product (I_C = 10 mAdc, V_{CE} = 20 Vdc, f = 100 MHz)	2N3903 2N3904		f_T	250 300	– –	MHz
Output Capacitance (V_{CB} = 5.0 Vdc, I_E = 0, f = 100 kHz)		3	C_{ob}	–	4.0	pF
Input Capacitance (V_{BE} = 0.5 Vdc, I_C = 0, f = 100 kHz)		3	C_{ib}	–	8.0	pF
Input Impedance (I_C = 1.0 mAdc, V_{CE} = 10 Vdc, f = 1.0 kHz)	2N3903 2N3904	13	h_{ie}	0.5 1.0	8.0 10	k ohms
Voltage Feedback Ratio (I_C = 1.0 mAdc, V_{CE} = 10 Vdc, f = 1.0 kHz)	2N3903 2N3904	14	h_{re}	0.1 0.5	5.0 8.0	X 10-4
Small-Signal Current Gain (I_C = 1.0 mAdc, V_{CE} = 10 Vdc, f = 1.0 kHz)	2N3903 2N3904	11	h_{fe}	50 100	200 400	–
Output Admittance (I_C = 1.0 mAdc, V_{CE} = 10 Vdc, f = 1.0 kHz)	2N3903 2N3904	12	h_{oe}	1.0	40	μmhos
Noise Figure (I_C = 100 μAdc, V_{CE} = 5.0 Vdc, R_S = 1.0 k ohms, f = 10 Hz to 15.7 kHz)	2N3903 2N3904	9, 10	NF	– –	6.0 5.0	dB

SWITCHING CHARACTERISTICS

Characteristic			Symbol	Min	Max	Unit
Delay Time	(V_{CC} = 3.0 Vdc, $V_{BE(off)}$ = 0.5 Vdc,	1, 5	t_d	–	35	ns
Rise Time	I_C = 10 mAdc, I_{B1} = 1.0 mAdc)	1, 5, 6	t_r	–	35	ns
Storage Time	(V_{CC} = 3.0 Vdc, I_C = 10 mAdc,	2, 7	t_s	2N3903 2N3904	175 200	ns
Fall Time	I_{B1} = I_{B2} = 1.0 mAdc)	2, 8	t_f	–	50	ns

* Pulse Test: Pulse Width = 300 μs, Duty Cycle = 2.0%.

Figure C.2-1 (*Continued*)

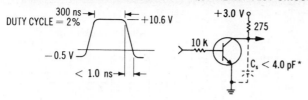

FIGURE 1 — DELAY AND RISE TIME EQUIVALENT TEST CIRCUIT

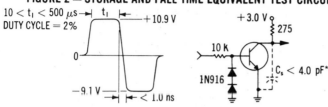

FIGURE 2 — STORAGE AND FALL TIME EQUIVALENT TEST CIRCUIT

*Total shunt capacitance of test jig and connectors

Figure C.2-1 *(Continued)*

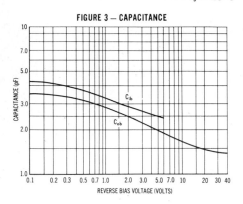

FIGURE 3 — CAPACITANCE

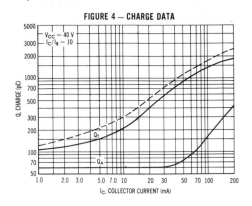

FIGURE 4 — CHARGE DATA

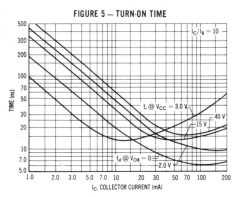

FIGURE 5 — TURN-ON TIME

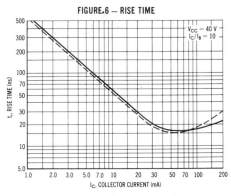

FIGURE.6 — RISE TIME

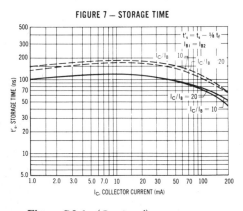

FIGURE 7 — STORAGE TIME

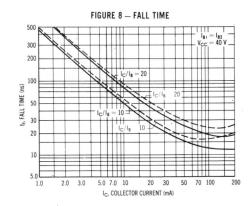

FIGURE 8 — FALL TIME

Figure C.2-1 *(Continued)*

AUDIO SMALL SIGNAL CHARACTERISTICS

NOISE FIGURE VARIATIONS
$V_{CE} = 5.0$ Vdc, $T_A = 25°C$

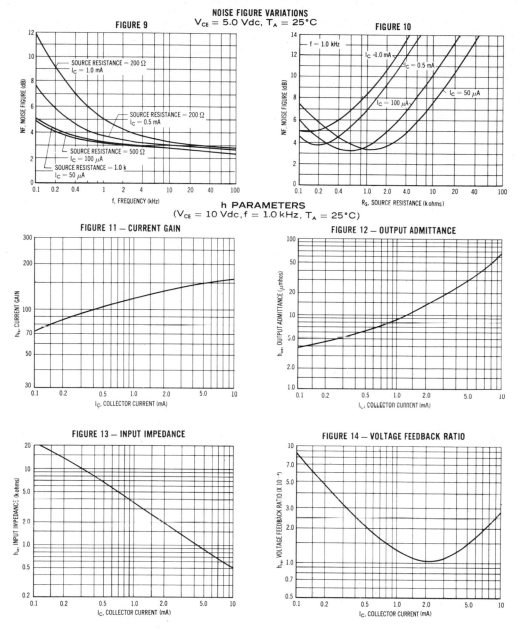

FIGURE 9

FIGURE 10

h PARAMETERS
$(V_{CE} = 10$ Vdc, $f = 1.0$ kHz, $T_A = 25°C)$

FIGURE 11 — CURRENT GAIN

FIGURE 12 — OUTPUT ADMITTANCE

FIGURE 13 — INPUT IMPEDANCE

FIGURE 14 — VOLTAGE FEEDBACK RATIO

Figure C.2-1 *(Continued)*

FIGURE 15 — NORMALIZED CURRENT GAIN

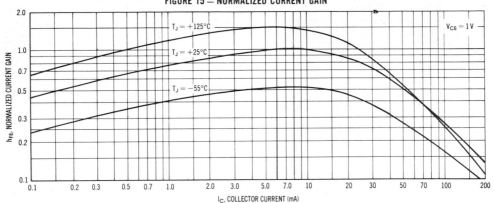

FIGURE 16 — COLLECTOR SATURATION REGION

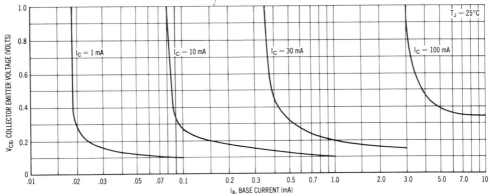

FIGURE 17 — "ON" VOLTAGES

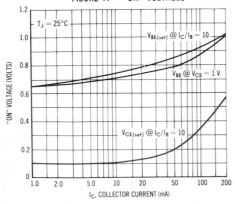

FIGURE 18 — TEMPERATURE COEFFICIENTS

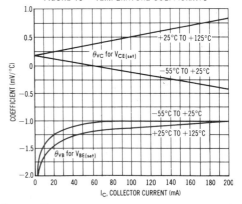

Figure C.2-1 (*Continued*)

MAXIMUM RATINGS

Characteristic	Symbol	Rating	Unit
Collector-Base Voltage	V_{CB}	30	Volts
Collector-Emitter Voltage	V_{CEO}	20	Volts
Emitter-Base Voltage	V_{EB}	3	Volts
Collector Current	I_C	100	mA
Total Device Dissipation @ T_A = 60°C @ T_A = 25°C	P_D	210 310	mW
Thermal Resistance, Junction to Ambient	θ_{JA}	0.357	°C/mW
Junction Temperature	T_J	135	°C

Figure C.2-2 Characteristics of the MPS6507 *npn* silicon transistor. Designed as a VHF mixer in TV applications. (*Motorola, Inc.*)

ELECTRICAL CHARACTERISTICS ($T_A = 25°C$ unless otherwise noted)

Characteristic	Symbol	Min	Typ	Max	Unit
Collector-Emitter Breakdown Voltage (I_C = 1 mAdc, I_B = 0)	BV_{CEO}	20	—	—	Vdc
Collector-Emitter Breakdown Voltage* (I_C = 10 mAdc, V_{EB} = 0)	BV_{CES}*	30	—	—	Vdc
Collector Cutoff Current (V_{CB} = 15 Vdc, I_E = 0) (V_{CB} = 15 Vdc, I_E = 0, T_A = 60 °C)	I_{CBO}	— —	— —	0.05 1.0	μAdc
DC Current Gain (I_C = 2 mAdc, V_{CE} = 10 Vdc)	h_{FE}	25	—	—	—
High Frequency Current Gain (I_C = 2 mAdc, V_{CE} = 10 Vdc, f = 44 mc)	$|h_{fe}|$	20	—	—	db
Output Capacitance (V_{CB} = 10 Vdc, I_E = 0, f = 100 kc)	C_{ob}	—	—	2.5	pf
Current-Gain – Bandwidth Product (I_C = 10 mAdc, V_{CE} = 10 Vdc)	f_T	700	—	—	mc

* Pulse Test: Pulse Width ≤ 300 μsec, Duty Cycle ≤ 2%

Figure C.2-2   (*Continued*)

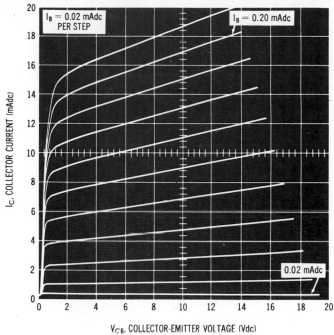

FIGURE 1 — COLLECTOR CURRENT versus COLLECTOR-EMITTER VOLTAGE

Figure C.2-2 (*Continued*)

FIGURE 2 — CONTOURS OF CONSTANT GAIN — BANDWIDTH PRODUCT

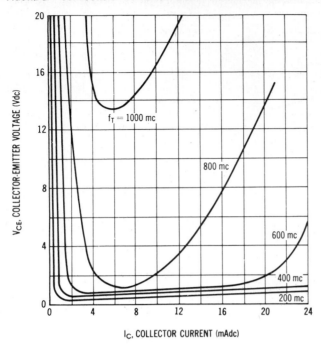

y PARAMETER VARIATIONS

($V_{CE} = 10\,Vdc$, $I_C = 3\,mAdc$, $T_A = 25°C$)

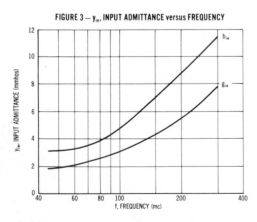

FIGURE 3 — y_{ie}, INPUT ADMITTANCE versus FREQUENCY

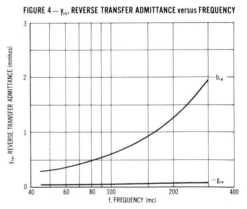

FIGURE 4 — y_{re}, REVERSE TRANSFER ADMITTANCE versus FREQUENCY

Figure C.2-2 (*Continued*)

774

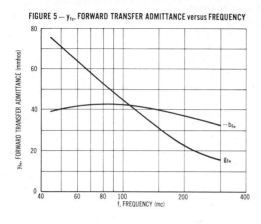

FIGURE 5 — y$_{fe}$, FORWARD TRANSFER ADMITTANCE versus FREQUENCY

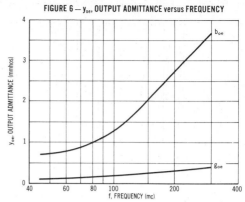

FIGURE 6 — y$_{oe}$, OUTPUT ADMITTANCE versus FREQUENCY

Figure C.2-2 (*Continued*)

C.3 FET CHARACTERISTICS

MAXIMUM RATINGS (T$_A$ = 25°C)

Characteristic	Symbol	Rating	Unit
Drain-Source Voltage	V_{DS}	30	Vdc
Drain-Gate Voltage	V_{DG}	30	Vdc
Gate-Source Voltage	V_{GS}	-30	Vdc
Drain Current	I_D	20	mAdc
Power Dissipation Derate above 25°C	P_D	300 2	mW mW/°C
Operating Junction Temperature	T_J	175	°C
Storage Temperature Range	T_{stg}	-65 to +200	°C

Figure C.3-1 Characteristics of the 2N4223 and 2N4224 silicon *n*-channel junction FETs, designed for VHF amplifier and mixer applications. (*Motorola, Inc.*)

775

ELECTRICAL CHARACTERISTICS $(T_A = 25°C$ unless otherwise noted)

OFF CHARACTERISTICS

Characteristic		Symbol	Min	Max	Unit
Gate-Source Breakdown Voltage $(I_G = -10 \ \mu Adc, V_{DS} = 0)$		$V_{(BR)GSS}$	-30	–	Vdc
Gate Reverse Current $(V_{GS} = -20 \ Vdc, V_{DS} = 0)$	2N4223 2N4224	I_{GSS}	– –	-0.25 -0.50	nAdc
$(V_{GS} = -20 \ Vdc, V_{DS} = 0, \ T_A = 100°C)$	2N4223 2N4224		– –	-250 -500	
Gate-Source Cutoff Voltage $(I_D = 0.25 \ nAdc, V_{DS} = 15 \ Vdc)$	2N4223	$V_{GS(off)}$	–	-8	Vdc
$(I_D = 0.50 \ nAdc, V_{DS} = 15 \ Vdc)$	2N4224		–	-8	
Gate-Source Voltage $(I_D = 0.3 \ mAdc, V_{DS} = 15 \ Vdc)$	2N4223	V_{GS}	-1.0	-7.0	Vdc
$(I_D = 0.2 \ mAdc, V_{DS} = 15 \ Vdc)$	2N4224		-1.0	-7.5	

ON CHARACTERISTICS

Characteristic		Symbol	Min	Max	Unit
Zero-Gate-Voltage Drain Current* $(V_{DS} = 15 \ Vdc, V_{GS} = 0)$	2N4223 2N4224	$I_{DSS}*$	3 2	18 20	mAdc

DYNAMIC CHARACTERISTICS

Characteristic	Symbol	Device	Min	Max	Unit
Forward Transfer Admittance ($V_{DS} = 15$ Vdc, $V_{GS} = 0$, $f = 1$ kHz)*	$|y_{fs}|$	2N4223	3000	7000	μmhos
		2N4224	2000	7500	
($V_{DS} = 15$ Vdc, $V_{GS} = 0$, $f = 200$ MHz)		2N4223	2700	–	
		2N4224	1700	–	
Input Conductance ($V_{DS} = 15$ Vdc, $V_{GS} = 0$, $f = 200$ MHz)	$\text{Re}(y_{is})$		–	800	μmhos
Output Conductance ($V_{DS} = 15$ Vdc, $V_{GS} = 0$, $f = 200$ MHz)	$\text{Re}(y_{os})$		–	200	μmhos
Input Capacitance ($V_{DS} = 15$ Vdc, $V_{GS} = 0$, $f = 1$ MHz)	C_{iss}		–	6	pF
Reverse Transfer Capacitance ($V_{DS} = 15$ Vdc, $V_{GS} = 0$, $f = 1$ MHz)	C_{rss}		–	2	pF
Noise Figure ($V_{DS} = 15$ Vdc, $V_{GS} = 0$, $R_S = 1$ kohm, $f = 200$ MHz)	NF	2N4223	–	5	dB
Small-Signal Power Gain ($V_{DS} = 15$ Vdc, $V_{GS} = 0$, $f = 200$ MHz)	G_{ps}	2N4223	10	–	dB

*Pulse Test: Pulse Width ≤ 630 ms, Duty Cycle $\leq 10\%$

Figure C.3-1 (*Continued*)

FIGURE 2 — INPUT IMPEDANCE

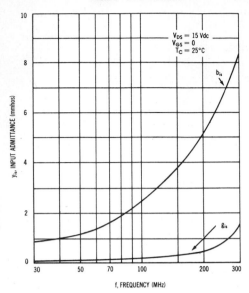

FIGURE 3 — REVERSE TRANSFER ADMITTANCE

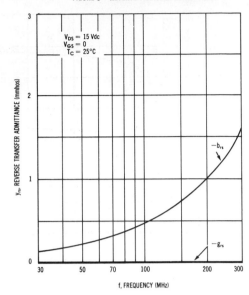

FIGURE 4 — FORWARD TRANSFER ADMITTANCE

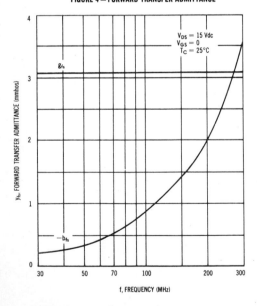

FIGURE 5 — OUTPUT ADMITTANCE

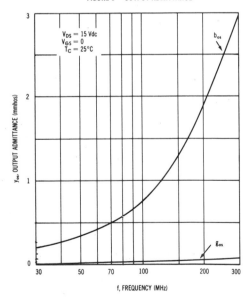

Figure C.3-1 (*Continued*)

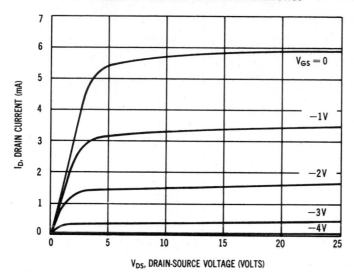

FIGURE 6 — TYPICAL DRAIN CHARACTERISTICS

Figure C.3-1   (*Continued*)

FIGURE 7 — COMMON SOURCE TRANSFER CHARACTERISTICS

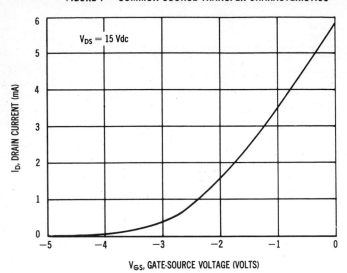

Figure C.3-1 (*Continued*)

MAXIMUM RATINGS (T_A = 25°C unless otherwise noted)

Rating	Symbol	Value	Unit

Drain–Source Voltage
  2N3796
  2N3797
	V_{DS}	
		25
		20
			Vdc

Gate–Source Voltage	V_{GS}	±10	Vdc
Drain Current	I_D	20	mAdc
Power Dissipation at $T_A = 25°C$ Derate above 25°C	P_D	200 1.14	mW mW/°C
Operating Junction Temperature	T_J	+200	°C
Storage Temperature	T_{stg}	−65 to +200	°C

Figure C.3-2 Characteristics of the 2N3796 and 2N3797 silicon n-channel MOSFETs, designed for low-power applications in the audio-frequency range. (*Motorola, Inc.*).

ELECTRICAL CHARACTERISTICS ($T_A = 25°C$ unless otherwise noted)

Characteristic		Symbol	Min	Typ	Max	Unit
Drain-Source Breakdown Voltage		BV_{DSX}				Vdc
($V_{GS} = -4.0$ V, $I_D = 5.0\ \mu A$)	2N3796		25	30	—	
($V_{GS} = -7.0$ V, $I_D = 5.0\ \mu A$)	2N3797		20	25	—	
Zero-Gate-Voltage Drain Current		I_{DSS}				mAdc
($V_{DS} = 10$ V, $V_{GS} = 0$)	2N3796		0.5	1.5	3.0	
	2N3797		2.0	2.9	6.0	
Gate-Source Voltage Cutoff		$V_{GS(off)}$				Vdc
($I_D = 0.5\ \mu A$, $V_{DS} = 10$ V)	2N3796		—	-3.0	-4.0	
($I_D = 2.0\ \mu A$, $V_{DS} = 10$ V)	2N3797		—	-5.0	-7.0	
"On" Drain Current		$I_{D(on)}$				mAdc
($V_{DS} = 10$ V, $V_{GS} = +3.5$ V)	2N3796		7.0	8.3	14	
	2N3797		9.0	14	18	
Drain-Gate Reverse Current *		I_{DGO} *				pAdc
($V_{DG} = 10$ V, $I_S = 0$)			—	—	1.0	
Gate-Reverse Current *		I_{GSS} *				pAdc
($V_{GS} = -10$ V, $V_{DS} = 0$)			—	—	1.0	
($V_{GS} = -10$ V, $V_{DS} = 0$, $T_A = 150°C$)			—	—	200	

Small-Signal, Common-Source Forward Transfer Admittance ($V_{DS} = 10$ V, $V_{GS} = 0$, f = 1.0 kHz) 2N3796 2N3797	$\|y_{fs}\|$	900 1500	1200 2300	1800 3000	μmhos
($V_{DS} = 10$ V, $V_{GS} = 0$, f = 1.0 MHz) 2N3796 2N3797		900 1500	— —	— —	
Small-Signal, Common-Source, Output Admittance ($V_{DS} = 10$ V, $V_{GS} = 0$, f = 1.0 kHz) 2N3796 2N3797	$\|y_{os}\|$	— —	12 27	25 60	μmhos
Small-Signal, Common-Source, Input Capacitance ($V_{DS} = 10$ V, $V_{GS} = 0$, f = 1.0 MHz) 2N3796 2N3797	C_{iss}	— —	5.0 6.0	7.0 8.0	pF
Small-Signal, Common-Source, Reverse Transfer Capacitance ($V_{DS} = 10$ V, $V_{GS} = 0$, f = 1.0 MHz)	C_{rss}	—	0.5	0.8	pF
Noise Figure ($V_{DS} = 10$ V, $V_{GS} = 0$, f = 1.0 kHz, $R_S = 3$ megohms)	NF	—	3.8	—	dB

*This value of current includes both the FET leakage current as well as the leakage current associated with the test socket and fixture when measured under best attainable conditions.

Figure C.3-2 *(Continued)*

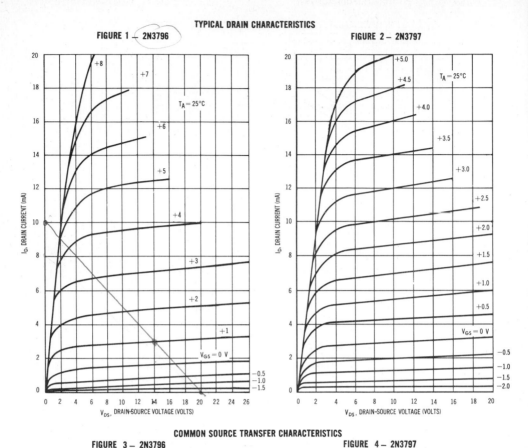

TYPICAL DRAIN CHARACTERISTICS

FIGURE 1 — 2N3796

FIGURE 2 — 2N3797

COMMON SOURCE TRANSFER CHARACTERISTICS

FIGURE 3 — 2N3796

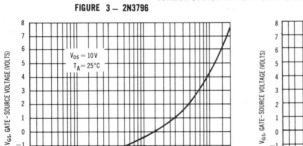

FIGURE 4 — 2N3797

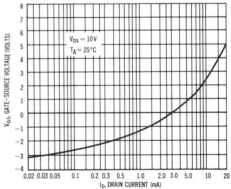

Figure C.3-2 *(Continued)*

FIGURE 5 — FORWARD TRANSFER ADMITTANCE

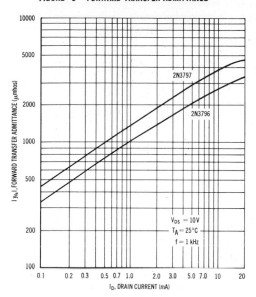

2N3797

2N3796

$V_{DS} = 10V$
$T_A = 25°C$
$f = 1 kHz$

I_D, DRAIN CURRENT (mA)

$|y_{fs}|$, FORWARD TRANSFER ADMITTANCE (μmhos)

FIGURE 6 — AMPLIFICATION FACTOR

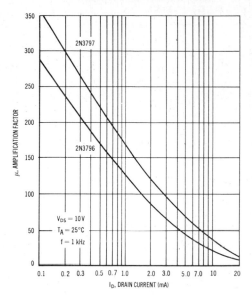

2N3797

2N3796

$V_{DS} = 10V$
$T_A = 25°C$
$f = 1 kHz$

μ, AMPLIFICATION FACTOR

I_D, DRAIN CURRENT (mA)

FIGURE 7 — OUTPUT ADMITTANCE

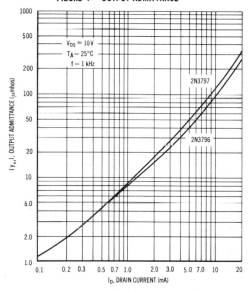

$V_{DS} = 10V$
$T_A = 25°C$
$f = 1 kHz$

2N3797

2N3796

$|y_{os}|$, OUTPUT ADMITTANCE (μmhos)

I_D, DRAIN CURRENT (mA)

FIGURE 8 — NOISE FIGURE

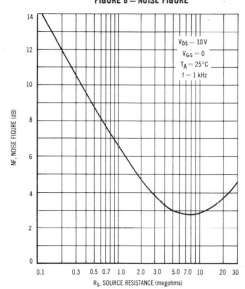

$V_{DS} - 10V$
$V_{GS} = 0$
$T_A = 25°C$
$f = 1 kHz$

NF, NOISE FIGURE (dB)

R_S, SOURCE RESISTANCE (megohms)

┌FLIP-FLOPS─────────────────────────

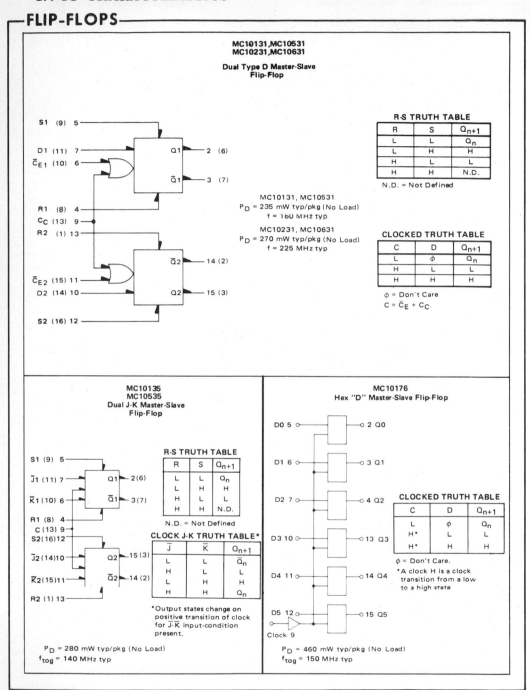

MC10131, MC10531
MC10231, MC10631

Dual Type D Master-Slave
Flip-Flop

S1 (9) 5

D1 (11) 7   Q1 2 (6)
$\bar{C}_{E1}$ (10) 6

$\bar{Q}1$ 3 (7)

R-S TRUTH TABLE

R	S	Q_{n+1}
L	L	Q_n
L	H	H
H	L	L
H	H	N.D.

N.D. = Not Defined

R1 (8) 4
C_C (13) 9
R2 (1) 13

$\bar{Q}2$ 14 (2)

$\bar{C}_{E2}$ (15) 11
D2 (14) 10   Q2 15 (3)

S2 (16) 12

MC10131, MC10531
P_D = 235 mW typ/pkg (No Load)
f = 160 MHz typ

MC10231, MC10631
P_D = 270 mW typ/pkg (No Load)
f = 225 MHz typ

CLOCKED TRUTH TABLE

C	D	Q_{n+1}
L	ϕ	Q_n
H	L	L
H	H	H

ϕ = Don't Care
$C = \bar{C}_E + C_C$.

MC10135
MC10535
Dual J-K Master-Slave
Flip-Flop

S1 (9) 5

$\bar{J}1$ (11) 7   Q1 2(6)
$\bar{K}1$ (10) 6 $\bar{Q}1$ 3(7)

R1 (8) 4
C (13) 9
S2 (16) 12

$\bar{J}2$ (14) 10   Q2 15 (3)

$\bar{K}2$ (15) 11 $\bar{Q}2$ 14 (2)

R2 (1) 13

R-S TRUTH TABLE

R	S	Q_{n+1}
L	L	Q_n
L	H	H
H	L	L
H	H	N.D.

N.D. = Not Defined

CLOCK J-K TRUTH TABLE*

$\bar{J}$	$\bar{K}$	Q_{n+1}
L	L	$\bar{Q}_n$
H	L	L
L	H	H
H	H	Q_n

*Output states change on
positive transition of clock
for $\bar{J}\cdot\bar{K}$ input-condition
present.

P_D = 280 mW typ/pkg (No Load)
f_{tog} = 140 MHz typ

MC10176
Hex "D" Master-Slave Flip-Flop

D0 5   2 Q0

D1 6   3 Q1

D2 7   4 Q2

D3 10   13 Q3

D4 11   14 Q4

D5 12   15 Q5

Clock 9

CLOCKED TRUTH TABLE

C	D	Q_{n+1}
L	ϕ	Q_n
H*	L	L
H*	H	H

ϕ = Don't Care.
*A clock H is a clock
transition from a low
to a high state

P_D = 460 mW typ/pkg (No Load)
f_{tog} = 150 MHz typ

Figure C.4-1 Characteristics of the MC10131 dual type-*D* master-slave flip-flop. *(Motorola, Inc.)*.

MC10131

ELECTRICAL CHARACTERISTICS

Each MECL 10,000 series circuit has been designed to meet the dc specifications shown in the test table, after thermal equilibrium has been established. The circuit is in a test socket or mounted on a printed circuit board and transverse air flow greater than 500 linear fpm is maintained. Outputs are terminated through a 50-ohm resistor to -2.0 volts. Test procedures are shown for only one input, or for one set of input conditions. Other inputs tested in the same manner.

L SUFFIX
CERAMIC PACKAGE
CASE 620

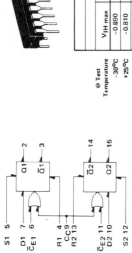

TEST VOLTAGE VALUES (Volts)

@Test Temperature	V$_{IH}$ max	V$_{IL}$ min	V$_{IHA}$ min	V$_{ILA}$ max	V$_{EE}$	Gnd (V$_{CC}$)
-30°C	-0.890	-1.890	-1.205	-1.500	-5.2	1, 16
+25°C	-0.810	-1.850	-1.105	-1.475	-5.2	1, 16
+85°C	-0.700	-1.825	-1.035	-1.440	-5.2	1, 16

VOLTAGE APPLIED TO PINS LISTED BELOW:

	V$_{IH}$ max	V$_{IL}$ min	V$_{IHA}$ min	V$_{ILA}$ max	V$_{EE}$	Gnd
	4	—	—	—	8	1, 16
	5	—	—	—	8	1, 16
	6	—	—	—	8	1, 16
	7	—	—	—	8	1, 16
	9	—	—	—	8	1, 16
	5	—	—	—	8	1, 16
	7	—	—	—	8	1, 16
	5	—	—	—	8	1, 16
	7	—	—	—	8	1, 16
	—	—	5	9	8	1, 16
	—	—	7	9	8	1, 16
	—	—	5	9	8	1, 16
	—	—	7	9	8	1, 16
	+1.11 Vdc		Pulse In	Pulse Out	-3.2 Vdc	+2.0 Vdc
	7	—	9	2	8	1, 16
	7	—	9	2	8	1, 16
	6	—	6	2	8	1, 16
	7	—	9	2	8	1, 16
	7	—	9	2	8	1, 16
	6	—	5	2	8	1, 16
	—	—	12	15	8	1, 16
	9	—	5	3	8	1, 16
	—	—	12	14	8	1, 16
	6	—	4	2	8	1, 16
	—	—	13	15	8	1, 16
	9	—	4	3	8	1, 16
	—	—	13	14	8	1, 16
	—	—	6.7	2	8	1, 16
	—	—	6.7	2	8	1, 16
	—	—	6	2	8	1, 16

MC10131L Test Limits

Characteristic	Symbol	Pin Under Test	-30°C Min	-30°C Max	+25°C Min	+25°C Typ	+25°C Max	+85°C Min	+85°C Max	Unit
Power Supply Drain Current	I$_E$	8	—	—	—	45	56	—	—	mAdc
Input Current	I$_{inH}$	4	—	—	—	—	330	—	—	µAdc
		5	—	—	—	—	330	—	—	
		6	—	—	—	—	220	—	—	
		7	—	—	—	—	245	—	—	
		9	—	—	—	—	265	—	—	
Input Leakage Current	I$_{inL}$	4,5,*	—	—	0.5	—	—	—	—	µAdc
		6,7,9*	—	—	0.5	—	—	—	—	
Logic "1" Output Voltage	V$_{OH}$	2	-1.060	-0.890	-0.960	—	-0.810	-0.890	-0.700	Vdc
		2†	-1.060	-0.890	-0.960	—	-0.810	-0.890	-0.700	Vdc
Logic "0" Output Voltage	V$_{OL}$	3	-1.890	-1.675	-1.850	—	-1.650	-1.825	-1.615	Vdc
		3†	-1.890	-1.675	-1.850	—	-1.650	-1.825	-1.615	Vdc
Logic "1" Threshold Voltage	V$_{OHA}$	2	-1.060	—	-0.980	—	—	-0.910	—	Vdc
		2†	-1.060	—	-0.980	—	—	-0.910	—	Vdc
Logic "0" Threshold Voltage	V$_{OLA}$	3	—	-1.655	—	—	-1.630	—	-1.535	Vdc
		3†	—	-1.655	—	—	-1.630	—	-1.535	Vdc
Switching Times Clock Input Propagation Delay	t$_{9+2-}$	2	1.4	4.6	1.5	3.0	4.5	1.5	5.0	ns
	t$_{9+2+}$	2								
	t$_{6+2-}$	2								
	t$_{6+2+}$	2								
Rise Time (20 to 80%)	t$_{2+}$	2	1.0		1.1	2.5	4.3	1.1	4.9	ns
Fall Time (20 to 80%)	t$_{2-}$	2	1.0		1.1	2.5	4.3	1.1	4.9	ns
Set Input Propagation Delay	t$_{5+2+}$	2	1.1	4.4	1.2	2.8	4.3	1.2	4.6	ns
	t$_{12+15+}$	15								
	t$_{5+3-}$	3								
	t$_{12+14-}$	14								
Reset Input Propagation Delay	t$_{4+2-}$	2	1.1	4.4	1.2	2.8	4.3	1.2	4.6	ns
	t$_{13+15-}$	15								
	t$_{4+3+}$	3								
	t$_{13+14+}$	14								
Setup Time	t$_{setup}$	7	—	—	2.5	—	—	—	—	ns
Hold Time	t$_{hold}$	7	—	—	1.5	—	—	—	—	ns
Toggle Frequency (Max)	f$_{Tog}$	2	125	—	125	160	—	125	—	MHz

*Individually test each input; apply V$_{IL}$ min to pin under test.

†Output level to be measured after a clock pulse has been applied to the $\overline{C}_E$ input (pin 6)

†Output level to be measured after a clock pulse has been applied to the $\overline{C}_E$ input (pin 6)

MC14042AL
MC14042CL
MC14042CP

QUAD LATCH

The MC14042 quad latch is constructed with MOS P-channel and N-channel enhancement mode devices in a single monolithic structure. Each latch has a separate data input, but all four latches share a common clock. The clock polarity (high or low) used to strobe data through the latches can be reversed using the polarity input. Information present at the data input is transferred to outputs Q and $\overline{Q}$ during the clock level which is determined by the polarity input. When the polarity input is in the logic "0" state, data is transferred during the low clock level, and when the polarity input is in the logic "1" state the transfer occurs during the high clock level. Additional characteristics can be found on the Family Data Sheet.

- Buffered Data Inputs
- Common Clock
- Positive or Negative Edge Clocked
- Q and $\overline{Q}$ Outputs
- Double Diode Input Protection
- No Limit on Clock Rise or Fall Times

McMOS
(LOW-POWER COMPLEMENTARY MOS)

QUAD LATCH

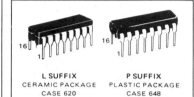

L SUFFIX
CERAMIC PACKAGE
CASE 620

P SUFFIX
PLASTIC PACKAGE
CASE 648

This device contains circuitry to protect the inputs against damage due to high static voltages or electric fields; however, it is advised that normal precautions be taken to avoid application of any voltage higher than maximum rated voltages to this high impedance circuit. For proper operation it is recommended that V_{in} and V_{out} be constrained to the range $V_{SS} \leqslant (V_{in}$ or $V_{out}) \leqslant V_{DD}$.
Unused inputs must always be tied to an appropriate logic voltage level (e.g., either V_{SS} or V_{DD}).

MAXIMUM RATINGS (Voltages referenced to V_{SS}, Pin 8.)

Rating		Symbol	Value	Unit
DC Supply Voltage	MC14042AL	V_{DD}	+18 to -0.5	Vdc
	MC14042CL/CP		+16 to -0.5	
Input Voltage, All Inputs		V_{in}	V_{DD} to -0.5	Vdc
DC Current Drain per Pin		I	10	mAdc
Operating Temperature Range	MC14042AL	T_A	-55 to +125	°C
	MC14042CL/CP		-40 to +85	
Storage Temperature Range		T_{stg}	-65 to +150	°C

PIN ASSIGNMENT

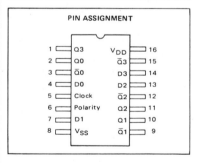

1	Q3	V_{DD}	16
2	Q0	$\overline{Q}3$	15
3	$\overline{Q}0$	D3	14
4	D0	D2	13
5	Clock	$\overline{Q}2$	12
6	Polarity	Q2	11
7	D1	Q1	10
8	V_{SS}	$\overline{Q}1$	9

LOGIC DIAGRAM

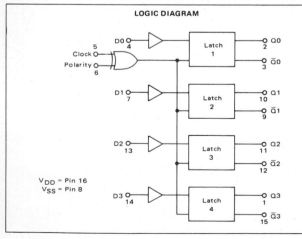

V_{DD} = Pin 16
V_{SS} = Pin 8

TRUTH TABLE

CLOCK	POLARITY	Q
0	0	Data
⌐	0	Latch
1	1	Data
⌐	1	Latch

Figure C.4-2 Characteristics of the MC14042 quad latch. (*Motorola, Inc.*).

MC14042

ELECTRICAL CHARACTERISTICS

Characteristic	Symbol	V_{DD} Vdc	T_{low}* Min	T_{low}* Max	25°C Min	25°C Typ	25°C Max	T_{high}* Min	T_{high}* Max	Unit
Output Voltage "0" Level	V_{out}	5.0	–	0.01	–	0	0.01	–	0.05	Vdc
		10	–	0.01	–	0	0.01	–	0.05	
		15	–	0.05	–	0	0.05	–	0.25	
"1" Level		5.0	4.99	–	4.99	5.0	–	4.95	–	Vdc
		10	9.99	–	9.99	10	–	9.95	–	
		15	14.95	–	14.95	15	–	14.75	–	
Noise Immunity #	V_{NL}									Vdc
($\triangle V_{out} \leqslant 0.8$ Vdc)		5.0	1.5	–	1.5	2.25	–	1.4	–	
($\triangle V_{out} \leqslant 1.0$ Vdc)		10	3.0	–	3.0	4.50	–	2.9	–	
($\triangle V_{out} \leqslant 1.5$ Vdc)		15	4.5	–	4.5	6.75	–	4.4	–	
($\triangle V_{out} \leqslant 0.8$ Vdc)	V_{NH}	5.0	1.4	–	1.5	2.25	–	1.5	–	Vdc
($\triangle V_{out} \leqslant 1.0$ Vdc)		10	2.9	–	3.0	4.50	–	3.0	–	
($\triangle V_{out} \leqslant 1.5$ Vdc)		15	4.4	–	4.5	6.75	–	4.5	–	
Output Drive Current (AL Device)	I_{OH}									mAdc
($V_{OH} = 2.5$ Vdc) Source		5.0	-0.62	–	-0.50	-1.7	–	-0.35	–	
($V_{OH} = 9.5$ Vdc)		10	-0.62	–	-0.50	-0.9	–	-0.35	–	
($V_{OH} = 13.5$ Vdc)		15	-1.8	–	-1.5	-3.5	–	-1.1	–	
($V_{OL} = 0.4$ Vdc) Sink	I_{OL}	5.0	0.50	–	0.40	0.78	–	0.28	–	mAdc
($V_{OL} = 0.5$ Vdc)		10	1.1	–	0.90	2.0	–	0.65	–	
($V_{OL} = 1.5$ Vdc)		15	4.2	–	3.4	7.8	–	2.4	–	
Output Drive Current (CL/CP Device)	I_{OH}									mAdc
($V_{OH} = 2.5$ Vdc) Source		5.0	-0.23	–	-0.20	-1.7	–	-0.16	–	
($V_{OH} = 9.5$ Vdc)		10	-0.23	–	-0.20	-0.9	–	-0.16	–	
($V_{OH} = 13.5$ Vdc)		15	-0.69	–	-0.60	-3.5	–	-0.48	–	
($V_{OL} = 0.4$ Vdc) Sink	I_{OL}	5.0	0.23	–	0.20	0.78	–	0.16	–	mAdc
($V_{OL} = 0.5$ Vdc)		10	0.60	–	0.50	2.0	–	0.40	–	
($V_{OL} = 1.5$ Vdc)		15	1.8	–	1.5	7.8	–	1.2	–	
Input Current	I_{in}	–	–	–	–	10	–	–	–	pAdc
Input Capacitance ($V_{in} = 0$)	C_{in}	–	–	–	–	5.0	–	–	–	pF
Quiescent Dissipation (AL Device)	P_Q	5.0	–	0.025	–	0.000025	0.025	–	1.5	mW
		10	–	0.01	–	0.00010	0.1	–	6.0	
		15	–	0.3	–	0.0003	0.3	–	18	
Quiescent Dissipation (CL/CP Device)	P_Q	5.0	–	0.25	–	0.000025	0.25	–	3.5	mW
		10	–	1.0	–	0.0001	1.0	–	14	
		15	–	3.0	–	0.0003	3.0	–	42	
Power Dissipation**† (Dynamic plus Quiescent) ($C_L = 15$ pF)	P_D	5.0 10 15		$P_D = (1.5$ mW/MHz$)$ f $+ P_Q$ $P_D = (6.0$ mW/MHz$)$ f $+ P_Q$ $P_D = (20$ mW/MHz$)$ f $+ P_Q$						mW

* T_{low} = -55°C for AL Device, -40°C for CL/CP Device.
 T_{high} = +125°C for AL Device, +85°C for CL/CP Device.
\# Noise immunity specified for worst-case input combination.
† For dissipation at different external load capacitance (C_L) use the formula:

$$P_T(C_L) = P_D + 8 \times 10^{-3} (C_L - 15 \text{ pF}) V_{DD}^2 f$$

where: P_T, P_D in mW (per package), C_L in pF, V_{DD} in Vdc, and f in MHz is input data frequency.
** The formula given is for the typical characteristics only.

SWITCHING CHARACTERISTICS* (C_L = 50 pF, T_A = 25°C)

Characteristic	Symbol	V_{DD}	All Types Typical	All Types Maximum	Unit
Output Rise Time t_r = (3.0 ns/pF) C_L + 30 ns t_r = (1.5 ns/pF) C_L + 15 ns t_r = (1.1 ns/pF) C_L + 10 ns	t_r	5.0 10 15	180 90 65	360 180 130	ns
Output Fall Time t_f = (1.5 ns/pF) C_L + 25 ns t_f = (0.75 ns/pF) C_L + 12.5 ns t_f = (0.55 ns/pF) C_L + 9.5 ns	t_f	5.0 10 15	100 50 40	200 100 80	ns
Propagation Delay Time, D to Q, $\bar{Q}$ t_{PLH}, t_{PHL} = (1.7 ns/pF) C_L + 135 ns t_{PLH}, t_{PHL} = (0.66 ns/pF) C_L + 57 ns t_{PLH}, t_{PHL} = (0.5 ns/pF) C_L + 35 ns	t_{PLH}, t_{PHL}	5.0 10 15	220 90 60	440 180 120	ns
Propagation Delay Time, Clock to Q, $\bar{Q}$ t_{PLH}, t_{PHL} = (1.7 ns/pF) C_L + 135 ns t_{PLH}, t_{PHL} = (0.66 ns/pF) C_L + 57 ns t_{PLH}, t_{PHL} = (0.5 ns/pF) C_L + 35 ns	t_{PLH}, t_{PHL}	5.0 10 25	220 90 60	440 180 120	ns
Minimum Clock Pulse Width	PW_C	5.0 10 15	150 50 40	300 100 80	ns
Maximum Clock Rise Time	t_r	5.0 10 15	No Limit		—
Hold Time	t_{hold}	5.0 10 15	50 25 20	100 50 40	ns
Setup Time	t_{setup}	5.0 10 15	0 0 0	50 30 25	ns

*The formula given is for the typical characteristics only.

FIGURE 1 – AC AND POWER DISSIPATION TEST CIRCUIT AND TIMING DIAGRAM
(Data to Output)

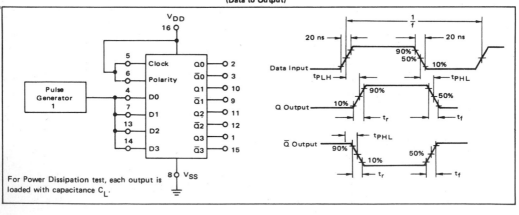

For Power Dissipation test, each output is loaded with capacitance C_L.

MOTOROLA Semiconductors
BOX 20912 • PHOENIX, ARIZONA 85036

MC14194B

4-BIT BIDIRECTIONAL UNIVERSAL SHIFT REGISTER

The MC14194B is a 4-bit static shift register capable of operating in the parallel load, serial shift left, serial shift right, or hold mode. The asynchronous Reset input, when at a low level, overrides all other inputs, resets all stages, and forces all outputs low. When Reset is at a logic 1 level, the two mode control inputs, S0 and S1, control the operating mode as shown in the truth table. Both serial and parallel operation are triggered on the positive-going transition of the Clock input. The Parallel Data, Data Shift, and mode control inputs must be stable for the specified setup and hold times before and after the positive-going Clock transition.

- Quiescent Current = 5.0 nA typ/pkg @ 5 Vdc
- Typical Shift Frequency = 9.0 MHz @ 10 Vdc
- Synchronous, Right/Left Serial Operation
- Synchronous Parallel Load
- Asynchronous Hold (Do Nothing) Mode
- Functional Pin for Pin Equivalent of 74194

McMOS MSI
(LOW-POWER COMPLEMENTARY MOS)

4-BIT BIDIRECTIONAL UNIVERSAL SHIFT REGISTER

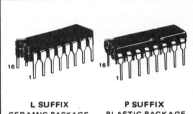

L SUFFIX	P SUFFIX
CERAMIC PACKAGE	PLASTIC PACKAGE
CASE 620	CASE 648

ORDERING INFORMATION

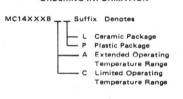

MC14XXXB — Suffix Denotes

- L Ceramic Package
- P Plastic Package
- A Extended Operating Temperature Range
- C Limited Operating Temperature Range

TRUTH TABLE

OPERATING MODE	INPUTS (Reset = 1)					OUTPUTS (@ t_{n+1})			
	S1	S0	DSR	DSL	D_{P0-3}	Q0	Q1	Q2	Q3
Hold	0	0	X	X	X	Q0	Q1	Q2	Q3
Shift Left	1	0	X	0	X	Q1	Q2	Q3	0
	1	0	X	1	X	Q1	Q2	Q3	1
Shift Right	0	1	0	X	X	0	Q0	Q1	Q2
	0	1	1	X	X	1	Q0	Q1	Q2
Parallel	1	1	X	X	0	0	0	0	0
	1	1	X	X	1	1	1	1	1

X = Don't Care
t_{n+1} = State after the next positive-going transition of the clock.

LOGIC DIAGRAM

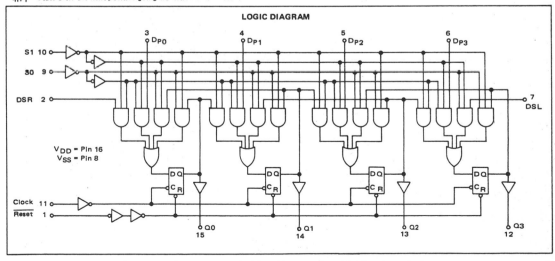

Figure C.4-3 A 4-bit bidirectional universal shift register. *(Motorola, Inc.)*.

MM54C192/MM74C192 synchronous 4-bit up/down decade counter
MM54C193/MM74C193 synchronous 4-bit up/down binary counter

general description

These up/down counters are monolithic complementary MOS (CMOS) integrated circuits. The MM54C192 and MM74C192 are BCD counters. While the MM54C193 and MM74C193 are binary counters.

Counting up and counting down is performed by two count inputs, one being held high while the other is clocked. The outputs change on the positive going transition of this clock.

These counters feature preset inputs that are set when load is a logical "0" and a clear which forces all outputs to "0" when it is at logical "1." The counters also have carry and borrow outputs so that they can be cascaded using no external circuitry.

features

- High noise margin   1V guaranteed
- Tenth power   drive 2 LPTTL
  TTL compatible   loads
- Wide supply range   3V to 15V
- Carry and borrow outputs for N-bit cascading
- Asynchronous clear
- High noise immunity   0.45 V$_{CC}$ typ

connection diagram

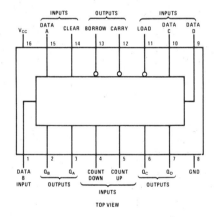

cascading packages

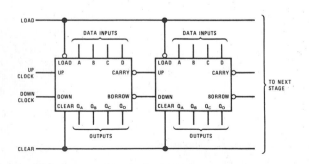

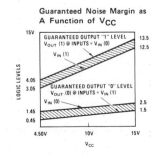

Figure C.4-4 Specifications of the MM74C192 and MM74C193 synchronous 4-bit up-down counters. *(National Semiconductor, Inc.).*

schematic diagrams

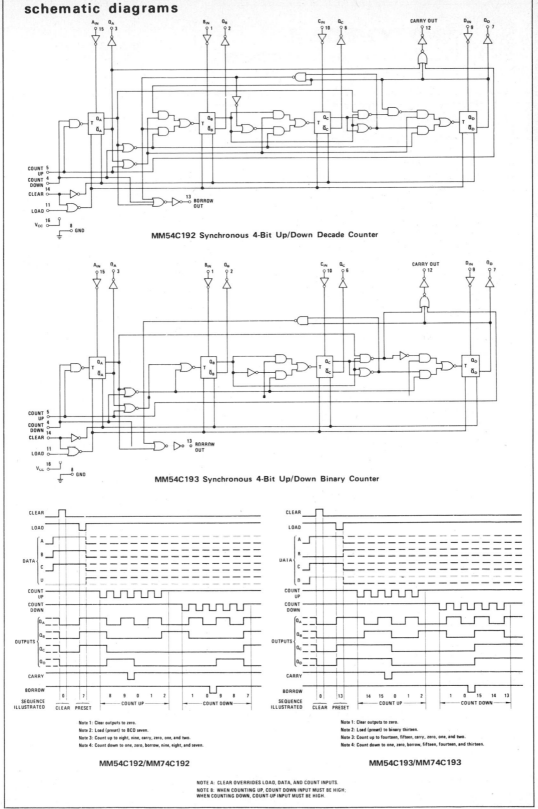

MM54C192 Synchronous 4-Bit Up/Down Decade Counter

MM54C193 Synchronous 4-Bit Up/Down Binary Counter

MM54C192/MM74C192

Note 1: Clear outputs to zero.
Note 2: Load (preset) to BCD seven.
Note 3: Count up to eight, nine, carry, zero, one, and two.
Note 4: Count down to one, zero, borrow, nine, eight, and seven.

MM54C193/MM74C193

Note 1: Clear outputs to zero.
Note 2: Load (preset) to binary thirteen.
Note 3: Count up to fourteen, fifteen, carry, zero, one, and two.
Note 4: Count down to one, zero, borrow, fifteen, fourteen, and thirteen.

NOTE A: CLEAR OVERRIDES LOAD, DATA, AND COUNT INPUTS.
NOTE B: WHEN COUNTING UP, COUNT DOWN INPUT MUST BE HIGH;
WHEN COUNTING DOWN, COUNT-UP INPUT MUST BE HIGH.

MOTOROLA
Semiconductors
BOX 20912 • PHOENIX, ARIZONA 85036

MC14582B

LOOK-AHEAD CARRY BLOCK

The MC14582B is a CMOS look-ahead carry generator capable of anticipating a carry across four binary adders or groups of adders. The device is cascadable to perform full look-ahead across n-bit adders. Carry, generate-carry, and propagate-carry functions are provided as enumerated in the pin designation table shown below.

- Quiescent Current = 5.0 nA/package typical @ 5 Vdc
- High Speed Operation — 140 ns typical @ V_{DD} = 10 Vdc (from Data-in to Carry-out)
- Expandable to any Number of Bits
- Noise Immunity = 45% of V_{DD} typical
- All Buffered Outputs
- Low Power Dissipation
- Diode Protection on All Inputs
- Supply Voltage Range = 3.0 Vdc to 18 Vdc
- Capable of Driving Two Low-Power TTL Loads, One Low-power Schottky TTL Load or Two HTL Loads Over the Rated Temperature Range

McMOS MSI

(LOW-POWER COMPLEMENTARY MOS)

LOOK-AHEAD CARRY BLOCK

L SUFFIX	P SUFFIX
CERAMIC PACKAGE	PLASTIC PACKAGE
CASE 620	CASE 648

ORDERING INFORMATION

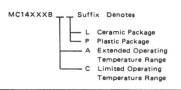

MC14XXXB ____ Suffix Denotes

- L   Ceramic Package
- P   Plastic Package
- A   Extended Operating Temperature Range
- C   Limited Operating Temperature Range

MAXIMUM RATINGS (Voltages referenced to V_{SS})

Rating	Symbol	Value	Unit
DC Supply Voltage	V_{DD}	–0.5 to +18	Vdc
Input Voltage, All Inputs	V_{in}	–0.5 to V_{DD} + 0.5	Vdc
DC Current Drain per Pin	I	10	mAdc
Operating Temperature Range — AL Device	T_A	–55 to +125	°C
CL/CP Device		–40 to +85	
Storage Temperature Range	T_{stg}	–65 to +150	°C

LOGIC EQUATIONS

$C_{n+x} = \overline{G0} + (\overline{P0} \bullet C_n)$

$C_{n+y} = \overline{G1} + (\overline{P1} \bullet \overline{G0}) + (\overline{P1} \bullet \overline{P0} \bullet C_n)$

$C_{n+z} = \overline{G2} + (\overline{P2} \bullet \overline{G1}) + (\overline{P2} \bullet \overline{P1} \bullet \overline{G0}) + (\overline{P2} \bullet \overline{P1} \bullet \overline{P0} \bullet C_n)$

$\overline{G} = \overline{G3} + (\overline{P3} \bullet \overline{G2}) + (\overline{P3} \bullet \overline{P2} \bullet \overline{G1}) + (\overline{P1} \bullet \overline{P2} \bullet \overline{P3} \bullet \overline{G0})$

$\overline{P} = \overline{P3} \bullet \overline{P2} \bullet \overline{P1} \bullet \overline{P0}$

PIN DESIGNATIONS

DESIGNATION	PIN NO's	FUNCTION
$\overline{G0}, \overline{G1}, \overline{G2}, \overline{G3}$	3,1,14,5	Active-Low Carry-Generate Inputs
$\overline{P0}, \overline{P1}, \overline{P2}, \overline{P3}$	4,2,15,6	Active-Low Carry-Propagate Inputs
C_n	13	Carry Input
C_{n+x}, C_{n+y} C_{n+z}	12,11,9	Carry Outputs
$\overline{G}$	10	Active-Low Group Carry-Generate Output
$\overline{P}$	7	Active-Low Group Carry-Propagate Output

BLOCK DIAGRAM

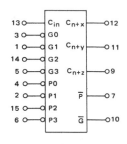

V_{DD} = Pin 16
V_{SS} = Pin 8

Figure C.4-5 Look-ahead carry circuits. (*Motorola, Inc.*).

FIGURE 4 – SWITCHING TIME TEST CIRCUIT AND WAVEFORMS

TEST TABLE

AC PATHS		DC DATA	
INPUT	OUTPUT	To V_{SS}	To V_{DD}
$\bar{P}0$	$\bar{P}$	Remaining $\bar{P}$'s, C_n	G's
G0	$\bar{G}$	P's, C_n	Remaining $\bar{G}$'s
C_n	$C_{n+x} \cdot C_{n+y}$ C_{n+z}	$\bar{P}$'s	$\bar{G}$'s

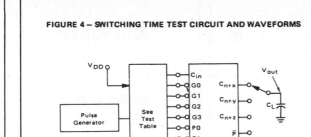

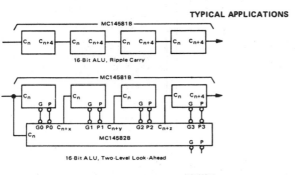

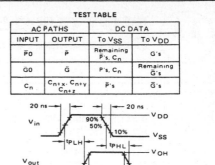

TYPICAL APPLICATIONS

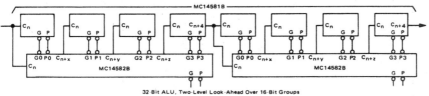

16-Bit ALU, Ripple Carry

16-Bit ALU, Two-Level Look-Ahead

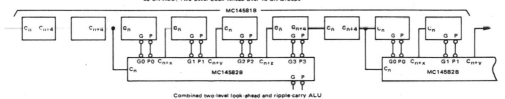

32-Bit ALU, Two-Level Look-Ahead Over 16-Bit Groups

Combined two-level look-ahead and ripple-carry ALU

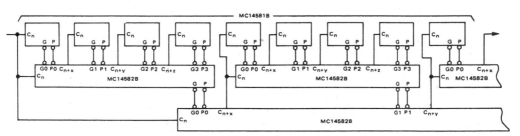

64-Bit ALU, Full-Carry Look-Ahead in Three Levels.

A and B inputs and F outputs are not shown (MC14581B).

 MOTOROLA *Semiconductor Products Inc.*

MOTOROLA Semiconductors

BOX 20912 • PHOENIX, ARIZONA 85036

MC14581B

McMOS MSI

(LOW-POWER COMPLEMENTARY MOS)

4-BIT ARITHMETIC LOGIC UNIT

4-BIT ARITHMETIC LOGIC UNIT

The MC14581B is a CMOS 4-bit ALU logic unit capable of providing 16 functions of two Boolean variables and 16 binary arithmetic operations on two 4-bit words. The level of the mode control input determines whether the output funciton is logic or arithmetic. The desired logic function is selected by applying the appropriate binary word to the select inputs (S0 thru S3) with the mode control input high, while the desired arithmetic operation is selected by applying a low voltage to the mode control input, the required level to carry in, and the appropriate word to the select inputs. The word inputs and function outputs can be operated with either active high or active low data.

Carry propagate ($\overline{P}$) and carry generate ($\overline{G}$) outputs are provided to allow a full look-ahead carry scheme for fast simultaneous carry generation for the four bits in the package. Fast arithmetic operations on long words are obtainable by using the MC14582B as a second order look ahead block. An inverted ripple carry input (C_n) and a ripple carry output (C_{n+4}) are included for ripple through operation.

When the device is in the subtract mode (LHHL), comparison of two 4-bit words present at the $\overline{A}$ and $\overline{B}$ inputs is provided using the A = B output. It assumes a high-level state when indicating equality. Also, when the ALU is in the subtract mode the C_{n+4} output can be used to indicate relative magnitude as shown in this table:

Data Level	C_n	C_{n+4}	Magnitude
	H	H	A ≤ B
Active High	L	H	A < B
	H	L	A > B
	L	L	A ≥ B
	L	L	A ≤ B
Active Low	H	L	A < B
	L	H	A > B
	H	H	A ≥ B

FEATURES:

- Functional and Pinout Equivalent to 74181.
- Quiescent Current = 5.0 nA/package typical @ 5 Vdc
- High Noise Immunity = 45% of V_{DD} typical
- Diode Protection on All Inputs
- Low Input Capacitance — 5.0 pF typical
- All Outputs Buffered
- Supply Voltage Range = 3.0 Vdc to 18 Vdc
- Capable of Driving Two Low-power TTL Load, One Low-power Schottky TTL Load or Two HTL Loads Over the Rated Temperature Range

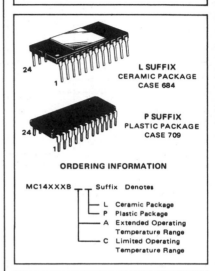

L SUFFIX
CERAMIC PACKAGE
CASE 684

P SUFFIX
PLASTIC PACKAGE
CASE 709

ORDERING INFORMATION

MC14XXXB ___ Suffix   Denotes

- L   Ceramic Package
- P   Plastic Package
- A   Extended Operating Temperature Range
- C   Limited Operating Temperature Range

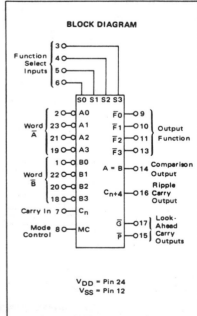

BLOCK DIAGRAM

V_{DD} = Pin 24
V_{SS} = Pin 12

MAXIMUM RATINGS (Voltages referenced to V_{SS})

Rating	Symbol	Value	Unit
DC Supply Voltage	V_{DD}	–0.5 to +18	Vdc
Input Voltage, All Inputs	V_{in}	–0.5 to V_{DD} + 0.5	Vdc
DC Current Drain per Pin	I	10	mAdc
Operating Temperature Range — AL Device CL/CP Device	T_A	–55 to +125 –40 to +85	°C
Storage Temperature Range	T_{stg}	–65 to +150	°C

Figure C.4-6 A 4-bit arithmetic logic unit. (*Motorola, Inc.*).

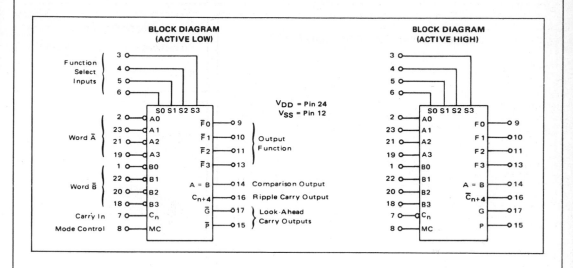

BLOCK DIAGRAM (ACTIVE LOW)

Function Select Inputs: 3, 4, 5, 6 → S0 S1 S2 S3

V_{DD} = Pin 24
V_{SS} = Pin 12

Word $\overline{A}$: 2 → A0, 23 → A1, 21 → A2, 19 → A3

Word $\overline{B}$: 1 → B0, 22 → B1, 20 → B2, 18 → B3

Carry In: 7 → C_n

Mode Control: 8 → MC

$\overline{F}0$ → 9, $\overline{F}1$ → 10, $\overline{F}2$ → 11, $\overline{F}3$ → 13 } Output Function

$A = B$ → 14 Comparison Output
C_{n+4} → 16 Ripple Carry Output
$\overline{G}$ → 17, $\overline{P}$ → 15 } Look-Ahead Carry Outputs

BLOCK DIAGRAM (ACTIVE HIGH)

3, 4, 5, 6 → S0 S1 S2 S3

2 → A0, 23 → A1, 21 → A2, 19 → A3
1 → B0, 22 → B1, 20 → B2, 18 → B3
7 → C_n
8 → MC

F0 → 9, F1 → 10, F2 → 11, F3 → 13
$A = B$ → 14
$\overline{C}_{n+4}$ → 16
G → 17, P → 15

TRUTH TABLE

FUNCTION SELECT				INPUTS/OUTPUTS ACTIVE LOW		INPUTS/OUTPUTS ACTIVE HIGH	
				LOGIC FUNCTION (MC = H)	ARITHMETIC* FUNCTION (MC = L, C_n = L)	LOGIC FUNCTION (MC = H)	ARITHMETIC* FUNCTION (MC = L, $\overline{C}_n$ = H)
S3	S2	S1	S0				
L	L	L	L	$\overline{A}$	A minus 1	$\overline{A}$	A
L	L	L	H	$\overline{AB}$	AB minus 1	$\overline{A+B}$	A+B
L	L	H	L	$\overline{A}+B$	$A\overline{B}$ minus 1	$\overline{A}B$	$A+\overline{B}$
L	L	H	H	Logic "1"	minus 1	Logic "0"	minus 1
L	H	L	L	$\overline{A+B}$	A plus $(A+\overline{B})$	$A\overline{B}$	A plus $A\overline{B}$
L	H	L	H	$\overline{B}$	AB plus $(A+\overline{B})$	$\overline{B}$	$(A+B)$ plus $A\overline{B}$
L	H	H	L	$A \odot B$	A minus B minus 1	$A \oplus B$	A minus B minus 1
L	H	H	H	$A+\overline{B}$	$A+\overline{B}$	$A\overline{B}$	$A\overline{B}$ minus 1
H	L	L	L	$\overline{A}B$	A plus $(A+B)$	$\overline{A}+B$	A plus AB
H	L	L	H	$A \oplus B$	A plus B	$A \oplus B$	A plus B
H	L	H	L	B	$A\overline{B}$ plus $(A+B)$	B	$(A+\overline{B})$ plus AB
H	L	H	H	$A+B$	$A+B$	AB	AB minus 1
H	H	L	L	Logic "0"	A plus A	Logic "1"	A plus A
H	H	L	H	$A\overline{B}$	AB plus A	$A+\overline{B}$	$(A+B)$ plus A
H	H	H	L	AB	$A\overline{B}$ plus A	$A+B$	$(A+\overline{B})$ plus A
H	H	H	H	A	A	A	A minus 1

* Expressed as two's complements. For arithmetic function with C_n in the opposite state, the resulting function is as shown plus 1.

 MOTOROLA *Semiconductor Products Inc.*

Device	Alternate	Temperature Range	Package
MC1455G	—	0°C to +70°C	Metal Can
MC1455P1	NE555V	0°C to +70°C	Plastic DIP
MC1455U	—	0°C to +70°C	Ceramic DIP
MC1555G	—	−55°C to +125°C	Metal Can
MC1555U	—	−55°C to +125°C	Ceramic DIP

MC1455
MC1555

Specifications and Applications Information

TIMING CIRCUIT

SILICON MONOLITHIC INTEGRATED CIRCUIT

TIMING CIRCUIT

The MC1555/MC1455 monolithic timing circuit is a highly stable controller capable of producing accurate time delays, or oscillation. Additional terminals are provided for triggering or resetting if desired. In the time delay mode of operation, the time is precisely controlled by one external resistor and capacitor. For astable operation as an oscillator, the free running frequency and the duty cycle are both accurately controlled with two external resistors and one capacitor. The circuit may be triggered and reset on falling waveforms, and the output structure can source or sink up to 200 mA or drive MTTL circuits.

- Direct Replacement for NE555/SE555 Timers
- Timing From Microseconds Through Hours
- Operates in Both Astable and Monostable Modes
- Adjustable Duty Cycle
- High Current Output Can Source or Sink 200 mA
- Output Can Drive MTTL
- Temperature Stability of 0.005% per °C
- Normally "On" or Normally "Off" Output

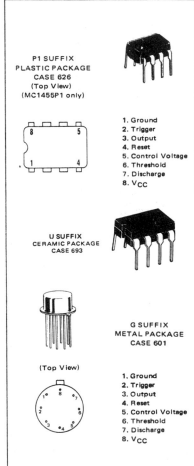

P1 SUFFIX
PLASTIC PACKAGE
CASE 626
(Top View)
(MC1455P1 only)

1. Ground
2. Trigger
3. Output
4. Reset
5. Control Voltage
6. Threshold
7. Discharge
8. V_CC

U SUFFIX
CERAMIC PACKAGE
CASE 693

G SUFFIX
METAL PACKAGE
CASE 601

(Top View)

1. Ground
2. Trigger
3. Output
4. Reset
5. Control Voltage
6. Threshold
7. Discharge
8. V_CC

FIGURE 1 – 22-SECOND SOLID-STATE TIME DELAY RELAY CIRCUIT

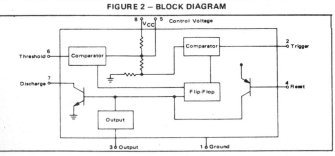

FIGURE 2 – BLOCK DIAGRAM

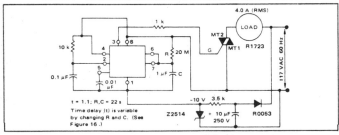

TYPICAL APPLICATIONS

- Time Delay Generation
- Sequential Timing
- Linear Sweep Generation
- Precision Timing
- Pulse Generation
- Pulse Shaping
- Missing Pulse Detection
- Pulse Width Modulation
- Pulse Position Modulation

Figure C.4-7 A 555 timer and some applications. (*Motorola, Inc.*).

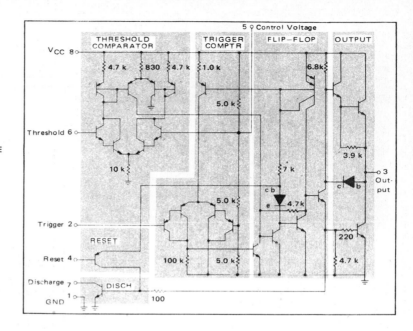

FIGURE 13 – REPRESENTATIVE
CIRCUIT SCHEMATIC

GENERAL OPERATION

The MC1555 is a monolithic timing circuit which uses as its timing elements an external resistor — capacitor network. It can be used in both the monostable (one-shot) and astable modes with frequency and duty cycle controlled by the capacitor and resistor values. While the timing is dependent upon the external passive components, the monolithic circuit provides the starting circuit, voltage comparison and other functions needed for a complete timing circuit. Internal to the integrated circuit are two comparators, one for the input signal and the other for capacitor voltage; also a flip-flop and digital output are included. The comparator reference voltages are always a fixed ratio of the supply voltage thus providing output timing independent of supply voltage.

Monostable Mode

In the monostable mode, a capacitor and a single resistor are used for the timing network. Both the threshold terminal and the discharge transistor terminal are connected together in this mode, refer to circuit Figure 14. When the input voltage to the trigger comparator falls below 1/3 V_{CC} the comparator output triggers the flip-flop so that it's output sets low. This turns the capacitor discharge transistor "off" and drives the digital output to the high state. This condition allows the capacitor to charge at an exponential rate which is set by the RC time constant. When the capacitor voltage reaches 2/3 V_{CC} the threshold comparator resets the flip-flop. This action discharges the timing capacitor and returns the digital output to the low state. Once the flip-flop has been triggered by an input signal, it cannot be retriggered until the present timing period has been completed. The time that the output is high is given by the equation $t = 1.1\,R_A C$. Various combinations of R and C and their associated times are shown in Figure 16. The trigger pulse width must be less than the timing period.

A reset pin is provided to discharge the capacitor thus interrupting the timing cycle. As long as the reset pin is low, the capacitor discharge transistor is turned "on" and prevents the capacitor from charging. While the reset voltage is applied the digital output will remain the same. The reset pin should be tied to the supply voltage when not in use.

FIGURE 14 — MONOSTABLE CIRCUIT

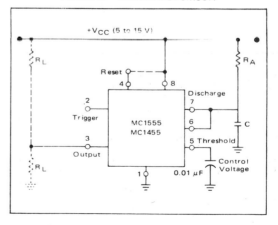

GENERAL OPERATION (continued)

FIGURE 15 – MONOSTABLE WAVEFORMS

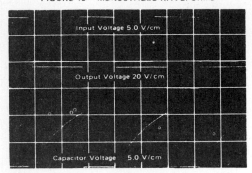

Input Voltage 5.0 V/cm

Output Voltage 20 V/cm

Capacitor Voltage 5.0 V/cm

t = 50 µs/cm
(R_A = 10 kΩ, C = 0.01 µF, R_L = 1.0 kΩ , V_{CC} = 15 V)

FIGURE 17 – ASTABLE CIRCUIT

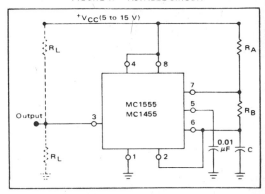

FIGURE 16 – TIME DELAY

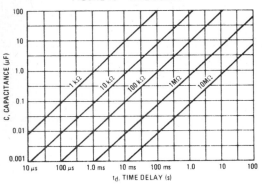

t_d, TIME DELAY (s)

FIGURE 18 – ASTABLE WAVEFORMS

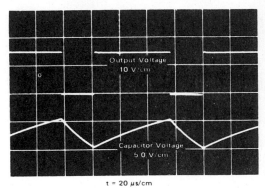

Output Voltage
10 V/cm

Capacitor Voltage
5.0 V/cm

t = 20 µs/cm

(R_A = 5.1 kΩ, C = 0.01 µF, R_L = 1.0 kΩ;
R_B = 3.9 kΩ, V_{CC} = 15 V)

Astable Mode

In the astable mode the timer is connected so that it will retrigger itself and cause the capacitor voltage to oscillate between 1/3 V_{CC} and 2/3 V_{CC}. See Figure 17.

The external capacitor charges to 2/3 V_{CC} through R_A and R_B and discharges to 1/3 V_{CC} through R_B. By varying the ratio of these resistors the duty cycle can be varied. The charge and discharge times are independent of the supply voltage.

The charge time (output high) is given by: t_1 = 0.695 (R_A+R_B) C

The discharge time (output low) by: t_2 = 0.695 (R_B) C

Thus the total period is given by: T = t_1 + t_2 = 0.695 (R_A+2R_B) C

The frequency of oscillation is then: $f = \dfrac{1}{T} = \dfrac{1.44}{(R_A+2R_B)\,C}$

and may be easily found as shown in Figure 19.

The duty cycle is given by: $DC = \dfrac{R_B}{R_A+2R_B}$

To obtain the maximum duty cycle R_A must be as small as possible; but it must also be large enough to limit the discharge current (pin 7 current) within the maximum rating of the discharge transistor (200 mA).

The minimum value of R_A is given by:

$$R_A \geq \dfrac{V_{CC}\,(Vdc)}{I_7\,(A)} \geq \dfrac{V_{CC}\,(Vdc)}{0.2}$$

FIGURE 19 – FREE-RUNNING FREQUENCY

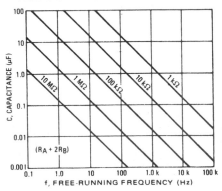

(R_A + 2R_B)

f, FREE-RUNNING FREQUENCY (Hz)

INDEX

INDEX